Hand Deburring:
Increasing Shop Productivity

Hand Deburring:
Increasing Shop Productivity

LaRoux K. Gillespie

Society of Manufacturing Engineers

Association for Finishing Processes/SME

Dearborn, Michigan

Copyright © 2003 by the Society of Manufacturing Engineers

987654321

All rights reserved, including those of translation. This book, or parts thereof, may not be reproduced by any means, including photocopying, recording or microfilming, or by any information storage and retrieval system, without permission in writing of the copyright owners.

No liability is assumed by the publisher with respect to use of information contained herein. While every precaution has been taken in the preparation of this book, the publisher assumes no responsibility for errors or omissions. Publication of any data in this book does not constitute a recommendation or endorsement of any patent, proprietary right, or product that may be involved.

Library of Congress Catalog Card Number: 2002112971

International Standard Book Number: 0-87263-642-9

Additional copies may be obtained by contacting:
Society of Manufacturing Engineers
Customer Service
One SME Drive, P.O. Box 930
Dearborn, Michigan 48121
1-800-733-4763
www.sme.org

SME staff who participated in producing this book:
Mike McLelland, Staff Editor
Rosemary Csizmadia, Production Supervisor
Frances Kania, Production Assistant
Kathye Quirk, Graphic Designer/Cover Design
Jon Newberg, Production Editor

Printed in the United States of America

*This book is dedicated to the thousands of manual deburring operators
who never received recognition for their contributions
and the impact of their work on part performance.
In some far recess of the plant is a soul who would be very surprised and grateful
if others honored his or her work.*

Table of Contents

Preface xv
Abbreviations xvii
Acknowledgments xxi

1 **Overview of Deburring and Hand Deburring Technology** 1
   What is Hand Deburring? 1
   Fundamentals of Burr Technology 2
   Why Use Hand Deburring? 2
   When to Use Hand Deburring 3
   Making Hand Deburring More Effective 4
   Keys to Success 7
   References 8

2 **What's Right about Hand Deburring?** 9
   Negative Views of Hand Deburring 9
   Toward a Positive View of Hand Deburring 11
   Selecting the Best Method 12
   Selection Approach 21
   Three Companies, Three Examples 21
   References 22
   Bibliography 22

3 **Edge Requirements and Standards** 25
   Edge Requirements: Issues and Problems 25
   Edge Standards 26
   Sample Standard for Edges and Burrs 29
   How Standards Apply to Hand Deburring 38
   Dictionaries for Edge and Burr Terminology 38
   Glossary 39
   References 41
   Bibliography 41

4 Common Hand Deburring Practices 43
- What Is Happening Today 43
- Good Management Practices 46
- Good Hands-on Operator Practices 47
- Lathe Parts 66
- Environmental, Health, and Safety Considerations 66
- Appendix 66
- References 68
- Bibliography 69
- Bibliography of Appropriate Drilling Conditions for Burr Minimization 69

5 Deburring Costs—A Primer 71
- First Estimate of Costs 71
- Estimate Based on Amount of Edge Deburred 71
- Broad View 71
- Economic Calculations 72
- Equipment Procurement Costs 74
- Helpful Facts 75
- References 80
- Bibliography 80

6 Manual Deburring—The Real Costs 81
- Deburring Versus Part Finishing 81
- What Increases Hand Deburring Costs? 82
- Time and Motion Studies 82
- Reducing Hand Deburring Costs 82
- Validating Relative Deburring Costs 84
- Calculating Costs 85
- Tooling Costs 85
- References 86
- Bibliography 87

7 Deburring with Knives 91
- Knife Styles 91
- Blade Materials 101
- Using Deburring Knives 101
- Knife Considerations 105
- Knives Make Burrs 106
- Relation between the Knife and Part 106
- Knife Sharpening 106
- Environmental, Health, and Safety Issues 106
- References 106
- Bibliography 106

8 Deburring with Scrapers 109
- Scraper Designs 109
- Using Deburring Scrapers 112
- Environmental, Health, and Safety Issues 113
- Reference 114

9 Countersinks and Chamfering Tools 115
Tool Types 115
Countersink Use 123
Environmental, Health, and Safety Issues 124
References 124

10 Rotary Burs 125
Rotary Bur Styles 125
Tooth Design 137
Materials 142
Applications 144
Environmental, Health, and Safety Issues 154
References 155

11 Reamers as Deburring Tools 157
Reamer Design 157
Unique Adaptations 157
Environmental, Health, and Safety Issues 158
References 159

12 Files 161
Types of Files 161
Selecting and Using Files 169
Environmental, Health, and Safety Issues 172
Appendix 173
References 175
Bibliography 175

13 Mounted Points 177
Types of Mounted Shapes 177
Types of Materials in Mounted Stones 178
Hardness of Mounted Stones 178
Applications for Mounted Stones 181
Environmental, Health, and Safety Issues 181
Bibliography 182

14 Hand Stones 187
Hand-stone Composition 187
Stone Bonds 189
Grit Size 189
Basic Shapes and Sizes 190
Basic Usage 191
Environmental, Health, and Safety Issues 191
Reference 191

15 Abrasive Wood Tools 197
Woods 197
Common Usage 198
Environmental, Health, and Safety Issues 198

16 Abrasive Cork Tools 201
Why Cork? 201
Cork Tool Forms 202
Typical Applications 202
Operating Recommendations 202
Resin/Rubber Cork Combination Wheels 202
Environmental, Health, and Safety Issues 202

17 Abrasive Cotton-fiber Products 205
Resin-bonded Tools 205
Sizes and Shapes 205
Product Materials 207
Usage 208

18 Abrasive Rubber Tools 211
Characteristics 211
Available Sizes 213
Shank Configurations 214
Potential Limitations 214
Tool Comparisons 214
Tool Usage 216
Environmental, Health, and Safety Issues 216

19 Bonded Abrasive Tools (Sanding Products) 221
Shapes and Sizes Available 221
Tool Materials 231
Tool Use 232
Environmental, Health, and Safety Issues 233
References 235

20 Brushes 239
Brush Types 239
Brush Fill Material 243
Brush Aggressiveness 246
Tool Wear 248
Power Requirements 249
Brush Selection 251
Brush Deburring Equipment 252
Brush Deburring Applications 257
Wire-filled, Elastomer-bonded Brushes 262
Troubleshooting Issues 263
Side Effects 263
Nontraditional Brushing 264
Combined Process Deburring 266
Environmental, Health, and Safety Issues 266
References 267
Bibliography 268

21 Ballizing 271
Side Effects 272
Environmental, Health, and Safety Issues 272
References 273

22 Miscellaneous Special Hand Tools 275
Tools 275
Other Tools 280
Environmental, Health, and Safety Issues 281
References 281

23 Handheld Motors 283
Types of Handheld Motors 283
Handheld Motors in Use 290
Environmental, Health, and Safety Issues 290
References 292

24 Bench and Pedestal Motors 299
Bench Motors 299
Speed Lathes 301
Bench Grinders 301
Pedestal-mounted Motors 302
Industrial Buffers 302
Environmental, Health, and Safety Issues 303
References 303

25 Hand-operated Mechanized Machines 305
Tool Types 305
Environmental, Health, and Safety Issues 311
References 312

26 Support Items 313
Support Devices 313
Environmental, Health, and Safety Issues 328
Reference 332

27 Tools that Beat Down Edges 333
Tool Types 333
Environmental, Health, and Safety Issues 334
References 334

28 Hot-wire Tools for Thermoplastic Parts 335
The Hot-tool concept 335
Suitable Plastic Materials 335
Environmental, Health, and Safety Issues 336
References 336

29 Mechanization and Automation 341
Mechanization to Replace Manual Deburring 341
Automation to Replace Manual Deburring 346
References 348

30 Environmental, Health, and Safety Issues 351
Knives 351
Files 354
Motorized Operations 354
Injury Statistics 355
Training for Safety 355
References 356

31 Ergonomics 357
Ergonomic Data 357
Cost of Poor Ergonomics 357
Improving Productivity 358
Legal Liabilities 358
Medical and Insurance Costs 358
Social Costs 358
Red Flags 358
Preventing Ergonomic Injuries 359
Cumulative Trauma Disorders 359
Hand Tool Design 363
Back Pain 366
High Stress 366
Illumination 366
Noise 367
Summary 368
References 369
Bibliography 369

32 Getting More from Your Operation 371
Reducing Deburring Costs 371
Problem Identification 371
Coordinating People 377
Social Aspects of Deburring 381
Equipment Allocation 382
Data-driven Solutions 383
Typical Problem Areas 405
Practical Hints 409
Appendix 410
Summary 419
Environmental, Health, and Safety Issues 419
References 420
Bibliography 421

33 Training for Manual Deburring ... 423
Training Elements ... 423
Training Goals ... 423
People for Deburring ... 424
Parts for Training ... 424
Identifying Tools ... 424
Hands-on Training Experiences ... 424
Quality Requirements ... 425
Deburring Economics ... 425
Plant Expectations for Deburring Workers ... 425
Case History ... 427
References ... 433

34 Inspecting for Burrs and Sharp Edges ... 435
Burr and Edge Standards ... 435
Inspection Approaches ... 435
Training Workers and Inspectors ... 456
Calibration of Burr Inspection Techniques ... 458
Which Edge Characteristics Should be Measured? ... 458
One Measurement is Not Enough ... 459
Measurement Capability Studies ... 459
Documentation ... 459
The Burr Inspector ... 459
References ... 459
Bibliography ... 461

Index ... 463

Preface

Burrs, perennial thorns in the manufacturing industry's paw, do not go away with the purchase of new and faster machines. Buying more machines to remove burrs is a good answer for high-volume work, but it may not be the most cost-effective answer for job shops and low-volume facilities. Hand deburring, also called manual deburring, often is the best choice for these situations. In many applications, hand deburring needs to be vastly improved. This book is the result of the author's 37 years of research, floor application, international observation, and teaching experience.

There are two keys to being more productive in the hand deburring field. First, know what your competitor already knows—the things that are in this book. Second, your people are the key to your success. Spurring their innovation and desire to learn and do more is the cheapest investment you can make to improve your operation. Help workers become world leaders in their operation—they can be in only two or three years.

This book begins with an overview of hand-deburring technology, explains why it is the right answer for many situations, then dwells on related issues that are often ignored—knowing the edge standards needed and calculating the real cost of deburring. If you have never calculated it, you do not understand the real cost to your operation, and the potential savings that are available with the right knowledge remain unknown.

Twenty chapters provide a wide perspective on the tens of thousands of tools used for deburring. The author's experience has come from precision, very close tolerance deburring, but he also has experience with commercial companies having less exacting requirements. A chapter on simple, low-cost mechanization also is presented.

Ergonomics is covered in most chapters, but Chapter 31 is specifically targeted toward preventing this subtle yet serious drain on your profits and your workers' health. Pain is hard to measure, but its effects on our lives can be documented. Why should workers have to suffer when prevention is available?

Chapter 32 (Getting More from Your Operation) could be 300 pages in length, but that is a bit impractical. This chapter may be the most helpful in the book, but it would not be as useful without all the detail provided in previous chapters.

The chapter on training (Chapter 33) is unique. Training workers on how to deburr parts simply is not described in literature. Commercial courses do not exist. In this chapter, you will read about how one company prepared some 700 workers, 100 of them to perform microscope-quality deburring.

The missing detail of how to inspect for burrs is covered in great detail in Chapter 34, but it should not be a chapter for inspectors—it is for those doing the deburring. If deburring operators inspect their own work well, inspectors are not needed.

While this is arguably the most complete work ever published on manual deburring, many tricks of the trade saving millions of dollars remain to be written down and shared. The author would like to help potential authors publish their experiences in this field so that everyone can be more productive. You can make a difference in world productivity if you will share your method of making hand deburring more effective. Drop the SME staff editor or the author a line with your improvements and we will find a way to tell the world.

Abbreviations

3-D three-dimensional

A
A ampere
AA arithmetic average
ABS acrylonitrile-butadiene-styrene
ADCI American Die Casting Institute
AFM abrasive flow machining
AISI American Iron and Steel Institute
AJM abrasive-jet machining
AMS aerospace material specification
AMSA American Metal Stamping Association
AMT Association for Manufacturing Technology
ANSI American National Standards Institute
aq. aqueous
AS acrylic styrene
ASM American Society for Metals
ASME American Society of Mechanical Engineers
ASTM American Society for Testing and Materials

B
Bhn Brinell hardness number

C
C Celsius
CAMI Coated Abrasive Manufacturers' Institute
CBD chronic beryllium disease (berylliosis)
CFR Code of Federal Regulations
CHM chemical machining
CLR circuit-loop resistance
cm centimeter
CNC computer numerical control

D
dB decibel
DC direct current
dia. diameter
DOE design of experiments
DOF depth of field
dm dyne per meter

E
EBM electron beam machining
ECD electrochemical deburring
ECG electrochemical grinding
ECH electrochemical honing
ECM electrochemical machining
EH&S environmental, health, and safety
ELP electropolishing
EPA Environmental Protection Agency
EPDM ethylene-propylene rubber
Eq. equation
ESM electrostream machining

F
F Fahrenheit
FEPA Federation of European Producers of Abrasives
fig. figure
FMS flexible manufacturing system
FOV field of view
ft foot/feet
ft^3/min cubic feet per minute

G
g gram
gal gallon
GP general-purpose plastic (PVC, polyethylene, polypropylene)

H
h helical
HCG hot chlorine gas
HDM hydrodynamic machining

HIPS	high-impact polystyrene		NEMA	National Electrical Manufacturers Association
hp	horsepower		NIST	National Institute for Science and Technology
hr	hour		nm	nanometer
HSS	high-speed steel		N/m	Newton/meter
Hz	hertz		NSMPA	National Screw Machine Products Association
			NTIS	National Technical Information Service

I

IBM	ion beam machining
IC	integrated circuit
ID	inside diameter
in.	inch
ipm	inches per minute
ipr	inches per revolution
ISO	International Organization for Standardization

J

J	joule

K

kg	kilogram
kHz	kilohertz
km	kilometer
kPa	kilo Pascal
ksi	1,000 pounds per square inch
kW	kilowatt

L

L	liter
lb	pound
lbf	pound force
LBM	laser beam machining
LED	light-emitting diode

M

m	meter
µin.	microinch
µm	micrometer
mg	milligram
MSDS	material safety data sheet
mi	mile
MIL	military specification
min	minute
mm	millimeter
MPa	mega Pascal
ms	millisecond

N

N	Newton
NC, N/C	numerical control

O

OD	outside diameter
OHD	overhead drive
ohm	unit of electrical resistance
OSHA	Occupational Safety and Health Administration
oz	ounce

P

PAM	plasma-arc machining
PBT	polybutylene terephthalate
PC	polycarbonate
PCM	photochemical machining
PH	precipitation hardening
PMMA	polymethyl methacrylate
ppm	parts per million
PPS	polyphenylene sulfide
psi	pounds per square inch
PVA	polyvinyl alcohol
PVC	polyvinyl chloride
PWB	printed wiring board

Q

qt	quart

R

r	rotation
R	radius
Ra	roughness average
RCRA	Resource Conservation and Recovery Act
rev	revolution
R_{max}	maximum roughness depth
RMS	root mean square
rpm	revolutions per minute

S

s, sec.	second
SAE	Society of Automotive Engineers
sfm	surface feet per minute
sfpm	surface feet per minute
SiC	silicon carbide
SME	Society of Manufacturing Engineers
SPC	statistical process control

T

TCLP	toxicity characteristic leaching procedure
TEM	thermal energy method
TiN	titanium nitride

U

U.K.	United Kingdom
UL	Underwriters' Laboratories
U.S.	United States
USM	ultrasonic machining

V

V	volt

W

W	watt
WAM	waterjet abrasive machining
WBTC	Worldwide Burr Technology Committee
WJM	waterjet machining

Y

yd	yard
yr	year

MATH SYMBOLS

~	about equal to
°	degree
/	divided by or per
>	greater than
≥	greater than or equal to
α	Greek alpha
β	Greek beta
Δ	Greek delta
ε	Greek epsilon
γ	Greek gamma
λ	Greek lambda
μ	Greek mu
Ω	Greek omega
φ	Greek phi
π	Greek pi
σ	Greek sigma
θ	Greek theta
ϑ	Greek theta (lower case)
<	less than
≤	less than or equal to
−	minus
%	percent
+	plus
±	plus or minus
×	times

Acknowledgments

Dozens of individuals have directly contributed to this book and, behind them, hundreds of practitioners and equipment manufacturers have generated material that helped provide original developments and ideas. Some of the material has previously been published in SME books and papers. I have brought much of my own experience to this handbook, and am indebted to the many persons who helped me develop my background in this international field.

A special acknowledgment goes to Mabel Taylor, my mentor in hand deburring. Mabel taught me the how and the why of her approaches. She was a teacher and innovator, a full-time manual burr remover, and expressive with her wealth of knowledge.

John Bolinger and Mark Mitchell brought new ideas and insights. Together, we learned how to expand our knowledge to others. Mark lugged a 70-pound microscope around the country for years, showing suppliers what they were missing, how to search for burrs, thus transferring our knowledge to them.

We have lived the ideas presented here, traveled the world looking for new ideas, and faithfully kept notes from every encounter for over 35 years. We have reached into blackened soot-filled rooms of machined gray-iron casting houses, stood next to hot plastic molds, and developed safety programs for the deflashing workers who have sliced their hands on scalpels or razor-sharp sheet-metal edges. The material in this book is based on the training of over 600 machinist and deburring workers on how to perform precision microscopic deburring. From rough deburring to investigating edges and surfaces at 10,000-power magnification, we have searched for the problem and solutions. Book material also is based on scraping the edges of parts weighing hundreds of pounds, and consulting with a variety of automotive, aerospace, engine, valve, and other companies. From a tableware factory in TinJing, China, to aircraft manufacturers in Canada, from Japan to Korea, Russia, Germany, and correspondence with engineers and students from around the world at many other sites, I have incorporated all that I have lived and others have breathed. I have reviewed 4,500 deburring, deflashing, and edge-finishing articles in many languages and incorporated the best of them in this work. I have taken my guidebook, *Guide to Deburring, Deflashing, and Trimming Equipment, Supplies, and Services,* which lists all the deburring products of over 2,000 companies, and blended the information into each chapter.

Many of the ideas and illustrations in this book are reprinted with permission from a variety of company and association publications. Grateful acknowledgments are extended for permission to use material from the following sources: 3M Corp.; ABB Flexible Automation; Abbeon Cal; Ace Manufacturing Co.; Acme Manufacturing Co.; Aerosharp Tool Co., Inc.; Allied Signal Aerospace, Inc.; Almco, Inc.; Aluminum-Verlag GmbH (Germany); AM Machinery Sales; American Heart Association; *American Machinist* magazine; American Rotary Tools Company, Inc.; Anderson Products Div. of Weiler Corp.; Baldor Electric; Basset Rotary Tool Co.; Bates Abrasive Products; Bendix; Boeing; Brasseler USA; Brush Research Manufacturing Co.; Burr King Manufacturing; Burrtec Co. (Japan); Cahners Business Information; Carl Hanser Verlag (Germany); Charles G. Allen Co.; CIMID Corporation; Cogsdill Tool Products; Comco, Inc.; Congress Tools; Craftics; Cratex Manufacturing Co.; D.C. Morrison Co.; Danmarks Tekniske Hooskole (Denmark); Dedeco International; Delta International Machinery Co.; Deprag Industrial; D.L. Rici Corp.; Dynabrade; Dynetics; Edmund Industrial Optics; Empire Abrasive Equipment Corp.; Eshenbach; Everede Tool; Extrude Hone Corporation; Federal File Co.; Field Tool Supply Co.; Flexarm/Midwest Specialties, Inc.; Flexbar Machine Co.; Foredom Electric Co.;

Fraunhofer-Institute fur Producktionstechnik und Automatisierung (Germany); Dr. Friedrich Shäfer (Germany); Fullerton Tool Co.; Gesswein; Hansco Enterprises; Honeywell; Hyde Manufacturing Company; IMCO Carbide Tool, Inc.; American Institute of Industrial Engineers; Jarvis Cutting Tools, Inc.; J.J. Claus; Kadia Corp.; Kent Corp./Tesgo, Inc.; Dr. Klaus Przyklenk (Germany); Kennametal, Inc.; Klingspor Abrasive; Dr. Koya Takazawa (Japan); Leica Microsystems, Ltd. (Germany); M.A. Ford Manufacturing Co.; Martindale Electric; Marvel Abrasive Products; Maxi-Blast, Inc.; Metal Cutting Tools, Inc.; Michigan Drill/APT; Midwest Specialties, Inc.; Morton Machine Works; Munson Machinery Co., Inc.; National Screw Machine Products Association; Nitto Kohki USA; Norton Coated Abrasives; NSK America; O.C. White Co.; Osborn Manufacturing; Pacer; Pferd, Inc.; Pfingst; Pines Manufacturing; Polygrat GmbH (Germany); Precision Machined Products Association; Quality Carbide Tool; Reid Tool Supply Co.; Rex-Cut Products; Rico Tool Co.; Robert W. Johannesen; Rovi Products, Inc.; Royal Products, Inc.; Senn Co.; Severance Tool Industries, Inc.; SGS Tool Co.; Shoreview Metalcraft, Inc.; Simco Automotive Division/UFP Industries; Somma Tool Co.; SPI; Springer-Verlag Corporation; Star Dental Manufacturing Co.; Sunnex, Inc.; Superior Abrasives; SurfTran; Tesco Corp.; Timesavers, Inc.; Turbocarver; Underwriters' Laboratories; Vargus Ltd. (Israel); Vernon Devices; Vision Engineering; Wallach Surgical Devices, Inc.; Waldman Lighting; Weiler Corp.; Weldon Tool Co.; Wyco/Veeco; Yasuo Kato (Japan).

Lastly, SME has continued its efforts to provide a top-quality publication by assigning an excellent book editor to improve style, format, and understanding. This joins the top-notch *Deburring and Edge Finishing Handbook* as a critical resource in a field for which there are no other major publications.

Thanks to all of those mentioned here and the many persons in the background who also helped put this book in print.

LaRoux K. Gillespie
Deburring Technology International
Kansas City, Missouri

Overview of Deburring and Hand Deburring Technology

Burr: "Illegitimate offspring of a machining process which grows just to frustrate inspectors and assemblers, and which puts itself where armies of patient workers are required to remove it with miniature knives, hatchets and axes."

Bob Hishaw, Process Engineer

This tongue-in-cheek definition of a burr aptly expresses the sentiment of many engineers, machinists, and deburring workers who have had to remove them. They exist; therefore, someone must remove them.

Throughout this book, *deburring* means the removal of burrs at the edges of parts. It does not mean finishing surfaces, doing paperwork, or cleaning parts. Deburring involves many individuals and groups, including machine operators, inspectors, maintenance workers, assemblers, and full-time deburring staffs. In later chapters, these and other topics are discussed—including tools, fettling (removal of sprues and gates), and deflashing; however, the emphasis is on burrs and edges.

WHAT IS HAND DEBURRING?

There are 109 deburring and edge-finishing processes. Of these, hand deburring, also known as hand or benchwork, benching, or manual deburring, is the most widely used process. It describes any deburring activity that requires a handheld deburring tool (Figure 1-1) or the placement of a handheld part against a fixed tool or machine (Figure 1-2). Hand deburring employs a variety of cutters and motorized tools. Because it uses these types of tools, hand deburring overlaps other processes, such as brush deburring and bonded-abrasive finishing. Just as hand deburring employs a variety of tools, it also involves (like deburring in general) a wide variety of workers, in-

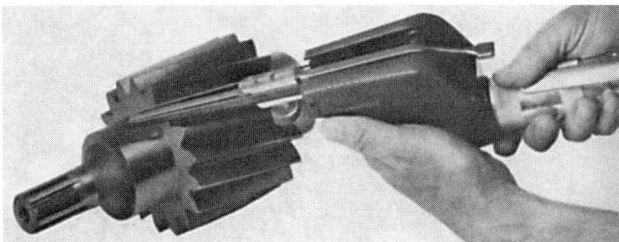

Figure 1-1. Deburring with a handheld deburring tool. (Courtesy Dynabrade)

Figure 1-2. Hand deburring using a mechanized machine. (Courtesy Senn Company/Reishauer Corporation)

cluding full-time deburring staffs, machine operators, inspectors, maintenance workers, and assemblers.

This book focuses on hand deburring and how to make it more effective. Chapters 1–6 concentrate on aspects of edge requirements, costs, and hand deburring that must be understood before hand deburring can be made more effective. Many of the ways in which hand deburring is misunderstood and why its application is often inefficient are also discussed. Chapters 7–28 focus on hand deburring tools and how to use them effectively. Chapters 29–34 focus on health and safety concerns and on optimizing deburring operations.

For a comprehensive look at the entire field of deburring, see the *Deburring and Edge Finishing Handbook* (Gillespie 1999), which is recommended as a companion to this book and will provide a better understanding of burr formation, burr prevention, and part geometry effects on burr size and deburring processes. It also includes an extensive bibliography on deburring, which can aid in further evaluation and comparison of the costs of hand deburring.

FUNDAMENTALS OF BURR TECHNOLOGY

Burr technology relies on five principles. The first principle is that cost-effective deburring requires knowledge of burr properties (for each edge), part design, part function, and deburring process capabilities. The following factors must be considered as a whole (rather than individually) because, in combination, they provide unique constraints that limit processes and define costs.

- Burr properties are a function of material properties, the machining and blanking process, and part configuration.
- Acceptable deburring is a function of burr properties, part configuration, acceptance standards, and deburring process parameters.
- Cost-effective deburring is a function of acceptable deburring quality and scheduled quantities; cycle time; environmental, safety, and health issues; and the side effects of deburring processes (Gillespie 1977).
- All deburring processes have side effects.

The second principle states that there are five basic approaches to reducing deburring costs. These approaches correspond to some of the goals discussed later in this chapter to:

- improve product design (goal 1);
- prevent burrs (goal 2);
- minimize burr properties (goal 3);
- remove burrs during the machining/blanking cycle (goal 6), and
- develop or obtain more effective deburring processes (goal 11).

The third principle maintains that *edge finishing* and *edge quality* are separate views of deburring.

The fourth principle states that there is a vast number of deburring innovations, deburring processes, and deburring process variations—the 109 deburring and edge finishing processes each have variations.

The fifth principle notes that even the subtlest tricks of the trade can generate substantial savings. They can eliminate, for example, the need for new machines, as well as high-tech training and maintenance. Innovation is the key to success.

WHY USE HAND DEBURRING?

Hand deburring is labor intensive and often yields inconsistent or undesirable results, but it offers several general advantages over other deburring processes:

- it is versatile;
- it requires minimal floor space, and
- it requires minimal capital investment.

Hand deburring, in fact, is the most versatile deburring process because it allows employees to use their creativity to continually find better approaches. Machines are not versatile. People are enormously versatile. They conceive new approaches to improve operations. They can quickly choose from among 10,000 different tools. They can reach areas that seem inaccessible, and they can identify and eliminate unique problems without affecting other part requirements. They can substitute for missing tools, take immediate action, and document problems.

The following story illustrates the importance of employee creativity. At the onset of the robotics era, a newspaper reporter was touring a plant in the heart of Japan's automation district. He noted the absence of robots and asked the owner, "Why don't you have any robots?" The owner responded, "Because robots do not give me ideas of how to improve my operation."

Trained, talented workers invent ways to improve operations when they are equipped with proper tools and a strong management team. These advantages explain why hand deburring is the most economical and effective deburring method for many operations. Apart from them, there are a number of more specific or technical reasons for using hand deburring. Likewise, there are a number of poor reasons for using hand deburring. The challenge arises in determining when to use it—the key to making hand deburring more effective is to reduce its improper application.

Good Reasons for Hand Deburring

Cost-effective reasons for using hand deburring fall into five categories:

1. Burr properties and location.
   - Burrs are hard to reach or remove by any other process.
2. Run considerations.
   - The small number of workpieces being processed makes manual approaches the most cost effective.
   - Highly variable burr sizes on short runs prohibit other techniques.
   - Short turnaround time is required for a few parts.
3. Limitations of other processes.
   - Small burr traces left by mechanized deburring must be removed (hand deburring finishes the part).
   - The deburring or finishing machine will not accept the workpiece size (hand deburring can be done on any size of part).
   - The required mechanized machine is out of commission (a hand deburring operator is available any day).
   - Prevention of media impregnation from loose abrasive processes is required (hand deburring does not impregnate media).
   - Prevention of oxide formation from other deburring processes is required (hand deburring leaves part clean).
   - Prevention of size change from other deburring processes is required (appropriate hand deburring does not change part dimensions).
   - Prevention of residual stress changes from other deburring processes is required (hand deburring does not change part stresses).
   - Prevention of undesirable side effects from other finishing processes is required.
   - Lead-in/lead-out threads cannot be removed by other approaches.
   - Workpieces are delicate or require a precision finish that would be destroyed by mechanized operations.
   - Workpieces require close tolerances that would be affected by mechanized deburring (for example, ±0.0001 in. [±2.5 μm] or less).
   - Required edge radii are large—0.060 in. (1.52 mm) or more; or allowable edge radii are small—0.002 in. (51 μm) or less (hand de-burring can handle these, while many mechanized processes cannot).
   - Part geometry cannot tolerate the edge-break variation resulting from a mechanized process. For example, vibratory finishing produces different radii on different edges because of part geometry (hand deburring can provide a uniform edge break, regardless of part geometry).
4. Machine cycle.
   - Hand deburring can be performed while other machining operations are performed (within cycle).
5. Process qualification.
   - The plant is unable to change the hand deburring process because it has been accepted in conformance with a military specification.

Poor Reasons for Hand Deburring

There are many less-than-admirable industry reasons for hand deburring, including the following:

- ignorance of other deburring processes;
- it is easier to call for hand deburring than to evaluate the best approach;
- no one cares;
- no time to think about deburring;
- no knowledge of where the burrs will appear, and
- other processes didn't work in the past.

Some of these explanations might seem ridiculous, yet casual conversations with those who specify manufacturing sequences confirm this type of reasoning is still found among engineers and is not limited to small companies or easily produced parts. Curbing this type of thinking is the first step toward improving hand deburring effectiveness because it will reduce its improper application. This is not easily accomplished, but it can be done by close and continuous scrutiny of deburring operations.

An important step in reducing the improper application of hand deburring is to remain aware of innovations. Notable advances occur in deburring every year. This means that whether or not a burr could be removed seven years ago by process A, B, or C is often irrelevant today. A few years ago, for example, vibratory deburring on large parts resulted in part-on-part impingement. When this was objectionable, these parts were often hand deburred. Today's vibratory machines, however, can virtually ensure such impingement does not happen. Similar improvements have occurred among many other deburring methods, including abrasive flow and thermal energy processes.

WHEN TO USE HAND DEBURRING

It is important to remember that hand deburring is not always the best choice. Thus, as previously stated, the challenge arises in determining when to use it. For example, hard-to-reach burrs are not always good candidates for hand deburring, because seven of the 109 known deburring processes (see Chapter 5) can access remote areas.

MAKING HAND DEBURRING MORE EFFECTIVE

The key to selecting the most effective process is to consider a number of factors, including burr size and accessibility, edge requirements, available equipment, and cost. These factors are part of the fundamentals of deburring discussed earlier. Careful consideration of these factors is a step toward making hand deburring more effective. Companies strive to reduce cost and increase effectiveness. The goal is to do whatever it takes to attain the most cost-effective plant operation possible, so companies should strive to make hand deburring more effective. Few companies aim to eliminate burrs or hand deburring, because that might increase costs. In fact, when plant managers choose to use traditional cutting or blanking operations, they choose to deal with burrs. Managers could have selected processes that do not produce burrs, but they are more costly. Hand deburring, therefore, is not a process to avoid but should be made more effective. To do this, companies must strive toward the following 12 goals:

1. improve product design (as it affects burrs);
2. prevent burrs;
3. minimize burrs;
4. control burr size and repeatability;
5. understand deburring economics;
6. remove burrs during the machine cycle;
7. provide proper tools;
8. use trained and skilled operators;
9. understand edge-quality issues;
10. understand customer expectations of edge quality;
11. find unique approaches to reduce deburring costs, and
12. use hand deburring where it is most effective and mechanized processes for other portions.

Improvements can be made in any of these areas, but the most effective program—a world-class deburring operation—strives toward each goal. Each of these is discussed briefly in the following sections. Remember, the goal is to maximize profit and production, not eliminate burrs.

Goals 1–4: Improve Product Design and Prevent, Minimize, and Control Burr Size

Goals (1–4) are important for a number of reasons, including the following:

- If you do not make a burr, you do not have to remove it.
- Small burrs can be readily and inexpensively removed.
- Consistent burrs allow for consistent deburring methods, affording more accurate costing and scheduling.

The recommended companion book, SME's *Deburring and Edge Finishing Handbook*, contains more than 125 pages showing users how to prevent burrs, minimize them, and make them readily removable. These four goals are brought to the reader's attention since there is important information available, but there simply is not enough space here to present all of it. Chapters 5 and 6 in this book provide some critical insight into where readers can concentrate their deburring expenses. That, in turn, can help focus efforts to reduce costs.

Goal 5: Understand Deburring Economics

Understanding shop economics is crucial to making the best decisions. Hand deburring is often a misnomer for the work performed by the deburring staff. For example, operators may be expected to smooth surfaces; remove broken tools; sand to a specific thickness; and inspect nicks, dings, scratches, pits, voids, and stains. When staff wants to reduce hand deburring costs, it often fails to recognize that these costs have nothing to do with burr removal. Hand deburring, which is often ignored, is frequently the most cost effective and timely process. Significant improvements can be made by being aware of the possibilities. Later in this book, chapters 5 and 6 explore shop economics in detail to demonstrate how to calculate the real costs of deburring and hand deburring.

Goal 6: Remove Burrs During the Machine Cycle

Removing burrs during the machine cycle is a popular method of reducing hand deburring costs. Either a part is physically hand deburred while another is machined, or machine deburring is performed as part of the cycle. Both methods are employed today, but most companies rely on numerical-controlled (NC) or computer-numerical-controlled (CNC) machines to remove all or most of a burr during the machine cycle. Chamfering tools and abrasive-filled nylon brushes, which are especially built for CNC machines, are typically used in the latter case (Gillespie 1999; Hettes 2000; Mahadev 1995).

Goal 7: Provide Proper Tools

This book provides depth and breadth in identifying the appropriate deburring tools because effective hand deburring is based on an understanding of which tools to use for which tasks and making them available to workers. An estimated 10,000 hand deburring tools are commercially available today—one catalog lists over 1,000 rotary burs alone! Table 1-1 contains a list identifying 23 effective hand-deburring tools.

The tools in Table 1-1 are discussed throughout this book, but most of them will be discussed individually and in greater detail in chapters 7–28. It is important to remember that an *effective* tool does not necessarily mean an *inexpensive* tool. Rather, it means having economical tools on hand when needed. Saving pennies is not advantageous if parts must await a tool's arrival. Often, success is measured by the rapidity with which parts move from machining to assembly.

Goal 8: Use Trained and Skilled Operators

Training is often overlooked and undervalued. Some sites consider deburring an entry-level job without skill requirements. Non-critical deburring requires minimal or on-the-job training, whereas critical and cost-effective deburring demand training that focuses on a desired outcome.

As stated in previous sections, people are the most important resource—or can be. Training on the following topics can enhance this resource: awareness of available tools, typical deburring practices, economics of hand deburring, inspection or acceptance requirements, skill enhancement, and the capabilities and economics of mechanized deburring. (See Chapter 33 for a detailed look at training for manual deburring.)

Goal 9: Understand Edge-quality Issues

Burrs and sharp edges create many problems. Sharp edges on finished parts can result from neglecting a sharp edge or by producing a burr with many

Table 1-1. List of effective hand-deburring tools and the chapters in which they are discussed

1. Abrasive-filled products	
a. Cork products (bullets, ball nose, cylinders)	Chapter 16
b. Cotton products (cylinders, balls, bars)	Chapter 17
c. Nylon synthetic products	Chapter 20
d. Rubber products (bullets, cylinders, flat bars, disks, cups, dental bullets)	Chapter 18
2. Abrasive wood tools	Chapter 15
3. Ballizing tools	Chapter 21
4. Bonded abrasive tools (abrasive paper products—disks, rolls, sheets, cord)	Chapter 19
5. Brushes (wheel, end and cup, tube, cross hole, side action)	Chapter 20
6. Burs, bur balls, rotary files	Chapter 10
7. Countersinks	Chapter 9
8. Drills and reamers	Chapter 11
9. Felt bobs (disks, bullets, cylinders)	Chapter 22
10. Files (large, miniature, round, half-round, triangular, curved, bent)	Chapter 12
11. Hand-operated mechanized machines	Chapter 25
12. Hand stones (bars, triangles, cones, points)	Chapter 14
13. Hot wire tools for thermoplastic parts	Chapter 28
14. Knives (triangular, oval, special shapes, scalpel blades)	Chapter 7
15. Lapping compounds	Chapters 15, 22
16. Mandrels for tools	Chapters 18, 19
17. Motorized tools (bench motors, air motors, dental tools, belt sanders, reciprocating files, jitterbug sanders)	Chapters 23, 24, 25
18. Mounted points (balls, disks, cylinders, cones, special shapes)	Chapters 13, 16
19. Peening tools (ball peening, blade peening)	Chapter 27
20. Picks	Chapter 8, 22
21. Pin vises (dog nose, collet)	Chapter 26
22. Scrapers	Chapter 8
23. Miscellaneous tools (back-side cutters, special designs, vacuum probes)	Chapters 9, 22, 31

sharp facets. Burrs and sharp edges are problematic and raise many issues. Burrs on sheet metal parts cause premature tearing during formation. Plating over burrs and sharp edges allows early corrosion attack or a poor fit in assembly. Fine burrs on automotive cylinders (grinding micro-slivers) can cause engine failure. Burrs on life safety devices can provide unrealistic levels of perceived safety. Automotive mechanics regularly sustain cuts and bruises from burred and sharp-edged components.

Edge-quality issues include safety, performance, appearance, and cost, as detailed in the following list. While the list is divided into several categories, note that when some products fail, the performance issue creates a life safety problem.

Safety issue:
- cut hands in assembly or disassembly.

Performance issues:
- interference fits (from burrs) in assemblies;
- jammed mechanisms (from burrs);
- scratched or scored mating surfaces (that allow seals to leak);
- friction increases or changes (disallowed in some assemblies);
- increased wear on moving or stressed parts;
- electrical short circuits (from loose burrs);
- cut wires from sharp edges and sharp burrs;
- unacceptable high-voltage breakdown of dielectric;
- irregular electrical and magnetic fields (from burrs);
- detuning of microwave systems (from burrs);
- metal contamination in unique aerospace assemblies;
- clogged filters and ports (from loose burr accumulation);
- cut rubber seals and O-rings;
- excessive stress concentrations;
- plating buildup at edges;
- paint thinout over sharp edges (from liquid paints);
- edge craters, fractures, or crumbling (from initially irregular edges);
- turbulence and nonlaminar flow;
- reduced sheet-metal formability;
- inaccurate dimensional measurements;
- microwave heating at edges;
- reduced fatigue limits;
- reduced volumetric efficiency of air compressors;
- reduced cleaning ability in clean-room applications, and/or
- reduced photo-resist adherence at edges.

Aesthetics:
- paint buildup at edges (from electrostatic spray over burrs).

To reduce the cost of edge finishing, companies must develop an awareness of these and other ways that burrs and sharp edges hinder production and increase its costs. Then, of course, they must actively try to minimize burrs and sharp edges.

Goal 10: Understand Customer Expectations for Finished Edge Quality

It is important to understand finished edge quality from the customer's perspective. Asking customers to clearly articulate their expectations is a low-cost strategy for making hand deburring more effective. Some customers state their expectations on drawings and purchase orders. Others, however, simply write "burr-free" on a drawing, assuming the manufacturer will finish the edges according to common practice. The problem is that the industry lacks common practice, which has been one of the major obstacles in the battle against burrs.

Companies and quality-control departments define "burr-free" differently, so what constitutes a "burr-free" edge to one company may not be to another. To some companies, it means the part has no loose materials at its edges. To others, it means there are no flaws visible to the naked eye. Others define it as an edge condition that will not cause a functional problem in the next assembly. To make matters more confusing, "burr" is also defined differently among companies. Some define a burr as absent material (a chamfer or broken edge, for example). To others it is a hump of rounded metal at an edge. Others define burrs as sharp projections at an edge.

Given the lack of industrywide standards concerning what constitutes "burr" and "burr-free," customer expectations should be met not only with face-to-face discussions but also with clearly written, detailed instructions. This is also important because customer expectations include far more than deburring. They include surface finish, rounded-edge uniformity, color, cleanliness, welding capability, and other aspects of the deburring process. See Chapter 3 for a more detailed discussion of the many issues surrounding edge requirements and standards. It presents several ways in which standards, edge requirements, and customer expectations can be documented, understood, and communicated.

Goal 11: Find Unique Approaches to Reduce Deburring Costs

Human innovation is the source of economic improvement. This applies as much to deburring as to any other industrial process. Unfortunately, hand-deburring innovations rarely make the news, unlike other high-tech innovations. Yet, unsung heroes have developed unheralded and generally unknown approaches that result in substantial savings. Later chapters (21, 22, 25, 27, 28, and 32) discuss some of these approaches. None is particularly earthshaking, but a single idea can reduce problems, delays, and waste. One innovation can generate others and each, in turn, can increase effectiveness.

Goal 12: Use Hand Deburring Where it is Most Effective

A combination of mechanized and hand deburring is often more economical than hand deburring alone, so companies must know the limitations of hand deburring (as discussed earlier in this chapter). For example, precision edge breaks are difficult to maintain consistently with hand deburring, especially when the workpieces have a number of edges that require breaking. It is almost impossible to consistently produce breaks of 0.002–0.003 in. (51–76 μm) on each portion of each edge of intricate work-pieces (Gillespie 1976). Larger breaks, typically 0.005–0.010 in. (0.13–0.25 mm), generally must be specified to ensure part acceptability. Nonetheless, at least one company has trained over 100 people to use microscopes to work to such exacting tolerances. It is time-consuming, but it can be done. Hand deburring can also be time-consuming when edge break or radii requirements are greater than 0.015 in. (0.38 mm). In the aerospace industry, for example, edge radii to 0.062 in. (1.57 mm) are sometimes specified.

KEYS TO SUCCESS

Deburring fundamentals and hand deburring goals have been discussed. Operations that use deburring should be aware of these fundamentals and goals, but if they remember nothing else, they should remember the following three keys to success.

- Minimize burr size.
- The machine and blanking process used determine burr size and deburring difficulty.
- People are integral to success.

Burr Size

Minimizing burr size is always the key to reducing edge issues. Typical burr size in many operations is 0.003 in. (76 μm) thick at the base by 0.010 in. (0.25 mm) high. Deburring a small burr requires little thought. If a burr is only 0.0001 in. thick × 0.0001 in. tall (2.5 × 2.5 μm), it can be removed in a matter of seconds from any part by any process. In contrast, if a burr is 0.005 in. thick × 0.005 in. tall (127 × 127 μm) and part tolerances are critical, carefully conceived approaches are required. Figure 1-3 illustrates how manual deburring time increased as burr thickness increased for one simple shape of a precision part.

Keep in mind, too, that dull tools significantly heat a part while it is machined. This affects burr properties, because normally small burrs become monstrous. The underlying cause of these monster burrs is poor control of machining, but dull tools increase part temperatures, which further increases ductility and, subsequently, burr thickness.

Machining and Blanking Process Effects

Typical burrs are a natural product of machining and blanking processes. They usually do not result from poor planning or engineering. Large burrs, however, may result from poor planning. Burr removal cost, in part, is an aspect of planning because it was incurred when individuals selected specific machining or blanking processes. Therefore, the cost is partially determined by the choices of other manufacturing operations.

Burr location and removal efforts are affected by the sequence in which dimensions are machined or blanked. Inattention to tool sharpness can make deburring by traditional means impossible. A number of other factors also affect burr size and removal costs, including feeds, speeds, depths of cut, cutter geometry, sequence of cutter paths, and even machine

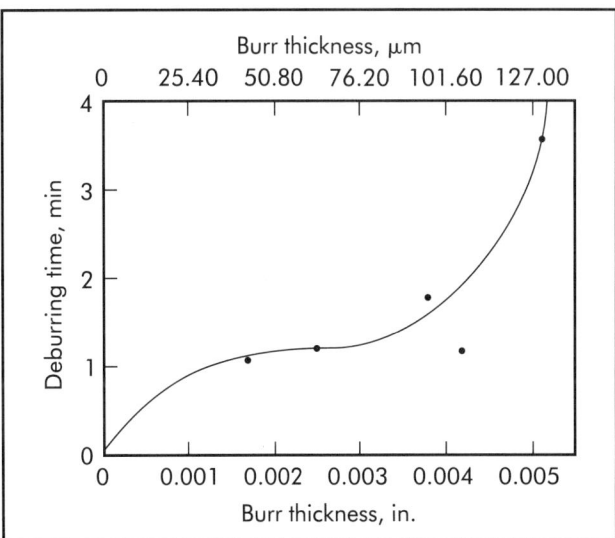

Figure 1-3. Effect of burr thickness on hand-deburring time (Gillespie 1999).

tool design and repair. (The *Deburring and Edge Finishing Handbook* provides detailed information about machining several products to reduce deburring costs. It also discusses the affect of process variables on burr size and provides many insightful pages on how to prevent burrs.) Clearly, high feed rates usually increase burr dimensions. Processes that normally employ rough and finish passes to hold tolerances also tend to result in smaller burrs, because the finish passes are taken at smaller tooth or chip loads. Process control is essential, and users must become accustomed to measuring burr size to make improvements.

The number of edges on a part and the total length of edges to be deburred on a single part may not be apparent. Both of these factors provide some measure of the difficulty of obtaining perfect edges when high precision is required. Figure 1-4 shows that a single gear tooth has 10 different line segments that must be burr-free. A miniature instrument gear that is small enough to fit under a fingernail and has 12 teeth requires 120 burr-free edges.

Figure 1-5 illustrates a simple fine-pitch 5.00 × 0.35 screw that is 0.320 in. (8.12 mm) long. It has 17 threads with 24.41 in. (620 mm) of burrs on its crests. In other words, this small part has over 2 ft (0.6 m) of thread to deburr! A burr left anywhere on its crest could jam its mating part. The burrs on a single part of this design could produce 32,632 particles if they were broken into loose particles of 0.00076 in. (19 μm) diameter or length. This would be unacceptable in precision applications, but it would not be an issue for many commercial parts.

The People Aspect of Hand Deburring

People are the key to success in any business. Chapter 32 discusses improvement through people. Improvements described in this book would be short-lived if workers functioned in dark, dusty, crowded conditions. People rise to the occasion. If treated as losers, they behave as losers. If treated as winners with valuable opinions, however, they respond in kind. This is not a philosophical assessment; it is the result of practice and an ongoing effort to learn new ways to demonstrate interest and recognition for excellence in the workplace. This book is one result of a deburring staff that was recognized for its excellence. Those workers dazzled admirals, generals, congressional representatives, and industry leaders with their abilities and performance. You and your staff can do the same.

REFERENCES

Gillespie, LaRoux K. 1976. "Deburring Capabilities for Miniature Precision Parts." Report BD X-613-1604 (July). Kansas City, MO: Bendix Corporation. (Available from National Technical Information Service [NTIS].)

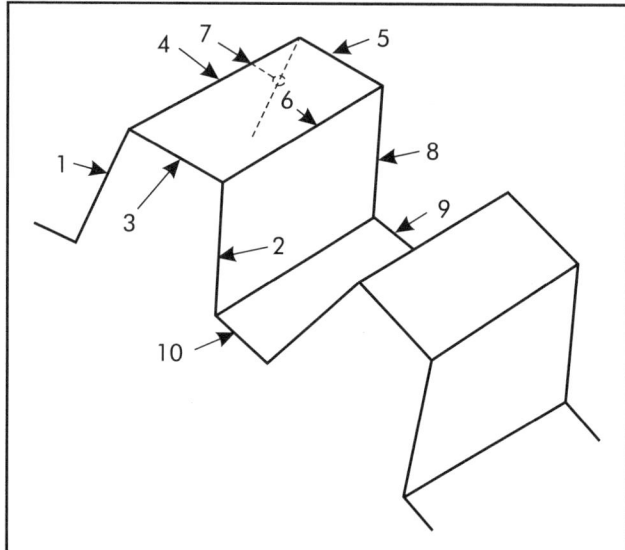

Figure 1-4. Line segments on a gear tooth (Gillespie 1999).

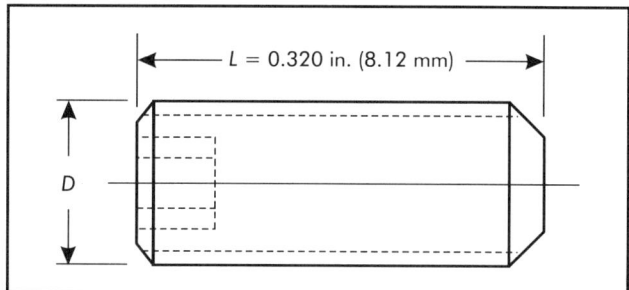

Figure 1-5. Fine-pitch screw (5.0 × 0.35 – 5h) has over 2 ft (0.6 m) of burrs on its crests (D = diameter, L = length) (Gillespie 1999).

———. 1977. "Side Effects of Deburring Processes." Technical Paper MRR77-17. Dearborn, MI: Society of Manufacturing Engineers (SME).

———. 1994. "Process Control for Burrs and Deburring." Report TR94-1. Kansas City, MO: Deburring Technology International.

———. 1999. *Deburring and Edge Finishing Handbook*. Dearborn, MI: Society of Manufacturing Engineers (SME).

Hettes, Frank J. 2000. "Advances in On-machine Deburring Using Abrasive Filament Brushes." In *Proceedings of the 6th International Conference on Precision Surface Finishing and Deburring Technology*, 96–108. St. Petersburg, Russia (September). (Available from Deburring Technology International, Kansas City, MO.)

Mahadev, Prasad S. 1995. "Nylon Abrasive-filament (NAF) Brushes—An Alternative to Deburring, Edge Radiusing, and Surface Finishing Problems." Technical Paper MRR95-02. Dearborn, MI: Society of Manufacturing Engineers (SME).

What's Right about Hand Deburring?

Manual deburring adds value to products. If it did not, burr removal would be unnecessary. Someone in your plant decided it is cheaper to machine parts than to use a burr-free process. Machines make unwanted burrs. The people who remove them add value to the part edges and product function. Unfortunately, hand deburring has had an image problem for many years, which has fed a general lack of concern for this process. Even when hand deburring is a concern, it fails to receive the attention it deserves. As this chapter shows, however, many things are "right" about hand deburring.

NEGATIVE VIEWS OF HAND DEBURRING

Traditionally, hand deburring has been considered a necessary evil that interferes with production and causes great annoyance. The following observations about deburring have been made in the past:

- it is a common problem;
- it is a new problem;
- it is a cover-up or cosmetic operation;
- it is a low-paying job;
- it is done by hand, so it cannot be too difficult;
- it is a mysterious art because little information was available in the past;
- its past contains little rationale for minimizing the problem, and
- it is considered a trivial problem by engineers and managers, so they assign the wrong people.

If we understand some of the reasons people feel this way, we can better understand how hand deburring can become an important profit center, rather than a wasteful cost.

Deburring versus Edge Finishing— A Matter of Perception

Managers and engineers may want to consider the real and perceived difference between deburring and edge finishing in their company presentations. Among some professionals, *edge finishing* is a more positive term than *deburring*. Some facilities see deburring as removing another's mistakes. Others consider it unworthy of resource or planning allocations. Wherever you go, engineers and managers still discuss deburring as a "wasted" effort, a process that does not add value, and one that should be eliminated. Yet, those same people are not heard saying that machining is a wasted effort, that holes in parts should be eliminated, or that good surface finishes are unnecessary on a sealing surface. One of the reasons hand deburring is perceived in a negative light is that its role and potential are seldom understood.

Many of today's hand deburring problems are rooted in a lack of understanding. Companies strive to reduce costs, some of which seem as if they could be easily reduced. The perception of easiness changes, however, when companies and individuals discover the absence of data on the real cost of deburring and when they realize they:

- lack understanding of actual work performed, the value of the work performed, and how hand deburring can be more effective;
- lack understanding of the industrial baseline for such work (there are an estimated 40,000 pages of information on deburring and over 800 just on hand deburring);

- lack trained resources to improve the situation (where is the expert who can heal the problem?), and
- fail to value the people doing the deburring.

Companies that do not understand these factors will fail to reduce costs as effectively as companies that truly understand the value of hand deburring and hand deburring workers.

Poor Image

One reason that hand deburring is poorly understood is that it has suffered image problems for most of its history. Hand deburring has traditionally suffered from a social stigma. Its poor image is related to perceptions of deburring workers and the simplicity of hand deburring tools.

Workers Undervalued

For most of its history, burr removal has not required any training or skill. The only concern was that a file, stone, or abrasive paper left a chamfer on a large, loosely toleranced part. Deburring was considered a mundane task performed by unskilled laborers. Because of this, deburring workers were usually undervalued, often overlooked or ignored, and sometimes wrongly perceived as unimportant by individuals who did not recognize the value of their work or understand the complexity of their assignments. Therefore, deburring needs were not seriously considered. Even today, unskilled workers can perform a large percentage of deburring. As a result, the deburring industry has remained, largely, an art lacking in engineering or research efforts. This image partly results from companies' failure to provide workers with information. For example, if you ask your staff, "How much does your labor contribute to the cost of our product on a per-part basis?" More than likely, none of your deburring staff will be able to answer that question, because no one thought to help them understand their impact on the product from either the cost or benefit perspective.

Tools are Simple

The undeservedly poor image of hand deburring is industry-wide. Often, deburring is considered a repair job or white elephant, rather than an integral, value-added part of the manufacturing process. Because the supposed "repair" job is done with simple tools and machines, hand deburring is often considered a mundane task that anyone can easily perform and master. This image is perpetuated when companies fail to support the deburring processes with teams of engineers and managers.

In fact, it is not true that anyone can easily perform and master all deburring tasks. Low-skilled workers can perform only some of today's deburring tasks. Many deburring workers are critically skilled, and many machinists cannot perform their work. The need is for workers to find ways to reduce costs and improve the product.

Lack of Support Team

Negative views of deburring influence many managers and engineers to disassociate themselves from deburring activities. Managers and engineers, like most people, are hesitant to enter an unfamiliar field of work. This is particularly true when the company pays little attention to deburring and its needs. No one wants to be associated with a department whose tasks are perceived as mundane, whose efforts are perceived as wasted, and whose workers are undervalued.

Complexity

One of the most significant, yet least appreciated, reasons deburring is shunned by both workers and engineers is that successful solutions often require more technical background on burrs and deburring than most engineers, managers, and workers possess. For example, burr formation is a plastic deformation process. A thorough understanding of burr formation practically requires a doctorate in stress analysis. Easily understood descriptions of burr formation exist. However, accurately predicting formation and defining deburring capabilities requires a technical background and considerable experience, astuteness, and cleverness. Staffs with deburring responsibilities must include professionals who possess a sophisticated understanding of material plasticity and a thorough knowledge of techniques for producing electrical, chemical, thermal, and mechanical erosion. They must interface daily with shop, management, and other engineers—people who may not appreciate or understand the complexities involved.

Other Factors

Another reason deburring is ignored is that it is often economically less significant than machining (3% versus 97% of the total spent on labor).

High turnover rates in deburring departments can further complicate efforts to implement improvements. It is also important to remember that, until recently, little useful information on deburring has been visible to the public.

TOWARD A POSITIVE VIEW OF HAND DEBURRING

Even when edge requirements and productivity became more critical, the status of deburring remained little improved. In the late 1950s, this began to change with the introduction of vibratory deburring, which heralded a decade of innovation in deburring processes. Now, if a company is serious about cost reduction, it must evaluate 109 deburring processes. Today's interest in burrs results, in part, from a number of factors, including:

- the existence of a basic deburring capability database;
- the urgent need to evaluate all contributions to productivity;
- want of quantifiable forms of deburring capabilities and needs;
- the existence of large amounts of precision or high-volume deburring, and/or
- worldwide recognition that deburring can be improved and has suffered from a lack of skilled, dependable labor.

All of these reasons have contributed to a more positive view of hand deburring.

Reasons to Use Hand Deburring

There are at least 14 reasons for using hand deburring, rather than automated approaches.

1. It is convenient.
2. It is the cheapest method.
3. It eliminates equipment costs.
4. It reduces facility costs (for equipment).
5. It eliminates the need for machine repairmen and the time lost while waiting for repairs.
6. It eliminates capital equipment energy costs.
7. It eliminates equipment waste disposal costs.
8. It is readily available.
9. It can provide extraordinary quality.
10. It is flexible (it can perform other tasks at any time).
    - It can instantly shift from gross to microscopic-quality finishing. Mechanized deburring, on the other hand, has only one quality level.
    - It can access parts too big for deburring machines.
    - It can be performed on assemblies, whereas mechanized processes damage the assembly.
11. It allows plant owners to benefit from workers' improvement ideas.
12. It reduces cycle time, in many instances, because machine availability is not an issue. (In other instances, for the same reason, it adds cycle time.)
13. It can accommodate any part material and configuration.
14. Used effectively, it eliminates the side effects of mechanized processes.

Each of these positive aspects of hand deburring will be discussed individually in the following sections, but special attention will be paid to deburring side effects and how they influence process selection.

Convenience

In job shops, "ready availability" means profit. Job delays delay income. Manual deburring, unlike mechanized, can instantly respond to any demand. (You can "start" a person in a few seconds.)

Least Expensive Process

Job-shop work demands flexibility and low cost. Mechanized systems simply cannot reach some burrs, so hand deburring is not only the only feasible way to finish the part, but also the least costly. Note, however, hand deburring is not the least expensive solution for advanced companies that daily produce thousands of identical parts.

It is the least expensive method in most developing countries, where wages are low and mechanization is expensive. Economic factors in these places, which greatly differ from those of high technology centers, make mechanization impractical.

Minimal Capital Cost

Each machine costs money. If kept busy, it can make money. If not, it loses money and can increase costs. It is taxed as an asset and it occupies space; consumes heat, light, and electricity; and requires maintenance. Moreover, it performs only one task.

New machines are more than a capital cost. They require trained operators. Training is good, but it increases cost and the higher the employee turnover rate, the higher the cost. A technical staff for some of the process planning may be lacking. One-of-a-kind machines add to the costs.

Minimal Facility Cost

The space a machine occupies costs money. The machine will also require space for spare parts, along with aisle and maintenance clearance.

Minimal Repair and Maintenance Cost

Some advanced deburring equipment requires available and skilled maintenance workers. Small shops do not want to maintain such equipment and, generally, do not have the staff to do so.

Usually Consumes Less Energy

Energy costs in the United States are among the lowest in the world and traditionally they have not greatly increased deburring costs. Some world events, however, can quickly double or triple energy costs. Therefore, energy requirements must be considered before investing in high technology. Some hand deburring processes also use energy, which contribute to cost.

Eliminates Waste Disposal Cost

Industrial waste disposal has become a major cost for some operations. Some deburring processes remove burrs rapidly, but they generate legacy wastes. The legacy may outlive the owner or stakeholder and transfer to new owners. Waste is expensive to remove and can be a clean-up quagmire. Conversely, waste is not an issue with properly managed hand deburring.

Instant Communication of Problems

Hand deburring allows workers to instantly detect and communicate problems and quickly resolve them. These benefits defy quantification but provide major benefits. Unrelated problems can be communicated immediately, including information about pits, dings, rust, broken tools left in parts, wrong sizes, and finish differences. Mechanized operations do not detect such things, so they cannot provide this level of information.

Improves Quality

"People skills" allow workers to consistently remove burrs and add quality to select areas of a part without affecting other critical areas. In other words, properly performed, manual deburring improves quality without changing part dimensions or finishes. Mass finishing, in contrast, changes all exposed edges and because of part geometry, some edges will be changed more than others.

Flexibility

People are flexible; machines are not. People can respond quickly to varying burr sizes, different materials, and nonsequential operations. They can respond differently at different times of the day. They can respond differently to different parts and specifications, and they can immediately respond to priority changes.

Innovative Solutions

Every plant using manual deburring can recount stories of how "Old Joe" in the back made a slick tool that solved the burr problem. The "Old Joes" solved burr problems, which reduced deburring costs. Joe was a specialist—he understood the real needs and knew what his tools could do.

Reduced Cycle Time

On-time delivery is crucial for most companies around the world. Time is money, and queue time is wasted money. Manual deburring allows plants to respond quickly to unusual situations. Certainly, mechanized equipment should be used, where possible, to reduce cycle time; but a capable staff should also be employed to respond quickly to sudden changes. Chapter 32 provides simple approaches to cycle time and cost reduction.

Part Configuration

Part shape and features are critical aspects of mechanized deburring. Many processes do not reach into many areas, or do so poorly. Combining mechanized and manual deburring is an effective strategy for maximizing profits. Use mechanized deburring for obvious areas and manual deburring for areas that are more difficult to reach. (The *Deburring and Edge Finishing Handbook* [Gillespie 1999] explains the major impact of part configuration on mechanized deburring effectiveness.)

SELECTING THE BEST METHOD

The challenge to manufacturing engineers is to efficiently produce the desired edge definition at every edge; therefore, they must select a method that can remove burrs quickly and at a reasonable cost, without adversely affecting part definition and function. Hand deburring is generally more efficient for small job shops because it is less likely to affect changes in part size, edge break, edge break repeatability, and surface finish.

Tables 2-1 through 2-4 provide details on the capabilities of various deburring processes. These tables are based on the best and most current data available. (Keep in mind, however, that results of additional research may necessitate changes in and rearrangement of some of the tabular material.) The information in these tables is important because process capabilities and limitations affect selection and, as the

Table 2-1. Mechanized deburring processes affect part dimensions and finishes

Process	Number of Key Variables	Typical Burr Thickness Removed (μm)	Typical Effect on Part Size (μm)	Typical Edge Break (μm)	Normal Best Repeatability of Edge Radii (μm)	Surface Finish Range (μm)	Typical Part Size (mm)	Typical Part Materials
Abrasive blasting	7–10	25–75	None	75–250	±75	1–7	up to 1–1,000	Zinc, steel, plastics; no soft metals
Abrasive flow	5–7	75	5	75–250	±38	0.4	20–500	All metals
Barrel tumbling	13–15	75	5	25–750	±25	0.2–0.8	20–300	All metals, plastics, rubber
Bonded abrasive (sanding)	8–12	25–75	2–50	75–500	—	0.8–1.6	50–500	All metals
Brushing	5–10	5–75	0–2	50–250	±50	0.4	20–500	All metals, ceramics, some plastics
Centrifugal barrel	14–15	5–75	0–12	75–500	—	0.05–0.8	1.5–200	All metals, ceramics, some plastics
Chemical	4–5	2–12	1–50	25	—	0.15–0.2	20–500	Some limit on metals
Cryogenic vibratory	15–16	—	—	—	—	—	20–200	—
Edge coining	4–5	2–12	0	—	—	—	10–100	—
Edge rolling/ burnishing/skiving	4–5	2–12	0	25–75	—	0.05–1.0	—	—
Electrochemical	5–6	7–127	0	50–250	±50	0.4	20–500	Metals other than titanium
Electropolish	5–7	2–25	5–12	25–125	±1	0.05	20–200	Many metals
Hand (manual)	20–50	5–75	0	50–250	±50	—	5–500	All metals, ceramics, plastics, rubber
Magnetic abrasive	5–10	5–25	1–12	5–40	—	0.1	50–200	Metals
Mechanized cutting	5–10	75–250	0	125–750	—	—	50–500	All metals, plastics
Robotic	5–10	75–250	0	125–750	—	—	50–500	All metals, plastics
Roll flow	10–15	25–125	1–12	75–750	±30	0.2–1.0	20–200	All metals
Spindle finishing	10–15	25–125	—	75–750	±30	0.2–1.0	20–200	All metals
Thermal energy	3–6	5–50	0–10	25–75	—	0.8–2.0	20–200	Many metals, a few plastics
Vibratory	12–15	25–125	1–12	75–750	±30	0.2–1.0	20–200	All metals, some plastics
Water jet	4–6	12–50	0	50–125	—	—	50–500	All metals, plastics

Dashes indicate a lack of published data. 1 μm = 39.37 μin.
(Gillespie 1999)

Table 2-2. Typical processes for precision deburring of external metal edges

Maximum Allowable Edge Break (μm)	Allowable Thickness Change on External Surfaces (μm)					Allowable Thickness Change Repeatability (μm)		
	1.25	2.50	25.0	125	0.5	5.0	50
2.5	—	—	—	—	—	—	—
25	—	—	AB3	AB3	—	AB3	AB3
50	—	M2, M5, EC	AB3, M2, M5, EC	AB3, M2, M5, EC	EC	AB3, EC, M2, M5	AB3, EC, M2, M5
125	EC, M1–3, M5–8	EC, M1–3, M5	A, AM, AB1, AB3, E1, M1–5, M7, T1, T2	Most	EC, M1–3, M5	A, AB3, EC, M1–5	Most
250	E1, EC, M1–8	E1, EC, M1–M10	A, M, AB1, AB3, E1, M1–10, T1, T2	All	E1, EC, M1–8	A, AB3, EC, M1–5, T2	All

Letters are codes for processes—see Table 2-4 for definitions. To use this table, select the maximum edge break that a drawing allows and then move horizontally across the row that has that edge break or the next largest value. Move until you reach a column that has an allowable stock loss (part-thickness change). The processes shown in that cell will most likely yield acceptable effects on stainless steel parts having a surface finish requirement of 32 μin. (0.8 μm) or better (1 μm = 39.37 μin.). Data are based on complete removal of burrs with an initial thickness and height of 0.003 in. (76 μm). The last three columns provide an estimate of size-change repeatability for the various processes and conditions stated. Dashes indicate that no process will produce the desired result for the specified burr size, edge break, surface finish, part material, or stock loss. In this case, the burr must be hand worked or machined to a smaller size before edge-finishing processes can provide complete burr removal.

(Gillespie 1999)

Table 2-3. Typical processes for precision deburring of intersecting holes

Maximum Allowable Edge Break (μm)	Allowable Hole Diameter Change Repeatability (μm)				Allowable Hole Diameter Change (μm)		
	1.25	2.50	25.0	125	0.5	5.0	50
2.5	—	—	—	—	—	—	—
25	—	—	—	—	—	—	—
50	EC1–2, T2	EC1–2, T2	EC1–2, T2	EC1–2, T2	—	EC1–2, T2	EC1–2, T2
125	EC1–2, T2	EC1–2, T2, M1	EC1–2, M1–2, T2, AB3, E1, M5	C1, EC1–2, EL3, M1–2, T2, AB3, E1, M5M2	EC1–2	EC1–2, T2, M1, M2	Most
250	EC1–2, M2, T2	E1, EC, M1–2	EC1–2, M1–2, T2, AB3, E1, M5	C1, EC1–2, 2, EL3, M1–2, 5, T2, AB3, E1, M5	EC1–2, M2	AB3, C1, EC1–2, EL3, M1–2, 5, T2	Most

To use this table, select the maximum edge break that a drawing allows and then move horizontally across the row that has that edge break or the next largest value. Move until you reach a column that has an allowable stock loss (hole-diameter change). The letters in that cell define the processes most likely to yield acceptable effects on stainless steel parts having a surface finish requirement of 32 μin. (0.8 μm) or better (1 μm = 39.37 μin.). The data are based on complete removal of burrs with an initial thickness and height of 0.003 in. (76 μm). Letters are codes for processes—see Table 2-4 for definitions. The last three columns provide an estimate of size-change repeatability for the various processes and conditions stated. Dashes indicate that no process will produce the desired result for the specified burr size, edge break, surface finish, part material, or stock loss. In this case, the burr must be hand worked or machined to a smaller size before edge finishing processes can provide complete burr removal.

(Gillespie 1999)

Table 2-4. Typical side effects of deburring processes

Process	Size	Finish	Stresses	Flatness	Plating	Soldering	Welding	Corrosion	Luster	Color
Abrasive Finishing (A)										
Barrel tumbling (A1)	Y	Y	Y	S	Y	Y	Y	Y	Y	Y
Vibratory finishing (A2)	Y	Y	Y	Y	Y	Y	Y	Y	Y	Y
Vibratory shaker mixer finishing (A2s)	Y	Y	Y	S	Y	Y	Y	Y	Y	Y
Tube flow through vibratory finishing (A2t)	Y	Y	Y	S	Y	Y	Y	Y	Y	Y
Roll-flow (centrifugal disc) finishing (A3)	Y	Y	Y	S	Y	Y	Y	Y	Y	Y
Centrifugal barrel finishing (A4)	Y	Y	Y	Y	Y	Y	Y	Y	Y	Y
Spindle finishing (A5)	Y	Y	Y	S	Y	Y	Y	Y	Y	Y
Fluidized bed spindle finishing (A5a)	Y	Y	Y	S	Y	Y	Y	Y	Y	Y
Recipro finishing (A6)	Y	Y	Y	S	Y	Y	Y	Y	Y	Y
Orboresonant finishing (A7)	Y	Y	Y	S	Y	Y	Y	Y	Y	Y
Flow finishing (A8)	Y	Y	Y	S	Y	Y	Y	Y	Y	Y
Cascading media (A9)	Y	Y	Y	S	Y	Y	Y	Y	Y	Y
Immersion lapping (A10)	Y	Y	Y	S	Y	Y	Y	Y	Y	Y
Screw rotor deburring (A11)	Y	Y	Y	S	Y	Y	Y	Y	Y	Y
Chemical Loose Abrasive Finishing (AC)										
Chemical barrel tumbling (AC1)	Y	Y	?	S	?	?	?	?	?	?
Chemical vibratory finishing (AC2)	Y	Y	?	S	?	?	?	?	?	?
Chemical roll-flow (centrifugal disc) finishing (AC3)	Y	Y	?	S	?	?	?	?	?	?
Chemical centrifugal barrel finishing (AC4)	Y	Y	?	S	?	?	?	?	?	?
Chemical spindle finishing (AC5)	Y	Y	?	S	?	?	?	?	?	?
Chemical fluidized bed spindle finishing (AC5a)	Y	Y	?	S	?	?	?	?	?	?
Chemical recipro finishing (AC6)	Y	Y	?	S	?	?	?	?	?	?
Chemical orboresonant finishing (AC7)	Y	Y	?	S	?	?	?	?	?	?
Chemical flow finishing (AC8)	Y	Y	?	S	?	?	?	?	?	?
Cryogenic Loose Abrasive Finishing (ACRY)										
Cryogenic barrel tumbling (ACRY1)	Y	Y	S	S	?	?	?	?	?	?
Cryogenic vibratory finishing (ACRY2)	Y	Y	S	S	?	?	?	?	?	?
Cryogenic vibratory shaker mixer finishing (ACRY2s)	Y	Y	S	S	?	?	?	?	?	?
Cryogenic roll-flow (centrifugal disc) finishing (ACRY3)	Y	Y	S	S	?	?	?	?	?	?
Cryogenic centrifugal barrel finishing (ACRY4)	Y	Y	S	S	?	?	?	?	?	?
Cryogenic spindle finishing (ACRY5)	Y	Y	S	S	?	?	?	?	?	?
Cryogenic fluidized bed spindle finishing (ACRY5a)	Y	Y	S	S	?	?	?	?	?	?
Cryogenic recipro finishing (ACRY6)	Y	Y	S	S	?	?	?	?	?	?
Cryogenic orboresonant finishing (ACRY7)	Y	Y	S	S	?	?	?	?	?	?
Cryogenic flow finishing (ACRY8)	Y	Y	S	S	?	?	?	?	?	?

Table 2-4. (continued)

Process	\multicolumn{10}{c}{Part Attributes Affected by Process}
	Size	Finish	Stresses	Flatness	Plating	Soldering	Welding	Corrosion	Luster	Color
Magnetic Loose Abrasive Finishing (AM)										
Magnetic abrasive barrel finishing (AM1)	Y	Y	?	?	?	?	?	?	?	?
Magnetic abrasive vibratory finishing (AM2)	Y	Y	?	?	?	?	?	?	?	?
Magnetic abrasive spindle finishing (AM5)	Y	Y	?	?	?	?	?	?	?	?
Magnetic abrasive cylindrical finishing (AM5a)	Y	Y	?	?	?	?	?	?	?	?
Magnetic abrasive tube-ID finishing (AM5b)	Y	Y	?	?	?	?	?	?	?	?
Magnetic abrasive ball finishing (AM5c)	Y	Y	?	?	?	?	?	?	?	?
Magnetic abrasive special shape finishing (AM5d)	Y	Y	?	?	?	?	?	?	?	?
Magnetic abrasive prismatic finishing (AM7)	Y	Y	?	?	?	?	?	?	?	?
Mixed metal fibers magnetic media (AM8)	Y	Y	?	?	?	?	?	?	?	?
Chemical Magnetic Loose Abrasive Finishing (AMC)										
Chemical magnetic abrasive barrel finishing (AMC1)	Y	Y	?	?	?	?	?	?	?	?
Chemical magnetic abrasive vibratory finishing (AMC2)	Y	Y	?	?	?	?	?	?	?	?
Chemical magnetic abrasive spindle finishing (AMC5)	Y	Y	?	?	?	?	?	?	?	?
Chemical magnetic abrasive cylindrical finishing (AMC5a)	Y	Y	?	?	?	?	?	?	?	?
Chemical magnetic abrasive tube-ID finishing (AMC5b)	Y	Y	?	?	?	?	?	?	?	?
Chemical magnetic abrasive ball finishing (AMC5c)	Y	Y	?	?	?	?	?	?	?	?
Chemical magnetic abrasive special shape finishing (AMC5d)	Y	Y	?	?	?	?	?	?	?	?
Chemical magnetic abrasive prismatic finishing (AMC7)	Y	Y	?	?	?	?	?	?	?	?
Electrochemical Loose Abrasive Finishing (EAC)										
Electrochemical barrel tumbling (EAC1)	Y	Y	?	?	?	?	?	?	?	?
Electrochemical vibratory finishing (EAC2)	Y	Y	?	?	?	?	?	?	?	?
Electrochemical roll-flow (centrifugal disc) finishing (EAC5)	Y	Y	?	?	?	?	?	?	?	?
Electrochemical centrifugal barrel finishing (EAC3)	Y	Y	?	?	?	?	?	?	?	?
Electrochemical spindle finishing (EAC4)	Y	Y	?	?	?	?	?	?	?	?
Electrochemical fluidized bed spindle finishing (EAC4a)	Y	Y	?	?	?	?	?	?	?	?
Electrochemical recipro finishing (EAC6)	Y	Y	?	?	?	?	?	?	?	?

Table 2-4. (continued)

Process	Part Attributes Affected by Process									
	Size	Finish	Stresses	Flatness	Plating	Soldering	Welding	Corrosion	Luster	Color
Electrochemical orboresonant finishing (EAC7)	Y	Y	?	?	?	?	?	?	?	?
Electrochemical flow finishing (EAC8)	Y	Y	?	?	?	?	?	?	?	?
Lapping (A12)	Y	Y	Y	Y	Y	Y	Y	Y	Y	?
Abrasive Jet (AB1)	Y	Y	Y	Y	Y	Y	Y	?	Y	Y
Cryogenic Abrasive Jet (ABC1)	?	Y	Y	S	?	?	?	?	?	?
Liquid Hone Abrasive Flow Deburring (AB2)	?	?	?	N	N	N	N	N	N	N
Abrasive Flow Finishing (AB3)	Y	Y	Y	S	?	?	?	?	?	?
Abrasive Flow Orbital Finishing (AB3o)	Y	Y	?	?	?	?	?	?	?	?
Abrasive Flow Stream Finishing (AB3s)	?	Y	?	?	?	?	?	?	?	?
Ultrasonic Abrasive Flow Finishing (AB3u)	?	Y	?	?	?	?	?	?	?	?
Ultrasonic Liquid Deburring (AU1)	Y	Y	?	?	N	?	?	?	?	?
Ultrasonic Slurry Finishing (AU2)	?	Y	?	?	?	?	?	?	?	?
Chemical Deburring (C1)	Y	Y	Y	?	Y	?	?	Y	Y	Y
EDM Deburring (E1)	N	?	?	N	?	N	N	N	N	N
Electrochemical Deburring Processes (EC)										
Electrochemical deburring (salt) (EC1)	N	N	N	N	?	N	N	Y	Y	N
Electrochemical rotary electrode (EC1R)	N	N	N	N	?	N	N	Y	Y	N
Electrochemical deburring (glycol) (EC2)	N	N	N	N	?	N	N	Y	Y	N
Electrochemical mesh deburring (EC5)	Y	?	?	?	?	?	?	Y	Y	?
Electrochemical moving electrode deburring (EC4)	?	?	?	?	?	?	?	Y	Y	?
Electrochemical brush deburring (ECM3)	?	?	?	?	?	?	?	Y	Y	?
Electrochemical orbital abrading (ECM5)	?	?	?	?	?	?	?	Y	?	?
Electrochemical nonwoven abrasive magnetic finishing (ECM6)	?	Y	?	?	?	?	?	?	?	?
Electropolish (EL3)	Y	Y	Y	S	Y	N	N	Y	Y	Y
Manual Deburring (M1)	N	N	N	S	N	N	N	N	N	N
Mechanized Cutting (M2)	N	N	N	N	N	N	N	N	N	N
Ballizing (M11)	Y	Y	Y	N	?	?	?	?	?	?
Tearing (M12)	?	N	N	N	N	N	N	N	N	N
Skiving (M2s)	N	N	N	?	N	N	N	N	N	N
Robotic Deburring (M2R)	N	N	N	N	N	N	N	N	N	N
Brushing (M3)	N	N	N	N	Y	?	?	?	?	?
Buffing (M3b)	?	Y	Y	?	Y	?	?	?	?	?
Wheel Blending (M4)	N	N	N	N	?	N	N	N	N	N
Sanding (M5)	Y	Y	Y	Y	?	?	?	?	?	?
Vibratory Conveyor Deburring (M5a)	Y	Y	Y	Y	?	?	?	?	?	?
Abrasive Chemical (Mechanochemical Polishing) (MC1)	Y	Y	?	?	?	?	?	?	?	?
Edge Rolling (M7)	N	N	N	N	N	N	N	N	N	N
Edge Peening (M13)	N	N	N	N	N	N	N	N	N	N
Edge Burnishing (M8)	N	N	N	N	N	N	N	N	N	N
Trimming (press work) (M9)	N	N	N	N	N	N	N	N	N	N

Table 2-4. (continued)

Process	Part Attributes Affected by Process									
	Size	Finish	Stresses	Flatness	Plating	Soldering	Welding	Corrosion	Luster	Color
Edge Coining (M10)	N	N	N	N	N	N	N	N	N	N
Water Jet Deburring (M6)	N	N	?	?	N	N	N	N	N	N
Torch Cut Deburring (T1)	?	?	?	Y	?	N	N	N	N	N
Plasma Deburring (T3)	?	?	?	S	N	?	?	?	?	?
Plasma Glow Deburring (T5)	N	N	N	N	?	?	?	?	?	?
Thermal Energy Method (T2)	?	?	?	?	?	?	?	?	?	?
Hot Wire Deburring (T4)	N	N	N	N	N	N	N	N	N	N
Hot Blade Deflashing (T4b)	N	N	N	N	N	N	N	N	N	N
Hot Tool Deflashing (MT4c)	N	N	N	N	N	N	N	N	N	N
Resistance Heat Deburring (T6)	?	?	?	?	?	?	?	?	?	?
Laser Deburring (T7)	N	N	N	?	?	N	N	N	N	N
Chlorine Deburring (TC1)	Y	Y	Y	?	?	?	?	Y	?	?

Y = yes
N = no
S = suspected
? = lack of data or experience

Data reflect the most probable results of normal industry use. Some effects are positive, whereas others should be avoided. All processes affect edges, but this table shows which processes affect part surfaces not necessarily near the edges being deburred. Electrochemical deburring will affect surfaces next to the edge. Question marks indicate lack of data or experience. An "S" indicates that the part is probably affected, or it is suspected that it would be affected. Many processes can be modified to prevent the side effects listed, but the operator must be aware of the issue to prevent it. For example, vibratory deburring in a water solution can cause parts to rust, but adding a rust inhibitor can prevent this side effect. Thus, one of the basic materials in the process (water) would affect corrosion unless it was considered in the process steps.

tables show, hand deburring is often the best selection.

Table 2-1 shows some of the common side effects of deburring. Tables 2-2 and 2-3 focus on precision deburring—Table 2-2 on external metal edges and Table 2-3 on intersecting holes. Table 2-4 provides a more extensive list of processes than Table 2-1. It lists and assigns a code letter to each known deburring process and identifies its typical side effects. The codes are used in tables 2-2 and 2-3 to identify applicable deburring processes.

Limitations

Table 2-1 lists a number of deburring processes and how they affect the dimension and finish of typical parts. Careful hand deburring, as the table shows, generates none of the listed side effects produced by other processes. Mechanized deburring can also affect a number of other part and part-related attributes, including cleanliness, flatness, plating, soldering, welding, residual stresses, surface imperfections, corrosion rates, luster, and color (see Table 2-4). It is important to know which processes generate which side effects, because this information will help you determine which process will produce the most desirable and efficient results.

The data of tables 2-1 through 2-4 are based on the principle that effective deburring requires knowledge of four factors:

- burr properties—thickness, height, and hardness relative to part hardness;
- allowable changes in part dimension resulting from the deburring operation;
- edge radius requirement, and
- final surface finish requirement, as well as knowledge of the part's initial surface finish.

For precision parts, knowledge of these factors is not just desirable, it is essential to selecting the appropriate deburring process. Other factors are also important, including burr location, part geometry, and part material, but these may not be critical to selection. The simple cube in Figure 2-1 illustrates these basic requirements.

Burr Size

Tables 2-2 and 2-3 show that a typical burr, by many precision processes, measures 0.002–0.003 in. (51–76 µm) thick. The significance of this information can be further appreciated by scanning the data on typical burr sizes in later chapters and elsewhere (Gillespie 1972, 1978). No process appears capable of

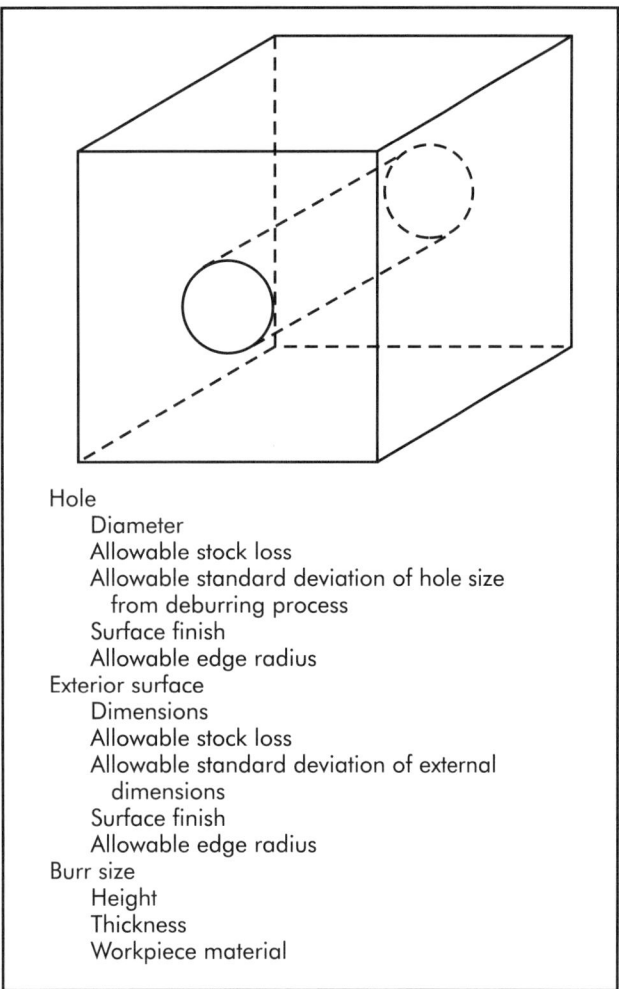

Figure 2-1. Considerations for selecting a deburring process for a workpiece with a hole (Gillespie 1999).

dii. As parts become smaller, however, some processes, such as centrifugal barrel finishing, begin to lose their effectiveness. For example, it may take twice as long to remove a burr from a 0.02 in. (0.5 mm) pin that is 0.125 in. (3.18 mm) long than it would to remove a burr of the same size from a 0.125 in. (3.18 mm) pin of the same length. If threaded parts are small and have close tolerances, they can present problems to many deburring processes.

External Edges

Table 2-2 focuses on processes used for precision deburring of external edges. If allowable stock loss must be kept below 0.0001 in. (2.5 μm), there is little margin for error; however, selecting a deburring process for a complex part—one with many edges, surfaces, and burr sizes—may require considerable thought. It is desirable to have the actual data available, but previous experience usually intuitively provides the necessary facts for selecting the best process.

Each of the processes in Table 2-2 can be used to deburr external edges while maintaining maximum allowable edge break, allowable stock loss, and a given stock-loss repeatability. The data in this table are valid, however, only for 0.003 in. thick × 0.003 in. high (76 × 76 μm) burrs on a stainless steel workpiece requiring a surface finish of 32 μin. (0.8 μm) or better. An initial surface finish of 32 μin. (0.8 μm) is assumed, although some deburring processes would improve a rougher finish.

Holes

Table 2-3 lists typical processes for precision deburring of holes. To accurately select a deburring process, all the following information must be known: approximate hole size, amount of hole stock-loss resulting from deburring, repeatability of stock loss, initial surface finish, final surface finish requirements, and edge radius requirements at the entrance and exit of the hole.

The size of the hole is important because some deburring processes work well on miniature holes, whereas others work well on large holes. For precision holes, you must know both the average and the repeatability of stock loss. Although the holes can be drilled smaller to accommodate stock loss from deburring, this action is pointless if the stock removal cannot be repeated.

When two holes intersect, the angles between the hole surfaces vary considerably around the intersection. Because edge angle affects many, if not all, deburring processes, deburring tolerances are also affected. For the burr size and surface finish cited in

removing this typical size burr while maintaining stock losses of less than 0.0005 in. (12.7 μm) and guaranteeing final edge radii of 0.002 in. (51 μm) or less. If edge radii of 0.005 in. (127 μm) are allowable, several processes can be used. However, when parts require precision small radii and a low stock-loss condition, only two possibilities exist:

- develop nonstandard variations of the deburring processes, or
- make smaller burrs.

When burr size is reduced to a thickness and height of 0.001 in. (25 μm), several processes can produce the desired result (Gillespie 1977). Precision parts from automatic screw machines typically have burrs of this size. As a result, they generally can be deburred by most processes without excessive loss in stock or ra-

Table 2-3, no process will maintain a 0.002 in. (51 μm) radius when allowable hole size repeatability must be held to 19.7 μin. (0.5 μm). In this instance, process inability is caused by the varying angles of intersection, as well as natural process capabilities.

SELECTION APPROACH

Table 2-4 provides an extensive list of processes and how they affect part attributes. This is the first table you will use in selecting the most appropriate deburring process. The following steps present a rational approach to selection (other variations also work).

1. Determine the typical burr sizes produced today or in the past.
2. Determine drawing edge requirements.
3. Using Table 2-4, cross out processes whose limitations make them unusable.
4. Using tables 2-1 through 2-4, determine which processes meet part drawing requirements.
5. With the list from step 4, the user is ready to look in much more depth at processes that have real merit. (*Deburring and Edge Finishing Handbook* [Gillespie 1999] is one source for that more in-depth information, along with discussions with equipment vendors.)
6. Estimate production rates and deburring costs of the processes remaining from step 5.
7. Select the best process from step 6, or continue obtaining more details from manufacturers, distributors, job shops, or other sources, such as those listed in the reference and bibliography sections of this book.
8. Verify the selection with a staff member who has previous knowledge.
9. Implement the best process.
10. Fine tune the experimental results (Gillespie 1994).

Table 2-4 data reflect the most probable results in normal industry use. Some effects are positive, whereas others, normally, should be avoided. All processes affect edges, but this table was created to show which processes affect part surfaces that are not necessarily near edges that are being deburred. Electrochemical Deburring (ECD) will affect surfaces next to the edge. Question marks indicate lack of data or experience. An "S" indicates that the part is probably affected, or it is suspected. Many processes can be modified to prevent the side effects listed, but the operator must be aware of the issue to prevent it. For example, vibratory deburring in a water solution can cause parts to rust, but adding a rust inhibitor can prevent this side effect. Thus, one of the basic materials in the process (water) would affect corrosion, unless it was considered in the process steps.

When using tables 2-1, 2-2, and 2-3, note that the data are for typical hand-sized and smaller parts, but the processes can be used for larger parts, nonprecision parts, and bigger burrs. As shown in tables 2-2 and 2-3, abnormal burr size, sharp edges, and near-zero part size severely limit the number of possible deburring processes. (Another title, *Advances in Deburring* [Gillespie 1978], contains almost 100 pages of discussion on the side effects and capabilities of deburring.)

Users can simplify the tables to fit their own applications and product needs. Tables such as these not only document capabilities but also provide a training aid for new workers (engineers, managers, and production workers alike). They also help in preparing finishing costs.

THREE COMPANIES, THREE EXAMPLES

As this chapter has shown, many things are "right" about hand deburring. The following true vignette can best illustrate the positive effects of hand deburring. Once upon a time, three companies machined and assembled metal parts. Important people visited each plant to see their impressive products and nice machines.

Company A

Company A was proud of its new machines that held tolerances of ±0.000050 in. (50 millionths of an inch [1.27 μm]) on some parts. Its assembly areas were clean and the parts worked exactly as they were designed. Quality was excellent and costs were reasonable. Cost could have been reduced a little if "we could just get rid of deburring." The flexible machining system (FMS) was exciting, unmanned, and sophisticated. The floor was clean, and employees were busy and smiling.

The burr bench was in a distant corner where no one ever stopped. Why stop there? The area contained files and sandpaper and was often dirty. The workers did not say much; perhaps because no one told them much about the product—after all, they were "just" deburring workers. What could any of these workers have to say that would be of interest to a visitor? Company management hated to admit that it still used hand deburring because all the modern plants, they thought, had switched to mechanical deburring processes. Maybe they had even advanced to a point where deburring was unnecessary. Company A was sure that other plants had found a better method, but there just was not enough time to investigate these methods.

There was not much money to go around, either, and hand deburring received none, because it was just deburring.

Company B

Company B was located in another country and was famous for its high automation and lights-out factories. Shuttles carried big parts from one machine to another without human assistance. It truly was a marvel of technology. Its magic was unforgettable. Hidden away in a back room, however, a man with a file was manually finishing edges that were inappropriate for automated equipment, either because of burr properties or equipment limitations. Company B—a showcase of automation—clearly understood the value of manual deburring. Unfortunately, this highly publicized plant never credited hand deburring in its press releases or interviews. The company appeared embarrassed to admit reliance on a method that seemed outdated in its high-tech plant. Its attitude was rooted in the negative views that have long been associated with hand deburring.

Company C

Company C was in the same town as Company A. When important visitors came to this plant, they first saw the deburring area—the showplace of the plant—because it demonstrated the company's standards for high quality and excellence. The clean, modern work area had modern hand-deburring tools, such as dental motors, precision tools, and fine microscopes. In fact, deburring received better microscopes than inspection, so workers could detect and remove minute burrs and particles, rather than have inspection reject parts for features that manufacturing could not see. The talented workers were well-paid, trained, experienced, and proud of their work. It was clear that complex parts would not work without them. It was impressive. Hand deburring at Company C was so impressive that generals, admirals, congressional representatives, CEOs from giant corporations, and other dignitaries visited it.

Views

How do these companies differ? Clearly, companies A and B could boast of automated facilities that were showplaces with standard production. Both companies had a vision that demanded automated rather than manual labor. Companies B and C held different visions of what could be done and how to treat employees. Company C encouraged the best from its employees, and it valued and recognized their contribution. Company B, however, seemed as if it would rather sweep its "back-room" employees under the carpet. Company C's job-shop environment allowed it to produce extraordinary edge quality. Company B, on the other hand, considered deburring "just another job" that was unworthy of recognition or visibility.

It is important to remember that burr problems are entwined with many plant choices and decisions, which means that, in practice, making improvements in hand deburring can be complex. You may never optimize deburring, but you can greatly improve the efficiency of many of these operations. The first step in doing this is to understand why companies have traditionally treated it as a castoff. Once companies understand the issues and images that affect hand deburring, they can make it more efficient with many small steps. It is not easy, but as Company C illustrates, it can be done, and in ways that will make your company a showplace.

REFERENCES

Gillespie, LaRoux K. 1976. *Deburring Capabilities and Limitations*. Dearborn, MI: Society of Manufacturing Engineers (SME).

———. 1977. "Side Effects of Deburring Processes." Technical Paper MRR77-17. Dearborn, MI: Society of Manufacturing Engineers (SME).

———. 1978. *Advances in Deburring*. Dearborn, MI: Society of Manufacturing Engineers (SME).

———. 1994. "Process Control for Burrs and Deburring." Report TR94-1. Kansas City, MO: Deburring Technology International.

———. 1999. *Deburring and Edge Finishing Handbook*. Dearborn, MI: Society of Manufacturing Engineers (SME).

BIBLIOGRAPHY

Alting, L., and F. Nygaard. 1980. "Systematic Generation of Deburring Processes." Technical Paper MR80-339. Dearborn, MI: Society of Manufacturing Engineers (SME).

Datsko, Joseph. 1966. *Material Properties and Manufacturing Processes*. New York: John Wiley and Sons.

Gillespie, LaRoux K. 1976. "Deburring Capabilities for Miniature Precision Parts." Report BDX-613-1604 (July). Kansas City, MO: Bendix Corporation. (Available from NTIS.)

———. 1977. "Machinability of Metals as Related to Precision Miniature Parts." Report BDX-613-1723. Kansas City, MO: Bendix Corporation. (Available from NTIS.)

———. 1982. "The Management of Deburring." Technical Paper MRR82-04. Dearborn, MI: Society of Manufacturing Engineers (SME).

Ioi, Toshihiro, Hisamine Kobayashi, and Masahisa Matsunaga. 1981. "Computer-aided Selection of Deburring Methods." Technical paper MR81-389. Dearborn, MI: Society of Manufacturing Engineers (SME).

Kittredge, John B. 1985. "Computer-aided Mass Finishing Process Development." Technical Paper MR85-836. Dearborn, MI: Society of Manufacturing Engineers (SME).

Narayanaswami, Rangarajan, and David A. Dornfeld. 1994. "Design and Process Planning Strategies for Burr Minimization and Deburring." In *Transactions of NAMRI/SME,* vol. 22, 313–22. Dearborn, MI: Society of Manufacturing Engineers (SME).

Takazawa, K., and Y. Kato. 1977. Edge Quality Classification and Grade." In *Proceedings of BEST-J,* vol. 3, no. 17. Tokyo, Japan.

Wick, Charles, and Raymond Veilleux, eds. 1982. *Tool and Manufacturing Engineers Handbook,* 4th ed., vol. 3, *Materials, Finishing, and Coating.* Dearborn, MI: Society of Manufacturing Engineers (SME).

Edge Requirements and Standards 3

Webster's Collegiate Dictionary defines *burr* as "a thin ridge or area of roughness produced in cutting or shaping metal." This definition provides a visual image that is adequate for most people's understanding of a burr. Unfortunately, it is an inadequate definition for engineers and other deburring professionals charged with removing them from parts. How do they define a burr? This presents a problem—the industry lacks a uniform standard for burrs and edge finishing because its definitions of burr and burr-free are numerous and varied.

As discussed in Chapter 1, burr and burr-free are defined differently among companies. To many, a burr-free part has no loose material at its edges. To some, it has no visible imperfections. To others, burr-free refers to an edge that will not cause problems in the next assembly. Companies also define a burr differently. Some engineers call missing material a burr. Others define a burr as a hump of rounded metal at an edge; some do not.

The idea that parts must be burr-free is meaningless without reference to inspection approaches—which may be as numerous as the definitions of burr and burr-free. For example, some inspection departments identify burrs as the microscopic slivers left on surfaces, rather than on edges. One company reportedly uses 400× magnification to check for burrs. Another may use 10×. Still another may use 40× to 100×, while another uses no magnification at all. If drawings or inspection procedures do not identify magnification levels, it does not necessarily mean that items unseen by the unaided eye are acceptable. Legal implications could arise if particles, regardless of size, remain on the part. Such issues can be a matter for courts to decide when companies lack explicit written standards.

Diverse definitions and approaches all contribute to the lack and necessity of edge requirements and standards, which is the focus of this chapter.

EDGE REQUIREMENTS: ISSUES AND PROBLEMS

Deburring departments, in general, exist to remove burrs and produce rounded edges on parts. Figure 3-1 illustrates a burr-laden edge and the desired and expected rounded edge that results from deburring. Table 3-1 shows there are many levels, or variations, of deburring. These many variations reflect the need for edge standards and some agreement about the definition of burr-free. The levels in Table 3-1 are real examples of product needs and illustrate why simply writing "burr-free" on a drawing inadequately reflects product needs. Standards provide a means of expressing which level of quality users and producers expect to receive or produce.

Clear communication of the desired edge condition is a low-cost method of reducing deburring and edge finishing costs. It is an easy thing to do, but it has been a major company omission in the battle against the burr. As customer expectations increase, however, so does the need for standards. Furthermore, costly scrap and functional problems arise from three aspects of the standards and edge requirements issue, which are discussed in the following sections.

Problem 1: Diverse Definitions of "Burr"

Because of varied and inexact definitions of burr, many parts are rejected that should be passed, and many parts are passed that should be rejected. These unwarranted actions can occur when inspectors, floor supervisors, or engineers each have a different notion of what

Hand Deburring: Increasing Shop Productivity

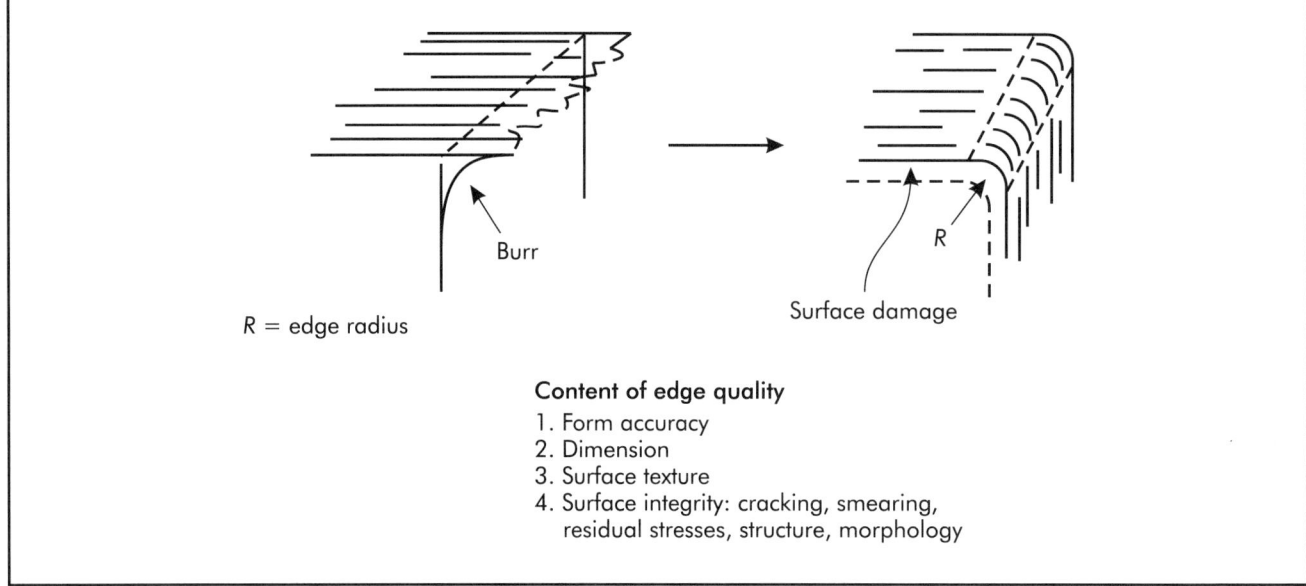

Figure 3-1. Definition of edge conditions (Takazawa and Kato 1997).

constitutes a burr and, thus, argue about whether the small speck on a part is a burr or raised material. The cost of either action is enormous, and virtually all plants have paid the price in lost time, unnecessary rework, and confusion.

Problem 2: Lack of Understanding

Failure to understand and meet real edge needs can cause overzealous deburring, which can waste tens of millions of dollars each year. The result can be scratches, cuts, tears, product failures, assembly line stoppages, excessive scrap, cycle-time delays, and frustration.

Problem 3: Oral Standards Change with Time and Observer

Even if the specifications do not change, the implementation of standards may change daily, which creates unnecessary expenses and delays. The inconsistent application of standards causes questions, discussions, or arguments among engineers and supervisors about whether a part detail is or is not a burr and how to handle it: "Is it a 'normal' burr?" "Is it bigger than yesterday's burr, or is it the same size?" "If we accepted the part last week, why is it unacceptable today?" How does this happen? When inspectors, engineers, or manufacturing supervisors resign or transfer to other departments, their knowledge goes with them. Over time, memories of some long-approved policies fade. Written standards, when consistently applied, prevent the confusion and delays that result from ambiguous oral standards.

EDGE STANDARDS

In all fields, companies create standards to ease work. Standards help ensure a common level of performance and understanding. It is important to remember, however, that standards do not necessarily restrict users to a single solution. Many standards define multiple allowable levels or states, which testifies to the tolerance of a wide range of alternatives when a single standard provides too little breadth.

Reasons to Adopt Edge Standards

The reasons to adopt edge standards parallel the three problems outlined in the previous section. They also include the increasing demands of the international marketplace. There are at least three reasons to adopt edge standards:

1. Effective plantwide communication—most manufacturing and design personnel have encountered instances in which their assumptions about edge quality differed markedly from those of other personnel. Such instances stress the need for a common language among plants, within plants, and among individuals.
2. Effective drawings—how do you interpret drawings that contain only a notation such as "deburr" or "burr-free"? Do they mean that a burr is invisible to the naked eye or under magnification? If a plant uses magnification to check some parts, is it legally required to do the same for all parts? Engineers often respond differently than lawyers to this question. The occurrence

Table 3-1. Overview of edge quality requirements (Takazawa and Kato 1997)

Class	Edge Grade	Drawing Definition	Target Radius Tolerance in.	Target Radius Tolerance μm or mm	Qualitative Evaluation	Quantitative Evaluation	Typical Application
E0	Exceptional	0.0002R	0.0000004–0.000008	0.01–0.02 μm		Interference microscope, scanning electron microscope (SEM)	Diamond microtome knife edge
E1	High quality	0.002R	0.000012–0.0002	0.3–5 μm	Cuts paper	Taper section metallography universal tool microscope, profile measuring machine, light section microscope, surface roughness tester	Edge of cutting tools, edge of dies
E2	Sharp	0.02R	0.000031–0.0012	8–30 μm	Cuts fingernail	Same as above	Hydraulic orifice edge
E3	Rounded	0.2R or chamfer	0.0031–0.012	0.08–0.30 mm	Will not cut finger	Stereo microscope, replica measurements	Mechanical parts, gyro pivots, piston rings, hydraulic spools
E4	Chamfered	0.5R or chamfer	0.015–0.024	0.4–0.6 mm	Naked eye, magnifying glass	Optical comparator	Mechanical parts
E5	Dull				No cut fingers	UL sharpness gage	Some automotive parts

of such questions would shrink if a national standard could be referenced, or even a comprehensive company standard.
3. Effective international communication—with the trend toward international manufacturing liaisons, combined with electronic communication and the need for rapid decisions, standards become even more important to meet deliveries, quality, and cost expectations. A standard would help diverse cultures with different languages and practices communicate more effectively, because many countries interpret words differently.

Existing Edge Standards

Every plant has at least one standard for burrs and edge finishing. As noted earlier, some plants may rely on oral, unwritten, standards. In these plants, standards can vary daily and among individuals, but there is some general understanding of allowable edge conditions. Other plants have comprehensive written standards. Large companies are more likely than small companies to rely on detailed written standards because consistency is important and written standardization is the only method of providing consistency throughout the plant. A few companies may have such consistent process controls that the control provides the consistency.

Often, standards are not exciting enough for organizations to continually update. Leaders come and go in the field of standards and interests change. As a result, some standards disappear with time. Others change and are improved. One standard disappeared due to lack of use and problems during implementation, and emerged later with great vigor. The following paragraph lists some of the older standards that are useful for companies wishing to establish their own. These provide a pattern and a rationale for company choice.

The ideal standard is one that everyone understands—a single document that every worker is comfortable using. Currently, however, industry remains diverse enough that many companies need only simple standards for discussions of edge conditions. Existing specialty standards in the United States include one by the Precision Metalforming Association (PMA) for allowable burrs on stamped parts (PMA 1976). The Washer Division of this association has yet another standard for stamped washers (PMA 1977). The American Iron and Steel Institute (AISI) defines allowable edge conditions of sheared sheets (ASTM 1993). The American Die Casting Institute (ADCI) has two standards for allowable flash on die-cast parts (ADCI 1955a, b). Underwriter's Laboratories (UL 1973, 1976) and the Consumer Product Safety Commission (Consumer Product Safety Commission, 1975, 1976) also define allowable edge sharpness. The American Society for Testing and Materials (ASTM) defines powder-coated edge quality (ASTM 1991). The National Institute for Science and Technology (formerly the National Bureau of Standards) has also proposed a standard for detecting hazardous edges (Sorrells and Berger 1974). (*Deburring Technology for Improved Manufacturing* [Gillespie 1981] summarizes each of the listed standards.) Need, however, far exceeds the listed standards.

The Worldwide Burr Technology Committee (WBTC), a group of leading deburring authorities from the major industrial countries, provides seven different standards for use by small and large shops around the world. The simplest standard is two pages long and is suitable for small shops (Gillespie 1978; 1995; 1996; Gillespie 1996a, b, c, d). The German DIN 6784 standard (DIN 1982), which is used by some companies around the world, also defines edge conditions.

Requirements for Good Edge Standards

A standard must include the following information:

- a clear definition of when the standard applies;
- a clear definition of how to specify the standard;
- a clear narration of the standard;
- a business reason for the specifications within the standard;
- industry awareness of the standard;
- periodic review;
- a mechanism for incorporating user comments, and
- unambiguous language and terminology.

In the administration of national standards, some countries require the governing agency that generates the standard to ensure that people affected by the standards have a valid voice in their establishment.

Content of Good Edge Standards

At least three levels, or forms, of standards are required by most design and manufacturing firms. With them, a company can ensure complete conformity to edge expectations; communicate clearly with customers, vendors, and in-house staff; and consistently guarantee conformity, despite changes in personnel. The three levels include:

- design specifications,
- manufacturing specifications, and
- inspection specifications and practices.

Design Specifications

Companies can specify design requirements on notes, icons, or textual references to other documents.

Drawings should include all requirements for the part or its assembly, both functionally and aesthetically, including clear expectations of edge configuration. Unfortunately, current U.S. standards for drawings inadequately define edge quality needs.

Drafting personnel and designers typically focus on design and function. Standards exist for making drawings that visually or textually detail edge requirements in clear, unambiguous language. Special symbols for edges are either included or referenced in the standard for making drawings.

Manufacturing Specifications

Manufacturers transform drawing requirements into real products by one or more processes, each of which can produce different edge characteristics. To convey clear instructions to the machine operators, various forms of further instructions and specifications are required. The manufacturing specification for adequate edge quality may state: "Blast surfaces and edges with 600 mesh aluminum oxide to remove burrs and loose particles." Some plants use even more detail for edge finishing expectations. Typically, several illustrations are included so operators clearly understand the needs of each edge. Manufacturing specifications differ from drawing specifications because operators require more detail than drawings contain; in some instances, they require different information.

Operators sometimes produce edges that later operations machine away. Drawings, however, do not address those temporary edges, which may be important to those who manufacture the parts. Thus, manufacturing standards define how to produce the expected edge standards. Subsequent operations may remove edges, so the staff that is responsible for manufacturing sequences and operations must recognize not only final but intermediate edge needs. Edge standards, specified and applied, may provide several quality levels and may be higher than the drawing specifies and the plant requires. Manufacturing edge standards include the following: process routing instructions, manufacturing specifications (departmental, product-related, or plantwide procedures), edge-finishing machine operation instructions, and machine calibration instructions.

Inspection Specifications and Practices

Inspection standards answer such questions as: "Exactly how will the inspector check for the presence or absence of burrs?" One company might use the naked eye. Another may use a number two pencil to probe edges. Some will paint part edges to illustrate areas to check more closely. Each approach can produce either the same or widely differing results. A company may produce and inspect edges at a higher quality level than that defined in drawings to set itself apart as a precision facility capable of the highest quality. In some cases, a unique device, such as a borescope, must monitor edges. Inspection standards or specifications help ensure compliance of both the drawing intent and company policy.

Inspection standards include instructions for inspectors. They describe what to inspect, acceptable quality levels, how to inspect, tools and equipment to use, calibration instructions, and record-keeping details. For example, there are over 40 different ways to measure burrs and their existence. Different edges, different products, and different materials may dictate using several of these approaches.

SAMPLE STANDARD FOR EDGES AND BURRS

Companies can use the following sample standard as a framework for integrating their respective edge and burr requirements. The Worldwide Burr Technology Committee's Standard (Gillespie 1998) provides the basis of the standard. It is compatible with ISO 9000 requirements for clear definition of expectations. Table 3-2 provides an overview of the standard. Numbered sections divide it: Section 01.01 is subordinate to section 01, etc. Companies can reproduce this standard as is, or they can copy individual sections as needed into an existing standard. Companies may also modify the title, section numbers and titles, and section details to suit their respective needs.

Section 01. Corporate Quality Expectations

This company shall consider the effects of burrs and edges on the products it designs and produces. This standard shall provide appropriate actions and controls to detect burrs and finish edges. Product designs shall identify edge requirements. Technical staff shall be aware of edge requirements.

Section 01.01. Design Requirements

The company's quality expectations motivate design requirements for edges and burrs. To ensure compliance with these requirements, the company, herein, provides a comprehensive system, or standard, to define edge conditions on product drawings. All new and existing products shall conform to this requirement. Manufacturing

Table 3-2. Outline of burr standards using an ISO 9000 approach

Standard No.	Directed at the Corporate Level	Directed at Design Engineering	Directed at Manufacturing Engineering	Directed at the Shop Floor
01	Quality expectations			
01.01		Design requirements		
01.01.01			Manufacturing requirements	
01.01.01.01				Work instruction details
01.01.01.02				Machine operation details
01.01.01.03				Training details
01.01.01.04				Calibration details
01.01.01.05				Documentation details
01.01.02			Inspection requirements	
01.01.02.01				Inspection instruction details
01.01.02.02				Inspection techniques and equipment details
01.01.02.03				Training details
01.01.02.04				Calibration details
01.01.02.05				Documentation details
01.01.03			Applied edge and burr research requirements	
01.02		Microscopic and scientific considerations		

and inspection practices will ensure compliance of design requirements.

Definitions of burr and edge terms are included at the end of this standard. (Definitions appear in the glossary at the end of this chapter. Companies should copy this glossary and append it to the standard.)

Edge-finish Levels

Edge finishing has seven levels (A–G). When a design drawing or purchase document specifies an edge-finishing level for a part or assembly, the edge quality must conform to the requirements of that level. The specified level applies to the entire part, unless the drawing or document states otherwise. If a drawing or document does not specify a level, it implies "deburring is not required." The following list details the levels and their respective requirements.

A. **Deburring not required.** If a drawing, purchase order, or specification fails to specify edge-finish requirements, this level is implied. All edges can remain as is; that is, as produced by the sequence of production processes. Burrs or similar protrusions produced at edges may cause some dimensions to fall outside their normal limits. Producers may remove burrs and finish edges if they desire.

B. **Remove sharp edges.** Edges defined by this level will be smoothed to the extent that they will not cut hands, electrical wires, or mating parts. Burrs may remain on the product; they may be beaten over, flattened, completely removed, or rounded. Any remaining material at an edge shall not cause the material dimensions to exceed drawing limits.

C. **Remove visible burrs.** Projections visible to the naked eye are not permitted beyond the plane of adjacent surfaces. Small projections may exist at an edge, if they cannot be detected by the naked eye. Remaining material should not cause dimensions to exceed drawing specifications. This level requires removal of sharp edges to the extent that they cannot cut hands, wiring cables, or mating parts.

D. Remove all burrs visible at __ × magnification. Projections visible at the specified level of magnification (user inserts magnification level on the drawing) are not allowed beyond the normal plane of adjacent surfaces. Small projections may exist at an edge if they cannot be detected by the specified power of magnification. Remaining material should not cause dimensions to exceed drawing specifications. This level allows for inspection by any quality of optical instrument and in any form of lighting. It is recognized that the quality of the optic and the form of lighting can greatly affect detection of minute particles and burrs. This level of deburring does not allow tactile and other non-optical forms of inspection, and it requires removal of sharp edges to the extent that they cannot cut hands, wiring cables, or mating parts.

E. Break edges __ × __ in. (__ × __ mm) minimum. Edges shall be chamfered, blunted, or smoothed so that no material falls above a chamfer of the indicated minimum dimensions (user must state on drawing the values for the indicated level). Small burrs may remain on the edges of the chamfers, and some raised material may exist near the edges. Any material left at the edges shall not cause product dimensions to fall outside their tolerances.

F. Round edges __ to __ in. (__ to __ mm) radius. Edges shall have a curvature falling within the indicated limits. Chamfers are not acceptable (user must state on drawing the values for the indicated level). Projections at the edge, regardless of radius, are not allowed when viewed by the naked eye.

G. Do not deburr. Edges shall be left as produced by the sequence that produced them. This statement, in contrast to level A, explicitly prohibits deburring.

Edge-finish Specifications

Normal practice requires mention of this standard on designs or other specifications. For example, the following statement adequately defines expectations: "Break all edges 0.004–0.010 in. (0.10–0.25 mm) as specified in WBTC STD-14, 1998 draft." Designers may also list the standard from the following list in the "bill of materials," and note it on the face of the drawing, "Note 1. Break all edges 0.004–0.010 in. (0.10–0.25 mm)." It is unnecessary to use a level identifier (A–G) on the drawings. (Company inserts its own standard number here, in place of the WBTC reference, if it has one.)

Specification	Title
WBTC STD-14.1998	The integrated international standard for burrs and edge finishing
WBTC STD-01.1996	Burr and edge terminology: definitions

Other methods of specifying edge-finishing levels are acceptable, as long as the intent of the requirement is clear.

Components requiring multiple edge-finishing levels should clearly indicate the requirement that corresponds to each edge. An explicit direction with arrows drawn to the edges is the most common approach, although other approaches are acceptable.

Figures 3-2 through 3-5 illustrate common burr-related notes when deviation from the standard note is required. These notes also illustrate approaches to identify requirements unique to specific edges.

Inadequate notes. Avoid inexact specifications for parts made by outside vendors. For parts made within the plant, the following types of notes may do: "small burr satisfactory," or

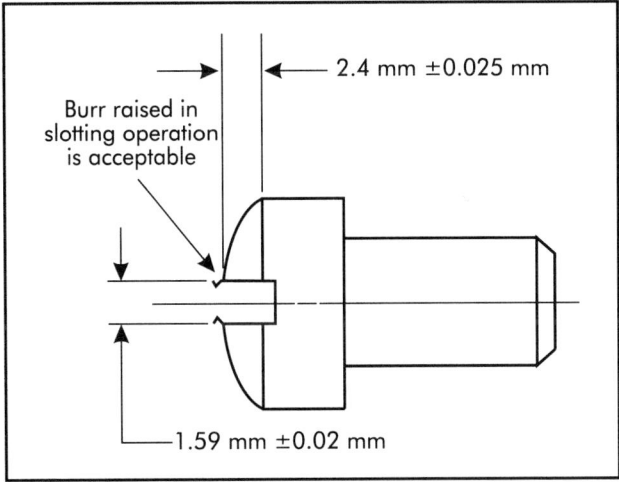

Figure 3-2. Example of note for slotting burr (Precision Machined Products Association 1955).

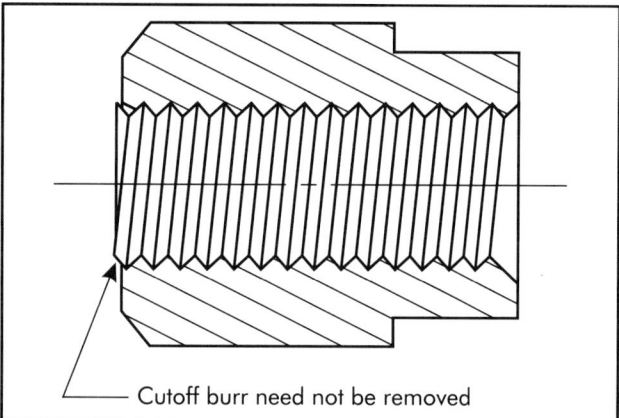

Figure 3-3. Example of note for cutoff burr (Precision Machined Products Association 1955).

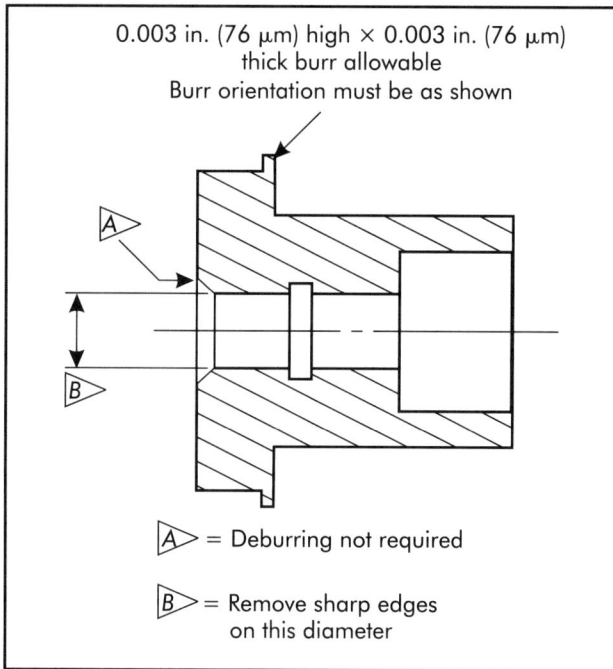

Figure 3-4. Allowable burr size on one part (Precision Machined Products Association 1955).

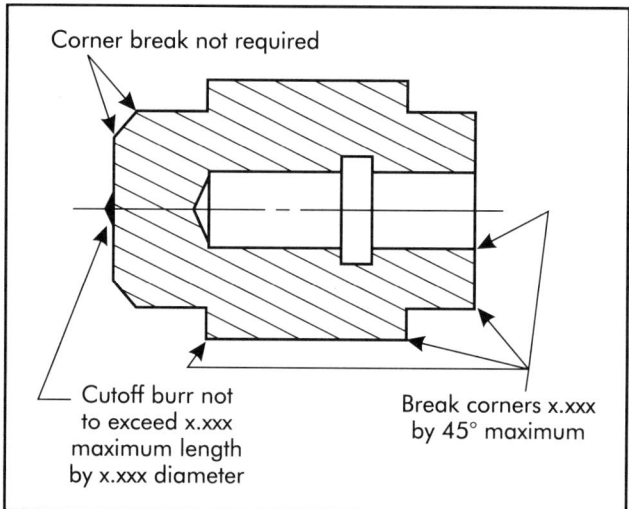

Figure 3-5. Typical burr notes for external edges (Precision Machined Products Association 1965).

"burr raised in slotting operation is acceptable." Eventually, however, someone will ask the product designer to define "small burr." One company faced a $9M lawsuit over lack of detail.

Preferred notes. Figure 3-4 illustrates the preferred practice of edge quality specification for notes that describe allowable burr sizes. It is preferable to use drawing notes or an in-plant standard to indicate whether chamfering adequately deburrs. For parts that require a smooth edge, specify the edge with a radius.

Edge breaks. Companies should specify edge breaks so that either a chamfered or a radiused condition is acceptable. This specification permits the manufacturing engineer to determine whether a machining or a deburring process will provide the most economical edge condition. Typical minimum corner breaks are 0.010 × 0.010 in. (0.25 × 0.25 mm). Normally, radii on precision parts should neither be specified larger than 0.010 in. (0.25 mm) nor smaller than 0.003 in. (76 μm), unless required by function or stress concentrations. Normal deburring processes readily provide these median value edges. Chamfering or corner-rounding operations using machine tools may require larger values.

Chamfers. When machined chamfers are required, they can be defined as chamfers by drawing notes (see figures 3-3 and 3-5). Companies usually specify chamfers as a dimension and an angle. A chamfer of 0.010–0.015 in. (0.25–0.38 mm) × 45° is typical of small parts, whereas a chamfer of 0.020–0.050 in. (0.50–1.27 mm) × 45° may be common for earth-moving equipment. A 45° angle is implied, but good practice requires inclusion of the angle. Inspection of the angle usually does not occur, unless it is obviously wrong. Small and large chamfers can significantly increase machining times.

Burr direction. A burr's direction is sometimes more critical than its size. In these instances, the part drawing should note the burr's orientation (see Figure 3-4). With symmetrical threaded parts, the designer can aid the manufacturer by indicating at which end the thread begins, thus eliminating the need to deburr both ends.

Edge breakout. Typically, fractured edges have a rough surface with variable edge geometry. Companies can state allowable conditions in several ways—by the standard edge break, special notes, or separate specification.

Specific notes. A burr always forms at the intersection of two holes. Note on the part if a burr is intolerable at one hole but not another. Defining the places burrs can exist on formed parts can help eliminate the need to deburr the whole stock. With thoughtfulness and cooperation among the product designer, tool designer, and manufacturing engineer, die forming can be designed so that burrs on the blank will be created in an insignificant location on the finished part.

Remove sharp edges. Many parts have only one edge requirement: "Remove all sharp edges." In such cases, beating out burrs and dulling edges is adequate. Designers can handle this requirement in one of two ways:

- by specifying the process, or
- by defining the necessary edge quality.

When an edge-finishing process specification—rather than edge requirement—is desirable, documents must clearly indicate the process and related details. Relatively minor process changes can greatly affect some components. If the documentation fails to define the process, the producer selects it.

It is important to define *sharp*. The glossary at the end of this chapter includes three definitions. If none of these apply, clearly state on the drawing your definition of sharp and your sharpness expectations.

Section 01.01.01.
Manufacturing Requirements

Design requirements drive manufacturing requirements, which completely define edge conditions for manufacturing operations. All new manufacturing instructions and practices should conform to this requirement.

Section 01.01.01.01.
Work Instruction Details

Work instructions for burrs, deburring, and edge finishing should accurately describe the operations so those who perform the operations can understand the requirements and methods for completing the work. Work instructions should ensure that manufacturing complies with drawing specifications.

When work instructions refer to machine operations, the company should provide a written description. Machine operators shall follow work instructions. Operators who deburr or finish edges shall receive appropriate training, and the company shall keep a record of operator training.

If unspecified by drawings or manufacturing instructions, all edges entering inspection should be broken to a maximum of 0.015 in. (0.38 mm), and neither burrs nor sharp edges shall be visible to the naked eye. Instructions such as "deburr complete" are adequate for many operations and shall imply a maximum break of 0.015 in. (0.38 mm) and no burrs or sharp edges visible to the naked eye. (Inspection can accept or reject based only on drawing requirements.)

Figure 3-6 illustrates one acceptable approach to define edge requirements in work instructions. The approach provides precise instructions about which edges to deburr, the size of the final edge configuration, and which equipment to use. Figure 3-7 shows another acceptable approach. To indicate which deburring tool to use, paint a sample part white and mark each feature with a different color. Desired radii or chamfers may be hand-painted if the requirements differ from the plant standard.

Section 01.01.01.02.
Machine Operation Details

If a specific deburring process is critical, or its costs are clearly advantageous, specify it in the work instructions. If, on the other hand, manufacturing instructions do not include a specific deburring process, use hand deburring. Table 3-3 defines this plant's requirements for processes, calibration, and training. Specify the process by name, or use the codes in Table 3-3. The following list describes process requirements (meaning that operators must verify their work).

A. Hand deburring—An operator may use any available tool, material, and technique, unless these are specified in the work instructions.
B. Vibratory deburring—(Company inserts machine operation descriptions here.)
C. Blasting—(Company inserts machine operation descriptions here.)
D. Sanding—(Company inserts machine operation descriptions here.)

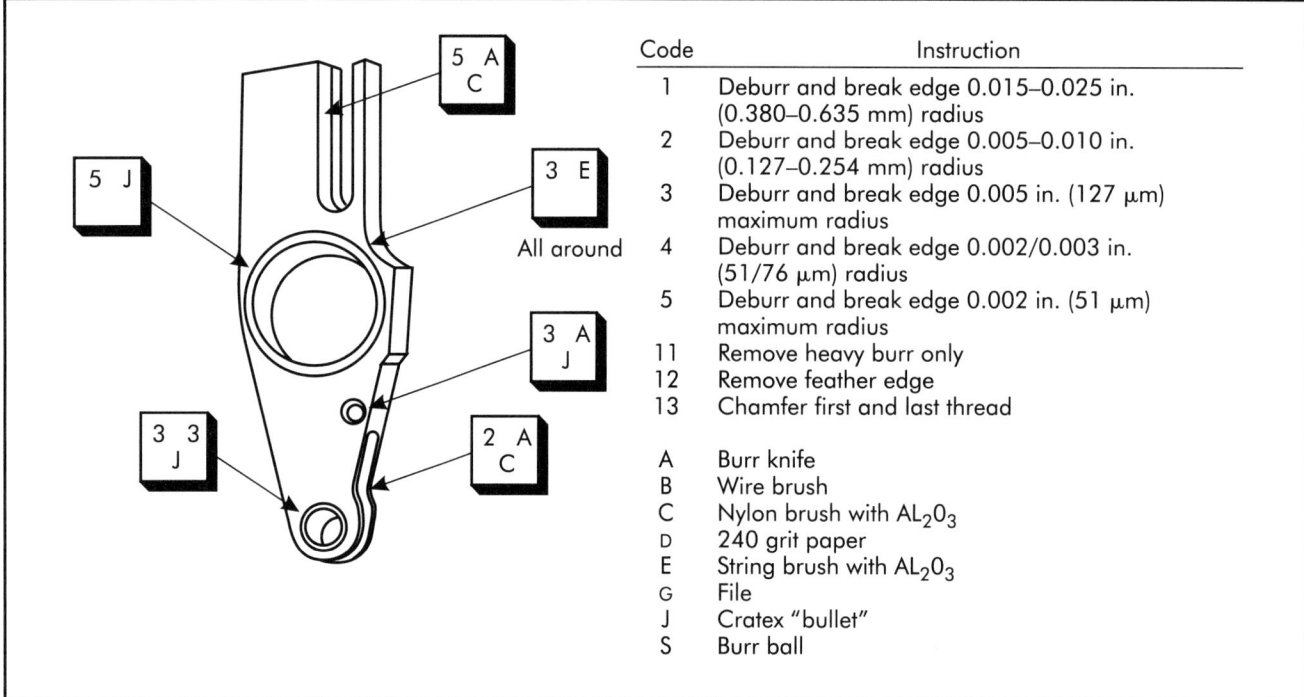

Figure 3-6. Example of work instructions that define edge requirements.

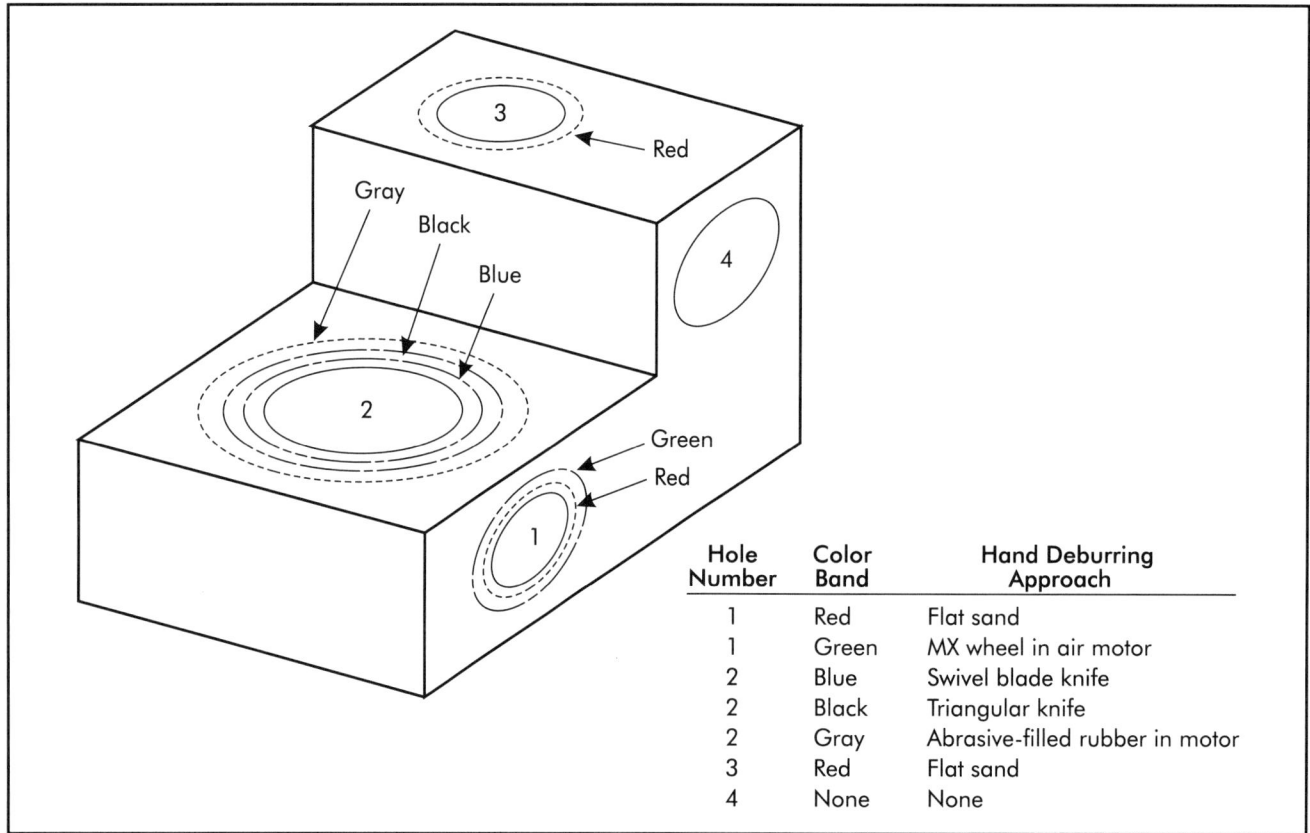

Figure 3-7. Example of work instructions using color-coded approach to define edge requirements.

Table 3-3. Summary of manufacturing process approaches for deburring and edge finishing*

Code	Process Name	Process Spec. No.	Calibration Type	Calibration Spec. No.	Training Name	Training Spec. No.
A	Hand deburring	01.01.01.01A	None	N/A	Hand deburring	01.01.01.01 t-1
B	Vibratory finishing	01.01.01.01B	Vibratory machine cal-1 & cal-17	01.01.01.01 basics	Vibratory finishing	01.01.01.01 t-2
C	Blasting	01.01.01.01C	Abrasive blasting	01.01.01.01 cal-2	Blasting basics	01.01.01.01 t-3
D	Sanding	01.01.01.01D	Sanding machines	01.01.01.01 cal-3 sander	Principles of flat belt	01.01.01.01 t-4
E	Brushing	01.01.01.01E	Brushing machine no. 1	01.01.01.01 cal-4 machine operation	Three-headed brush	01.01.01.01 t-5
F	ECD	01.01.01.01F	ECD machine	01.01.01.01 cal-5 ECD machine	Operating the dynetics	01.01.01.01 t-6
G	Centrifugal barrel	01.01.01.01G	Barrel finisher	01.01.01.01 cal-6	Harperizer operation	01.01.01.01 t-7

*Company inserts its own specification numbers in this table.

E. Brushing—(Company inserts machine operation descriptions here.)

F. Electrochemical deburring (ECD) —(Company inserts machine operation descriptions here.)

G. Centrifugal barrel—(Company inserts machine operation descriptions here.)

Section 01.01.01.03.
Training Details

Training should include all elements specified on work instructions, machine operation procedures, plant general specifications, expectations, and requirements.

Section 01.01.01.04.
Calibration Details

To ensure equipment is working as expected, calibrate all equipment dials, gages, measuring devices, and controls to a set of written standards and keep a record of the calibrations.

Section 01.01.01.05.
Documentation Details

Document work instruction-related issues and maintain the documents.

Section 01.01.02.
Inspection Requirements

Design requirements drive inspection requirements, which completely define edge conditions for inspection. All current and new inspection instructions and practices should conform to this requirement.

Section 01.01.02.01.
Inspection Instruction Details

Instructions for burr inspection, deburring, and edge finishing should be written and provide adequate detail for inspectors to understand requirements, measuring and recording devices, and methods for completing work. Inspection instructions ensure compliance with drawing specifications.

If inspection instructions reference equipment operations, provide a written description, as indicated by the specification number. Inspectors should follow inspection instructions. The company shall train deburring and edge-finishing inspectors, and it shall maintain training records.

Section 01.01.02.02. Inspection
Techniques and Equipment Details

Use any of several approaches to document the presence of burrs on parts, their measurements, edge configuration, etc. The approaches in Table 3-4 can be used, but the inspection instructions must specify the approach for edges or parts. If it is unspecified, use approach "A." The indicated procedures, or their respective subsections, describe the operational details, calibration, and training requirements for each piece of equipment. Unless otherwise specified, inspect

Table 3-4. Summary of inspection approaches for edges and burrs[a]

Code	Process Name	Spec. No.	Calibration Machine/Gage No.	Calibration Spec. No.	Type	Training Spec. No.
A	Visual (unaided)	01.01.02.01A	None	None	Burr inspection	01.01.02.01.01
B	Toothpick	01.01.02.01B	None	N/A	Burr inspection	01.01.02.01.01
C	No. 2 pencil point	01.01.02.01C	None	N/A	Burr inspection	01.01.02.01.01
D	Fingernail	01.01.02.01D	None	N/A	Burr inspection	01.01.02.01.01
E	4× glass	01.01.02.01E	None	N/A	Burr inspection	01.01.02.01.01
F	Microscope	01.01.02.01F	M1, M2, M3s	cal821	Microscope	01.01.02.01.02
G	Borescope	01.01.02.01G	Bore 1, 2, 3	cal986	Borescope	01.01.02.01.03
H	Metallurgical mount	01.01.02.01H	LTR 347	cal555	Mount preparation	01.01.02.01.04
J	Height gage	01.01.02.01J	CE4235	cal345	Height gage use	01.01.02.01.05
K	Profilometer	01.01.02.01K	CE3981	cal941	Profilometer use	01.01.02.01.06
L	3-D comparator[b]	01.01.02.01L	CE125	cal767	Laser comparator use	01.01.02.01.07
M	Light comparator	01.01.02.01M	CE45	cal4532	Comparator use	01.01.02.01.08
N	Toolmaker's microscope	01.01.02.01N	CE3247	N/A	Toolmaker's microscope use	01.01.02.01.09
P	Light section	01.01.02.01P	CE127	cal 887	Light section gage use	01.01.02.01.10
Q	Diode scan	01.01.02.01Q	CE128	cal 969	Diode scan gage use	01.01.02.01.11
R	Taper section mount	01.01.02.01R	None	N/A	Mount preparation	01.01.02.01.04
S	UL sharpness gage	01.01.02.01S	ga76	Ol-ga76	UL sharpness gage use	01.01.02.01.12
T	Capacitance gage	01.01.02.01T	ga120	Ol-ga73i	Capacitance gage	01.01.02.01.14
U	Water path	01.01.02.01U	FM125342	Ol-ga43	Water flow use	01.01.02.01.13
V	Laser line vision system[c]	01.01.02.01V	CE321	Ol-ga435	Disruption in line	01.01.02.01.15

[a] Company inserts its own equipment, specification, and document numbers in this table.
[b] Flores 1992.
[c] Stein et al. 1993.

all parts and assemblies. Inspect all features on all parts from lots chosen for inspection.

A. Visual (unaided). Unaided visual inspection includes inspection by workers who normally wear eyeglasses. This method cannot guarantee burr-free conditions, because visual acuity varies among individuals.

B. Toothpick (tactile) inspection. Wooden toothpicks can generally detect burrs and steps as small as 0.0005 in. (13 µm). Move the toothpick end slowly across the edge (or edges) in question. Slight upturns and momentary stoppages are indicative of a burr. Sharpness is detected by lightly pressing the edge with the toothpick body. If shavings appear, the edge is sharp. Shavings generally indicate edge radii 0.001 in. (25 µm) or smaller.

C. No. 2 pencil-point inspection. Jab at the burr with a sharpened No. 2 pencil point. If the point breaks before the burr breaks, you can assume that the burr is rigidly adhered.

D. Fingernail inspection. Draw a fingernail across the edge (or edges) in question. A burr, or raised metal, is indicated if the fingernail catches, slows, or moves slightly upward. This technique can often detect burrs as small as 0.0005 in. (13 µm) high. Sharpness is detected by lightly moving the top surface of the fingernail across the edge. If shavings appear, the edge is sharp. Shavings would generally indicate edge radii of 0.001 in. (25 µm) or smaller.

E. 4× magnification inspection. Use a 4× magnifying glass, with or without accompanying lighting. Inspectors may wear eyeglasses, if they normally wear them. Visual acuity varies significantly among individuals, so a burr-free condition is not guaranteed.

F. Microscope inspection. Use a stereo-zoom microscope with any type of lighting. If the power of magnification is unstated, use 10×. High optic quality requires that microscopes are in working order and maintained according to a defined maintenance plan.

G. Borescope inspection. Use a borescope to inspect cavity interiors. Operators must be qualified to detect burrs and associated edge conditions. Unless otherwise specified, acceptance may be made by borescope optic eyepiece or borescope video screen.

H. Metallurgical mounts inspection. Use metallurgical cross-sectional mounts only when specified, or if other approaches provide inconclusive results. This approach destroys the specimen.

J. Height gage inspection. Pass a height gage and indicator across the edge (or edges) in question. This approach requires several measurements along each edge. Carefully used, it can detect burrs as small as 0.0002 in. (5 µm) high, but it cannot determine thickness.

(Company: insert here definitions of any additional techniques used.)

Section 01.01.02.03.
Training Details

Training must cover all specifications that relate to the following: inspection instruction, equipment operation procedures, plant general specifications, and expectations and requirements.

Section 01.01.02.04.
Calibration Details

Calibrate all equipment dials, gages, measuring devices, and controls to a set of written standards, and maintain the calibration records.

Section 01.01.02.05.
Documentation Details

Maintain documentation of part conformity to edge and deburring requirements, along with any deviations from the requirements.

Section 01.01.03.
Standards for Applied Edge and Burr Research

Corporate requirements drive research standards, which provide a means to better define edge conditions for design, manufacturing, and inspection. All new research efforts should conform to this requirement.

Applied research terminology relating to burrs, deburring, and allied edge conditions must be consistent with that of the plant and with international practice. Clear definitions related to existing terms must accompany new terms, because those who use the research must fully understand the meaning of each new term. Where possible, plant practice uses existing terminology. For research, measure and identify burr dimensions in a manner that allows company personnel to analyze the data. Maintain research results records.

Section 01.02. Microscopic and Scientific Considerations

Some scientific studies provide a way to better understand and define the minute details of edges, burrs, and related effects. This specification concerns edge effects seen, or believed to exist, when viewed at magnification powers ranging from 400× to 100,000×. New scientific and theoretical research efforts shall conform to this specification.

Terminology related to advanced scientific research on burrs, deburring, and allied edge conditions must be consistent with both plant terminology and international practice. As with applied research, new terms require clear definitions that relate to existing terms, so users of the research can fully understand them. Where possible, plant practice uses existing terminology.

HOW STANDARDS APPLY TO HAND DEBURRING

Sections 01.01.03 and 01.03 of the sample standard do not directly affect hand deburring, but they do provide a complete company view of standard needs. In short, they set the stage with examples for more effective hand deburring.

The people who perform hand deburring must know exactly what the product and the plant manager or owner require. Clearly written requirements better ensure that each person has the same understanding of quality expectations. Written requirements also ensure consistency of part quality, because written documents remain in the plant, whereas employees may transfer to another department or leave the company altogether.

Inspectors, as well as the operators, must know the production standard. Inspection techniques are discussed in Chapter 34, but the standard used to accept parts, in general, differs from how edges are inspected—one is conceptual and one is detailed.

DICTIONARIES FOR EDGE AND BURR TERMINOLOGY

Large companies may require three different dictionaries of burr-related terms: one for manufacturing staffs, one for researchers, and one for scientists.

The important thing to remember about this chapter is that those who deburr must know the edge-quality definitions used by their respective companies or familiarize themselves with those cited in the previous sample standard. A single page might adequately define plant requirements. Special needs for specific parts or specific features on a part may require additional, clearly written or illustrated notes, as exemplified in figures 3-1 through 3-5. Similarly, the inspection approach requires clear definition.

For the Shop

Manufacturing staffs need a common-use dictionary, which provides clear definitions for the kinds of burr and edge conditions found on most commercial parts. The glossary to this chapter is this type of dictionary.

For Researchers

Researchers require a dictionary of burr taxonomies, which is more detailed and sophisticated than the common-use dictionary recommended for manufacturing staffs. They require this type of dictionary because their research covers vast numbers of burr types, material at edges, and edge configurations; and they must communicate research results to a wide audience. The beginnings of such a dictionary already exist (Gillespie 1996a).

Machinists, inspectors, and engineers may use a researcher's dictionary, but their requirements are not as sophisticated. For example, researchers need to know that a single-face-milling cut can generate six different edges in a straight pass, and it can produce more edges on complex parts.

For Scientists and Theorists

Individuals who study the scientific, theoretical, and microscopic (nanometer) levels of edges and surfaces often work with equations and advanced analytical lab equipment. Their universe is different from that of applied researchers and shop staffs; therefore, they require a different dictionary. Scientists and theorists study the physics of materials; thus, the production floor cannot readily use their dictionaries. For example, the surface of an atomic-force microscope probe literally scrapes across molecules of material. The probe end may require description in normal vectors, by B-splines, or other definitions. Because the minute end of the tool is important, the end edge is critical and may be defined in the same highly mathematical formats, rather than by images.

GLOSSARY

Standard terminology is necessary to ensure better communication among individuals involved in the deburring process. This glossary provides the terms and definitions with which those individuals should be familiar. It can be appended to the sample standard included in this chapter or it can be appended to a company's existing standard.

Burr and Edge Terminology

The following are four categories of key terms:

- edge characteristic terms,
- burr-related terms,
- flash-related terms, and
- miscellaneous edge-related terms.

The following paragraphs define the terms associated with each category.

Edge Characteristic Terms

Burr: Plastically deformed material that is produced, because of machining or shearing, at workpiece edges. A burr includes any metal that extends beyond the intersection of the two surfaces surrounding the burr (Figure 3-8a). Sometimes, a burr is inside the intersection (Figure 3-8b). A burr can be a firmly adhered, sharp, ragged projection, or it can be a loosely hanging projection. It can be a swell of material at an edge.

In Germany and other European countries, burr also means flash from castings and sintered metal operations; however, the definition in this standard is limited to cutting and shearing processes.

Chamfer: A machined, inclined surface cut at an edge, which is typically in the order of 0.005–0.050 in. (0.13–1.27 mm). Some chamfers are used solely to remove burrs. A chamfer may have small burrs at its edges.

Corner: A corner is the intersection of three or more edges.

Edge: An edge is the intersection of two surfaces.

Edge break: Edge break refers to the amount of material removed from the theoretical intersection of two surfaces. Typically, an edge break is identified as if it were a uniform chamfer or radius, such as "0.004 in. (0.10 mm) minimum edge-break requirement."

Edge breakout: Edge breakout is the condition left when a brittle material fractures at the edge and leaves an irregular, fractured surface. Cast iron and ceramics particularly display this condition. It is an absence of material at an edge, whereas burrs and raised material are extra material at an edge.

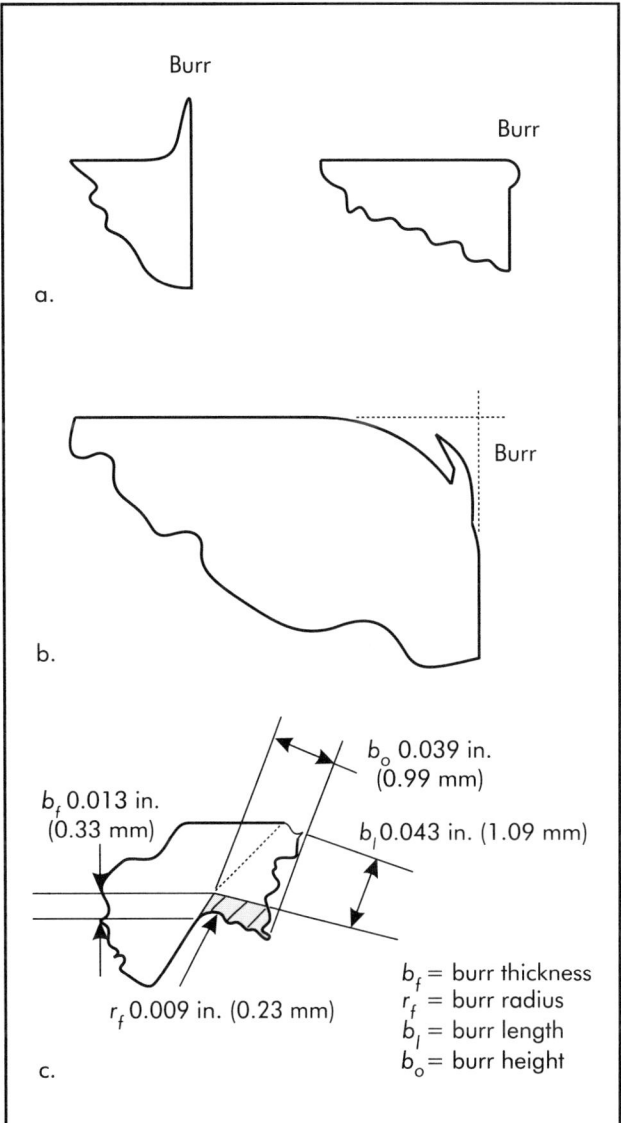

Figure 3-8. Examples of burrs and burr properties. (a) Metal extending outside the theoretical intersections of two surfaces is a burr; (b) example of a burr that lies inside the theoretical intersection of two surfaces; (c) definition of burr length, thickness, and radius, and actual measurements.

Edge sharpness: See "Sharp edges."

Sharp edges: Sharp edges indicate that an edge can readily cut or tear. There are three classifications of sharp edges:

- *Class 1.* For thin materials—0.010 in. (0.25 mm) thick or less—*sharp* means a burr-free edge (when viewed at 5× magnification) with a radius, chamfer, or edge break less than $t/10$, where t = workpiece thickness at the edge to be measured.

- *Class 2.* For a workpiece thickness greater than 0.010 in. (0.25 mm) at the edge of interest, *sharp* is a burr-free edge (when viewed without magnification) with a radius, chamfer, or edge break of 0.001 in. (25 μm) or less.
- *Class 3.* For normal commercial usage, *sharp* is any edge that would cut hands, containers, or nearby components in normal handling, installation, or repair. This class includes burr-laden edges, as well as burr-free edges that can cut. Cutting ability is determined by experiments, or by use of the Underwriters' Laboratories sharpness testing unit.

Burr-related Terms

Burr hardness: Burr hardness refers to the hardness at the root of the burr—at the theoretical intersection of the two surfaces adjoining the burr.

Burr height: Burr height refers to the dimension measured perpendicular to the edge from the theoretical intersection of surfaces to the most distant point of the burr in its cross section.

Burr length: Burr length is the longitudinal length of a burr along an edge. (See Figure 3-8c.)

Burr radius: Burr radius is the radius of the burr profile at its root. For a rollover burr, the radius occurs on the side of the burr, opposite that touched last by the cutting or shearing tool.

Burr thickness: Burr thickness is the width or thickness of the burr at its root, as measured along one surface.

Cutoff projection: Cutoff projection is the material that remains on the workpiece when the workpiece is allowed to fall before a cutting tool completely cuts through the material. On a turned workpiece, the cutoff projection is normally short, has a small diameter, and sticks out from the last surface machined (for example, Figure 3-5). On milled stock, it may be a thin rectangular piece attached to one edge. It is not plastically deformed material or material raised above a surface because of cutting. It is material left uncut.

Entrance burr: An entrance burr forms on the surface at which the cutting tool enters the workpiece.

Exit burr: An exit burr forms on the surface at which the cutting tool leaves the workpiece.

Feather burr: A feather burr refers to a fine or thin burr. It can look like a feather with many fine spokes.

Feather edge: Feather edge sometimes indicates a feather burr, but it can also refer to the ends of a lead-in or lead-out thread (a thin machined ridge that is not a burr).

Hanging burr: Hanging burr refers to loose portions of a burr (portions that are not firmly attached to a workpiece); that is, a hanging burr "hangs" from the workpiece.

Poisson burr: A poisson burr forms primarily by the phenomena responsible for Poisson's ratio. This type of burr occurs whenever the cutting edge extends past an edge of the workpiece.

Rollover burr: Cutters form rollover burrs when they exit a surface and allow a chip to roll away, instead of shearing it.

Tear burr: Cutter sides form tear burrs as they tear chips from workpieces.

Tensile burr: A tensile burr forms in blanking or piercing operations. It is caused by slug separation from the stock because of tensile stresses.

Tensile plus compressive burr: A tensile plus compressive burr forms in blanking or piercing operations in which the slug separates from the stock in a stress field that is initially tensile but changes to compressive at actual separation. This burr is usually associated with small die clearances and short die life.

Wire edge: A wire edge refers to a fine or thin burr that, typically, is sharp, looks like a wire, and is parallel to an edge.

Flash-related Terms

Fin: Fin refers to the portion of flash that flows into the small gap between the movable parts of the mold.

Flash: Flash refers to the excess material squeezed from the mold cavity as a compression mold closes, or as a transfer or injection mold receives pressure. Flash includes both the fin and fragments that remain in the mold.

Flash extension: Flash extension refers to the portion of flash that remains on the workpiece after it is trimmed. Measure it from the intersection of the draft and flash at the body of the forged or molded workpiece to the trimmed edge of the stock.

Miscellaneous Edge-related Terms

Dross: Dross is the molten metal that resolidifies on the workpiece. It occurs at the edges of material cut by thermal processes, such as plasma arc, oxyacetylene, and heliarc cutting.

Raised metal: Raised metal is material raised above the surrounding surface. It may be the root of a partially removed burr, or the material may have risen when something bumped against the part or edge. Raised material or raised metal is generally a small hump or bump of material, as opposed to a sharp burr. In some instances, raised material may be as objectionable as burrs.

REFERENCES

American Die Casting Institute (ADCI). 1955a. "Flash Removal." Product Standard ADCI-ElO-65. New York: American Die Casting Institute.

——. 1955b. "Flash Formed by Ejector Pins." Product Standard ADCI-E9. New York: American Die Casting Institute.

American Society for Testing and Materials (ASTM). 1993. "General Requirements for Steel Sheet and Strip, Alloy, Hot Rolled, and Cold Rolled." ASTM Standard A505-7. In *1998 Annual Book of ASTM Standards,* vol. 01.03. West Conshohocken, PA: American Society for Testing and Materials.

——. 1991. "Standard Method of Test for Edge Coverage of Coating Powders." ASTM Standard D-2967. In *1998 Annual Book of ASTM Standards,* section 06.02. West Conshohocken, PA: American Society for Testing and Materials.

Consumer Product Safety Commission. 1975. "Bicycles, Toys, and Other Children's Articles." Proposed Rules. *Federal Register* 49, no. 4 (7 January): 372 (to be codified at 16 CFR 1500). Washington, D.C.: Consumer Product Safety Commission.

——. 1976. "Reproposed Rule for Sharp Edges: Working Draft of Revised Proposed Rules for Allowable Sharp Edges on Children's Toys or Related Articles." Draft no. 1, 17 May. Washington, D.C.: Consumer Product Safety Commission.

Deutsches Institut für Normung e.V. (DIN). 1982. "DIN 6784" (German standard). Berlin, Germany: Deutsches Institut für Normung. (Available from Document Center, 1504 Industrial Way, Unit 9, Belmont, CA.)

Flores, Gerhard. 1992. "Mechanical Deburring: Process, Tools, Machines and Applications." Technical Paper MR95-272. Dearborn, MI: Society of Manufacturing Engineers (SME).

Gillespie, LaRoux K. 1978. *Advances in Deburring.* Chap. 2, 21–29. Dearborn, MI: Society of Manufacturing Engineers (SME).

——. 1981. *Deburring Technology for Improved Manufacturing.* Chapter 3, 65–109, 605–646. Dearborn, MI: Society of Manufacturing Engineers (SME).

——. 1995. "The Small Firm Standard for Burrs and Edge Finishing." WBTC-STD 11. Draft. Kansas City, MO: Deburring Technology International.

——. 1996. "Burr and Edge Terminology: An International Dictionary." WBTC-STD 3. Draft. Kansas City, MO: Deburring Technology International.

Gillespie, LaRoux K., ed. 1996a. "Burr and Edge Terminology: Definitions." WBTC-STD 01. Draft. Kansas City, MO: Deburring Technology International.

——. 1996b. "Comments on Use of the DIN 6784 Standard for Burrs and Edge Conditions." WBTC-MS96-1. Kansas City, MO: Deburring Technology International.

——. 1996c. "Standard Nomenclature for Researchers in Burrs, Deburring, and Edge Finishing." WBTC-STD 2. Draft. Kansas City, MO: Deburring Technology International.

——. 1996d. "The International Standard for Burrs and Edge Finishing." WBTC-STD 13. Draft. Kansas City, MO: Deburring Technology International.

——. 1998. "An Integrated International Standard for Burrs and Edge Conditions." WBTC-STD 14. Draft. Kansas City, MO: Deburring Technology International.

——. 1999. *Deburring and Edge Finishing Handbook.* Dearborn, MI: Society of Manufacturing Engineers (SME).

Precision Machined Products Association (PMPA). 1955. *Buyers Guide for Design of Screw Machine Products.* Cleveland, OH: Precision Machined Products Association.

——. 1965. *Draftsman's Guide for Screw Machine Products.* Cleveland, OH: Precision Machined Products Association.

Precision Metalforming Association (PMA). 1976. "Design Data Sheets." In *Stamping Buyer's Guide.* Independence, OH: Precision Metalforming Association.

——. 1977. *Facts About Washers.* Independence, OH: Precision Metalforming Association.

Sorrells, John R., and Robert E. Berger. 1974. "An Inspection Procedure for Detecting Hazardous Edges." National Bureau of Standards Report NBSIR 74-428 (April). Washington D.C.: National Bureau of Standards.

Stein, Julie M., Ranga Narayanaswami, Samuel Ho, Anselm Y. Lam, Ilwhan Park, Madhu Babu, Asim Afzal, and David Dornfeld. 1993. "Intelligent Deburring of Precision Components." Technical Paper MR93-319. Dearborn, MI: Society of Manufacturing Engineers (SME).

Takazawa, K., and Y. Kato. 1997. "Edge Quality Classification and Grade." In *Proceedings of BEST-J,* vol. 3, no. 17:4. Toyko, Japan. Kansas City, MO: Deburring Technology International, Inc.

Underwriters' Laboratories (UL). 1976. "Subject 1410: Proposed Revisions of the Standard for Television Receivers and Video Products—Proposed Effective Dates." UL Bulletin, 26 Oct. Melville, NY: Underwriters' Laboratories.

——. 1993. "Subject 1439: Standard for Determination of Sharpness of Edges of Electrical Equipment." UL Bulletin, 19 Feb. Northbrook, IL: Underwriters' Laboratories.

BIBLIOGRAPHY

Gillespie, LaRoux K. 1977. "An Extension of Proposed Definitions for Burrs and Related Edge Condition." Technical Paper MRR77-09. Dearborn, MI: Society of Manufacturing Engineers (SME).

———. 1981. "Progress in the Battle With the Burr." Technical Paper MRR81-07. Dearborn, MI: Society of Manufacturing Engineers (SME).

Przyklenk, Klaus and M. Schlatter. 1987. *Entgraten von Werkstucken aus Aluminum,* 5–9. Dusseldorf, Germany: Aluminum-Verlag.

Seimens, C. 1982. "Prufeinrichtung fur die Mechanische Messung Dunner und Labiler Metallteile." Brochure. Munchen, Germany: Siemens.

Common Hand Deburring Practices 4

Every company has efficient and cost-reducing deburring solutions, but most also have a number of poor practices. This chapter details the poor common practices happening today. Then, it discusses good management and hands-on operator practices.

WHAT IS HAPPENING TODAY

Most poor practices are the result of lack of management attention and priority. For example, discussions with a major transmission manufacturer revealed that some of its parts were on the burr benches for nine months. At that time, a six-month backlog awaited hand deburring. It is difficult to make money with backlogs like that.

Deburring simply has not been given the emphasis necessary to overcome past cultures and practices. The good news is that correcting the problems, while daunting at first, does not require major costs—but it does require commitment.

Lack of a Deburring System

Operators often lack a system to ensure that the edges of each part have been deburred. One of the more disruptive problems is that a deburring operator, while taking a break, or if distracted, can accidentally place parts to be deburred in the "finished" basket. There is no guarantee that a deburring tool ever touched a particular part. Someone other than the operator can also, absent-mindedly, pick up a burr-laden part to look at it through a microscope and never give a thought to which pile it came from or where it should be returned. Engineers, managers, machinists, and inspectors all have done this without thinking. A system can be as simple as two boxes—one for unfinished parts and one for finished parts. A system can also use three boxes, with the third one for parts in-work. Some operators will finish all of one edge before working other edges, since that one edge requires a unique tool. Each operator can have his or her own system for deburring parts, but until that system is developed and practiced consistently, parts with burrs will go to inspection. In a mistake-proof system, such as the Kaizen system, every box or position on the table will be marked, so everyone knows exactly where to place parts.

Another system to ensure that parts are burr-free before they leave the burr bench is to place all the parts on the worktable and look at all of them together. Inspection finds burrs, partly because they look at all the parts at one time. Anything different stands out. Deburring workers typically deburr one part at a time, and there is nothing comparative for them to judge quality. It is amazing to see what people can find when they compare one part against all others.

Another key system approach is to just have the deburring operator check over each part before sending the group to inspection. That last-minute review, whether done in a group or part-by-part, often finds the little things missed.

Unique Requirements Demand Unique Approaches

A recent study of the ability of precision hand-deburring operators to produce edge radii less than 0.004 in. (102 µm) indicates that, while possible, it is not often achieved. Such small radii require special approaches that are significantly different from normal deburring. These needs require measurement of results, and most deburring does not involve measuring edge results. Engineering development and close attention are required at the beginning of the effort, which are often overlooked. Many of the computerized cost-estimating programs fail to acknowledge or account for unique edge needs. Unique needs must be identified before quoting a job, and before parts are on hand to deburr.

Lack of Attention Adds Hidden Costs

Inspectors establish a pseudo-edge standard in their minds. When several inspectors review the same parts, some accept parts that others reject.

Companies reject parts every day. This raises costs substantially because they must re-examine, deburr, and reinspect the parts. Operators must complete reject forms, as well as resubmit forms. The problem must be shown to the inspection department, the supervisor, and the deburring staff. Someone must select a rework technique and evaluate how to prevent the problem in the future. In some cases, the quality engineer must make a weekly written report of rejection reasons and corrective actions taken. Other engineers and managers then review this report. Each of these activities costs money and disrupts the plant's mission of meeting schedules and making a profit. Every unessential piece of paper, meeting, discussion, and rework step wastes money and builds hostility. Money is wasted every day.

High Turnover Rate

Sudden personnel changes affect hand-deburring operations. Deburring departments experience a high turnover rate, as operators quit or transfer to other departments. New employees mean less experience, slower production, and more problems. These operators receive little, if any, formal training, which means that quality or deburring time, or both, suffer. During a layoff, machine operators with more seniority often "bump" down into deburring. Suddenly, an operator used to machining large parts is expected, overnight, to become skilled at deburring miniature parts under a microscope. It does not have to be that way. In some plants, deburring is a respected position with turnover rates so low as to be an asset to the plant, rather than an anchor slowing down progress.

Lack of Consistency

Other than "excessive" time required for deburring, the major complaint about hand deburring is lack of consistency. Consistency can be a problem lot-to-lot, part-to-part, edge-to-edge, and person-to-person. The following list identifies some reasons why part deburring lacks consistency:

- operator attitude (happy, preoccupied, frustrated, sad, motivated, mad, afraid, can-do, do not know how, no one cares, etc.);
- comments of people around him or her (Jo cannot do them right, she always has to have people help her; this is a lousy job, no one cares; do not worry about what the boss says—all are comments that degrade performance);
- lack of operator education (education of the basic process, plant requirements, specific part needs, job performance expectations, inability to read, inability to comprehend, inconsistency among operators);
- status within working family (deburring operators are treated as the lowest-class worker in the plant);
- time available to do parts (end-of-month shipments force deburring workers to double output; absent workers cause others to have to pick up the slack, but no additional time is available);
- rated time (some parts are not time-rated correctly for the quality required, so there is never enough time to deburr them adequately);
- peer pressure (do not be different, or we will be unhappy with you; you better not produce more than we do, or we will look bad);
- supervisor pressure (get it done in the next hour—or else);
- difficulty of edge access (each edge has its own level of difficulty, which forces unique small increments of care—there has to be an understanding of what each edge requires), and
- availability of hand tools (tool substitutions can result in lower-quality edges and almost always increase deburring time).

Four other deburring consistency issues affect quality and costs. They include:

- high turnover of engineers who are responsible for solving burr problems;
- the inspector has a "bad" day;
- high turnover of inspectors, and
- operators in many shops are not following work instructions.

Lack of Operator Reading Skills

The inability of operators to read and follow directions plays a major role in today's poor practice. Deburring operators in successful shops read and follow work instructions; but in many U.S. shops today, operators cannot read. They cannot read English. They cannot read Spanish. They cannot read technical instructions, and they receive no written instructions, which they might not be able to read anyway. Process instructions and actual practice do not agree in these shops. Deburring is no exception. These shops, in practice, seldom follow available printed instructions. Part of this results from one or more of four reasons:

- instruction details are overlooked;
- instructions are misinterpreted;
- the shop has better solutions to the situation than engineers wrote, and/or
- people are resistant to change.

Poor Communication

Poor communication, in general, creates problems that affect quality and costs. The following list is by no means exhaustive, but it includes some of the most significant problems resulting from poor communication.

- Workers develop their own tooling, but because of large plant size and lack of communication among three shifts, these tools are unknown to two shifts and the engineering staff. This results in one efficient and two less efficient shifts.
- Operators consistently fail to meet production-rate expectations because they have no knowledge of expected rates.
- Supervision does not emphasize the importance of meeting production rates.
- Supervision instructs operators to deburr more than what is specified on the written work instructions.
- Work instructions (routing sheet) are unavailable to the worker, so he or she is unaware of them. One example would be where several people in a large department share the same copy of instructions.
- Machinists do not identify scrap parts, so deburring operators unknowingly deburr them.
- Inspection personnel do not communicate with deburring personnel.
- Engineering deletes illustrations showing where and how to deburr, therefore the time it takes to hand deburr increases.

Other poor practices and their causes include the following:

- One manufacturing department routinely forwards its parts without deburring them, even though instructions state they should deburr parts for credit. The receiving department shows a low efficiency rate because it deburrs previously deburred (supposedly) parts.
- Machining rework after final deburring causes rework in the deburring process.
- Inadequate in-process deburring causes a need for more deburring in the final deburring operation, which may appear as reduced efficiency in final deburring.
- Running parts out of normal machining sequence increases the deburring workload.

- Some workers are only one-third as efficient as others.
- Deburring tool quality is inadequate (for example, heavy grinding burrs remain on deburring knives, portions are broken off deburring tools, or edges become soft from too much grinding heat).
- Operators are unaccountable for their work, unless a method is in place to trace parts to the individual who performed the work.
- Tool inspection or monitoring is insufficient (for example, workers can no longer use a certain tool to look into small, deep holes).
- When deburring during the machine cycle (prior to a final hand deburr), there is no way to control how much deburring occurs in the machine cycle (breaking edges too much or too little, as well as the percentage of edges touched).
- Actual hand-deburring times are frequently twice the rated standard hour time (50% efficiency). Mechanized deburring is closer to 80–90% efficient.
- Vendors do not deburr parts fully, even though it is specified in the contract, and management does not require the vendor to honor the contract.
- The production department lives with "little" problems year after year, because it does not believe the engineering or methods department will solve the problem. Production, engineering, and methods departments do not communicate burr problems. The supervisor believes it is easier to rework parts every month, rather than work with the engineering and methods departments.
- The company institutes 100% inspection for burrs because it failed to take appropriate deburring actions.

Deburring Staff can Provide Insight

Some of the most revealing aspects of ineffective deburring come from those who are actually doing the deburring. The appendix to this chapter contains comments made by several deburring workers who participated in a plant survey. The following are some of their more prevalent concerns:

- The right tools are not available.
- Operators do not like one another (interpersonal relationships).
- Rumors are rampant about operator status within the company.
- Engineers are not interested in helping with such a mundane assignment.
- Plants send equipment vendors sample parts for review with unrealistic burrs and without facts

about the parts, schedules, or costs. When the vendor notifies plant personnel that the parts have been processed and they later turn out to be unsatisfactory, plant management quickly drops the subject.

It is unrealistic to expect wonderful results with a "one-shot" evaluation, no dialogue, unrealistic parts, and no cost statements. Successful solutions to many burr problems require a continuing dialogue with facts. Some vendors have not developed an appreciation for a customer's burr needs, but many vendors have a good understanding of several processes—as well as burr formation and minimization—because of extensive recent publications in this field.

In one observer's company, NC machine operators refused to partially hand deburr parts because when they had once before it generated too large of an edge break. The high expense of these parts convinced them that, despite the available time and management's wishes, they were not going to repeat the mistake. In this instance, failure to educate workers and provide appropriate tools, along with the stubbornness of a few individuals, contributed to higher than necessary hand-deburring costs.

Engineers who like to over-specify (to get higher quality) dislike adding burr notes. Machining operators search for ways to shortcut standard practice. Therefore, they are not interested in adding deburring work. Deburring, they believe, is beneath their standard. Quality control people are reluctant to accept new tasks, such as burr control—especially since scientific measurement of burrs is difficult, if not impossible. Purchasing staff does not want to criticize vendors whose parts do not meet burr standards. Industrial Relations staff will strongly object, since the reallocation, elimination, or new control of the work effort can create grievances.

Operators believe that large burrs make work for the deburring department, which is good for keeping people employed. The implication that machine operators do not worry about burr minimization because their friends might be laid-off if the burr bench ever caught up is a reality, although seldom mentioned.

GOOD MANAGEMENT PRACTICES

Hand deburring does not lend itself to ready analysis. If it did, it would not be an area of mystery and high cost. However, today's cost-effective plants use some basic approaches and considerations to improve manual-deburring costs. They include:

- analyzing actual hand-deburring costs;
- determining actual in-house burr needs;
- determining the key problem areas of in-house hand deburring (Pareto principle—35% of rejected part numbers account for 70–90% of rejected parts);
- tentatively evaluating whether potential benefits outweigh the cost of improvements, and
- seeking solutions.

Analysis of actual manual-deburring costs is relatively straightforward. These costs are discussed in chapters 5 and 6. Costs alone, however, are not the only criteria used to determine whether manual efforts should receive special consideration.

Considerations and Solutions

For each part with burrs, there is a shop record of actual deburring time spent to remove them. This provides evidence of improvement and performance.

Shop supervisors, owners, and engineers search the Internet and published literature for better deburring approaches. Solutions to burr problems are easier to find than they were in the past. The *Deburring and Edge Finishing Handbook* (Gillespie 1999) provides in-depth insight into most deburring processes, as well as ways to minimize and prevent burrs. Another publication, *Guide to Deburring, Deflashing, and Trimming Equipment, Supplies and Services* (Gillespie 2000), provides 150 tables of deburring specialty machines, tools, and services, and lists 2,500 suppliers of these items. Deburring leaders are taking advantage of *Deburring: A 70 Year Bibliography* (Gillespie and Repnikova 2001), which is a convenient listing of 4,500 deburring reports and documents for immediate insight into available publications. *Deburring Capabilities and Limitations* (Gillespie 1976) and *Advances in Deburring* (Gillespie 1978) are two other sources of extensive deburring details. Explore published research results, attend deburring conferences, and ask questions of online experts.

Successful plants follow some or all of these solutions to hand-deburring problems:

- Define in-house standards on burrs and deburring.
- Provide the time and money to truly solve burr-related problems.
- Involve people who are willing to solve the problems.
- Buy and read the literature on the subject.
- Combine two types of deburring whenever possible.
- Determine exactly which tools, machines, and processes are available.
- Copy effective methods used by others.

- Provide worker training on effective deburring tools and processes.
- Establish multiple grades of deburring classification.
- Use consultants when necessary.
- Outsource portions to job-shop deburring facilities.
- Work with deburring machine vendors.
- Document results.
- Maintain a continuing awareness of edge-finishing technology and plant needs.
- Machine heavy burrs, rather than hand deburr them.
- Use a tool to push the burr into a more accessible location to reduce deburr time.

The Hamilton Standard Division of United Aircraft Corporation related one well-publicized case of minimizing hand deburring costs (Miller 1965). The company performed an in-depth study similar to that proposed in the previous list and elsewhere in this chapter. After evaluating approaches and problems, the company found it could save 30% of its deburring costs by developing a simple in-house approach to defining deburring needs by implementing what was then a new process. Since then, others have focused on standards as one means to reduce costs (Gillespie and Repnikova 2001; Gillespie 1980; Stein 1995; Adachi et al. 1987).

Several large companies have been able to dramatically improve deburring by implementing intensive one-year or six-month reviews. In almost every case, a portion of this improvement resulted in some form of standardization. Unfortunately, in many cases, reassignment of the individuals responsible for making the changes followed this intensive effort, resulting in the loss of some of this knowledge.

Training

Pratt & Whitney's slide-tape presentation, *Clean Engines,* is an example of one form of training aid that is essential for effective education of deburring workers and engineers throughout the world. Vendors and other sources now offer a few videos useful for hand deburring, but most are focused on mechanized processes (SME 2000). As will be discussed later, however, a key to success is blending manual operations with low-cost mechanized approaches. Many videos, while not directly focused on manual methods, provide background to help identify which parts should be moved from manual-only approaches. Even though other deburring topics are under consideration for packaged audio and visual presentations, do not wait for industry to solve specific company burr problems, because industry moves relatively slowly. Generally, by the time industry solves a problem, a company has already spent unnecessary deburring dollars for several years awaiting the solution. With today's low-cost digital systems, companies of any size can readily make their own training films.

Image

As a part of the total program, plantwide improvement efforts include an investigation of the facility's deburring image. The company can take the following steps to improve a poor image:

- Emphasize that deburring is a challenging job.
- Demonstrate that management is willing to support changes.
- Make funds available for buying needed equipment.
- Recognize individuals who make improvements.
- Ask everyone to help minimize deburring problems.
- Provide well-lighted worktables, microscopes, and special hand tools required to work efficiently.
- Provide remuneration that is commensurate with work efficiency and quality.

GOOD HANDS-ON OPERATOR PRACTICES

Hand deburring has several technical approaches, including:

- deburring the entire part manually;
- deburring the heavy burrs by hand, then using mechanized processes to finish the part, and
- using mechanized processes to remove most of the burrs, and using hand deburring to finish hard-to-reach areas.

For effective deburring, leaders must determine for each department and each part which of these approaches is most cost and schedule effective. In addition, they must consider that:

- a single tool can complete a few applications, and
- most applications rely on two or more tool types to provide a finished part.

Use hand deburring to provide rough or in-process deburring, and completely finish deburring the part. In some instances, tool specification is critical. In others, the operator should select the tools.

Published solutions to most manual deburring problems are rarely found. One deburring conference attendee spent three days listening to authorities on various deburring processes. While traveling home, he casually mentioned to another attendee that he increased the life of his deburring files up to five times by the simple addition of chrome flashing. Since he was working on very tough Inconel® and related high-nickel alloys, this was a significant observation; but because he thought it would not be of interest, this simple solution was never mentioned during the conference. Unfortunately, too many individuals have assumed their solutions are too basic to warrant public notice.

As an example of the types of questions users must answer and for which there are no published guidelines, consider the following questions.

- When should you use a bur ball or abrasive-filled rubber?
- Should you deburr before or after boring a hole? (On small precision holes, do it before to prevent boring bar deflection by heavy work-hardened burrs; then deburr the boring burrs.)
- Is it faster to rotate a knife around a hole; sand the surface then use a knife around the hole; rotate a Shavit tool around the edge; or sand, ream, and use a bullet?
- Is it easier to rotate the tool or the part?

Some General "Rules"

Generally, the fastest manual-deburring processes for job-shop operations include sanding-like operations, brushing and countersinking, and tube or bar-end chamfering. Hand-fed mechanized machines provide low cost and relatively rapid deburring and finishing with little investment and floor space. Hand scrapers are fast, but precision knife work is relatively slow. Swivel-blade tools are among the fastest and lowest-cost methods of deburring straight edges and long contours. They are fast hole-deburring tools when the holes are large enough to allow use. As shown in Chapter 32, while sanding is one of the cheapest operations, costs can still be reduced by some simple tooling, or by changing to one of the many new abrasive paper products.

Intersecting Holes

Deburring intersecting holes is one of the most difficult deburring tasks faced by many industries. Effectively deburring them is a function of many variables. For an example of normal practice, consider the intersecting holes in Figure 4-1—a stainless steel, handheld-sized part. The intersecting edges must be sharp, with no burr. In other words, the edges can have an edge break of 0.0005–0.0020 in. (13–51 µm). The long holes are about 0.125 in. (3.18 mm) in diameter and, by scaling the figure, it is clear that the intersection is about seven diameters deep from the top surface. In this example, there are 100 of these to make. Any change in material properties at the intersecting edges is reason to reject the part (meaning that electrical discharge machining [EDM] cannot be used to make either hole). In short, economic factors force hand deburring of these parts.

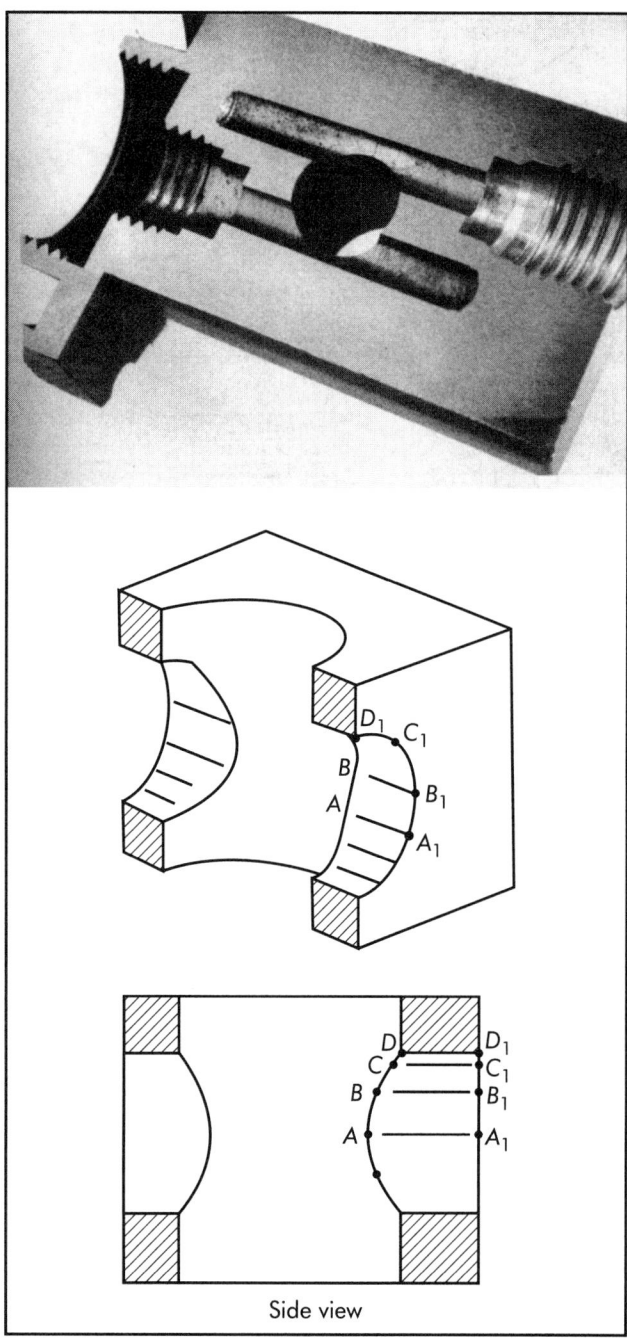

Figure 4-1. Three-dimensional configuration of hole intersections.

Deburring considerations for intersecting holes include:

- hole size;
- hole tolerance;
- hole surface finish;
- depth of intersection;
- type of hole (blind or through);
- wall thickness;
- desired edge condition;
- initial edge condition;
- workpiece material;
- material compatibility (deburring side effects);
- quantity of intersections;
- part schedules, and
- accessibility of intersection.

When users have the difficult requirements of removing 0.003-in. (76-μm) thick burrs, not changing hole size more than 0.0001 in. (2.5 μm), and not exceeding edge chamfers of 0.001 in. (25 μm), deburring by either manual or mechanized methods is challenging. *Deburring Small Intersecting Holes* (Gillespie 1980) defines the capabilities of a variety of individual processes, which should be considered when manual means are insufficient. In difficult situations, manufacturers, by necessity, must combine processes. Thus, by first reaming the hole to make the burrs smaller and then using one of the mechanized deburring processes, it is possible to meet the indicated constraints.

One subtle point on intersecting holes is that the resulting edge breaks or chamfers are in part a function of edge geometry. Consider Figure 4-1: When two holes intersect, the angles formed by the intersection vary from acute to obtuse. The diameter and nature of the intersection (centerline to centerline or just slight breakthrough) dictate the variation in angles (Stein 1995).

To appreciate the effect of angle variations, see Figure 4-2. To produce a 0.005 in. (127 μm) edge break on a 90° edge, 0.0021 in. (53.3 μm) of stock material must be removed. On a 120° edge, the same stock removal would produce a 0.015 in. (381 μm) radius, and on a 30° edge, the resulting radius would be approximately 0.0005 in. (13 μm).

Hand-deburring Tools

Because of the geometry of intersecting holes, many hand-deburring tools are not useful. Normally, the following tools are used:

- drills;
- reamers;
- bur balls;
- cross-hole deburring brushes;
- mounted dental points;
- abrasive-filled rubber dental points;
- abrasive cord and tape;
- abrasive-filled nylon fibers;
- abrasive silicon-carbide balls on brush fibers;
- miniature hand stones;
- pin vises, and
- dental motors.

Drills and reamers. In most applications, use reamers first to cut out the large burr and, insofar as possible, leave a sharp edge. When burrs are too thick for reamers to remove by hand, use a drill. The helix angle on drills reduces the initial torque to shear the burrs.

Bur balls. If the largest hole is 0.0625 in. (1.588 mm) or larger, a bur ball is normally used to chamfer or break the edges of the intersection. Bur balls are available as small as 0.004 in. (102 μm), but the shanks on these very small tools are normally much larger, which may present a problem when two very small holes intersect. As Figure 4-3 illustrates, the relation between the absolute minimum size of the larger hole must be

$$D_4 \geq \frac{d_3}{2} + L_2 \tan\theta + \frac{d_4}{2\cos\theta} \qquad (4\text{-}1)$$

For a standard commercial bur ball (d_2) of 0.040 in. (1.02 mm) diameter, using the tabled data from Figure 4-3 and an intersection depth (L_2) of 0.50 in. (12.7 mm), the minimum large-hole size (D_4) would be 0.101 in. (2.57 mm). This calculated size is considerably smaller than actually required, because the tooth pattern on most balls, by design, cuts with the end of the ball rather than the side (Figure 4-4). In practice, the ball diameter should be twice the diameter of the hole to be deburred to provide an approximate 45° chamfer. In addition, the number of teeth of these tools can vary significantly among suppliers. When the orientation of tooth cut is a problem, some users resort to ball-shaped mounted dental stones. These are commercially available in several grades of hardness in sizes down to 0.094 in. (2.39 mm) diameter.

Bur balls and miniature mounted stones produce burrs themselves or, at the minimum, leave chamfered but sharp edges. When these are not allowable, standard practice is to use miniature cross-hole deburring brushes, abrasive-filled rubber dental points, abrasive cord or tape, or abrasive-filled nylon fibers to provide a radius or blend at the intersection.

Cross-hole deburring brushes and mounted dental points. Stainless-steel tube brushes are available in sizes down to 0.024 in. (0.61 mm). They do not provide a smooth blend, but they remove fine burrs

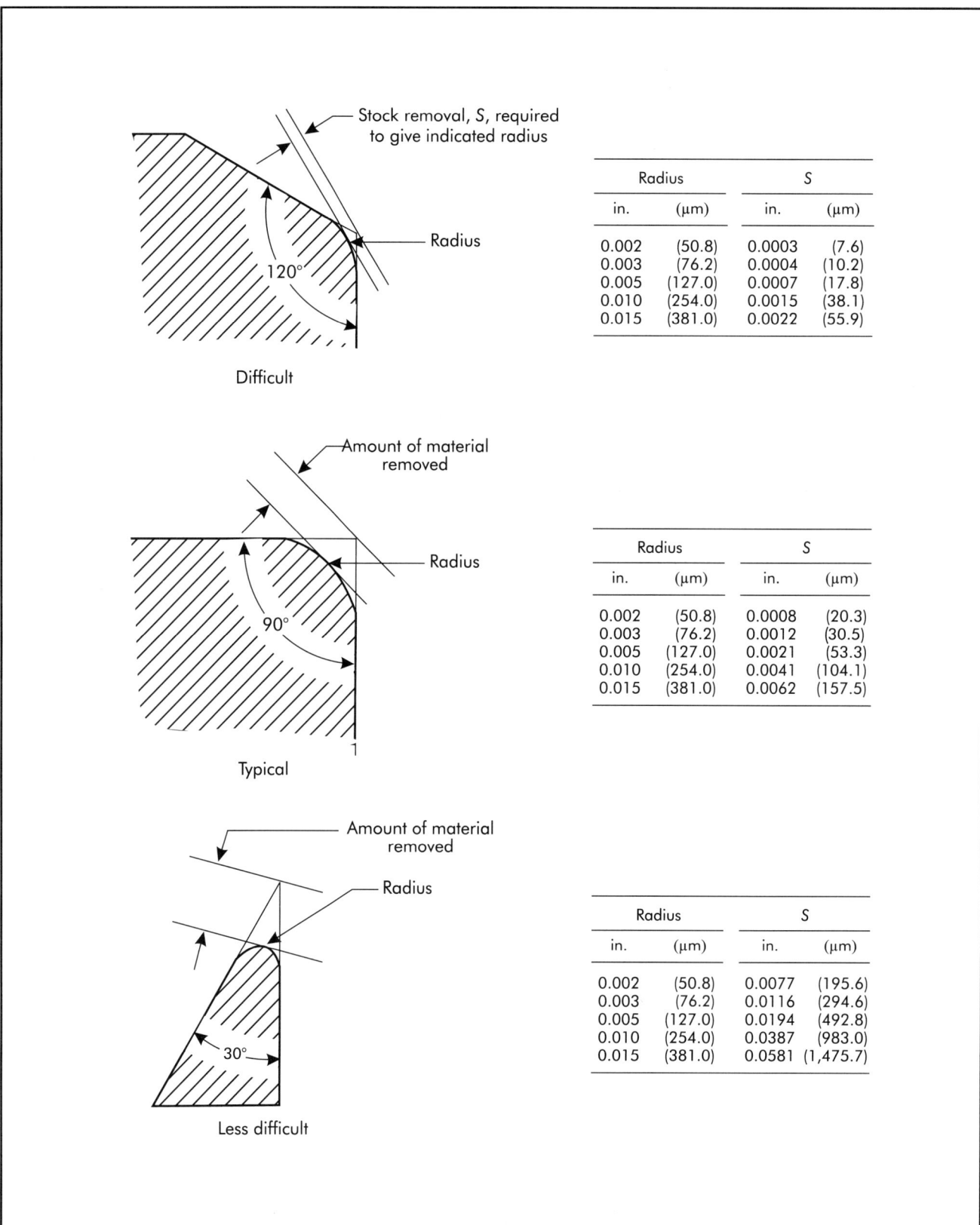

Figure 4-2. Effect of geometry on edge radiusing.

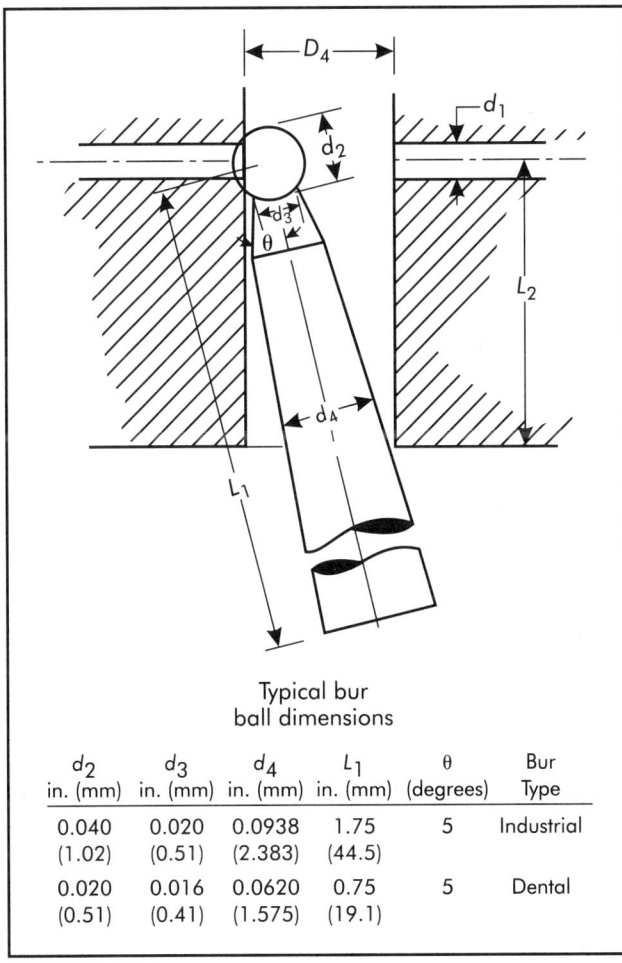

Figure 4-3. Bur ball dimensions related to intersecting hole location and size.

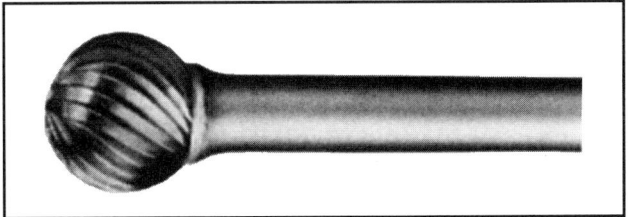

Figure 4-4. Bur ball is designed to cut with the end of the ball.

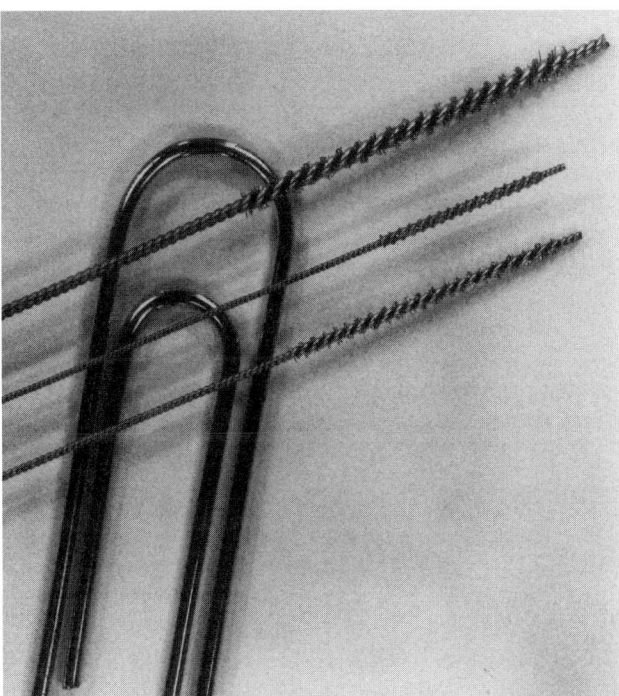

Figure 4-5. Cross-hole deburring brushes.

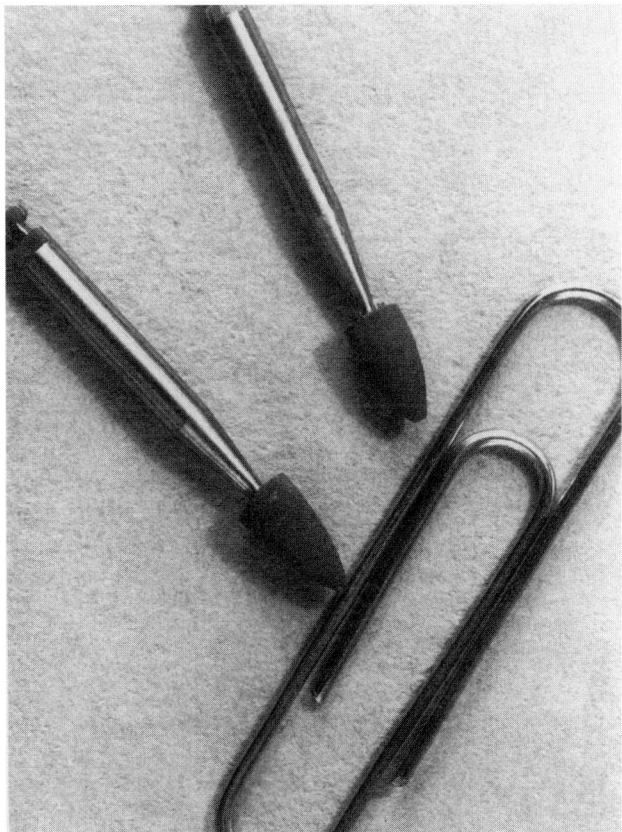

Figure 4-6. Abrasive-filled rubber dental points.

and most loose particles. Figure 4-5 illustrates the size of these brushes. The crosshole deburring brushes and the mounted points scratch hole walls. In most situations, these should not worsen finishes of 16 μin. (0.41 μm).

Abrasive-filled rubber dental points. This group of tools represents the fastest manual approach to radius edges of small holes (Figure 4-6). These are available in at least two grades of aggressiveness and three shank configurations. Because of the limited shank length, they cannot be used on intersections deeper than 0.50 in. (12.7 mm) unless the access hole is 0.875 in. (22.23 mm) or larger (Figure 4-7). If a sharp burr exists when these are used, it cuts the rubber point, destroying its usefulness in a matter of a few revolutions. These tools are available from most dental supply houses, as are a number of mounted dental stones, diamond-plated dental stones, and dental rotary burs.

Abrasive cord and tape. In some situations, the use of an abrasive string or cord provides satisfactory radiusing. These products are available in diameters down to 0.012 in. (0.31 mm), and their limpness allows use for a variety of unusual geometries. In use, the string is simply drawn back and forth over the hole intersection to blend the edge (Figure 4-8). The abrasive does not last long in this application, however, when used on stainless-steel parts, and it is an "if all else fails" application.

Adhesive-filled nylon fibers. Some individuals who perform deburring have noticed that the fibers in abrasive-filled nylon brushes have some excellent characteristics for deburring holes. In this application, a radial brush is disassembled to yield 50 or 100 fibers, each 0.50–1.00 in. (12.7–25.4 mm) long. One of these fibers is trimmed to provide a pointed end. Then, it is inserted in a bench motor to provide a long-wearing and abrasive chamfering tool. This approach can provide radii on holes 0.007–0.030 in. (0.18–0.76 mm). These same fibers can be used to polish 0.020-in. (0.51-mm) diameter holes. By using a long fiber, extending it through the hole intersection, and rotating the motor fast enough, the fiber whips, which increases the abrading force. Several adaptations of this

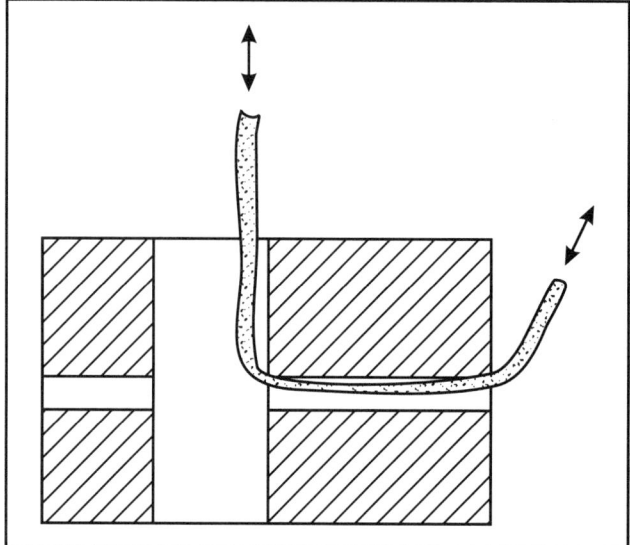

Figure 4-8. Abrasive-coated cord used to deburr intersecting holes.

approach have been used on holes 0.50 in. (12.7 mm) or larger. The same result can be obtained by using a cluster of long steel wires instead of a single nylon fiber.

One company makes a similar tool with a cluster of fibers, each of which has a ball of silicon carbide bonded to the fiber end. This ball adds weight (in other words, force) to the abrasion by fiber, and the carbide ball, in itself, is an abrasive material. By replacing the carbide ball with hard metal stars, a much more aggressive action is possible (Figure 4-9). Abrasive-filled rubber dental tools and abrasive-filled nylon fibers polish to less than 8 μin. (0.20 μm).

When large radii are not required, nonwoven abrasive-filled nylon wheels can be used, which can be cut with scissors to provide any diameter "brush" desired (Figure 4-10). If the access hole is 0.50 in. (12.7 mm) diameter or larger, pencil-thin air motors and dental motors with a variety of abrasive rubber wheels or stones can be employed (figures 4-11 and 4-12).

Figure 4-7. Rubber dental point in right-angle dental motor.

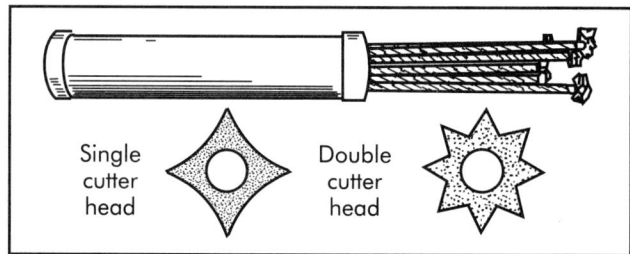

Figure 4-9. Sintered tungsten-carbide stars on stainless-steel cable for heavy-duty cutting.

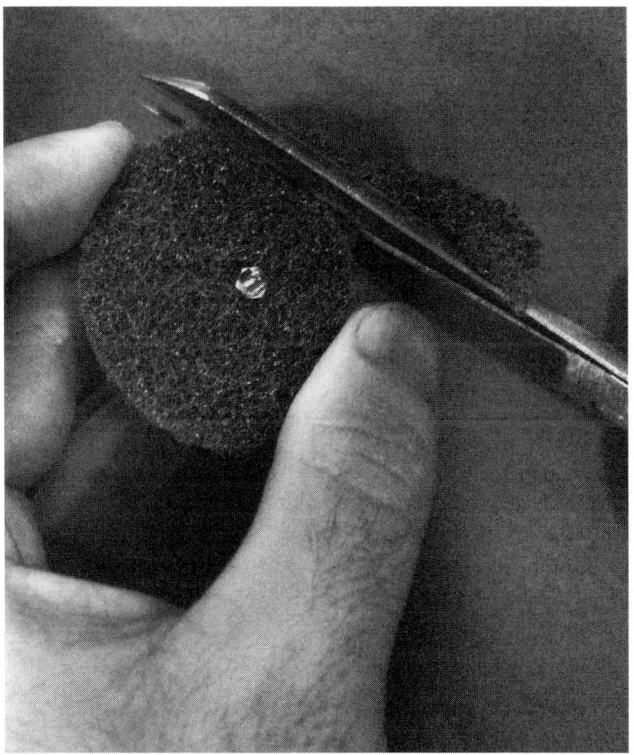

Figure 4-10. Cut abrasive-coated nylon to fit hole size.

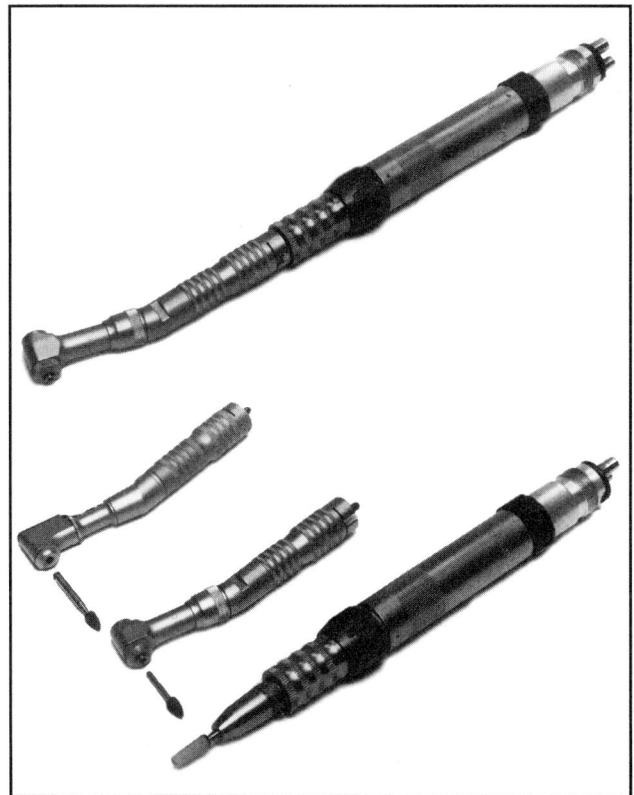

Figure 4-11. Heads and shanks used with dental air motors.

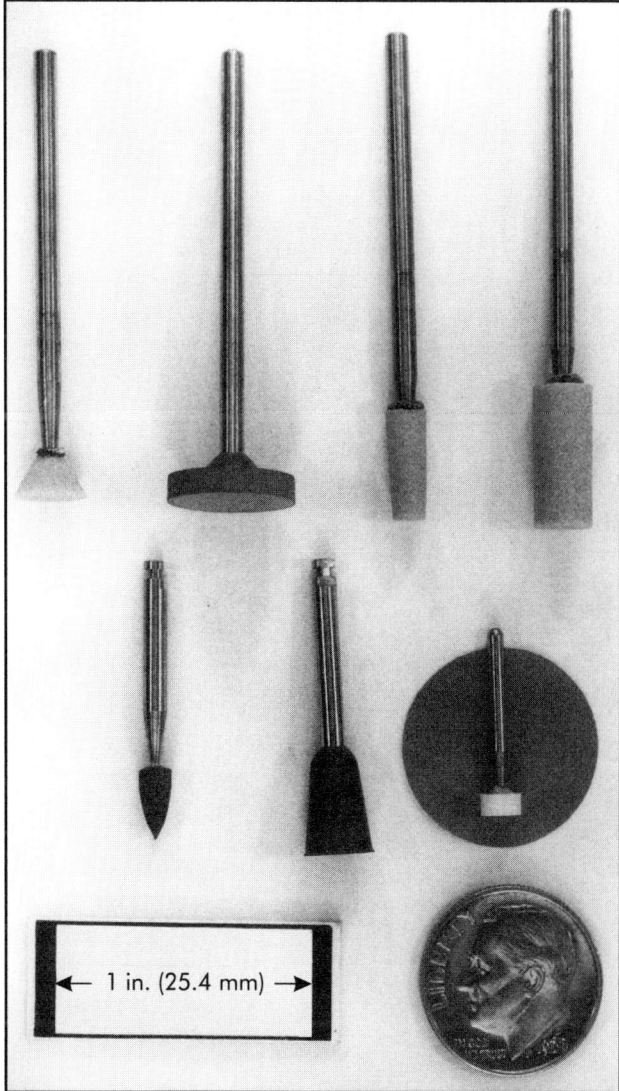

Figure 4-12. Dental tools used in dental air motor.

Abrasive silicon-carbide balls. In some cases, the same result can be more quickly achieved by using a tube-type brush with abrasive silicon-carbide balls on the fiber ends (Figure 4-13). This type of tool is commonly used to hone automotive engine and brake cylinder walls. These brush-like tools are available in standard diameters as low as 0.25 in. (6.4 mm). A simple and similar approach uses narrow strips of abrasive paper looped through a slotted mandrel (Figure 4-14). You can individually tailor this pseudo flap wheel to each part at little cost. Some countries use miniature eggbeater-style brush tools, shown in Chapter 20, for intersecting holes.

Miniature hand stones. It is not generally known, but hand stones are available in diameters down to 0.094 in. (2.39 mm). These stones are excellent for

Figure 4-13. Abrasive balls on end of nylon fibers hone and deburr. (Courtesy Brush Research Manufacturing)

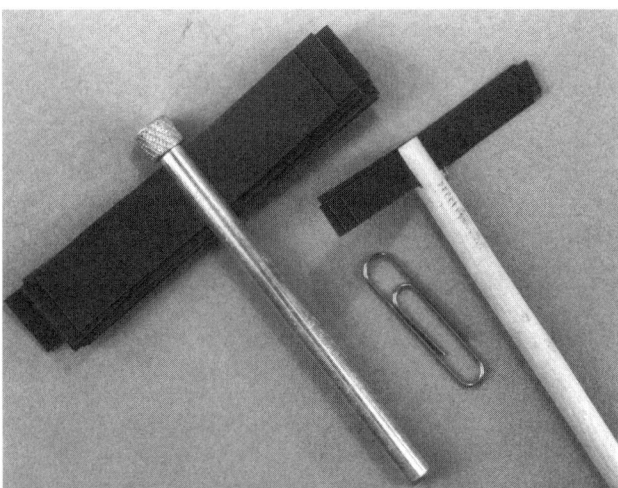

Figure 4-14. Pseudo flat wheel.

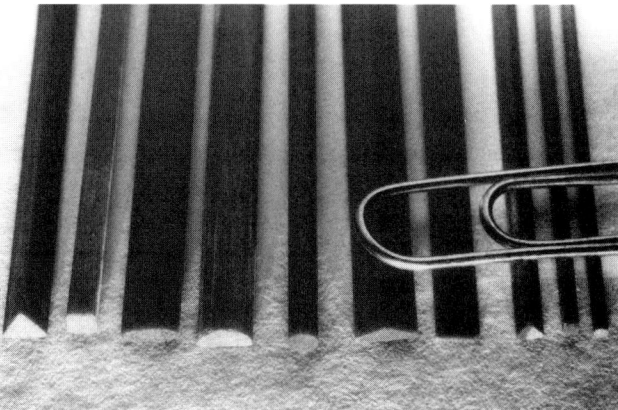

Figure 4-15. Miniature hand stones.

many hard-to-reach places on small parts. As shown in Figure 4-15, they are available in a variety of shapes, and one is only 0.010 in. (0.25 mm) thick.

Dental motors. Economic deburring almost demands motorized tools. For miniature parts and intricate or hard-to-reach areas, this normally necessitates the use of dental or dental-like motors. As Figure 4-7 illustrates, you can insert these tools in holes of $\frac{7}{8}$ in. (22.2 mm) diameter. They are available in speeds of 6,000–400,000 rpm and with integral air or water coolant. Some even have a built-in light aimed at the tool. A variety of interchangeable handpieces is available. An estimated 500 different dental cutters or stones are used in these tools.

Inspection

Verification that intersections are burr-free can be as difficult as the deburring itself. One company reportedly spends $500 monthly on miniature light bulbs just to inspect for burrs in hard-to-see areas. Table 4-1 lists some size limitations of many of the more commonly used deburring and inspection tools.

Abrasive Jet Deburring

Miniature abrasive jets are frequently used to manually deburr complex intersecting holes. These miniature blasters use nozzle openings as small as 0.003 in. (76 μm) diameter. By varying the nozzle diameter, nozzle configuration, and nozzle tip-to-part distance, a wide variety of deburring conditions can be met.

Abrasive jet deburring has been used to deburr hypodermic needles with 0.028 in. (0.71 mm) openings, intersections of miniature slots and threads, and intersecting holes and external features. In most applications, deburring by this process requires less than one minute per intersecting hole. Several case histories of this process are discussed elsewhere (Gillespie 1978).

Intersecting holes and holes intersecting threads are particularly noteworthy applications of abrasive jet deburring. Threads that have been machined through are a particularly difficult deburring problem. Because the blasting process conforms to feature contours, this process is particularly applicable to such features.

The depth of the intersection and the size of the holes to be deburred limit abrasive-jet deburring. Table 4-2 illustrates some of these limitations, based on existing nozzle size. In this instance, assume that deburring action is ineffective on metals when the nozzle is more than 0.50 in. (12.7 mm) from the in-

Table 4-1. Dimensional limitations of some deburring and inspection tools

Tool	Diameter of Large Hole			Maximum Depth Tool/ Press is Effective	
	0.010 in. (0.25 mm)	0.100 in. (2.54 mm)	1.00 in. (25.4 mm)	in.	(mm)
Dental motors			———	4.0	(101.6)
Handheld tools		———————		6.0	(152.4)
Abrasive string	———————			36.0	(914.4)
Dental mirrors (standard)		———		6.0	(152.4)
Sight pipes		———		12.0	(304.8)
Borescope		———		18.0	(457.2)
Otoscope		———		6.0	(152.4)
Fiberoptic display		———————		36.0	(914.4)

Table 4-2. Approximate hole diameter and depth limitations of miniature blasting nozzles*

Nozzle Approach**	Smallest Access Hole Diameter Allowable		Maximum Depth of Cross Hole	
	in.	(mm)	in.	(mm)
Standard nozzle in biggest hole	0.040	(1.02)	0.250	(6.35)
End of head in biggest hole	0.155	(3.94)	0.575	(14.61)
Long nozzle in hole	0.040	(1.02)	2.000***	(50.80)
Right angle head in hole	0.400	(10.16)	1.000***	(25.40)

* This table assumes that the maximum effectiveness of the process is lost when the nozzle is more than 0.50 in. (12.7 mm) from the area to be deburred.
** The nozzle is the thin tube at the end of the tool head (Gillespie 1980).
*** Nozzles this size are in use; actual maximum is probably larger.

tersection. If burrs are thin enough, this assumption might be overly restrictive.

In addition to the process variables already mentioned, the aggressiveness of the process can be increased by changing the abrasive material, abrasive size, the ratio of abrasive to air, and the air pressure. Some miniature blasters use pressures up to 190 psi (1.31 MPa). This is twice the pressure used by some larger units. This 2-to-1 difference results in four times faster cutting.

Blasting processes affect surface finish; but in most situations, texture—rather than average roughness—changes. Blasting processes can readily deburr many parts without worsening a 16 μin. (0.41 μm) finish. Complete burr removal is a function of time, the variables described, and initial burr size. In most instances, however, burr size can be controlled by a reaming operation to remove heavy burrs before blasting. While the process is most applicable to hard metals, 303 Se stainless steel (R_C 30) can be easily deburred by this approach. Since media size is typically 0.001 in. (25 μm) or larger, the process probably would not be used on cross holes smaller than 0.005 in. (127 μm).

This is normally a handheld operation, but it is possible to fixture it or combine it with one of the miniature robots.

Burr Minimization

Several approaches are used to minimize the size of burrs at hole intersections. This minimization may eliminate the need to perform deburring, or at least reduce deburring time. The approaches include:

- using appropriate drill geometry, feed rate, sharpness, and sequence;
- using removable backup material to fill the first hole;
- reaming the holes;
- drilling holes through, then plugging to provide blind holes;
- designing a large counterbore in the blind end for access, then plugging it in assembly;
- using the largest hole size possible;
- using the largest hole tolerances possible, and
- using nonductile workpiece materials when possible.

Appropriate drilling conditions for burr minimization have been the subject of many research studies. (See "Bibliography of Appropriate Drilling Conditions for Burr Minimization" at the end of this chapter.) The following factors always play a major and controllable role in burr size: helix angle, point angle, feed rate, and sharpness. In any cross-hole situation, the manufacturing engineer needs to practice the concepts outlined in the indicated references.

Sacrificial backup material has often been used for minimizing burrs. In cross-hole situations, its use is somewhat limited because its removal may be more difficult than burr removal. Low-melting metal alloys, plastic alloys, and special metal pins or tubes are used in these applications. Low-melting-point alloys have been highly effective in some applications, but useless in others. In some instances, these materials adhere too tightly to allow complete removal. In other cases, they are simply too soft to provide adequate support. Harder backup minimizes burr size.

Frequently, plants drill the larger hole first. In this case, it is possible to plug this hole with a solid pin, then drill the cross hole. If the pin fits tightly in the hole and is as hard as the workpiece, then the resulting burrs must be very small. Large burrs can occur if the pin is softer or significantly smaller than its hole (more than 0.002 in. [51 μm] difference, for example). When large burrs form, it may be difficult to withdraw the pin, because the burr acts as a locking feature. If the intersecting holes are the same size, the drill severs the sacrificial rod, leaving part of it in the hole. In many cases, acid etching is the only process that can remove this piece.

Burrs are the result of ductile deformations. When the workpiece is not ductile, only small burrs can form. In many cases, a permanent or temporary heat-treat operation reduces the ductility.

Threads

Four areas of threads require special attention:

- burrs on thread crests;
- raised metal at threaded hole entrances;
- burrs in threaded hole exits, and
- incomplete first thread.

On high-precision parts, plants may also have to:

- remove micro-burrs on surfaces;
- remove loose particles in threaded holes;
- remove incomplete chips at the bottom of tapped holes, and
- prevent torn thread surfaces.

Figure 4-16 illustrates some burrs on screws, and Figure 4-17 illustrates the types of issues found on

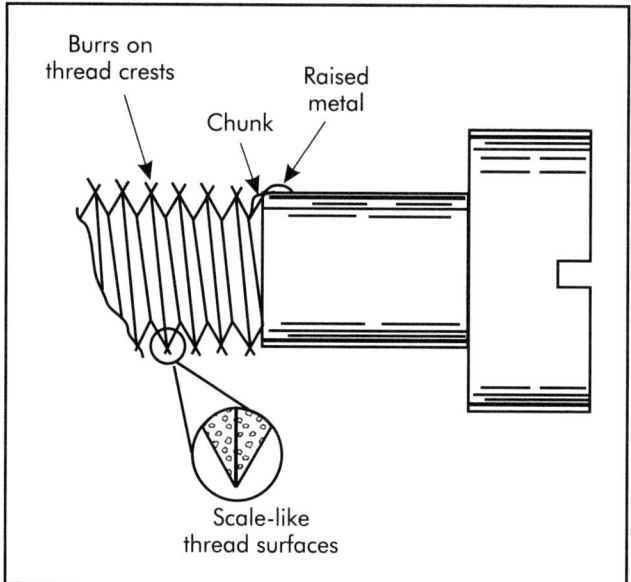

Figure 4-16. Schematic of burr-related features on threaded shafts.

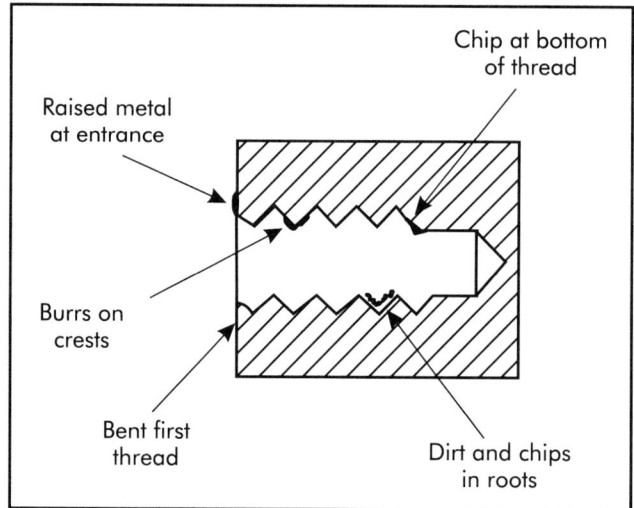

Figure 4-17. Schematic of burr-related features on threaded holes.

threaded holes. Many more issues require consideration for high-reliability screws (Gillespie 1981).

A burr remains on the crests of threads, regardless of whether it is produced by tapping or cold forming (figures 4-18 and 4-19). On external threads, you can easily remove these fine burrs from most parts by sanding, tumbling, or brushing. Internal thread burrs are more difficult to remove, but after they have been removed they too can appear as burr-free as external threads, particularly in the larger thread sizes. On the smaller threads, such as 0.0394, 0.0315, and 0.0236

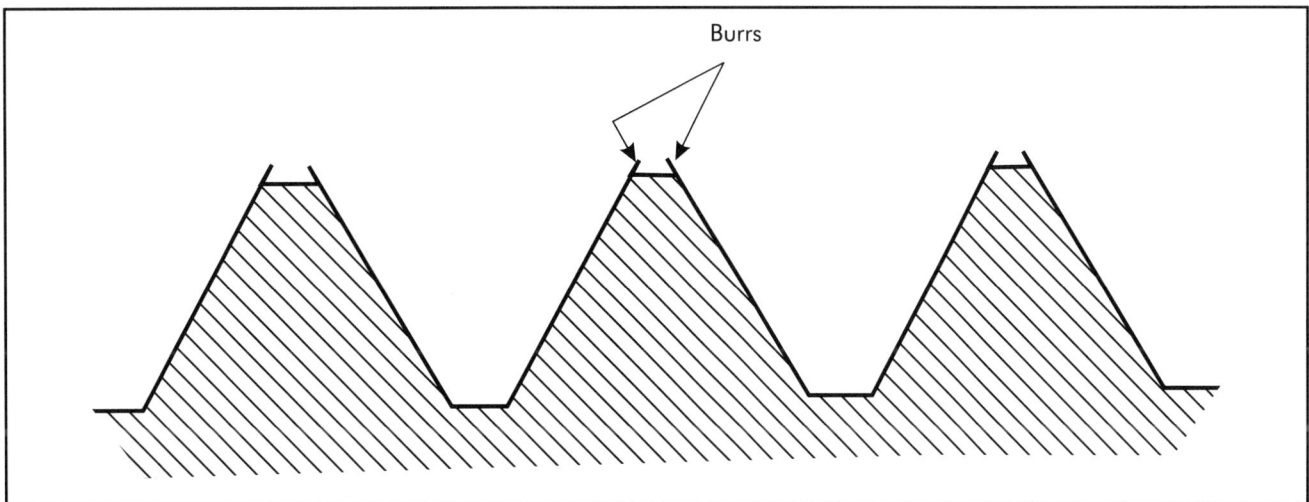

Figure 4-18. Burrs produced by conventional taps.

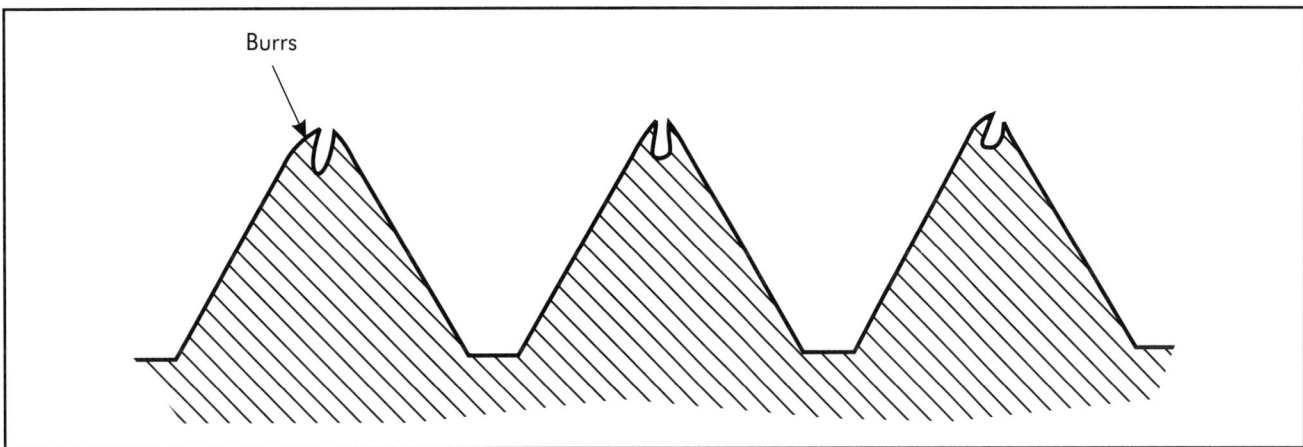

Figure 4-19. Realistic schematic of thread cross-section for cold-formed internal threads.

in. (1.000, 0.800, and 0.599 mm) internal threads, it is more difficult to provide such high-quality surfaces.

The naked eye cannot detect crest burrs on a 0-80 Unified National Fine (UNF) thread. They are only obvious when viewed at 10–20× magnification. One must wonder if such small burrs actually create problems in assembly. The answer is unclear; but it is important to realize that a small metal fragment, such as one of these hair-like thread crest burrs, can keep a miniature electrical contact from closing and, thus, cause a failure. Larger metal fragments or burrs can jam close-fitting components. Although the probability of such events is small, considering the many times that some parts are assembled, disassembled, and tested, the actual possibility is real for many situations. In high-reliability applications, such concerns are real. If an airbag failed to deploy because of such burrs, the concern would be valid. Someone's life would be at stake. If a billion-dollar Mars probe failed because of such a burr, the result would be a decade of wasted work, massive wasted funds, and a lifetime of engineering regret.

Most plants require burr-free parts. Since they do not specify the use of magnification, we will assume (for the purpose of this discussion) that parts must be burr-free to the following extent:

- The unaided eye cannot see burrs.
- Fingernails or toothpicks cannot detect burrs.
- Burrs will not cause cross-threading during gaging.
- Part dimensions are not to be exceeded because of measurements over burrs.

Fingernails can detect burrs only 0.0005 in. (12.7 μm) high. Drawing wooden toothpicks over edges also reveals the presence of burrs this size. On miniature threads, these methods are impossible, because the threads are not greatly larger than burrs on normal parts. Both techniques may be useful on larger threads.

Rolled Threads, Threaded Shafts

External threads formed by thread rolling demonstrate raised metal projecting from the end of the shaft and small burrs on thread crests.

While thread-cutting tools throw material up on shaft diameters, thread rolling extrudes metal axially out the end of the thread (Figure 4-20). The crests on rolled threads can have a smooth, full radius or some small burrs. One can easily tumble these external threads to remove the burrs.

Figure 4-21 shows a small turning burr thrown out into the threaded area. This may be allowable in some cases and not in others. To reduce wasted deburring effort, workers should have clear instruction on whether to remove or leave them.

Finishing Threaded Shafts

Typically, plants use the following procedure to deburr small threaded screws. A tumbling process is preferable, but it may not always provide the necessary quality.

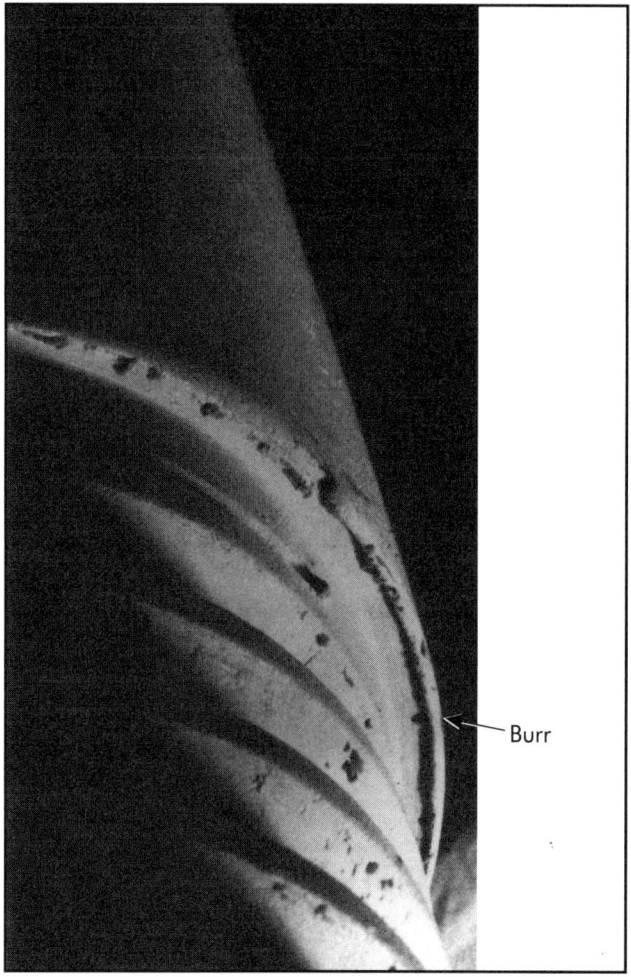

Figure 4-21. Burr on shoulder at end of thread.

1. Use a small knife to cut away any flimsy or damaged first and last thread.
2. Use a knife to cut away any metal thrown on the shaft diameter or past the end of the part.
3. With the point of the knife, dig out any loose chunks of metal at the end of the last thread.
4. Brush the part with a small abrasive-filled nylon wheel to remove crest burrs.
5. Swish the part in a cleaner and air-blow dry.
6. Inspect the part (person performing the deburring).
7. Repeat steps 1–6 as required.

It is more difficult to remove crest burrs from cold-formed threads and threads in aluminum.

Cold-formed Threads, Internal Threads

Internal threads created by cold forming or flowing metal have burrs similar to those produced by cutting taps. Burrs formed on thread crests, however,

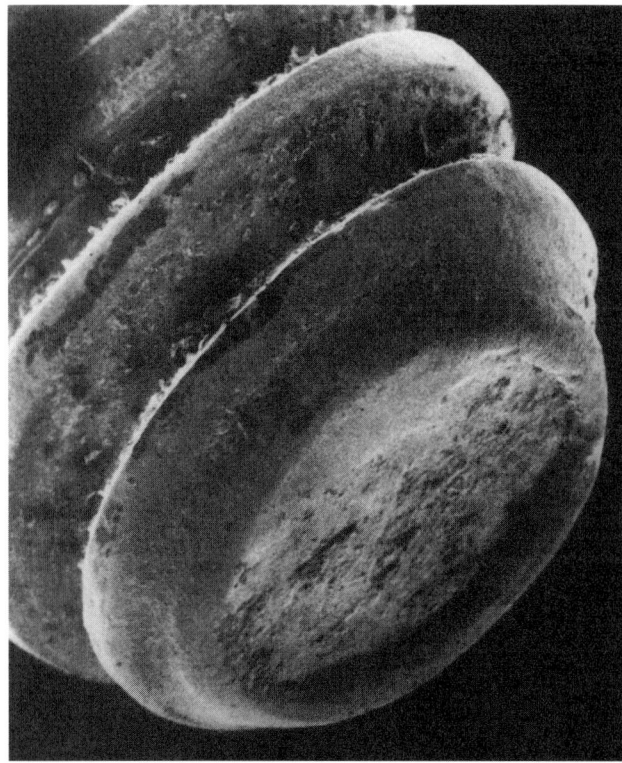

Figure 4-20. Metal extruded past end of shaft by thread rolling.

are significantly different. A cross section of one of these threads reveals a trough in the center of each thread crest. Figure 4-22 illustrates one plant's specification for cold-formed threads. Unfortunately, the actual cross-sectional shape of miniature threads looks more like that illustrated in figures 4-19 and 4-23 than tapped threads. As seen in Figure 4-24, the two sides of the crests are folded over and become the minor diameter. In some instances, they actually close. In many others, however, the cold-formed extruded metal looks like a burr—it does not close.

Normally, the cross section has sharp spikes (Figure 4-23) and many loose particles. In some cases, the trough area is closed over, resulting in a small cavern underneath the thread crest. With this type of thread, it can be difficult to determine how much of this fragmented material should be removed. In many cases, particles are temporarily trapped within the troughs. If the fragmented material were removed, the minor diameter of the thread would be too large (Figure 4-24).

Raised Metal at Hole Entrances and Exits

As shown in Figure 4-25, raised metal forms at threaded hole entrances. Regardless of whether it is considered a burr, it needs to be removed to allow proper seating of mating parts, normally by chamfering the holes. You can also sand surfaces to accomplish this task. Figure 4-26 illustrates one plant's definition for this chamfer. Those performing the drilling and tapping are responsible for chamfering the holes. Although those performing deburring may occasionally have to perform such chamfering, the absence of chamfers should be noted to supervision to ensure that deburring workers are not considered the cause of lower efficiency.

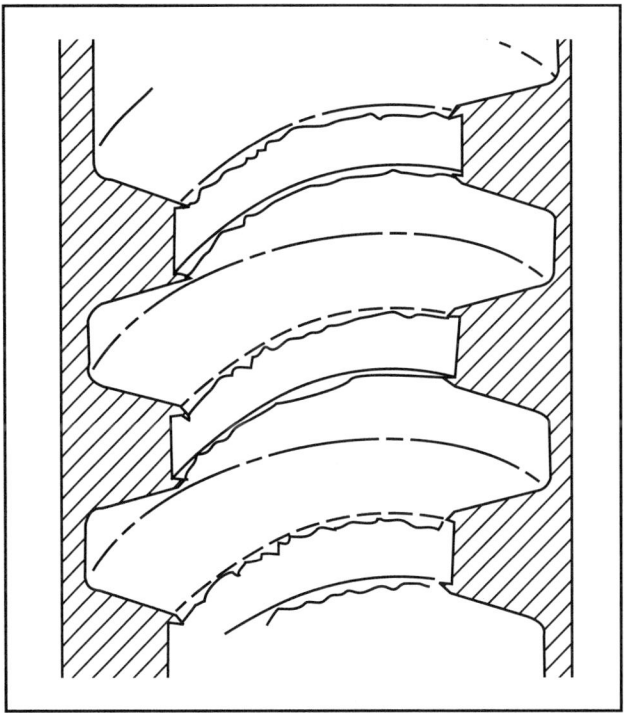

Figure 4-23. Cross section of cold-formed internal threads with irregular crests.

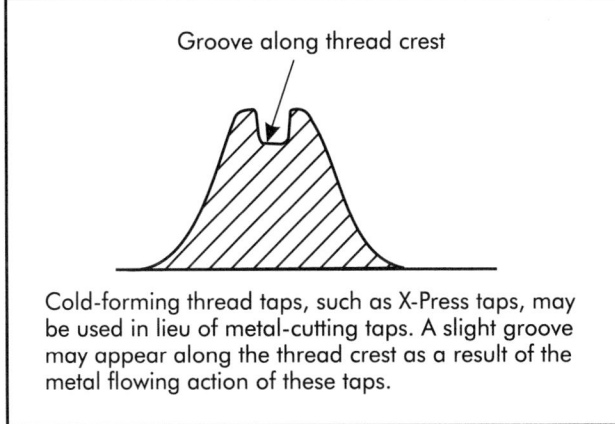

Figure 4-22. Condition allowed by one company's plant workmanship specification for cold-formed threads.

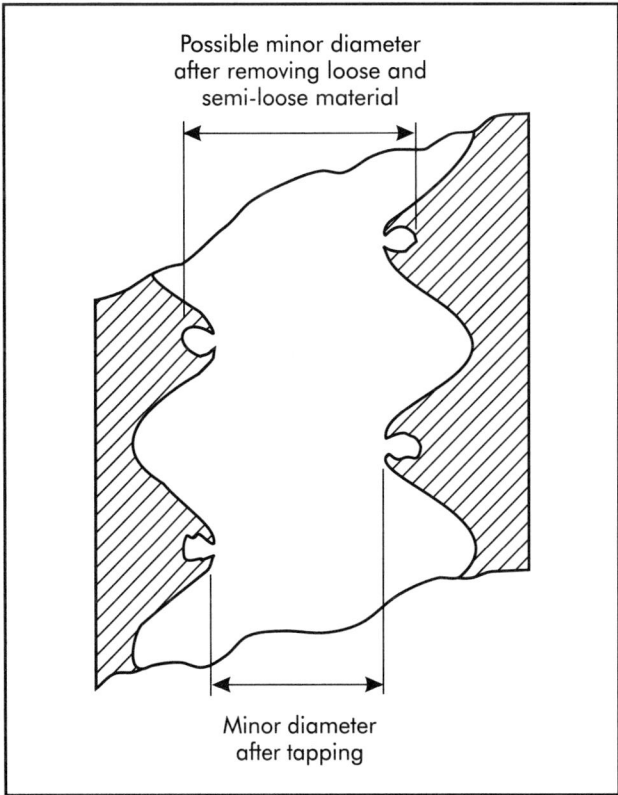

Figure 4-24. Schematic of cold-formed thread dimensions.

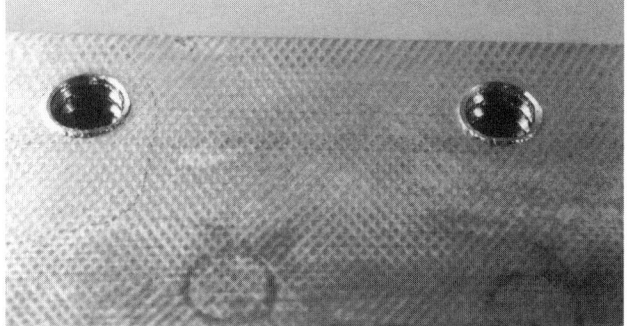

Figure 4-25. Raised metal at threaded hole entrance.

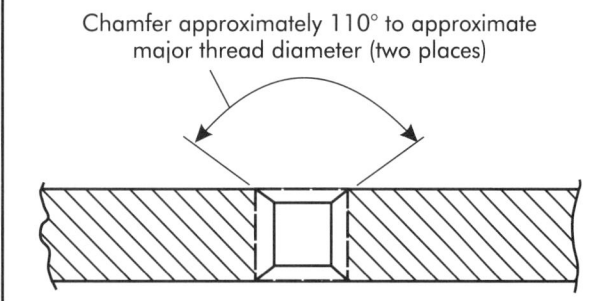

Figure 4-26. Internal thread chamfer required by one plant.

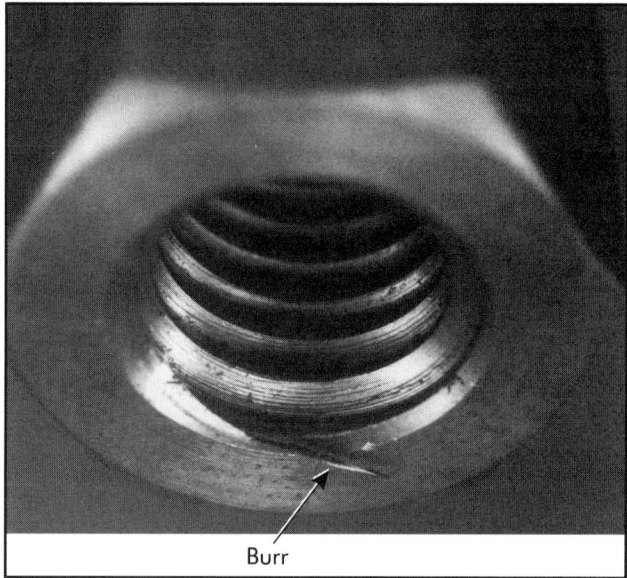

Figure 4-27. Portion of first thread extends outside part geometry.

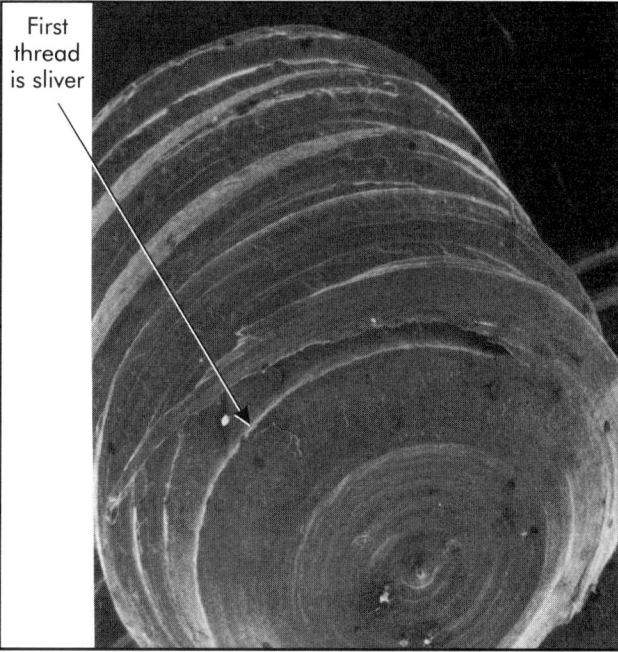

Figure 4-28. First thread on purchased 0.0112-80 "burr-free" screw.

Lead-in Thread

Some plants require partial removal of the lead-in or lead-out thread (referred to in national thread standards as the "incomplete thread" or "blunt start"); others do not, except for Acme-style threads. In practice, it is essential to remove enough of these thread ends to ensure that the thin thread portion cannot be bent over or otherwise damaged. Inspection practice is to reject any threads with this thin fin bent or those damaged in any way (figures 4-27 and 4-28). Many specifications, however, require the leading edges of external threads to be chamfered, which minimizes some of this fin. Figure 4-29 shows a similar incomplete thread on a tumbled part. Since this thread cannot cause a cross threading or break loose, it may be a nonfunctional issue. If that is the case, plant standards should define it that way, and deburr workers should know, through training, that removal of these incomplete threads is not required.

Blind tapped holes frequently have hard-to-reach burrs or chips at the end of the bottom thread (figures 4-30 and 4-31). Though difficult, in some instances these chips may have to be removed since mating screws that are inserted and removed frequently during assembly can dislodge them in the next assemblies. If these chips or burrs are allowable, they also need to be defined. Figure 4-31 shows the use of a modified reamer to dig out these chunks.

Microburrs

The actual thread surface appearance can be quite unsettling when viewed closely. Figure 4-32, for example, shows the surface of a small, roll-formed thread. In this instance, there are many small, semi-loose particles, some loose fibers, and some adhered burrs. These particles, sometimes called microburrs,

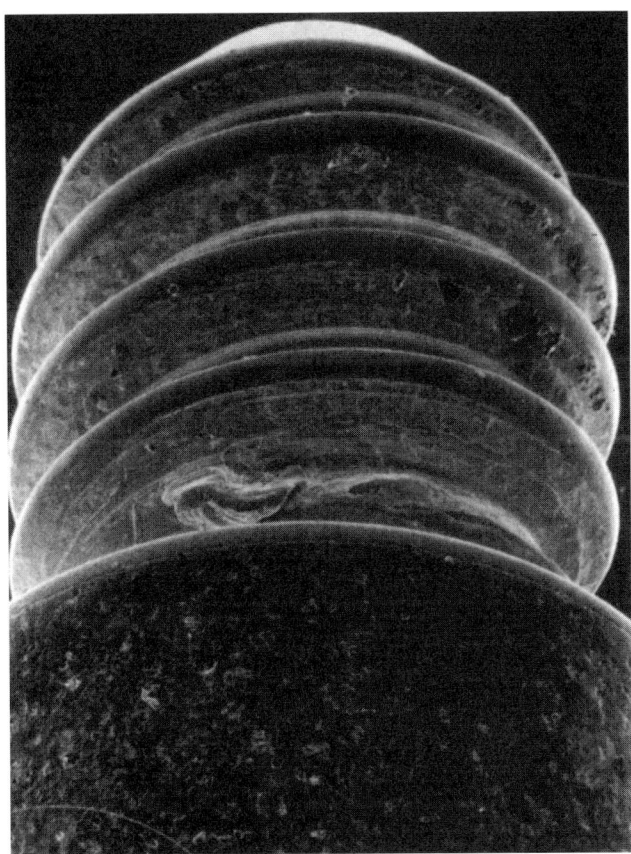

Figure 4-29. Damaged lead-out thread.

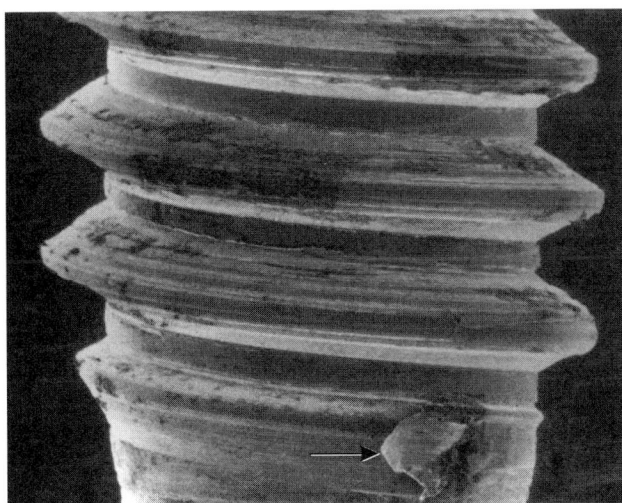

Figure 4-30. Chip in bottom of blind hole.

can only be seen at 200× magnification or higher. A similar burr can be found on ground surfaces when examined closely. Unlike common burrs, microburrs are found on part surfaces, rather than edges. (Actually, most of these occur at microscopic edges.)

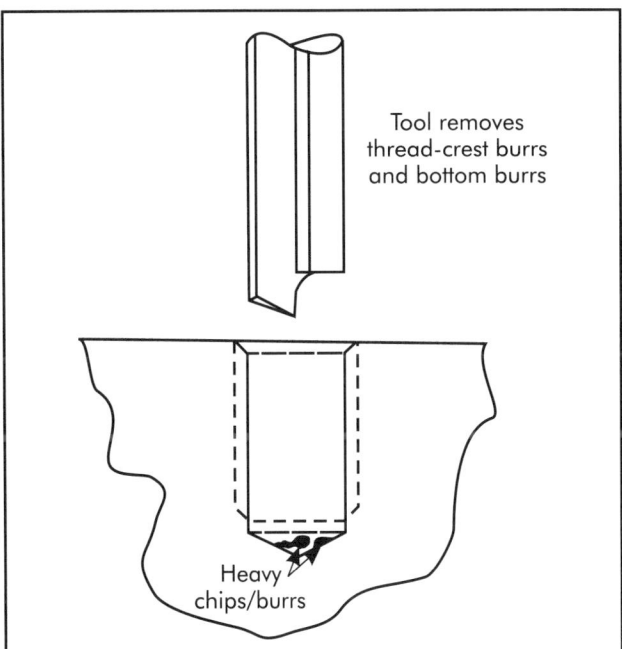

Figure 4-31. Use of chip-removal tool in bottom of holes.

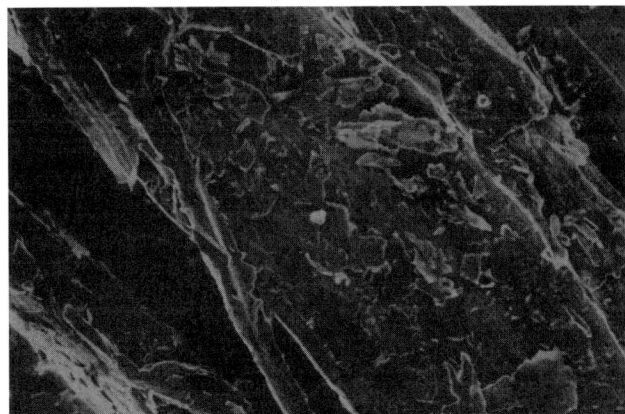

Figure 4-32. Highly magnified view of threads.

Deburring and Finishing Threaded Holes

The following typical procedure is used to deburr small threaded holes.

1. Stone and sand the top flat surface to remove raised metal.
2. Use a knife to cut away any bent portion of the first thread.
3. Use a knife, bur ball, or countersink to provide a chamfer (if one does not exist) or enlarge an existing chamfer. (Ideally, this would have been done in a mechanized operation.)
4. Radius the hole entrance with a rubber dental point.

5. On critical holes, ream the thread crests with an undersize (0.0005 in. [12.7 μm] diameter) reamer.
6. Use a cross-hole deburring brush to clean the threads.
7. Swish the part in a cleaner and air-blow dry.
8. Inspect the part visually (person performing the deburring).
9. Verify that thread gages go as required (to ensure that burrs are not causing gage failure).
10. Repeat steps 1 through 9 as required.
11. Flush holes with small hole-flushing unit.

One plant noted that every time it ran a thread gage or screw into threaded holes, the gage or screw generated fine, hair-like burrs on the crests of threads. In these instances, manufacturing, engineering, and inspection must work together to determine whether those fine hairs of metal really affect function and, if they do, how to remove them after inspection.

In some cases, frequent use of a gage is sufficient to make miniature threads fail "go" gages. Gages are specified in some thread deburring operations to define when adequate deburring quality has been achieved. In this case, if a "go" gage is used, by definition (or agreement with the using department), no further deburring effort is required in the thread area, even if minor burrs exist.

Assume that a number 0.112-40 Unified National Coarse (UNC) thread requires deburring and, in this case, the thread is in a blind hole. Note that a 0.112-40 UNC thread is approximately the same diameter as a normal wooden pencil. Although production workers use their own deburring techniques, all would probably use the following sequence to deburr inside the hole.

1. Stone or power sand the flat surface.
2. Use a countersink or bur ball.
3. Use a knife to remove the bent first thread.
4. Use a knife to wipe away any burrs produced by the bur ball or countersink tool.
5. Use a reamer to remove burrs on crests.
6. Use a cross-hole deburring brush.
7. Radius the hole entrance with a rubber dental point.
8. Use a "go" gage.
9. Dig out any chips in the bottom of the threaded hole.
10. Swish cleaners inside the hole or clean it ultrasonically.
11. Air-blow dry.
12. Scrutinize the hole using gooseneck, fiberoptic, or microscope coaxial lighting.
13. Repeat steps 1 through 12 as required.

Note that, in this instance, a knife removes the burrs produced by the bur ball. Also, a "go" gage verifies adequate burr removal from the crest of the threads. One of many tools can be used to dig the chip from the bottom of the hole, but Figure 4-31 illustrates the most frequently used tool. A reamer is also used in some holes to shave burrs from thread crests during the chip-removal operation. While these steps are relatively time-consuming, they are necessary in some cases to remove each chip. In a few cases, a threaded hole may be entirely packed with wire-like chips. These chips may have to be removed with tweezers or a small knife before using other methods.

One plant has designed miniature knives for miniature threads. Their points and edges must be sharp to reach into the threads. The total thread depth on a 0.02 in. (0.5 mm) thread, for example, can be as small as 0.004 in. (0.10 mm), and the knives must operate within these valleys without damaging other threads.

Figures 4-33 and 4-34 summarize most of the major steps and concerns in thread deburring. In the figures, the starting condition begins at the left and progresses right to the finished condition. You can stop deburring at any point in the progression to the right. As seen in the raised metal series, sanding the raised metal leaves a sanding burr.

Chamfering provides the necessary chamfer, but it leaves a chamfering burr. Rubber dental points provide a radiused entry and remove the chamfering burr. Some of the steps shown in figures 4-33 and 4-34 are applicable only to large-diameter threads.

It is easy to miss burrs in blind threaded holes. In particular, it is difficult for light to penetrate small holes. Gooseneck lamps and coaxial lighting provide the only ways to see inside these holes.

Consider a 0.23 in. (5.8 mm) thread. In this case, the hole is one-fifth the diameter of the 0.112-40 UNC thread described earlier. It is permissible, in a few instances, to use a small reamer to remove burrs on crests. In the last example, the threads are so small that it is very easy for a reamer to remove too much material and, thus, scrap the part. For this reason, if the production work instructions do not specify a reamer, the individual performing the deburring will not use it unless authorized by a supervisor.

Cross-hole deburring brushes cannot remove each crest burr, because the brush fibers do not extend to the end of the brush. A small length of the brush lacks bristles. The depth of a shallow hole may be no deeper than this nonbristle area. The cross-hole deburring brushes also wear very quickly. As they wear, the bristle area tapers, which means that neither the tapered portion of the fiber area nor the nonfiber portion can adequately remove material from the lower

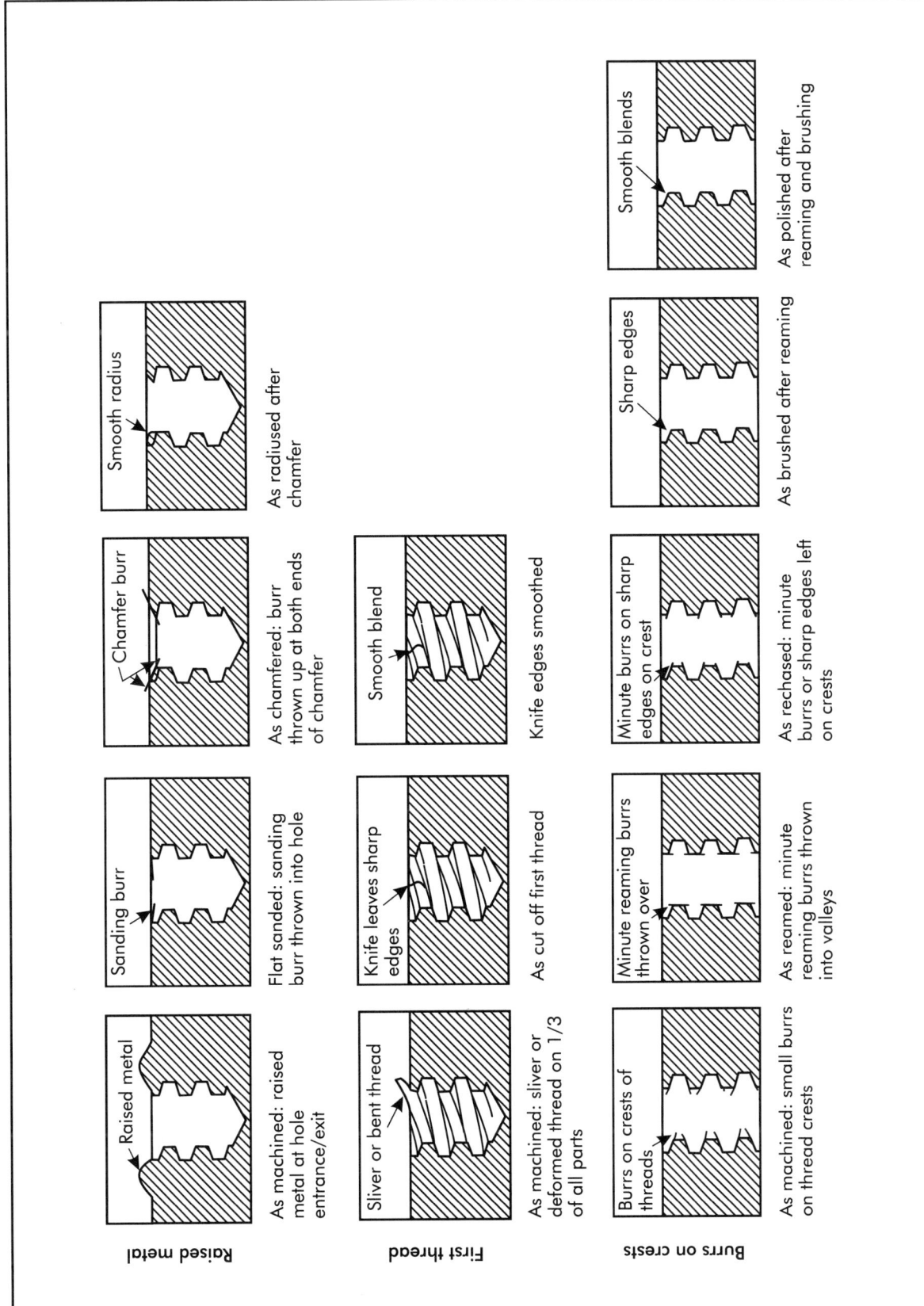

Figure 4-33. Methods of thread deburring.

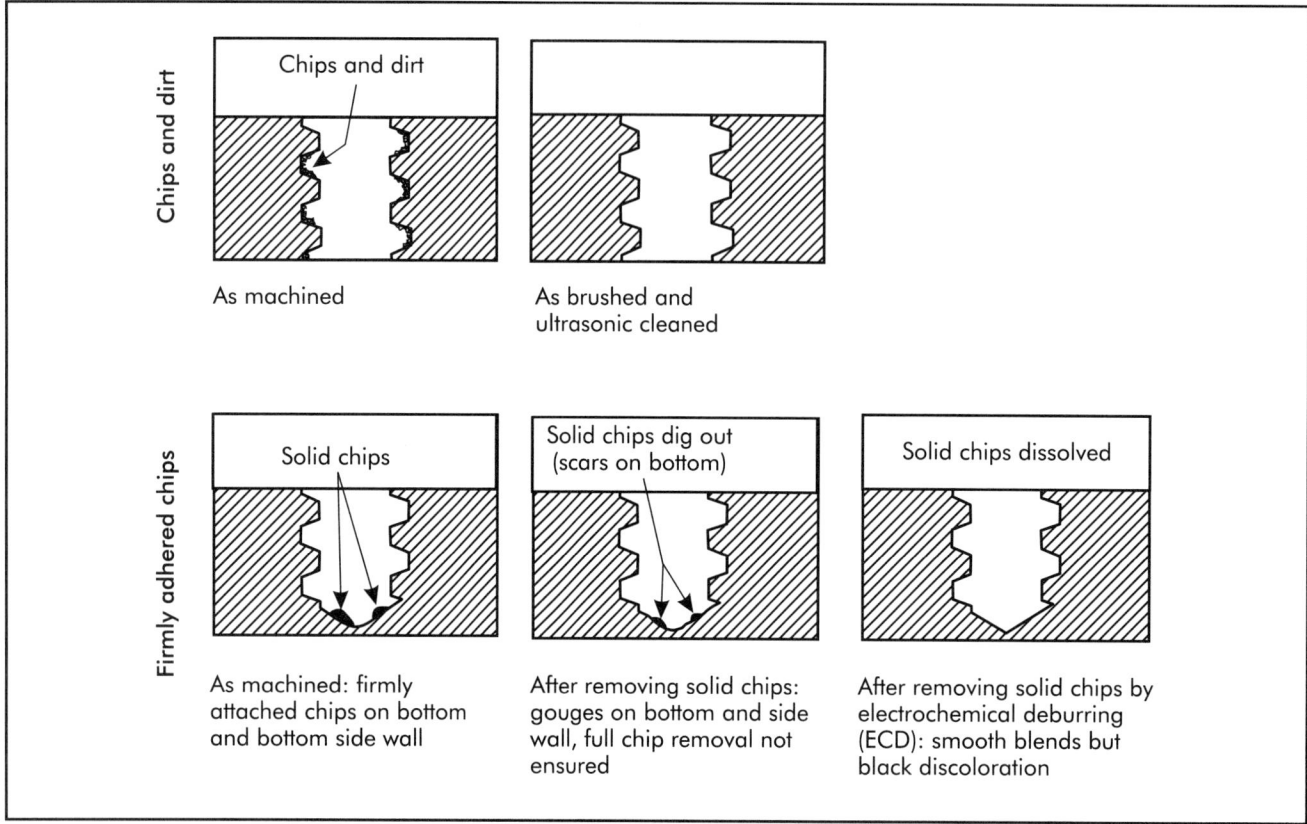

Figure 4-34. Methods of removing chips from holes.

portion of the threads. Table 4-3 lists the brushes used by one plant for threaded holes.

Despite the procedures described thus far, because of the small size of the miniature threads, it is impossible in many cases to get the aesthetic thread quality obtainable on larger threads. It takes longer to hand deburr miniature threads than it takes machining to make the thread. Table 4-4 lists standard hour estimates for various deburring steps. Actual deburring times can be two to 10 times longer, depending on thread condition and operator efficiency.

In thread deburring, avoid these common bad practices:

- Do not use rat-tail files to deburr thread crests—they cut the minor diameter oversize.
- Do not use commercial screws as a cleaning tool to remove fine burrs—many have burrs and sharp edges, which create more burrs.
- Do not use commercial screws to "crunch" particles in the bottom of holes by tightening the screws down or breaking loose contamination in the holes—some of the small screws can be broken.
- Do not run thread gages in miniature threads any more than necessary, because it makes precision miniature threads oversize.

Flash Removal

Flash can be easily removed when it is thin and when parts are designed to minimize flash problems. When inserts are molded, one source recommends lubricating them beforehand to prevent flash adhesion. Polishing and plating the inserts also helps. For insert applications, the flash can be cut, then peeled from the part in many instances. Usually, a mild solution of caustic soda loosens the flash for easy removal (Society of the Plastics Industry 1954). However, plastic surfaces can be harmed if they remain in contact with the solution too long.

Files are often used to remove flash and smooth edges. Select a file based on plastic hardness and the size, shape, and contour of the part. Use sharp, thin-topped teeth to hold cutting-edge sharpness. Well-rounded gullets minimize clogging.

Three-square scrapers are often used more than files today, but many soft materials are suitable for

Table 4-3. Cross-hole deburring brush sizes used by one plant

Thread Size	Maximum Allowable Minor Diameter in.	(mm)	Brush Diameter in.	(mm)	Reamer Diameter for Thread Crests, if Needed in.	(mm)
0.3 mm						
0.5 mm			0.024	(0.61)		
0.6 mm	0.0198	(0.503)	0.024	(0.61)	0.0193	(0.490)
0.8 mm	0.0263	(0.668)	0.032	(0.81)	0.0258	(0.655)
1.0 mm	0.0327	(0.831)	0.032	(0.81)	0.0322	(0.818)
1.2 mm	0.0406	(1.031)	0.047	(1.19)	0.0401	(1.019)
#0-80	0.0514	(1.305)	0.047	(1.19)	0.0509	(1.293)
#2-56	0.0737	(1.872)	0.079	(2.01)	0.0732	(1.859)
#4-40	0.0939	(2.385)	0.097	(2.46)	0.0934	(2.372)
#5-40			0.109	(2.77)		
#6-32			0.125	(3.18)		
#6-40			0.125	(3.18)		
#8-32			0.142	(3.61)		
#10-32			0.156	(3.96)		
#10-24			0.156	(3.96)		
#1/4-20			0.250	(6.35)		

Note: the life of these brushes may be short, and it is essential to search for brush fibers in the holes before using the parts.

Table 4-4. Time required to deburr various threaded holes

Deburring Task	Time Required (Standard Hours per Part) Thread Size					
	0.5 mm	0-80	2-56	4-40	10-32	1/4-20
No deburring	0	0	0	0	0	0
Flat sand only	0.00500	0.00525	0.00875	0.00890	0.00925	0.01250
Chamfer only	0.01642	0.01875	0.01993	0.02111	0.02987	0.03331
Rechase only	0.00467	0.00467	0.00838	0.00999	0.01375	0.01625
Blend entrance only	0.02085	0.02099	0.02940	0.03139	0.03289	0.03687
Cut off first damaged thread only	0.00427	0.00427	0.00787	0.00879	0.00979	0.01439
Ream only	0.00420	0.00420	0.00753	0.00898	0.01236	0.01461
Brush only	0.00350	0.00350	0.00558	0.00678	0.00785	0.00895
Dig out chips only	*	*	0.06700	0.07258	0.07817	0.08933
Electrochemically deburr chips	*	*	0.01117	0.01338	0.01389	0.02010

*Currently not feasible

files, including cellulose acetate, acetate butyrate, and polyethylene, vinyl, acrylics, polystyrene, and copolymers of styrene. Nylon, on the other hand, is not easily filed because it is tough and abrasion resistant.

Teeth cut on a 45° angle are preferred with a coarse, single-cut shear-tooth form. This combination promotes self-cleaning of chips. Milled-tooth files are used for filing sheet edges. File toward a strong section of the part to prevent chipping. Mill files in bastard and western cuts are used to remove flash from flat or convex surfaces of molded articles and for removing saw burrs from edges of sheets. Swiss-pattern files are also used for fine finishes. Semi-automatic filers may be an improvement when significant filing must be performed. Piercing tools or punches may be the fastest way to remove flash from holes.

For Japanese readers, *The Illustrated Deburring Technique for Automation—152 Concrete Machining Examples* (Society of Cutting Fluids and Cutting Technology 1985) contains mostly illustrations—175 pages depicting deburring approaches used on about 500 products. This text is in Japanese.

LATHE PARTS

CNC screw machine parts can be quickly chamfered (beveled) or rounded to remove burrs, rather than manually deburred—which takes longer (Figure 4-35). Most CNC machines have built-in, canned cycles to do this. One plant saved one hour per assembly by changing to this type of chamfering. Chamfer sizes can be easily changed, and tool nose radii can also be adjusted for wear or size. When the drawing allows up to a 0.003 in. (76 μm) chamfer, a 0.002 in. (51 μm) chamfer should be programmed. For one plant, the actual programming was chosen as 0.001 in. (25 μm) less than the allowable break. Groove tools also can be programmed to perform the same operation, which removes or prevents the tear burr that results from grooving. For ongoing success, operators must carefully watch tool tip wear and nose radii variations for these tiny break improvements.

ENVIRONMENTAL, HEALTH, AND SAFETY CONSIDERATIONS

Some materials have safety or health considerations that require different or more controlled approaches to hand deburring. For example, beryllium and its compounds are carcinogenic materials, and the fine particles produced in deburring, if inhaled, can cause chronic beryllium disease. This disease is not detected until years after the work has been performed. In addition, some techniques can result in parts being thrown from the operator's hands, which is an unallowable safety issue.

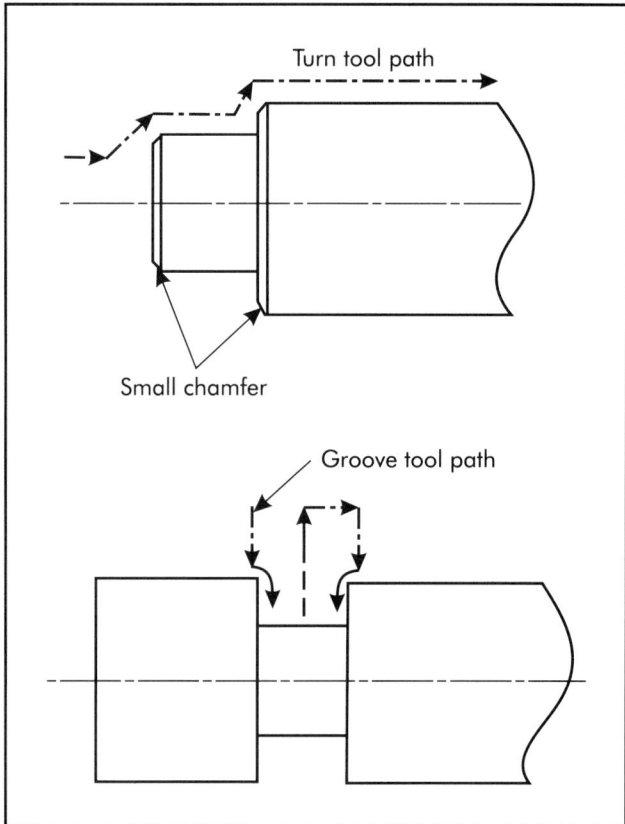

Figure 4-35. Screw machine and lathe CNC chamfering.

APPENDIX

Comments and Issues from People Who Hand Deburr

Many of those who perform hand deburring know painfully well the trials and tribulations that come from their duties. The following comments from a variety of personnel in the deburring field illustrate their observations, needs, desires, and suggestions. Every comment is a statement showing how the company can save money.

Subject 1

Subject 1 is a female with 18 years of experience using a microscope to deburr. She is extremely capable and efficient on precision miniature parts.

"We can't get any tools made to hold parts by the inside diameter of the hole."

"I have to use a dog-nose pin vise to hold the parts, and that takes longer on every part."

"The crib continually runs out of the small nylon brushes. Someone should increase the quantities for these brushes."

Subject 2

Subject 2 is a male with 15 years of experience. He began on large parts and spent 10 years on "scope" work.

"Machinists leave too much burr. They don't change tools often enough. They don't keep sharp tools in the machine."

"Methods engineering pushes machinists to increase feed rates." (Inference is that this increases burr size.)

"We don't have enough brass-jaw pin vises." (Normal pin vises distort and mark precision miniature parts.)

"Methods engineering's time studies and estimates grossly underrate the time required to deburr parts."

"You just can't work eight hours looking through a microscope."

"You can't take shortcuts on some of these extremely miniature parts with threads."

"We have to ultrasonically clean some parts to see the burrs, but 'methods' doesn't include that time in their allowance."

"We get some parts that will be scrapped later because they have scratches or tool marks on them, but we have to deburr them because we don't know what inspection will buy."

"If they would inspect parts for scratches before we deburr, we wouldn't waste our time on these parts." (The inference is that scratches should cause the part to be scrapped before any deburring is begun.)

"We deburr the parts as best we can and just ignore our efficiency rating because the foremen don't pressure us for efficiency. They understand the problem."

"No one makes their rated time."

"Yesterday, I deburred 18 miniature parts (pieces)—that was 40 percent efficiency. (Methods' rating was not applicable. Burr was heavier than normal.)

"We finally have some good microscopes."

"The magnification we use on the scope influences our results." (Some operators can work at higher magnifications.)

"Even our best women have parts rejected."

Subject 3

Subject 3 is a male who has spent 20 years as an electroplater and has spent three months as a deburr operator in a sheet metal department. He does some file and scope work.

"It would help if they would change the height of the fluorescent light above our table to prevent glare."

"I can't get some of the rotary cutters from the crib that used to be available" (rotary files and fine white emery wheels).

"I would refuse to deburr in the heavy machining area because the noise in there is too high. It's not bad here. That sound enclosure for the vibratory machine helps." (Hand deburring people work near vibratory finishing equipment.)

"The detail they put on our production work sheets is good. We don't have to guess what they (engineering) want."

"Part drawings aren't available in the department any more, so I have to walk up front to get a copy any time I have a question." (Implication—I lose 20 minutes of productive work every time I have to get a drawing.)

Subject 4

Subject 4 is a male who has spent 10 years operating vibratory and centrifugal barrel tumbling machinery with hand work to fill in.

"I need a three-fourths-inch water line to get enough flushing action when I wash off parts and media" (one-half-inch water line is available now).

"We need to put a wall up around the 'rock' (vibratory) area because when I rinse media, water splashes onto the slick floor around my area. I almost fell yesterday."

"I need a new oven to dry media and parts. This one doesn't hold the heat even though maintenance just fixed it."

"The special collars you made for the centrifugal barrels are too high to use." (Barrels were not used for two years because of this.)

"The hold-down clamps for the special centrifugal barrel lids don't lock in, and with the screen in the lid I can't reach under to lock them." (The screen is used to keep thin parts from floating out during the cycle.)

"We keep production traveler sheets for centrifugal barrel tumblers and vibratory machines in our area instead of in the file cabinet with the other operation sheets. It's not official, but I don't have to go looking in four foremen's file cabinets every time I have a part to run."

"Most operation sheets have the correct information. Occasionally, I find an operation that calls out a media but specifies a different media code number."

"I need some pans made so I can lay my arms over the edges without cutting them. These plastic pan lids hurt my arms after five minutes, and they warp easily, too." (The problem is that many parts must be separated by hand from the media because of part-to-media size ratio.)

"The plastic lid on the sound enclosure is a pain to use. Somebody should redesign it so that it won't crack and will slide freely."

"I need a light above my work table. Now, I have to adjust three ring lights to get enough light."

"The woman deburring through the scope says she wants the ceiling light above me turned off because it glares in her scope."

Subject 5

Subject 5 is a female with 13 years of experience using the scope. She is one of the fastest and best workers on scope parts but is perennially caught in layoffs because of low plant seniority.

"Several of our special design knives are not set up correctly to keep them in the system. When a certain part goes out of production, stores cancel the requirement for that tool." (Engineering needs to make some of these a general use tool, or list them as a requirement against parts having a long production schedule.)

"My foreman says he can't find the blueprint on two of these tools, and now he's lost the only tools I have of this design."

"I have our men make me some special tools when I need them. Only one of these men can make what I need when I tell him what I want. I keep the tools in my tool box, and they don't have any tool number."

"I can't get any more of these special-design tools."

"The new people (hand deburr operators) don't know about the discontinued tools. When I come back from layoff, I remember what I used to use and I scrounge around until I find one."

"We need a special classification for deburr people who have to use scopes."

"Every time there's a layoff, the skilled deburr scope people get bumped out by machinists with more plant seniority."

"Most of these machinists never had to use a scope; they come from a department where they used files and Red Devils to deburr. These people hack up miniature parts, and what isn't scrapped by inspection must be reworked by someone with experience."

"It takes at least six months for a person who is capable to learn how to do scope work correctly."

"The new people don't get any training; they are given parts and a set of knives and told to deburr."

"The new people don't get any help from anyone in the department."

"The people who can do the deburring get all the rework of those who can't. It's frustrating to get the same pay as someone who can't do the work and have to clean up his mess, too."

"Most of the people who can't do the job don't even know they are not doing it good enough. The foreman gives us the rework and doesn't say anything to the people who hacked them up."

"I hear the company wants to put all burr people in the same classification as the broom men (floor sweepers), oilers, and packaging men. Then they'll have a pool of people they can draw on for any need. Can you imagine a broom man trying to work under a scope on these parts?"

"The bumped-down machinists will leave the bench as soon as they can because of the work and the money."

"We have burr people just waiting for retirement. No one presses them. We get the hard parts, and they get the easy parts for the same pay."

"Some burr people have bad nerves and bad eyes. That limits the parts they can do."

"I get fifty cents less than a machinist, but most machinists can't do the burr work that I have to."

Subject 6

Subject 6 is an inspection foreman.

"We get too many parts with burrs." (The quality report shows at least one group of parts is rejected every week from this department because of burrs.)

"We get some parts month after month that we reject for burrs because the centrifugal barrel tumbling machine doesn't remove all the burrs, and no hand operation is called out." (Apparently, the responsible engineer is not advised. The foreman does the rework without notifying the engineer.)

"We used to have a lot of parts get into stores with minute burrs on them on two particular assemblies."

"We had to go to 100% inspection for burrs on all the parts that go into these assemblies."

"If we don't see the burrs, they are picked up on the precision gages and gouge the next five or six parts."

REFERENCES

Adachi, Katushige, Norihisa Arai, Shoichi Wakisaka, Susumu Harada, and Hidetoshi Hatutori. 1987. "A Study on Burr in Low-frequency Vibratory Drilling." *Bulletin of the Japan Society of Precision Engineering,* 21 (4): 258–64.

Gillespie, LaRoux K. 1976. *Deburring Capabilities and Limitations.* Dearborn, MI: Society of Manufacturing Engineers.

——. 1978. *Advances in Deburring.* Dearborn, MI: Society of Manufacturing Engineers.

——. 1980. *Deburring Small Intersecting Holes.* Kansas City, MO: Deburring Technology International.

——. 1981. "Visual Thread Quality for Precision Miniature Mechanisms." Report BDX-613-2600 (February). Kansas City, MO: Bendix Corporation. (Available from NTIS.)

——. 1999. *Deburring and Edge Finishing Handbook.* Dearborn, MI: Society of Manufacturing Engineers (SME).

——. 2000. *Guide to Deburring, Deflashing, and Trimming Equipment, Supplies and Services.* Kansas City, MO: Deburring Technology International.

Gillespie, LaRoux K., and Elena Repnikova. 2001. *Deburring: A 70 Year Bibliography.* Kansas City, MO: Deburring Technology International.

Miller, Bernard S. 1965. "Control Deburring Quality." *Metalworking.* Nov., pp. 140–148.

Society of Cutting Fluids and Cutting Technology. 1985. *The Illustrated Deburring Technique for Automation—152 Concrete Machining Examples* (in Japanese). Japan: Kogyochosaka Publishing Company Ltd.

Society of Manufacturing Engineers (SME). 2000. *Fundamentals of Manufacturing Processes* videotape, *Deburring Processes.* Product ID VT694. Dearborn, MI: Society of Manufacturing Engineers.

Society of the Plastics Industry. 1954. *Plastics Engineering Handbook.* New York: Reinhold Publishing.

Stein, Julie M. 1995. *The Burrs from Drilling.* Kansas City, MO: Deburring Technology International.

BIBLIOGRAPHY

Blount, Ezra. 1977. "A Review of Industry Standards for Burrs and Related Edge Conditions." Technical Paper 77-471. Dearborn, MI: Society of Manufacturing Engineers (SME).

Drozda, T. 1977. "Deburring: The Common Cold of Industry." *Production,* November: 62–75.

Gillespie, LaRoux K. 1974. "Deburring: An Annotated Bibliography." Technical Paper (no number). Dearborn, MI: Society of Manufacturing Engineers (SME).

——. 1975. "Deburring: an Annotated Bibliography." Vol. 2. Technical Paper MRR75-04. Dearborn, MI: Society of Manufacturing Engineers (SME).

——. 1976. "Deburring: an Annotated Bibliography." Vol. 3. Technical Paper MRR76-07. Dearborn, MI: Society of Manufacturing Engineers (SME).

——. 1976. "Economic Tradeoffs in Deburring." Technical Paper MRR76-17. Dearborn, MI: Society of Manufacturing Engineers (SME).

——. 1977. "The Burr: A 1977 Report on the Technology of Reducing its Cost." Technical Paper MRR77-11. Dearborn, MI: Society of Manufacturing Engineers (SME).

——. 1977. "Deburring: An Annotated Bibliography." Vol. 4. Technical Paper MRR77-04. Dearborn, MI: Society of Manufacturing Engineers (SME).

——. 1977. "An Extension of Proposed Definitions for Burrs and Related Edge Conditions." Technical Paper MR77-09. Dearborn, MI: Society of Manufacturing Engineers (SME).

——. 1977. "Hand Deburring of Precision Miniature Parts." Technical Paper MRR77-08. Dearborn, MI: Society of Manufacturing Engineers (SME); Topical Report BDX-613-1443, unclassified (October). Kansas City, MO: Bendix Corporation. (Available from SME or NTIS.)

——. 1977. "Plantwide Control of Deburring Costs." Technical Paper MR77-05 (March). Dearborn, MI: Society of Manufacturing Engineers (SME).

——. 1978. "Deburring Precision Miniature Parts." *Journal of Japan Society of Precision Engineering,* 1 (4): 89–198.

Kerr, G., 1977. "Phase I Report—AIAC Deburring Programs." Report RAM-000-121 (April). Montreal, Canada: Canadiar Limited.

Tresselt, A. R. 1977. "An In-plant Standard on Burrs." Technical Paper MR77-467. Dearborn, MI: Society of Manufacturing Engineers (SME).

BIBLIOGRAPHY OF APPROPRIATE DRILLING CONDITIONS FOR BURR MINIMIZATION

Fleming, C. M. 1973. "Precision Hole Generation Methods." Report AFML-TR-73-135 (March). St. Louis, MO: McDonnell Aircraft Company.

Gillespie, LaRoux K. 1974. "The Effects of Reaming Variables on Burr Properties." Report BDX-613-1083 (rev. September). Kansas City, MO: Bendix Corporation. (Available from NTIS.)

——. 1975. "The Burrs Produced by Drilling." Report BDX-613-1248 (December). Kansas City, MO: Bendix Corporation. (Available from NTIS.)

——. 1978. *Advances in Deburring.* Dearborn, MI: Society of Manufacturing Engineers (SME).

Gillespie, LaRoux K., and Elena Repnikova. 2001. *Deburring: A 70 Year Bibliography.* Kansas City, MO: Deburring Technology International.

Hasegawa, Yoshio, Shigeo Zaima, and Akiyasu, Yuki. 1975. "Burr Drilling Aluminum and Prevention of It." Technical Paper MR75-480. Dearborn, MI: Society of Manufacturing Engineers.

Kanai, M., and Y. Kanda. 1978. "Statistical Characteristics of Drill Wear and Drill Life for the Standardized Performance Tests." *CIRP Annals* 27:61–66.

Kent, Brian. 1976. "Case Histories of Burr Minimization by Radial Lip Drills." Technical Paper MR76-833. Dearborn, MI: Society of Manufacturing Engineers (SME).

Kitajima, Koiche, Koya Takazawa, Teruaki Miyake, Akihiro Yamamoto, and Yukoio Tanaya. 1990. "Study on Mechanism and Similarity of Burr Formation in Face Milling and Drilling." Technology Report 32 (March). Osaka, Japan: Kansai University.

Lee, G. B. 1989. "Digital Control for Burr Minimization in Drilling." Ph.D. dissertation. Berkeley, CA: University of California.

Lenz, E. 1978. "Investigation in Drilling." *CIRP Annals,* 27 (1): 49–53.

Link, R. 1992. "Gratbildung und Strategien zur Gratreduzierung." Ph.D. dissertation. Aachen, Germany: Rheinisch Westfalischen Technischen Hochschule.

Nakayama, K., and M. Arai. 1987. "Burr Formation in Metal Cutting." *CIRP Annals,* 36 (1): 33–36.

Ohshima, Ikuya, Katsuhiro Maekawa, and Ryoji Murata. 1993. "Burr Formation and Deburring in Drilling Cross Holes." *Journal of Japan Society of Precision Engineering,* vol. 59, 1:155–160.

Pande, S., and H. Relekar. 1986. "Investigations on Reducing Burr Formation in Drilling." *International Journal of Machine Tool Design and Research,* 26 (3): 339–48.

Pekelharing, A. J. 1978. "The Exit Failure in Interrupted Cutting." *CIRP Annals,* 27 (1): 5–10.

Phillips, Joseph L. 1976. "Multi-layer Fastener Systems." Final Report AFML TR-76-76, vols. 1–4 (June). Seattle, WA: Boeing Commercial Airplane Company.

Shikata, H., M. F. DeVries, and S. M. Wu. 1980. "Experimental Investigation of Sheet Metal Drilling." *CIRP Annals,* 29 (1): 85.

Sofronas, Anthony. 1975. "The Formation and Control of Drilling Burrs." Ph.D. dissertation. Detroit, MI: University of Detroit.

——. 1976. "The Effect of System Stiffness, Workpiece Hardness, and Spindle Speed on Drilling Burr Thickness." Technical Paper MR75-132. Dearborn, MI: Society of Manufacturing Engineers (SME).

——. 1976. "Reduction of Burr Formation in Drilling." Technical Paper MR75-376. Dearborn, MI: Society of Manufacturing Engineers (SME).

Sofronas, Anthony, and K. Taraman. 1976. "Model Development for Exit Burr Thickness as a Function of Drill Geometry and Feed." Technical Paper MR76-253. Dearborn, MI: Society of Manufacturing Engineers (SME).

Spur, O., and I. R. Masuha. 1981. "Drilling with Twist Drills of Different Cross Section Profiles." *CIRP Annals,* 30 (1): 31–35.

Stein, Julie M., and David A. Dornfeld. 1995. "An Analysis of the Burrs in Drilling Precision Miniature Holes." In *Manufacturing Science and Engineering,* E. Kannetey-Asibu, ed. MED-vol. 2-1-MH-vol. 3-1, 127–48. New York: American Society of Manufacturing Engineers (ASME).

Stumpf, William R. 1983. "Drilling for a Consistent Burr in AISI O1 Oil-hardening Tool Steel." Technical Paper MR83-676. Dearborn, MI: Society of Manufacturing Engineers (SME).

Subramanian, K. 1977. "Sensing of Drill Wear and Drill Prediction of Drill Life." Master's thesis. Cambridge, MA: Massachusetts Institute of Technology.

Sugawara, A., and K. Inagaki. 1978. "Effect of Shape of Tool Point with Dwindling of Drill Diameter on Drilling: Burr in Case of 0.02% C Steel." *Journal of Japan Society of Precision Engineering,* 44 (2): 179–84.

——. 1982. "Effect of Workpiece Structure on Burr Formation in Micro-drilling." *Journal of Japan Society of Precision Engineering,* 4 (1): 9–14.

Takeyania, H., Shunji Kato, Shoichi Ishiwata, and Hiroyuki Takeji. 1993. "Study on Oscillatory Drilling Aiming at Prevention of Burr." *Journal of Japan Society of Precision Engineering,* 59 (10): 1719–24.

Zaima, A. Yuki, and S. Kamo. 1967–68. "Drilling of Aluminum Alloy Plates with Special Type Point Drill." *Journal of Japan Institute of Light Metals,* 18 (5): 269–76; 18 (6): 307–13.

Deburring Costs—A Primer 5

Few published examples exist of comparative deburring economics. A 1981 book provides estimates of equipment and operation costs for a few popular deburring processes, but it lacks comparative data and fails to account for burr size, part finishes, and tolerances—all of which affect cost (Rhoades 1981). We can provide, however, some cost-based decision-making approaches.

FIRST ESTIMATE OF COSTS

One of the most effective methods of estimating costs is to make the following calculations:

- count the machinists and related machine operators within the company;
- multiply this sum by 0.03, and
- multiply the product by the yearly salary average of machinists.

This calculation provides a rough estimate of in-house hand-deburring costs. It is based on the estimation that the equivalent of 3% of machinists and related workers perform hand deburring and finishing. While this is an easy method of estimating hand deburring costs, it may be grossly in error. According to one survey, many companies devote 10% of their piece-part cost to burr removal (Drozda 1977). A Canadian study reported similar 10% results (Kerr 1977). In an unreported 1979 study by the Society of Manufacturing Engineers (SME), only 10% of the companies polled used fully automatic deburring equipment, and 73% used some hand deburring. Therefore, if a shop employs 10 machinists who each earn (or cost the owner) $60,000/yr, the estimated cost of deburring is 10 × 0.03 × $60,000 = $18,000/yr.

ESTIMATE BASED ON AMOUNT OF EDGE DEBURRED

Another simple approach is to estimate the number of inches (millimeters) of edge to be deburred and multiply by an average plant cost/in. (cost/mm) of edge. For example, a simple rectangular plate that is 8 × 0.5 × 5 in. (203 × 13 × 127 mm) has four edges that are 8 in. (203 mm) long; four that are 5 in. (127 mm) long; and four that are 0.5 in. (13 mm). So, the formula for rectangles would be:

$$4 \times (8 + 5 + 0.5) = 54 \text{ in.} \qquad (5\text{-}1)$$
$$(1{,}372 \text{ mm}) \text{ of edges}$$

At a cost of $0.05/in. ($0.002/mm), each part would cost $0.05 × 54 = $2.70.

Estimating the inches (millimeters) of edges deburred can be relatively deceptive on precision miniature parts. For example, consider the fine-pitch injector screw shown in Figure 1-5. At first glance, readers may estimate that this part has 0.50 in. (12.7 mm) of edges to deburr, but because the thread is fine, and there are two thread crests at the tip of each thread that must be burr-free, the number of inches (millimeters) of edge on this part is 24.8 in. (630 mm).

BROAD VIEW

Table 5-1 provides a simple way to view plant deburring costs. It includes most costs associated with this hand process. One can readily obtain a usable estimate of costs and, at the same time, see how they are apportioned.

Machinists deburr during the machine cycle and, while that time may be internal to the processing cycle

Table 5-1. Elements of deburring costs

Source of Cost	Estimated hrs/yr on Deburring	Salary ($/hr)	Overhead ($/hr)	Yearly Cost ($)
Hand deburr people				
Deburring machine operators				
Machinists deburring at machine				
Deburr foreman				
Engineering or management support				
Inspections for burrs				
Deburr equipment depreciation		n/a	n/a	
Equipment maintenance				
Sharpen/repair deburr tools				
Deburr supplies			n/a	
Energy costs			n/a	
Water and other utilities				
Deburr on machine cycle				
Cost of scratches from burrs on parts bumping together				
Scrap due to inadequate deburr		n/a		
Warranty work caused by burrs				
Floor space costs		n/a	n/a	
Overhead cost				
Property taxes				
Waste removal				
Environmental legacy costs				
Total cost				

Note: hr × salary = yearly salary; overhead × salary = yearly overhead; yearly costs = yearly salary and yearly overhead; depreciation only goes in yearly costs; energy costs = hr × overhead cost = yearly costs; deburring supplies = hr × overhead cost = yearly costs; floor space only goes in yearly costs.

time, it is time that might be better applied to another task. It is a deburring cost. When machinists remove burrs to allow themselves to check a dimension, a deburring cost is incurred. It may not be a cost that management chooses to reduce, but it is a real cost.

ECONOMIC CALCULATIONS

Few people ever really look at the details of costs. By comparing hand deburring with other processes, a user should be able to find the most economical combination of deburring that fits the operation (which is why readers should know something about other deburring processes as well as hand deburring). The following equations provide a reasonable estimate of costs by various processes.

For vibratory or other loose abrasive processes, use

$$C = [(C_D + C_M + WC_Pt + C_B + C_E + C_C + C_W)/N] + [C_L (1 + D_O) (K_1 + K_2)/N] \quad (5\text{-}2)$$

For the thermal energy method (TEM), use

$$C = [C_D + C_M + C_L (1 + D_O) + WC_pt + C_A]/N + C_g/n + C_t/N_P \quad (5\text{-}3)$$

For brush deburring, use

$$C = [C_D + C_M + C_L(1 + D_O) + WC_p t + C_A]/N + C_b/N_{pl} \quad (5\text{-}4)$$

For flame deburring, use

$$C = [C_D + C_M + C_L(1 + D_O) + WC_p t + C_A]/N + C_g/n \quad (5\text{-}5)$$

For manual deburring, use

$$C = [C_L(1 + D_O) + C_A]/N + C_t/N_P \quad (5\text{-}6)$$

For mechanical deburring, use

$$C = [C_D + C_M + WC_p t + C_A + C_L(1 + D_O)]/N + C_S/N_P \quad (5\text{-}7)$$

For chemical deburring, use

$$C = [C_D + C_M + WC_p t + C_A + C_L(1 + D_O)]/N + C_S/N_P \quad (5\text{-}8)$$

For electrochemical deburring, use

$$C = [C_D + C_M + WC_p t + C_A + C_L(1 + D_O)]/N + C_t/N_P + C_S/N_{pl} \quad (5\text{-}9)$$

For electropolish deburring, use

$$C = [C_D + C_M + WC_p t + C_A + C_L(1 + D_O)]/N + C_t/N_p + C_S/N_{pl} \quad (5\text{-}10)$$

where:

C = deburring per part
C_A = cost of cleaning/hr after deburring (labor + material)
C_b = cost of brush
C_B = cost of cleaning materials/hr
C_C = cost of compound/hr
C_D = depreciation cost/hr = machine cost/operating hours
C_E = cost of media/hr = media cost × percent hourly attrition
C_g = cost of gas/cycle
C_L = labor cost/hr to run machine
C_M = maintenance cost/hr of operation
C_P = cost of power used ($/kWh)
C_S = total cost of solution
C_t = total tool cost
C_W = cost of water/hr
D_O = overhead as percentage of labor rate
K_1 = percent of cycle time operator actually spends controlling deburring operation
K_2 = percent of cycle time operator spends cleaning parts
n = number of parts run per cycle
N = number of parts run/hr = n/t
N_P = total number of parts run
N_{pl} = number of parts run for a given quantity of solution or tool life
t = time (hrs) per cycle
W = power used, in kW (1 hp = 0.75 kW)

These equations have some important limitations. First, they assume use of a conventional form of the deburring process. As mentioned elsewhere, it is frequently possible to alter the process slightly to obtain faster or better results. Such alterations may require insertion of another cost term to the equation. Unless the conventional approach is used, these equations provide only initial cost estimates.

A second limitation is that all the equations assume knowledge of the value of each individual component and the time required to remove the burr. While it is possible to use "rule of thumb" costs for media, compounds, and the like, only a few publications provide any information on the time required to remove specific burr sizes. As additional research is reported, this will become less of a limitation. In the interim, analogies can be drawn between other parts subjected to the same process.

A third limitation is that the equations ignore the costs of floor space, area heating, lighting, maintenance, insurance, and supervision. These costs can add $1–2 more per hour than indicated by Equations 5-2 through 5-10. For example, assume that 400,000 parts are to be deburred, and the machine used must have a life that will accommodate that many parts. Assume, too, that the values in Table 5-2 are representative. For the values shown, the thermal energy method (TEM) would be the least expensive process (calculated C = $0.02/part), while manual deburring would obviously be an undesirable approach ($1.21/part). These calculations are predicated on the fact that the processes, in fact, will remove the burrs without adversely affecting parts.

Table 5-2 data is believed to be a reasonable estimate of costs for precision aerospace components. Before decisions are made, however, the values should be discussed with knowledgeable vendors or users. The number of parts per cycle and the cycle duration are functions of part size, burr size, and other variables.

It is also necessary to point out that vibratory deburring is typically one of the least expensive processes if burrs are accessible (keeping in mind that the key to effective hand deburring is to quit using it when a better approach is available). In this example, three factors contribute to the high cost of this process:

1. After deburring, special cleaning is required to remove all traces of compound and loose particles (common on precision aerospace parts, but often not on commercial parts).

Table 5-2. Process used and cost values in dollars per unit

Cost Item	Process			
	Vibratory	Thermal Energy Method (TEM)	Manual	Chemical
C_D	0.40	5.00		0.20
C_M	0.04	1.00	—	0.02
C_L	5.00	5.00	5.00	5.00
C_P	0.04	0.04	0.04	0.04
C_A	—	5.80	5.50	5.50
C_E	0.60	—	—	—
C_C	0.30	—	—	—
C_W	0.15	—	—	—
D_O	0.8	0.8	0.8	0.8
N	50	1500	12	400
n	100	6	1	100
t	2	0.004	0.08	0.25
C_g	—	0.024	—	—
W	4	4	0	0
C_t	—	1,000	1,000	—
N_P	400,000	400,000	400,000	400,000
C_S	—	—	—	6,000
Calculated C	0.106	0.020	1.210	0.052
K_1	0.17	—	—	—
K_2	0.17	—	—	—
C_B	0.60	0.80	0.60	0.50

Note: This table assumes that no automatic load/unload is used.

2. The run time is relatively long, and the quantity of parts per cycle is relatively small.
3. The TEM and chemical processes are basically automated, whereas the vibratory method is not.

In most cases, these three factors are not as pronounced. In all equations except 5-2, it is assumed that the operator is devoting full time to the deburring operation.

A computer program to estimate finishing costs using loose abrasive processes is available and has been described in literature (Kittredge 1985). More in-depth cost calculations for mass finishing are available in *Deburring and Edge Finishing Handbook* (Gillespie 1999).

As mentioned earlier, almost anyone can make an existing process more effective, which means that valid comparisons of the output or costs of different machines depends on a clear understanding of how the comparison is being made. For example, is it a comparison of how you use the machine or how another expert would use it? If more knowledgeable workers are hired, would the same process result in lower costs? Also, there are several accounting procedures today that provide conflicting results, so users should understand their company's preferred practice for establishing process and part costs. For example, one accounting procedure of some companies assigns costs simply by the number of days a part or assembly is in house. It ignores all other variables.

Industrial engineering standard methods data exists to help companies estimate some deburring and finishing costs (Ostwald 1983). The data provide good first-cut information and can be refined with specific standard data for groups of parts or materials.

EQUIPMENT PROCUREMENT COSTS

The average of 20 industries indicates that 5–10% of capital equipment dollars are spent on equipment that cleans or finishes parts. Table 5-3 provides some insight into the costs of common mechanized deburring

Table 5-3. Economic characteristics of deburring processes (relative)

Process	Labor Time for Each Part (min.)	Machine Cost ($)	Tool Cost
Hand deburring	3–60	None	None
Vibratory deburring	0.1–1.0	30,000	None
Barrel deburring	0.1–1.0	1,500	None
Spindle finishing	0.1–1.0	15,000	None
Centrifugal barrel	0.1–1.0	25,000	None
Blasting	1–5	2,000	None
Brushing	0.5–2.0	9,000	Low
Sanding	0.1–1.0	5,500	None
Abrasive flow	1–3	50,000	Medium
Mechanical	1–3	8,000	Medium
Thermal energy	0.01–0.5	100,000	Low
Chemical	0.001–1.0	2,000	None
Electrochemical	1–3	35,000	High
Electropolish	0.01–3.0	5,000	Low

Data are for processes used on small lots of small parts. Machine costs are approximations; cheaper machines can often be obtained. For larger parts, equipment may be twice these values. Most processes also require some expendable items, which add cost to each part.

equipment. Table 5-4 lists 109 known edge finishing processes. Each of these processes can lead operators to create derivative processes, depending on what is needed. One source notes that over 400 applications exist in Russia for magnetic abrasive finishing. Another source reports that 1,500 robots are used for deburring in Germany. No separate verification exists for either number.

The general field of equipment justification is described in many books, and several different systems are used (see Bibliography). Three criteria are used to justify equipment procurement:

1. maximum profit,
2. maximum production, and
3. minimum cost.

The costs, or break-even points, calculated from these three approaches will each be noticeably different.

HELPFUL FACTS

Figure 5-1 provides useful geometry insight for readers who need to consider the volume of material removed. Earlier in this chapter, the approach for evaluating cost per inch (millimeter) of edge removed was discussed. A similar effort may be appropriate on parts that require large edge breaks. In the latter, the volume of material removed may be a more applicable measure. For example, referring to row 1 in Figure 5-1, if a bar of diameter

$2(R + b)$

is chamfered, the area (A) removed in any cross section is

$$A = \frac{ab}{2} \tag{5-11}$$

The volume (V) of material removed from the bar end would be

$$V = \pi Rab + \frac{ab^2\pi}{3} \tag{5-12}$$

Chamfering a hole of radius R to $R + b$ would result in the same values just described (row 2). When radii are generated, rather than chamfers, the values in rows 3 and 4 result. From row 1, for a bar end of radius

$(R + b) = 0.250$ in. (0.50 in. diameter)
(6.35 mm radius [12.7 mm diameter])

With 0.030 in. (0.76 mm) chamfer, using Eq. 5-11, the area of removed material is

$A = \dfrac{ab}{2}$ or $(0.030 \times 0.030)/2 = 0.00045$ in.^{2}/bar

(0.2890 mm^2/bar)

Table 5-4. Deburring processes known in 2001[a]

Process (Code)	Quantity in U.S.	Quantity in World
Abrasive finishing (A)		
Barrel tumbling (A1)	8,000	16,000
Vibratory finishing (A2)	12,000	30,000
Vibratory shaker mixer finishing (A2s)	15	30
Tube flow through vibratory finishing (A2t)	4	5
Roll-flow (centrifugal disc) finishing (A3)	300	500
Centrifugal barrel finishing (A4)	1,000	2,000
Spindle finishing (A5)	500	1,200
Fluidized bed spindle finishing (A5a)	20	40
Recipro finishing (A6)	0	5
Orboresonant finishing (A7)	5	10
Flow finishing (A8)	5	10
Cascading media (A9)	35	50
Immersion lapping (A10)	10	50
Screw rotor deburring (A11)	0	10
Lapping (A12)	10	20
Cryogenic abrasive jet (ABC1)	100	300
Chemical loose abrasive finishing (AC)		
Chemical barrel tumbling (AC1)	100	200
Chemical vibratory finishing (AC2)	200	400
Chemical roll-flow (centrifugal disc) finishing (AC3)	5	10
Chemical centrifugal barrel finishing (AC4)	10	20
Chemical spindle finishing (AC5)	0	0
Chemical fluidized bed spindle finishing (AC5a)	0	0
Chemical recipro finishing (AC6)	0	0
Chemical orboresonant finishing (AC7)	0	0
Chemical flow finishing (AC8)	0	0
Abrasive jet (AB1)	20,000	40,000
Liquid hone abrasive flow deburring (AB2)	1	5
Abrasive flow finishing (AB3)	500	700
Abrasive flow orbital finishing (AB3o)	2	4
Abrasive flow stream finishing (AB3s)	1	2
Ultrasonic abrasive flow finishing (AB3u)	1	2
Cryogenic loose abrasive finishing (ACRY)		
Cryogenic barrel tumbling (ACRY1)	50	90
Cryogenic vibratory finishing (ACRY2)	50	90
Cryogenic vibratory shaker mixer finishing (ACRY2s)	0	0
Cryogenic roll-flow (centrifugal disc) finishing (ACRY3)	0	0
Cryogenic centrifugal barrel finishing (ACRY4)	0	0
Cryogenic spindle finishing (ACRY5)	0	0
Cryogenic fluidized bed spindle finishing (ACRY5a)	0	0
Cryogenic recipro finishing (ACRY6)	0	0
Cryogenic orboresonant finishing (ACRY7)	0	0
Cryogenic flow finishing (ACRY8)	0	0

Table 5-4. (continued)

Process (Code)	Quantity in U.S.	Quantity in World
Magnetic loose abrasive finishing (AM)		
Magnetic abrasive barrel finishing (AM1)	5	50
Magnetic abrasive vibratory finishing (AM2)	0	0
Magnetic abrasive spindle finishing (AM5)	0	5
Magnetic abrasive cylindrical finishing (AM5a)	0	5
Magnetic abrasive tube-ID finishing (AM5b)	0	5
Magnetic abrasive ball finishing (AM5c)	0	5
Magnetic abrasive special shape finishing (AM5d)	0	5
Magnetic abrasive prismatic finishing (AM7)	0	5
Mixed metal fibers magnetic media (AM8)	0	1
Chemical magnetic loose abrasive finishing (AMC)		
Chemical magnetic abrasive barrel finishing (AMC1)	0	0
Chemical magnetic abrasive vibratory finishing (AMC2)	0	0
Chemical magnetic abrasive spindle finishing (AMC5)	0	0
Chemical magnetic abrasive cylindrical finishing (AMC5a)	0	0
Chemical magnetic abrasive tube-ID finishing (AMC5b)	0	0
Chemical magnetic abrasive ball finishing (AMC5c)	0	0
Chemical magnetic abrasive special shape finishing (AMC5d)	0	0
Chemical magnetic abrasive prismatic finishing (AMC7)	0	0
Ultrasonic liquid deburring (AU1)	20	30
Ultrasonic slurry finishing (AU2)	10	15
Electrochemical loose abrasive finishing (EAC)		
Electrochemical barrel tumbling (EAC1)	0	0
Electrochemical vibratory finishing (EAC2)	1	1
Electrochemical centrifugal barrel finishing (EAC3)	0	0
Electrochemical spindle finishing (EAC4)	0	0
Electrochemical fluidized bed spindle finishing (EAC4a)	0	0
Electrochemical roll-flow (centrifugal disc) finishing (EAC5)	0	0
Electrochemical recipro finishing (EAC6)	0	0
Electrochemical orboresonant finishing (EAC7)	0	0
Electrochemical flow finishing (EAC8)	10	20
Chemical deburring (C1)	20	40
EDM deburring (E1)	1	2
Electrochemical deburring processes (EC)		
Electrochemical deburring (salt) (EC1)	600	1,200
Electrochemical rotary electrode (EC1R)	1	2
Electrochemical deburring (glycol) (EC2)	50	90
Electrochemical moving electrode deburring (EC4)	0	0
Electrochemical mesh deburring (EC5)	0	0
Electrochemical brush deburring (ECM3)	0	0
Electrochemical orbital abrading (ECM5)	1	1
Electrochemical nonwoven abrasive magnetic finishing (ECM6)	0	6

Table 5-4. (continued)

Process (Code)	Quantity in U.S.	Quantity in World
Electropolish (EL3)	50	100
Manual deburring (M1)	300,000	800,000
Mechanized cutting (M2)	10,000	20,000
Skiving (M2s)	50	100
Robotic deburring (M2R)	300	500
Brushing (M3)	50,000	100,000
Buffing (M3b)	1,000	2,000
Wheel blending (M4)	300	600
Sanding (M5)	20,000	50,000
Vibratory conveyor deburring (M5a)	1	1
Waterjet deburring (M6)	500	800
Edge rolling (M7)	400	600
Edge burnishing (M8)	300	300
Trimming (press work) (M9)	800	1,600
Edge coining (M10)	100	150
Ballizing (M11)	4	10
Tearing (M12)	10	50
Edge peening (M13)	1,000	2,000
Abrasive chemical (mechanochemical polishing) (MC1)	5	20
Hot tool deflashing (MT4c)	100	300
Hot tool deflashing (MT4L)	5	20
Torch cut deburring (T1)	200	400
Thermal energy method (T2)	500	700
Plasma deburring (T3)	50	60
Hot wire deburring (T4)	1	1
Hot blade deflashing (T4b)	100	200
Plasma glow deburring (T5)	5	5
Resistance heat deburring (T6)	1	1
Laser deburring (T7)	1	5
Chlorine deburring (TC1)	0	0

Note: Process codes are in parentheses.

[a] These estimates are based on the intentional use of the process identified for deburring. Several of these processes are also used in other finishing functions. Several processes have been studied in the laboratory, but have not been used industrially. Manual deburring refers to the estimated number of persons who perform deburring with some form of handheld tool.

[b] All processes in this category, except fluidized bed spindle finishing, can use steel or metal media, which peens rather than abrades edges.

and the volume, using Eq. 5-12, is

$$V = \pi(0.25 - 0.03)(0.03 \times 0.03) + \frac{(0.03 \times 0.03 \times 0.03\pi)}{3}$$

$$A = \frac{ab}{2} = 0.00023 \text{ in.}^3 \ (0.0038 \text{ cm}^3)$$

The length of edge deburred in this example is

$V/A = 0.00070/0.00045$ in. $= 1.76$ in. (39.6 mm)

It is also the circumference, which is

$$2\pi(R + b) = 2 \times 3.14159 (0.250 + 0.03) = 1.76 \text{ in. (39.6 mm)} \quad (5\text{-}13)$$

	Type	A	V
1	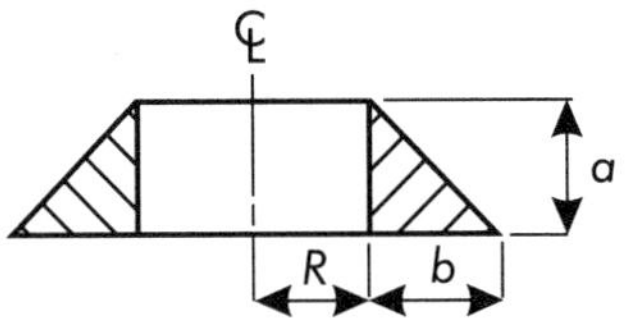 (diagram with dimensions R, b, a, centerline)	$\dfrac{ab}{2}$	$\pi Rab + \dfrac{ab^2\pi}{3}$
2	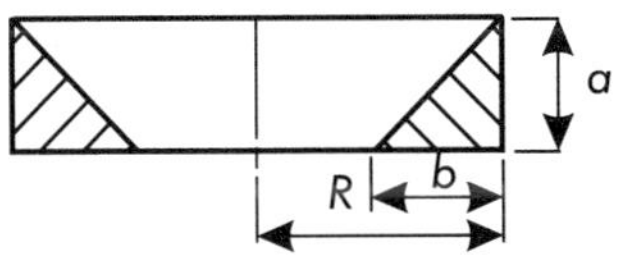 (diagram with dimensions R, b, a)	$\dfrac{ab}{2}$	$\pi Rab - \dfrac{ab2\pi}{3}$
3	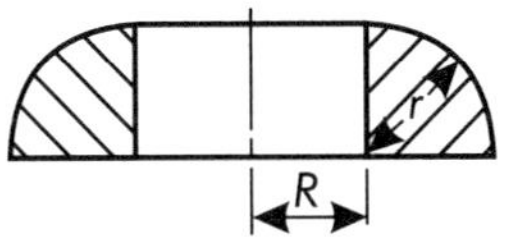 (diagram with dimensions R, r)	$\dfrac{\pi r^2}{4}$	$\pi^2 r^2 R + \dfrac{2\pi r^3}{3} = 1.5708 r^2 R + 2.0944 r^3$
4	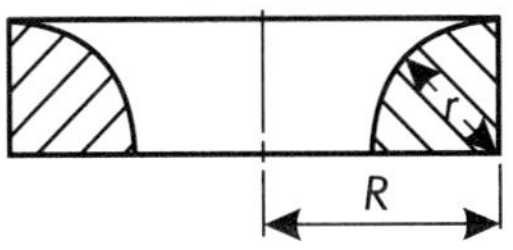 (diagram with dimensions R, r)	$\dfrac{\pi r^2}{4}$	$\pi^2 r^2 R - \dfrac{2\pi r^3}{3} = 1.5708 r^2 R - 2.0944 r^3$
5	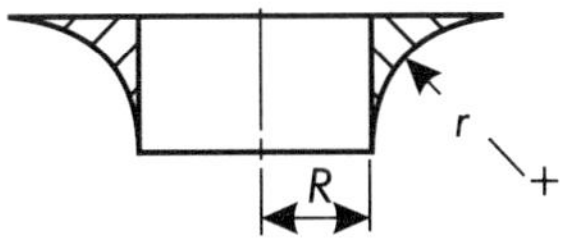 (diagram with dimensions R, r)	$0.2146 r^2$	$1.3486 R r^2 + 0.0476 r^3$
6	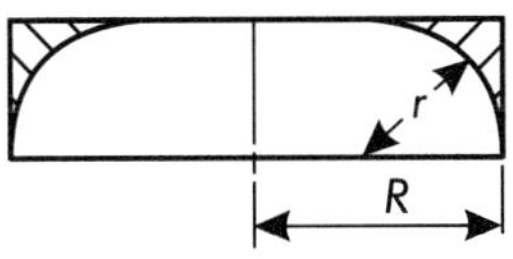 (diagram with dimensions R, r)	$0.2146 r^2$	$1.3486 R r^2 - 0.0476 r^3$

Figure 5-1. Area and volumes of material removed from edges. (Courtesy Cahners Publishing Co.)

If the product were a straight rectangular bar 10 in. (254 mm) long, and a 0.030 in. (0.76 mm) chamfer was desired, the volume of material removed would be

$$\dfrac{(0.030 \times 0.030 \times 10)}{2} = 0.00450 \text{ in.}^3 \; (0.0737 \text{ cm}^3) \quad (5\text{-}14)$$

for each of the four long edges, or 0.01800 in.³ (0.2950 cm³). The short ends of the bar would add a little more. It is uncommon to talk about deburring costs per cubic inch of material removed, but it can be a highly effective calculation to make the point that better approaches are needed. Milling costs might be rated at 10 in.³/min (163.9 cm³/min) while manual deburring might be as low as 0.03 in.^{3}/min (0.5 cm^3/min). With labor rates roughly equal per minute, the boss will become very interested in making manual operations more effective.

REFERENCES

Drozda, T. 1977. "Deburring, the Common Cold of Industry." *Production,* November: 62–75.

Kerr, G. 1977. "Phase I Report—AIAC Deburring Program." Report RAM-000-121 (April). Montreal: Canadair Limited.

Kittredge, John B. 1985. "Computer-aided Mass Finishing Process Development." Technical Paper MR85-836. Dearborn, MI: Society of Manufacturing Engineers.

Ostwald, Phillip F. 1983. *American Machinist Manufacturing Cost Estimating Guide.* New York: American Machinist.

Rhoades, Lawrence J. 1981. *Cost Guide for Automatic Finishing Processes.* Dearborn, MI: Society of Manufacturing Engineers.

BIBLIOGRAPHY

Association for Manufacturing Technology (AMT). 2001. *Economic Handbook of The Machine Tool Industry—2000–2001 edition.* McLean, VA: AMT.

Carpenter, Richard W. 1976. "Cost Control Program for Job Shops." *Finishing Highlights,* January/February: 16–30.

Clarke, P. K. 1969. "Determining Finishing Costs Accurately." *Metal Finishing Journal,* November: 396–98.

Conn, Harry C. 1979. "Equipment Justification—Based on Earnings, Not Savings." Technical Paper 520B. W.A. Whitney Corporation.

"Cost Reduction—Your No. 1 Capital Spending Objective." 1975. *Production's Manufacturing Planbook* (Production special issue), 24–29.

Gillespie, LaRoux. 1976. "What is the Actual Cost of a Burr?" *Abrasive Methods,* May/June: 4–7.

———. 1976. "What is the Actual Cost of a Burr—II?" *Abrasive Methods,* July/August: 4–6.

———. 1977. "How to Estimate Your Deburring Costs." *Machine and Tool Blue Book,* August: 62–68.

———. 1999. *Deburring and Edge Finishing Handbook.* Dearborn, MI: Society of Manufacturing Engineers.

"Hidden Savings Brought to Light." 1976. *American Machinist,* March: 65–74.

Humphreys, Kenneth K., and Paul Williams. 1996. *Basic Cost Engineering.* New York: Marcel Dekker.

Lang, Hans J. 1989. *Cost Analysis for Capital Investment Decisions.* New York: Marcel Dekker.

"Machine Justification and Manufacturing Productivity." 1978. *Machine and Tool Blue Book,* August: 110–32.

Rampe Manufacturing Company. 1970. "How to Determine Exact Finishing Cost Per Piece." Brochure. Cleveland, OH: Rampe Manufacturing Company.

Rosenau, Milton D. Jr. 1979. "Probability and Internal Rate of Return." *Industrial Research/Development,* January: 100–105.

Sims, E. Ralph Jr. 1995. *Precision Manufacturing Costing.* New York: Marcel Dekker.

Steffy, Wilbert. 1977. *Economics of Machine Tool Procurement.* Dearborn, MI: Society of Manufacturing Engineers (SME).

Winchell, W. 1989. *Realistic Cost Estimating,* Second Edition. Dearborn, MI: Society of Manufacturing Engineers.

Wise, Robert L. 1966. "What Comes After the Make or Buy Decision?" *Machine and Tool Blue Book,* October: 106–109.

Manual Deburring—The Real Costs 6

This chapter explores four aspects of calculating hand deburring costs in the shop, including: determining how much hand deburring actually is performed and what is causing the extra costs; determining how costs can be reduced; and validating the cost.

DEBURRING VERSUS PART FINISHING

Most individuals assume that manual deburring workers spend their time deburring. Thus, when it is time to "reduce deburring costs," the target is to eliminate deburring workers. However, if you bought a new piece of equipment or were to eliminate deburring, would you eliminate the need for deburring workers? Table 6-1 provides some insight into what manual deburr operators actually do during their workday.

It is important to recognize that, in many plants, deburring typically encompasses not only removing burrs, but also:

- taking out broken taps and drills;
- improving finishes;
- hiding scratches, nicks, and dings;
- completing paperwork;
- removing stains and discoloration, and
- other assorted functions.

These additional, and often unspecified, functions are clearly beyond the capabilities of many deburring processes, whereas other processes can accommodate some of these added functions. When the intent is to reduce operators' deburring time, it is essential to recognize the complete breadth of their work. Many so-called cost savings never happened after equipment was bought because users never took into account that the person only spent a small part of his or her time performing actual deburring.

Observations of operators at work, or discussions with them, reveal the additional work elements appli-

Table 6-1. Tasks commonly performed by manual deburring operators

Task	Hours/Day	$/Hr	# of Days × $/Day
Deburr parts			
Polish to remove surface defects			
Sand surfaces			
Remove broken drills or otherwise rework parts			
Correct dimensional defects			
Complete paperwork			
Inspect			
Total			

Note: Only one row of this table refers to actual deburring.

Hand Deburring: Increasing Shop Productivity

cable to a specific plant. From Table 6-1, or an extended table based on a specific plant, one can readily estimate the cost of hand deburring and other associated elements.

Table 6-1 provides key decision-making information for individuals considering purchasing mechanized deburring approaches. It focuses attention on the actual time that a person deburrs, and it allows owners and engineers to see how much non-deburring time is spent by the individual performing ancillary tasks. In addition, the table prevents the mistaken assumption that the entire job currently performed will disappear with the introduction of mechanization.

WHAT INCREASES HAND DEBURRING COSTS?

There can be a significant amount of wasted effort in manual deburring, in the sense that the operation may not have been evaluated carefully. So, although deburring is truly required, the operators may be wasting effort. They may be deburring to microscope quality when only naked-eye quality is needed. They may have to wander through the shop to get the necessary tooling. In a large shop, this can cost 30 minutes or more per day per person. Untrained operators will be much slower. Also, better manual tooling can cut costs greatly.

Table 6-2 provides insight into the causes of wasted time that unnecessarily increase hand deburring costs.

In addition to the shop issues described in Table 6-2, it is critical to realize that hand-deburring costs are also affected by the following issues:

- burr size (keep them small to reduce deburr costs),
- burr location (put them where they are easily removed),
- adjacent critical surfaces (do not force heavy deburring next to critical surfaces),
- edge radii (change edge requirements from very small or very large to more normal radii), and
- total length of edges to be deburred.

The number of edges on a part and the total length of edges to be deburred on a single part may not be apparent. Both factors indicate the difficulty of obtaining perfect edges when high precision is required. In Chapter 1, Figure 1-4 shows that a single gear tooth has 10 different line segments that must be burr-free. For a miniature instrument gear that has 12 teeth and is small enough to fit under a fingernail, 120 edges must be burr-free. Figure 1-5 illustrates a simple, fine pitch 5.0 × 0.35 screw that is 0.32 in. (8.2 mm) long. It has 17 turns with 24.4 in. (620 mm) of burrs on its crests. A burr left anywhere on its crest could jam its mating part. The burrs on a single part of this small screw could produce 32,632 particles, if broken into loose particles of 0.00076 in. (19.3 μm) diameter or length.

TIME AND MOTION STUDIES

A simple industrial engineering time study can reveal the potential in your company for reducing costs. These studies detail every second of an operator's time and the associated tasks. Most organizations are not as concerned about this level of detail, but that detail reveals wasted motion and time, and the potential for reducing costs, which are real factors in standard production operations. Such detail is not as useful in job shops where jobs change every day. Industrial engineering studies can be procured from several types of sources:

- university students and faculty in industrial engineering programs;
- commercial firms, and/or
- texts that provide details to enable untrained individuals to perform their own studies.

Industrial engineering approaches also provide cost estimating time standards that allow users to estimate time before actual work is available to monitor (Maynard 1971).

REDUCING HAND DEBURRING COSTS

Table 6-2 provides some insight into what a company can do to reduce hand deburring costs. Are these significant enough to implement an improvement effort? Table 6-3 provides a clear view of the cost of wasted time, based on only 15 wasted minutes per day. Typically, the actual wasted effort in hand deburring in a job shop may approach four times that level.

Table 6-2 entries can generate "buckets" of time savings. They represent a class of issues that, when looked at more closely, reveal other, similar areas of potential improvement. The entry, "use staff that can see burrs," is a technical issue. How do you evaluate the visual acuity of operators and inspectors? It is a fact of life that our visual abilities decline as we age. Shops have operators and inspectors who cannot see the fine burrs that must be removed. Rejects for burrs increase the cost of inspection (they find things to reject, they fill out the paper, they explain what they saw, then they re-inspect the parts). The operator spends time understanding what and why things were rejected, then must remove burrs a second time. Typically, the shop supervisor has to understand the problem and dedicate time to fixing it. Between five and

Table 6-2. Improving hand-deburring efficiency

Issue	Possible Cost Savings (%)	Daily $ Savings	Time Savings (Min/Day)	Labor Rate ($/Min)	Cost Savings ($/Min × Min/Day)
Keep deburring tools at operator's workstation so workers do not have to walk to the crib	5				
Reduce fatigue	10				
Provide optimum tools	5				
Complete hand deburring on machine cycle	5				
Reduce rework that causes extra hand deburring	1				
Use trained operators	15				
Use staff that can see burrs	4				
Clarify what is to be deburred	5				
Provide team environment	10				
Reduce over-deburring	5				
Validate that inspection and manufacturing are working to same quality expectation	5				
Identify scrap parts so they are not deburred	1				
Total					

Table 6-3. The yearly cost of apathy (cost of 15 wasted min/day)

Hourly Rate ($/Hr)	$ Lost/Yr Number of Employees				
	1	5	10	50	100
8	510	2,550	5,100	25,500	51,000
10	638	3,188	6,375	31,875	63,750
15	956	4,781	9,562	47,810	95,620
20	1,275	6,375	12,750	63,750	127,500
25	1,594	7,969	15,938	79,688	159,376
30	1,912	9,562	19,125	95,625	191,250

Note: Data are based on an 8-hour day, 255 working days/yr, no overhead.

10 people could work on a burr reject, wasting time that would otherwise be used to perform work that is more useful.

The "provide team environment" entry in Table 6-2 is an approach to get better solutions by involving all the appropriate staff to solve underlying problems. There are some excellent hand-deburring operations in the United States. They get to that level by empowering teams of workers and engineers to improve operations. That, in turn, builds pride, reduces direct costs, and develops a staff that is more capable and better understands the economics of the company. The lack of such effective teams is one indicator that an individual plant's deburring costs may be above the appropriate level.

In the 1960s, the Hamilton Standard Division of United Aircraft Corp. performed an in-depth study similar to that recommended in Chapter 32 and other places in this book. After evaluating its approaches and problems, it found it could save 30% of its deburring costs by developing a simple in-house approach to defining deburring needs, and by implementing electrochemical deburring.

In Chapter 3, Figure 3-6 illustrates the method used to define the approach. The approach was successful because every worker knew which edges to deburr, the required edge break, and the tools and sequence to use. In other words, lack of specific instructions had been one of the major causes of unnecessary hand-deburring costs.

VALIDATING RELATIVE DEBURRING COSTS

The analysis of actual manual-deburring costs is relatively straightforward (Koenig 1987; Gillespie 1994). Compared to equipment analysis, it is simple (Sims 1995; Humphreys and Wellman 1996; Lang 1989). Costs alone, however, should not be the only criteria for determining whether special effort should be concentrated on manual deburring. After costs are obtained, they must be compared to a base value, such as any one of the following:

- total annual dollar cost;
- cumulative dollar total (over several years or the life of the program);
- percentage of shop operator hours;
- percentage of machining costs;
- percentage of sales dollars;
- industry cost comparison, and/or
- specific industry percentage cost comparison.

While a cost reduction of any kind is desirable, these simple comparisons or evaluations provide a relative measure of long-range potential benefits. They may also be useful for job-cost estimating, since such factors as total shop or machining employee hours are required in all costing and long-range planning.

Some manufacturers send a few of their parts to job-shop deburring houses to meet overloads. This prevents the problem of having to use untrained personnel. It also provides a relative measure of the effectiveness of plant deburring. If the job shop quotes deburr time and costs, the plant can compare its time to that of the job shop. If they are close, then the plant is performing well. If the difference is major, clearly there is a better way. An analysis of the edges and surfaces of job-shop-finished parts can provide many clues about how parts were deburred.

Several keys, or red flags, signal that hand deburring costs are out of line. A flag should go up when deburring operator hours or costs exceed a specified percentage of total machining time or costs. When all the work is held up at the deburring bench, another flag should go up. A third flag involves capital equipment expenditures. When an industry as a whole spends 5–7% of its equipment dollars on finishing equipment, and the plant in question spends considerably less, someone needs to consider if the right balance exists. A tractor company had a nine-month backlog at the hand deburr bench before it realized the seriousness of its problem.

Another simple method of determining if hand deburring is excessive is to ask those who perform the work. One of the perennial complaints is that machinists do not change dull tools often enough, so "burr hands" have to work unnecessarily hard, which adds to the cost and slows production.

One of the interesting facets of deburring is that few companies are willing to invest significant time or money to solving burr-related problems. For example, if deburring requires three people at an annual salary of $30,000, few companies would invest $3,000 a year to search for better methods. In fact, some companies shudder when someone suggests spending $200 to buy the latest literature on how to reduce costs. Similarly, few companies assign their best engineers to deburring improvement because they do not believe it deserves such high-quality effort.

If, however, one views the subject from a 10-year vantage point, the picture looks more dramatic. Over a 10-year period, the three workers will have consumed $300,000 for a problem "unworthy of solution!" Yet, you cannot expect to reduce hand-deburring costs in a one-shot effort, without dialogue, without using real parts, and without treating the topic as a truly significant problem. Many helpful aids are available, including the following:

- There are over 2,500 manufacturers of deburring equipment and supplies (Gillespie 2001).
- There are an estimated 50,000 or more pages of burr-related publications.
- Most of the world's published literature on burrs and deburring has been listed in one convenient bibliography (Gillespie and Repnikova 2001).
- An international group coordinates and reports on worldwide deburring efforts (Gillespie 1996).
- There are regular U.S. and international conferences on deburring and edge-finishing technology.

- Deburring handbooks, including this one, summarize much of the known knowledge in a convenient-to-use format (Gillespie 1976; Gillespie 1977; Gillespie 1999; Wick and Veilleux 1982).
- There are videotapes available on deburring.

Generally, consideration of schedule, workforce, and equipment constraints, combined with some previous experience, results in acceptable costs from the first iteration. The result may not be optimum, but the user will be able to determine if the cost is in the correct range of allowable costs. Purchases of new types of equipment and new processes require experimentation and data-driven solutions—solutions that document the size of burrs removed, the numerical value of edge condition produced, the time required to attain it, and associated indirect costs.

CALCULATING COSTS

There are no explicit guidelines for calculating hand-deburring costs, but some of the industrial engineering approaches provide a method. Table 6-4 provides a starting point for calculating actual deburring costs.

Figure 6-1 illustrates a simple, small part to explain Table 6-4. It depicts a 1 in. (25.4 mm) cube of stainless steel. The edge requirement is to remove all burrs and smooth the edges, but the chamfer left must not exceed 0.003 in. (76 μm). The parts will be inspected at 20× magnification. Because of the near-sharp edge requirement, swivel-bladed scrapers or knives cannot be used. Instead, a triangular knife and an abrasive-filled nylon brush on a bench motor will be used. Then the tops of the four holes will be deburred with a countersink, followed by a rubber dental bullet. A reamer in the big hole is to be used in the intersecting holes, followed by a reamer in the small holes, then a small burr ball and rubber bullet. Table 6-5 illustrates the calculations. Note that the table treats the tops of the intersecting holes (the easy-to-deburr areas) as blind holes. The table lists the intersections as a separate challenge.

It takes 10 minutes to select the best from among the operator's tools and make replacements in the two handheld dental motors used to polish the holes. The inspection time shown in Table 6-5 is for the operator to validate his or her work. An inspector, rather than a deburring worker, would normally do the actual acceptance of these features. While the values shown may seem long, recognize that a single scratch on the surface of this part rejects the part, and a single area exceeding the 0.003 in. (76 μm) maximum edge break scraps the part.

Many modern plants spend 33% of their labor rates for employee medical and dental insurance, worker's compensation, and other worker costs. Thus, the true cost to the company for a $16/hr worker is $16 × 1.33 = $21.28/hr.

The example shown in Figure 6-1 took almost 11 minutes per part to produce, and it was only a 1-in.^{2} (6.4-cm^2) cube! When looking at the cost elements, however, it can be seen that the hole intersections required the most time. If they have to be smooth and burr-free at 20× magnification, skill and time are required to reach that level of quality. If the holes were larger, it could be done faster. If there were many parts, then a more mechanized effort would be appropriate. Note also that this example includes surface finishing costs, as well as deburring costs. Table 6-5 does not include paperwork or other costs that also may be needed.

Phillip Ostwald presents another, somewhat simpler, approach (Ostwald 1983). He makes a standard cost for each type of deburring, whereas the approach illustrated in tables 6-4 and 6-5 requires estimations of each feature to be deburred. Table 6-6 illustrates the Ostwald approach. Use it for quick-costing. It takes a basic box shape, for example, and recognizes that the total length of edges = 4(length) + 4(width) + 4(height). The average box-shaped part using a bench motor requires about $0.505/in. for each edge deburred for 100 parts. Stated differently, each edge of each part would require about $0.005/in. to deburr. This would include cutting off the burr and then smoothing the edge. There are some basic handling and minor setup costs.

The data shown in Table 6-6 are for reference purposes only to explain the approach. Each company would create a similar table to allow estimators to make quick calculations of deburring costs. Costs may vary or remain the same, regardless of which piece of motorized equipment is used.

Overtime cost has been ignored up to this point, but it greatly increases costs, so it should be considered in any cost reduction evaluation. Also, include in deburring costs all the rework costs resulting from inadequate manual deburring. These costs can be eliminated with a good hand-deburring program.

TOOLING COSTS

Several problem and expense areas exist that have not yet been discussed. They include:

- designing hand tools;
- sharpening hand tools;
- storing hand tools;
- retrieving hand tools, and
- maintaining adequate hand-tool inventory.

Table 6-4. Calculating hand-deburring costs

Element	Quantity	Rate of Work	Tools Used	Total Time	Cost/Min	Cost/Part
Deburr	___ in. of edge	At ___ in./min	Using a ___	Takes ___ min	And at ___ $/min	Costs $___
Smooth edges						
Deburr holes	___ number of blind holes					
	___ number of through holes × 2					
Smooth hole edges	___ number of blind holes					
	___ number of through holes × 2					
Deburr hole intersections	___ number of hole intersections					
Smooth hole intersections	___ number of hole intersections					
Deburr threads	___ number of threaded features					
Smooth threads	___ number of threaded features					
Deburr intricate areas	___ number of areas					
Smooth intricate area edges	___ number of areas					
Inspect work						
Total deburr costs						
Improve surface finish	___ in.^{2} of surface					
Inspect surface finish work						
Clean part						
Inspect cleaning						
Total cost						

Large facilities of 25–100 deburring workers can spend several hundred dollars each week keeping up with these items. Some large companies have a $250,000 inventory of hand-deburring tooling and related aids.

REFERENCES

Gillespie, LaRoux K. 1976. "Deburring Capabilities for Miniature Precision Parts." Report BDX-613-1604 (July). Kansas City, MO: Bendix Corporation. (Available from NTIS.)

——. 1977. "Side Effects of Deburring Processes." Technical Paper MRR77-17. Dearborn, MI: Society of Manufacturing Engineers (SME).

——. 1994. "Process Control for Burrs and Deburring." Report TR94-1. Kansas City, MO: Deburring Technology International.

——. 1996. "State of the Art of Deburring in the U.S." In *Proceedings of the 4th International Deburring and Edge Finishing Conference*. Seoul, Korea (September). Kansas City, MO: Deburring Technology International.

——. 1999. *Deburring and Edge Finishing Handbook*. Dearborn, MI: Society of Manufacturing Engineers (SME).

——. 2001. *Guide to Deburring, Deflashing, and Trimming Equipment, Supplies and Services*. Kansas City, MO: Deburring Technology International.

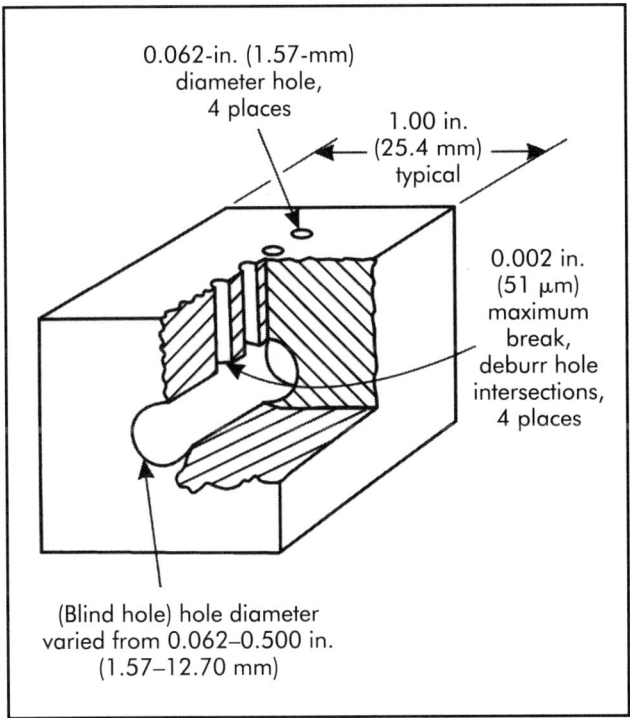

Figure 6-1. Geometry of example part.

Gillespie, LaRoux K., and Elena Repnikova. 2001. *Deburring: A 70 Year Bibliography*. Kansas City, MO: Deburring Technology International.

Humphreys, Kenneth K., and Paul Wellman. 1996. *Basic Cost Engineering*. New York: Marcel Dekker.

Koenig, Daniel T. 1987. "Capital Equipment Programs." *Manufacturing Engineering*, pp. 77–97. New York: Hemisphere Publishing Corporation.

Lang, Hans J. 1989. *Cost Analysis for Capital Investment Decisions*. New York: Marcel Dekker.

Maynard, H. B. 1971. *Industrial Engineering Handbook*, 3rd ed. New York: McGraw-Hill.

Ostwald, Phillip F. 1983. *American Machinist Manufacturing Cost Estimating Guide*. New York: American Machinist.

Sims, E. Ralph, Jr. 1995. *Precision Manufacturing Costing*. New York: Marcel Dekker.

Wick, Charles, and Raymond Veilleux, eds. 1982. *Tool and Manufacturing Engineers Handbook*. 4th ed., Vol. 3, *Materials, Finishing, and Coating*, pp. 10-1–17-19. Dearborn, MI: Society of Manufacturing Engineers (SME).

BIBLIOGRAPHY

Gillespie, LaRoux K. "The Management of Deburring." Technical Paper MRR82-04. Dearborn, MI: Society of Manufacturing Engineers (SME).

Ioi, Toshihiro, Hisamine Kobayashi, and Masahisa Matsunaga. 1981. "Computer-aided Selection of Deburring Methods." Technical Paper MR81-389. Dearborn, MI: Society of Manufacturing Engineers (SME).

Table 6-5. Hand deburring calculations

Element	Quantity	Rate of Work	Tools Used	Total Time (Min)	Cost/Min ($)	Cost/Part ($)
Obtain the correct tools and make them readily available				10	0.25 ($15/hr)	2.50
Deburr	12 in. (30.5 cm) of edges	24 in./min	Triangular knife	0.5	0.25 ($15/hr)	0.125
Smooth edges	12 in. (30.5 cm) of edges	10 in./min	Nylon brush	1.2	0.25	0.30
Deburr holes	4 blind holes	16 holes/min	Countersink	0.25	0.25	0.062
	1 through hole × 2	16 holes/min	Countersink	0.0625	0.25	0.016
Smooth hole edges	4 blind holes	8 holes/min	Dental bullet	0.5	0.25	0.125
	1 through hole × 2	10 holes/min	Commercial rubber bullet	0.10	0.25	0.25
Deburr hole intersections	4 hole intersections	2 min/part	Reamer (1 large, 1 small)	2	0.25	0.50
Smooth hole intersections	4 hole intersections	1 hole/min	Rubber dental tool	4	0.25	1.00
Deburr threads	0 threaded features					
Smooth threads	0 threaded features					
Deburr intricate areas	0 areas					
Smooth intricate area edges	0 areas					
Inspect work		2 min total	Microscope	2	0.25	0.50
Total deburr labor costs				10.61		5.49
Total deburr cost of tools and supplies			1. Knife 2. Dental bullet 3. Commercial bullet 4. Countersinks (2) 5. Reamers (2)		1. $7/100 parts 2. $1.25/100 parts 3. $0.25/100 parts 4. $0.50/100 parts 5. $2.00/100 parts	0.11
Total direct deburr costs						
Improve surface finish	6 in.² (38.7 cm²) of surface	3 in.²/min		2	0.25	0.50
Inspect surface finish	6 in.² (38.7 cm²)	6 in.²/min		1	0.25	0.25
Clean part	12 per 3 min load	4/min		0.25	0.25	0.062
Inspect cleaning		0.1/min		0.1	0.25	0.025
Total non-deburr time				3.35		0.84
Lost time				15 per 100 parts	0.25	0.038
Total operations per part				3.50		6.33
Overhead						
Total part finishing cost						
Profit (@ 15%)						
Cost to customer						
Tax to customer (@ 7%)						
Total bill						

Table 6-6. Cost using standard element costs (Ostwald 1983)

Machine or Device	Setup Cost* ($)	Variable Cost	Cost/100 Units ($)
Bench motor	0.75	Box shape (brush)	2.16 + 0.220 (L + W + H)
Drill press	1.30	Number of holes (countersink)	0.736 (number of holes)
		Number of holes (brush)	0.900 (number of holes)
		Number of holes (buff)	7.32/in.^{2} (1.13 cm^2) of buffed area
		Number of rehandle times	3.24 (number of rehandles)
Abrasive belt	1.88	With conveyor feed	6.90 + 0.244 (in. of conveyor length)
		With manual handling	6.90 + 2.74 (in. of edge length)
Pedestal motorized equipment	1.92	Brush (in. of edge)	0.326 (L + W + H)
		Buff (surface area)	0.772 (number of in.^{2})
Handheld motors	1.30	Profile edges	0.866 + 0.220 (L + W + H)
		Deburr holes	1.874 (number of holes)
Hand tools	0.90	Swivel tools	0.90 (L + W + H)
		Triangular knives	2.90 (L + W + H)
		Scrapers	1.20 (L + W + H)
		Files	2.95 (L + W + H)
Plastic finishing	0.64	Hand degate	2.80 + 3.64 (number of gates)
		Saw degate	0.426 (number of gates)
		Deflash rim	0.426 (number of gates)
		Deflash holes	1.292 (number of holes)

*Data are based on $16/hr wages.

Example: Handheld motor cost to profile edges is $0.866 + $0.220 (length + width + height of part)

Deburring with Knives 7

Knives are one of the oldest known tool types used for deburring. They provide a quick edge-cutting action, and can accommodate almost any part configuration and material. Knives can be sharpened easily so they will last for many years, and they can be made in many different configurations to meet product needs.

While it may not be obvious, expert use of knives requires some skill. For precision work, users treat the knife like a surgeon's scalpel. In fact, scalpels are excellent tools for some plastics deflashing. Knives come in many forms as described in this chapter. In one plant, these tools are the predominant burr removal tool, used by each worker as the natural tool for precision work. At another plant, they are the natural tool to use for fast non-precision work. There is no similarity in the knives used at these two plants, but both use them. The versatile knife is ubiquitous!

KNIFE STYLES

As shown in Table 7-1, 14 different deburring and deflashing knife styles are commonly used. Some individuals use common pocketknives for deburring, but blade size and design make them a poor choice. One source lists 48 manufacturers and suppliers of knife and scraper tools for deburring (Gillespie 1996). Each has its own area of excellence and limitations. Thin scalpels, for example, are not normally a good choice for metals because they deflect too much and dig into metal edges. They are widely used for deflashing plastic parts and removing fine hair-like projections. Triangular cross-section knife blades do not cut plastic well because they are too thick.

Swivel-blade Knives

Today, swivel-type tools are the standard in most plants for rapid deburring of general features because

Table 7-1. Common deburring and deflashing knife styles

Type	Typical Use
Swivel tool	All metal contours and holes
Triangular cross section	Part contours
Oval blade	Burrs on holes
Surgical or hobby knife	Deflash plastics
Bench knife	Scrape metals or cut plastics
Utility knife	Deflash plastic
Hook knife	Thin flash
Cone end	Deburr holes in hard-to-reach areas
Chisel knife	Hard-to-reach areas
Hot blade	Electric heater systems to cut or melt plastic flash, sprues, or runners
Commutator slot shaver	Pull knife to chamfer edges by shearing
Special knives	Deburr unusual locations and unusual features
Pocketknife	Not a good choice for deburring
Razor blade	Deflashing and nonmetals

of their modular, throwaway design and smooth cutting action. Several forms are available, and they each have a variety of replaceable blades.

Figure 7-1 illustrates one swivel-type tool chamfering the inside diameter of a large tube. The tool bit swivels within the handle so the operator just makes a rotary motion to complete the deburring. This tool does not require a fixed wrist position, which might irritate repetitive motion injuries. A similar tool has an extendable length from 0.50–5.00 in. (12.7–127 mm) for hard-to-reach areas. These tools come in replaceable and disposable styles with a multitude of colored handles for ready identification of the blade, its application, or its owner. Some are designed for light deburring and some for thick metal burr removal. Some have very small diameter handles and others are designed more ergonomically (see Figure 7-2). They can deburr holes as small as 0.0625 in. (1.588 mm) in diameter. A complete set of all the available combinations of blades, handles, and associated tooling includes more than 50 elements. Blades are ground to deburr either soft metals, such as copper and brass, or steels and aluminum (see Table 7-2 and Figure 7-3). For longer life, blades can be titanium-nitride (TiN) coated. The same handles also house scrapers, which are discussed in Chapter 8.

Triangular Knife Design

Triangular knives are used as scrapers and true cutting knives. Scraper use is discussed in Chapter 8.

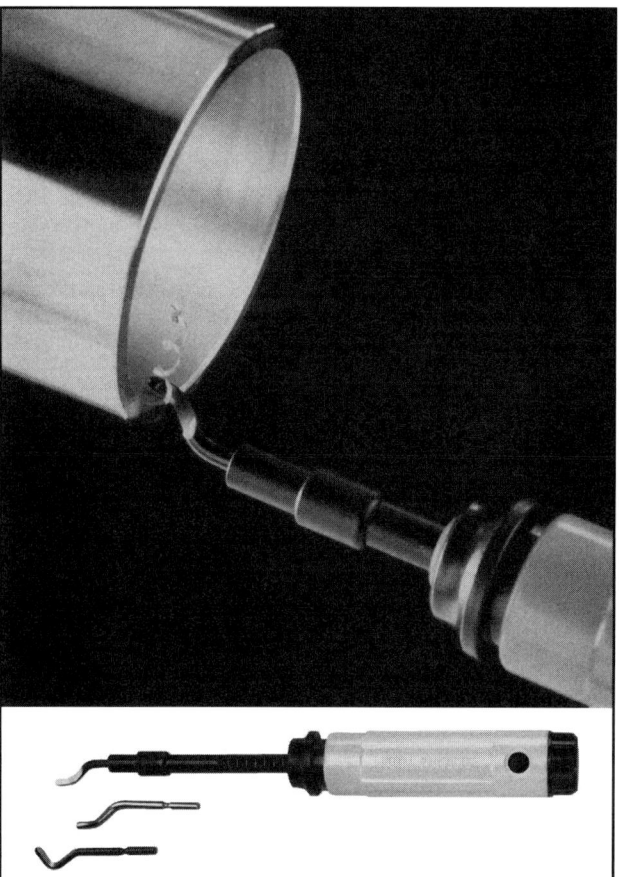

Figure 7-1. Swivel-blade knife removes burrs from a tube (Gillespie 1999). (Courtesy Royal Products)

Figure 7-4 illustrates a triangular knife used on a typical part. The same tool is used to deburr tubing. In this instance, the tool is actually cutting—just as a knife would—rather than pulling the burr off, as a scraping operation does.

Some triangular knives are not designed for easy cutting, rather for rough operations or scraping. They are called knives, but are really scrapers and cannot be used for precision deburring. The knives shown thus far are designed for commercial edge break and deburring practice. They can remove burrs at least 0.005 in. (0.13 mm) thick and produce edge chamfers of 0.005–0.025 in. (0.13–0.64 mm), depending on material, burr size, and other variables. In general, they are unsuitable for microscopic finishing when burrs must be removed and edges broken to tolerances of 0.0005–0.005 in. (0.013–0.13 mm).

Figure 7-5 illustrates one detail design for a precision triangular knife used to produce minute edge breaks (as low as 0.001 in. [25 µm]). These are tools made from Gorton cutter blank K1704. They are packaged in 0.25 in. (6.4 mm) diameter clear plastic tubes to protect the tool sharpness and protect workers and store personnel from injury. One company has designed 15 of these tools, each with different working diameters, holding diameters, and lengths. Some similar scraper tools use a knurled handle, which may be easier to hold. One tool uses a pin-vise holder, which allows adjustable tool length. In this particular design, the blade is double-ended, which enables the user to switch it when it becomes dull. The handle also accommodates scraper tools.

As implied by its name, the triangular knife has three cutting edges (Figure 7-6). These knives can be ground either concave or convex. Concave- or hollow-ground knives have a very sharp edge, but they dull quicker and are more likely to chip. Convex-ground tools have a more rigid body on each edge, thus they can last longer. In addition, they can be reground much faster using existing equipment than the concave tools. Both convex- and concave-ground tools are used. In many cases, both types of tools can be found in the same tool storage bin. Whether one gets convex- or concave-ground tools depends on how they were ground and what the tool design specified. One manufacturer, in addition to providing concave grinds, provides a deep hollow spot in the center of the concavity for additional clearance.

Oval Knives

Figure 7-7 illustrates an oval knife. Figure 7-8 provides the design for an oval-blade deburring knife. In this instance, a round or tapered Gorton cutter blank is ground at an angle to the axis to generate an oval

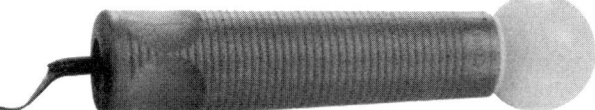
- Ergonomically designed handle provides precise fingertip blade control
- Deburrs small holes down to 1/16 in. (1.588 mm) diameter

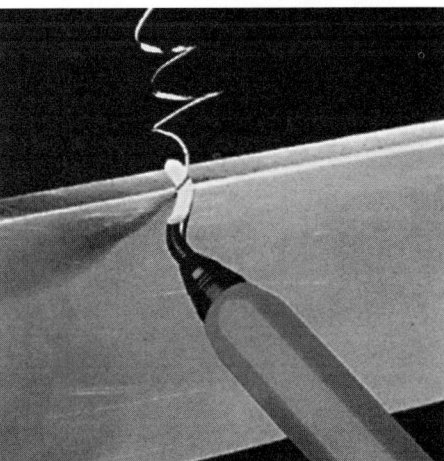

- Similar to disposable bur but blade can be changed and tool takes any E series blade
- Has a clip for pocket or pocket protector

Figure 7-2. Two different handle designs for swivel-type deburring knives. (Courtesy Royal Products)

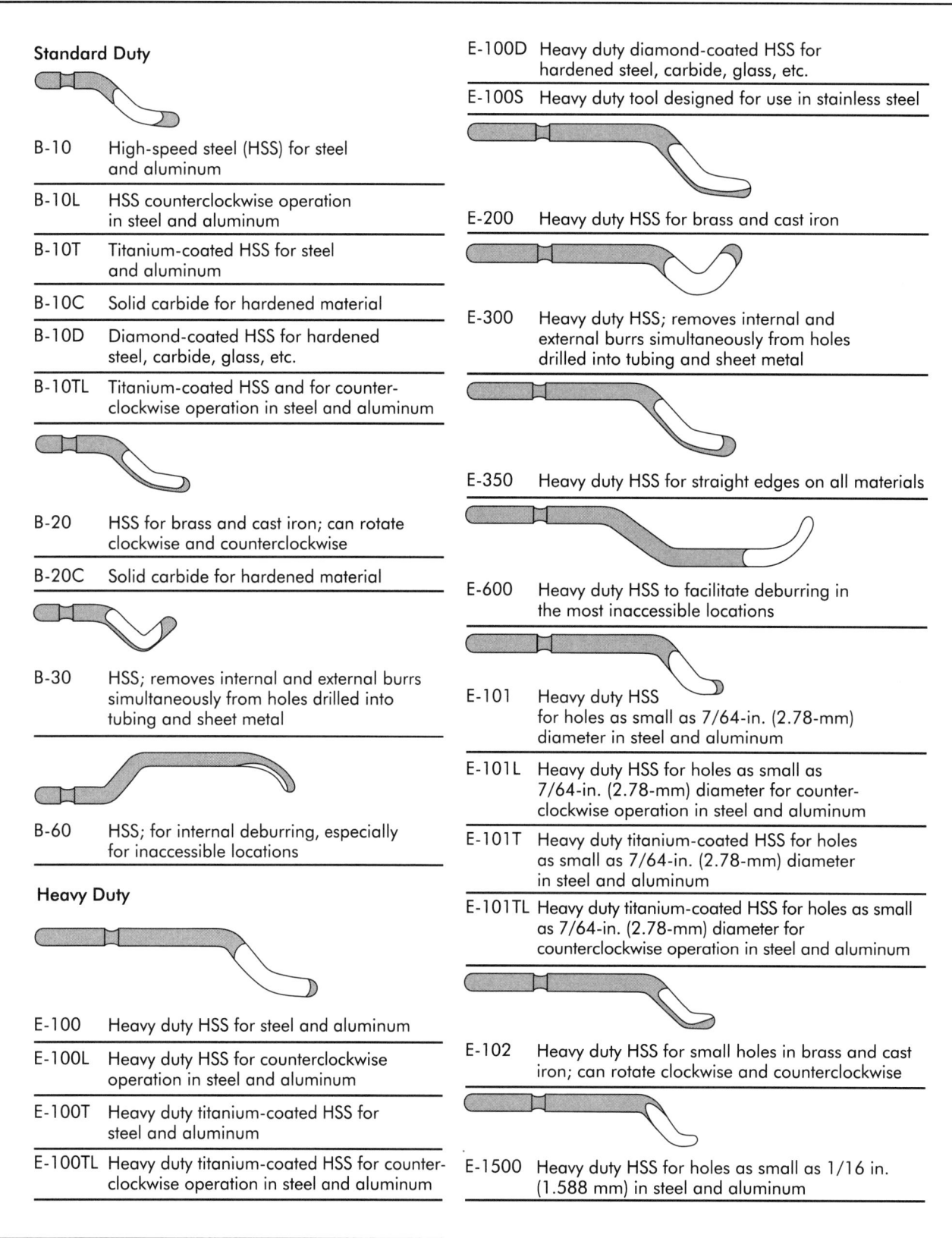

Figure 7-3. Various blade designs for swivel tools. (Courtesy Royal Products)

Table 7-2. Swivel blade variations

Item	Application	Illustration
Standard high-speed steel (HSS) blade	Steel, aluminum, and plastics	See Figure 7-3, illustration B-10
TiN-coated blade	Long life in steel or aluminum	
HSS blade for soft materials	Brass and copper	
Solid carbide blade	Hard steel and plastic	
HSS blade	Plastics	
TiN-coated blade	Long life in plastics	
HSS blade for internal holes	Internal holes	See Figure 7-3, illustration B-60
Heavy duty blade	Steel and aluminum	See Figure 7-3, illustration E-100
Heavy duty blade (R_c 68)	Stainless steel	
Heavy duty blade	Brass and cast iron	See Figure 7-3, illustration E-200
Bi-directional rotation blade	Brass and cast iron; cuts when swiveled clockwise or counterclockwise	See Figure 7-3, illustration B-20
TiN-coated blade	Long life in brass and cast iron	
Tube wall deburr blade	Deburr inner and outer diameters simultaneously	See Figure 7-3, illustration B-30
Heavy duty tube wall deburr blade	Deburr inner and outer diameters simultaneously	See Figure 7-3, illustration E-300
Straight-edge blade	Deburr straight edges and keyways	
Heavy duty straight edge blade	Heavy duty straight work on all materials	See Figure 7-3, illustration E-350
Heavy duty HSS internal blade	Hard-to-reach internal areas	See Figure 7-3, illustration E-600
Heavy duty HSS for 7/64 in. (2.778 mm) holes	Holes as small as 7/64 in. (2.778 mm) in steel and aluminum	See Figure 7-3, illustration E-101
Heavy duty HSS counterclockwise blade for 7/64 in. (2.778 mm) holes	Holes as small as 7/64 in. (2.778 mm) in steel and aluminum for left-handed operators	
Heavy duty TiN-coated HSS for 7/64 in. (2.778 mm) holes	Holes as small as 7/64 in. (2.778 mm) in steel and aluminum	
Heavy duty TiN-coated HSS counterclockwise for 7/64 in. (2.778 mm) holes	Holes as small as 7/64 in. (2.778 mm) in steel and aluminum for left-handed operators	
Bi-directional rotation heavy duty blade for small holes	Brass and cast iron; cuts when swiveled clockwise or counterclockwise, but heavy duty design for small holes	See Figure 7-3, illustration E-102
Heavy duty HSS for 1/16 in. (1.588 mm) holes	Holes 1/16 in. (1.588 mm) in steel or aluminum	See Figure 7-3, illustration E-1500
Counterclockwise HSS blade	Steel and aluminum for left-handed operators	
Heavy duty TiN-coated HSS for counterclockwise application	Heavy left-handed work in steel and aluminum	
Heavy duty diamond-coated blade	Hardened steel, carbide, glass, etc.	

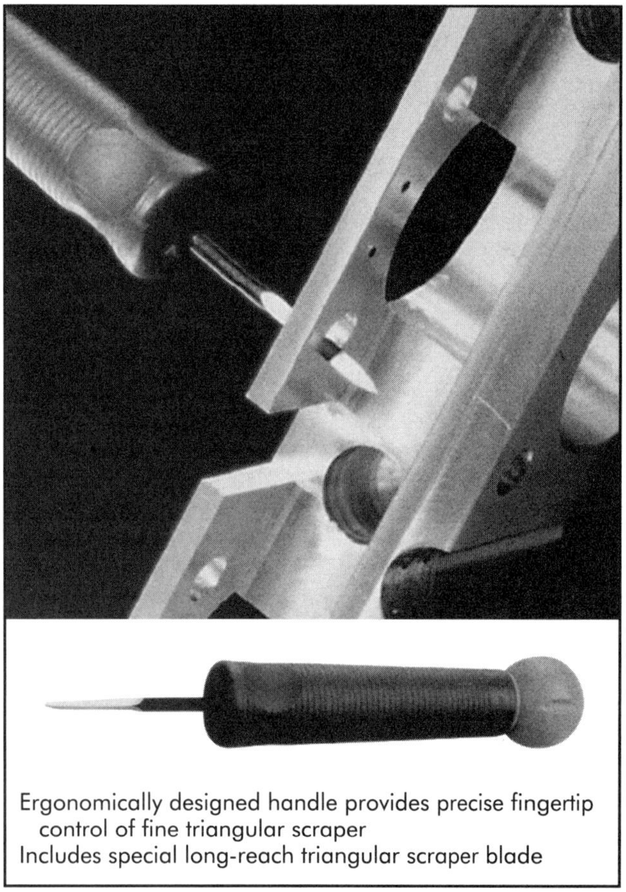

Figure 7-4. Small triangular knife cuts burr from a machined part. (Courtesy Royal Products)

Ergonomically designed handle provides precise fingertip control of fine triangular scraper
Includes special long-reach triangular scraper blade

surface, which becomes the cutting edge for deburring holes. Since the tool can be made very small, it can readily reach into small holes. While oval knives are principally designed for removing burrs in holes, they are often used both to chisel out burrs and reach difficult areas.

Surgical and Hobby Knives

Surgical and hobby knives are widely used to remove fine flash from molded rubber and plastic parts. They are unsuitable for metal parts or burrs. They come in at least 12 handle configurations (Figure 7-9) and at least 24 different blade designs. Some handles have a nonslip, soft grip. Others have plastic, stainless steel, nickel, or chrome-plated steel handles. The key attributes of these knives are their thin blade and very sharp cutting edge. Almost all are throwaway-blade design tools. Many of these are scalpel or surgical tools. High-carbon steel blades reportedly last longer than those of stainless steel. The medical knives provide many more variations than those manufactured by traditional supply houses. As with many hand tools, manufacturers also provide special design blades.

Surgical and hobby knives also may be called:

- surgical blades;
- heavy-duty steel knife blades;
- dissecting scalpels;
- microdissecting scalpels;
- microtechnique scalpels;
- cartilage knives;
- hobby knives;
- frisket knives;
- flat-blade carving knives;
- laboratory knives, and/or
- graphic arts knives.

Bench Knives

Bench knives are relatively large, all-purpose knives with steel blades mounted in a hardwood or plastic handle (Figure 7-10). They are not useful for metals, but can trim plastics and rubber parts.

Utility Knives

The retractable, all-purpose utility knives found on many home workbenches are used for plastics deflashing and trimming. These are not generally applicable for metal burrs. Some have rounded leading edges to minimize product damage and prevent puncture-type wounds from pointed ends.

Large Hook Knives

Figure 7-11 illustrates hook-shaped knives and other general knife types. These are used in limited applications where the edge would be drawn toward the operator. Generally, they would not be used for metals, but can be used for plastic and rubber parts.

Cone-end Tools

Figure 7-12 illustrates a cone-end tool. It is made by grinding two tapers on the end of the tool, then grinding through those surfaces at 90°. These tools let you reach to the bottom side of holes and slice away metal burrs.

Chisel Knives

The chisel, or flat-pointed, knife is generally a cylinder-shaped tool with a flat ground on each side. It leaves a flat, sharp edge, much like a sharpened screwdriver (Figure 7-13).

Hot Blade Knives

Hot blade knives are somewhat unusual because they are designed to cut with both sharp edges and

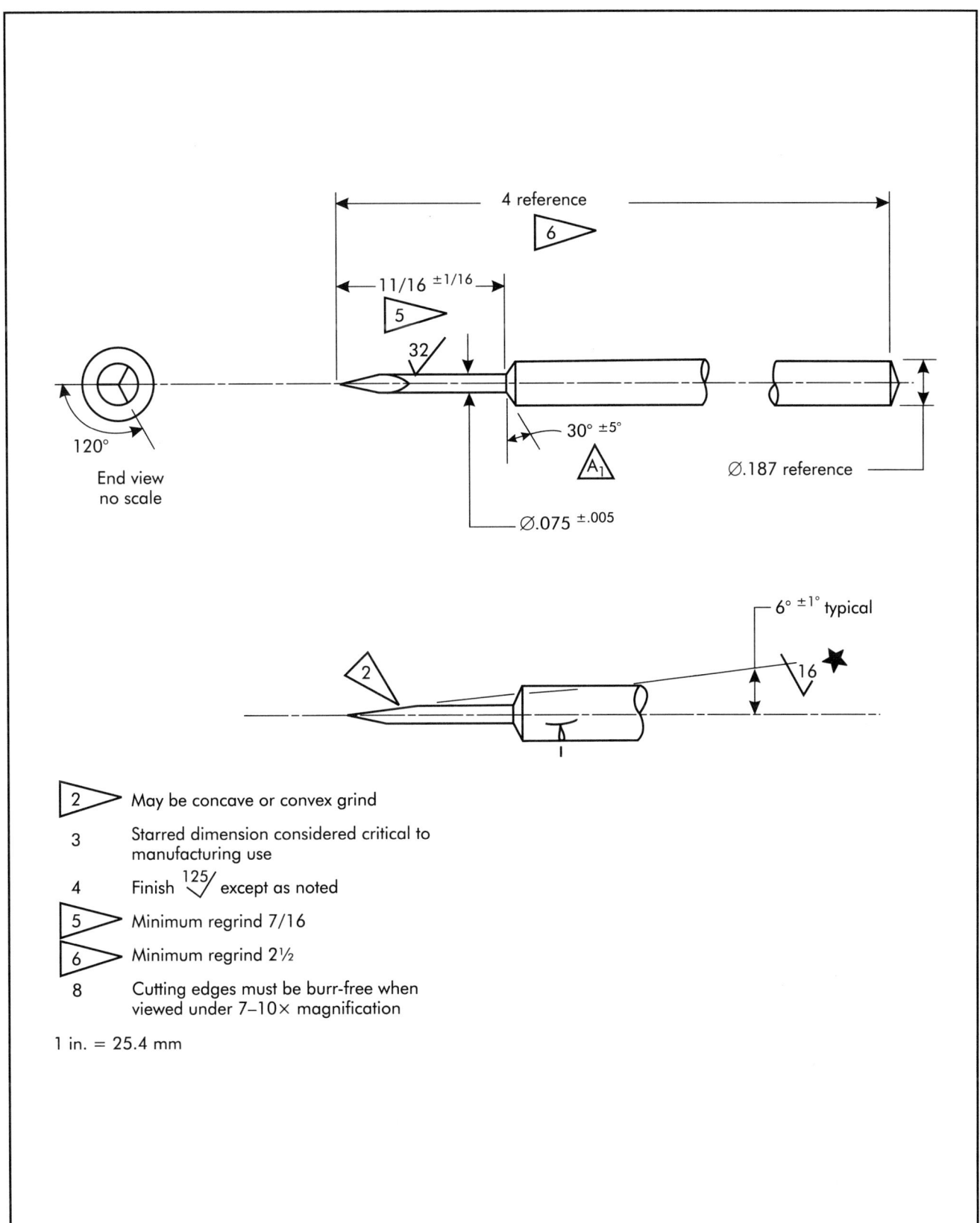

Figure 7-5. Precision triangular knife dimensions. (Courtesy Honeywell)

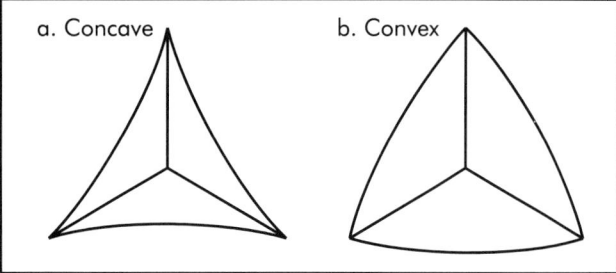

Figure 7-6. Concave and convex grinds on a triangular knife (Gillespie 1999).

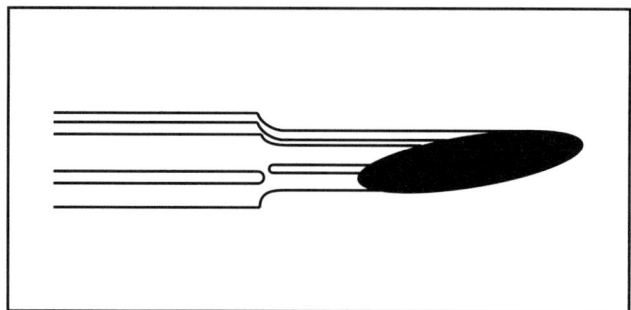

Figure 7-7. Oval knife (Gillespie 1999).

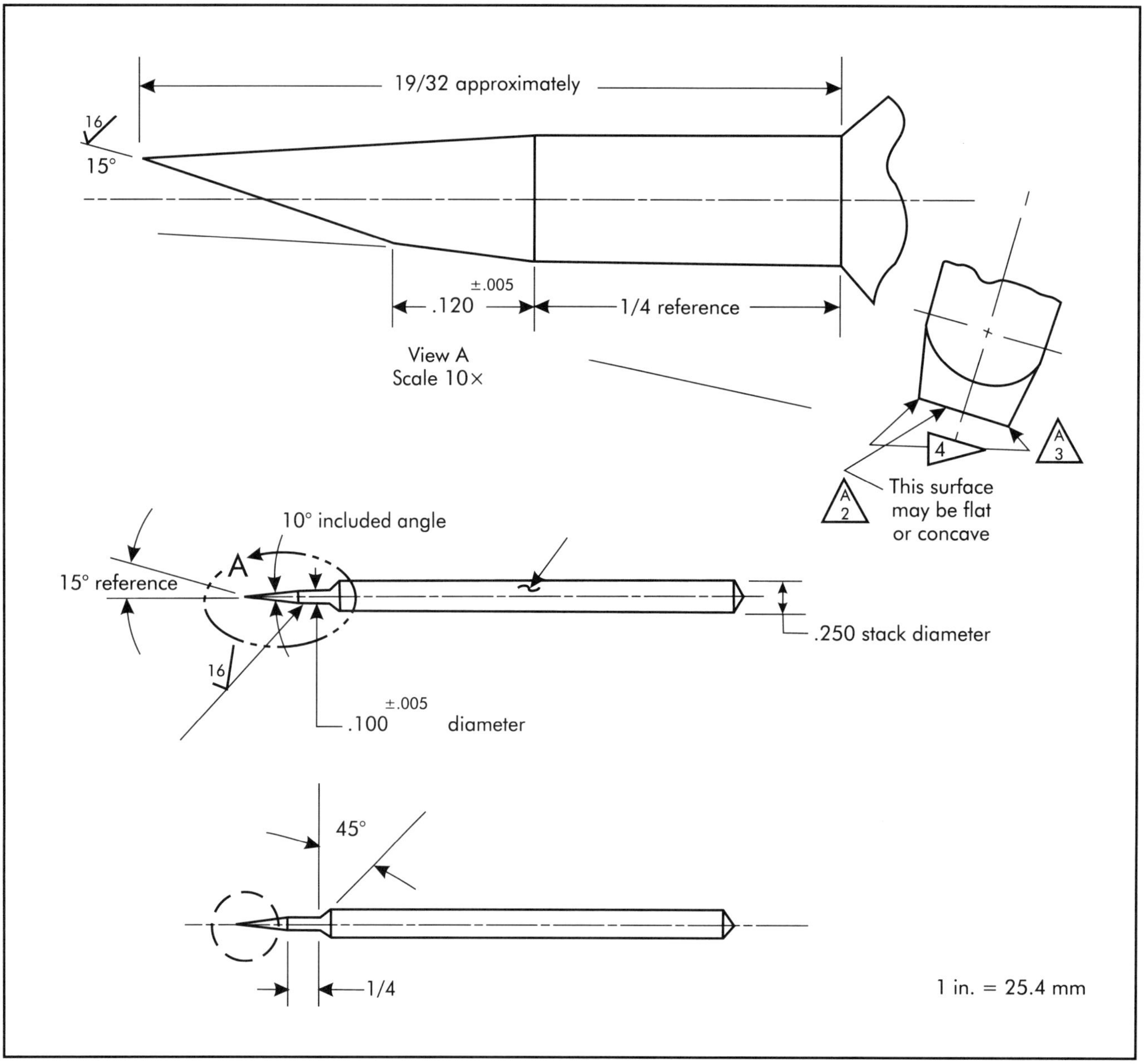

Figure 7-8. Oval deburring knife for precision hole deburring. (Courtesy Honeywell)

Replaceable blade scalpel, plastic. Handle patterned after surgeon's handle.

Utility scalpel. Stainless-steel blade embedded in black plastic handle. Total length, 6¼ in. (158.8 mm), blade length, 1¾ in. (44.5 mm).

Utility scalpel, carbon steel. Handle patterned after surgeon's handle. Total length 6 in. (152.4 mm), blade length 1½ in. (38.1 mm).

Screw-lock scalpel handle, stainless steel. Screw lock holds blade securely, yet allows for safe and easy blade changes.

Screw-lock scalpel, nickel. Similar to screw-lock scalpel handle.

Surgeon's scalpel, chrome. Quality German-made handle.

Student's scalpel. One-piece nickel-plated scalpel with polished carbon steel blade. Total length, 6 in. (152.4 mm); blade length, 1½ in. (38.1 mm).

Dissecting scalpel, chrome. Handle 0.16 in. (4 mm) thick with frosted finish; polished 1½ in. (38.1 mm) blade.

Dissecting scalpel, stainless steel. One-piece scalpel with frosted 0.16 in. (4 mm) thick handle and 1½ in. (38.1 mm) polished blade.

French pattern scalpel, stainless steel. Extra-large hollow handle with satin finish and a 1½ in. (38.1 mm) blade.

Cartilage knife, nickel. Made in USA. Heavy, serrated 6¼ in. (158.8 mm) handle with 1¾ in. (44.5 mm) blade. Equipped with scraper on handle end.

Microdissecting scalpel, stainless steel. For fine dissection. Total length 5½ in. (139.7 mm), blade length ¾ in. (19.05 mm).

Microdissecting scalpel, stainless steel. Total length, 5¼ in. (133.4 mm); 0.71 × 0.12 in. (18 × 3 mm) blade.

Microtechnique scalpel, stainless steel. Extra-fine scalpel for very delicate work. Total length, 5 in. (127 mm); 1.18 × 0.08 in. (30 × 2 mm) blade.

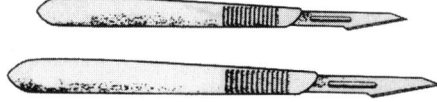
Flat blade carving knives. Stainless-steel handles with replaceable surgical blades.

Scalpel handle. Stainless-steel handle with satin finish.

Figure 7-9. Scalpel and hobby knives. (Courtesy Carolina Science Materials)

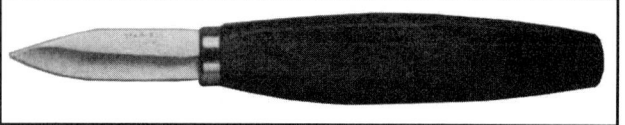

Figure 7-10. Bench knife. (Courtesy Gesswein)

heat. For use on plastic or foam, they come in many shapes. Figure 7-14 illustrates those provided by one manufacturer. They are used on natural and synthetic rubber, rubber sheets, rubber with textile, polyvinyl chloride (PVC) materials, fabric plies, industrial tubing, profile gaskets, silicone gaskets, bituminous ma-

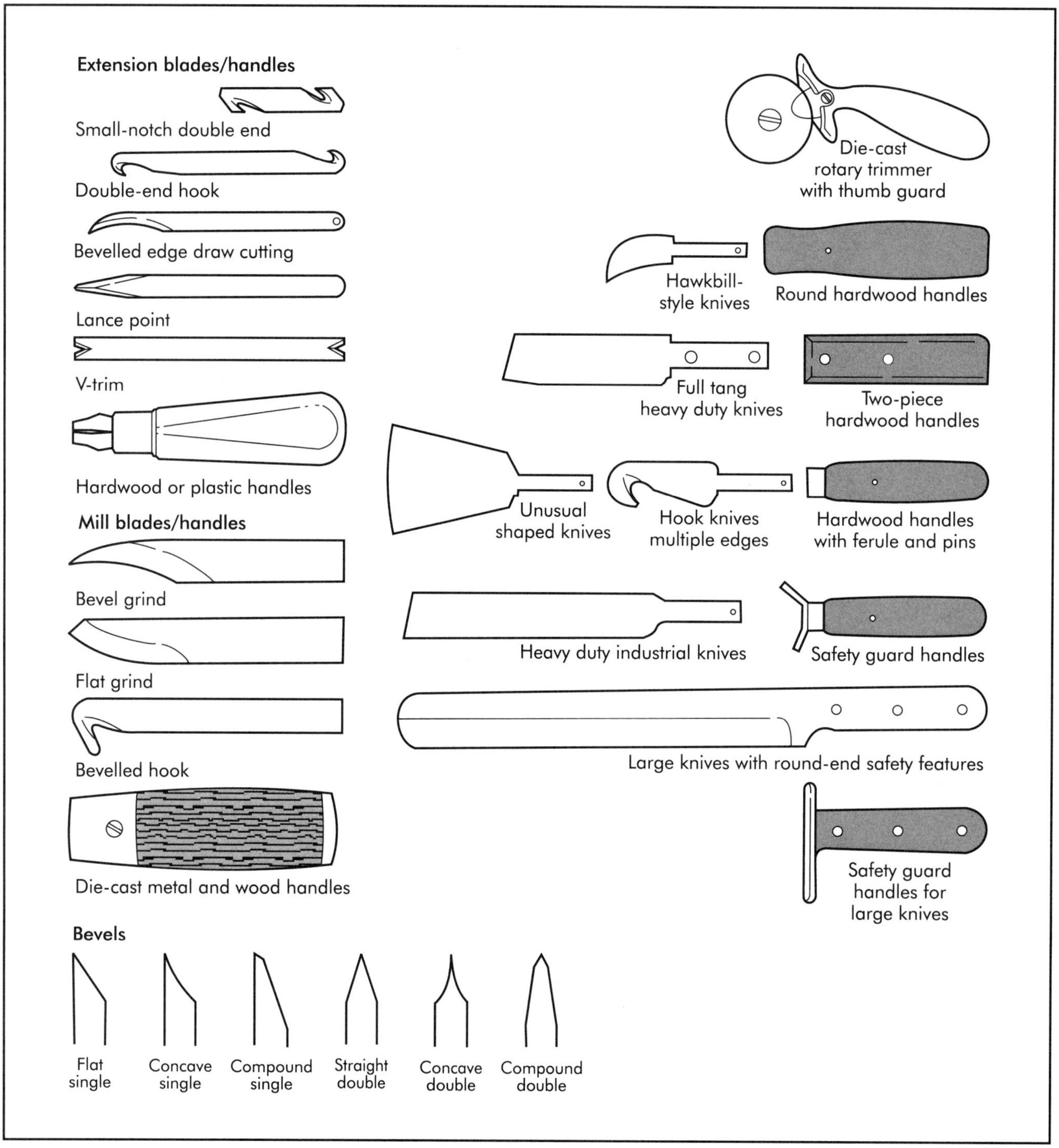

Figure 7-11. Hook-shaped knives and other general-purpose knives. (Courtesy Hyde)

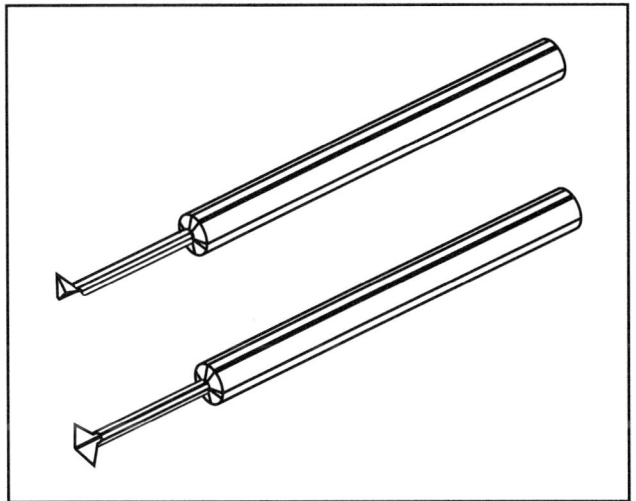

Figure 7-12. Cone-end deburring knife (Gillespie 1999).

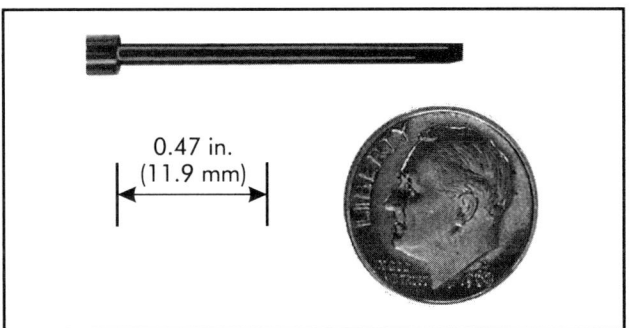

Figure 7-13. Chisel knife. (Courtesy Honeywell)

terials, neoprene rubber, and other materials and applications.

Razor Blades

Razor blades present clear personnel hazards, but some companies use them to remove thin flash and trim plastic and rubber parts.

Special Knives

Most knife manufacturers and many companies design special knives for unique situations. As previously noted, one company has designed over 40 such knives to meet the needs of its products. For example, of the specials that can be made, consider the dental and medical tools shown in Figure 7-15. These are just a few of the 100–200 shapes available today. They may not be designed as knives or scrapers, but with suitable modification they can be used as such—at least for a short time. Wax carvers can be altered, but they do not have high strength. For strong cutting edges, users alter pillar and Swiss files.

The special shapes category also includes angle-hooked tools, which look very similar to golf clubs. Figure 7-16 illustrates an angle-hooked knife used for microscopic deburring of miniature parts.

Commutator Slot Shavers

Figure 7-17 illustrates a slot shaver in use. Note that in contrast to most knives, it is pulled toward the operator. Each side (top and bottom) has a sharp cutting edge.

BLADE MATERIALS

Knife blades no longer have to be made from metal. Ceramic blade knives are now in use, which provide a longer life (Gillespie 1996). Space-age ceramics are used to deflash plastics, particularly the ultra-hard resins that resist carbon-steel tools. They have been used for glass-filled nylon, carbon fiber, talc-filled plastics, polyphenylene sulfide (PPS), polybutylene terephthalate (PBT), polycarbonate (PC), polymethyl methacrylate (PMMA), acrylonitrile butadiene styrene (ABS), high-impact polystyrene (HIPS), melamine, phenol, and 30% glass-filled polypropylene (PP). They have been used on high-density polyethylene (HDPE); however, they are not judged to be as good for that application. Reportedly, they also can be used on soft materials, such as aluminum, copper, and brass. One advantage of these tools is that they reduce the potential for cut hands and fingers.

USING DEBURRING KNIVES

There are two basic approaches to using knives (Gillespie 1979). They can be used as fist tools or as finger tools. Fist tools, scrapers, large knives, and many large tools are powered by the whole hand and arm. An example is a paint scraper, where the fist is clutched around the tool, holding it tightly, and applying a great deal of pressure. Figure 7-18 illustrates the fist position. It is the position used for the swivel tools, bench, hook, hot blades, and utility knives. While the fist position is used with triangular knives on large parts, it is not a successful approach on miniature or precision parts because the burrs cannot be felt.

On precision parts, the finger is used as a tactile indicator of the actual condition. In the fist position, the knife is essentially scraping through and past the edge being deburred. Considerable momentum is developed, and it is difficult to stop or control the amount of edge break applied to a part. Using the fist position to pull a knife through the part can cause cuts on the thumb, gouges on the part, or cuts to the hand that is holding the part. This is an awkward use of a precision tool, which is inherently dangerous (that is, the

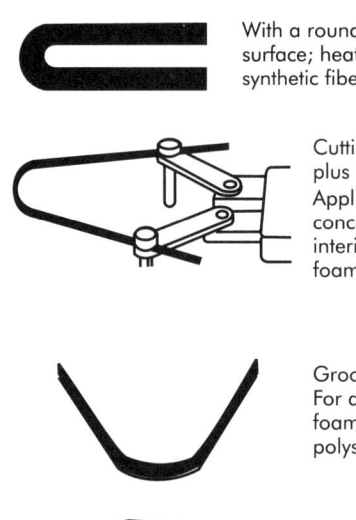 With a rounded edge to glide along cutting surface; heat seals the edges to prevent fraying of synthetic fibers and textiles.

Cutting sling consists of adjustable arms plus blade, which can be reshaped once. Applications: for rounded, angular, concave, and other shaped cuts, shaping interior spaces, and cutting hollows into foam materials.

Grooving blade (for use with cutting sling). For quick-cutting of grooves and flutes in foam materials made of polyethylene, polystyrene, and polyurethane.

 Blade is sharpened on two sides to prevent distortion and allows cutting in two directions; ideal for cutting foam.

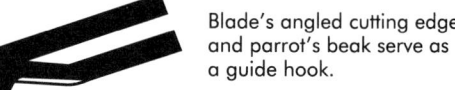 Blade's angled cutting edge and parrot's beak serve as a guide hook.

 With curved cutting edge and thick blade, the high-temperature blade is suitable for long cuts under pressure.

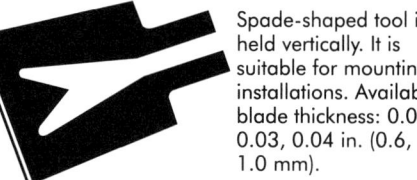 Spade-shaped tool is held vertically. It is suitable for mounting installations. Available blade thickness: 0.02 0.03, 0.04 in. (0.6, 0.8, 1.0 mm).

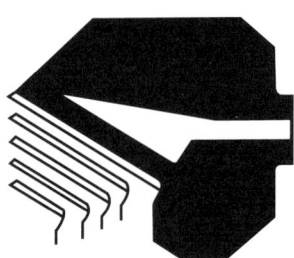 Blades are designed for rapid heating to maximum temperature. The entire cutting edge must be used in the material to obtain full benefit of the heat and to maximize cutting efficiency. Blades are available in five cutting lengths and should be chosen according to thickness of material.

Suitable for thermoplastics with higher melting points such as rubber, synthetic rubber, and synthetic resins.
- Cutting rubber up to 90 Shore combustible, oil resistant rubber.
- Cutting out rubber linings on containers, mixers, vibration, and conveyor chutes and troughs.
- Angled cuts at joints on conveyor belts, camelbacks, and rubber anti-abrasion coatings.
- Cutting and peeling rubber sectioning belts and flexible polyvinyl chloride (supported or unsupported).

 Blade hooks into material and glides along cutting base; offset at 90° angle.

Blade has short, obtuse angled cutting edges and both sides ground.

 Wide blade back and reinforced cutting flange provides stable cutting. The high-temperature blade is designed for heat-sealing the cutting edges.

With evenly heated cutting edge and thick blade, from 2 in. (50.8 mm) cutting length upward, blade is suitable for thermoplastics with low melting points such as foam polyethylene, polystyrene, and polyurethane.

Blade gives a clean cut that allows seamless joining of material and subsequent sealing or gluing operations.
Cuts profile gaskets of polyvinyl chloride and neoprene rubber for windows and door frames without crushing the material.

Figure 7-14. Hot-blade knives. (Courtesy Abbeon Cal)

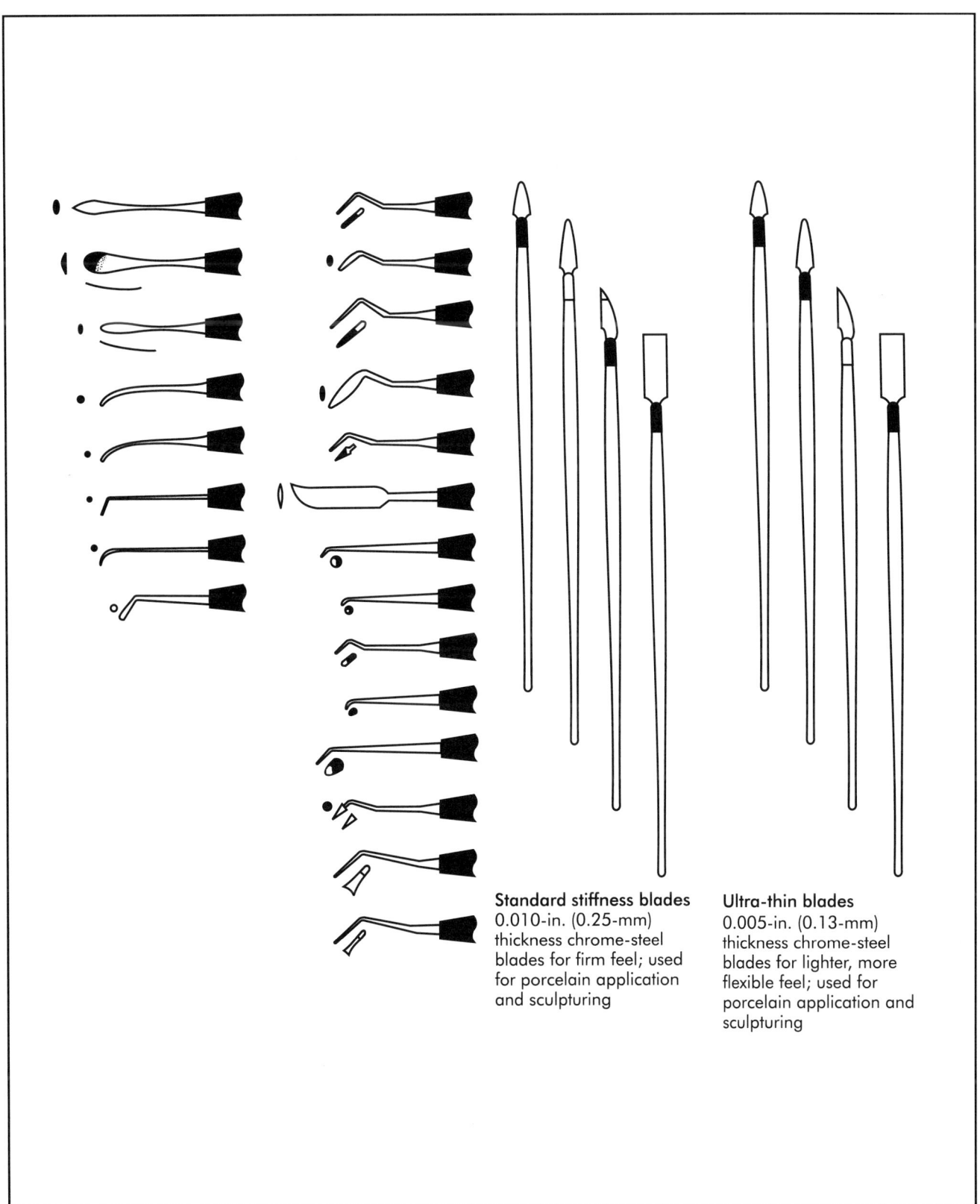

Figure 7-15. Dental and medical tool shapes that can be used, with some modification, as knives or scrapers. (Courtesy Belle de St. Claire)

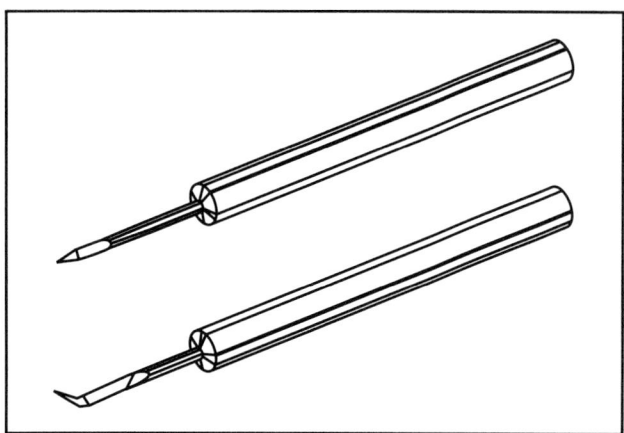

Figure 7-16. Top and side views of angle-hooked knife often used for deburring slots (Gillespie 1999).

The commutator slot shaver is a simple little hand tool to lightly chamfer the edges of commutator bars after undercutting. Pull it along the copper, shaving off the burrs—flip it over and do the other edge. Made of hardened high-speed steel, it can be quickly resharpened on a grinding wheel.

Figure 7-17. Commutator slot shaver. (Courtesy Martindale Electric)

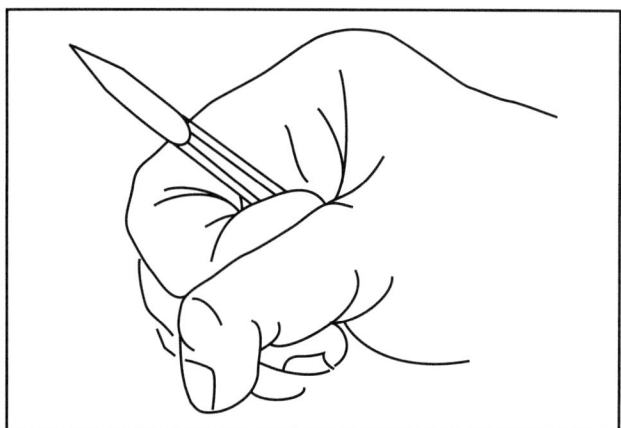

Figure 7-18. Fist position for knife deburring (Gillespie 1999).

knife is designed to cut, and it will do just that—to hand or part).

Figure 7-19 illustrates one of the two approaches to the finger position. In this case, the knife is held with two fingers and the thumb, almost like holding a pencil. This is called the pencil position. The second position is the underhand position (Figure 7-20). In the pencil position (Figure 7-19), the knife can be rotated to accommodate changes in intersecting planes on the part. The middle finger can be used as a surface feeling probe. By using this method under a microscope, a small burr can be removed as easily as writing with a pencil.

In the underhand position (Figure 7-20), the knife can also be rotated for easy burr removal. In this position, the ring and small fingers can be used as a surface support. The index finger can be used as a surface

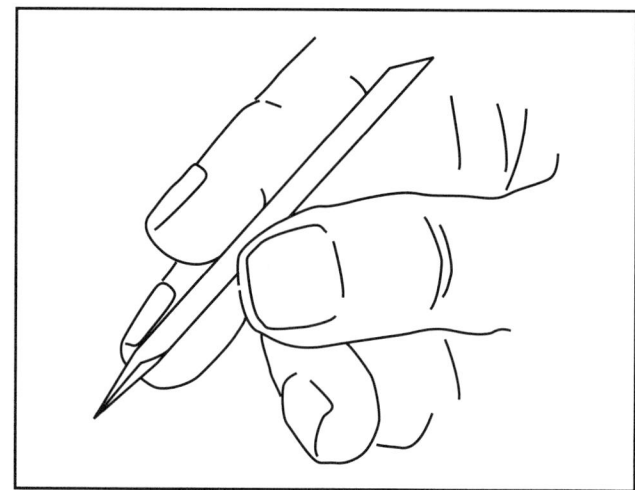

Figure 7-19. Overhand, or pencil, position for deburring with precision knives (Gillespie 1999).

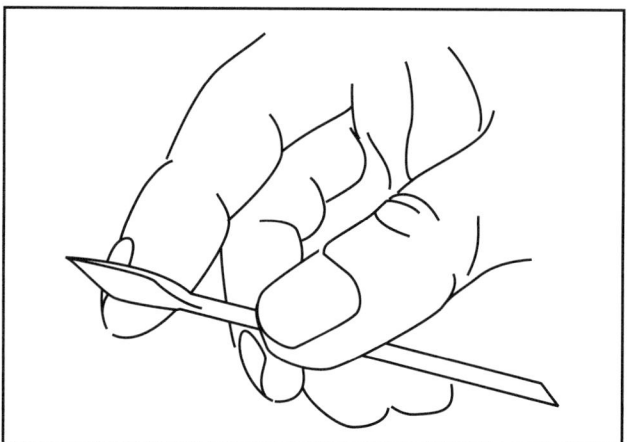

Figure 7-20. Underhand finger position for deburring with precision knives (Gillespie 1999).

feeling probe. This is a particularly useful approach for deburring holes. It permits rotating the tool around the contour, providing uniform force against the edge.

In deburring, both the underhand and finger positions are widely used. While it takes a certain amount of practice to develop skill, in just a few minutes one can develop a general appreciation for handling knives in this manner. In addition to greater safety for the part, these approaches minimize the danger of cutting oneself or losing control near the edge of a part. Precision deburring requires precision movements and skill. Using these approaches is not greatly different from free-flowing handwriting. Smooth, even strokes around the contour of the parts are required for precision and minimum fatigue.

Knives used for deburring come in a variety of shapes and sizes. Each shape was developed to do a specific job, but many shapes and sizes can be used on many different parts. The actual use of these knives will vary slightly among individuals. Some will find that a tool is particularly adaptable to a left-handed worker, whereas others will find that it works equally well for right-handed workers in certain applications. Personal preference plays a major role in the use of tools, just as it does in the brush of a painter. The worker needs to practice and observe those around him.

Triangular knives can be used in the pencil or underhand position. Using these knives with the fingers enables the user to rotate the knife to suit the edge or corner to be deburred. Typically, a precision triangular knife is used on a straight edge about 1 in. (25.4 mm) long. In contrast, oval knives are used in round holes. These distinctions, while generally true, are not always followed in practice and should be used only as general guidelines.

The angle-hooked knife (Figure 7-16) has two cutting edges at 90° to the shank. This allows the user to remove the edge burrs along a slotted groove (Figure 7-21). This tool will fit into most slots to remove edges, and sometimes it can be used to remove burrs from both edges of the slot at the same time. Special angle-hook knives are designed to clean out and deburr threads or to scrape along a single edge. These tools can be used both as knife and scraper.

The chisel is used as a knife and is not used with pounding motion; rather, it is used with a gentle pushing motion to scrape loose burrs. A typical application would be to clean out chips in the bottom of a blind tapped hole.

The cone-end tool is often used to clean out and deburr threads. This tool has a double cone back-to-back with a flat surface ground to the centerline (Figure 7-12). This enables the tool to remove burrs from hard-to-reach areas, such as blind holes. The cone-end tool is also used on a slot cut circumferentially into the wall of a tube, like a retaining ring groove. It can deburr most of the inside edges. Some individuals prefer this tool to an angle-hook tool for deburring threads or tapped holes, particularly for the larger sizes (for example, 0.25 in. [6.4 mm] diameter).

A scalpel (surgeon's scalpel or sharp-bladed hobby knife) is used almost exclusively to remove flash from plastic molded parts. Because of its thinness, this type of knife is not well-suited for removing chips or burrs from metal.

KNIFE CONSIDERATIONS

Knives cost from $3–60 each. Swivel-blade and oval knives, which are the easiest to make, are the least expensive. In many cases, if the order quantities are high enough, the cost of these tools is only $2–3 each. Triangular knives require more grinding so they are more expensive. Precision angle-hook knives may cost $10–50 each because they require considerable machining. Chisels cost from $3–10, and commercial scrapers cost $1–2 each. Any specially designed tool becomes expensive when small quantities must be made.

Individuals who deburr on a daily basis must have the proper tools on hand when they are needed, so most will have 20–50 knives in a variety of sizes and shapes in their toolbox at any one time. There is a natural tendency to use some tools too long, allowing them to become dull before they are returned for regrinding. Dull tools require considerably more strength when deburring and do not work as well as sharp tools. Whenever tools are dull they should be returned to the tool crib and exchanged for new ones.

Before using a deburring knife, the deburrer should investigate the edge under a microscope. It is easy to grind deburring knives so that the edges are ragged or chipped. In some cases, these tools are reground (1,000 at a time), and it is easy for those doing the

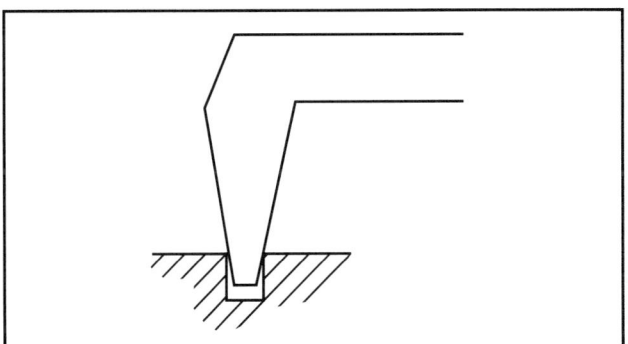

Figure 7-21. Use of angle-hooked tool to deburr a slot. (Courtesy Honeywell)

deburring to overlook those that are improperly ground. Improperly ground tools do not last. In some cases, when looking at newly ground tools one will find a burr on the tool. In this situation, it is important to inform the supervisor of the problem so he or she can correct the situation. Deburring knives that have burrs on them will not successfully deburr parts.

Table 7-3 lists the characteristics of precision knives currently being used at one company. These are in addition to the swivel-blade tools and scrapers that are also common.

KNIVES MAKE BURRS

Knives are widely used for deburring, but it should be noted that, while removing a burr, they typically make two more small burrs (Figure 7-22). Even these small burrs are objectionable on many parts. As a result, one must either scrape off (as opposed to cut off) these small burrs or use abrasive-filled rubber products, brushes, or related tools to remove knife-made burrs.

RELATION BETWEEN THE KNIFE AND PART

When deburring with knives under a microscope, the worker should not hold the part down on the microscope base or table. It is important for both part quality and personal safety to hold the part above a solid surface (Figure 7-23). If necessary, wrists or arms may rest on the tabletop. If the worker is using the table to support the part, any sudden lunge with the hand that is holding the knife will gouge the part or cut the other hand. By holding the part in the air, the grasping hand easily and quickly responds to sudden motions by the knife-wielding hand. Abrupt changes in burr sizes or location can easily cause minute lunges.

KNIFE SHARPENING

Normally, regrinding knife edges resharpens them when they become dull. Some knives can be sharpened with normal knife-sharpening techniques (Juranitch 1985; Lee 1995). In less industrialized nations, where labor is inexpensive, resharpening minimizes costs. In highly industrialized countries, throwaway tools are often the most economical.

ENVIRONMENTAL, HEALTH, AND SAFETY ISSUES

Knives are inherently dangerous. They are designed to cut. In normal use, with skilled operators, knives do not present a safety problem. Any time a knife is used in a fist position, however, the possibility is ever present for cutting oneself or injuring nearby workers. For this reason, workers should use knives in the fist position only for special circumstances. Knives should never be used for anything but their intended purpose. Because of their danger, knives should only be used when operators are mentally alert. Drowsiness or medicines may impair the user's normal mental sharpness.

To replace scalpel blades safely, grab the blade with forceps or special blade removal tools. Do not use fingers. Razor blades should be stored in razor-blade safety dispensers and never thrown into trash receptacles, which would expose others to cuts.

REFERENCES

Gillespie, LaRoux K. 1979. "A Training Manual for Precision Hand Deburring." Pt. 1. Report BDX-613-2245 (November), pp. 69–90. Kansas City, MO: Bendix Corporation. (Available from National Technical Information Service [NTIS].)

———. 1996. *Guide to Deburring, Deflashing, and Trimming Equipment, Supplies and Services*. Kansas City, MO: Deburring Technology International.

———. 1999. *Deburring and Edge Finishing Handbook*. Dearborn, MI: Society of Manufacturing Engineers.

Juranitch, John. 1985. *The Razor Edge Book of Sharpening*. New York: Warner Books.

Lee, Leonard. 1995. *The Complete Guide to Sharpening*. Newtown, CT: Taunton Press.

BIBLIOGRAPHY

Juranitch, John. 1977. "Sharpening Secrets of a Pro." *Popular Science*, February: pp. 118–121.

Kingshott, Jim. 1994. *Sharpening—The Complete Guide*. Lewes, East Sussex, England: Guild of Master Craftsman.

Spielman, Patrick. 1991. *Sharpening Basics*. New York: Sterling Publishing Company.

Table 7-3. Dimensions of one company's precision deburring knives (see figures 7-6 through 7-8)

Tool Style	Shank Diameter in.	(mm)	Working Diameter in.	(mm)
Triangular	3/16	(4.76)	0.188	(4.76)
Triangular	1/4	(6.35)	0.250	(6.35)
Triangular	5/32	(3.97)	0.156	(3.97)
Triangular	1/8	(3.18)	0.125	(3.18)
Triangular	1/8	(3.18)	0.062	(1.57)
Triangular	1/4	(6.35)	0.250	(6.35)
Triangular	3/16	(4.76)	0.188	(4.76)
Triangular	1/4	(6.35)	0.156	(3.97)
Triangular	3/16	(4.76)	0.075	(1.91)
Triangular	5/32	(3.97)	0.037	(0.94)
Oval	1/8	(3.18)	0.060	(1.52)
Oval	1/4	(6.35)	0.100	(2.54)
Oval	1/4	(6.35)	0.093	(2.36)
Oval	1/4	(6.35)	0.156	(3.96)
Oval	1/8	(3.18)	0.125	(3.18)
Oval	3/16	(4.76)	0.096	(2.44)
Oval	1/4	(6.35)	0.156	(3.97)
Oval	1/4	(6.35)	0.020	(0.51)
Angle hooked	1/4	(6.35)	0.063	(1.60)
Angle hooked	1/4	(6.35)	0.031	(0.79)
Angle hooked	1/4	(6.35)	0.100	(2.54)
Angle hooked	1/4	(6.35)	0.200	(5.08)
Angle hooked	1/4	(6.35)	0.005	(0.13)
Angle hooked	1/4	(6.35)	0.096	(2.44)
Angle hooked	1/8	(3.18)	0.062	(1.57)
Angle hooked	1/8	(3.18)	0.091	(2.31)
Cone end	1/4	(6.35)	0.100	(2.54)
Chisel	1/8	(3.18)	0.096	(2.44)
Chisel	1/8	(3.18)	0.065	(1.65)
Chisel	1/8	(3.18)	0.030	(0.76)
Chisel	1/8	(3.18)	0.310	(7.87)
Chisel	1/8	(3.18)	0.125	(3.18)
Chisel	1/8	(3.18)	0.156	(3.97)
Chisel	5/32	(3.97)	0.084	(2.13)

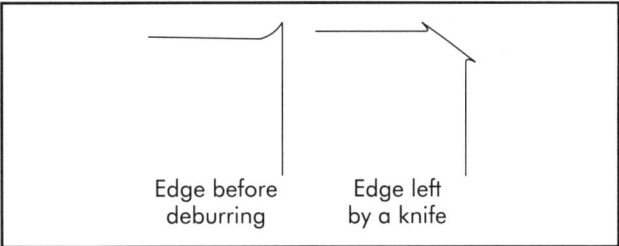

Figure 7-22. Knives generate burrs while they remove burrs. (Courtesy Honeywell)

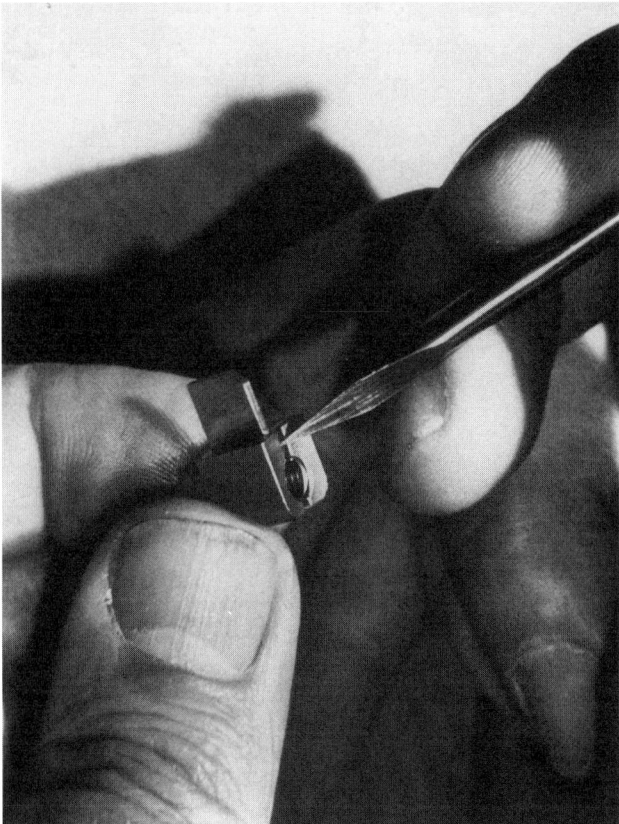

Figure 7-23. Correct way to hold parts while hand deburring with precision knives. (Courtesy Honeywell)

Deburring with Scrapers 8

Scrapers, as implied by the name, are tools designed to be drawn or pulled over a workpiece, as opposed to being pushed into it. One of today's popular scrapers is used for larger edges, typically on parts that are 5–10 in. (127–254 mm) in diameter or length. Figure 8-1 illustrates two scrapers. The long scraper is a triangular cross-section knife used as a scraper, rather than as a knife. Small scrapers have also been designed for use in areas where other tools do not work satisfactorily.

SCRAPER DESIGNS

As shown in Table 8-1, there are 15 different deburring and deflashing scraper styles in common use.

Pocketknives

Pocketknives are not good scrapers, but they are used for large, open edges.

Triangular Cross-section Scrapers

Triangular cross-section scrapers are basically the same tools described as knives in Chapter 7, but they are more often used to scrape the edge, rather than provide accurate edge cutting. They usually have large wooden handles, as shown in Figure 8-1, although plastic handles and replaceable blades are also available. Some companies call large scrapers "machinist scrapers," others call them "wood-handle scrapers." They may have a straight 120° surface or a relieved or hollow ground surface, such as the one shown in Figure 8-2. These also may be called "three-square" shapes. Chrome alloy steel is the common blade material. Some approach 8 in. (203 mm) in overall length. These, of course, would be used on relatively long, open surfaces. Most scrapers can be resharpened by a bench stone.

Thinner tools of similar design are used to remove burrs from long channels and holes in machined castings. Some of these are identified as "channel tools." Others are known simply as "straight scraping tools."

Keyway Scrapers

Keyway scrapers, as the name indicates, are designed to scrape burrs from long, narrow slots, such as keyways. Figure 8-3 illustrates one of these tools. They can be made with round-, square-, and V-shaped inserts to scrape both sides of the slot at the same time. One company manufactures slotting files with 60° V-shaped surfaces. They are neither knife nor scraper, but they are used in the same applications to deburr narrow slots.

Edge Shavers

Figure 8-4 illustrates an edge shaver, or chamfering tool, designed to provide a uniform chamfer on a straight edge. The illustrated tool can chamfer up to 0.040 in. (1.02 mm) wide. It can be used on lathe parts, as well as straight edges, and it can be safer than using a file on a rotating lathe part because it is used while the part is stationary. Different blades are provided for steel and aluminum and brass and cast iron. One edge shaver design is similar to a cheese cutter. The wide carbide blade is dragged over metal, plastic surfaces, or edges. It is designed for surfaces, rather than edges; with care, however, it can be used for both.

Sheet Metal Skivers and Tube-end Tools

Sheet metal skivers and tube-end tools employ two replaceable inserts on the end of the tool. In one design, the two inserts form a V-notch that rides over the edge of sheet-metal strips or parts. The "V" feature allows one to deburr and chamfer both sides of

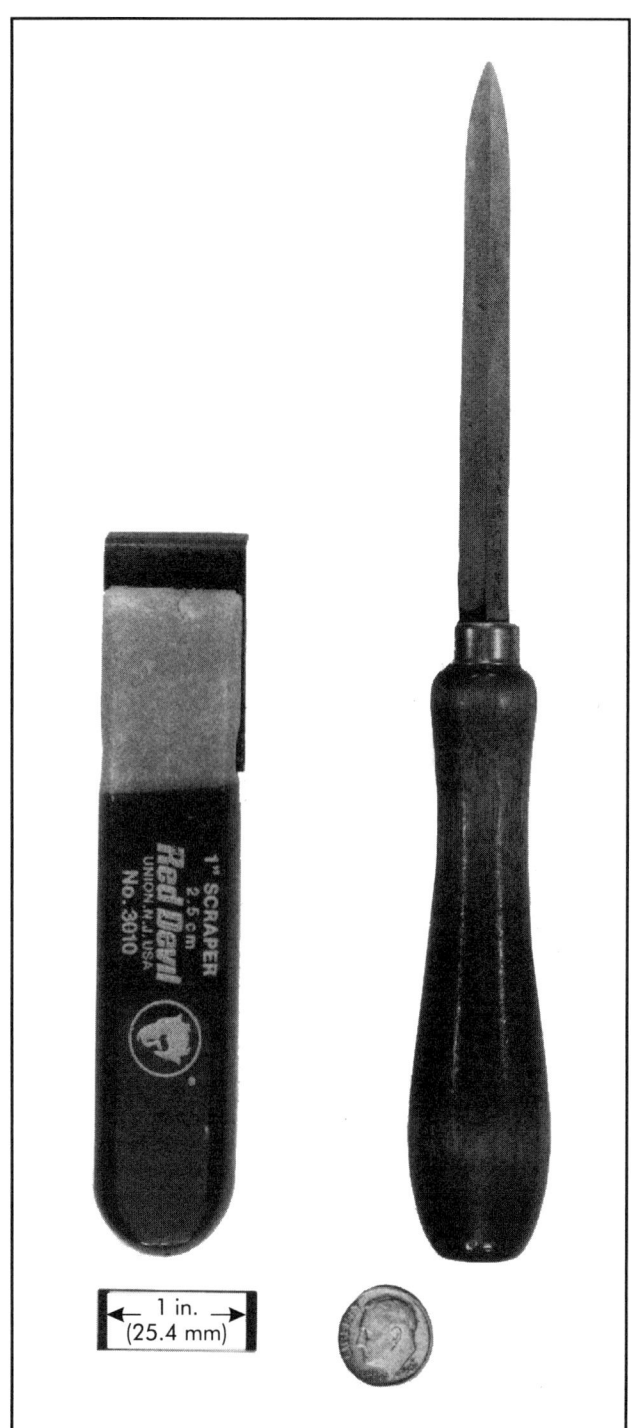

Figure 8-1. Two common large scrapers used for deburring.

employs a hand guard to prevent slicing hands on sharp edges before the tool removes the burr. This is one of the few knife or scraper tools with an integral hand protector. Commercial tools can accommodate sheets from 0.031–0.500 in. (0.79–12.70 mm) thick.

The V-notch, in addition to scraping both sides, provides a guide to keep the tool on the edge. Users can fashion a similar feature on the paint scraper, shown in Figure 8-1, by grinding a V-notch in the center of the cutting edge. The same "V" feature that works for sheet metal is also employed for large-diameter tube ends.

Bearing Scrapers

Some bearing scrapers are made from tool steel. Their unique cross-sectional shape facilitates deburring cylindrical edges.

Gravers

Gravers are tools used primarily for engraving. They can be found in toolmaker supply catalogs. Gravers come in several shapes and are used in hard-to-reach areas for small lengths of surface.

Angle-hooked Tools

Angle-hooked tools were described in Chapter 7, where they were treated as knives. They are also used as scrapers, and were first used to reach down into deep intersecting holes and cross channels in machined castings. Figure 8-6 illustrates some precision miniature shapes made by one plant for its employees, while Figure 8-7 illustrates a small commercial replaceable blade unit and its application. Each different tool design in Figure 8-6 represents not only different sizes of tools and shapes, but also different levels of tool quality and product needs.

Buttonhook Tools

Figure 8-8 illustrates buttonhook tools. They are triangular cross-section tools that are bent into a circle and used for rough deburring or to deflash holes.

Wide, Flat Scrapers

Producers of large parts use a kind of tool often called a wide, flat scraper, or hand scraper. They are used to scrape machine ways' flat surfaces to free them from sludge, buildup, paint, or other accumulations. They are also used for rough deburring of large part edges. The flat model looks like the ground-down end of a file. A half-round model allows scraping inner diameters, and the triangular model is a large triangular knife shape. These are big tools.

the strip in one pass of the tool. The inserts may be made of carbide or M-2 tool-steel hardened to R_C 64. Another design utilizes a single insert that has V-notches ground into it. The tool shown in Figure 8-5

Table 8-1. Common deburring and deflashing scraper styles

Type	Typical Use
Pocketknife	Not a good choice for deburring, but can act as a scraper on metal
Triangular cross-section scraper	Large parts, generally
Keyway scraper	Burrs on slots and keyways
Edge shaver	Long, straight edges (design is similar to a cheese cutter)
Sheet-metal skiver	Scrape both sides of sheet-metal strips simultaneously
Bearing scrapers	Long, open edges
Graver	Engraving cutters have a variety of shapes, useful for specific parts.
Angle-hook tools	Hard-to-reach areas, bottoms of hard-to-reach areas
Buttonhook tools	Chamfering holes
Wide, flat scrapers	Long, open edges
Chisel-like scrapers	Hole bottoms or hard-to-reach areas
Cone end	Holes in hard-to-reach areas
Rolling scraper	Slots on wide surfaces
Scribers/engraver/deburring tools	Also called pocket scribers, slim-grip scrapers, and adjustable scrapers
Special scrapers	Deburr unusual locations and unusual features

Shape: Three square in shape with handle. Hollow ground with 60° angles. Blade length is approximately 3.25 in. (82.6 mm) with overall length of about 8 in. (203.2 mm).
Use: Great for hand finishing bearings, casting joints, etc.

Figure 8-2. Hollow-ground triangular scraper. (Courtesy Federal File)

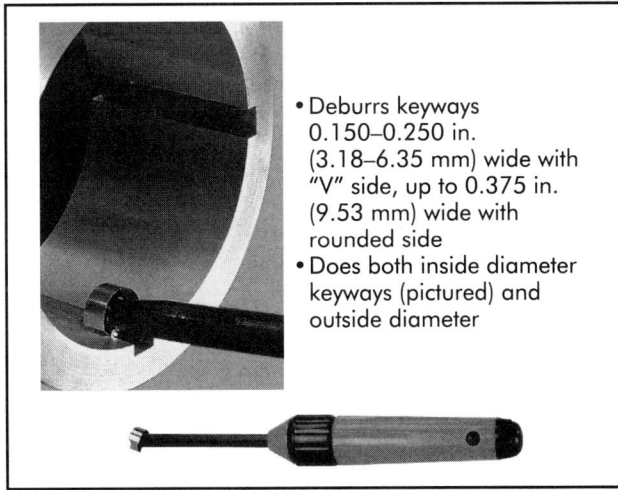

- Deburrs keyways 0.150–0.250 in. (3.18–6.35 mm) wide with "V" side, up to 0.375 in. (9.53 mm) wide with rounded side
- Does both inside diameter keyways (pictured) and outside diameter

Figure 8-3. Keyway scraper. (Courtesy Martindale Electric)

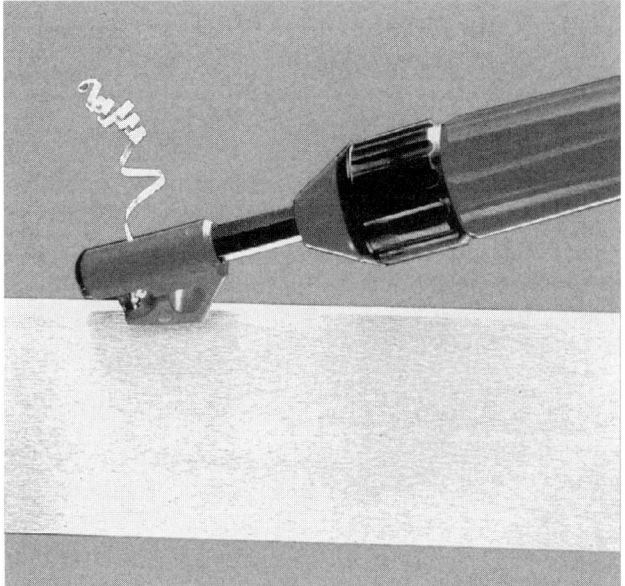

Figure 8-4. An edge shaver chamfering tool. (Courtesy Royal Products)

Chisel-like Scrapers

A variety of special chisels are used to scrape burrs from out-of-the-way places. They are inefficient deburring tools, but they can effectively remove relatively brittle sprues or similar plastic projections. They are also useful for removing chips and projections at the bottom of blind holes. Figure 7-13 in Chapter 7

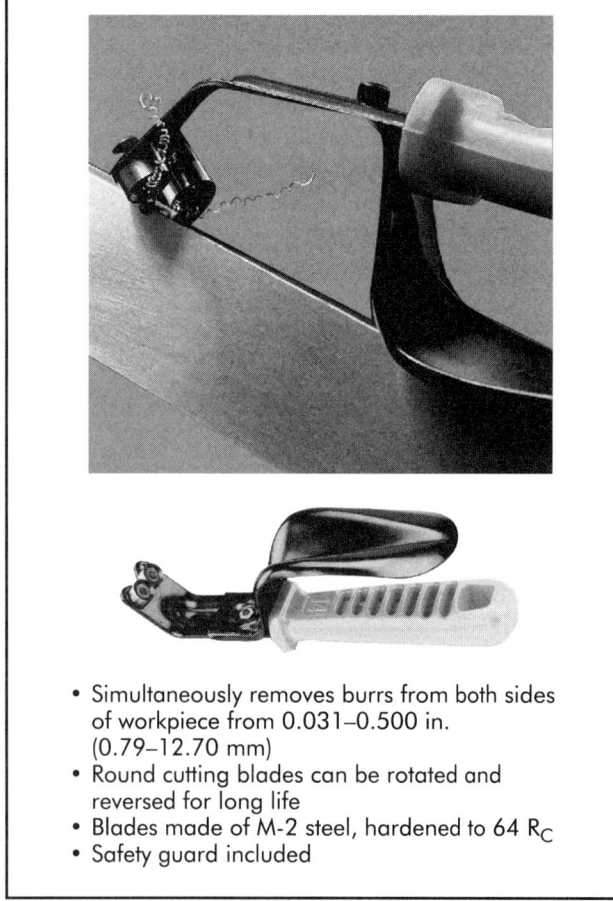

- Simultaneously removes burrs from both sides of workpiece from 0.031–0.500 in. (0.79–12.70 mm)
- Round cutting blades can be rotated and reversed for long life
- Blades made of M-2 steel, hardened to 64 R_C
- Safety guard included

Figure 8-5. Sheet-metal skiver. (Courtesy Royal Products)

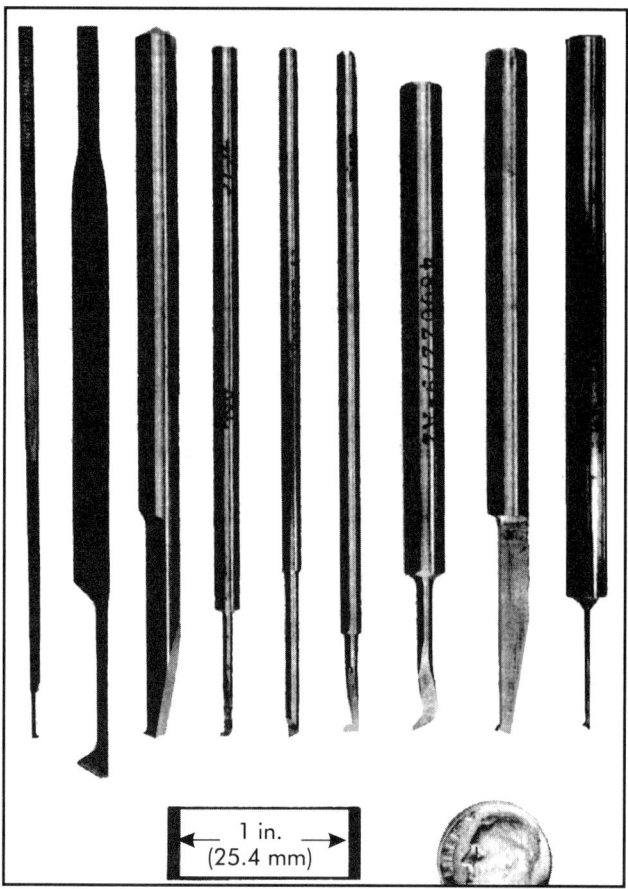

Figure 8-6. Precision angle-hooked scrapers made from Gorten cutter blanks.

illustrates one of these tools. Others have a "T" shape, while still others have an angle cut like a scalpel blade (Gillespie 1979).

Cone-end Tools

Cone-end tools are described in Chapter 7. They are used as knives in some instances, as scrapers in others.

Rolling Scrapers

Users who do not have a straight-line scraping motion can move up to wheels. Two different straight-line deburring and scribing systems are available. Two steel wheels guide and support the deburring scraper for straight-line deburring. These are used for slots on wide surfaces.

Scribers/Engravers/Deburring Tools

Scribers/engravers/deburring tools represent a loose class of tools that provides several functions. One series is a solid-carbide tool mounted in a steel collet and held in an aluminum pin vise. A variety of solid-carbide inserts are used with these tools. None of them are exceptional for deburring, but their convenience and small size make them useful for hard-to-reach areas.

Special Scrapers

Some large machinist scrapers have half-round shapes; others have a chisel edge on a long, flat blade. The flat-blade scrapers are used mostly to scrape surfaces, as opposed to edges, but they also can be used on many edges. Hand scrapers are another variation of machinist scrapers.

USING DEBURRING SCRAPERS

Most deburring scrapers are used as fist tools with the power supplied by the whole hand and arm. As such, many scrapers provide commercial or coarse edge quality. That satisfies many of the commercial product needs. Although it takes a certain amount of practice to develop skill, in just a matter of a few minutes it is

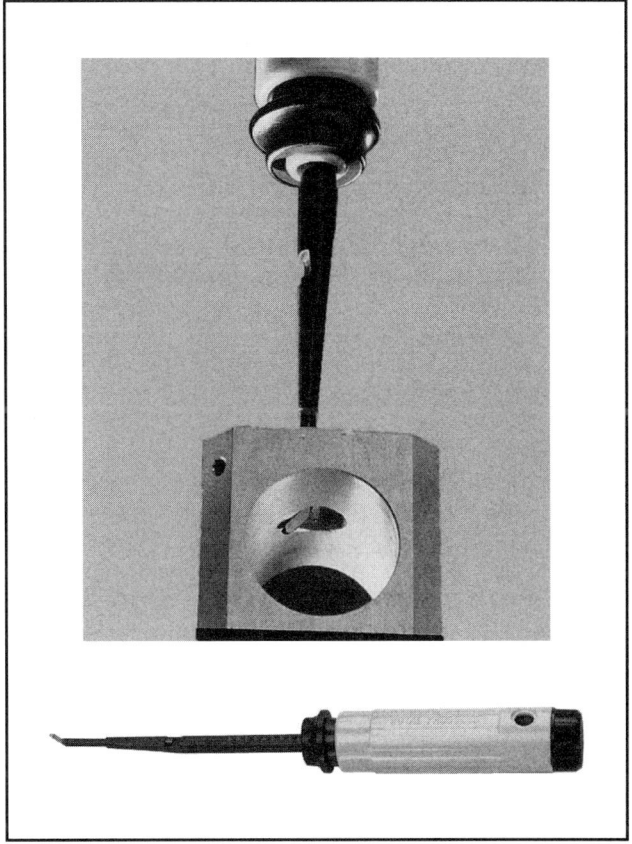

Figure 8-7. Adjustable scraper with triangular blade and angle-hooked blades. (Courtesy Royal Products)

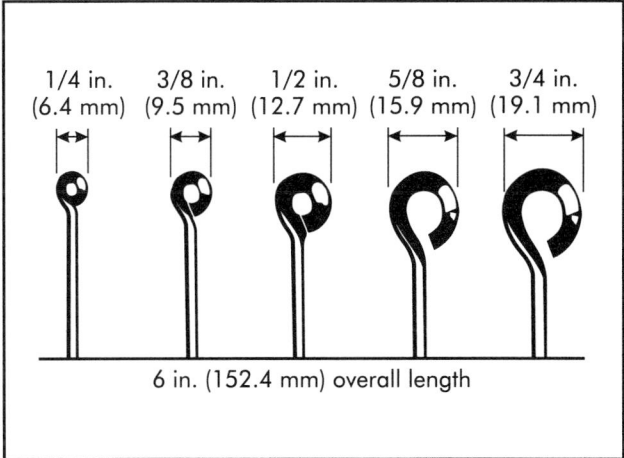

Figure 8-8. Buttonhook scrapers.

possible to gain a general appreciation for handling scrapers. In addition to greater safety for the part, these tools minimize the danger of cutting oneself or losing control near the part's edge. Commercial scrapers cost $1–6 each. Any special-design tool becomes expensive when ordered in small quantities.

There is a natural tendency to use some tools too long, allowing them to become dull before they are returned for regrinding. Dull tools require considerably more strength to use for deburring, and they do not work as well as sharp tools. Thus, whenever tools become dull, they should be returned to the crib and exchanged for new ones.

Figure 8-7 illustrates a typical scraper application. In this instance, the scraper scrapes the burr from the bottom of the hole. There is some likelihood that the scraper's edges will scratch the hole surface. That may be a quality problem for some holes. For precision deburring, the tool would be used as a knife, the part would be turned over, and the cutting edges would slice the burr away without touching hole walls.

In Chapter 4, Figure 4-31 illustrates a reamer that has been modified to scrape away an adhered chip in the bottom of a threaded hole. In most companies, this chip is not a concern, but for some high-precision applications in which the chip can fall into assemblies, it must be removed. The key for deburring personnel is to know exactly what does and does not have to be removed. If you must perform this operation, keep in mind that it is time-consuming.

Scrapers Make Burrs

Scrapers are widely used for deburring, but they also create burrs. Typically, scrapers (like knives) make two more small burrs (Chapter 7, Figure 7-22). Even these small burrs are objectionable on many parts. As a result, one must either scrape off (as opposed to cut off) these small burrs or use abrasive-filled rubber products, brushes, or related tools to remove them.

Scrapers are perfectly acceptable as the final deburring tool for many applications. They can generate chamfers up to 0.040 in. (1.02 mm), as noted previously; however, 0.030 in. (0.76 mm) would be a more normal high limit on many materials. Scrapers can be used on all metals, as well as many plastics. They may leave chattered edges on some plastics and metals until operators find the best technique.

ENVIRONMENTAL, HEALTH, AND SAFETY ISSUES

Some scrapers are heavy; therefore, they should be applied to burrs that are larger than those on which some of the knives would be used. After a full day of applying heavy pressure, operators may exhibit repetitive motion syndrome.

REFERENCE

Gillespie, LaRoux. 1979. "A Training Manual for Precision Hand Deburring, Part 1." Report BDX-613-2245. November 1979, pp. 76-81. Kansas City, MO: Bendix.

Countersinks and Chamfering Tools 9

Most holes are deburred, in part, with countersinks or chamfering tools. There are at least 100 different designs of countersinks and chamfers. They come in high-speed steel and various carbide grades, and most are available in 60, 82, and 90° included angles. This chapter describes their most common attributes. A complete presentation of each design can be found in other sources (Gillespie 2000).

The advantage of countersinks is that they can be applied on-machine, as well as in a hand-deburring operation. It is a quick operation and, sometimes, it is all the deburring required. In many instances, the small burrs made by the chamfer tool must be polished off.

TOOL TYPES

Table 9-1 provides some insight into the variety of tool types available. Some of the more common types used for deburring are discussed and illustrated in the following sections. The complete list is shown elsewhere (Gillespie 2000). Countersinks and chamfering tools are used on live spindles in a manual mode or in pin vises. Pin-vise work provides a more careful feel for cutting and can be performed under a microscope if needed.

Traditional Countersink Designs

Traditional countersinks, as shown in Figure 9-1, come in 1-, 2-, 3-, 4-, 5-, 6-, 8-, and 12-flute configurations. They can be made into any chamfer angle desired, although 60, 82, and 90° angles are standard. They are made from high-speed steel and carbide, and can be coated with diamond or other materials for finishing abrasive materials. Although any two tools may have the same principal design, they may function significantly different because of clearance angles and ground surfaces.

There seems to be neither rhyme nor reason as to which tool is chosen for deburring work. "If it works, use it," is probably as sage advice as one can find. If it does not work, try one of the 99 other designs or the same design from another manufacturer.

Carbide Insert Countersinks

Countersinks with carbide inserts (Figure 9-2) are typically used on larger holes. Generally, they are used in a drill press or other motorized device.

Flat Blade Insert

A flat blade insert is a simple deburr tool for soft materials. This tool is used in a motorized application.

Cylindrical Insert Countersinks

Cylindrical insert countersinks (Figure 9-3) are typically used in a drill press or similar motorized device.

Spring-loaded Countersink

Spring-loaded tools are typically used in motorized applications. The spring loading limits unintentional chamfer depth.

The spring-loaded countersink is loaded into a drill press or similar spindle and placed at the hole entrance. When force is applied and the tool rotates, the blade cuts off the burr and it moves up or down, depending on the three-dimensional shape of the hole surface. These tools can be used in any rotating spindle, such as low-speed portable drills, drill presses, or automatic equipment. Light contact is sufficient

Table 9-1. Chamfer tool designs used for deburring and chamfering

Style	Application	Comment
Single-flute countersink		Can countersink smaller holes than multi-flute tools
2-flute countersink		
3-flute countersink	Odd number of flutes minimizes chatter	Run at low speeds
4-flute countersink		
5-flute countersink		
6-flute countersink	Six flutes minimizes chip load and chatter	
8-flute countersink		
12-flute countersink		
Carbide insert countersink	Use in high-wear applications	Comes in many design variations
Flat blade insert		
Cylindrical insert		
Spring-loaded countersink		
Rubber spring countersink	Reduces chatter	
Elliptical hole countersink	Will cut some materials cleaner	
Corner rounding end mill/countersink	Produces a radius rather than a chamfer	
Quick-change countersink	Allows quick cutter change	
Retractable blade		Not available in very small sizes. Not for precision hole finishes
Clothespin tool	Used for fast deburring	Success depends on burr properties and hole size
Back countersinks	Used when other tools cannot get to bottom of hole	Several designs are available
Disk-in-pin vise	Simple design scrapes off burrs	
Copper-tube-blade countersink	Low-cost design still works	
Pipe-burring reamer	Still used in plumbing industry	
Diamond-coated clothespin	Allows many shapes to be made and many levels of deburring quality	
Rotary bur	Widely used for finishing	
Tube-end tools	Fast finishing for both inside and outside of tubes	Generally used in conjunction with special tube end manual and automated finishing machines
Back chamfer rotary bur		
Outside chamfering mills	Chamfers the ends of bars	

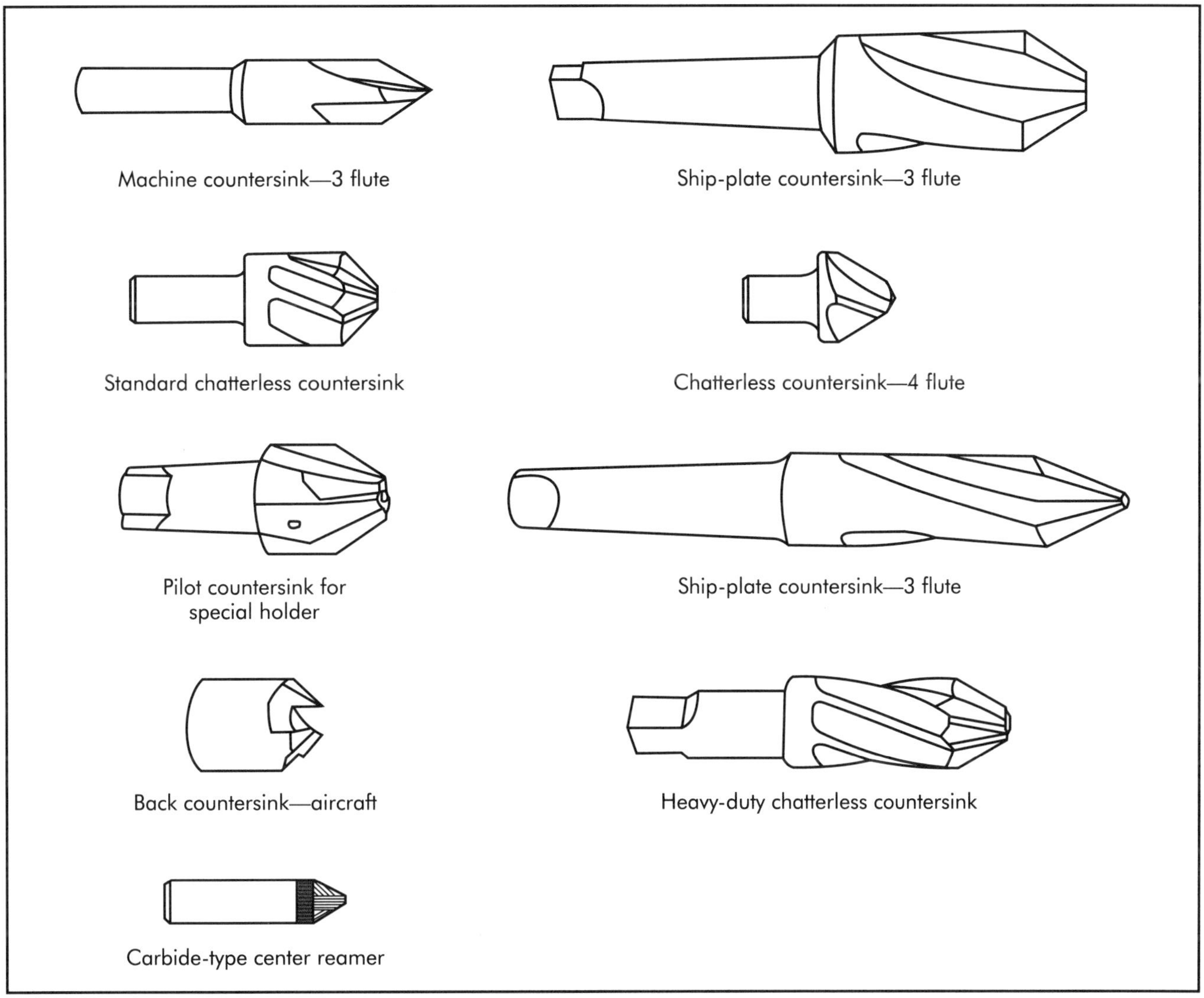

Figure 9-1. Common countersinks (Drozda and Wick 1983).

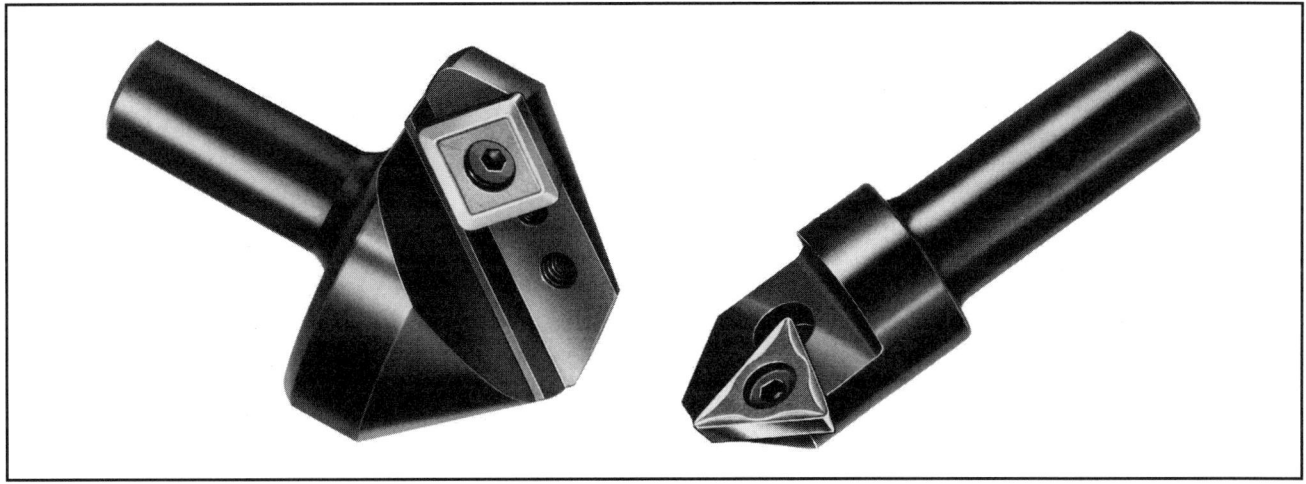

Figure 9-2. Carbide insert countersinks. (Courtesy Everede Tool and Metal Cutting Tools)

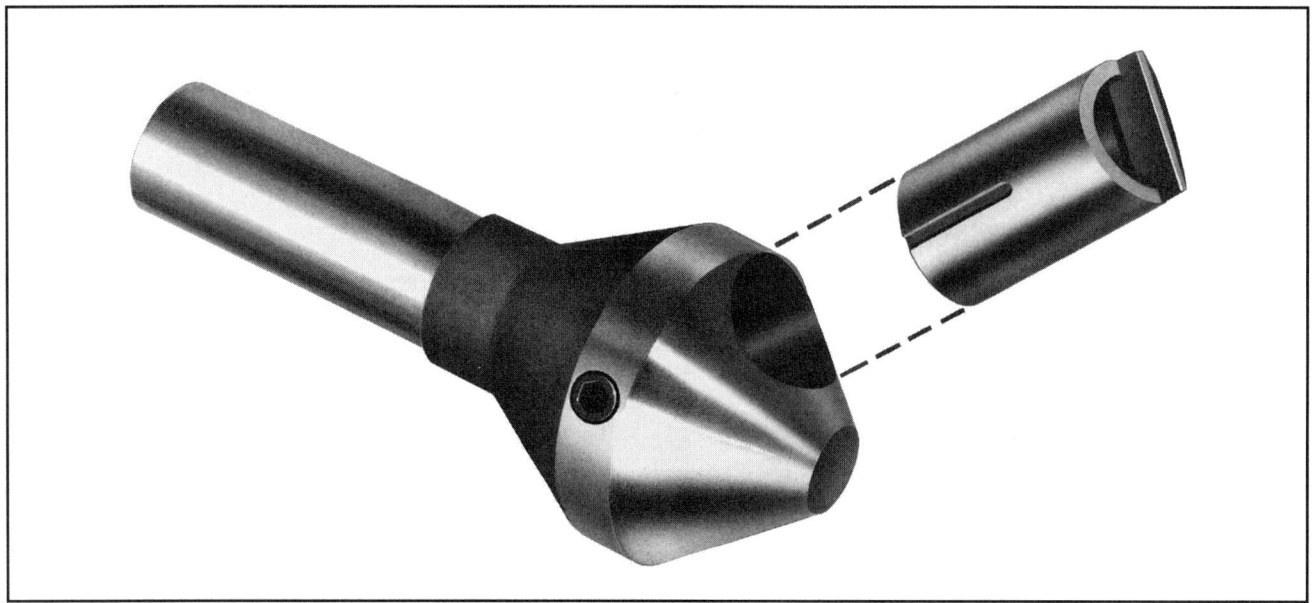

Figure 9-3. Cylindrical insert countersink. (Courtesy Field Tool)

for most deburring. They provide a minimum cut diameter of 0.1875 in. (4.763 mm) and a maximum of 0.75–2.00 in. (19.1–50.8 mm). Similar tools are available that use a pilot to keep the tool in the hole.

Rubber-spring Countersink

Rubber-spring countersinks (Figure 9-4) are used in operations involving larger holes, and require the use of a spindle motor.

Elliptical Hole Countersink

Elliptical hole countersinks come in sizes that enable their use as a handheld tool, as well as in larger sizes suitable for drill press or other motors. In some materials, they provide a cleaner cut than traditional tools. Some are designed to deburr the outside edge of curved and elliptical hole openings, such as those produced by drilling through a curved or cylindrical surface, or holes drilled at an angle in flat parts.

Figure 9-5 illustrates an elliptical hole countersink. A hole produced at an angle through the tool provides a slicing action to help cutting. These come with a plain or pilot nose to center on the hole.

Corner-rounding End Mills

Figure 9-6a provides an illustration of a corner-rounding rotary bur. Figure 9-6b shows a corner-rounding end mill, which has fewer teeth but performs the same kind of action. These tools are designed for larger holes, and a handheld holder is also illustrated for this application. The radius action not only provides a radiused edge, but, in many materials, prevents the formation of the little burrs produced by a typical chamfering tool. These tools are normally used on machining centers in automatic operations, but they work just as well in hand operations, as shown in Figure 9-6a with the wooden handle. Several designs are made as standards, and others can be made as special designs. The many small teeth on this design minimize the tendency for the tool to grab and throw the part. These tools may also be called "radius deburring cutters." They can be made of high-speed steel or carbide.

Quick-change Countersink

The uniqueness of the quick-change countersink is not on the cutting end, but that it can be used in a quick-change toolholder to allow rapid change-out (Figure 9-7a). Figure 9-7b shows a quick-change tapered hand reamer used to deburr.

Retractable-blade Countersink/Deburring Tool

In the retracted position, the retractable-blade countersink allows the tool to pass through the hole. With a simple hand motion, the blade projects outward, which allows the user to perform a chamfer on the backside of a hole or on the top inside of a slot milled through a hole (Figure 9-8). It is particularly suitable for holes larger than 0.250 in. (6.35 mm) in diameter.

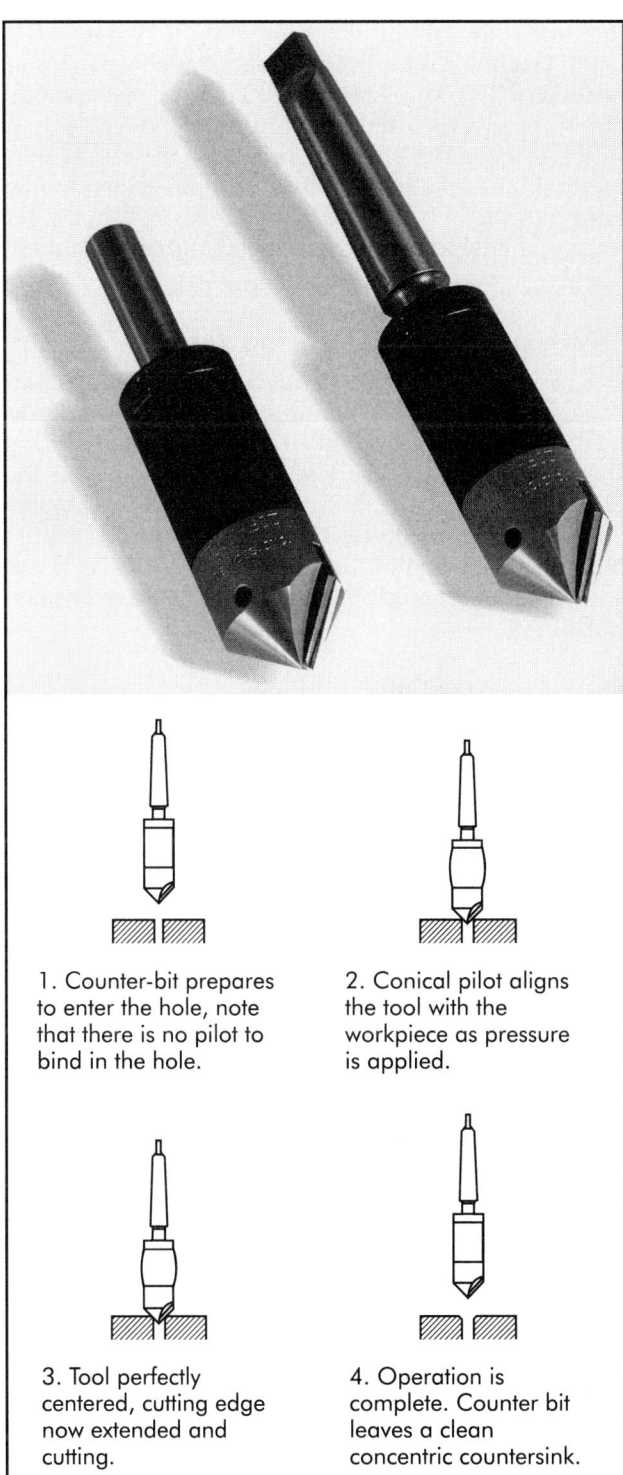

Figure 9-4. Rubber spring conical countersinks. (Courtesy Vernon Devices)

1. Counter-bit prepares to enter the hole, note that there is no pilot to bind in the hole.

2. Conical pilot aligns the tool with the workpiece as pressure is applied.

3. Tool perfectly centered, cutting edge now extended and cutting.

4. Operation is complete. Counter bit leaves a clean concentric countersink.

Figure 9-5. Elliptical hole countersink. (Courtesy Weldon Tool)

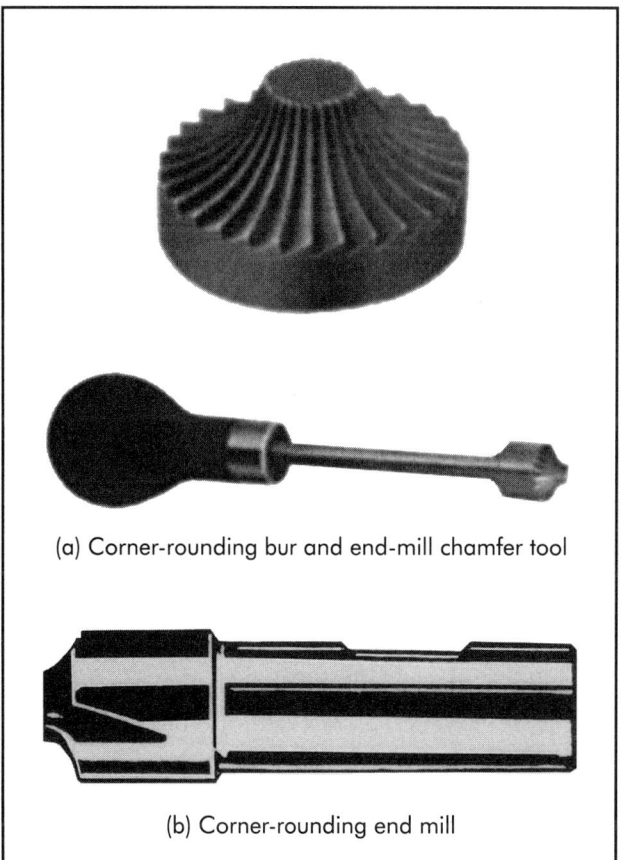

(a) Corner-rounding bur and end-mill chamfer tool

(b) Corner-rounding end mill

Figure 9-6. Corner-rounding bur and end mill. (Courtesy Severance Tool Industries)

As the retractable-blade countersink moves into the hole, spring tension pushes the blade from its housing. As the feed force increases (because the tool is being forced down into the part), a force will be reached that overcomes the spring force, allowing the blade to slip into the housing. The housing and blade

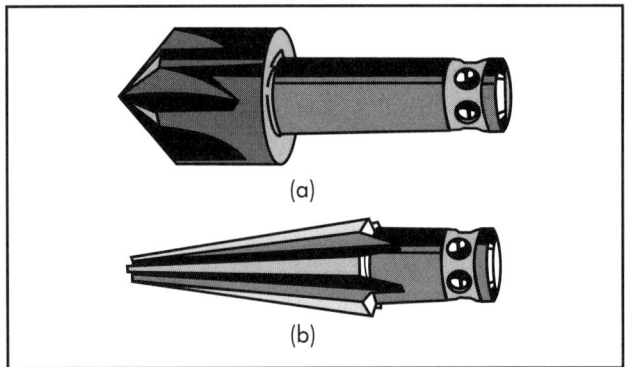

Figure 9-7. (a) Quick-change countersink and (b) taper hand reamer. (Courtesy Morton Machine Works)

facturer claims it will not mar a 30 μin. (0.8 μm) finish in the hole. Tools are available to deburr holes as small as 0.0787 in. (1.999 mm). They are also used on holes as large as 2 in. (50.8 mm) in diameter.

Reportedly, the chamfer produced in steel is normally 0.004–0.016 in. (0.10–0.41 mm) wide, and in aluminum it is 0.006–0.040 in. (0.15–1.02 mm), yet burr size has little effect on the chamfer (Gustafson 1983).

Clothespin-style Deburring Tool

The clothespin-style deburring tool (Figure 9-9) deburrs the topside of a hole and, with continued downward pressure, collapses to a smaller size sliding through the hole, which allows the tool to deburr the bottom side of the hole as well. Some users, apparently, can deburr 5,000–15,000 holes per resharpening. Since the tool expands and contracts, each size of the tool will accommodate a range of hole sizes and variations.

Back Countersinks

Back countersinks, as the name implies, are designed to chamfer the bottom or backside of holes.

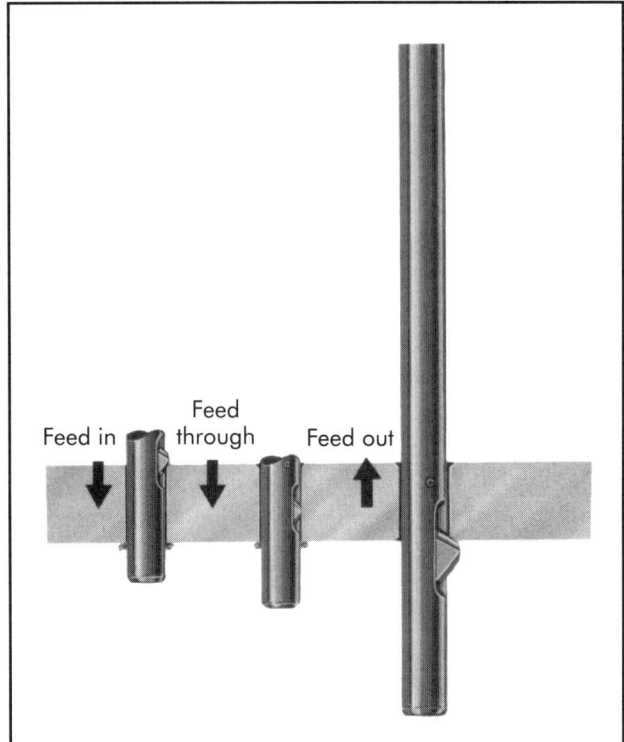

Figure 9-8. Retractable-blade deburring cutter. (Courtesy Cogsdill Tool Products)

slip through the hole until the blade reaches the backside of the hole, where no force is pushing it into the housing. The spring tension pushes the blade out. When the tool is retracted upward, the blade cuts against the bottom surface of the hole until this force again exceeds the spring force, then the blade slips back into the housing, and the housing is pulled out of the hole. A set screw adjusts spring force, which, in turn, increases or decreases stock removal. Since the crest of the blade is polished and noncutting, the manu-

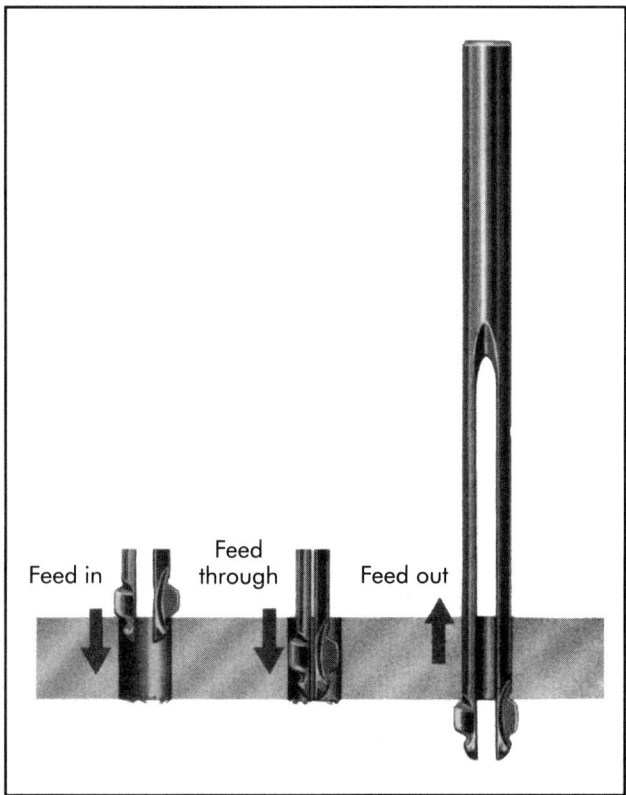

Figure 9-9. Clothespin-style deburring tool. (Courtesy Cogsdill Tool Products)

Two or three designs exist, but special designs can be made for any application. The challenge with some of these is to provide the correct upward force to remove the desired material. Figure 9-10 illustrates flat-face tools, but countersinks can be made in the same manner. In this design, the cutter is engaged to the driving shaft by a simple quarter-turn operation. Other designs use different engagements.

Figure 9-10. Back spot face/counterbore cutter. (Courtesy Metal Cutting Tools)

Disk-in-pin Vise

The disk-in-pin vise, an unusual approach, uses a thin metal disk, which is held in pin-vise jaws or other holders. The circular contour allows the tool to drop in the hole, and a little hand pressure and rotation scrapes off the burr and provides a small chamfer.

Copper-tube-blade Countersink

The copper-tube-blade countersink is a simple, triangular blade, and has been used to remove burrs from copper tubing for decades. It is widely used for home repair on copper water lines.

Pipe-burring Reamer

The pipe-burring reamer (Figure 9-11) has been used for decades. It is used for removing burrs from water and gas pipes as they are prepared. Most pipe-burring reamers are made from carbon steel and they can accommodate 1–2 in. (25–51 mm) pipe.

Diamond-coated Clothespin Tools

Almost any shape can be diamond-coated to cut or polish (Gustafson 1983). The clothespin configuration allows the tool to slip in and out of a hole with no

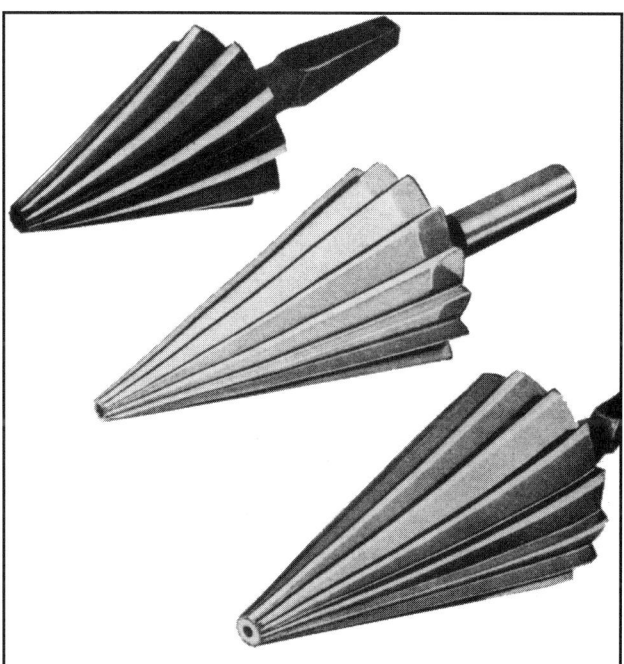

Figure 9-11. Pipe-burring reamers. (Courtesy Field Tool)

special mechanism to cause the action (Yinsen et al. 1992).

Rotary Bur

Rotary burs come in hundreds of configurations. They are described in Chapter 10 in more detail.

Tube-end Finishing Tools

Figure 9-12 illustrates two designs of tube-end finishing tools. Like some of the others mentioned in this chapter, tube-end tools come in many design variations. They are usually used in conjunction with powered motors. Automated systems as well as manual-fed machines are used (they can be used without power feed). The tools shown use a two-piece construction and accommodate only one size tube, but one company has almost 20 sizes for light deburring and other sizes for heavy deburring.

Cutting teeth are designed to give a shearing cut. In the designs illustrated in Figure 9-12, they curl the fine chips away from the cutter to avoid loading. The inside member produces a chamfer at a 30° angle with the inside and the outside member.

The tube-end deburring cutters in Figure 9-12 are intended for light deburring only and will quickly deburr tubes of almost any machinable material. For tougher and harder materials, they are available in high-speed steel or carbide. The tooth arrangement on these cutters has been adopted to cover a wide

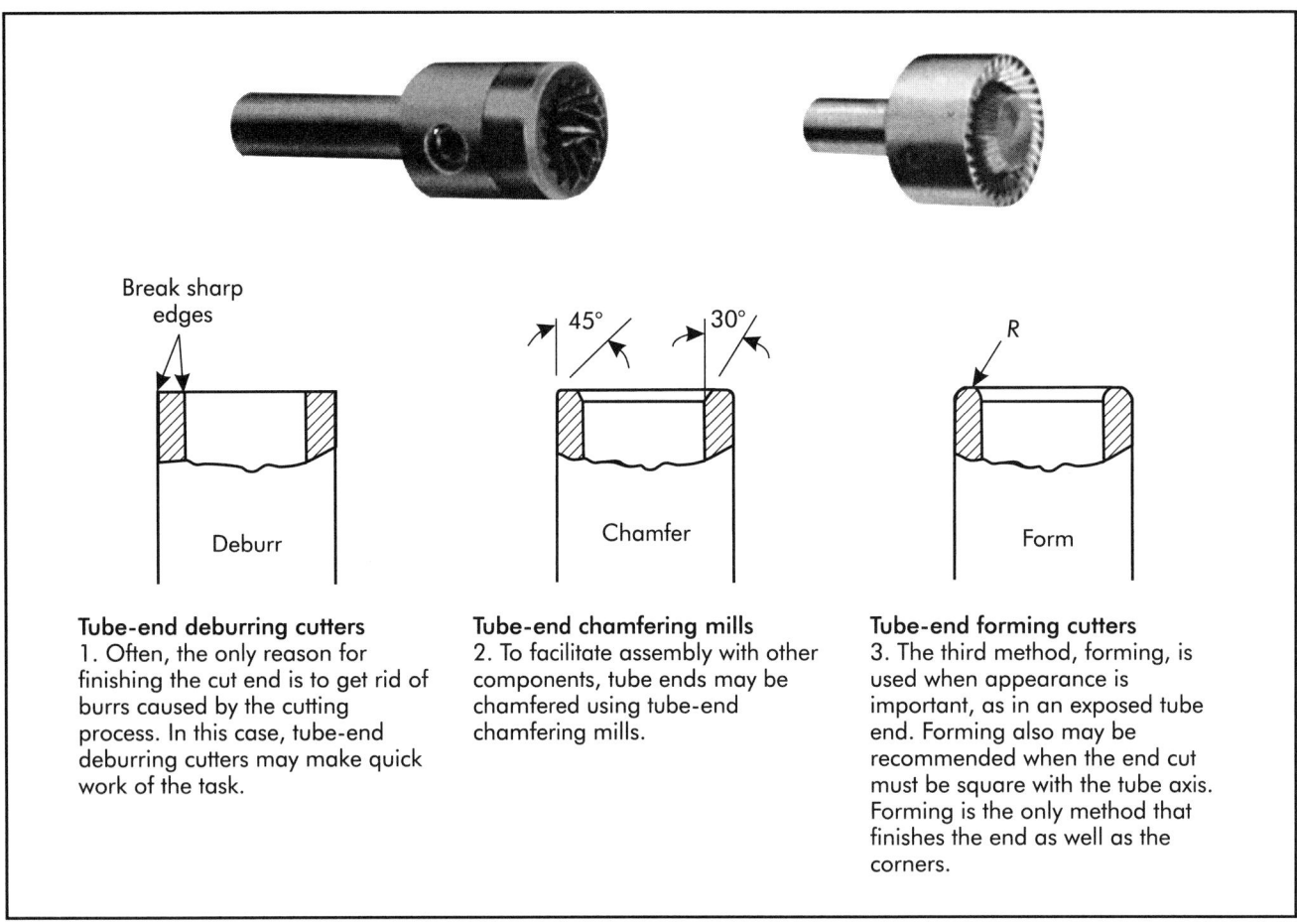

Figure 9-12. Tube-end deburring tools. (Courtesy Severance Tool Industries)

range of the most commonly encountered materials. Special designs are regularly made to accommodate unique end configurations or part material problems. Simple guides, such as V-blocks (correctly positioned), are recommended to locate the part under the rotating tool. The ring, or outside member, is adjustable for relative chamfer on the outside edge. It is secured after adjustment by socket and screws. Operating speeds range from 50–200 rpm, depending on the size of the cutter, material, and work condition. There are at least three forms of these tools:

- tube-end cutters;
- tube-end chamfering mills, and
- tube-end forming cutters.

Tube-end Deburring Cutters

Inside, outside, and tube-end deburring cutters are available in carbide and high-speed steel (HSS). Deburring cutters are identified as having smaller and more numerous cutting teeth than chamfering or forming tools. They can be operated over a wide speed range (slower is better in nonrigid setups) and take light cuts very quickly. Tube-end deburring cutters are available in sizes ranging from 0.125–2.250 (3.18–57.15 mm) outside diameter (OD). Each is adjustable for wall thickness. Standard cutters feature 30 and 45° centerline angles.

Tube-end Chamfering Mills

Tube-end chamfering mills have cutting edges designed to provide a shearing action, yielding a smoothly machined surface. The standard tool produces a 30° angle on the tube inside diameter (ID) and 45° on the OD. Other angles are available as special designs. This series of tools is offered in a range of sizes to accommodate pipe and tubing in 0.1875–2.5000 in. (4.763–63.500 mm) outside diameters. The models shown in Figure 9-12 are adjustable for different wall thicknesses.

Separate chamfering mills for inside and outside cutting are available. Inside chamfering mills are obtainable in 0.50–1.50 in. (12.7–38.1 mm) sizes, with

30° or 45° centerline angles. Outside chamfering mills, for working diameters from 0.125–3.000 in. (3.18–76.20 mm), are also stocked in 30 or 45° models.

Tube-end Forming Cutters

As the name implies, tube-end forming cutters completely machine the cut ends of tubular products. They produce a smoothly rounded surface that is both attractive and functional. Because they are of solid construction, a specific tool is required for each different workpiece diameter and wall thickness. Custom tube-end forming cutters are often made for nonstandard sizes and machining profiles other than blended radii. Carbide is available on sizes of 0.375 in. (9.53 mm) OD and larger.

Outside Chamfering Mills

Outside chamfering mills chamfer the ends of bar stock. They come in many designs and some can be used as a two-piece construction for finishing tube ends. Figure 9-13 illustrates a range of sizes of one design. This design accommodates bars as small as 0.25 in. (6.4 mm).

Figure 9-13. Outside chamfering mills. (Courtesy Rico)

Tap-chamfer Combinations

Figure 9-14 shows a combination tap and chamfering tool. This removes the burr raised by the tap in the same operation that performs the tapping.

COUNTERSINK USE

Normally, one desires to have machine tools apply countersinks and remove any large burrs, but when precision chamfers are required it may be necessary to resort to hand chamfers. One company has utilized hand chamfering for holes 0.016–0.125 in. (0.41–3.18 mm) for decades when edges must not be broken more than 0.001–0.003 in. (25–76 μm). A precision chamfer tool is needed, along with a good pin vise to hold it. Formal training and practice, as described in Chapter 33, are needed to develop the feel to accomplish such precision while removing burrs.

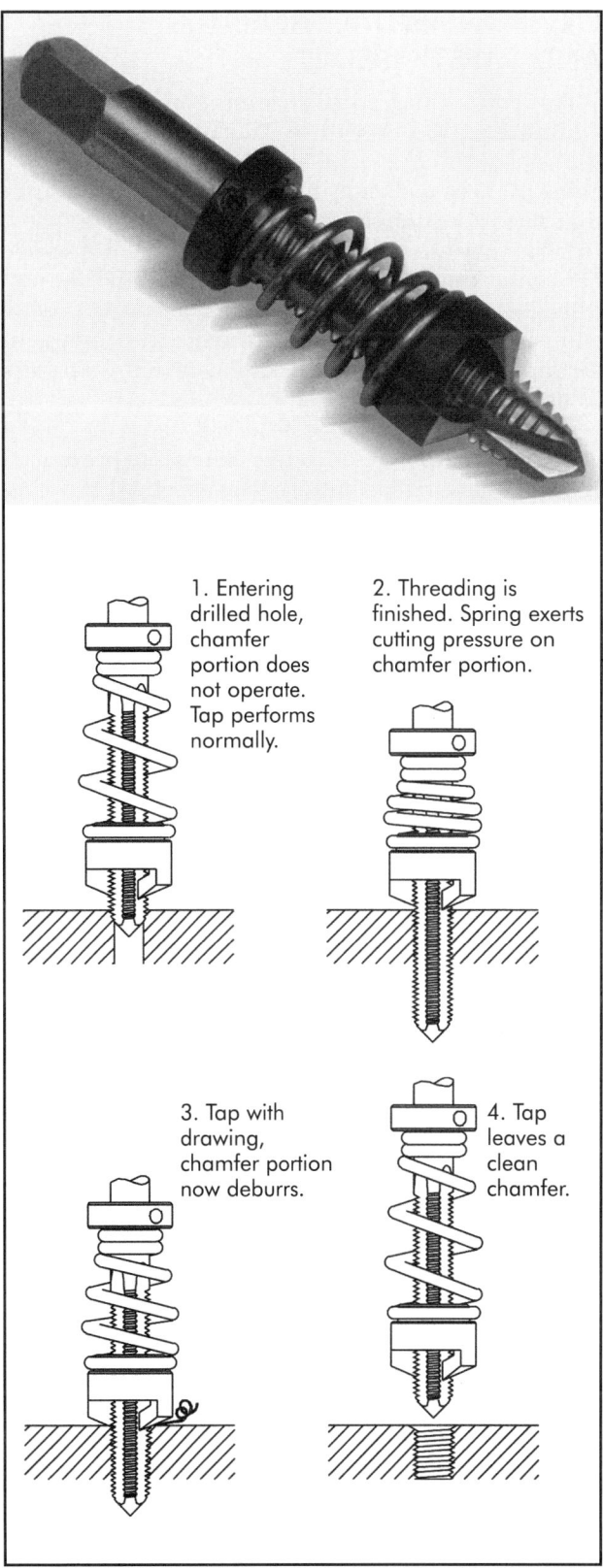

Figure 9-14. Chamfer tool combined with tap. (Courtesy Vernon Devices)

ENVIRONMENTAL, HEALTH, AND SAFETY ISSUES

Any cutter can grab loose clothing and quickly wrap it around the tool or spindle. This represents a potential injury to hands and fingers and even to arms. Some of these tools may have small-diameter shanks that bend if operated at too high of a spindle speed. In turn, they can be thrown from chucks or can scrape hands and knuckles. When performed by hand, the tool can also grab parts and throw them across the room. Some of these safety issues can be life-threatening in certain situations. Each can seriously injure workers or nearby observers.

Whenever a problem is observed, there are many opportunities to make effective changes to prevent it. Hand chamfering can be a safe and effective deburring process. Users can minimize risk by limiting the exposure of cutting and grabbing edges. To increase safety, some hand-fed bench-top machines expose only a few thousandths of an inch (hundredths of a millimeter) of the cutter end. Always use only sharp cutters and wear eye protection. Most shops will also require the use of safety shoes to prevent injury to toes if heavy parts or fixtures fall.

REFERENCES

Drozda, Thomas and Charles Wick, eds. 1983. *Tool and Manufacturing Engineers Handbook,* 4th ed., Volume 1, *Machining.* Dearborn, MI: Society of Manufacturing Engineers.

Gillespie, LaRoux K. 2000. *Guide to Deburring, Deflashing, and Trimming Equipment, Supplies and Services,* 2nd ed. Kansas City, MO: Deburring Technology International.

Gustafson, Lars. 1983. "Deburring with an Industrial Robot." Technical Paper MR83-677. Dearborn, MI: Society of Manufacturing Engineers.

Yinsen, Ye, Guo Hongdi, Huang Yaowen, Lin Feng, and Chen Qili. 1992. "The Technique of Finishing and Deburring Parts for Hydraulic Valves." In *Proceedings of the 2nd International Conference on Precision Surface Finishing and Burr Technology.* Dalian, China (September), pp. 363–67. Kansas City, MO: Deburring Technology International.

Rotary Burs 10

Rotary burs are widely used to deburr holes, as well as precision edges. They are used to remove heavy stock, flash, and risers from castings. In some plants, these tools are used more for surface material removal than for deburring, but they are found in almost every plant's arsenal of burr-fighting weapons. As this chapter shows, there are literally hundreds of tools from which to choose.

Although the industry literature lacks uniformity, most manufacturers use *bur* to indicate the tool that removes burrs. It becomes confusing in reading the literature if one uses a *burr* to remove a burr. For this reason, our discussion uses the most widely accepted nomenclature, which refers to bur as a tool that is often used to remove a burr.

Bur tools have a variety of more specific names, as shown in Table 10-1. They are known not only as bur tools but also as rotary files, mills, cutters, and combinations of these and many other names. As Figure 10-1 illustrates, one of the principal differences between a bur and a conventional countersink, or similar cutter, is that a burr has a relatively large number of small teeth. In many cases, a rotary bur has 2–5 times more teeth than a countersink tool. These many small teeth give the bur its smooth finishing action. The small teeth prevent excessive cutting into edges and reduce chatter that, in turn, improves surface finish. These tools can be used at high cutting speeds, which increase the number of part edges that can be finished at a given time.

There are two types of burs: commercial burs, typically, are relatively large (up to 2 in. [50.8 mm] in diameter), and dental burs are very small. Dentists use them to work on teeth. These distinctions are not hard and fast, and many catalogs intermix these tools or do not identify them by typical usage. Miniature bur balls only 0.004 in. (0.10 mm) in diameter are available. The smallest standard ball diameter, however, is 0.020 in. (0.51 mm).

ROTARY BUR STYLES

Rotary burs are named as such because early, finely serrated tools were made on cylinders and acted like rotary files. Burs and related tools are available in a wide variety of diameters, shapes, tooth coarseness, and materials. At least 80 national firms provide these tools (Gillespie 2000), and one company lists 1,028 different burs and rotary files in its catalog. Although bur balls are used most frequently, pointed cones and flame-like shapes are also used. Typically, these tools come in both high-speed steel and carbide, but today

Table 10-1. Common nomenclature for rotary burs

Rotary Files	Finish Mills	Burs
Disc cutters	Die mills	Rotary burs
Rotary cutters	Lab mills	Midget burs
Deburring cutters	Chamfering mills	Midget rotary burs
Micro center reamers	Edging mills	Miniature burs
Routers	Grinding mills	Dental burs
Rotary files	Junior burs	Surgical burs

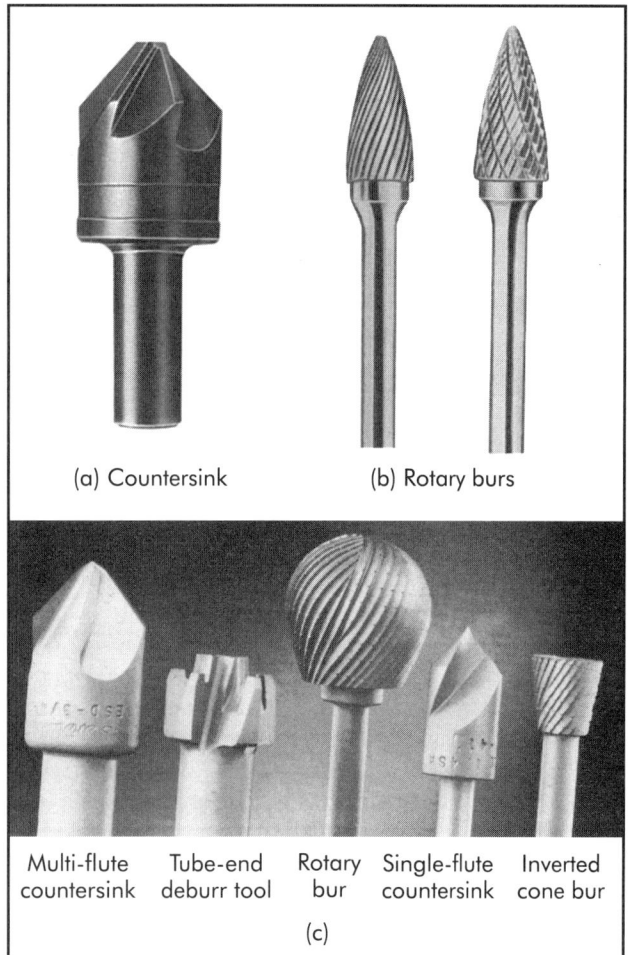

Figure 10-1. Typical countersink and rotary burs. (Courtesy Severance Tool Industries)

they are also produced as solid shapes coated with tungsten carbide or diamond particles, and are sold as industrial-grade tools. Many firms are turning to dental and medical burs for precision work. The nomenclature and family of tools contained within this broad heading include:

- rotary burs,
- midget burs,
- dental and jeweler's burs, and
- medical burs.

Table 10-2 presents a taxonomy of burs, which can be used to quickly understand the variations.

Rotary Burs

Figures 10-2 through 10-6 illustrate the standard geometry and sizes available from one manufacturer. As illustrated, this manufacturer provides both fine and coarse cuts for many of the shapes. The shapes are similar, but there are differences between the high-speed steel bur geometry and the carbide bur welded to steel shanks. These are representative of standard products, but many variations exist.

A typical angle range for cone tools is from 10–31°. The chamfer tools have 60, 82, or 90° angles. *Cylindrical ball nose* and *radius end* are nomenclature for the same basic design. Figure 10-7 illustrates additional geometries for these tools. The sizes shown are representative, but many other sizes are available or can be produced by other firms. Almost any shape can be purchased in sizes up to 1-in. (25.4-mm) diameter. A typical shank size is 0.25 in. (6.4 mm) in diameter, although 0.24 in. (6.1 mm) is also a standard. Tool length, overall, is roughly 1.5–3.0 in. (38–76 mm). The tools shown here are standard sizes, implied when one discusses burs. Miniature, or midget, burs are discussed in a later section. Their typical maximum size is 2 in. (51 mm) in diameter, but special sizes can be made to almost any diameter, and some suppliers make a living from making special sizes.

Router Cutters

Figure 10-8 illustrates fiberglass router cutters, also known as routing mills. They are used to trim and cut edges of printed circuit board and other fiberglass-filled plastics. Carbide withstands abrasion much better than high-speed steel, so these tools are widely used in the plastics industry.

Inside Deburring Cutters

Inside deburring cutters are designed for the inside diameter of tubing. High-speed steel is used for light deburring, while carbide is required for heavy-duty work. These cone-shaped tools employ angles of 30 and 45°. Standard sizes begin at 0.25 in. (6.4 mm) in diameter and go to 3 in. (76.2 mm) in diameter. Shank sizes of 0.25 in. (6.4 mm) are typical. Inside deburring cutters look something like the inner piece of the cutter shown in Figure 9-12.

Outside Deburring Cutters

Outside deburring cutters are used to deburr bar ends. They are designed for use on screw machines and computer-numerical control (CNC) machines, as well as for hand use. They are produced in both high-speed steel and carbide, and they come in sizes as small as 0.125 in. (3.18 mm) and as large as 4 in. (101.6 mm) in diameter and have included angles of 30 or 45°. Tube-end and bar-end cutters are produced in many variations—probably 15 different designs exist—and the bur variety is just one of those. These tools look something like the outer piece shown in Figure 9-12.

Table 10-2. Rotary bur taxonomy

Style/Shape	Variation	Figure No.
Rotary burs		10-2
Standard burs		10-2 through 10-7
Routing mills		10-8
Inside chamfer mills		
Bar-end chamfer tools (outside chamfer)		
Tube-end deburring tools		10-9
Tube-hole deburring cutters		10-34
Rivet shavers		
Radius deburring cutters		9-8
Disc cutters		10-10
Micro center laps		10-11
Die mills		
Sheet metal edge-finishing tools		10-12
Ball nose with guide pin		
Side-cutting disk		
Disk with radius	Full radius, truncated radius	
Midget burs		
Cylindrical	Side cutting, side and end cutting	
Cylindrical with ball nose		
Cylindrical with corner radius		
Cylindrical cutting only on end		
Ball "bur ball"	Full ball, blunted nose	10-4, 10-26
Cone	Sharp pointed nose, rounded nose, small radius at nose	10-27
Inverted cone		10-7, 10-30, 10-40
Pear shaped		10-15
Olive shaped		10-4
Tree	Radius nose, rounded nose	10-5, 10-6
Concave (U-shaped)	Single groove Multiple grooves	10-13 10-13
Blunt radius nose		
Convex-concave		10-14
Disk		10-28
Disk with radius	Full radius, truncated radius	
Disk with double taper sides		
Flame		
Ball-nose deburring cutters	Plain nose, guide on nose	
Midgets with long shanks	Extra long (>2 in. [50.8 mm]) 6 or 8 in. (152.4 or 203.2 mm) long 18 in. (457.2 mm) long 24 in. (609.6 mm) long 36 in. (914.4 mm) long	10-19
Cup		10-33

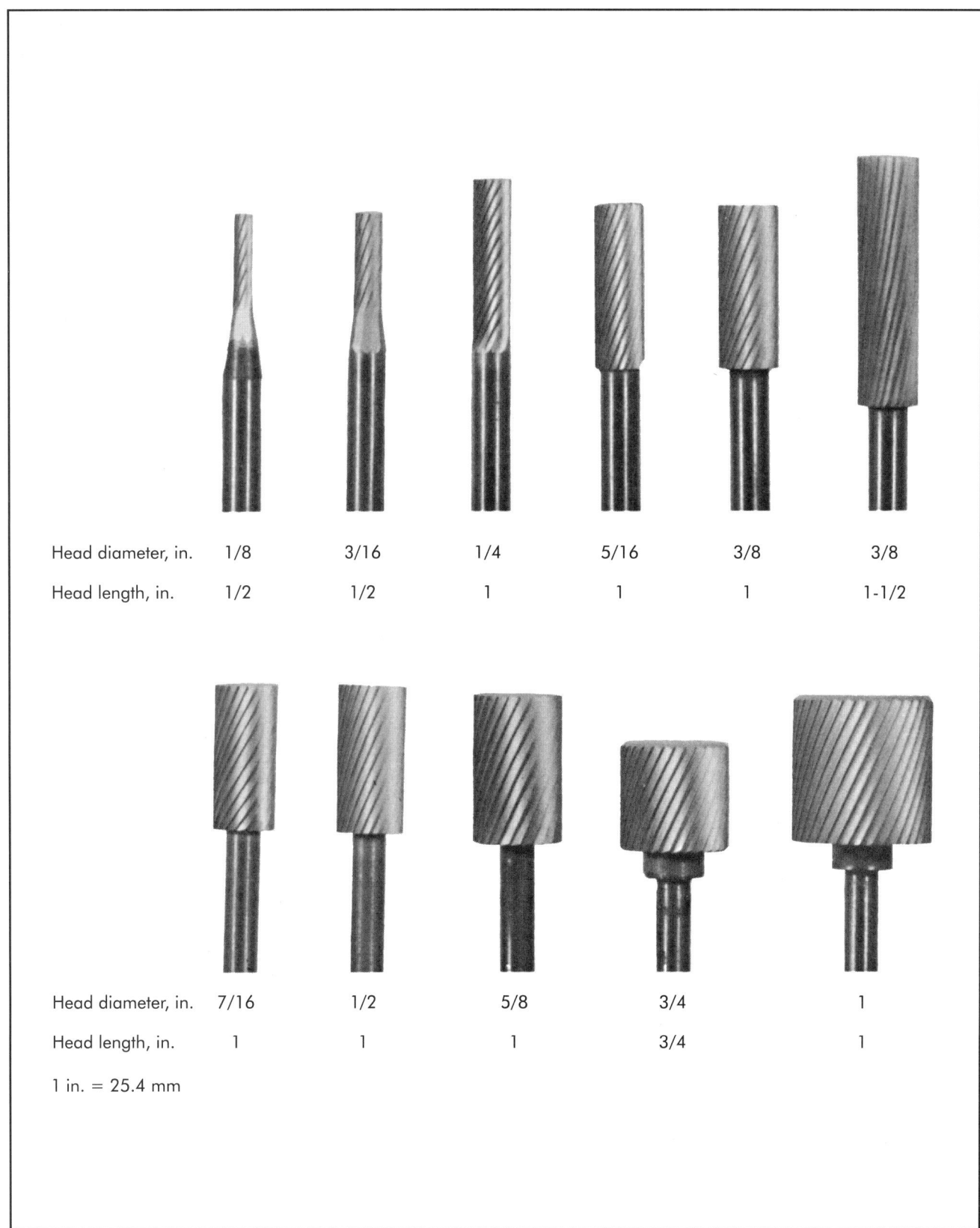

Figure 10-2. Cylindrical, high-speed steel burs (0.25 in. [6.4 mm] shank). (Courtesy Jarvis Cutting Tools)

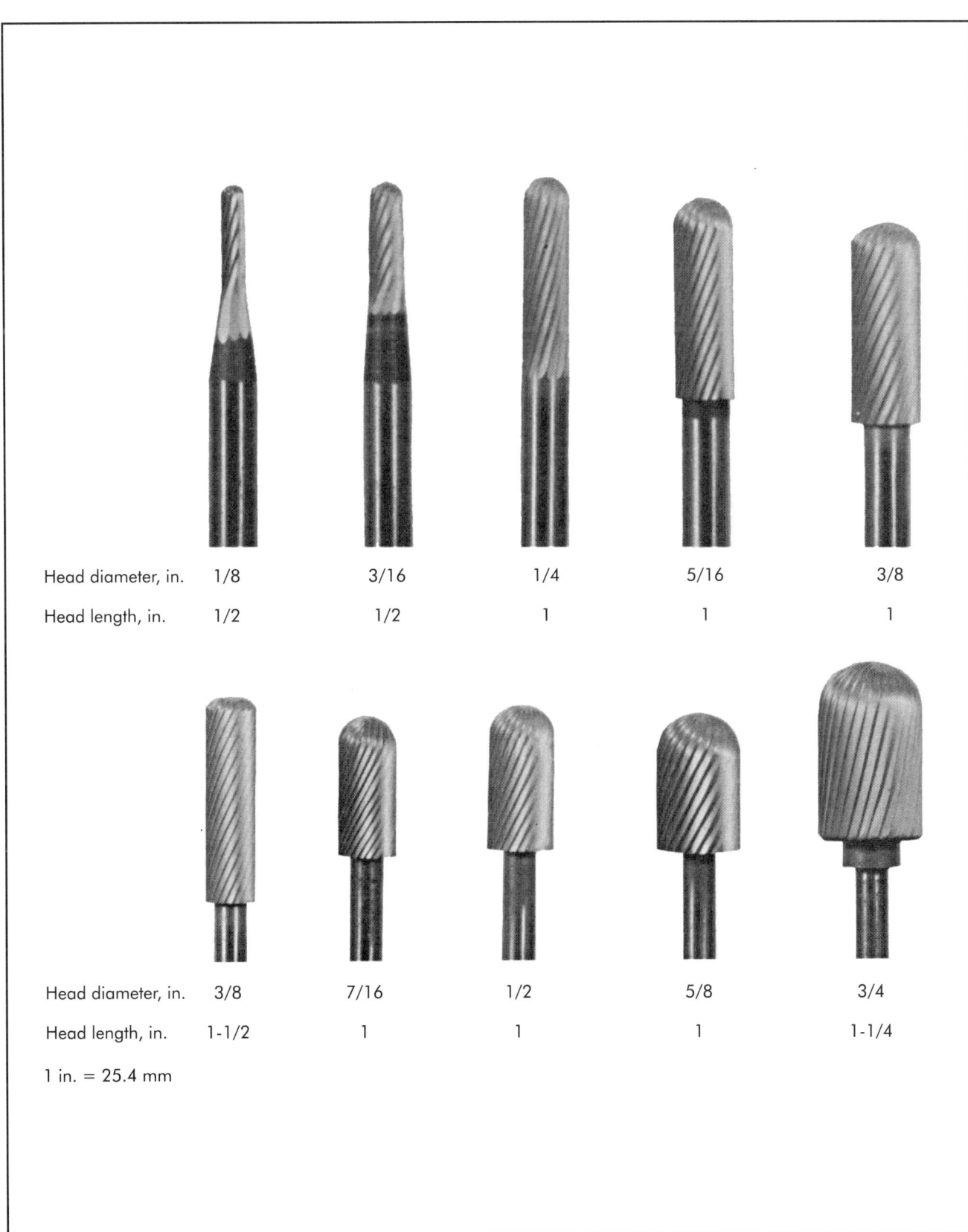

Figure 10-3. Cylindrical, radius-end, high-speed steel burs (0.25 in. [6.4 mm] shank). (Courtesy Jarvis Cutting Tools)

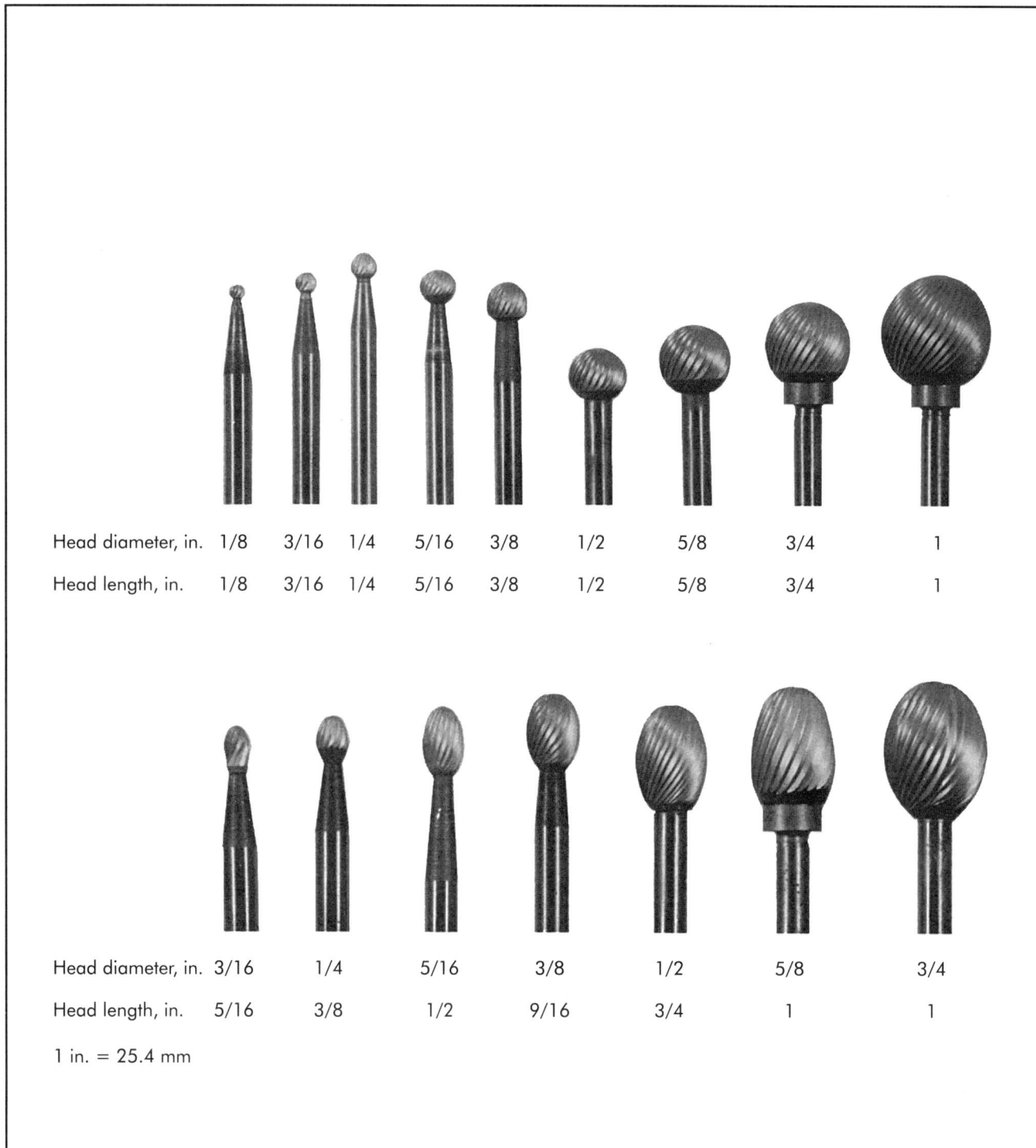

Figure 10-4. Ball and oval high-speed steel burs (0.25 in. [6.4 mm] shank). (Courtesy Jarvis Cutting Tools)

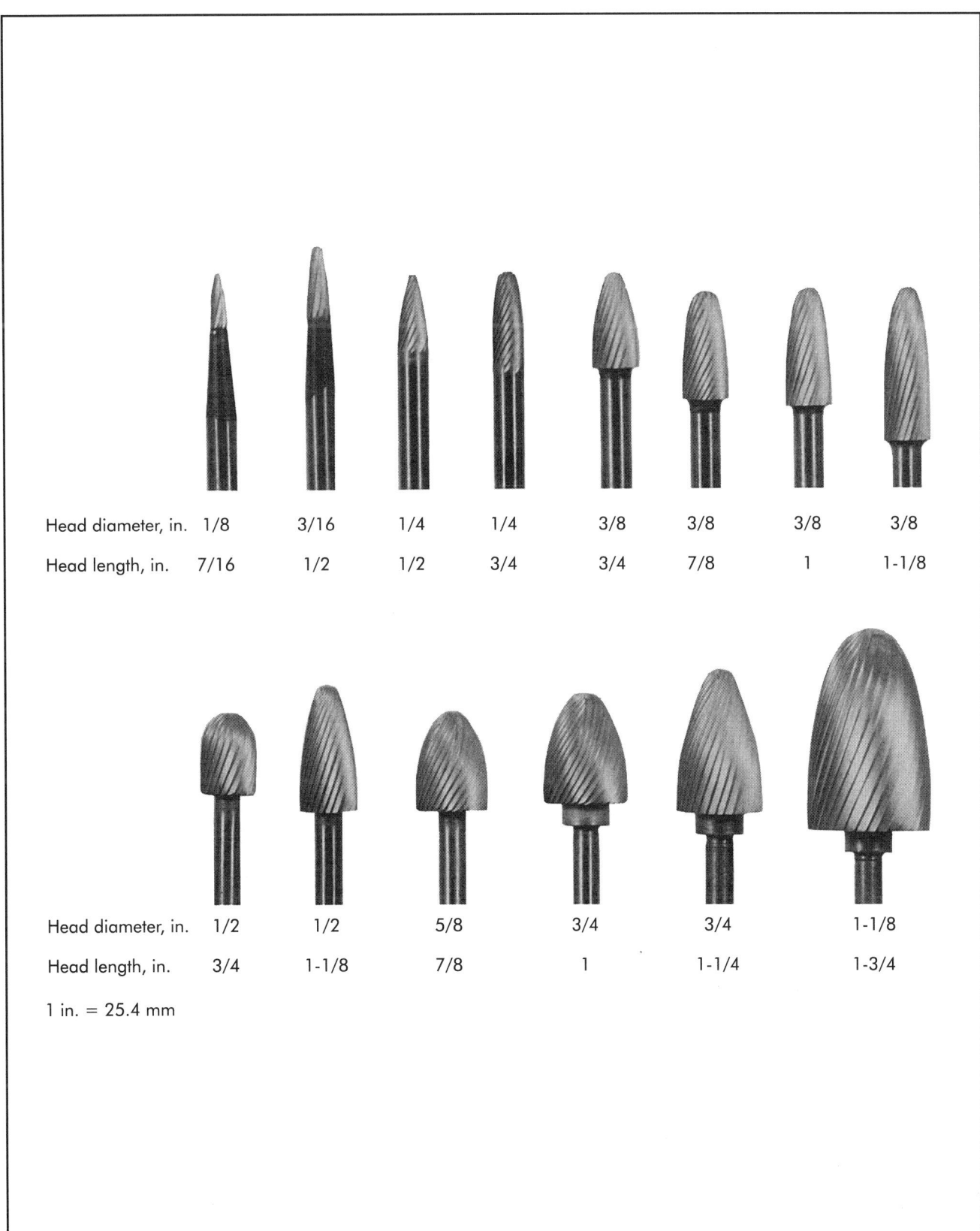

Figure 10-5. Tree-radius end, high-speed steel burs (0.25 in. [6.4 mm] shank). (Courtesy Jarvis Cutting Tools)

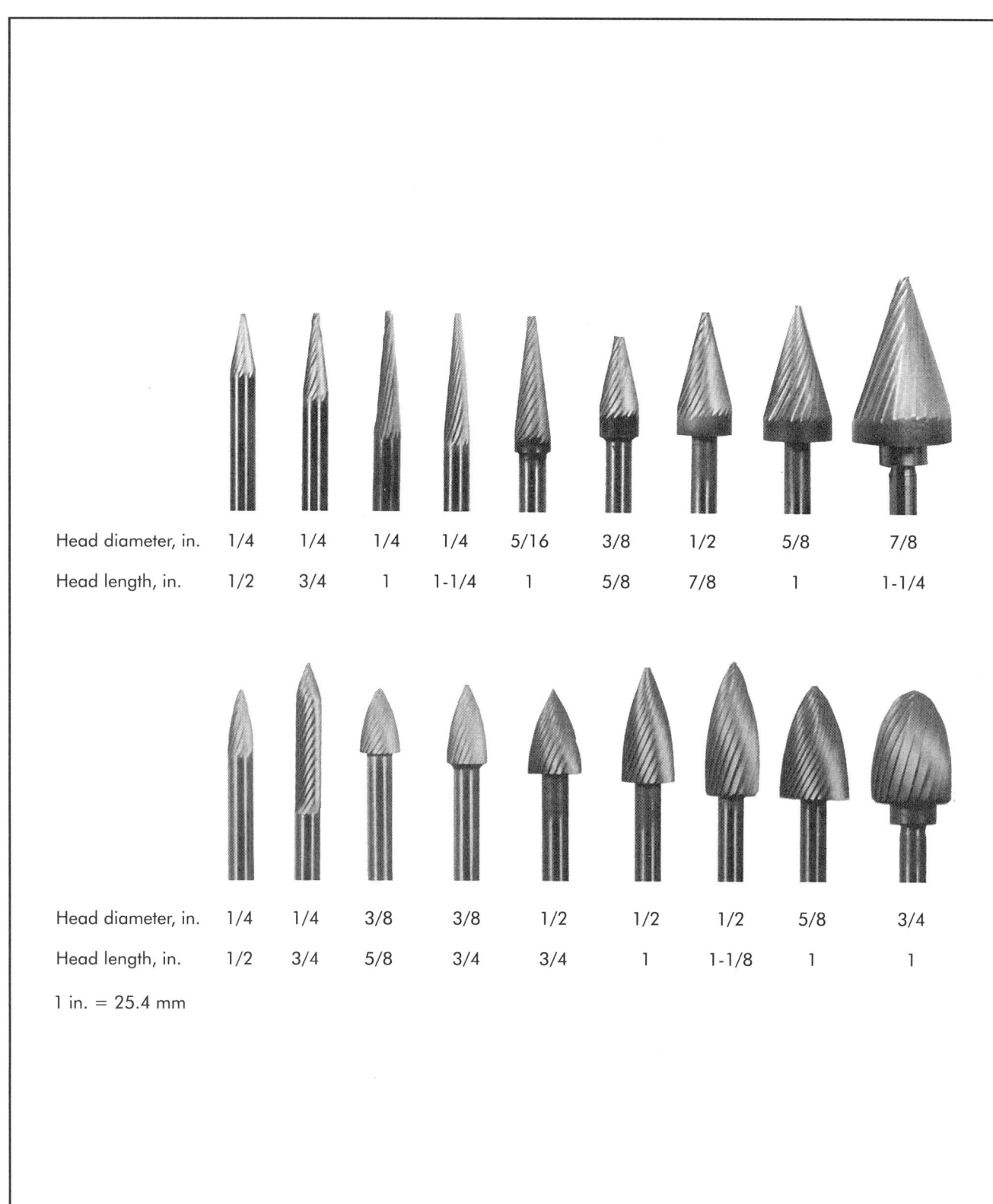

Figure 10-6. Cone high-speed steel burs (0.25 in. [6.4 mm] shank). (Courtesy Jarvis Cutting Tools)

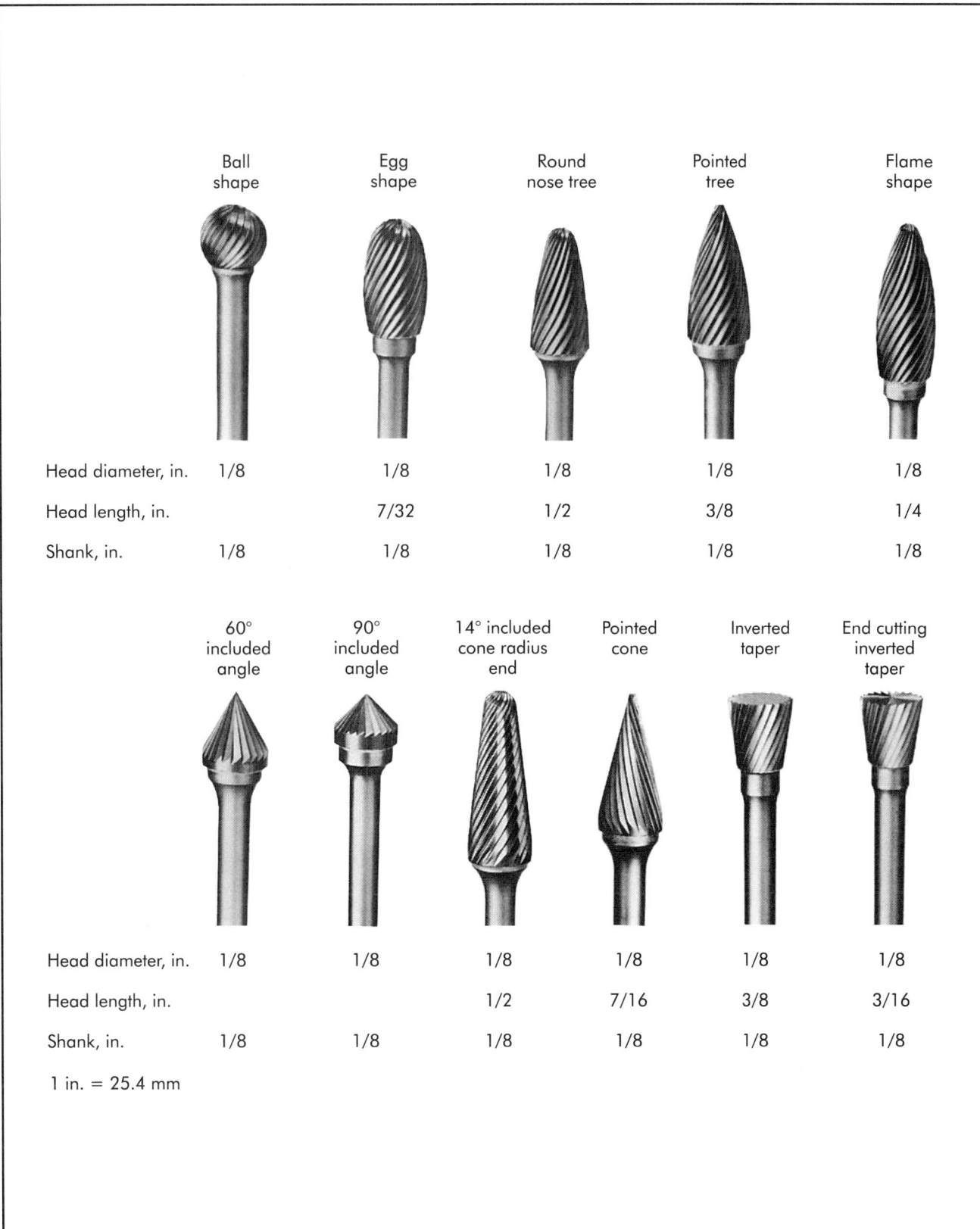

Figure 10-7. Additional standard bur geometries. (Courtesy Bassett Rotary Tool Company)

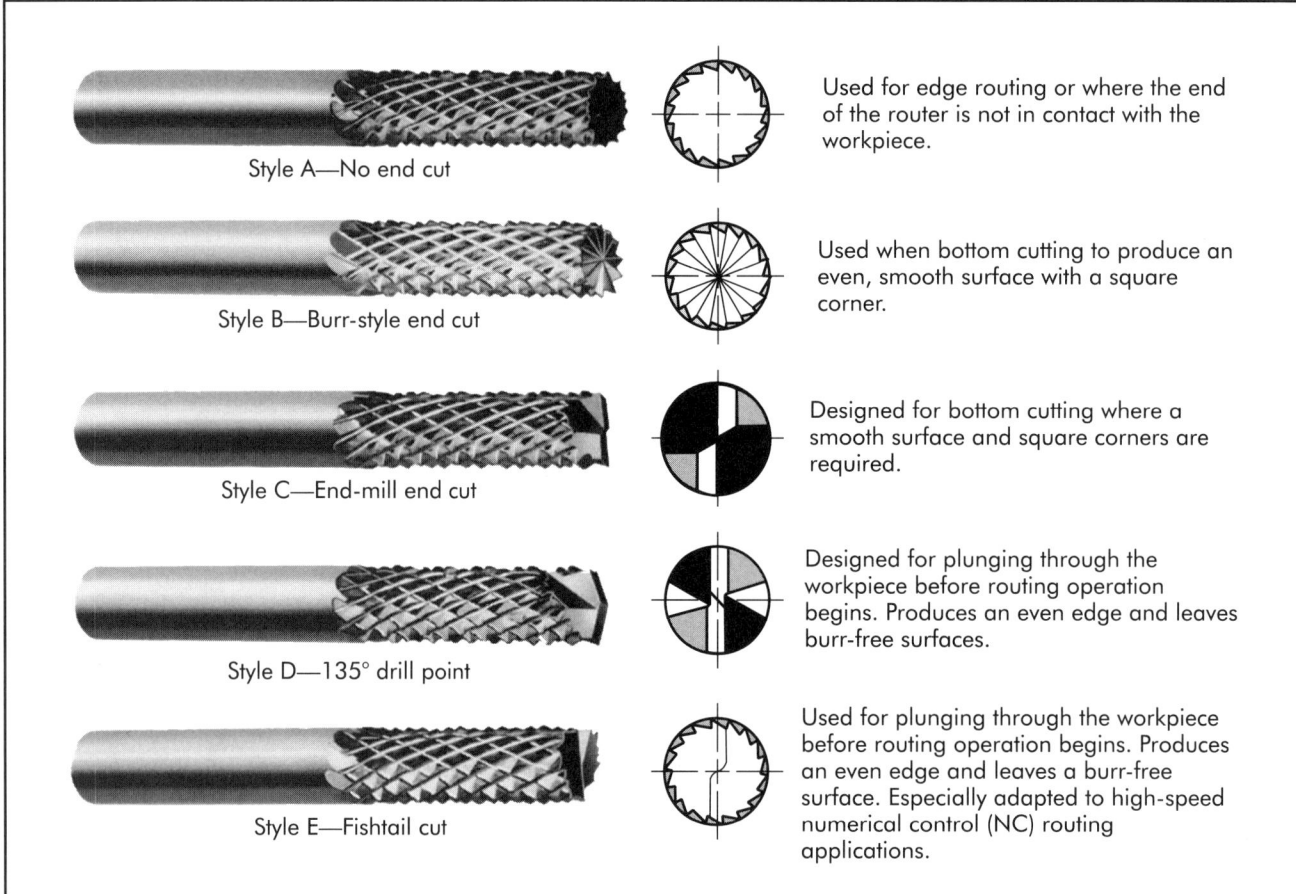

Figure 10-8. Fiberglass router cutters. (Courtesy Fullerton Tool Co.)

Tube-end Chamfering Cutters

Tube-end chamfering tools and their use were discussed in Chapter 9. Chamfered ends are produced by both conventional cutters and rotary burs. Figure 10-9 shows several sizes of one design. This tool simultaneously deburrs both inner and outer diameters. Also, tools are available that deburr just the inner or outer diameter. This particular design allows the outer piece to move relative to the inner piece. This results in either more or less chamfer than the inside portion provides.

Rivet Shavers

Rivet shavers, as the name implies, are traditionally used to cut off the heads of rivets. They have a solid cylinder with an end cut. This end-cutting ability is convenient for handwork in the bottom of flat bottom holes and in hard-to-reach areas. They range in size from 0.3125–1.0000 in. (7.938–25.400 mm) in diameter. They can be combined with micro-stop countersink units to provide accurate heights of material above or below the surrounding surface.

Radius Deburring Cutters

Figure 9-6 showed a radius deburring cutter based on rotary bur design. Similar integral shank tools are also provided. They can accommodate holes as small as 0.0938 in. (2.383 mm) and as large as 2 in. (50.8 mm).

Disc Cutters

Figure 10-10 illustrates a disc that allows users to cut with its face. This tool can replace sanding disks, snagging wheels, and milling cutters. It comes in a 0.375 in. (9.53 mm) diameter. This is a very aggressive cutter.

Micro-center Laps

Micro-center laps (Figure 10-11) are most frequently used to provide a smooth center in the ends

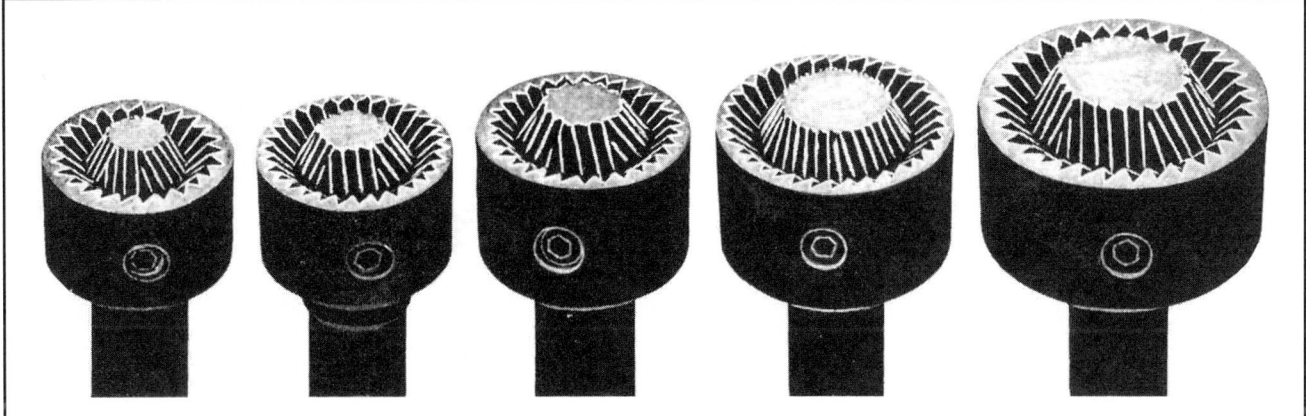

Figure 10-9. Tube-end deburring cutters.

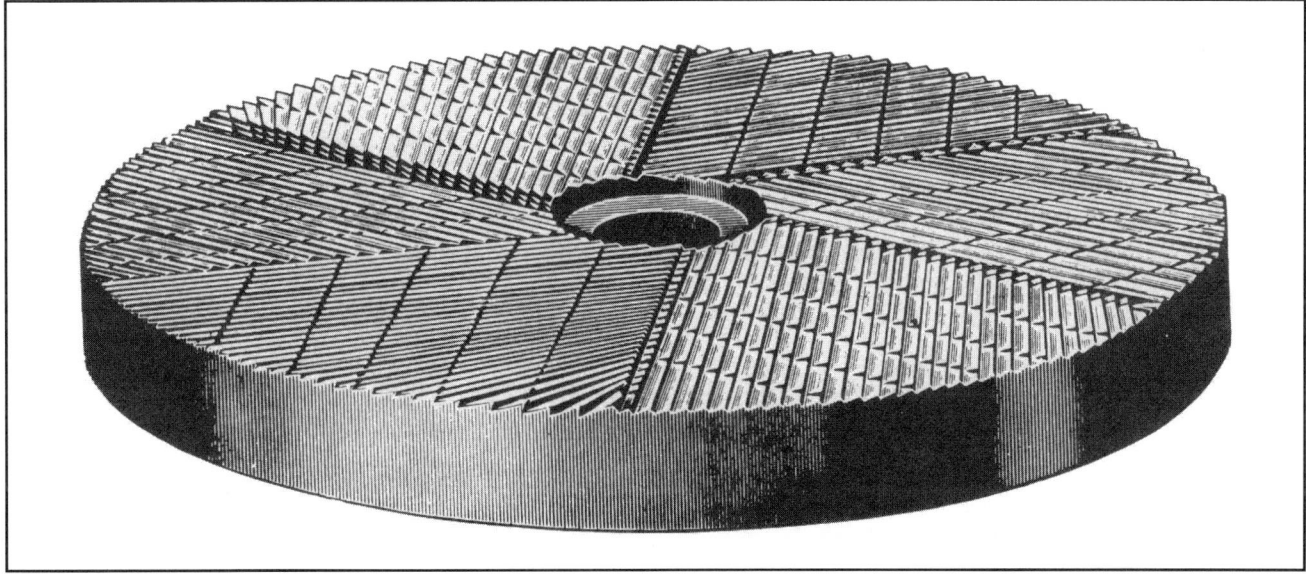

Figure 10-10. Disc cutter. (Courtesy Severance Tool Industries)

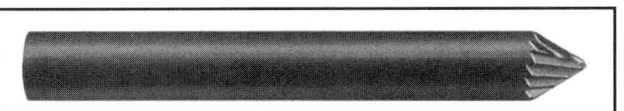

Figure 10-11. Micro-center lap. (Courtesy Severance Tool Industries)

of barstock. They come in a variety of tooth spacings (number of teeth), and there is also some variety in tooth configuration.

Die Mills

Die mills and cylindrical burs both have teeth on the diameter. Die mills, however, have the same measurement for shank and cutting diameter. The shank can serve as a guide to facilitate profiling. They come in 0.125–0.500 in. (3.18–12.70 mm) diameters in either high-speed steel or carbide. Like many burs, they can be reground to a smaller size.

Sheet-metal Edgers

Sheet-metal edgers have bur cutters facing each other (Figure 10-12). They are adjusted for a specific sheet-stock thickness. The complete tool is 0.75 or 1.00 in. (19.1 or 25.4 mm) in diameter. Also, the tool can be adjusted to have one noncutting and one cutting surface so the tool only trims one edge while the other side of the tool guides on the back of the workpiece strip.

Figure 10-12. Sheet-metal edger. (Courtesy Severance Tool Industries)

Ball Nose with Guide Pin

A built-in guide pin projects from the ball nose of this tool, which is used to deburr oil holes in crankshafts. The pin keeps the tool engaged in the hole. Tools can accommodate hole sizes of 0.125–0.438 in. (3.18–11.13 mm) in diameter.

Side-cutting Disks

Side-cutting disks are small saws that cut on the outer periphery. They come in flat faces, full radius faces, and truncated or angular-tapered radius configurations. Typically, they are 1 or 2 in. (25.4 or 50.8 mm) in diameter and have a thickness of 0.125–0.375 in. (3.18–9.53 mm). While a saw has normal teeth, a side-cutting disk bur has many slashes that act like teeth.

Midget Burs

Midget burs are also known as "miniature" or "junior" burs. For practical purposes, they are differentiated from standard burs only by their small size. They typically have 0.125 in. (3.18 mm) diameter shanks and a tool length of 1.5 in. (38 mm), but a shank size of 0.28 in. (3 mm) is also available. The head of the bur has a cut length of about 0.25 in. (6.4 mm). Tapers come in 7, 8, 10, 14, 16, 60, and 90° angles, and are based on inch or metric sizes. Other than for chucking, the differences between metric and English sizes are inconsequential.

U-shaped Midget Bur

U-shaped burs are useful for chamfering the sides of thin materials (Figure 10-13). They clean castings and smooth plastic edges, as well as deburr machined part flanges. Figure 10-13 also shows a modified tool that produces four edges at a time. Such tools accommodate a stock thickness thinner than 0.25 in. (6.4 mm) or as thick as 0.75 in. (19.1 mm).

Blunt Radius-nose Bur

Blunt radius-nose burs have teeth only on the radius, and they have a smooth, unfluted sidewall. This

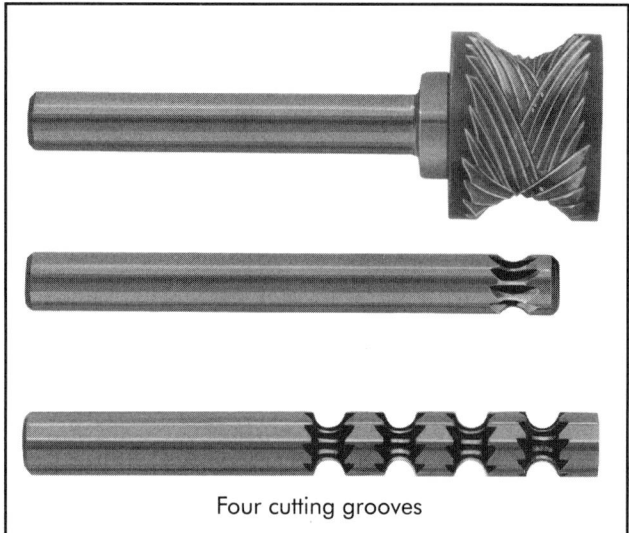

Figure 10-13. U-shaped midget burs. (Courtesy Severance Tool Industries)

allows them to be used down in fillet radii without marring walls. They have diameters of 0.25 in. (6.4 mm) or larger.

Convex-concave Midget Bur

The novel shape of the convex-concave midget bur allows users to reach inside small cavities to deburr with a ball- or pear-shaped nose (Figure 10-14). A typical small cutting diameter for this tool is 0.1875 in. (4.763 mm) with a total head length of 1.00 in. (25.4 mm).

Midget Side-cutting Disks

Midget side-cutting disks, like their bigger counterparts, are essentially miniature saws. They have cutting diameters as small as 0.25 in. (6.4 mm) and a thickness of 0.375 in (9.53 mm). Some have diameters of 1.50 in. (38.1 mm) and 0.25 in. (6.4 mm) thickness. They come in straight-sided, full-radius, and V-shaped contours, and can be produced as thin as 0.012 in. (0.31 mm). Note, when used as a slot cutter, the full-radius tool may produce a much larger burr at the bottom of the slot than any of the other shapes (Gillespie 1975).

Dental and Jeweler's Burs

The dental and jewelry industries use some of the smallest burs made. While they are designed to work on soft metals or human teeth, rather than hard metal, many are used on false teeth or other ceramics and plastics. The small sizes allow precision users to reach into the smallest of crevices.

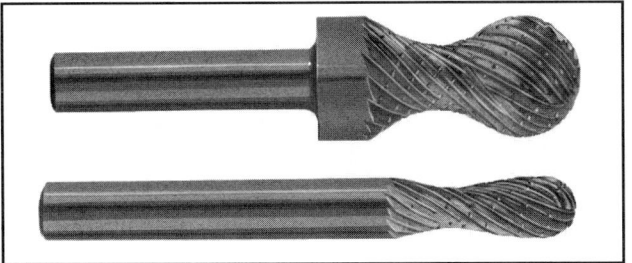

Figure 10-14. Convex-concave midget burs. (Courtesy Severance Tool Industries)

Figures 10-15, 10-16, and 10-17 illustrate some of the almost 200 shapes produced for the dental and jewelry industries. They are made in steel, carbide, and plated-diamond versions. Table 10-3 summarizes the size codes for the cutting diameter of these minute burs. The variety available is too extensive to display in this book. One manufacturer's catalog, for example, includes 25 pages of dental and jeweler's burs. Figure 10-17 illustrates the different shank sizes of dental and jeweler's burs. Figure 10-18 illustrates sizes that come in friction grip (FG) and handpiece (HP) shapes.

Medical Burs

Medical burs are designed, primarily, to cut bone. They are similar to dental burs, but larger. Because of limited demand, they tend to be more expensive.

Unique Variations

Burs are relatively easy to produce; as a result, several ingenious adaptations exist. The variations described in the following sections are believed to be some of the more useful design variations.

Long-stem Burs

While typical industrial burs have an overall length of 3 in. (76.2 mm) or less, some are made with 6, 8, 12, 18, 24, or even 36 in. (152.4, 203.2, 304.8, 457.2, 609.6, or 914.4 mm) long shanks for hard-to-reach areas (Figure 10-19). As with all the burs discussed so far, suppliers will produce any variation in shank or cutting end that users require, and these can be produced relatively quickly and inexpensively. Suppliers are particularly interested in exploring new shapes because someone's specific problem is another potential market.

Flexible Stems

Typically, users need a stiff stem to perform deburring. There are times however, when a long, straight stem will not meet the need. At least two firms provide long, flexible stems with a bur welded to a coiled spring similar to a screen door spring. When that is too limber, one firm runs the coiled spring shaft inside a copper sleeve, which allows the user to bend the sleeve into the required shape (Figure 10-20). Reportedly, tools like this can be run in an electric drill or air motor and have been used in sleeves up to 36 in. (914.4 mm) long. Shafts are made in 0.1875, 0.2500, or 0.3750 in. (4.763, 6.350, or 9.525 mm) diameters.

Handles for Burs

Many burs are used in handheld pin vises. As the burs get larger and more force is needed to cut, handles—such as shown in Figure 10-21—may be necessary. Note that if large forces are required and the workers must use these tools all day long, plants need to consider ergonomic handles and stress-reducing efforts. Chapter 31 covers this topic.

TOOTH DESIGN

Just as there is a wide variety of sizes and shapes of rotary burs, there is a wide variety in the configurations of the teeth found on these burs:

- the fineness of cut;
- the way the teeth were produced;
- the angles of cut;
- the rake angle on the teeth, and
- the basic tooth shape.

For an example of fineness of cut ranges, see Figure 10-22. Note that the number one cut in this instance has 62 teeth/in. (2.4 teeth/mm). The coarse number 14 cut has only 5 teeth/in. (0.2 teeth/mm). In many instances, manufacturers do not describe the number of teeth per inch. Rather, they merely designate *fine* or *finish* cut, versus *normal* or *coarse* cut. Fine, however, often does not imply the same cut if manufactured by different companies. In many cases, fine for one type of bur is not the same fineness of cut as that of another shape or style, even if the bur manufacturer is the same. Table 10-4 illustrates the fluting specifications of one company. A number of manufacturers also provide special fluting or fineness of cut for specific applications. The helix angle of cut also can be varied on these tools.

Figure 10-23 illustrates the rake angle on teeth. As illustrated, radial-cut teeth have a leading edge, which is the extension of a straight line passing through the center of the cutter. Positive-cut teeth have a small included angle at the tips. They are similar to the teeth found on the cutters of a radial arm or table saw. Negative-cut teeth have a large included

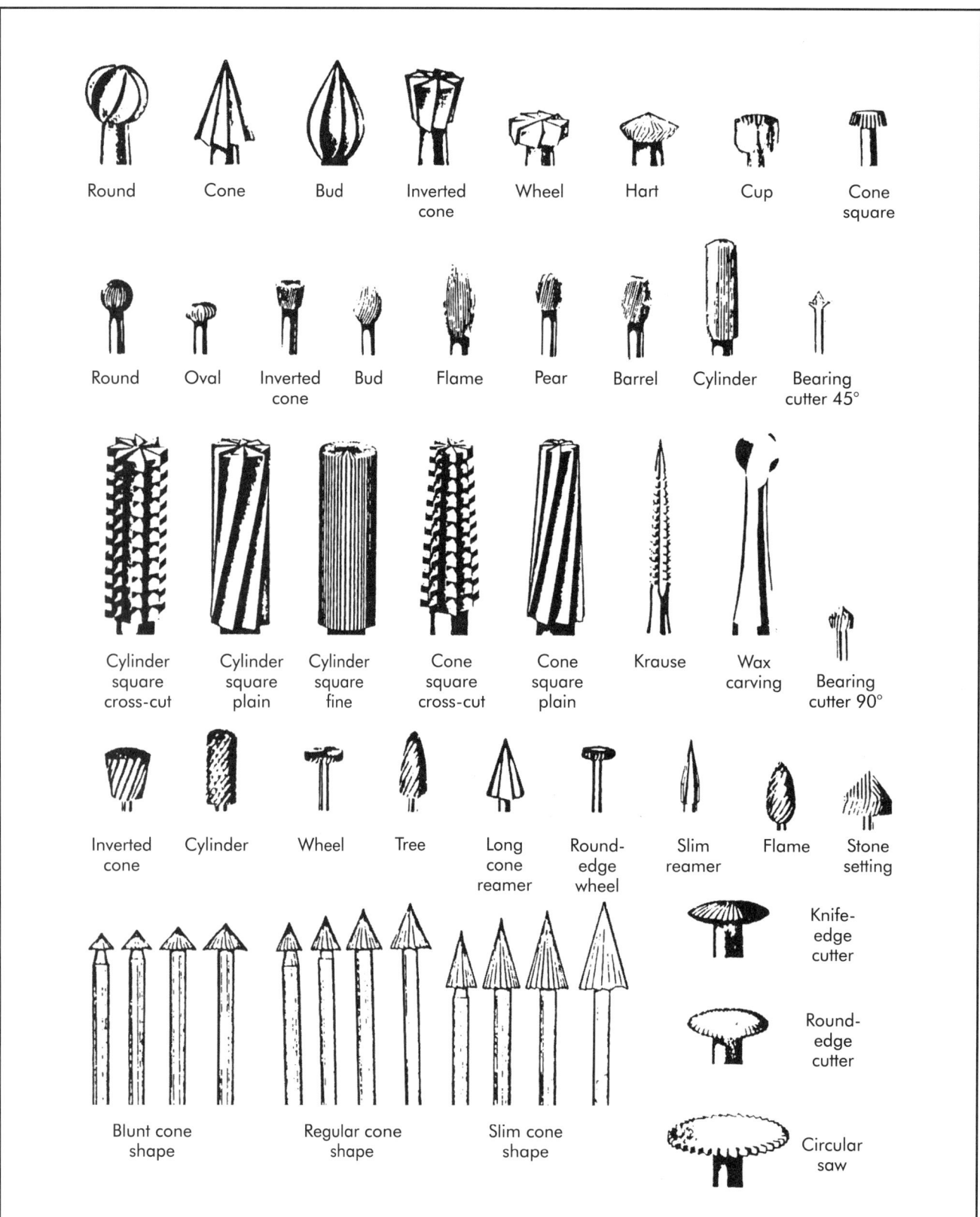

Figure 10-15. Common shapes of dental and jewelry burs.

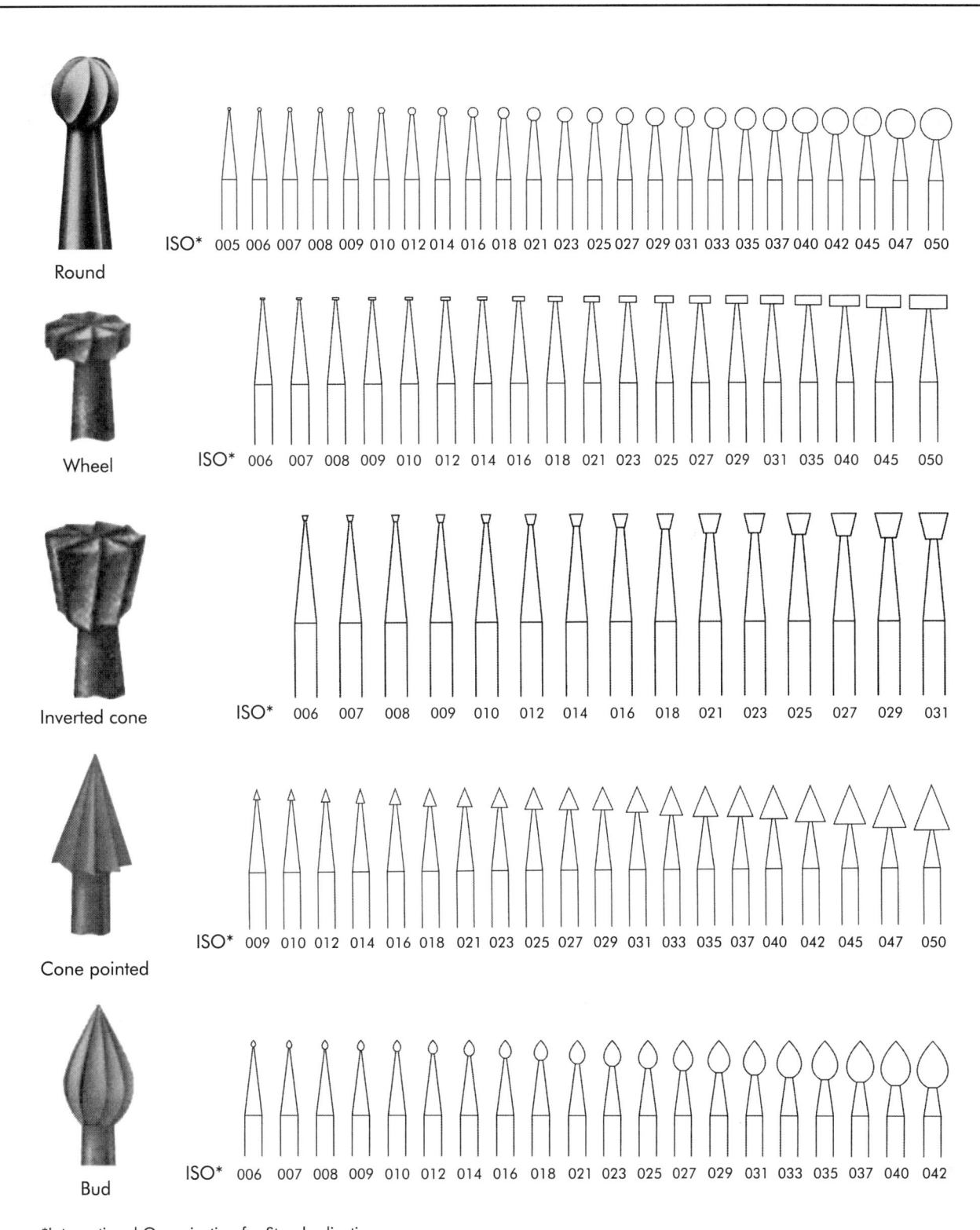

Figure 10-16. Size range available for some common shapes of jeweler's burs. (Courtesy Pfingst and Company, Inc.)

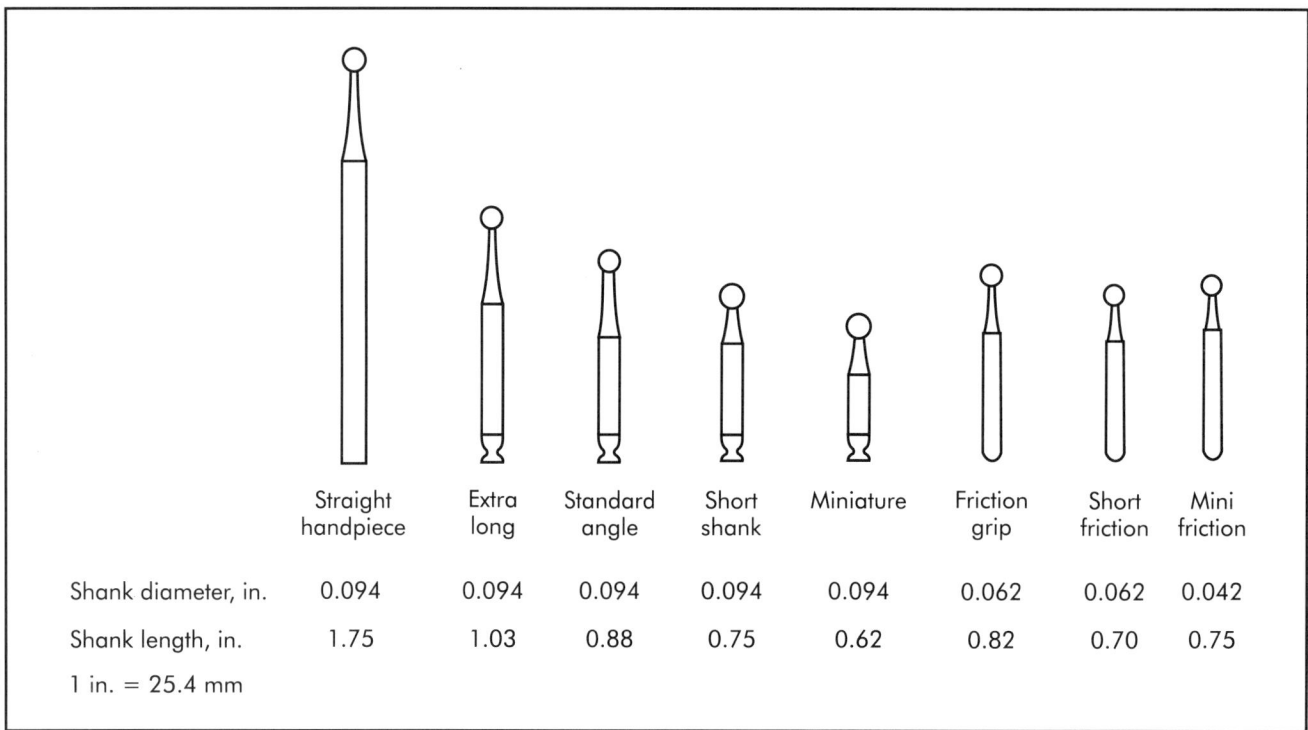

Figure 10-17. Shank sizes for dental tools.

Table 10-3. Dental and jewelry bur cutting diameters

ISO* Number	005	006	007	008	009	010	011	012	013
mm	0.500	0.600	0.700	0.800	0.900	1.000	1.100	1.200	1.300
in.	0.0197	0.0236	0.0276	0.0315	0.0354	0.0393	0.0432	0.0472	0.0511
ISO* Number	014	015	016	017	018	019	020	021	022
mm	1.400	1.500	1.600	1.700	1.800	1.900	2.000	2.100	2.220
in.	0.0551	0.0590	0.0630	0.0668	0.0709	0.0747	0.0786	0.0827	0.0872
ISO* Number	023	024	025	026	027	029	031	033	035
mm	2.300	2.400	2.500	2.600	2.700	2.900	3.100	3.300	3.500
in.	0.0906	0.0942	0.0984	0.1022	0.1063	0.1141	0.1220	0.1299	0.1378
ISO* Number	037	040	042	045	047	050	055	060	070
mm	3.700	4.000	4.200	4.500	4.700	5.000	5.500	6.000	7.000
in.	0.1457	0.1575	0.1654	0.1772	0.1850	0.1968	0.2165	0.2362	0.2756

*International Organization for Standardization—the ISO number defines a specific bur size.

angle. This provides a stronger tooth, but it also increases the amount of force required to cut metal. Figure 10-24 illustrates some of the basic styles of cuts available. The following sections discuss six standard configurations ground into the tool.

An *aluminum cut* (Figure 10-25) is used for deburring aluminum, soft steels, reinforced plastics, and nonferrous metals. The wide clearance and end-mill type geometry of the flutes allow fast stock removal with minimal loading. Because the flutes are

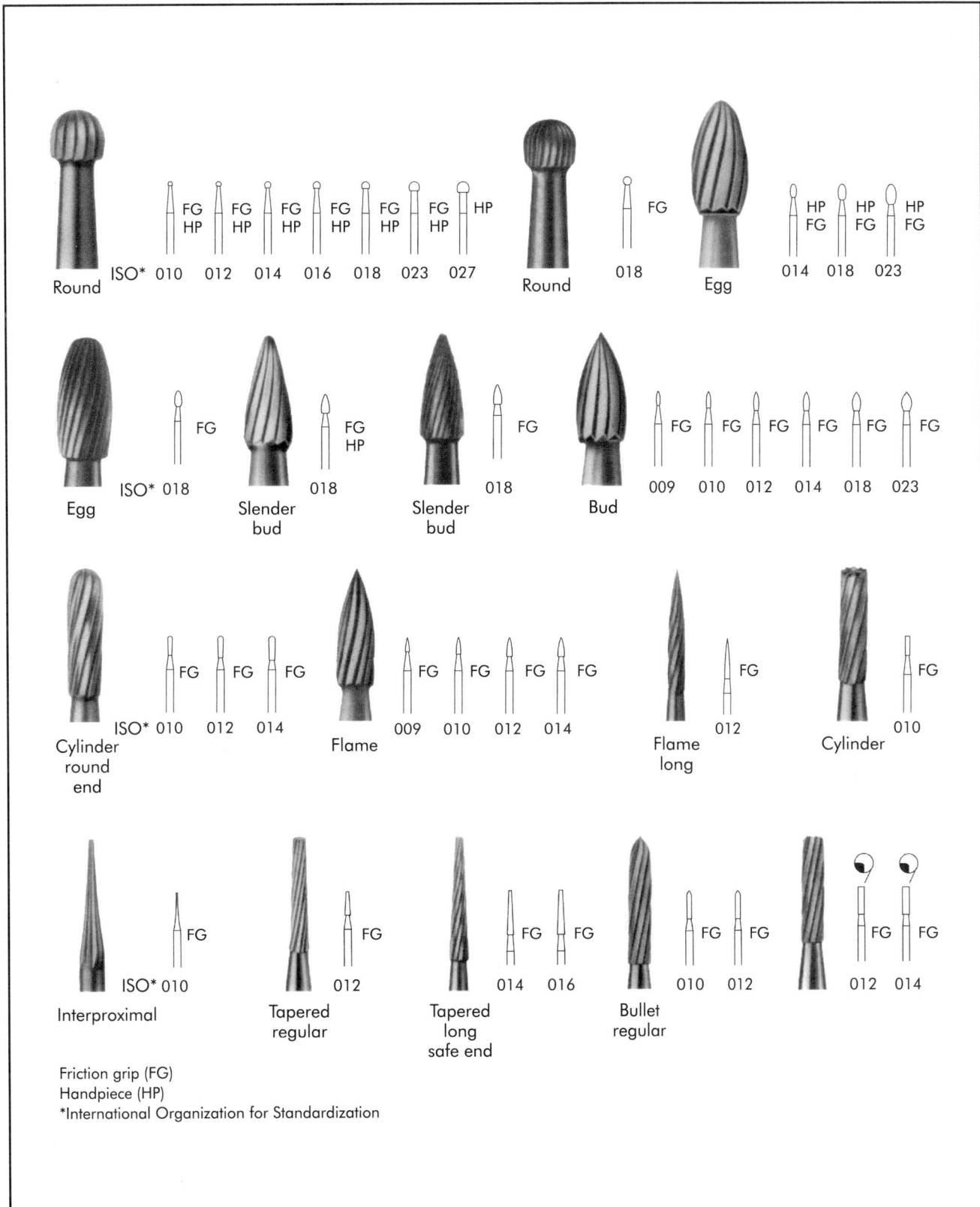

Figure 10-18. Jeweler's burs. (Courtesy Pfingst and Company, Inc.)

Figure 10-19. Extra-long shank burs. (Courtesy SGS Tool Co.)

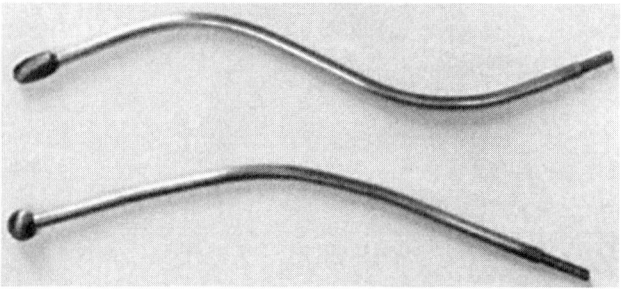

Figure 10-20. Snake bur for working around bends. (Courtesy O. G. Bell)

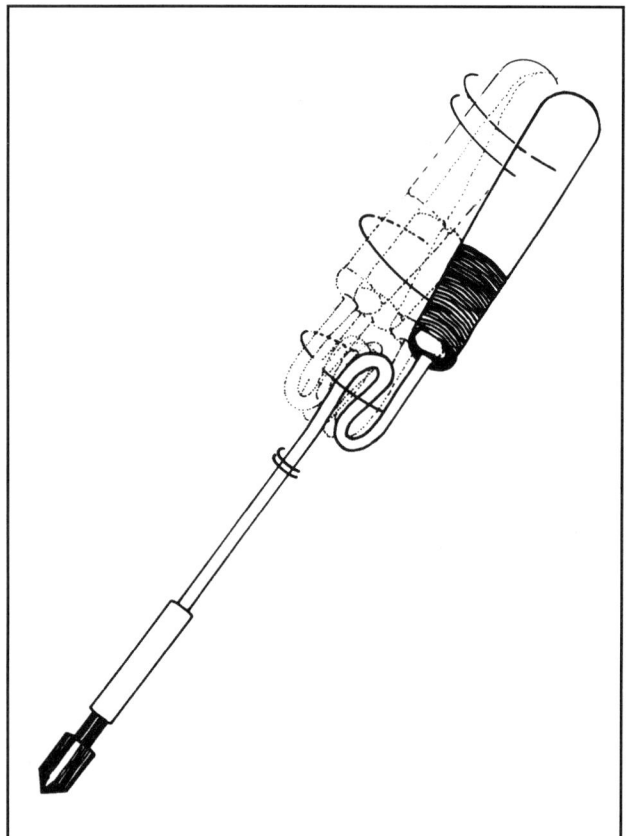

Figure 10-21. Quick-action handle for burs. (Courtesy Duke Machine Corporation)

highly polished, this style is often used on soft, sticky materials.

Chisel-cut (or hand-cut) rotary files have staggered teeth, in contrast to the smooth flutes of the ground tools. Therefore, they are particularly well-suited for work on dense or tough materials. Ferrous metals, such as die steels and steel forgings, are good materials for the chisel files. These are best-suited for work at lower speeds, such as work with handheld flexible-shaft machines, drill presses, and lathes. Chisel-cut tools are coarser, so they are generally not well-suited for working on precision miniature parts.

Additional nomenclature is used for burs produced for the dental industry, as follows:

- plain cut;
- cross-cut fissure, straight cut;
- cross-cut fissure, spiral cut, and
- fine cut.

These are not widely used for deburring metals, but they provide yet another alternative for difficult operations or materials.

MATERIALS

Rotary burs are made from a variety of tool materials, including:

- high-speed steel;
- high-speed vanadium steel;
- tungsten vanadium steel;
- tungsten carbide;
- titanium-nitride (TiN)-coated tungsten carbide;
- tungsten-carbide-plated materials;
- diamond-plated or brazed materials, and
- titanium-nitride (TiN)-coated diamond materials.

High-speed steel tools are adequate in many situations that call for only handheld pin vises to rotate the tools. They also can be used with low-speed, handheld motors. The addition of vanadium and small quantities of tungsten provide tougher tools that can be used at higher speeds. Tungsten-carbide tools can be operated at speeds considerably higher than the steel burs, and they will not rust when used in areas of high humidity or moisture.

Tungsten-carbide particles of relatively coarse size are plated or brazed onto a steel shank (Figure 10-26) to produce a rough cutting tool that has some of the same properties as burs made from solid metal. Both come in the same basic industrial shapes and sizes. They are not mounted points, by traditional terms, since they are not a bonded abrasive wheel in the typical grinding wheel construction. They are a new form of bur.

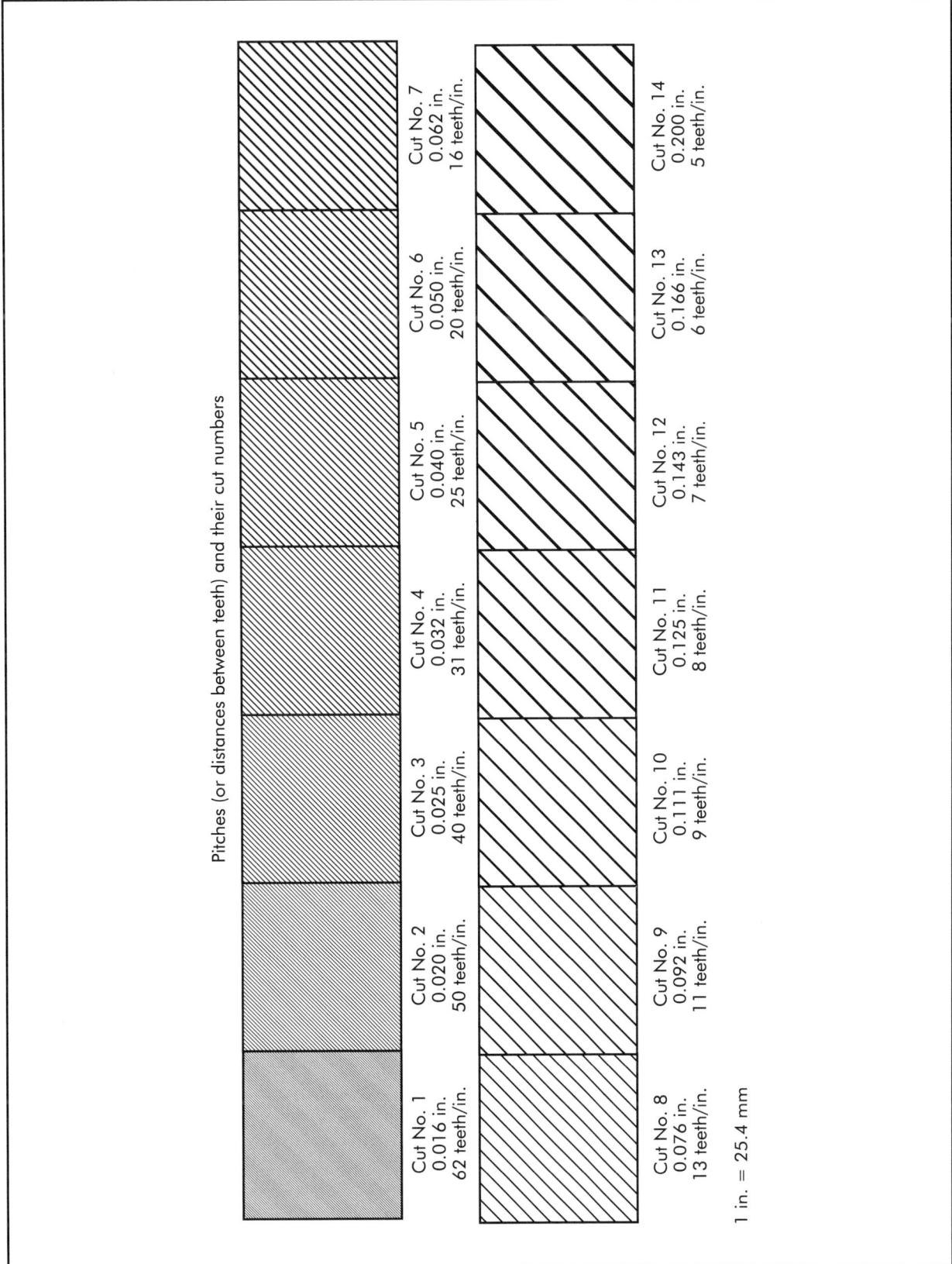

Figure 10-22. Standard cut sizes for rotary industrial burs.

Table 10-4. Fluting specification by bur diameter (Courtesy Aerosharp Tool Co.)

Diameter, in.	Fine Cut Number (teeth/in.)	Standard Cut Number (teeth/in.)	Coarse Cut Number (teeth/in.)
3/32	2 (50)	3 (40)	5 (25)
1/8	3 (40)	4 (31)	5 (25)
3/16	3 (40)	5 (25)	6 (20)
1/4	4 (31)	5 (25)	7 (16)
5/16	4 (31)	6 (20)	7 (16)
3/8	4 (31)	6 (20)	8 (13)
7/16	5 (25)	6 (20)	8 (13)
1/2	5 (25)	7 (16)	9 (11)
9/16	5 (25)	7 (16)	9 (11)
5/8	5 (25)	7 (16)	9 (11)
3/4	6 (20)	8 (13)	10 (9)
7/8	6 (20)	8 (13)	10 (9)
1	6 (20)	8 (13)	10 (9)

Note: 1 in. = 25.4 mm

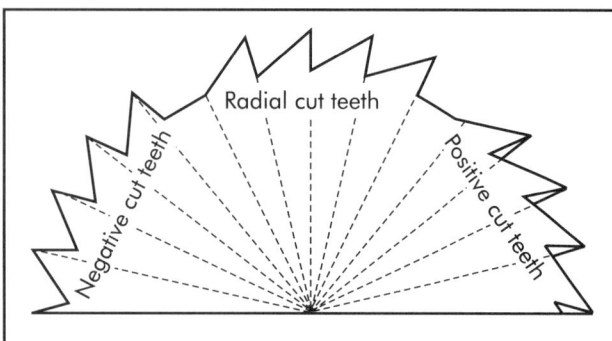

Figure 10-23. Rake angle variations used on rotary burs.

Diamond particles are plated or brazed today on almost any tool shape imaginable. Diamond dental instruments (burs) are widely used and available in various dental sizes (Figure 10-27). They come with nickel plating or titanium-nitride (TiN) coating. The diamond can either be applied in a coarse size for typical cutting or in fine abrasives for polishing. These diamond-coated tools are also available in thin, flexible discs with the top and bottom coated (or both) to reach into narrow slots (Figure 10-28).

APPLICATIONS

Some manufacturers indicate that a break-in or pre-honing step is required before metal burs are used at full operation. Pre-honing consists of several slow-speed cuts to remove the grinding burr on the bur's cutting edges. At least one manufacturer provides a version of a bur that is pre-honed at the factory, saving users the time it takes to perform this task. The manufacturer's pre-honing step leaves the surface a dull matte gray, in contrast to the bright finish of most of these tools. That surface contrasts significantly with the bright surface produced as the tool wears. This makes it easier for users to detect tool change needs.

Tools ground after the tool shank has been hardened are used for finishing ductile and stringy materials, particularly nonferrous metals, such as aluminum, brass, magnesium, and others. They are also effective on some plastics. These tools are used at medium speeds in flexible shaft machines and air motors, as well as in handheld applications.

For hand deburring of high-precision miniature workpieces, a bur is placed in pin vises and rotated by hand. Faster action, for parts having more chamfer tolerance, relies on tools inserted in bench-mounted electric motors, air motors, and dental motors. Burs are also used on various machine tools.

Table 10-5 and Figure 10-29 present recommendations for industrial rotary bur use. Tables 10-6 and 10-7 describe recommended cutting speeds for dental burs and diamond dental burs. As the tables indicate, the dental tools are operated at up to 450,000 rpm. The dental tools, particularly, require a light touch—

Standard Cut
This fluting pattern is considered standard. Flute structure is designed for both superior material removal and a smooth finish. The common use of the "standard cut" is on steel, steel alloys, cast iron, some copper and brass, and generally where the material is relatively hard and will not load the flutes.

Chip Breaker
The addition of a chip breaker on the three single spiral flute patterns—standard, coarse, fine—improves tool control. Chips will be broken up to reduce sliver size. Surface finish will be slightly reduced due to the chip breaker pattern.

Coarse Cut
This style of flute structure is recommended for use on soft materials that would normally load the standard cut. With this design, rapid stock removal in materials such as copper, brass, aluminum, plastics, and rubber is achieved. If loading is excessive, then the aluminum-cut style is recommended. Extremely fast stock removal is possible on the softer nonferrous materials. The extra-deep flute structure should eliminate bur loading.

Fine Cut
Selection of fine cut should be based on the required finish. While stock removal is reduced, the finish is greatly improved. This cut is recommended only on materials having a Rockwell "C" hardness in the range of 55–60. Use on softer materials will cause loading.

Double Cut
The double cut bur allows for rapid stock removal on harder materials. The chisel-tooth pattern not only minimizes tool chatter, it reduces the chip to a granular shape in most materials, thereby reducing or eliminating the sharp sliver chips normally experienced. Chip reduction also helps to eliminate loading of the flutes. An improvement in tool control will be realized as the double cut tends to reduce the pulling action of the main flute pattern. Although some finish reduction may be experienced, improvement in material removal and, therefore, increased production will be realized.

Diamond Cut
The diamond-cut-flute pattern produces teeth with a triangular style of point. This greater number of teeth will produce extremely small chips. The diamond cut virtually eliminates the pulling action of the main cut, thereby greatly improving tool control. With this cut, stock removal will be greatly increased at the sacrifice of finish.

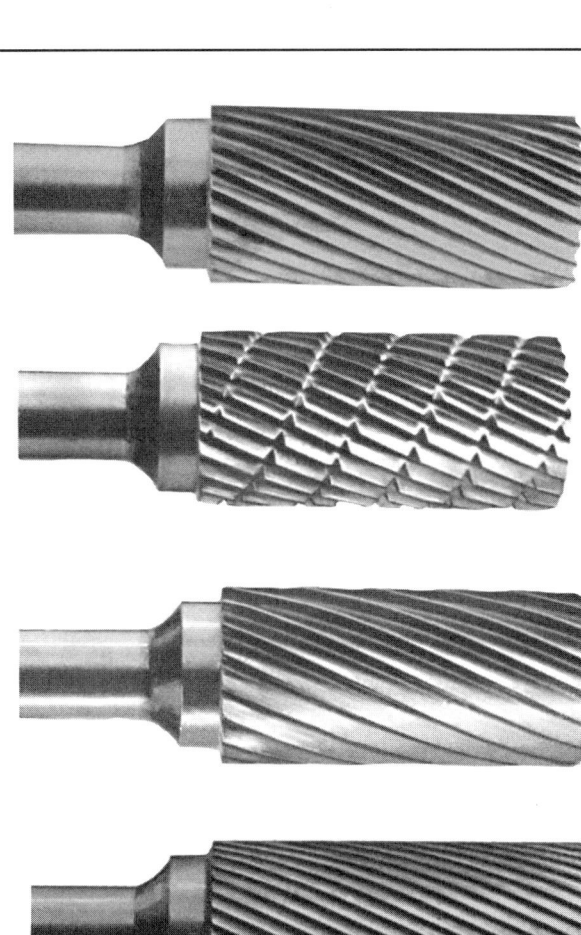

Figure 10-24. Basic cuts for burs. (Courtesy Menlo Tool Co.)

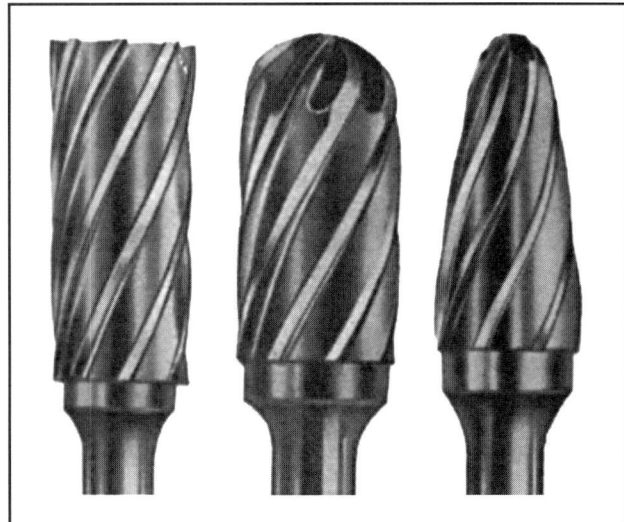

Figure 10-25. Aluminum cut for burs that cut aluminum and soft metals. (Courtesy Quality Carbide Tool)

Figure 10-26. Tungsten carbide particle-coated bur. (Courtesy Brasseler)

maximum contact pressure is 0.067–0.450 lbf (0.3–2 N). In contrast, heavy finishing work with industrial burs can require enough force to quickly tire hands and arms. Deburring using industrial burs can be performed with just a light touch, although many users force the cutter such that the air motors almost stall.

The smallest commercially available rotary bur has a 0.004 in. (0.10 mm) diameter ball, but the shanks are normally much larger, which may present a problem for internal deburring. Rotary burs tend to chatter or cut too deeply unless very fine teeth are used. Several shapes of these tools are used for deburring the backsides of holes that may be inaccessible by other means.

Bur balls are probably the most extensively used tool for chamfering and finishing hole entrances and exits. Because of the wide variety of available bur balls and the relative ease with which they may be ground, they are available in a tremendous variety of sizes.

Figure 10-30 illustrates the use of an inverted cone bur, a useful tool for deburring hard-to-reach undercuts. Sometimes, end-cut burs may be suitable for otherwise difficult-to-reach areas.

In a few instances, bur balls may be less desirable than the cone- or heart-shaped burs. Even small precision tools can leave fine burrs when they are used (Figure 10-31). In some cases, the use of a small-angle cone followed by a flat or heart-shaped bur will give the necessary edge break and provide a nearly burr-free edge (Figure 10-32). Some individuals rely exclusively on the use of these latter two types of tools rather than bur balls. Others follow the bur with a knife or swivel tool to wipe out any burrs created by the bur.

Concave cutters (cup cutters) can be used to deburr the ends of small pins. It is difficult to cut (make) inside cups, because they require such fine tooth spacing (Figure 10-33).

There are special cutters to deburr and finish both the inside and outside of tubing (Figure 10-34). These tube-hole cutters (as opposed to tube-end cutters), available for larger sizes, provide a chamfer and remove the burr produced when the tube is cross-drilled.

It may be difficult to use a standard bur ball in shallow counterbored holes and for holes next to shoulders; however, balls can be easily altered, as shown in figures 10-35 and 10-36, to meet these types of conditions. This simple alteration can frequently save hours of digging or cutting with knives. Figure 10-37 illustrates how one of these balls has been altered to deburr a counterbore or spotface around a shaft. The hole in the bur ball allows the tool to slip over the pin or shaft.

Figure 10-38 illustrates how users must consider the dimensions of the chamfer and counterbore in deciding which tool to use. The ends of the tree- or cone-shaped tools can be ground off, as shown in Figure 10-35, for the hole application.

Miniature rotary burs and bur balls should be used with even more care than precision files. Because the teeth are small, they are susceptible to breakage. Normally, carbide burs should not be cleaned in an ultrasonic cleaner, because the ultrasonic action fractures teeth ends. The large sizes are used in very aggressive cutting conditions.

The choice of fine, medium, or coarse flute spacing depends on the material to be deburred. The medium cut is used for general-purpose deburring of steel, cast iron, and other iron-base materials. Finer burs are used to provide a finer surface finish. The diamond-cut flute configuration will cut fast and allow better control when the tool is used in a handheld motor.

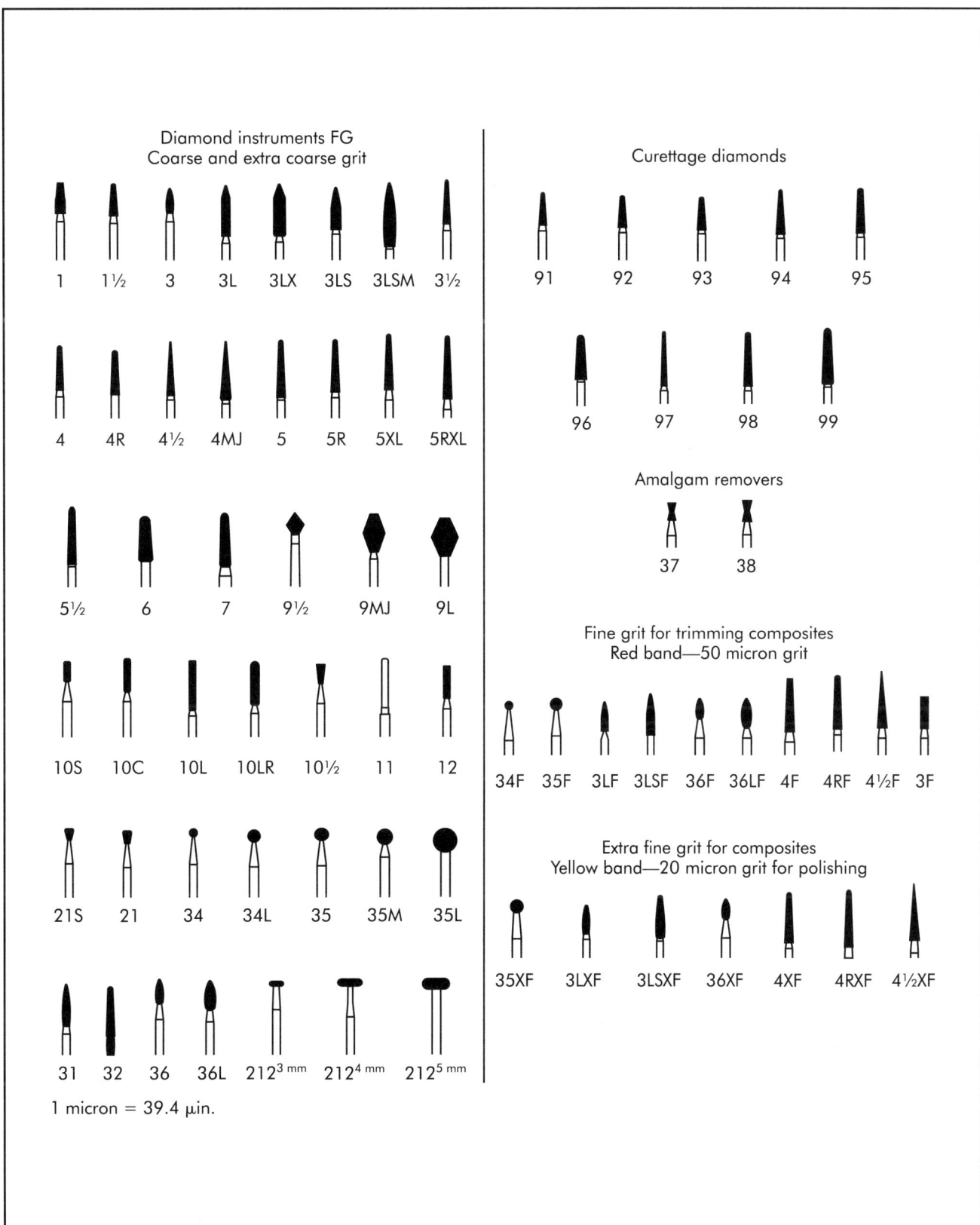

Figure 10-27. Diamond-coated dental burs. (Courtesy Charles W. Rode)

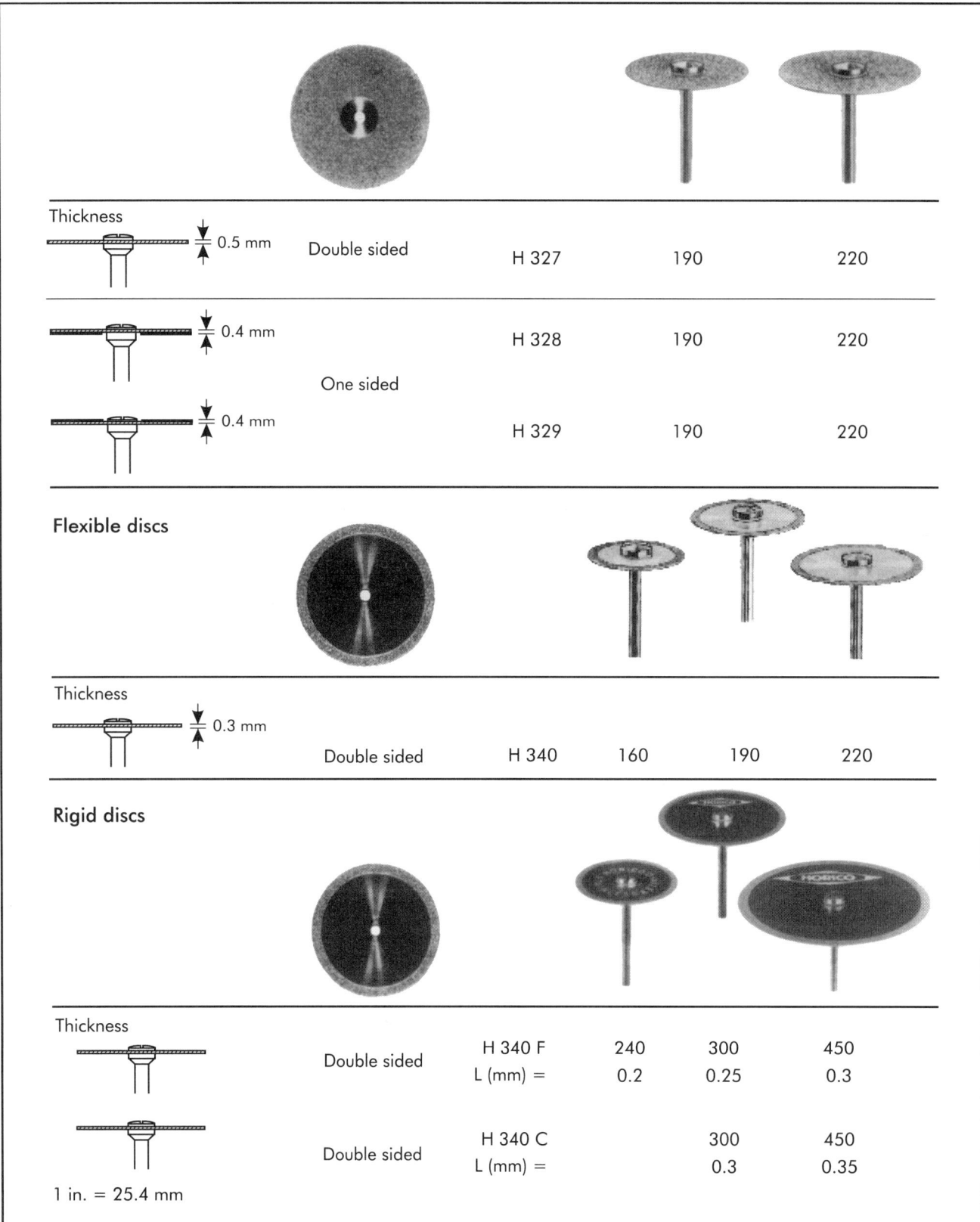

Figure 10-28. Diamond-coated flexible discs. (Courtesy Pfingst and Company, Inc.)

Table 10-5. Fluting style recommendations for burs (Courtesy SGS Tool Co.)

Materials	Single Cut	Double Cut	Chip Breaker	Diamond Cut	Nonferrous
Aluminum					•
Brass, bronze, copper	•	•	•		
Fiberglass				•	
Cast iron	•	•	•		
Plastics					•
Steel, 40–55 R_c	•	•	•	•	
Steel, 55–60 R_c	•	•	•	•	
Steel, carbon	•	•	•		
Steel, nickel chrome	•	•	•	•	
Steel, stainless	•	•	•		
Steel, weldments	•	•	•		
Titanium	•	•	•		
Zinc					•

Basic Fluting Styles for Carbide Burs

Standard Styles		Optional Styles		
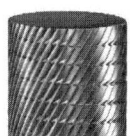				
Single Cut General-purpose fluting. The second most popular fluting style.	**Double Cut** The most popular of all fluting styles. Very efficient stock removal. Creates a small chip.	**Chip Breaker** Provides breakdown of chips. Chip size will be smaller than single cut.	**Diamond Cut** For use on heat-treated and tough alloy steels where control is important. Creates a powder-like chip.	**Nonferrous** Provides larger flute area and higher relief angles for nonferrous materials.

Stringy materials cut easily with this diamond pattern because it produces a powder-like chip with its hundreds of chisel-like edges.

Excessive pressure on bur tools will slow the spindle speed of the motor and damage the tool's cutting edges. For deburring, the rotary bur should be kept moving at all times to prevent it from digging into the work. Stainless-steel parts can take burs up to 0.25 in. (6.4 mm) in diameter at speeds up to 22,000 rpm without danger of tool breakage. Tools 1 in. (25.4 mm) in diameter may be operated at speeds of up to 12,000 rpm. On softer workpieces, the speed can be considerably higher. Tables 10-6 and 10-7 provide some of the working speeds for these types of tools recommended by one manufacturer. As with diamond-plated files, almost any bur shape can be coated with diamond particles. This permits reaching otherwise inaccessible places with conventional tools. These tools are slightly more expensive than steel or carbide rotary burs in the smaller sizes, but in many cases they also outperform the less-expensive rotary tools.

Rotary burs come in a variety of shank configurations and sizes. Commercial industrial tools typically have standard shank sizes of 0.125 in. (3.18 mm), 0.25 in. (6.4 mm), or multiples of 0.125 in. (3.18 mm). As previously discussed, there are eight basic shanks used in the dental industry (Figure 10-17). Typically, they are 0.0625 in. (1.588 mm) or 0.0937 in. (2.380 mm) in diameter.

Although most deburring is done with tapered tools, effective deburring demands a variety of tool configurations, tool axes, and motions. These include

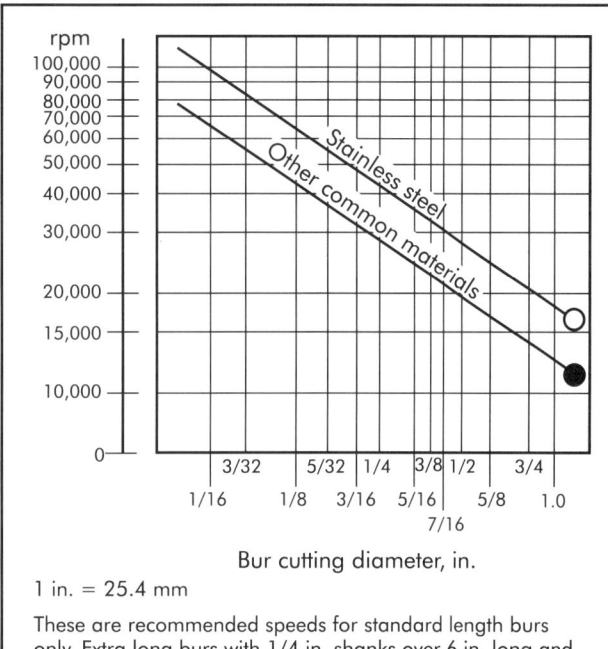

Figure 10-29. Bur speed recommendations. (Courtesy SGS Tool Co.)

Table 10-7. Recommended cutting speeds for dental diamond burs (Courtesy Pfingst and Company, Inc.)

ISO No. (1/10 mm)	Friction Grip, Right Angle (rpm)	Handpiece (rpm)
007–014	450,000	250,000
016–023	300,000	120,000
025–045	120,000	80,000
047–065	80,000	60,000
066–093	60,000	40,000
100–127		30,000
130–240		25,000
450		20,000

The optimal rpm recommended is approximately 50% of the maximum permissible velocity.

deburring with the flat end of some tools, the cylindrical portion in either of two axes, or brushing (Figure 10-39).

Special motions can be used on some features. The top surface of a hart bur or bur ball can be used to deburr the bottom of a hard-to-reach hole (Figure 10-40).

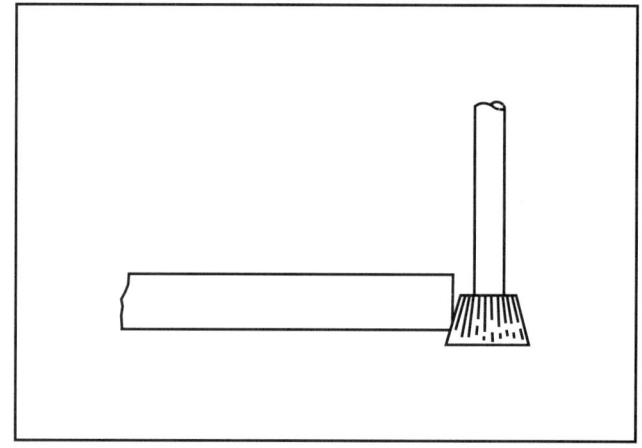

Figure 10-30. Inverted cone bur in use.

Table 10-6. Recommended speeds for carbide dental burs (Courtesy Pfingst and Company, Inc.)

	Dental Practice		Dental Laboratory	
ISO No.	Friction Grip (rpm)	Handpiece and Right Angle (rpm)	Metals (rpm)	Nonmetals (Acrylics, etc.) (rpm)
007 + 008	100,000–350,000	45,000–90,000	120,000–200,000	120,000–160,000
009 + 010	100,000–350,000	35,000–70,000	120,000–180,000	110,000–140,000
012 + 014	100,000–350,000	25,000–53,000	120,000–160,000	100,000–130,000
016 + 018	100,000–350,000	20,000–40,000	80,000–140,000	80,000–110,000
021 + 023	60,000–120,000	15,000–30,000	60,000–120,000	60,000–90,000
027 + 029		12,000–25,000	60,000–90,000	40,000–70,000
040			50,000–60,000	30,000–50,000
050	Friction-grip shank burs not over 120,000 rpm		30,000–50,000	20,000–40,000
060			30,000–40,000	20,000–30,000

Speeds are for guidance only. Variations above or below these speeds according to manner of use and particular material also may be satisfactory. Always use safety glasses.

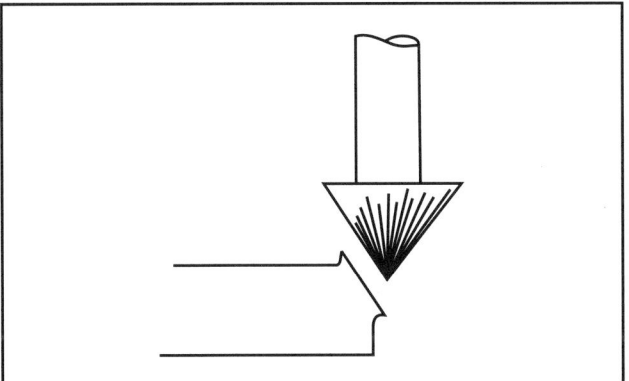

Figure 10-31. Rotary burs produce small burrs in most materials.

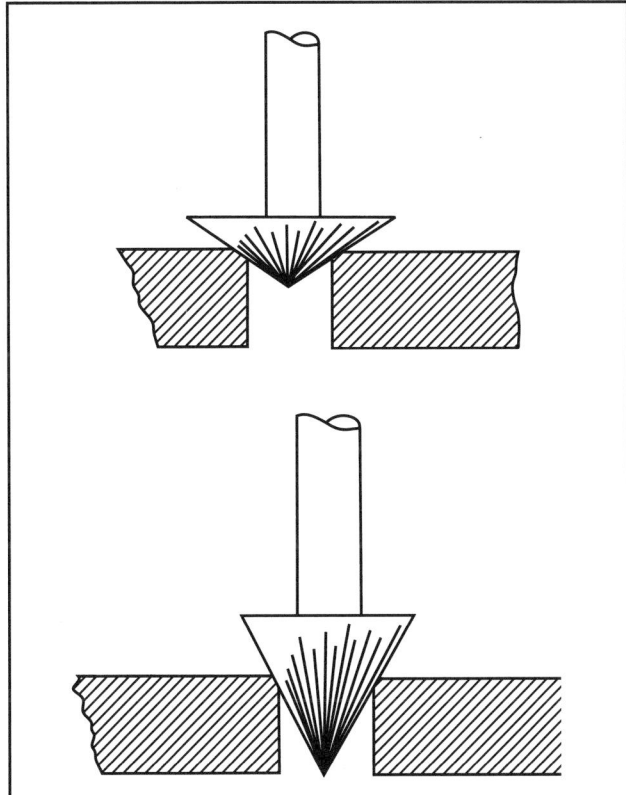

Figure 10-32. Using two cone-shaped burs to produce nearly burr-free holes.

Figure 10-33. Cup bur for finishing ends of bars.

In this instance, the tool must be used as a profiling tool for the hole edge—not as a countersink tool.

Double-sided chamfering tools are available to chamfer both the top and bottom of rectangular stock or related features (figures 10-12 and 10-40). The smallest standard cutters of this design are 0.76 in. (19.3 mm) in diameter.

Deburring small holes of 0.60 in. (15.2 mm) in a gearbox housing of cast iron is a good example of the accessibility problem. Both ends of the holes can be deburred by using a rotary bur ball with 0.48 in. (12.2 mm) diameter. The forward edge of the ball is used for the outer side of the hole, and the rear edge is used for the inner side (Figure 10-41).

An edge that flares to a wall is common on machined castings. With a small cutter, an operator can chamfer about one-half the edge before the tool hits the wall. If the burrs are small enough or only major burrs require removal, it may be feasible to use a stiff brush and completely deburr such areas.

Motor and tool interference are common potential geometry problems with mechanized deburring. The same issues occur for hand deburring. You can resolve some of these situations by using different motors, mounting arrangements, and cutters. Unless many changes are required, changing these aspects of tooling is feasible. Each tool, by necessity, must be set in its own motor and mount.

For aluminum alloys, generally it is important to use a tool with large flutes, deep and rounded gullets, small primary tooth clearance angle, large secondary clearance angle, and a positive rake angle (Figure 10-42) (Gillespie 1975).

In manual deburring, tool lives are hard to predict. Rotary burs used in robot deburring have a life of 30–300 hours, depending on the application. It is significant to note, however, that most carbide cutters generate two smaller, secondary burrs. By appropriate geometry selection, it is possible to minimize the size of these burrs.

The recommended optimum cutting velocity for solid carbide milling cutters (rotary burs) in 15-5 PH

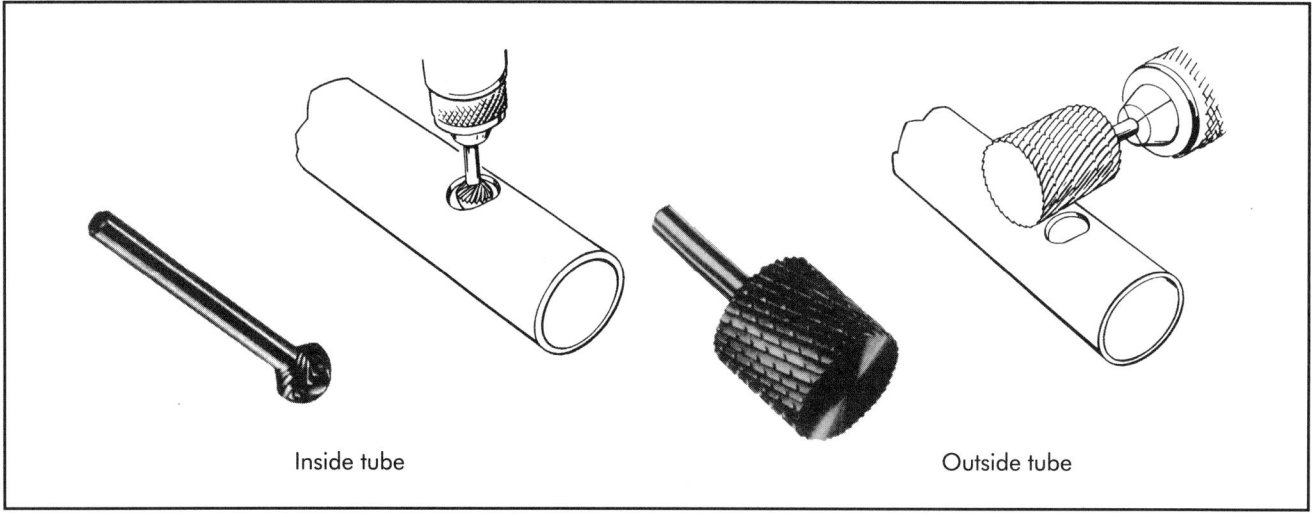

Figure 10-34. Finishing intersecting holes with burs. (Courtesy Severance Tool Industries)

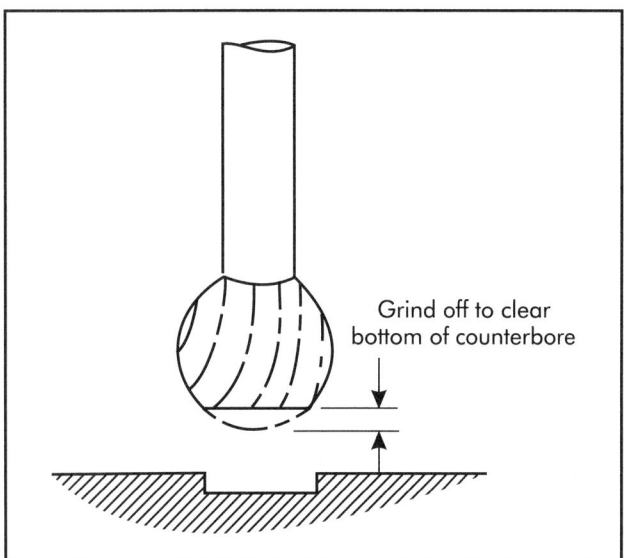

Figure 10-35. Bur ball altered to deburr shallow counterbored diameter.

stainless steel is 50 ft/min (15.2 m/min) with tooth loads of 0.0005 in. (0.013 mm) per tooth/revolution. Optimum, in this case, means acceptable production rates while providing adequate cutting, and maybe reducing chattering.

$$f_m = f_t \times N \times R_p \qquad (10\text{-}1)$$

where:

f_m = feed rate, in./min (mm/min)
f_t = feed per tooth, in. (mm)
N = number of teeth in the cutter
R_p = rpm of machine

Using a 30,000 rpm spindle and an 8-tooth cutter, the feed should be:

f_m = (0.0005) (8) (30,000) or [(0.0127) (8) (30,000)]
 = 120 in./min (3,048 mm/min)
 = 2 in./sec (50.8 mm/sec)

When milling cutters are used for fettling, the cutter diameter should be at least five times larger than the flash width to be removed. Because of the geometry of intersecting holes, many hand deburring tools are not useful in this application (Gillespie 1980). Chapter 4 includes a list of tools normally used for fettling. In most applications, reamers are used first to cut out the inside-diameter burr and, insofar as possible, leave a sharp edge.

For intersecting holes, if the largest hole is 0.0625 in. (1.588 mm) or larger, a bur ball is normally used to chamfer or break the edges of the intersecting hole. While bur balls can be manufactured as small as 0.004 in. (0.10 mm), shanks are normally larger, which may present a problem when two very small holes intersect. As Figure 4-3 illustrated, the relation between absolute minimum size of the larger hole must be:

$$D_4 \geq \frac{d_3}{2} + L_2 \tan\theta + \frac{d_4}{2\cos\theta} \qquad (10\text{-}2)$$

For a standard commercial bur ball of 0.040 in. (1.02 mm) in diameter and an intersection depth of 0.50 in. (12.7 mm), the minimum large hole size would be 0.101 in. (2.57 mm). This calculated size is considerably smaller than actually required, because the tooth pattern of most balls is designed to cut with the end of the ball, rather than the side (Figure 4-4). In practice, the ball diameter should be twice the diameter

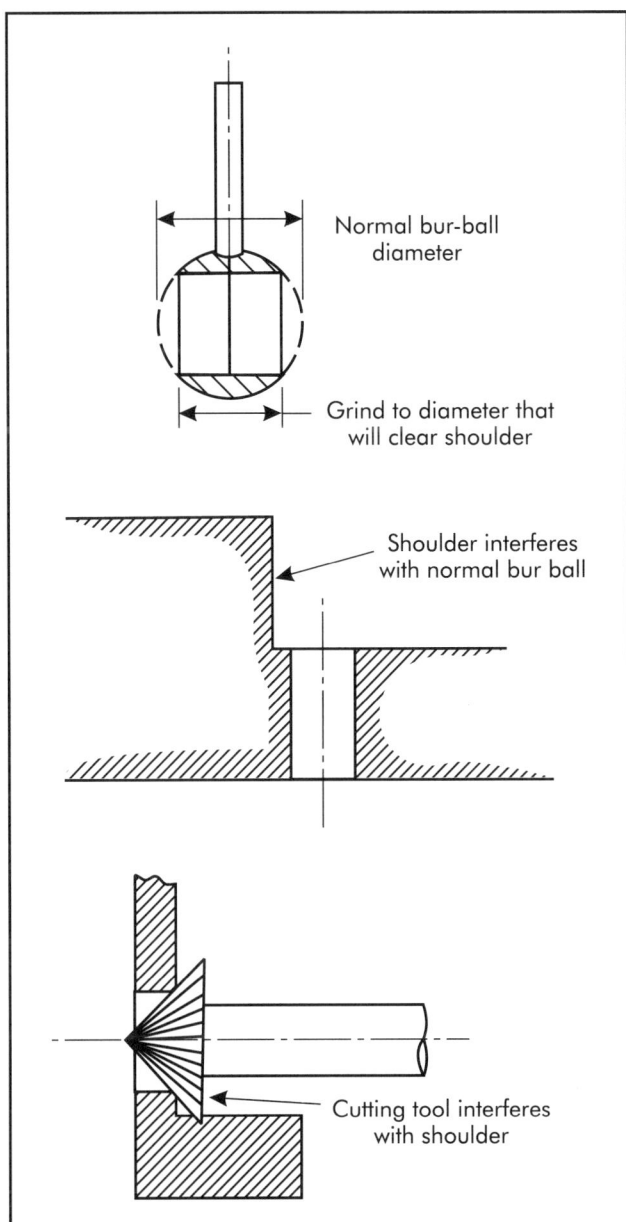

Figure 10-36. Bur ball altered to finish hole next to shoulder.

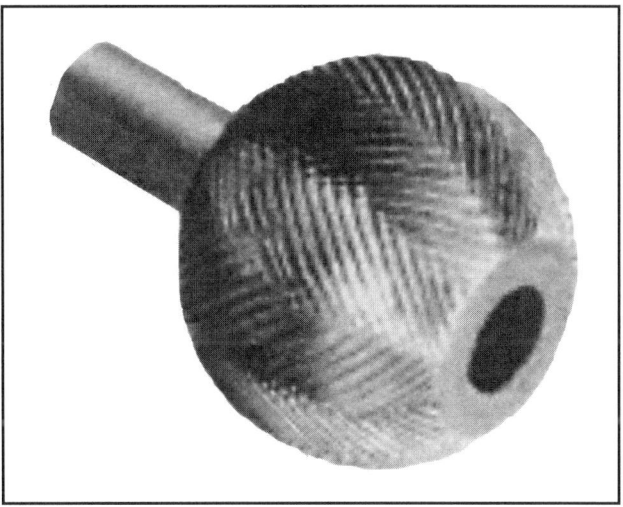

Figure 10-37. Bur ball modified to slip over a stud for counterbore deburring. (Courtesy Severance Tool Industries)

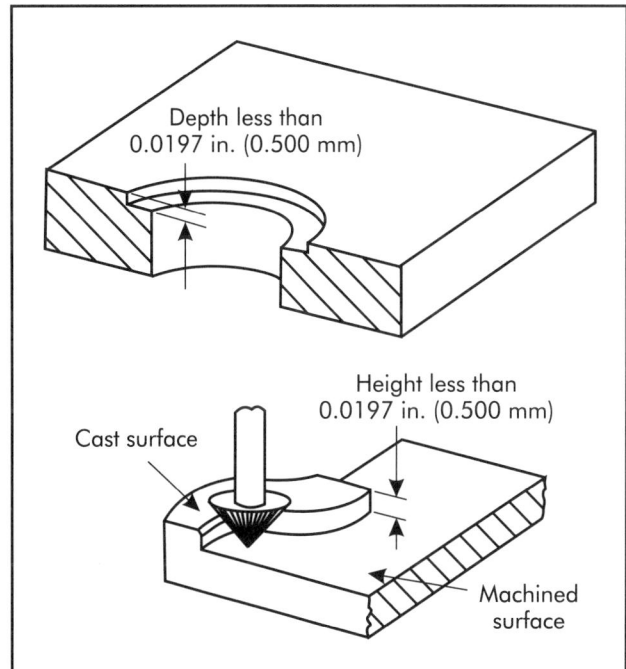

Figure 10-38. Effect of short step heights when picking rotary burs.

of the hole to be deburred to provide an approximate 45° chamfer. In addition, the number of teeth on bur balls can vary significantly among suppliers. When tooth-cut orientation is a problem, some users resort to ball-shaped mounted dental stones. These are produced commercially in several grades of hardness in sizes down to 0.0938 in. (2.383 mm) in diameter.

Abrasive-filled rubber dental points represent the fastest manual approach to radius edges of small holes after using bur balls (Figure 4-6). These are available in at least two grades of aggressiveness and three shank configurations. Because of the limited shank length, they cannot be used on intersections deeper than 0.50 in. (12.7 mm), unless the access hole allows the correct angle head to be used (Figure 4-7). Many of the dental tools may not be fully appropriate for use on hard metals. The cross-cut fissure tools, for example, may grab very thin workpieces. Dental tools, however, in addition to providing a wide variety of shapes, also afford a similar wide variety of flute configurations. Published literature does not indicate a

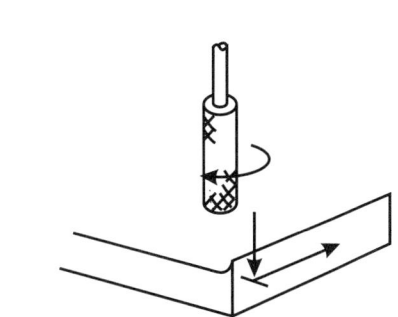
a. Use end of mounted point flush with top surface

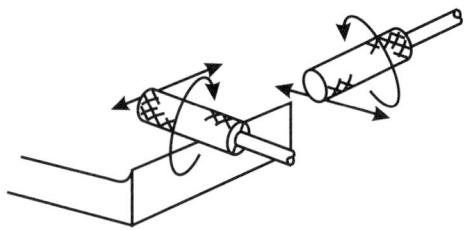
b. Use side of mounted point, cylindrical bur, or muslin-filled cylinder

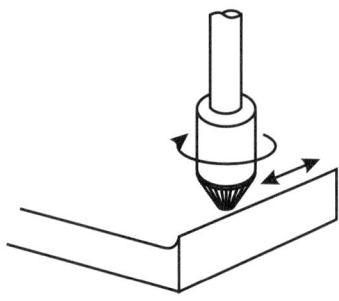
c. Use tapered cone to chamfer (rotary bur, countersink, or mounted point)

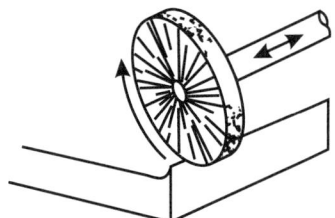
d. Brush edges to radius

Figure 10-39. Tool orientation for deburring with burs.

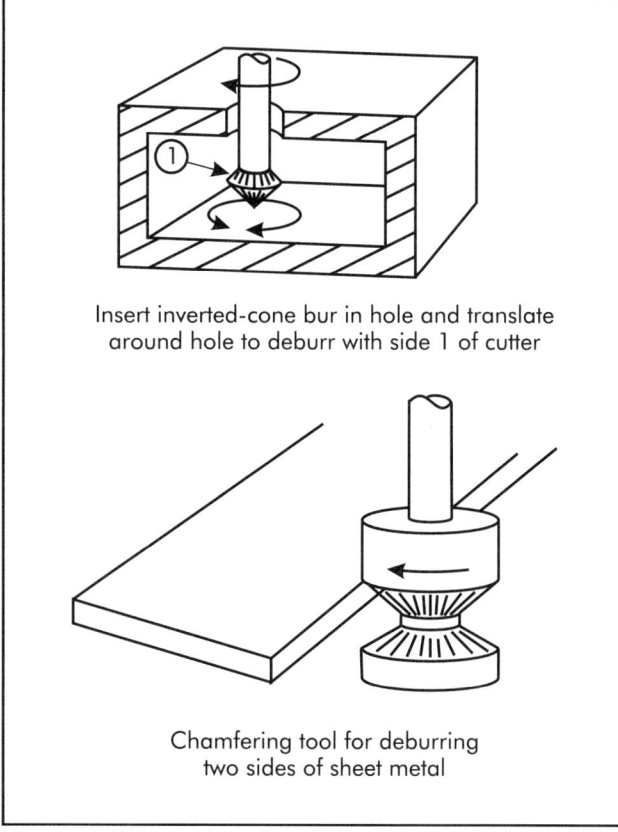

Figure 10-40. Deburring with inverted cones and sheet-metal chamfer burs.

uniform nomenclature for fineness of flute for many of the special dental tools.

ENVIRONMENTAL, HEALTH, AND SAFETY ISSUES

Environmental, health, and safety concerns for rotary burs include the following:

- they can catch and throw parts;
- chips from fast-moving cutters can hit eyes;
- heavy work creates heat in parts that can burn hands, and
- long thin stems bend when running at speeds that are too fast.

These tools are sharp, but their relatively shallow tooth depth makes them less likely than many other cutters to injure hands.

The fine dust that rotary burs make in some materials can create a health hazard. Fiberglass fibers can fill the air when these tools are used to trim glass fibers. Rotary burs may also place enough metal dust in the air to represent a hazard. Nickel dust is a suspected carcinogen, and dust from beryllium alloys and

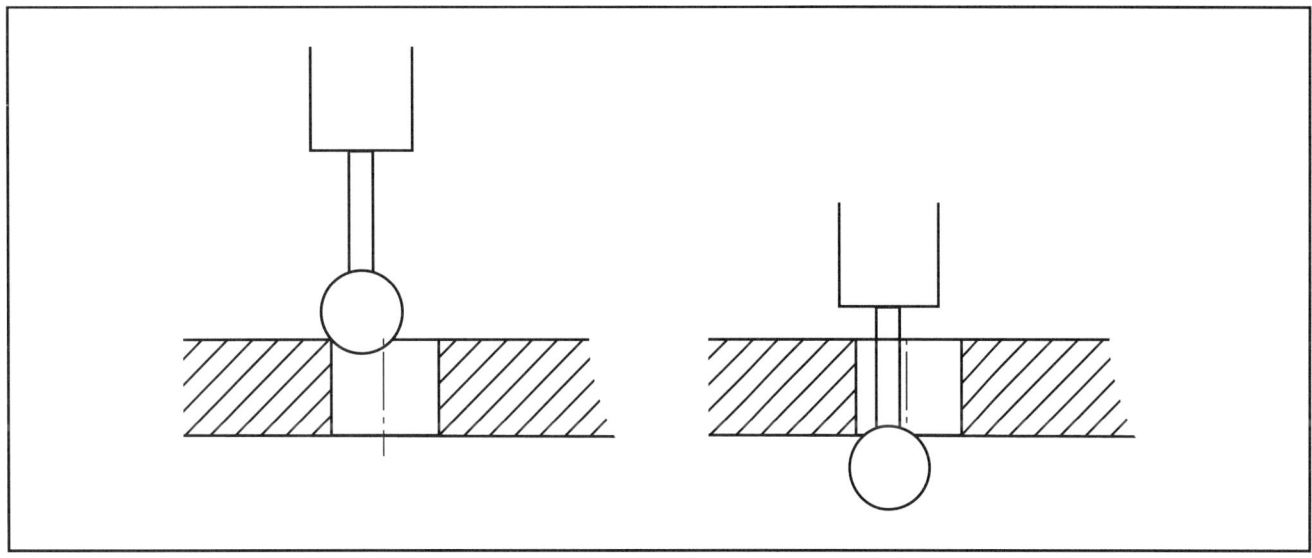

Figure 10-41. Bur ball deburrs top and bottom of holes.

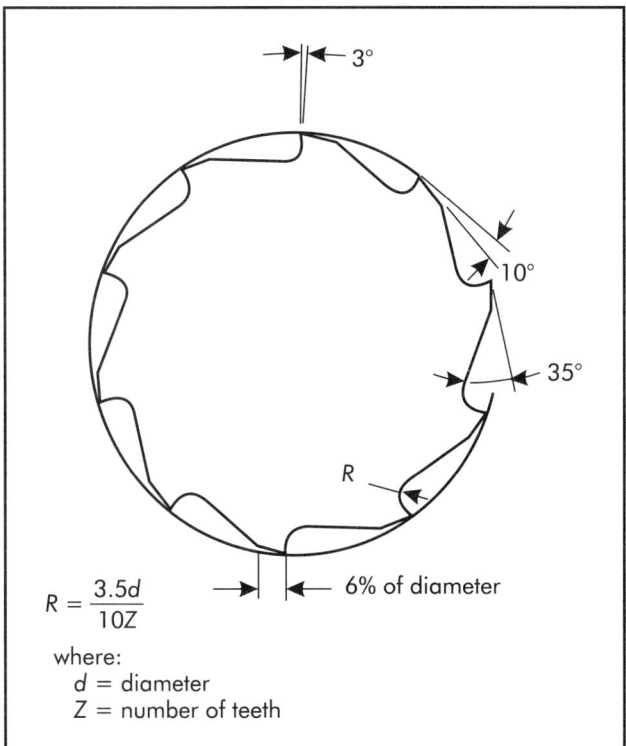

$$R = \frac{3.5d}{10Z}$$

where:
 d = diameter
 Z = number of teeth

Figure 10-42. Recommended cutter geometry for aluminum alloys. (Courtesy ABB ASEA)

compounds can cause chronic beryllium disease. Other hazardous materials can be also thrown into the air. Small magnesium and titanium chips may be fire hazards. Extensive amounts of fine dust of any material may be an explosion hazard, or at the least aggravate an asthma condition.

REFERENCES

Gillespie, LaRoux K. 1975. "Burrs Produced by Side Milling Cutters." Report BDX-613-1303 (August). Kansas City, MO: Bendix Corporation. (Available from NTIS.)

——. 1980. "Deburring Small Intersecting Holes." Technical Paper MRR080-007. Dearborn, MI: Society of Manufacturing Engineers.

——. 2000. *Guide to Deburring, Deflashing, and Trimming Equipment, Supplies, and Services,* 2nd ed. Kansas City, MO: Deburring Technology International.

Gustafson, Lars. 1983. "Deburring with an Industrial Robot." Technical Paper MR83-677. Dearborn, MI: Society of Manufacturing Engineers.

Reamers as Deburring Tools 11

Reamers are typically used to produce precision hole sizes. In the world of deburring, they are also widely used to deburr intersecting holes. When one hole breaks into another, a burr forms. Burr size depends on feeds, speeds, and drill geometry, as well as the relative sizes of the two holes, the angles at which they intersect, and the offset dimension between the axes of the two holes. The holes generally intersect well below the outer edges, so it is difficult to reach the burrs.

Mechanized processes, such as abrasive flow machine (AFM) deburring, thermal energy method (TEM), and electrochemical deburring (ECD), readily remove these burrs, but they require equipment. For manual operations, reamers provide a fast way to remove the major portion of the burr.

Reamers are not widely used for threads, but they can cut off the burrs on the crests of threaded holes (see Chapter 4). Care is required to prevent enlarging the minor diameter when using reamers. In some instances, this approach just pushes the burr back into the threads.

REAMER DESIGN

Figure 11-1 illustrates some key attributes of reamers that affect their ease of use in hand operations. For deburring, it is important to have as high a shear angle as possible to reduce torque. Small-hole reaming can be done with reamers held in pin vises. Larger holes typically require drill-press-type operations to produce the torque required to twist the reamer. The helix angle and the tooth radial rake angles, ideally, should be positive to reduce the torque.

The intent, using the reamer, is not to open the hole size, but to guide closely on the drilled hole diameter so the burrs are sheared off, leaving no material projecting into the diameter.

Figure 11-2 illustrates the problem faced by one engineer (Burnham 1981). In this instance, the holes are so small that few tools can reach into the intersection. The requirement is for the part to be burr-free when viewed under 15× magnification. The bores must not have visible scratches, and some surfaces have a 32 μin. (0.8 μm) finish. One hole breaks into the bigger bore at an angle. The drills used on this 300-series stainless steel make relatively large burrs. Holes A and C must have a sharp intersection (less than 0.001 in. [25 μm] edge break) and remain burr-free. Mechanized processes, as noted above, can remove the burrs from these holes; but removing the burrs, maintaining the hole size, and the 0.001 in. (25 μm) edge break are not yet possible for this configuration and low production quantities using mechanized processes. In this instance, reamers are used to cut the burrs in the large bore, then small reamers cut the small burrs and raised material is pushed back into the small holes by the large bore reamer.

Figure 11-3 illustrates the thin feather burr found on some intersecting holes that makes it difficult to completely remove by reamer (these burrs just push back and forth). In many situations, the burr is "bolder" and, thus, easier to remove.

Typically, most parts with intersecting holes do not have the sharp-edge requirement of the example just described. In these instances, reamers would be followed or preceded with bur balls to provide a chamfer at the internal edge (Gillespie 1980). Like all deburring cutting tools, success depends on having sharp reamers.

UNIQUE ADAPTATIONS

Some reamers can be converted to end scrapers, but it is not the reamer's action that removes burrs in this application; it is the convenient modification to make the bottom scraper.

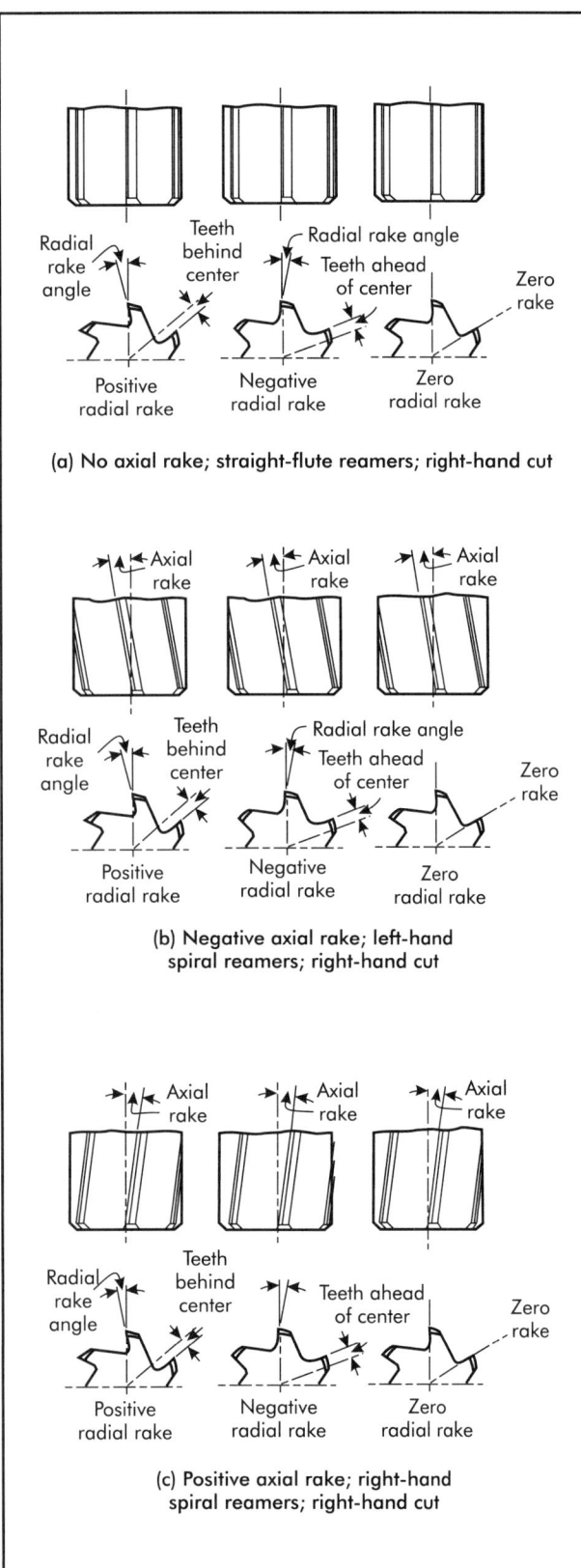

Figure 11-1. Reamer geometry.

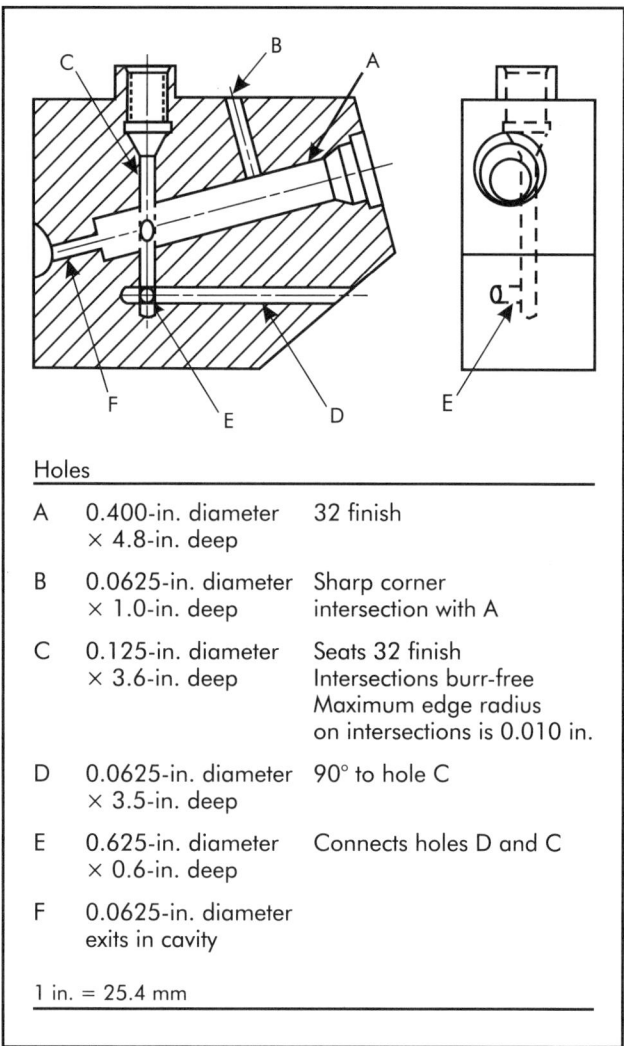

Figure 11-2. Stainless-steel part with intersecting holes (Burnham 1981).

Modified, high-speed steel reamers are good tools to begin with (if their design is applicable) because they sharpen easily and maintain their edge in deburring. Drills can also be used, and their helix angle helps shear material better, but reamers have a stiffer cross section.

ENVIRONMENTAL, HEALTH, AND SAFETY ISSUES

There are no obvious safety issues associated with the use of reamers. The size of the reamer must match the hole size or the reamer can grab, causing the part to twist in hands. This could cause the part to be thrown from the drill press or spindle, or tear hand skin. For the same safety reasons, handheld motors are not advised for use with reamers.

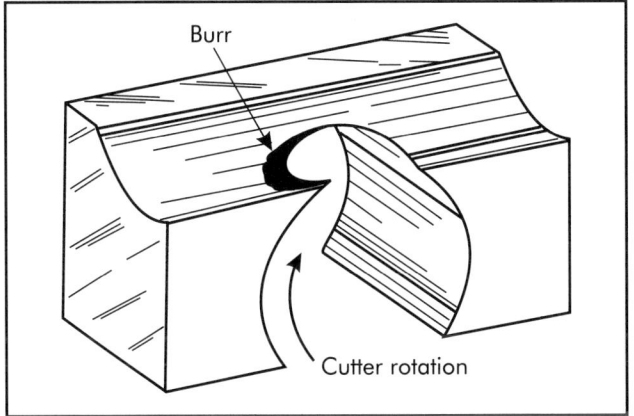

Figure 11-3. Thin burr on hole intersection (Burnham 1981).

For many small-hole applications, the reamers are rotated by hand. If prolonged work of this nature is required, hand reaming can cause extreme wrist and hand fatigue, which could easily lead to carpal tunnel or similar ergonomically related damage.

REFERENCES

Burnham, Marvin W. 1981. "Burrs Formed in Producing Small Holes." Technical paper MR81-218. Dearborn, MI: Society of Manufacturing Engineers.

Gillespie, LaRoux K. 1980. "Deburring Small Intersecting Holes." Technical paper MRR080-007. Dearborn, MI: Society of Manufacturing Engineers.

Files 12

Despite the presence of many sophisticated machining and finishing processes, filing remains one of the necessary hand operations for many parts. Even on precision parts and in many automotive assemblies, hand filing is required on edges with close tolerances. Files are one of the oldest metal cutting and stoneworking tools and one of the earliest known deburring tools. The oldest known metallic file, for example, is 3,400 years old. The early Egyptians used files and rasps as early as 3200 BC. This chapter describes some available files, the differences among them, and techniques for using them.

TYPES OF FILES

There are at least 2,300 different types of files. Despite efforts to standardize them among manufacturers and countries, considerable differences exist; therefore, it is difficult to categorize and describe all the commercially available variations. Basically, however, there are five general types of files:

- Swiss precision files,
- Swiss precision rifflers,
- American pattern files,
- millenicut files, and
- diamond-plated files.

Basic differences among these files include:

- variations in shape,
- variations in size, and/or
- variations in tooth configuration.

Two more distinctions must be considered. A file may be either single cut or double cut, as shown in Figure 12-1. As illustrated, the teeth of a single-cut file run at an angle to the edge of the file. The double-

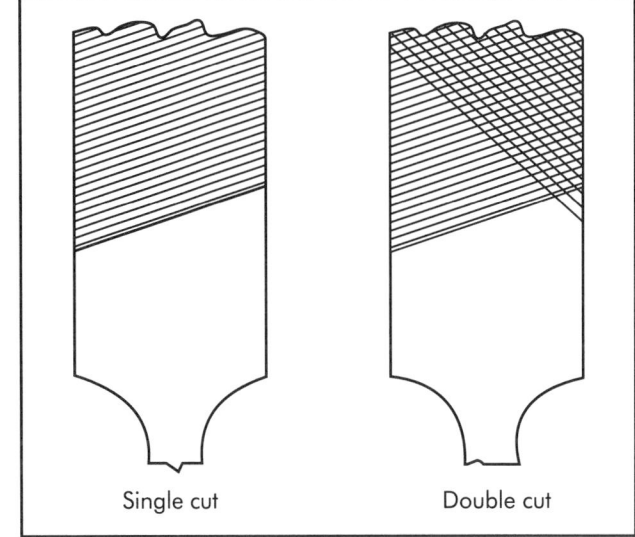

Figure 12-1. Single- and double-cut files.

cut file has another series of cuts running across the first series. Most deburring files are double-cut files. Unless otherwise indicated, assume that the files discussed in this chapter are double cut.

One of the interesting aspects of files is that they are formed by a tool that raises a sharp edge. A close look at many new files reveals a burr on the tips of the teeth. The burr (Figure 12-2) and sharp tips perform the deburring of parts. In this case, the burr is an advantage, but the orientation of the burr must be noted before using the file to obtain optimum cutting. Table 12-1 provides some comparisons between the number of teeth/inch on Swiss, American, and British files. In general, the American files are seldom

Hand Deburring: Increasing Shop Productivity 161

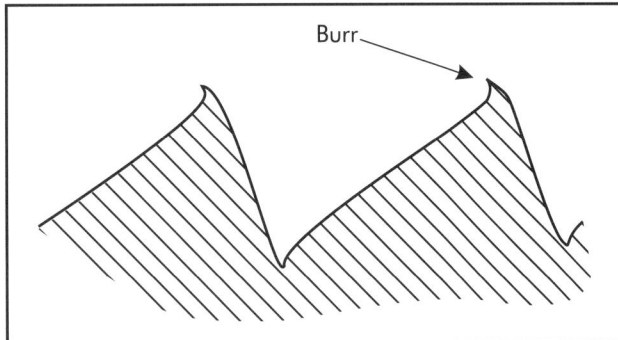

Figure 12-2. Burr on tips of file teeth.

finer than a #2 cut, although exceptions can be found. As Table 12-1 illustrates, the small numbers indicate coarse files.

Swiss Files

During precision deburring, most individuals use Swiss precision files or, in a few instances, diamond-plated miniature files. The primary difference between the Swiss precision and American file is that the Swiss file generally has a finer taper, and its teeth extend to the very ends and edges of the file. Generally, it is also narrower and smaller than an ordinary file. There are several types of precision files, that can be classified in three principal groups:

- tanged-needle files,
- escapement files, and
- riffler files.

Needle files, which are sometimes called *wire* or *French files*, have round handles (an integral part of the file). Note that to hold most American files, a wooden handle is required. Originally, French files were made for the European jewelry, watch, and clock trades. Escapement files are also self-handled, but the handle is square. As the name implies, they were made primarily for the delicate work of filing watch escapements and clock interiors.

The term *Swiss precision file*, or *Swiss file*, designates a group of shapes and scales of cuts that was originally developed nearly two centuries ago in Switzerland. These shapes and cuts, generally known as *Swiss pattern files*, have been widely imitated by many companies outside Switzerland. As several illustrations in this chapter demonstrate, they are quite different from other common types of files known as hardware, commercial, or American pattern files. Swiss precision files and rifflers are made in over 700 shapes. They are available in a wide range of sizes. Rifflers alone are available in more than 600 variations. At least 50 national firms manufacture or distribute files (Gillespie 2000).

Table 12-2 illustrates the wide variety of Swiss precision files. The majority range in size from 2–6 in. (50.8–152.4 mm) in length and are very narrow. In some cases, they are only 0.25 in. (6.4 mm) wide and 0.0625 in. (1.588 mm) thick. They are designed to work in very small areas to deburr small slots, holes, and hard-to-reach crevices. Figures 12-3 and 12-4 illustrate many precision Swiss files. As previously mentioned, escapement files are essentially square-handled versions of round-handled needle files. Escapement files and needle files are both Swiss files.

Swiss precision rifflers are bent files and form a unique series. As shown in Figure 12-5, they take a wide variety of configurations. Individuals who have

Table 12-1. Definition of file cuts

File No.	Name	Teeth/in.		
		Swiss	American	British
000	Rough cut			
00	Bastard		25–70	
0	Between bastard and second cut	40–70	35–60	42–65
1	Second cut	75–88	55–75	57–80
2	Smooth	88–104	80–95	72–95
3	Dead smooth	100–130	80–120	87–110
4	Dead smooth	120–160	125–135	102–125
6	No equivalent names	180–200	160–200	132–155
7	No equivalent names	213		147–170
8	No equivalent names	295		162–185

Note: There may be a wide variety in actual numbers of teeth/in. among manufacturers. 1 in. = 25.4 mm

Table 12-2. Types of precision files

Type of File	Length in.	Length mm	Width in.	Width mm	Thickness in.	Thickness mm
Hand or pottance	2–10	50.8–254	9/32–1-1/32	7.1–26.2	5/64–9/32	2.0–7.1
Pillar	2–10	50.8–254	3/16–45/64	4.8–17.9	5/64–15/64	2.0–6.0
Narrow pillar	2–10	50.8–254	1/8–25/64	3.2–9.9	5/64–13/64	2.0–5.2
Extra-narrow pillar	2–10	50.8–254	3/32–1-1/32	2.4–26.2	5/64–11/64	2.0–4.4
Half round	2–10	50.8–254	5/32–1-1/32	4.0–26.2	1/16–5/16	1.6–7.9
Half-round ring (high back)	6	152.4	1/2	12.7	11/64	4.4
Crossing	2–10	50.8–254	5/32–1-1/32	4.0–26.2	1/16–5/16	1.6–7.9
Three square (rectangular and thin)	2–10	50.8–254	1/8–7/16 (thin) 1/8–5/8 (normal)	3.2–11.1 3.2–15.9		
Knife	2–10	50.8–254	3/16–63/64	4.8–25.0	1/16–13/64	1.6–5.2
Round edge	4	101.6			0.016–0.065	0.4–1.7
Square edge joint	4	101.6				
Slitting	3–8	76.2–203.2	15/32–5/8	11.9–15.9	5/64–9/64	2.0–3.6
Taper flat	2–10	50.8–254	13/64–63/64	5.2–25.0	1/32–15/64	0.8–6.0
Entering	2–10	50.8–254	13/64–63/64	5.2–25.0	1/32–15/64	0.8–6.0
Warding	2–10	50.8–254	13/64–45/64	5.2–17.9	19G–12G*	
Equalling	2–6	50.8–152.4	1/4–1/2	6.4–12.7	0.24–0.79	6.1–20.1
Joint	4–6	101.6–152.4	7/16–9/16	11.1–14.3	0.05–0.18	1.3–4.6
Barrette	2–10	50.8–254	13/64–61/64	5.2–24.2	1/16–7/32	1.6–5.6

Table 12-2. (continued)

Type of File	Length in.	Length mm	Width in.	Width mm	Thickness in.	Thickness mm
Cant (or ridge-back nicking)	3-8	76.2-203.2	11/32-23/32	8.7-18.3	3/32-5/32	2.4-4.0
Round (or rattail)	2-10	50.8-254	Diameter 1/16-25/64	1.6-9.9		
Parallel round	2-10	50.8-254	Diameter 1/16-25/64	1.6-9.9		
Square taper (four sides cut)	2-10	50.8-254	Square 1/16-25/64	1.6-9.9		
Square taper	6	152.4	3/16	4.8		
Crochet (or hook files)	3-10	76.2-254	9/32-5/8	7.1-15.9	3/32-13/64	2.4-5.2
Pippin	3-8	76.2-203.2	7/32-15/32	5.6-11.9	5/64-11/64	2.0-4.4
Pivot	2-8	50.8-203.2	9/32-33/64	7.1-13.1	1/8-9/32	3.2-7.1
Double-ended pivot file and burnisher in tube	7	177.8	17/64	6.8	1/8	3.2
Screw head	2-6	50.8-152.4	23/64-9/16	9.1-14.3	17-28*	431.8-711.2
Checking	6-8	152.4-203.2	1/2-29/32	12.7-23.0	5/32-3/16	4.0-4.8

* Stubs iron wire gage.

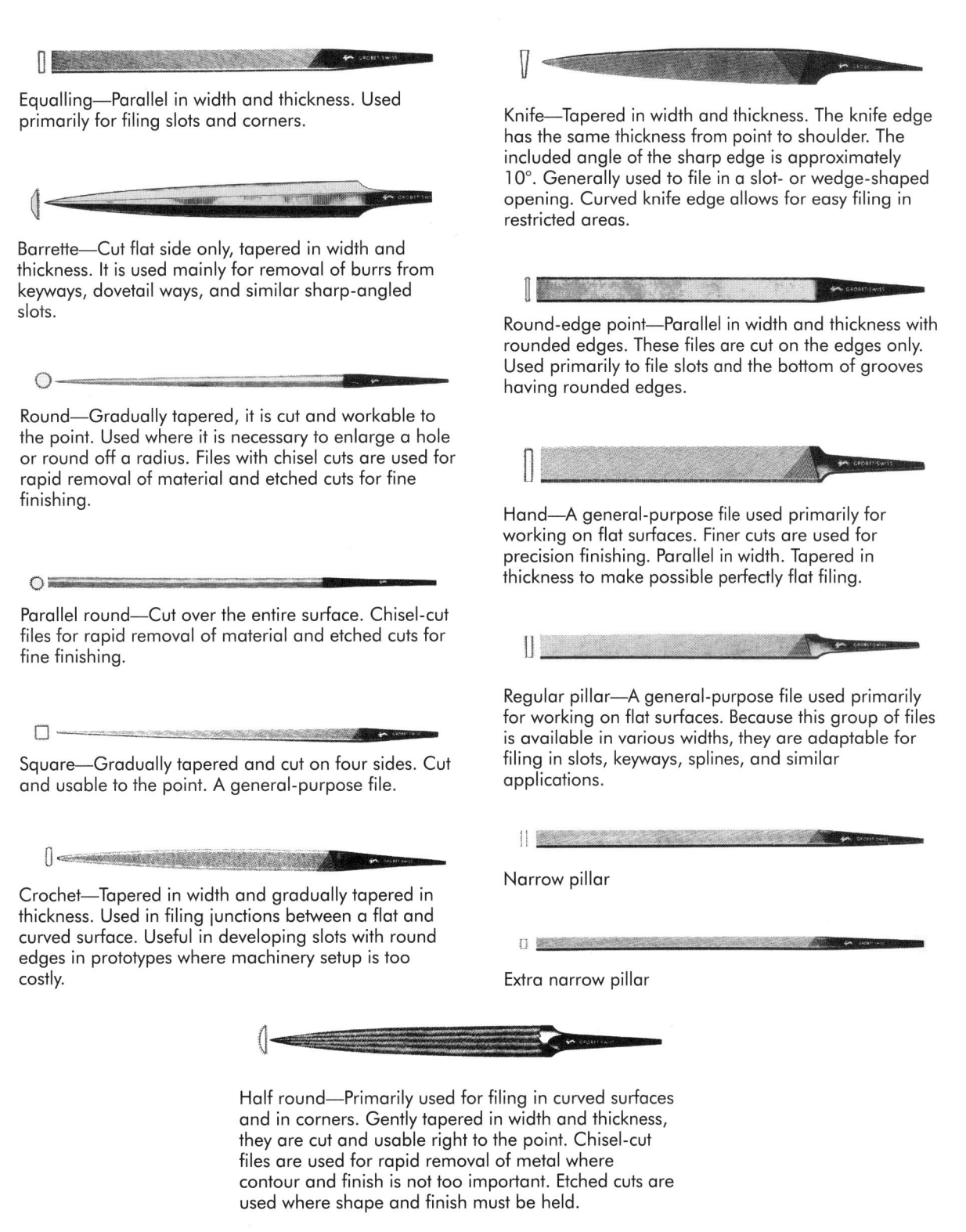

Figure 12-3. Swiss files used for deburring. (Courtesy Congress Tools)

Half round slim or ring—For applications similar to half-round files. Thinner in width and tapered in both width and thickness. The half-round side is on a smaller radius. Cut and usable right to the point. Half-round files and half-round slim files are the most versatile of files. They are found on almost every bench.

Crossing—Half round on two sides with one side having a larger radius than the other. Tapered in width and thickness. Cut and usable to the point. Used primarily for filing interior curved surfaces such as in dies. The double radius makes possible the filing at the junction of two curved surfaces or a straight and a curved surface.

Three square—Gradually tapered and cut and workable right to the point. Primary use is in filing corners and edges, such as in extrusion dies, where sharp edges must be held.

Three square slim—Same application and use as three square except thinner tapered shape permits working in smaller areas.

Parallel square—Parallel in width and thickness and cut on four sides. For general-purpose use.

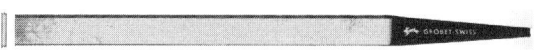

Square edge joint—parallel in width and thickness with square edges, these files are cut on edges only. Used primarily to file slots and the bottom of grooves having square edges.

Slitting—Parallel in width and thickness. While formerly used for repairing and finishing gears, their main use is now for fitting in slots too thin for knife files because of their finer edges.

Warding—Parallel in thickness and tapered in width. Used for precision removal of burrs after milling operations on office machines and other similar parts. Standard sizes are made to close tolerances.

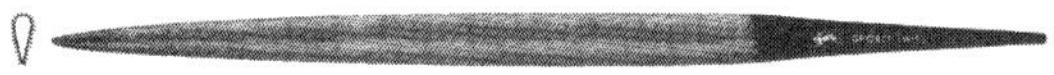

Pippin—Tapered in width and thickness. Chisel cuts for use in rapid removal of material; etched cuts for fine finishing. Combines the cross sections of the round with the crossing file. It has the edge of a knife file for finishing the junction of two different curved surfaces and for opening slots when a "V" shape is required.

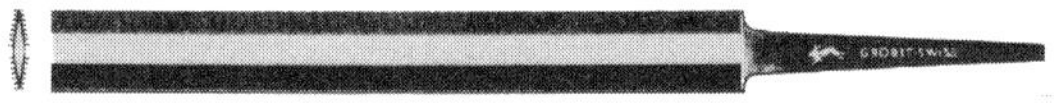

Screw head—Available in thicknesses ranging from No.1 (thickest) to No. 8 (thinnest). Used for repairing the heads of screws and filing in fine slots.

Checkering—Parallel in width and gently tapered in thickness. Overcut is made parallel to file edges and up cut is made at 90° to overcut. Used by cutters to put serrations on the edge of knives after regrinding and by gunsmiths to put a checkered area on a gun to make a firm hand grip. Also made in the form of a riffler for working in small areas.

Figure 12-3. (continued).

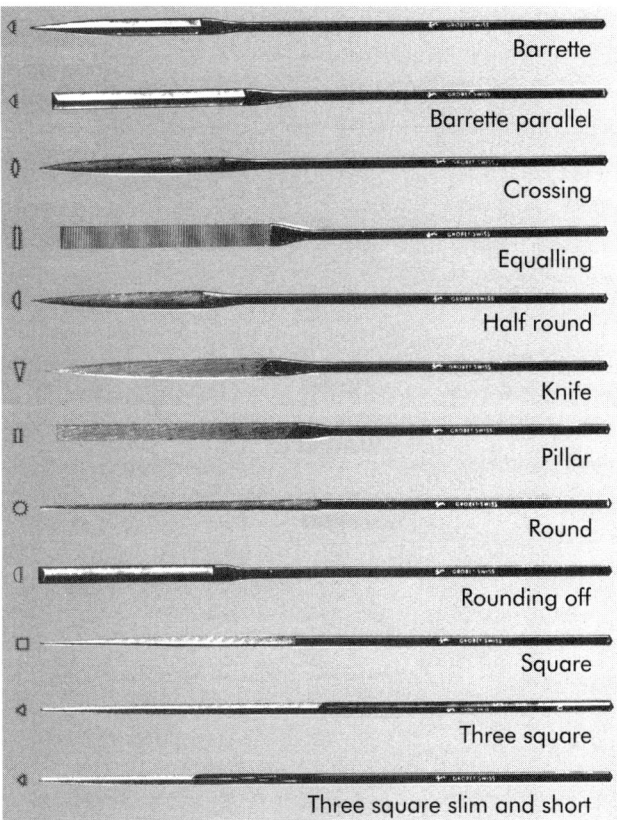

Figure 12-4. Escapement files (square-handle needle files). (Courtesy Congress Tools)

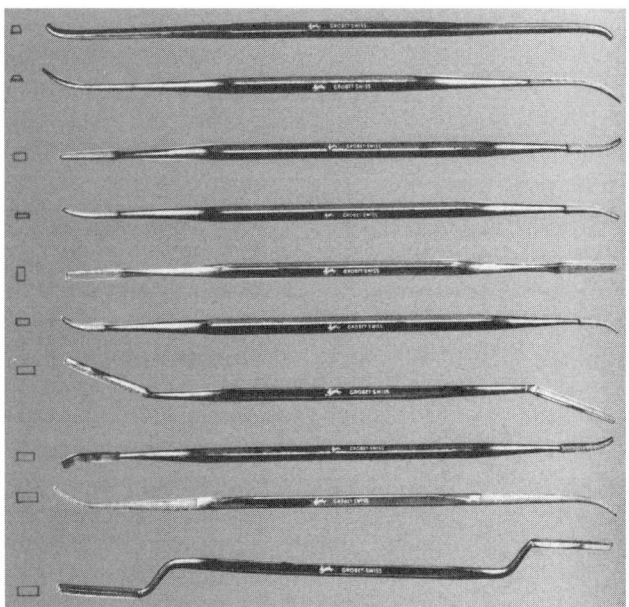

Figure 12-5. Die sinker regular riffler files. (Courtesy Congress Tools)

worked with wax carving tools will note that these are similar to those tools in configuration, but they have teeth on the ends rather than on the smooth surfaces. Many are shaped like buttonhooks or trowels, and some have gentle or sharp curves with needle or bayonet points. Some are extremely narrow and delicate, whereas others are relatively heavy. Each has different profiles and contours. They range in lengths from 6–12 in. (152.4–304.8 mm) and have teeth as fine as a #6 cut. They are sometimes called die sinkers, rifflers, die maker's rifflers, silversmiths, or toolmaker's rifflers, because they were made and used by the craftsmen with those titles. Others are designed for pattern making in the mold industry or for cabinet work.

Large Files

The long-angle lathe file cut is designed to provide smooth-finish lathe work. Curved-tooth or *vixen files* have innumerable variations in design, cut, form of the teeth, and the methods by which they are made. Every file-producing country appears to have its own variety of these types of files. The main purpose of a curved-tooth form is to provide a shearing cut, which means that each tooth automatically clears itself as it cuts, so filings do not build up between the teeth. The teeth are similar to those of milling cutters, which clear themselves in machining operations. In general, the radius of the arcs made by the teeth will be nearly equal to the width of the file. The curved-tooth file is generally used on soft materials, such as wrought iron, low-carbon steel, brass, and aluminum. Such files are also used for fibrous or tough materials.

The *millenicut file*, initially produced in England, has a pronounced undercut and intermittent interruptions in each file tooth. The teeth may be clean and un-notched, or they may have wavy interruptions (Figure 12-6). These tools do not clog on soft materials. They are particularly suitable for filing stainless steel, tough alloy steels, copper, brass, and aluminum.

Diamond-plated Files

A number of files today are made in an entirely different manner than normal steel files. They use particles of tungsten carbide, Borazon®, or diamond, which are plated onto a smooth surface. Their shapes are the same as the Swiss files previously discussed. The advantage of these files is that they are relatively inexpensive and long lasting, and they can be applied to any basic shape. An additional advantage is that any grit size can be plated onto these files. For example, for the file shown in Figure 12-7, the following grit sizes are obtainable: 100–150 ($\cong$ 0.006 in. [150 μm]), 140–200 ($\cong$ 0.004 in. [100 μm]), or 270–325

Figure 12-6. Millenicut file.

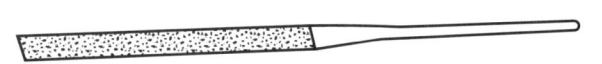

Figure 12-7. Diamond-plated file.

($\cong$0.002 in. [50 μm]). Other sizes are available as well. Since these files are harder than conventional steel files, in many cases they will last considerably longer, so they are economical choices for many applications. For files that can be placed in a reciprocating motorized file, these plated files can be used at faster speeds and higher temperatures than steel files. Figure 12-8 illustrates a miniature diamond-coated file in use.

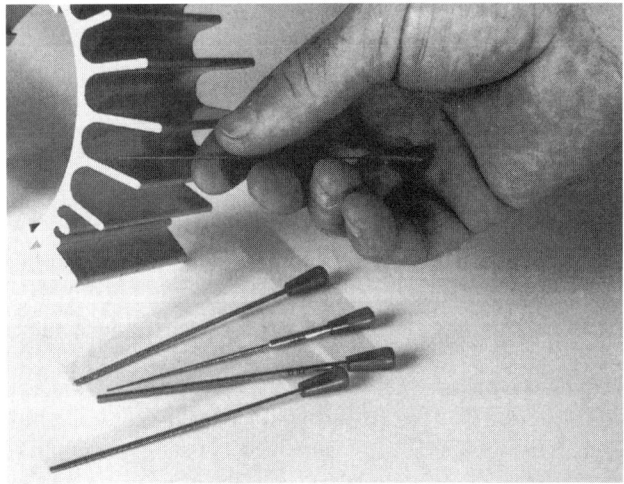

Figure 12-8. Miniature diamond-coated files. (Courtesy Gesswein)

Flexible Files

Steel needle files are available in bendable shapes. These have hardness values of 66 R_C, but they can be bent to conform to the shape being worked (Figure 12-9).

Diamond-plated foil is used as a file as well. This file is made from 0.010-in. (0.25-mm) thick nickel steel foil and is plated on one side with a heavy concentration of diamond particles (Figure 12-10). This strip can be glued with epoxy to many different shapes of tools. It is used extensively on plastic molds and die-casting dies, for polishing in thin ribs, and after electrical discharge machining (EDM) operations. It comes with grit sizes of 80–1,100 diamond. The foil allows the tool to exactly follow part contours.

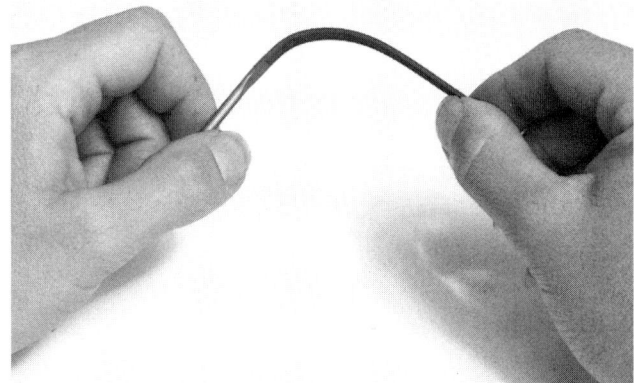

Figure 12-9. Flexible file. (Courtesy Gesswein)

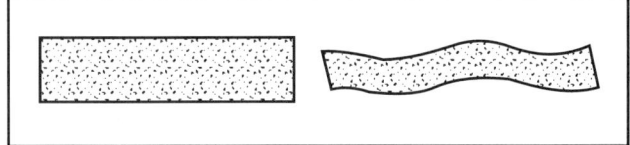

Figure 12-10. Diamond-plated foil. (Courtesy Congress Tools)

Contact Cleaners

One category of file is not normally termed a "file." Electrical contact cleaners, essentially, are very-fine-grit files. To the fingernail, they are entirely smooth; yet there is enough roughness to remove oxide surfaces from between electrical contacts. For a few miniature parts, electrical contact files are used to provide an extremely fine finish that will not exceed a critical edge break. Electrical contact cleaners are obtainable in thicknesses as small as 0.008 in. (0.20 mm). Although they are not called files, they are used in such a manner that they should be considered as a variation of commercially available files.

Folding Files and Whetstones

Some diamond-coated tools can be obtained in a pocketknife-like folding form. The whetstones used to sharpen tools also can be obtained in folding form. While whetstones are used for sharpening, they also can be useful for edge finishing of glass, ceramics, stone, and steel.

SELECTING AND USING FILES

Because of the wide variety of available files, it is difficult to define the proper one for each job. While the illustrations in this chapter provide usage guidelines, individuals must develop their own storehouses of knowledge in selecting which tool to use for which job. Note, however, that there is more to file selection than just finding a suitable shape. The type and form of material to be finished is one determining factor; the coarseness of cut to be used is another. In general, for rapid removal of stock, a coarse #0 cut might be indicated, while working on narrow surfaces requires a #2 cut file. Final finishing of precision miniature parts might call for a #4 or #6 file. Table 12-3 provides a general guideline for potential uses of each file. Table 12-4 lists properties of those most commonly used on precision miniature parts.

Materials Used in Files

Typical Swiss precision files are made from tough, chrome-alloy steel. Files are heat-treated for necessary hardness. The material used and the heat treatment can dramatically influence the life and usefulness of many of these files. Not all Swiss files are made from the same alloy, however. It is possible that some manufacturers use materials that are less conducive to long life. A few materials are highly susceptible to moisture and will rust immediately upon contact.

Filing Techniques

There is little published information on how to use precision miniature files. However, some basic criteria must be considered:

- the workpiece must be properly supported;
- the file must be correctly held;
- the proper pressure must be used;
- the file must be cleaned, and
- the correct cut and size of file must be used.

In contrast to knives, the workpiece is typically held against a solid surface with one hand and a precision file used with the other. To prevent marring small parts, it may be necessary to hold the part in the jaws of a vise or some other device covered with soft material.

There are four basic types of filing operations, but this discussion centers on the use of precision files. In these operations, the small files are held in much the same manner as a pen or pencil. Only enough pressure should be applied on the file during its forward motion to keep it cutting throughout its entire stroke. The file should be lifted during the return stroke. Too little pressure on the cutting stroke, especially when working with tool and chrome-alloy steels, quickly dulls the teeth of the file. Similarly, too much pressure will result in excess metal being removed and will cause the teeth of the file to become filled with

Table 12-3. Selection of appropriate file

File Name	Basic Application
Hand	Flat surfaces
Pillar	Flat surfaces, slots
Half round	Curved surfaces, corners, holes
Crossing	Curved surfaces, junctures of curved and flat surfaces, corners, holes
Three square	Corners, holes, edges
Knife	Slots, wedge-shaped openings
Sitting	Corners, slots
Warding	Slots
Equalling	Corners, slots
Joint	Edges, joints
Barrette	Flat surfaces, corners, keyways, dovetail ways, gear teeth, deburring
Cant corners	Corners, slots
Round	Rounded inside corners, holes
Square	Corners, holes
Crochet	Rounded corners, slots, flat surfaces, junctures between curved and flat surfaces
Pippin	Rounded corners, holes, "V" slots
Checkering	Roughening surfaces for hand grips, etc.
Screw head	Slots

Table 12-4. Files commonly used on precision miniature parts

Description	Cut Number	Thinnest File, in. (mm)
Screw head file	8	0.014 (0.36)
Screw head file	8	0.024 (0.61)
Screw head file	8	0.036 (0.91)
Narrow pillar	4	0.070 (1.78)
Equalling	2	0.014 (0.36)
Equalling	0	0.014 (0.36)
Equalling	2	0.018 (0.46)
Equalling	0	0.018 (0.46)
Equalling	4	0.029 (0.74)
Equalling	6	0.041 (1.04)
Round needle	0	0.032 (0.81)
Square-edge joint	2	0.028 (0.71)
Pattern pillar	6	0.076 (1.93)
Set needle/jewelers	4	Varies
Square needle	4	0.025 (0.64)
Set needle	2	Varies
Set needle	0	Varies
Three-square needle	2	0.019 (0.48)
Square needle	2	0.024 (0.61)
Barrette	2	0.028 (0.71)
Equalling	4	0.050 (1.27)
Equalling	6	0.014 (0.36)
Pillar escapement	4	0.041 (1.04)
Knife	8	0.036 (0.91)
Pillar	8	0.035 (0.89)
Half round	8	0.020 (0.51)
Round escapement	8	0.018 (0.46)

small chips. This will prevent effective cutting (Grobet File Company of America 1977).

Filing should always be carried out on the forward, never on the backward, stroke. No saw user would ever dream of expecting efficient cutting by running a circular saw backwards. Likewise, it is absurd to assume that a file will cut efficiently on the return stroke. The reason so many users continually file both ways is probably that the large number of teeth per inch on a file appears to render damage to a few of them insignificant. This, however, is a fallacy. Every tooth is of value and importance to the work.

Although all filing should be done on the forward stroke, it is unnecessary to lift the file entirely from the work during the return stroke, except when special files are used. The point is that no pressure must be exerted on the backward movement. The best results from the forward stroke will be obtained by giving the file a slight sideways motion, alternating to right and left with every few strokes. Narrow surfaces should be filed in the direction of length, not across. Sharp edges should not be filed with an upward stroke. Single-cut files are best for thin work, as their teeth are stronger and less likely to break than those of the double-cut file, and they also cut more smoothly.

Frequently, the inexperienced file user employs a file too large or too small for the work. For example, one might select a single file and do all deburring when a faster and more efficient method would be to use a coarse one first, followed by a smoother one. This enables the filer to remove the metal quickly and provide a good finish on the workpiece.

New files should not have too much pressure exerted on them on each stroke because some of the teeth may break off at the base. The proper method is to work lightly, preferably on brass or bronze until the fine tooth points are slightly worn, then employ more pressure as the teeth become more blunt. The file used for iron or steel is often unsuitable for brass. In filing steel, it is better to use second-cut files than those with coarser teeth.

A sharper file is needed for non-fibrous metals, such as brass, copper, aluminum, zinc, and cast iron, than for wrought iron or hard steel. A sharper file is also needed for a broad surface rather than for a narrow one. There are files specifically made for working on plastic-type materials, such as Vulcanite® or Bakelite®. When filing aluminum, better results may be obtained by using a lubricant, such as lard oil. While this may not be feasible on miniature parts, it is effective on hand-sized parts.

Some file users have noted that filing crosswise to the direction in which steel has been rolled is more effective than filing in other directions.

Best Practices

Filing is one of those arts handed down generation to generation among craftsmen. Few articles discuss best practices in this area. The following paragraphs provide some additional ideas for effective use.

Work determines the choice of file. The type of metal to be filed is the important factor in deciding which file cut to use. Cast iron, especially if the scale has not been removed, is hard on a new file. The glass-like scale tends to dull the cutting edges. New files should not be used on such a surface.

Two general rules for selecting the proper coarseness of file cut are:

1. when filing any hard metal, such as steel, use a second-cut file, and
2. when filing soft metals, such as brass, bronze, or copper, use a coarse or rough-cut file.

Nearly all large files used in the machine shop are double cut. The 10 in. (254 mm) or 12 in. (304.8 mm) bastard file is generally used for rough filing in bench work. The second-cut file gives a smooth finish. A finer file also provides a smoother finish, as does rubbing chalk on the file teeth.

How to file. First, select the proper file for the job. Make sure it is fitted with a good handle. Stand in a comfortable position, especially if the job will require considerable time. It is nearly impossible to file accurately in an awkward position. The worker (if right-handed) should stand left foot forward with feet slightly apart (Figure 12-11). The weight of the body should be balanced, allowing the arms to move easily forward and backward. The file should be regarded as an extension of the arm. That is, file and arm, from elbow to wrist, should be in line.

Grasp the file handle in the right hand, thumb on top. For light work, hold the point (tip) of the file with the thumb and first two fingers of the left hand, the thumb on top and the fingers below. For heavy cuts, the heel of the left hand may be placed on top and the fingers closed on the underside of the file (Figure 12-12).

Do not drag a file back over the work or attempt to cut on the return stroke. This will dull the file. Experienced workers know you should never rub your hand or fingers over the filed surface of cast iron or brass. Oil from the hand may be deposited on the metal. This will cause the file to slide over the work, instead of cutting it.

Rough filing. When rough filing, cross the stroke at short intervals. This will help keep the surfaces flat and straight while you are learning to file. Bear down only on the forward stroke.

Rounding a corner. To round a corner, rough file across the workpiece. Reduce the corner by filing a series of angles until the proper radius is secured (Figure 12-13). Finish the corner by following along the rounded corner with a fine-cut file.

Figure 12-12. Correct method of holding large file on a fixtured part.

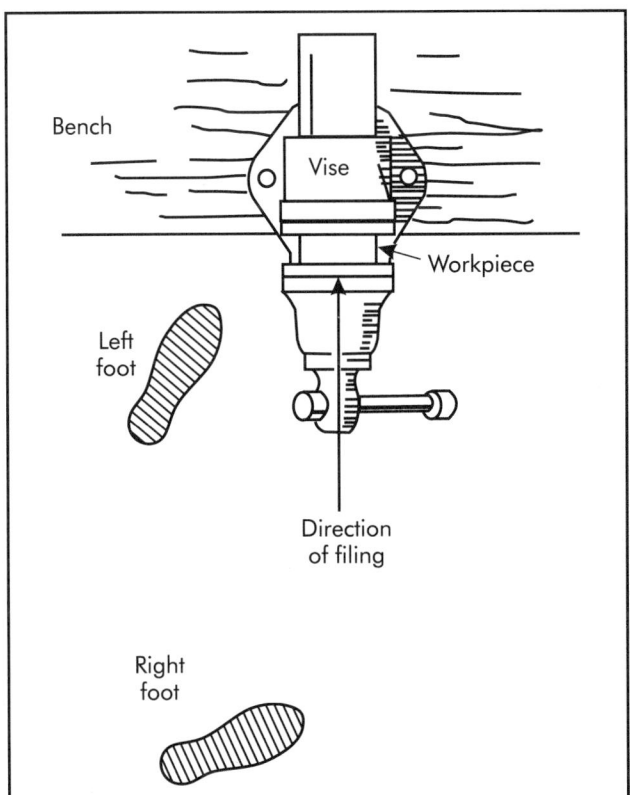

Figure 12-11. Deburring stance when filing.

Figure 12-13. Providing a large radius with a file. (Courtesy Honeywell)

Draw filing. Draw filing consists of grasping the file firmly at both ends and alternately pushing and pulling the file sideways across the work. The file is held at right angles to the line of stroke. The pressure should be the same for both forward and return strokes. Very little stock is removed in draw filing—the purpose of this operation is to produce a straight or square surface. Draw filing gives a smoother finish on edges and narrow surfaces than obtained by "straight" filing. For draw filing, single-cut files are preferred to double-cut files for producing a better finish.

General use. There are a number of ways to modify or select files for appropriate tasks. For example, a file only 0.014 in. (0.36 mm) thick cannot be used in a slot that is only 0.010 in. (0.25 mm) thick. Steel files, however, can be readily electropolished or chemically polished to reduce their size and still maintain most of the sharpness of the teeth. These operations can be accomplished in only a minute or two per file. A number of companies resharpen used files, but that may not be economical in industrialized countries.

One worker bonds adhesive-backed abrasive paper to shim stock to make a pseudo file, allowing production of a file of any thickness within a minute. Files work extremely fast in some applications, such as finishing high-nickel aerospace materials. Some companies have found that applying a thin, chrome-flash plating before use will extend the life of these files by three or four times.

While the thrust of this section is on handheld file use, many of the files discussed are commonly used in reciprocating, motorized handheld machines. Handheld reciprocating files greatly speed the finishing task and use the same tools one would use by hand. Both the plated and steel files are available for use with such equipment. It should be noted, however, for general use on precision miniature parts, motorized files may be too big and lack the necessary delicate touch.

Cleaning and Care of Files

Precision Swiss files, as the name states, are precision tools. They require care. Metal chips can easily clog spaces between the teeth. This prevents effective cutting, and clogging or "pinning" leaves deep scratches in the workpiece. If pinning cannot be avoided, its effects can be reduced by thoroughly cleaning the file at frequent intervals.

In some cases, the file may be cleaned by gently rapping its edge against a hard surface. If this is ineffective, one may use a fine wire brush or a file card. For working on wrought iron and steel, a soft iron or copper "scorer" may be used. For filing aluminum, one may need to use turpentine or paraffin on file surfaces to prevent pinning. When the files are clogged with dirt and grease, either a wire brush or an ultrasonic cleaning treatment may be necessary. If ultrasonic cleaning is used, however, make sure all water is removed from all surfaces, as it will rust the tools and completely erode many of the teeth.

Files exposed to magnets become magnetized. This will cause many of the chips to remain in the spaces between the teeth. As a result, the file may be unusable until it is demagnetized.

For prolonged life, particularly in coarser cut files, do not merely toss them into a drawer or pile them on the back of a bench, as such treatment will damage teeth and, in many cases, allow the files to rust. Rust eats the teeth. Eventually, they crumble to fine dust. Instead, place the files in a rack after use.

ENVIRONMENTAL, HEALTH, AND SAFETY ISSUES

File tangs (the part of the file that tapers from the shoulder that is intended to be fitted with a handle) are dangerous. Many workers have jammed the relatively sharp tang into the palm of their hand while filing. Several companies manufacture handles that should be used to avoid this injury. Figure 12-14 illustrates an ergonomically designed handle to reduce risk of injury. The operator can rotate the arm slightly to change the angle of the file. The manufacturer notes that it incorporates the natural wrist angle of 35°, so the tool's axis is aligned with the forearm, significantly reducing muscle fatigue.

Figure 12-15 illustrates another holder that allows two-handed usage. Figure 12-16 illustrates a universal edge sharpener that protects the hands while deburring sheet edges. Yet another approach is shown in Figure 12-17, where the tool is a cylindrical file

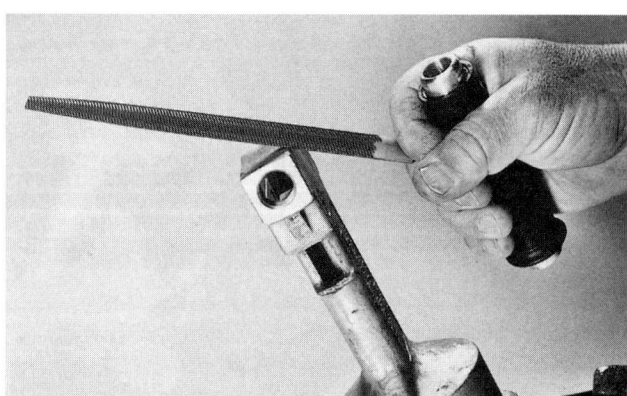

Figure 12-14. Ergonomic file handle. (Courtesy Gesswein)

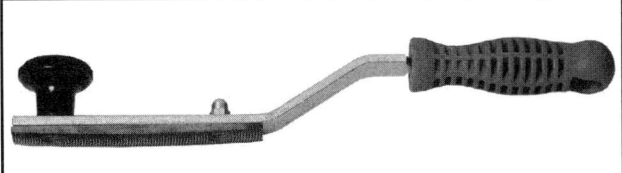

Figure 12-15. File holder requires two-hand use. (Courtesy Pferd)

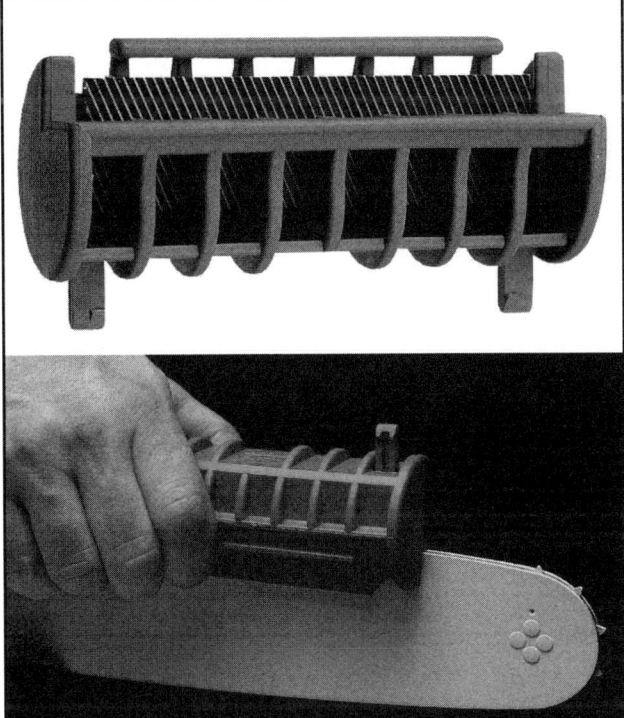

Figure 12-16. Universal edge sharpener. (Courtesy Pferd)

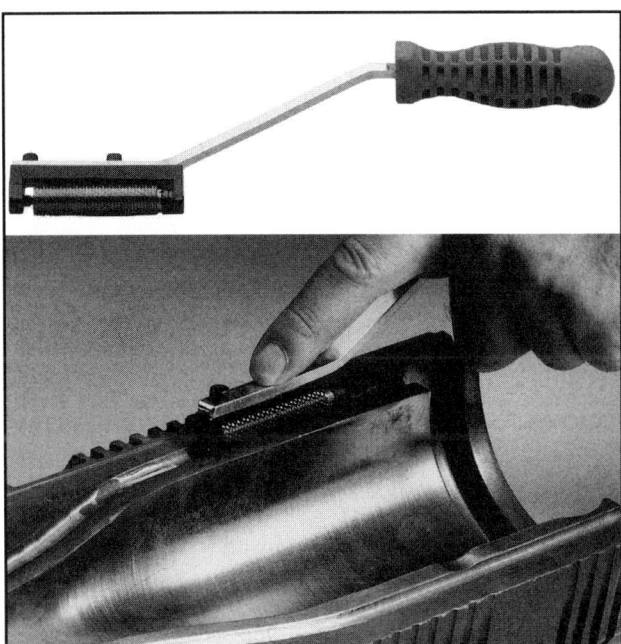

Figure 12-17. Round file in special holder for rounded grooves. (Courtesy Pferd)

that sits inside a frame. This tool is used to file in rounded grooves and concave surfaces.

Filing on lathes while the part is rotating has been performed for decades. The rotating spindle, however, presents several safety risks. Getting sleeves caught in the spindle or on the part will allow the machine to quickly draw the operator into the spindle. He or she will not have time to escape serious injury. If the operator slips while filing the end or near other edges of the part, his or her hands or arms can be forced into the sharp edges that have not yet been deburred. This will also result in serious injury. It is essential to hold the file soundly so it will not be thrown back at the operator if it hits an unexpected object. With modern machine tools, the program will chamfer the edges and manual deburring on the lathe is not required.

APPENDIX

The following glossary of Swiss file terms is adapted from a catalog of Swiss files (Grobet File Company of America 1977). It may prove useful for those wishing to further study or select appropriate files.

Glossary of Swiss File Terms

Auriform file. A die sinkers' file, it has a cross section that combines one half of a pippin file with one half of a crossing file.

Back. In a half round, barrette, cant, or other files of similar cross section, this is the convex side.

Barrette file. Cut on wide flat face and safe on sides and back, it is tapered in width and thickness.

Bench filing machine file. A parallel file, it is available in various cross sections for use in filing machines.

Blank. A steel forging from which a file is made, it is the basic shape of a file before teeth are cut or etched.

Cant file. Triangular in cross section, with one side wider than the other two, it is cut on three sides and tapered.

Checkering file. Rectangular in cross section and parallel in width and thickness, it has teeth cut at a 90° angle with the edge. The edges are safe.

Chisel cut. This is a method of cutting teeth into the surface of an annealed file blank by striking it with a series of repeated blows as the blank is moved beneath a chisel at a uniform speed. In the cutting operation, the chisel is placed obliquely to the length and is inclined to the surface of the file. This is done either by hand or machine. It is generally used to produce files of #2 cut and coarser.

Crochet file. Rectangular in cross section with rounded edges, it is cut on both faces and edges. It is tapered in length and slightly tapered in thickness.

Crossing file. A file with an oval cross section, with same radius as half-round files on one side and the other side is curved to a larger radius. It is cut on both sides and tapered in width and thickness.

Cut. Term used to describe the number of teeth/in. (teeth/mm) and the degree of coarseness of a file's teeth, which is from #00–8 for Swiss precision files. This term is also used to describe the type of file, such as single cut, double cut, etc.

Die maker's rifflers. These come in various cross-sectional shapes. The teeth cut on a small area of each end, leaving a long, middle portion as a handle. The cut ends are of various designs and the length is overall. Originally designed and hand-forged by die makers for specific purposes, it is now a generic term for this particular group of rifflers.

Die sinker's files. See also "die maker's rifflers." This group of rifflers has smaller cross-sectional shapes.

Double cut. A file on which the arrangement of teeth is formed by two series of cuts. The first is the overcut, which is followed by the upcut at an angle to the overcut.

Edge. The narrow cross section, or side, of a file.

Equalling file. A file of thin, rectangular cross section, parallel in width and thickness, and cut on both faces and edges.

Escapement file (also called square-handled needle file). A group of files of various cross-sectional shapes with a length of cut varying from 0.75–2.50 in. (19–64 mm) long with square handles. They are widely used by jewelers, watchmakers, die makers, and fine mechanics.

Etched cut. A method of cutting teeth into the surface of a file blank by drawing an etching tool, under sustained pressure, obliquely across an annealed file blank in a series of cuts. This may be done either by hand or machine. This method of cutting is used where it is necessary to retain the true cross section of a file. Generally, it is used to manufacture files finer than a #2 cut.

Face. The working surface of a file on which teeth are cut.

Filing block. Block of wood, soft metal, or other material used to protect the material being filed from damage from the jaws of a vise or other holding device. It may contain a series of grooves to hold work securely.

Flat file (also called a warding file). A form of escapement or square-handled needle file, it is parallel in thickness, cut on four sides, and has a tapered width.

Handle. Wood or plastic piece placed over the tang of a file to protect the hand of the user.

Half-round file. Cross section that is flat on one side and has a radius (not half circle) on the other side, and is cut on both sides. The width and thickness taper.

Half-round slim file (also called ring file). Same as half round, except thinner in width.

Heel (also called shoulder). The end of the file at a location where the body ends and the taper leading into the tang begins.

Joint file, round edge. Rectangular cross section with rounded edges, it is cut on the edges only, and is parallel in width and thickness.

Joint file, square edge. A rectangular cross section cut on the edges only. It is parallel in thickness and width.

Knife file. A file with a knife-shaped cross section that is tapered in width and thickness. Its edge has the same thickness from point to shoulder.

Length of cut. The length of a file measured between the shoulder or heel and the point.

Lozenge file. A file with a diamond-shaped cross section, parallel in width and thickness.

Machine file. A file made specifically for use in a filing machine, available in various cross-sectional shapes, and parallel in width and thickness.

Needle file, round-handled. Group of files of various cross sections with a knurled, round handle. Knurling gives the file a positive, nonslip grip for precision filing.

Needle file, square-handled (also called an escapement file). Group of files of various cross-sectional shapes with a length of cut varying from 0.75–2.50 in. (19.1–63.5 mm) and a long, square handle.

Oval file. This file has an oval cross section that tapers in width and thickness.

Overcut. First of a series of cuts in a double-cut file, its function is to act as a chip breaker. The second cut, or upcut, is made after the overcut.

Parallel machine file. Group of parallel files of varying cross-sectional shapes made specifically for use in reciprocating filing machines.

Parallel round file. A file of round cross section, parallel in width.

Parallel square file. This file has a square cross section, parallel in width and thickness.

Pillar file. A file of rectangular cross section with thickness greater, relative to width, than in other types. It is cut on its face or flat sides only, parallel in width, and tapered in thickness. It is also available in semi-narrow, narrow, and extra-narrow widths.

Pin or pinning. The term describes the tendency of small particles of materials to fill or clog the gullets between the teeth of a file. When the teeth become clogged, the file causes scratches on the work. When this occurs, the file is pinned.

Pippin file. Section that combines the cross section of a round file with that of an equalling file. It is tapered in thickness and width.

Point. Front end of a file, as contrasted with the tang end.

Pointed back barrette file. A file of triangular cross section with one side wider than the other two sides. It is cut on the wide or face side only, and is tapered in width and length.

Rasp cut. Cut used on wood rifflers made by a punch raising a series of individual cutting teeth.

Rifflers. The term originates from the German word *riefeln,* which means to channel, chaufer, flute, or groove. The tools were originally used and hand-forged by die sinkers, die makers, silversmiths, and other skilled artisans in shapes and cross sections appropriate to their work. Teeth are cut on small areas on each end that can be formed into any shape, from trowels to buttonhooks. A long, middle portion serves as a handle.

Ring file (same as half-round slim file). See "half-round slim file."

Round file. A file round in cross section and tapered in width.

Rounding-off file. Escapement or square-handle needle file that is half round in cross section, cut on the flat side, and parallel in width.

Safe. Side or edge of a file that has no teeth cut in it, so it will not mar a work surface that does not require filing.

Screw-head file. A file with a narrow, diamond-shaped section with short bevels to form sharp edges. It has cuts on beveled edges, but is safe on flat sides and parallel in width and thickness.

Section. Cross section or end view of a file cut squarely at the place of greatest width and thickness from the tang.

Silversmith's rifflers. Group of various cross-sectioned shapes originally designed for use by silversmiths. Teeth are cut on small areas of each, leaving a long, middle portion as a handle. The cut ends are of varied design.

Single cut. Teeth formed on a file by a single series of cuts.

Slitting file. A file with a flat, diamond-shaped cross section and cut on all sides. It is parallel in width and thickness.

Square file. A file square in cross section and tapered. It is cut on all sides.

Swiss pattern files. Files made to the same shape and cut as the files originated by F. L. Grobet in Switzerland over 150 years ago, they are made in cuts from #00–6.

Swiss precision files. Original Grobet Swiss files are made in hundreds of sizes and shapes and in cuts from #00–8. They are made to more exacting measurements and in much finer cuts than American pattern files.

Tang. The part of the file that tapers from the shoulder and is intended to be fitted with a handle.

Three-square file. Tapered and equilaterally triangular in cross section, the file is cut on all sides with sharp corners.

Toolmaker's rifflers. Various cross-sectional shapes with teeth cut on a small area at each end, they have a long, middle portion as a handle. The cut ends are of various designs to meet the needs of toolmakers.

Upcut. The second series of teeth cut in double-cut files made over the first series of cuts, called the overcut. This cut is made of an angle to the overcut.

Warding file. A file of rectangular cross section with teeth cut on all sides up to 4 in. (101.6 mm) in length and on three sides with one safe edge on files 6 in. (152.4 mm) and longer. It is tapered in width and parallel in thickness.

Wood rifflers. Various cross-sectional shapes cut with rasp teeth on both ends, leaving a long, middle portion as a handle. They are used by cabinet and pattern makers.

REFERENCES

Gillespie, LaRoux K. 2000. *Guide to Deburring, Deflashing and Trimming Equipment, Supplies, and Services.* Kansas City, MO: Deburring Technology International.

Grobet File Company of America. 1977. *Files.* Catalog 56C. Carlstadt, NJ: Grobet File Company of America, Inc.

BIBLIOGRAPHY

Simons, Eric N. 1947. *Steel Files.* London, England: Pitman and Sons, Ltd.

Mounted Points 13

The family of products known as mounted stones or mounted points is another type of efficient tool in burr removal. Mounted stones are actually miniature grinding wheels on small shanks. They are used much like rotary burrs or rotary brushes for deburring and edge finishing. This chapter describes the available industrial and dental mounted stones.

Although the focus of this chapter is the design and use of mounted stones or mounted points, it is important to recognize that a variety of abrasive-filled products are commonly used for similar purposes. These materials include:

- resin-bonded mounted stones;
- vitrified-bonded mounted stones;
- resin-impregnated cotton or muslin fiber;
- hard rubber bonded mounted wheels;
- soft rubber polishing wheels, and
- treated mounted wheels.

Resin-bonded mounted stones, essentially, are miniature grinding wheels. The resin bonding provides a wheel unusually strong and resistant to shock. The resin is malleable and adapted to operations where the wheel may be subjected to rougher usage than the more brittle vitrified bonds. This wheel is used in many companies for such things as grinding stainless-steel welds. Most mounted wheels discussed in this chapter are vitrified, rather than resin, bond.

Treated mounted wheels in vitrified and resin bonds can be purchased. The treatment on these wheels melts under the heat of grinding action and carries away metal material, which might otherwise stick to and clog the wheel grinding surface. Treated wheels are used chiefly for grinding aluminum, brass, plastics, or similar soft materials.

The impregnated cotton fiber reinforced wheel (muslin is one form of cotton fiber) is especially suitable where fast metal removal is not a prime concern. These wheels provide good action and finishing characteristics on the deburring and blending operations, and they are resilient. They are widely used in the aircraft industry for grinding aluminum and similar soft materials because of their ability to shed soft metal particles from the surface, which keeps tool surfaces clear and load-free. They are relatively soft, compared to the vitrified mounted points. These products are discussed in Chapter 17.

Hard rubber products are some of the most resistant products for shock and hard usage. They provide fine finishes because of the burnishing action of the rubber, and they are used frequently for deburring thin sheet steels. Because of their hardness, they are more resistant to wear than soft rubber wheels. These are used for small- and medium-size burrs. Large burrs will tear or cut the tools quickly. Chapter 18 discusses abrasive-filled rubber tools.

Soft rubber polishing wheels provide a slight degree of resiliency that results in excellent finishes. The wheels are used primarily for polishing operations in which almost no metal removal is required. They are frequently used after a mounted point, bur ball, or knife. The wheels will not grind out deep scratches, but they will polish out minor tool marks. In addition to deburring, this product has been used successfully for weld polishing of stainless-steel components.

TYPES OF MOUNTED SHAPES

Figures 13-1 through 13-3 illustrate the predominant shapes and sizes of mounted stone material (the code letters beneath each tool represent an industry standard designation for each size and shape). "A" shapes are primarily for offhand grinding, "B" shapes are used for light deburring, and "W" shapes for offhand and precision grinding for medium to heavy stock removal. Offhand grinding, as used here, means removing stock from the surface of the part, rather than just the edges. Despite the designation of A, B, or W, all of these tools can be and have been used for deburring.

Figures 13-4 and 13-5 illustrate some of the mounted stone shapes and sizes available in dental stones. As implied by the name, these stones are used to polish teeth or dentures. The photographs indicate that dental stones are typically smaller than commercial industrial stones.

Larger industrial tools and materials are also widely used in the casting industry to remove remnants of gates, runners, and flash. The open pore tools can be used for deflashing plastics, but mounted stones may not be the best tools for plastics.

TYPES OF MATERIALS IN MOUNTED STONES

Conventional mounted stones come in a variety of hardness and durability, and basically consist of two types of material:

1. Silicon carbide, a green or black abrasive, is found in many tools used on plastics. It has sharp edges that fracture, leaving even more sharp edges. Tools made from these materials will not glaze or clog, and they are exceptionally long-wearing in many situations.
2. Aluminum oxide particles are used for most mounted points. Aluminum oxide is inexpensive and long-lasting. It can be obtained in a wide variety of sizes and degrees of wear resistance.

Diamond particles brazed or plated onto shanks can also be considered mounted points. Particles were discussed with rotary burs in Chapter 10, but the similarity of diamond particles to silicon carbide and aluminum oxide allows them to also be categorized as mounted points.

Many manufacturers color-code their industrial stones so that users can quickly determine which tool is applicable to a specific situation. For example, a 120-grit aluminum oxide blue stone is designed for use on hard steel in a hard-bond fast-cutting tool. A red or reddish-brown color code indicates that the stone is 60-, 80- (medium), or 100-grit (fine) and should be used on softer steels or metals. For this application, the stones will wear rather quickly. Some companies distribute for several manufacturers, with each using various color codes. As a result, red and blue color codes may imply a specific amount of aggressiveness for one manufacturer, but something entirely different from another.

HARDNESS OF MOUNTED STONES

There is an established classification system to define the degrees of hardness and aggressiveness of mounted stones. Three basic factors influence the hardness and aggressiveness of stones:

Figure 13-1. Coarse mounted points. (Courtesy Abrasive Industries)

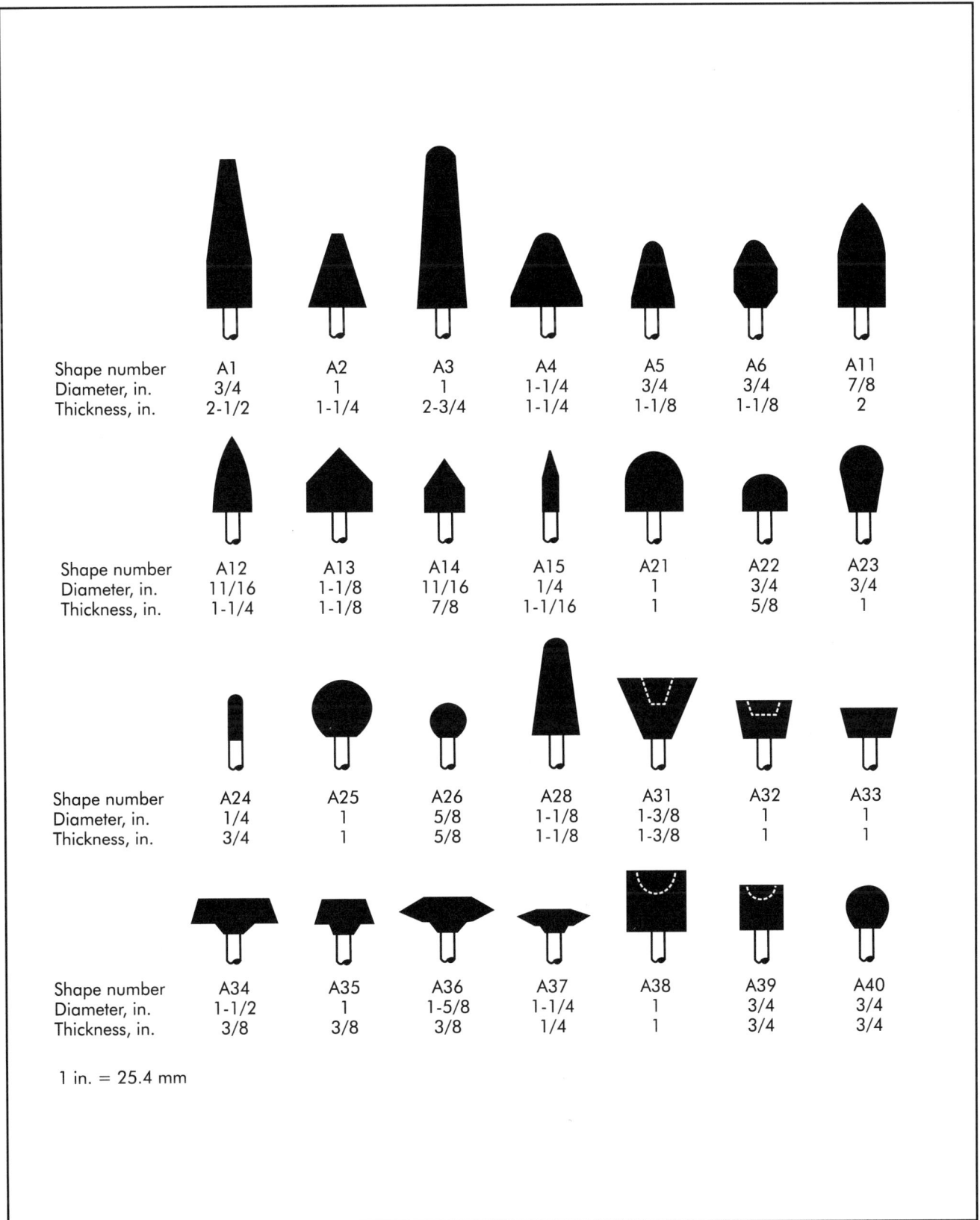

Figure 13-2. Standard shapes and sizes of "A" configuration mounted points. (Courtesy Abrasive Industries)

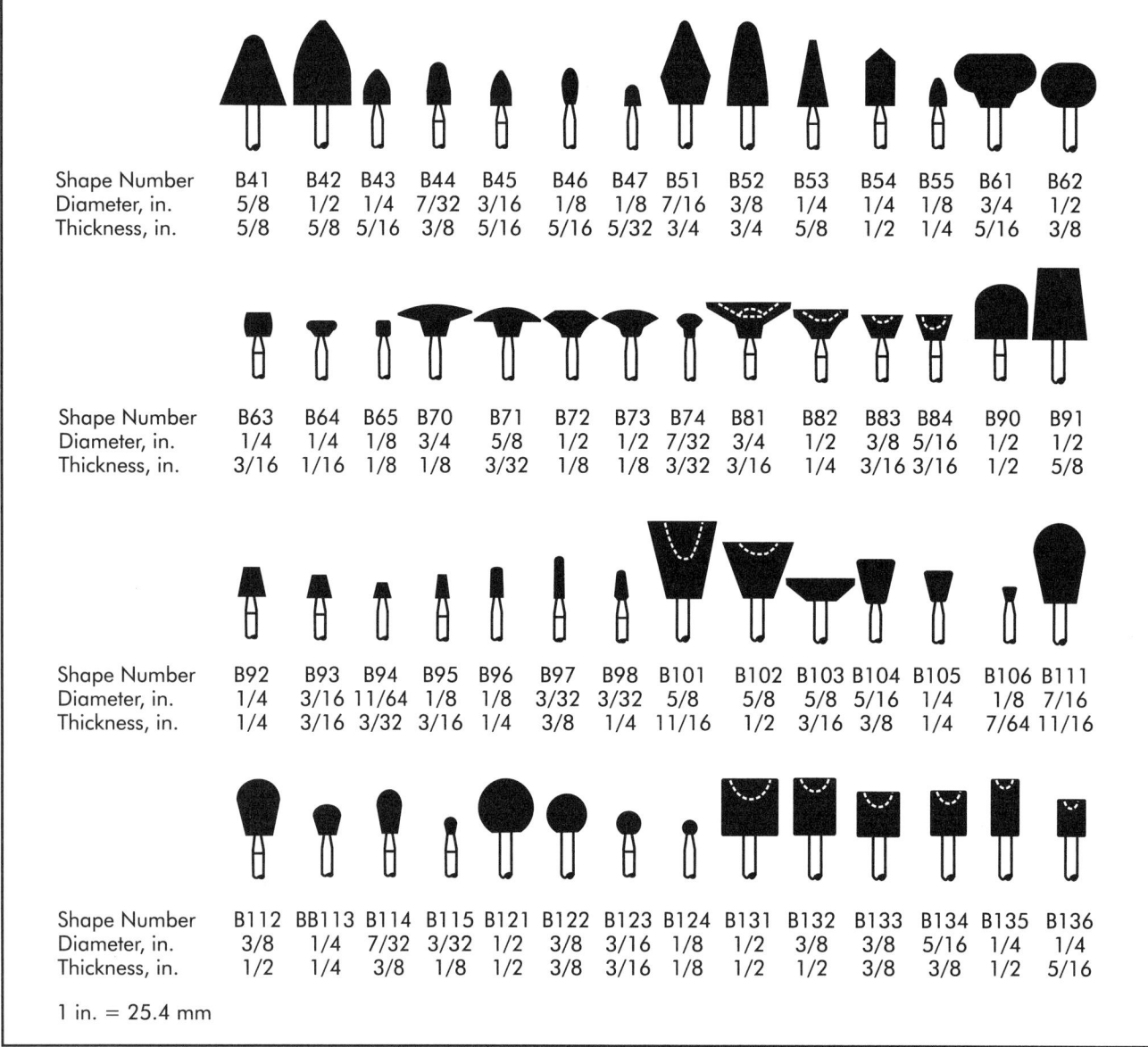

Figure 13-3. Standard shapes and sizes of "B" configuration mounted points. (Courtesy Abrasive Industries)

1. the size of abrasive particles bonded together to make the stone,
2. the type of bonding, and
3. the way in which the particles were manufactured.

Many industrial stones use a 120-grit particle bonded with a shellac-like resin. Others range from 60- to 600-grit, the higher number indicating smaller abrasive particles.

The second factor in the degree of stone aggressiveness is the type of bonding used to hold the particles together. It is possible to make a stone very porous, even though it may have 120-grit particles. A similar effect can be observed in the kitchen when a cake is made. Using the same cake mix, one can end up hard and rock-like while another is spongy and light. For aggressive cutting, a porous surface that will take large, discarded metal particles into the pores is desirable. For finishing however, a very tight, nonporous surface is desirable, because smooth, or nonporous surfaces, produce better finishes in this application. Only very small particles will be removed, so only small pores are required.

The third factor in the hardness and aggressiveness of stones depends on how particles are made,

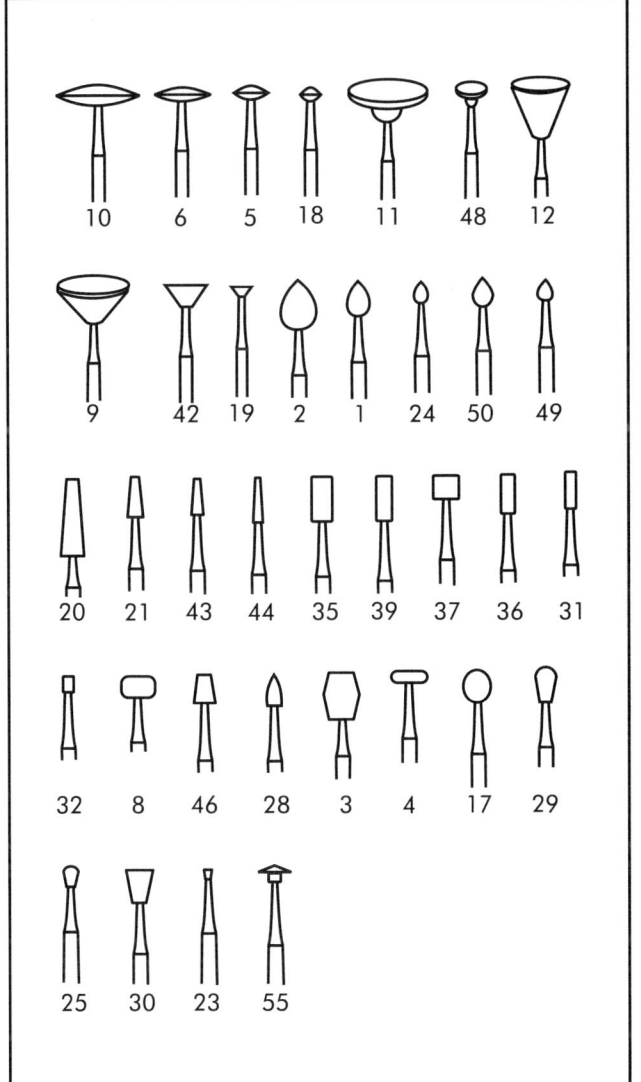

Figure 13-4. Dental mounted points.

typically in furnaces at a variety of temperatures and cooling rates. The matter in which these particles are made determines their individual aggressiveness. Unfortunately, many manufacturers or distributors do not indicate the full technical nomenclature for the characteristics of their mounted stones. Thus, selecting them for relative aggressiveness of stone life is difficult.

One manufacturer of dental stones produces a wide variety of stone-like materials. The color codes used on white dental stones indicate a soft, fast-wearing tool that provides an excellent finish and does not damage the workpiece, if properly handled. The same stone in green is a porous, fast-cutting tool that can quickly remove enough metal to take parts out of tolerance if it is not handled carefully.

The same manufacturer makes slightly larger stones in coral (brown) and pink. Pink stones are used for finishing and have a slightly different composition than the green ones. Coral stones are aggressive tools used to remove material quickly on surfaces. These tools are now widely used for deburring and edge finishing hand-size and larger parts. The white stones are used extensively for deburring and edge finishing miniature parts.

APPLICATIONS FOR MOUNTED STONES

Because of the wide variety of shapes and sizes available, defining the full capabilities and applications for mounted stones is difficult. The aggressive stones provide quick deburring of large burrs on parts having no close tolerance features. Miniature white stones remove small burrs and provide small edge breaks relatively quickly, even on the smallest parts. These are now being used with significant timesaving on many precision parts.

These tools are often used to scrape on the surface instead of on the edge, so they render a different surface texture than what was originally on the part. In many cases, however, the surface texture is only different—it is not necessarily a lesser finish. For example, with the white stones, a 16 μin. (0.41 μm) surface finish can be maintained even though the surface appears scuffed. As with rotary bur tools, they must be continuously moved over an edge or they leave a deep gouge. Tables 13-1 and 13-2 illustrate the characteristics of the mounted stones most commonly used for deburring miniature hand-size parts at one plant.

ENVIRONMENTAL, HEALTH, AND SAFETY ISSUES

Miniature grinding wheels must be accorded the same safety precautions as those used on large wheels. In general, they require two things:

1. never use them if they are cracked or greatly chipped, and
2. never use them at speeds higher than indicated in Table 13-3.

Rotated at high speeds, any object wants to throw itself apart. Because these grinding wheels are made of small particles that are essentially glued together, it is easier for them to come apart than for a solid metal piece. High speeds tremendously accelerate this tendency to come apart. For that reason, speeds should never exceed those shown in Table 13-3. Most mounted points are only 0.25-in. (6.4-mm) diameter or smaller, and large tools do not operate above 50,000 rpm. Allowable speeds vary with these conditions:

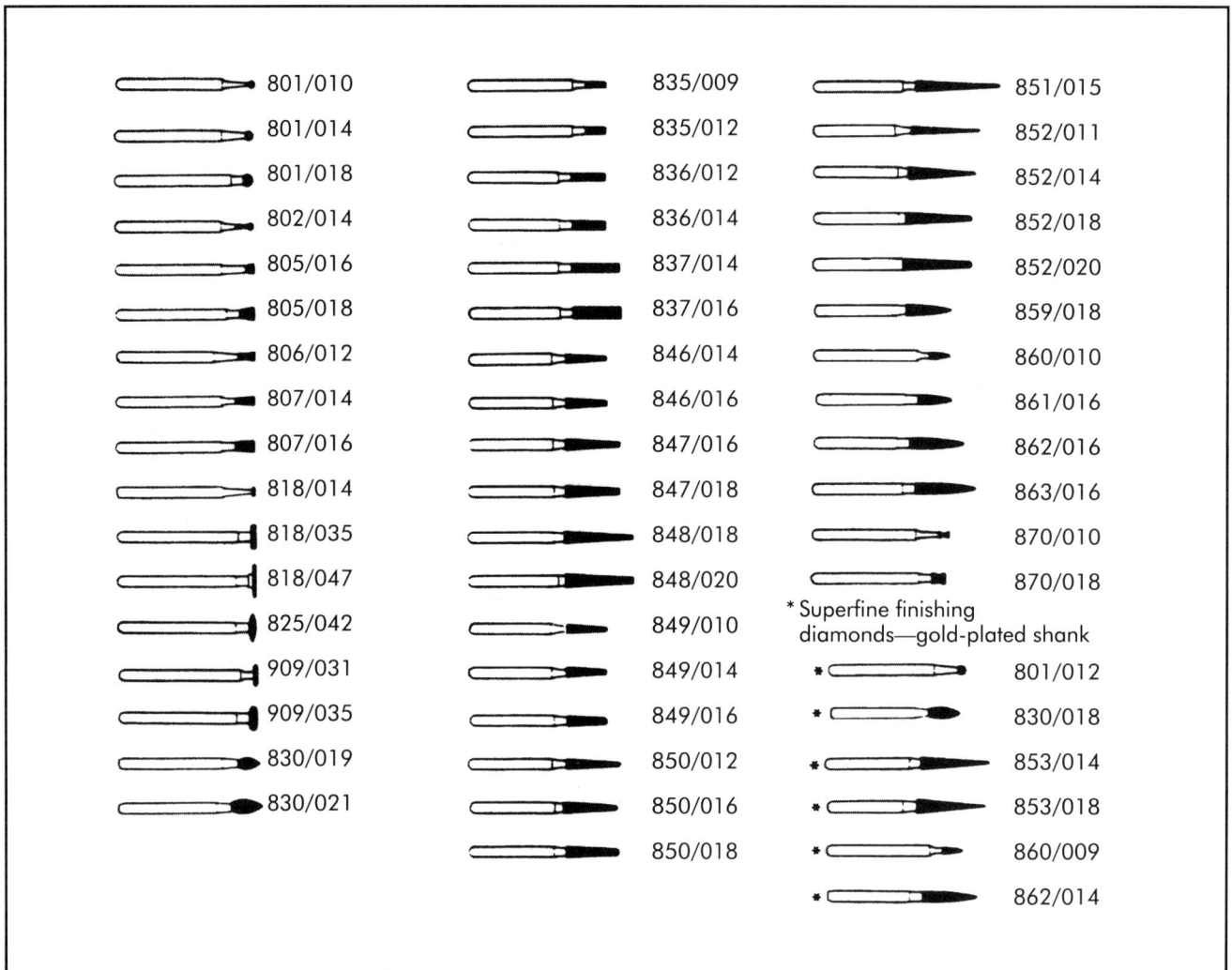

Figure 13-5. Mounted dental stones.

- wheel dimension,
- spindle diameter,
- overhang—distance from support to base of wheel, and
- wheel specification.

Some other safety considerations include:

a. Increasing the size of the wheel, either in diameter or thickness, generally reduces the maximum allowable speed. Reducing the size of the wheel, either in diameter or thickness, generally raises the maximum speed allowable.
b. Increasing the diameter of the spindle raises the maximum allowable speed. Reducing the spindle diameter reduces the maximum speed.
c. Increasing the overhang (the distance from end of support to base of wheel) reduces the maximum allowable speed. Reducing the overhang (pushing the spindle back into the chuck) raises the maximum allowable speed.

The tendency to fly apart is tremendously increased in cracked or chipped mounted points or stones. For this reason, do not try to get extra life out of these tools. Throw away cracked tools immediately, and do not allow others to use them.

BIBLIOGRAPHY

Abrasives Industries, 1991. *Grinding Wheel Application and Specification Manual*. Westboro, MA: Abrasive Industries.

Gitzen, Walter H. 1970. *Alumina as a Ceramic Material*. Columbus, OH: American Ceramic Society.

Weill, Gerald F. 1950. *Industrial Polishing of Metals*. London, England: Iliffe and Sons, Ltd.

Table 13-1. Characteristics of relatively aggressive mounted points typically used for hand-size or larger parts

Shape	Sizes in.	mm	Spindle Diameter in.	mm
A1	3/4 × 2-1/2	19.05 × 63.5	1/4	6.35
A3	1 × 2-3/4	25.4 × 69.85	1/4	6.35
A4	1-1/4 × 1-1/4	31.75 × 31.75	1/4	6.35
A5	3/4 × 1-1/8	19.05 × 28.56	1/4	6.35
A11	7/8 × 2	22.23 × 50.8	1/4	6.35
A12	11/16 × 1-1/4	17.46 × 31.75	1/4	6.35
A13	1-1/8 × 1-1/8	28.56 × 28.56	1/4	6.35
A14	11/16 × 7/8	17.64 × 22.23	1/4	6.35
A15	1/4 × 1-1/16	6.35 × 26.99	1/4	6.35
A21	1 × 1	25.4 × 25.4	1/4	6.35
A23	3/4 × 1	19.05 × 25.4	1/4	6.35
A24	1/4 × 3/4	6.35 × 19.05	1/4	6.35
A25	1 ball	25.4	1/4	6.35
A26	5/8 ball	15.88	1/4	6.35
A31	1-3/8 × 1	34.93 × 25.4	1/4	6.35
A32	1 × 5/8	25.4 × 15.88	1/4	6.35
A34	1-1/2 × 3/8	38.1 × 9.53	1/4	6.35
A36	1-5/8 × 3/8	41.28 × 9.53	1/4	6.35
A37	1-1/4 × 1/4	31.75 × 6.35	1/4	6.35
A38	1 × 1	25.4 × 25.4	1/4	6.35
A39	3/4 × 3/4	19.05 × 19.05	1/4	6.35
A40	3/4 ball	19.05	1/4	6.35
B41	5/8 × 5/8	15.88 × 15.88	1/8	3.18
B42	1/2 × 3/4	12.7 × 19.05	1/8	3.18
B43	1/4 × 5/16	6.35 × 7.94	1/8	3.18
B44	7/32 × 3/8	5.57 × 9.53	1/8	3.18
B45	3/16 × 5/16	4.76 × 7.94	1/8	3.18
B46	1/8 × 5/16	3.175 × 7.94	1/8	3.18
B52	3/8 × 3/4	9.53 × 19.05	1/8	3.18
B53	5/16 × 5/8	7.94 × 15.88	1/8	3.18
B54	1/4 × 1/2	6.35 × 12.7	1/8	3.18
B55	1/8 × 1/4	3.175 × 6.35	1/8	3.18
B61	3/4 × 5/16	19.05 × 7.94	1/4	6.35
B73	1/2 × 3/32	12.7 × 2.38	1/8	3.18
B81	3/4 × 3/16	19.05 × 4.76	1/8	3.18
B82	1/2 × 1/4	12.7 × 6.35	1/8	3.18
B91	1/2 × 5/8	12.7 × 15.88	1/8	3.18
B92	1/4 × 1/4	6.35 × 6.35	1/8	3.18
B95	1/8 × 3/16	3.175 × 4.76	1/8	3.18
B96	1/8 × 1/4	3.175 × 6.35	1/8	3.18

Table 13-1. (continued)

Shape	Sizes in.	Sizes mm	Spindle Diameter in.	Spindle Diameter mm
B97	1/8 × 3/8	3.175 × 9.53	1/8	3.18
B98	3/32 × 1/4	2.38 × 6.35	1/8	3.18
B102	5/8 × 1/2	15.88 × 12.7	1/8	3.18
B103	5/8 × 3/16	15.88 × 4.76	1/8	3.18
B111	7/16 × 11/16	11.11 × 17.46	1/4	6.35
B121	1/2 ball	12.7	1/8	3.18
B122	3/8 ball	9.53	1/8	3.18
B123	3/16 ball	4.76	1/8	3.18
B124	1/8 ball	3.175	1/8	3.18
B131	1/2 × 1/2	12.7 × 12.7	1/8	3.18
B132	3/8 × 1/2	9.53 × 12.7	1/8	3.18
B133	3/8 × 3/8	9.53 × 9.53	1/8	3.18
B134	5/16 × 3/8	7.94 × 9.53	1/8	3.18
B135	1/4 × 1/2	6.35 × 12.7	1/8	3.18
B136	1/4 × 5/16	6.35 × 7.94	1/8	3.18
W141	3/32 × 5/32	2.38 × 3.97	1/8	3.18
W142	3/32 × 1/4	2.38 × 6.35	1/8	3.18
W144	1/8 × 1/4	3.175 × 6.35	1/8	3.18
W145	1/8 × 3/8	3.175 × 9.53	1/8	3.18
W146	1/8 × 1/2	3.175 × 12.7	1/8	3.18
W152	3/16 × 1/4	4.76 × 6.35	1/8	3.18
W153	3/16 × 3/8	4.76 × 9.53	1/8	3.18
W154	3/16 × 1/2	4.76 × 12.7	1/8	3.18
W160	1/4 × 1/4	6.35 × 6.35	1/8	3.18
W162	1/4 × 3/8	6.35 × 9.53	1/8	3.18
W163	1/4 × 1/2	6.35 × 12.7	1/8	3.18
W164	1/4 × 3/4	6.35 × 12.7	1/8	3.18
W169	5/16 × 3/8	7.94 × 9.53	1/8	3.18
W170	5/16 × 1/2	7.94 × 12.7	1/8	3.18
W175	3/8 × 3/8	9.53 × 9.53	1/8	3.18
W176	3/8 × 1/2	9.53 × 12.7	1/8	3.18
W177	3/8 × 3/4	9.53 × 19.05	1/8	3.18
W178	3/8 × 1	9.53 × 25.4	1/8	3.18
W182	1/2 × 1/8	12.7 × 3.175	1/8	3.18
W184	1/2 × 3/8	12.7 × 9.53	1/8	3.18
W185	1/2 × 1/2	12.7 × 12.7	1/8	3.18
W186	1/2 × 3/4	12.7 × 19.05	1/4	6.35
W187	1/2 × 1	12.7 × 25.4	1/8	3.18
W188	1/2 × 1-1/2	12.7 × 38.1	1/4	6.35
W189	1/2 × 2	12.7 × 50.8	1/4	6.35
W194	5/8 × 1/2	15.88 × 12.7	1/4	6.35

Table 13-1. (continued)

Shape	Sizes in.	mm	Spindle Diameter in.	mm
W196	5/8 × 1	15.88 × 25.4	1/4	6.35
W197	5/8 × 2	15.88 × 50.8	1/4	6.35
W200	3/4 × 1/8	19.05 × 3.175	1/8	3.18
W203	3/4 × 1/2	19.05 × 12.7	1/4	6.35
W204	3/4 × 3/4	19.05 × 19.05	1/4	6.35
W205	3/4 × 1	19.05 × 25.4	1/4	6.35
W206	3/4 × 1-1/4	19.05 × 31.75	1/4	6.35
W207	3/4 × 1-1/2	19.05 × 38.1	1/4	6.35
W208	3/4 × 2	19.05 × 50.8	1/4	6.35
W215	1 × 1/8	25.4 × 3.175	1/8	3.18
W216	1 × 1/4	25.4 × 6.35	1/8	3.18
W217	1 × 3/8	25.4 × 9.53	1/4	6.35
W218	1 × 1/2	25.4 × 12.7	1/4	6.35
W220	1 × 1	25.4 × 25.4	1/4	6.35
W221	1 × 1/2	25.4 × 12.7	1/4	6.35
W222	1 × 2	25.4 × 50.8	1/4	6.35
W225	1-1/4 × 1/4	31.75 × 6.35	1/4	6.35
W226	1-1/4 × 3/8	31.75 × 9.53	1/4	6.35
W227	1-1/4 × 1/2	31.75 × 12.7	1/4	6.35
W230	1-1/4 × 1-1/4	31.75 × 31.75	1/4	6.35
W235	1-1/2 × 1/4	38.1 × 6.35	1/4	6.35
W236	1-1/2 × 1/2	38.1 × 12.7	1/4	6.35
W237	1-1/2 × 1	38.1 × 25.4	1/4	6.35
W238	1-1/2 × 1-1/2	38.1 × 38.1	1/4	6.35
W242	2 × 1	50.8 × 25.4	1/4	6.35

Table 13-2. Characteristics of mounted dental points typically used on miniature parts

Shape	Maximum Diameter in.	mm	Stone Length in.	mm	Shank Diameter in.	mm
Flame	0.062	1.58	0.125	3.18	0.062	1.58
Cylinder	0.080	2.03	0.140	3.56	0.062	1.58
Tapered cone	0.080	2.03	0.250	6.35	0.062	1.58
Cylinder	0.100	2.54	0.175	4.45	0.062	1.58
Wheel	0.160	4.06	0.080	2.03	0.062	1.58
Inverted cone	0.130	3.30	0.100	2.54	0.062	1.58
Flat cone	0.160	4.06	0.060	1.52	0.062	1.58
Cylinder	0.100	2.54	0.175	4.45	0.062	1.58
Inverted cone	0.130	3.30	0.100	2.54	0.062	1.58
Tapered cone	0.080	2.03	0.250	6.35	0.062	1.58
Wheel	0.160	4.06	0.080	2.03	0.062	1.58

The overall length of some of these stones is 0.750 in. (19.05 mm). (See Figure 10-17 for dimensions of other dental shank tools.)

Table 13-3. Maximum operating speed for miniature mounted points. (See manufacturer's recommended speeds for each tool and material.)

Wheel Diameter in.	mm	Shank Diameter in.	mm	Maximum Operating Speed* (rpm)
0.094	2.39	0.094	2.39	93,750
0.125	3.18	0.094	2.39	93,750
0.250	6.35	0.094	2.39	93,750
0.375	9.53	0.094	2.39	41,250
0.094	2.39	0.125	3.18	105,000
0.125	3.18	0.125	3.18	105,000
0.250	6.35	0.125	3.18	81,370
0.375	9.53	0.125	3.18	54,000
0.500	12.70	0.125	3.18	34,500

* Assumes that overhang length is 0.50 in. (12.7 mm) or less and wheel thickness is equal to diameter.

Hand Stones 14

Hand stones are normally referred to as *stones*. For some users, the same term also refers to grinding wheels, but it is used here to refer to hand stones or bench stones. Some companies also refer to hand stones as files, abrasive files, stone files, or abrasive stones. Stones come in a wide variety of shapes and sizes, as illustrated in Figure 14-1. They are classified by material, grit, shape, size, or application.

HAND-STONE COMPOSITION

Like grinding wheels, hand stones come in a variety of materials and structures. For example, hand stones can be made of silicon carbide (frequently abbreviated as SiC), aluminum oxide (frequently abbreviated as AlO, even though the correct formulation is Al_2O_3), or borozon (a synthetic material). Borozon (cubic boron nitride) is a tough, long-lasting manmade abrasive.

Stones are available with diamond particles in or on them. Other common minerals also can be used for stones, such as emery, garnet, quartz, and pumice. For general use, aluminum oxide and silicon carbide are by far the most commonly used materials.

The listing later in this chapter, however, does not adequately classify the materials found in stones. For example, some stones are made from corundum, a naturally occurring form of aluminum oxide that contains an impure form of ruby and sapphire gems in large crystalline form. Some manufacturers contend that this is better than synthetic aluminum oxides, which are available in degrees of toughness. Toughness and sharpness of many aluminum oxide stones are determined by additives placed in with the aluminum oxide before it is placed in a furnace, and by selecting specific cooling rates. Levigated alumina is a very fine, soft form of aluminum oxide.

Even silicon carbide, a synthetic product, comes in two forms: gray (or black) and green. The green material is a pure form and slightly harder than black silicon carbide. The actual difference between them is small, but users have found noticeable variations when in use. The synthetic stones consist of small particles of abrasive that are bonded with another material, frequently much like shellac. Because they are bonded, properties of the stone can be altered by either using larger particles or varying the amount of open spaces between each particle.

Nomenclature

In common usage, stones are designated by other nomenclature. The following list describes the nomenclature used to define the makeup of some stones:

Figure 14-1. Hand-stone variety. (Courtesy Norton Abrasives)

- aluminum oxide,
- aluminum oxide resin-bonded,
- white aluminum oxide,
- regular aluminum oxide,
- levigated aluminum oxide,
- silicon carbide,
- boron carbide,
- green silicon carbide,
- borozon,
- diamond,
- Arkansas,
- novaculite (silicon quartz),
- India,
- ruby, and
- ceramic fiber.

Common names include:

- electrical discharge machining (EDM) stone;
- oil-treated stone;
- mold-maker finishing stone;
- die-maker stone;
- tool dressing stick;
- resinoid stick;
- rough-out stone, and
- white polishing stone.

Aluminum Oxide

Probably the majority of synthetic stones are made from aluminum oxide. It comes in white, orange-brown, and other colors.

Silicon Carbide

Silicon carbide is the choice where speed of sharpening, rather than the fineness of the cutting edge, is most important. It is available in coarse, medium, and fine grits. One company calls its silicon carbide stones Crystolon®. They are produced by electric-furnace processes and result in an abrasive, gray tool. The fast-cutting abrasive is harder than any natural abrasive, except diamond, and does an outstanding job where moderate tolerances are desired.

Boron Carbide

Boron carbide is the hardest synthetic material. It is made into stick-form stones. This material is particularly recommended for cutting tungsten carbide and oxide ceramics. Stones of boron carbide are only available in a few sizes and shapes.

Diamond

Diamond stones are not usually beneficial in stoning applications. The use of diamond-plated or brazed files is discussed more fully in Chapter 12. Diamond lapping compound is also used in mold making, but it has few uses in deburring and edge finishing.

Arkansas

Arkansas stones are hard, natural stones often used when a fine finish surface is required and burrs are relatively small. Users can find "soft," "hard select," and "hard" Arkansas stones. The soft version is the least dense of the natural abrasive stones and occurs as extra-fine grit. It is often an off-white color. Arkansas stones are made from quarried novaculite (silicon quartz).

Hard select Arkansas is denser than the soft variety, has a super-fine grit, and is off-white. Hard Arkansas is very dense and can be described as an ultra-fine abrasive. Typically, it is white.

Arkansas natural abrasive stones are white to black in color. Hard Arkansas (ultra-fine grit), with its very dense, close construction, is recommended for final honing. It is used for sharpening tools requiring the sharpest precision edges possible. Soft Arkansas (super-fine grit), with its less dense and more open construction, is ideal for producing final finishes.

India

India stones are very soft and wear quickly when used for deburring and polishing. They will not leave as fine a finish as the Arkansas stone, although either stone should be able to provide a 10 µin. (0.254 µm) surface or better. All India stones have an orange-brown color.

The most noticeable difference between India and Arkansas stones is color. India stones are tan or orange; Arkansas stones are grayish white in most instances (one company sells a black Arkansas stone).

India stones are made of aluminum oxide electric-furnace abrasive. This abrasive is preferred for producing exceptionally keen, long-lasting edges, and for high-quality steel tool work. Generally, they are used where close tolerances and smooth cutting edges are required. This material is available in coarse, medium, and fine grits.

Ruby

Several suppliers provide ruby stones. They are made from a bond-free polycrystalline ruby. No binding agents are used. They have a hardness of nine on the Moh scale. Their very fine textures make them excellent for precision work. They are available in the following grades: very fine, fine, and medium. Porous and coarser grades are also available.

Ceramic Fiber

Fine-grit finishing stones are available in composite material that contains ceramic fibers in a very hard thermoset resin.

STONE BONDS

Synthetic stones are bonded with a variety of agents that provide different cutting characteristics. Resinoid stones are made from grit that is bonded with a resin. These tend to have a soft, resilient bond. They are used for light deburring and on difficult materials, such as stainless steel. They consist of laminates of cotton impregnated with aluminum-oxide abrasive. Vitrified stones have a harder bond, such as that found in most grinding wheels. They can be made of silicon carbide, aluminum oxide, or any other abrasive material.

GRIT SIZE

Table 14-1 illustrates that commonly used pencil stone kits (which are described later) consist of a wide variety of different grit sizes. Some stones are made of 90-grit material and others a fine particle size of 900. Table 14-2 indicates relationships between grit size and average particle size. As shown, 90-grit has an average particle size of 0.0057 in. (0.148 mm), about twice the diameter of human hair. In contrast, 800-grit has a particle size of approximately 0.00044 in. (0.0112 mm), approximately one-tenth the size of a human hair.

One manufacturer identifies the following stone grits:

- coarse (100 grit),
- medium (150–240 grit),
- fine (280–320 grit),

Table 14-1. Content of pencil-stone kit

Manufacturer's Description	Description	Grit
2	Diemaker semi-hard	220
3	Diemaker semi-hard	320
4	Diemaker semi-hard	400
6	Diemaker semi-hard	600
8	Diemaker semi-hard	800
9	Diemaker semi-hard	900
10	Diemaker semi-hard	India medium
12	Diemaker semi-hard	Super fine
51	Oil-treated	120
52	Oil-treated	220
53	Oil-treated	320
61	Electrical discharge machining (EDM)	120
62	EDM	180
RA490	Resin-bonded	90
RA40	Resin-bonded	120
RA40 1/2	Resin-bonded	150
RA41	Resin-bonded	180
RA42	Resin-bonded	220
RA43	Resin-bonded	320
RA46	Resin-bonded	600

Table 14-2. Grit/size conversion

Grit	Average Size in. (mm)
8	0.0870 (2.210)
10	0.0730 (1.854)
12	0.0630 (1.600)
14	0.0530 (1.346)
16	0.0430 (1.092)
20	0.0370 (0.940)
24	0.0270 (0.686)
30	0.0220 (0.559)
36	0.0190 (0.483)
46	0.0140 (0.356)
54	0.0120 (0.305)
60	0.0100 (0.254)
70	0.0080 (0.203)
80	0.0065 (0.165)
90	0.0057 (0.145)
100	0.0048 (0.122)
120	0.0040 (0.102)
150	0.0035 (0.090)
180	0.0030 (0.076)
220	0.0025 (0.064)
240	0.0020 (0.051)
280	0.00169 (0.0429)
320	0.00137 (0.0348)
360	0.00109 (0.0277)
400	0.00088 (0.0224)
500	0.00073 (0.0185)
600	0.00057 (0.0145)
800	0.00044 (0.0112)
1000	0.00032 (0.0081)
1200	0.00022 (0.0056)

- extra fine (400–600 grit),
- super fine (700 grit), and
- ultra fine (800–900 grit).

BASIC SHAPES AND SIZES

In general, the smallest dimensions of most stones commercially available are 0.25 in. (6.4 mm), but many stones are much larger. Figures 14-2 and 14-3 illustrate some of the common small hand stones. Figure 14-3 shows some of the smallest ones. Significantly smaller than a paperclip, some of these stones are extremely useful on precision miniature parts. Table 14-3 illustrates the sizes and shapes of commonly available handheld stones. This is not a complete list but a reasonable one for the types of products most commonly used. Table 14-4 defines several commercial hand-stone shapes. Figure 14-4 illustrates some of the cross-sectional geometries, or configurations, which Table 14-3 references.

Pencil stones are 0.156–0.500 in. (3.96–12.70 mm) square stones used in pencil-like holders. Sets of pencil stones contain 36 or more different stones of the same size. One company provides almost 100 different grit sizes and types of small pencil stones. The sets include India, EDM, oil-soaked, and others (listed individually in Table 14-1), all 0.125 in. (3.18 mm) square, in a variety of hardness values and compositions, which permit production of rapid deburring and smooth finishes.

Figure 14-3 illustrates the sizes, shapes, and dimensions of some miniature stones commonly referred to as "ruby stones." These red stones are among the most practical of those used for precision miniature features. To further appreciate the minute size of these tools, the complete set of stones listed in the table

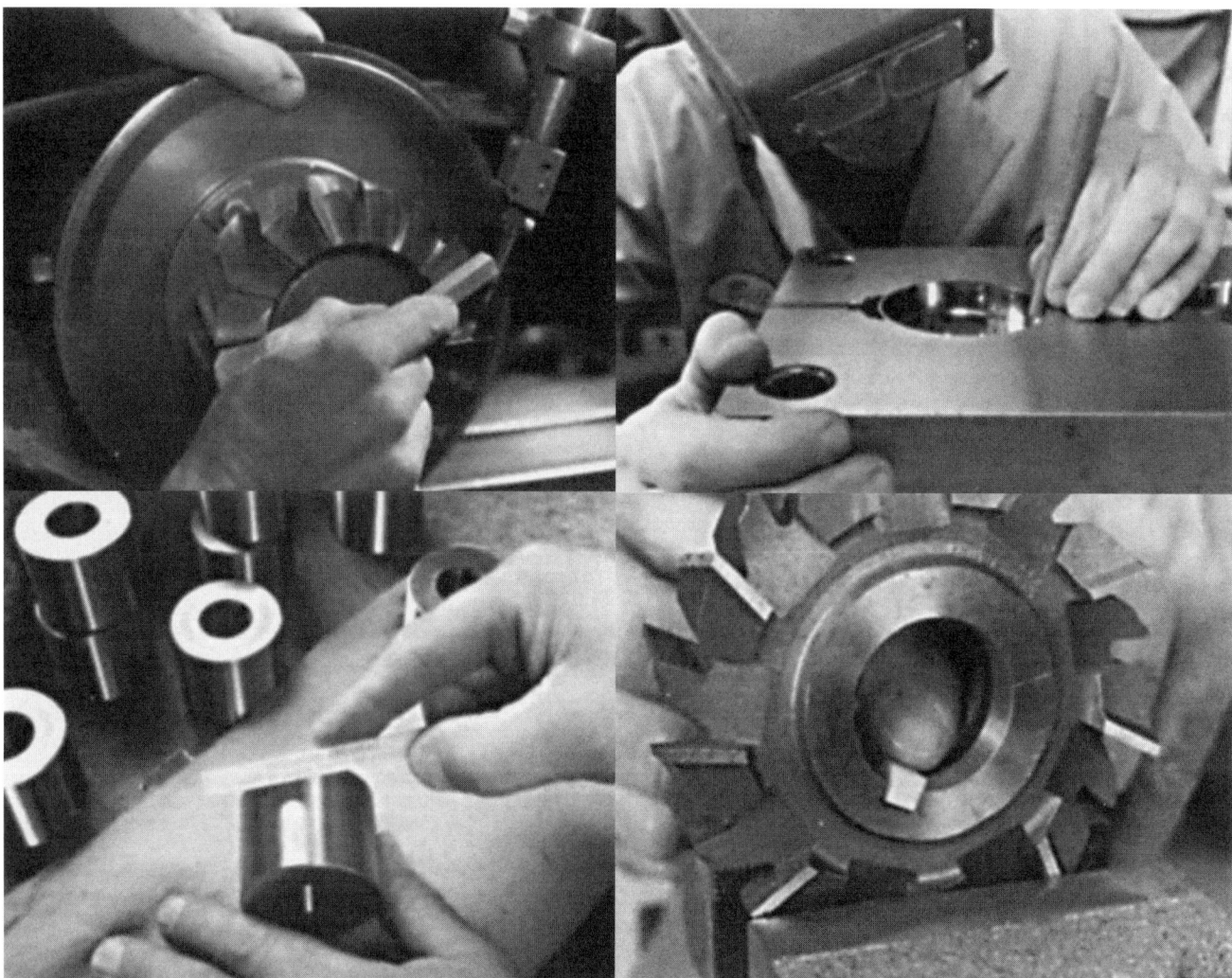

Figure 14-2. Small, handheld stones in use. (Courtesy Norton Abrasives)

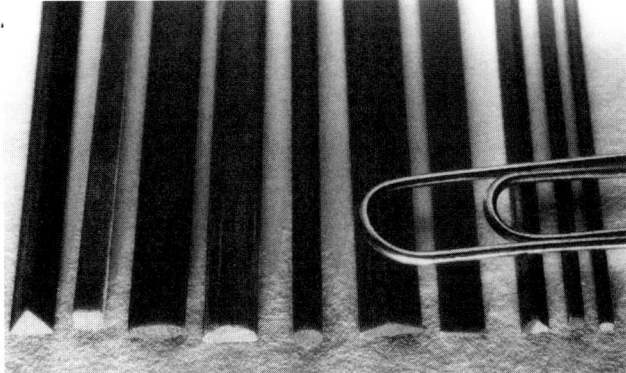

Figure 14-3. Midget ruby stone kit. (Courtesy Honeywell)

Table 14-3. Midget stone sizes and shapes

Shape	Available Size, in. (mm)
Square	0.047 × 2 (1.19 × 50.8)
Round	0.047 × 2 (1.19 × 50.8)
Triangular	0.094 × 2 (2.39 × 50.8)
Square	0.078 × 4 (1.98 × 101.6)
Flat	0.010 × 0.125 × 2 (0.254 × 3.18 × 50.8)
Half round	0.156 × 4 (3.96 × 101.6)
Barrette	0.188 × 0.063 × 4 (4.78 × 1.6 × 101.6)
Crossing	0.188 × 0.063 × 4 (4.78 × 1.6 × 101.6)

and the brass stone holder weighs only 1 oz (28 g). Table 14-5 provides the dimensions of some small ceramic-fiber stones.

BASIC USAGE

As with the use of any tool on precision parts, light pressure is necessary for successful finishing. Aggressive use can result in too large of an edge break and entrapment of large particles beneath the stone, which would scratch the workpiece. For general use, stones should be stored in oil to maintain necessary lubrication. On precision miniature parts, which are hard to hold, oil is seldom used. Figure 14-5 illustrates two typical hand-stone uses.

Stoning soft material, such as aluminum, is not recommended because the stones soon load or clog with aluminum and become glazed. When this happens, they are unsuitable for use.

Ultrasonic cleaners are frequently used to clean stones. The cleaner the stone, the better it will cut. In general, a very fine grit is desirable (about 320 or finer) for working on precision miniature parts. Extremely fine grit, such as 900, will not perform any significant cutting action; rather, it will merely remove the high spots on the surface. Aggressive work on big burrs will require a coarser grit and larger stones.

EDM stones are very hard—they were developed to remove hard surface scale left by the EDM process. Some of these are used with electronic reciprocating profilers.

Stoning Oil

Many users apply oil that is designed to prevent stone loading. Others use water to finish materials such as kirksite.

Some stones are sold "oil-soaked," implying that they are pretreated with oil. In larger applications, stoning oil is desirable to extend the life of the stone. Ruby stones can be used wet or dry. Reportedly, gasoline or paraffin oil can be used to remove oil from these stones and they can be cleaned in appropriate acids (American Rotary Tools Co. 1995).

ENVIRONMENTAL, HEALTH, AND SAFETY ISSUES

The stoning process, in general, has not raised ergonomic issues. Health issues could arise, however, if the user remains in the same position for hours and exerts heavy pressure, or if he or she breathes inappropriate stone cleaning solutions. As with all types of manual deburring, heavy pressure is a sign that the wrong tool is being used for the task. Several stone holders are sold that allow users to better grasp small stones. Several have a push-button release to allow stone changes (Figure 14-6).

REFERENCE

American Rotary Tools Co., Inc. 1995. *Catalog*. Port Washington, NY: American Rotary Tools Co., Inc., p. 25.

Table 14-4. Sizes and shapes of common handheld stones

Shape	Typical Minimum Size, in. (mm)	Typical Maximum Size, in. (mm)	Material
Square	0.25 × 0.25 × 4 (6.4 × 6.4 × 101.6)	1 × 1 × 6 (25.4 × 25.4 × 152.4)	SiC, India, Arkansas
Rectangular	0.25 × 1.50 × 4 (6.4 × 38.1 × 101.6)	1 × 2 × 8 (25.4 × 50.8 × 203.2)	SiC, India, Arkansas, AlO
Triangular	0.25 × 0.25 × 4 (6.4 × 6.4 × 101.6)	1 × 1 × 6 (25.4 × 25.4 × 152.4)	SiC, India
Tapered triangular	0.25 × 0.50 × 4 (6.4 × 12.7 × 101.6)		SiC, India
Round	0.25 × 4 (6.4 × 101.6)	1 × 6 (25.4 × 152.4)	SiC, India, Arkansas
Tapered round	0.25 × 0.50 × 4 (6.4 × 12.7 × 101.6)		SiC, India, Arkansas
Point stones	0.25 × 3 (6.4 × 76.2)		India, Arkansas
Half rounds	0.25 × 4 (6.4 × 101.6)	0.50 × 4 (12.7 × 101.6)	India
Tapered slips	0.25 × 4 × 1.75 × 1.5 (6.4 × 101.6 × 44.5 × 38.1)	0.375 × 0.75 × 2.25 (9.53 × 19.1 × 57.2)	AlO, SiC
Diamond files (diamond cross section)	0.1875 × 0.5625 × 4 (4.763 × 14.288 × 101.6)	0.1875 × 0.5625 × 4 (4.763 × 14.288 × 101.6)	India, Arkansas
Bevel files	0.125 × 0.375 × 3 (3.18 × 9.53 × 76.2)		Arkansas
Flat stones	0.0625 × 0.25 × 6 (1.588 × 6.4 × 152.4)	1 × 2 × 8 (25.4 × 50.8 × 203.2)	Arkansas, EDM, SiC, boron carbide
Oval stones	0.1875 × 0.50 × 3 (4.763 × 12.7 × 76.2)		Arkansas
Points	0.3125 × 3 (7.938 × 76.2)		India
Knife blades	0.125 × 1 × 4 (3.18 × 25.4 × 101.6)		India, Arkansas
Bench stones	4 × 1.75 × 0.625 (101.6 × 44.5 × 15.88)	8 × 2 × 1 (203.2 × 50.8 × 25.4)	SiC, India
Silversmiths' stones	0.0625 × 4 (1.588 × 101.6)	0.1875 × 4 (4.763 × 101.6)	India
Carving-tool slips	0.1875 × 0.875 × 2.25 (4.763 × 22.23 × 57.2)		India, Arkansas
Gouge sharpening stones	0.50 × 0.375 × 6 × 2 × 1 (12.7 × 9.53 × 152.4 × 50.8 × 25.4)		India
Reamer stones	0.1875 × 0.75 × 3.50 (4.763 × 19.1 × 88.9)	0.25 × 1 × 6 (6.4 × 25.4 × 152.4)	India
Grooved surgical stones	0.50 × 1.50 × 4 (12.7 × 38.1 × 101.6)		Arkansas
Engravers' pencils	0.125 square × 7 (3.18 square × 177.8)		India, Arkansas
Engravers' points	0.125 × 1 (3.18 × 25.4)		India, Arkansas
Utility stones	1 × 1 × 14 (25.4 × 25.4 × 355.6)		SiC
Dresser sticks	0.50 round × 6 (12.7 round × 152.4)	1 round × 6 (25.4 × 152.4)	SiC
Resinoid core files (stones)	0.50 × 1 × 10 (12.7 × 25.4 × 254.0)		SiC

SiC = silicon carbide
AlO = aluminum oxide
EDM = electrical discharge machining

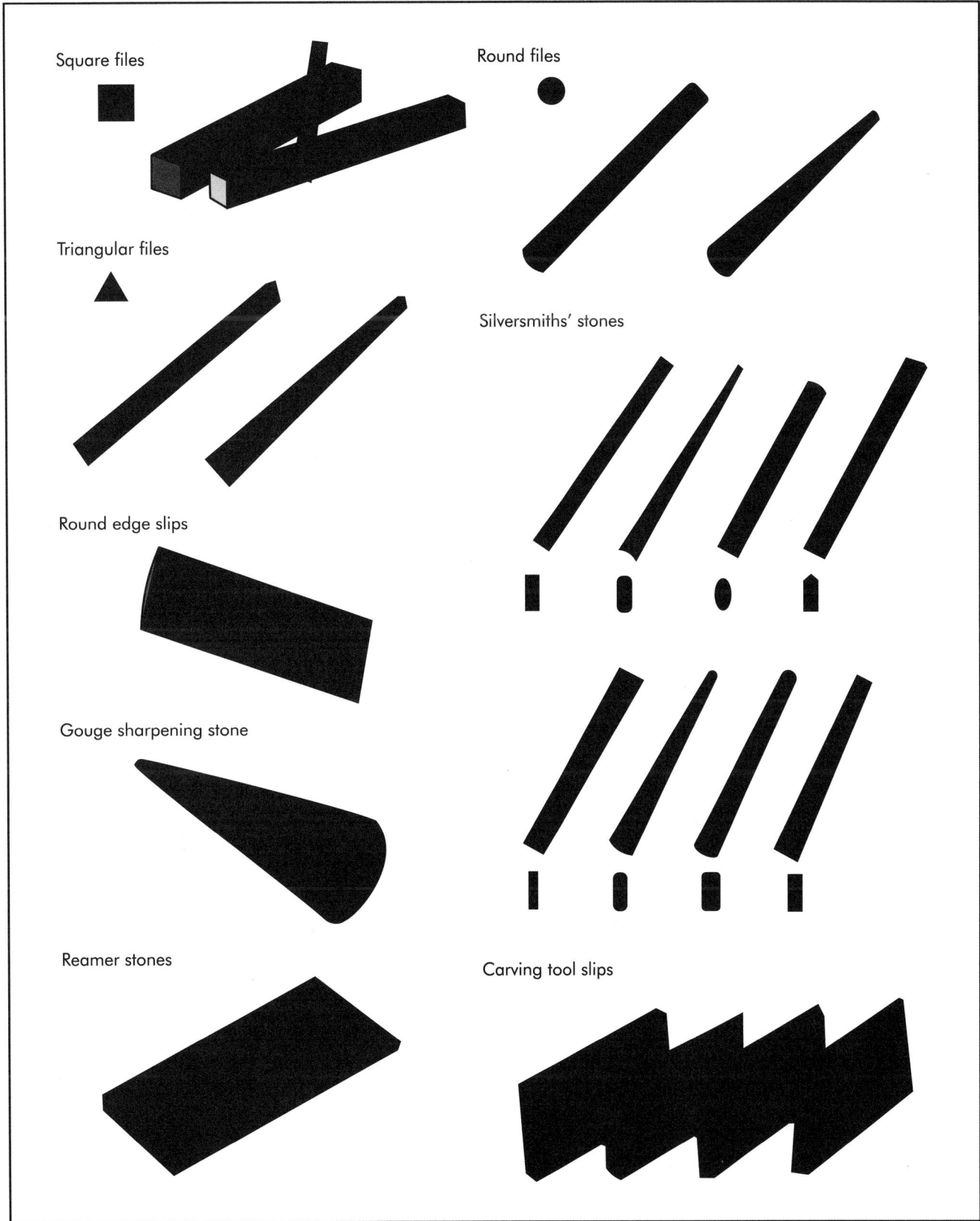

Figure 14-4. Common hand-stone geometries.

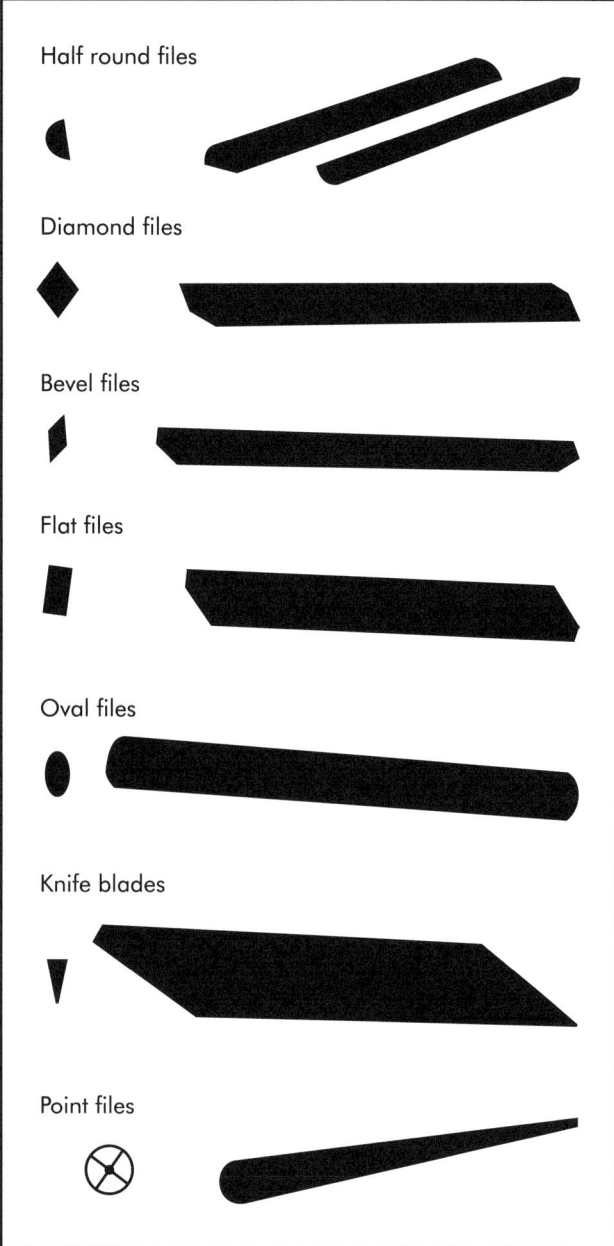

Figure 14-4. (continued).

Table 14-5. Ceramic-fiber stones available in 300, 400, 600, 800, 1000, and 1200 grit

Size, mm	Approximate Size, in.
0.50 × 4 × 100	0.020 × 0.157 × 3.94
0.50 × 6 × 100	0.020 × 0.236 × 3.94
1 × 2 × 100	0.039 × 0.079 × 3.94
1 × 4 × 100	0.039 × 0.157 × 3.94
1 × 6 × 100	0.039 × 0.236 × 3.94
1 × 10 × 100	0.039 × 0.394 × 3.94
1.5 × 4 × 100	0.059 × 0.157 × 3.94
1.5 × 6 × 100	0.059 × 0.236 × 3.94
1.5 × 10 × 100	0.059 × 0.394 × 3.94
3 × 6 × 100	0.118 × 0.236 × 3.94
3 × 6 × 30	0.118 × 0.236 × 1.18
3 round × 50	0.118 round × 1.97
3.2 round × 50	0.126 round × 1.97

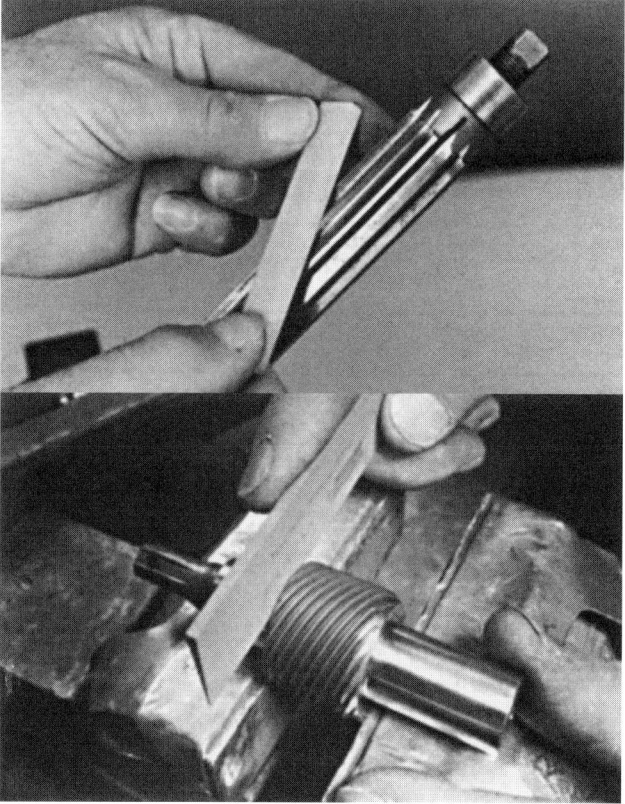

Figure 14-5. Flat file (stone) and knife blade (stone) in use. (Courtesy Norton Abrasives)

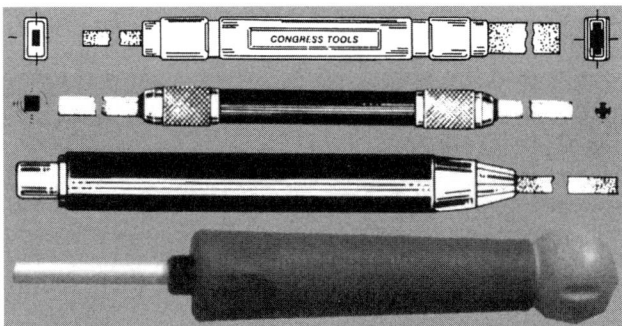

Figure 14-6. Stone holders for hand stoning. (Courtesy Congress Tools)

Abrasive Wood Tools 15

Wooden tools can be useful for the fine finishing of parts. Figure 15-1 illustrates some of the wooden tool shapes available for lapping and finishing. These tools are mounted to a 0.125-in. (3.18-mm) diameter shank and act like a soft mounted point. They are available in a hard dark wood or a soft white product. Typically, a 5.9 µin. (15 µm) or coarser diamond compound is used on hard forms and a compound smaller than 5.9 µin. (15 µm) is used with soft wood shapes. These can be used by hand or in rotary motors.

Figure 15-1. Set of wooden shapes. (Courtesy Gesswein)

Figure 15-2 illustrates some very hard laminated wood sticks. These are also used with diamond compounds. They all have blunt ends except for two—one with a pointed end and another with a beveled end.

WOODS

The woods used include:
- rock wood,
- white birch,
- balsa,
- poplar (used in the glass industry), and
- cork (described in Chapter 16).

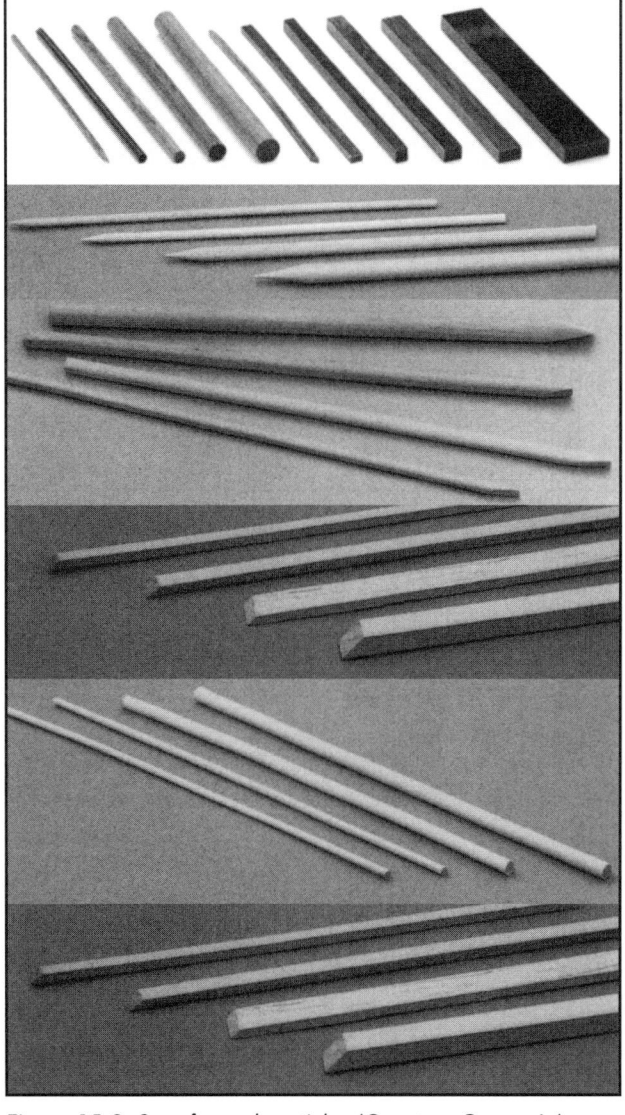

Figure 15-2. Set of wooden sticks. (Courtesy Gesswein)

Hand Deburring: Increasing Shop Productivity

COMMON USAGE

A general rule of thumb is that the softer the wood, the finer the finish obtained. Rock wood will cut the fastest of the woods. All can be dressed with a file or knife to fit the shape desired. Table 15-1 gives the sizes available of some common wooden tools. These tools would only be used for precision finishing, since lapping and the abrasiveness of the wood will only remove burrs on the order of tenths of a thousandth of an inch thick. While many users apply lapping compound, others use the bare wood tool.

Poplar-wood wheels are used to polish glass, but they are not available in the mounted-point configurations shown in Figure 15-1. They can be made from wedges glued together like wagon wheels. Pumice is used as an abrasive with this wood. One source notes that a wooden tool can be formed to the desired shape by using a sharp knife, woodworking lathe tools, or a piece of glass.

Wooden tools may be applicable to edges that contain plated gold or other unique materials. The wood has an abrasive action, but it is much less aggressive than stone or metal tools, and will not harm finishes unless they approach the 1–2 μin. (0.03–0.05 μm) level.

ENVIRONMENTAL, HEALTH, AND SAFETY ISSUES

Mounted tools all must be operated at safe rotary speeds. The manufacturer will provide recommendations. If abrasive compounds are used in conjunction

Table 15-1. Common wooden tools

Shape	Size, in. (mm)	Material		
		Rock Wood	Birch	Balsa
Round	0.080 (2.03)			Blunt end
	0.100 (2.54)			Blunt end
	0.120 (3.05)			Blunt end
	0.160 (4.06)			Blunt end
	0.240 (6.10)			Blunt end
Round	1/8 (3.18)	Blunt end	Pointed end or double beveled end	
	5/32 (3.97)		Pointed end or double beveled end	
	3/16 (4.76)	Blunt end	Pointed end or double beveled end	
	1/4 (6.35)	Blunt end	Pointed end or double beveled end	
	3/8 (9.53)	Blunt end		
	1/2 (12.70)	Blunt end		
Rectangular	1/8 × 1/8 (3.18 × 3.18)		Beveled end	
	3/16 × 3/16 (4.76 × 4.76)		Beveled end	
	1/8 × 1/4 (3.18 × 6.36)	Blunt end	Beveled end	
	1/4 × 1/4 (6.35 × 6.35)	Blunt end	Beveled end	
	1/4 × 3/8 (6.35 × 9.53)	Blunt end		
	1/4 × 1/2 (6.35 × 12.70)	Blunt end		
	5/16 × 1 (7.94 × 25.40)	Blunt end		
Cylinder on shank	1/4 × 5/8 (6.35 × 15.88)	X	X	
	3/8 × 3/4 (9.53 × 19.05)	X	X	
	1/2 × 3/4 (12.70 × 19.05)	X	X	
Short flame	3/8 × 5/8 (9.53 × 15.88)	X	X	
Long flame	3/8 × 1-1/4 (9.53 × 31.75)	X	X	
Ball	3/8 (9.53)	X	X	

with mounted tools, the compounds may fly off the wheels, so eye protection must be used.

Operations that involve glass finishing must consider the basic problem of glass sharpness and its inherent cutting ability if the operator slips and falls into the glass edge. If dust is produced, operators should wear protective facemasks.

Note that nickel dust is a suspected carcinogen and materials that contain nickel, such as stainless steel, may present the same nickel dust concerns as the nickel itself.

Abrasive Cork Tools 16

Chapter 15 provided examples of hardwood forms used for deburring and finishing. Cork is a wood product; however, it is discussed separately here because it is a unique material that may be overlooked if "hidden" in a chapter about wood.

WHY CORK?

Cork wheels can do work that may be difficult with other types of grinding wheels. They are made of a fine abrasive grain, a hard rubber bond, and a high concentration of small cork wood granules. The abrasive grain is evenly distributed throughout the wheel to promote consistent finishes.

Some cork wheels have a resin or epoxy bond. Others have a rubber bond. Rubber is best for resilience and smoothness. Other bonds work, but they do not polish like hard rubber-bond wheels. The combination of slightly resilient bond and unique cork material allows the abrasive grain to be cushioned during grinding. The effect is that the grain acts less aggressively on the work. This, plus a mild polishing action by the cork and bond combination, generates a bright finish of fine quality.

Depending on the choice of grain size, results vary from a mirror finish with virtually no removal to a very good finish with several thousandths of an inch removal. The actual results are determined by conditions, which vary among applications. Table 16-1 provides a guide to finishes.

The difficulty with fine finishing is that a very light cut produces little pressure on a grinding wheel. Wheels well-suited for heavy stock removal tend to glaze and load from light removal. The grinding pressure is insufficient to dislodge the flattened, worn abrasive grains. When the wheel glazes badly, it is ready to be dressed.

Microscopic analysis has shown that a wheel-burnished surface, although appearing bright and hav-

Table 16-1. Finish produced with various abrasive grains when using cork wheels

Abrasive Grain	Finish, μin. (μm) (Ra)	Maximum Stock Removal, in. (mm)
A120	16–20 (0.41–0.51)	0.002 (0.05)
A180	12–15 (0.31–0.38)	0.001 (0.03)
A240	8–11 (0.20–0.28)	0.0005 (0.013)
A320	5–7 (0.13–0.18)	0.0002 (0.005)
A500	2–4 (0.05–0.10)	0.0001 (0.003)

ing a low average microinch roughness (Ra) reading, may actually be unsatisfactory because the peaks have been forced into the valleys. Should the part then be chrome-plated, trapped chemicals and oxidation of folded but rough edges appear later in the job. Unplated parts have the same potential to stain and rust. Parts subjected to line wear may show premature surface degradation.

An advantage of the cork wheel over a harder grinding wheel is that the abrasive grains are not held after their useful cutting life. When grains dull and start to burnish, force dislodges them from the wheel surface. The finish results from fine cutting rather than burnishing. Additionally, cork wheels grind cooler.

Tube polishing or finishing other thin metals presents a special problem. Excessive wheel pressure tends to deflect a thin wall. A cork wheel, however, can cut with light pressure; therefore, it more easily generates proper dimensions and finish.

Stainless steel, because of its stringy chips, can cause troublesome wheel loading and workpiece galling. The cork wheel's free-cutting ability allows it to effectively polish stainless steel, and its low surface porosity prevents chip loading.

Aluminum and brass are even more difficult to polish than stainless steel. Their low tensile strength exerts little force on dull abrasive grain in a hard wheel. The wheel can become glazed and loaded. Rubber-bonded cork wheels have been used on both materials with excellent results.

CORK TOOL FORMS

Cork tools are difficult to find. They are or have been produced in:

- mounted-point shapes,
- centerless grinding wheels, and
- sanding belts.

Large flame shapes and round-end cylinders are two forms produced for manual mounted-point applications. With a 0.875 in. (22.23 mm) diameter and 1.5-in. (38-mm) long flame shape, users can attack many surfaces. The 0.375 in. (9.50 mm) cylinder enables usage on smaller features. Other shapes may be available. These shapes have a coarse appearance—cork particles are interspersed with both resin and resin voids. They can be formed to any shape with the use of a knife, lathe woodworking tool, or piece of glass.

One manufacturer produces large, centerless grinding wheels of cork materials. They can be used in an offhand mode with parts manually held against the wheels. Extra-large wheels are more typically used in a mechanized machine than in offhand operations.

Cork polishing belts, discussed in Chapter 19, provide some of the same advantages of cork, but in this instance, they can be applied to common belt sanding machines for a wide variety of applications.

TYPICAL APPLICATIONS

The growth of abrasive-filled cotton tools (described in Chapter 17) may have overtaken some of the previous applications for cork wheels on metal; but for glass, cork is still preferred. In the following list, note that cork is used widely for more delicate work, thin walls, and smooth finishes. In many applications, either material will produce similar results.

Some uses of cork include:

- hydraulic piston rods—pre- and post-chrome plating;
- tubing—0.010–10 in. (0.25–254 mm) in diameter;
- linear bearing bars;
- shock-absorber rods and struts;
- roller or needle bearings and races;
- automotive wrist pins, rocker-arm shafts, and valve stems;
- carbide rods and punches;
- Chrome, brass, and aluminum for decorative finish, and
- glass edges.

There is little literature about the use of cork or wood tools. The Internet is one source. For glass work, one must review both artistic glass work and commercial glass texts on glass cutting and polishing. Cork tools are not designed to remove large burrs; they are tools for precision and fine finishes.

OPERATING RECOMMENDATIONS

Cork wheels must be run wet. A cork wheel inadequately covered with coolant quickly loads, as evidenced by a noticeable black layer where it runs dry. When the wheel loads, the small particles of the workpiece fill (load) the gaps between the abrasive grains, resulting in a burnish rather than a cut on the surface. The type and concentration of coolant are important—too much oil loads cork wheels. Synthetics usually are best, but soluble oils are satisfactory if concentrations are 2%. Too much oil produces a thick black layer on the wheel face and the wheel will not cut properly. Heat, poor finishes, and erratic dimensions will result.

RESIN/RUBBER CORK COMBINATION WHEELS

A resin/rubber cork combination wheel consists of a resin section and a rubber cork section. The sections are separate wheels that mount on the spindle in a sandwich manner. For wide machines, there can be two resin sections and a rubber cork, or one resin and two rubber cork sections. The resin and cork sections are specially formulated to work together. They wear at the same rate to maintain, hour after hour, the same roughing and finishing performance.

ENVIRONMENTAL, HEALTH, AND SAFETY ISSUES

Mounted tools must be operated at safe rotary speeds, which the manufacturer provides. If abrasive compounds are used in conjunction with any cork tools, the compounds will fly off the wheels, so eye protection must be worn. Glass-finishing operations must consider the basic problem of glass sharpness and its inherent cutting ability to minimize the risk of an operator slipping and falling into the glass edge.

If dust is produced, operators should wear protective face masks. Note that nickel dust is a suspected carcinogen, and materials that contain nickel, such as stainless steel, may present the same nickel dust concerns as the nickel itself.

Abrasive Cotton-fiber Products 17

One of the newer materials gaining widespread use for deburring and blending surfaces is a soft wheel of cotton or related fibers impregnated with abrasives and a resin. These products, some of which are shown in Figure 17-1, are especially suitable for removing medium-size burrs from stainless steel or soft metals. Not only do they produce an aggressive deburring action, but when properly used, they do not damage surfaces. They will produce 16 μin. (0.4 μm) surface finishes and will make radii of up to 0.010 in. (0.25 mm) very quickly. In many applications, they are quicker than knives or other conventional deburring tools. They were developed for the aircraft industry, which requires fine finishes without burrs or sharp edges, so they are well-adapted to precision products.

RESIN-BONDED TOOLS

Resin-bonded tools are often identified by manufacturers with a "TX" or "G-Flex" prefix. Rubber- or latex-bonding tools are identified with letter codes from "FX" through "MX." Resin-bonded tools are used for coarser stock removal. For more aggressive cutting, additional abrasive material is added to the tool. Unlike grinding wheels, these tools do not load up or glaze over.

SIZES AND SHAPES

Table 17-1 describes available sizes and shapes of abrasive-filled cotton tools. The size codes are the same as those for mounted points (Table 13-1). At least two manufacturers make these products, but the tools are not necessarily identical in action. As Table 17-1 indicates, the smallest diameter wheel is 0.188 × 0.500 in. (4.78 × 12.70 mm). For a slightly smaller shape, put this tool on a lathe and either cut or grind the diam-

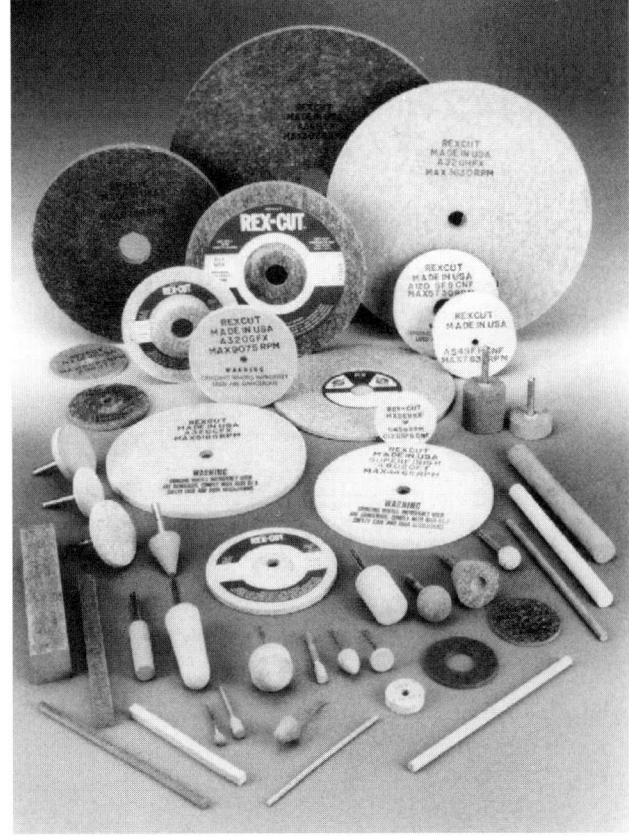

Figure 17-1. Abrasive cotton tools. (Courtesy Rex-Cut Products, Inc.)

eter. Wheels will adapt through use to a particular shape being deburred (they wear to match the contour).

In addition to the shapes identified in Table 17-1, materials can be produced in wheels of almost any diameter. One manufacturer lists 12-in. (304.8-mm)

Table 17-1. Commercially available abrasive-filled-cotton tools

Shape*	Size Code	Bond Hardness Grade Resinoid	Bond Hardness Grade Rubber	Size, in. (mm) Diameter	Size, in. (mm) Length	Availability of Bond and Grit** Resinoid	Availability of Bond and Grit** Rubber	Grit Size
Cone	A1			0.750 (19.05)	2.500 (63.50)	x	x	
Cone	A3			1.000 (25.40)	2.750 (69.85)	x	x	
Cone	A4			1.250 (31.75)	1.250 (31.75)	x	x	
Cone	A5			0.750 (19.05)	1.125 (28.58)	x	x	
Bullet	A11			0.875 (22.23)	2.000 (50.80)	x	x	
Bullet	A12			0.688 (17.48)	1.250 (31.75)	x	x	
Cone	A13			1.125 (28.58)	1.125 (28.58)	x	x	
Cone	A14			0.688 (17.48)	0.675 (17.15)	x	x	
Point	A15			0.250 (6.35)	1.063 (27.00)	x	x	
Round nose	A21			1.000 (25.40)	1.000 (25.40)	x	x	
Pear	A23			0.750 (19.05)	1.000 (25.40)	x	x	
Cylinder	A24			0.250 (6.35)	0.750 (19.05)	x	x	
Ball	A25			1.000 (25.40)	1.000 (25.40)	x	x	
Ball	A26			0.625 (15.88)	0.625 (15.88)	x	x	
Blunt cone	A34			1.500 (38.10)	0.375 (9.53)	x	x	
Blunt cone	A35	X		1.000 (25.40)	0.375 (9.53)	x	x	
Knife edge	A36	X		1.625 (41.28)	0.375 (9.53)	x	x	
Knife edge	A37	X		1.250 (31.75)	0.250 (6.35)	x	x	
Cone	B41	H, J	I–N	0.625 (15.88)	0.625 (15.88)	x	x	16–320
Bullet	B42	H, J	I–N	0.500 (12.70)	0.750 (19.05)	x	x	16–320
Cone	B51	H, J	I–N	0.434 (11.02)	0.750 (19.05)	x	x	16–320
Bullet	B52	H, J	I–N	0.375 (9.53)	0.750 (19.05)	x	x	24–320
Cone	B53	—	I–N	0.310 (7.87)	0.625 (15.88)	—	x	16–320
Flat ball	B61	G, H, I, J	I–N	0.750 (19.05)	0.310 (7.87)	x	x	16–320
Flat ball	B62	G, H, I, J	I–N	0.500 (12.70)	0.375 (9.53)	x	x	16–320
Cone	B91	H, J	I–N	0.500 (12.70)	0.625 (15.88)	x	x	24–320
Cone	B92	—	I–N	0.250 (6.35)	0.250 (6.35)	—	x	24–320
Inverted Cone	B104	H, J	I–N	0.310 (7.87)	0.375 (9.53)	x	x	16–320
Pear	B111	H, J	I–N	0.474 (12.04)	0.724 (18.39)	x	x	16–320
Pear	B112	H, J	I–N	0.375 (9.53)	0.500 (12.70)	x	x	16–320
Ball	B121	G, H, I, J	I–N	0.500 (12.70)	—	x	x	16–320
Ball	B122	G, H, I, J	I–N	0.375 (9.53)	—	x	x	16–320
Wheel	B131	G, H, I, J	I–N	0.500 (12.70)	0.500 (12.70)	x	x	16–320
Cylinder	B132	H, J	I–N	0.375 (9.53)	0.500 (12.70)	x	x	16–320
Cylinder	W160	H, J	I–N	0.250 (6.35)	0.250 (6.35)	x		16–320
Cylinder	W162	H, J	—	0.250 (6.35)	0.375 (9.53)	x		16–320
Cylinder	W163	H, J	—	0.250 (6.35)	0.500 (12.70)	x		16–320
Cylinder	W164	H, J	I–N	0.250 (6.35)	0.750 (19.05)	x		16–320
Cylinder	W175	G, H, I, J	I–N	0.375 (9.53)	0.375 (9.53)	x	x	16–320
Cylinder	W176	H, J	I–N	0.375 (9.53)	0.500 (12.70)	x	x	16–320
Cylinder	W177	H, J	I–N	0.375 (9.53)	0.750 (19.05)	x		16–320
Cylinder	W178	H, J	I–N	0.375 (9.53)	1.000 (25.40)	x	x	16–320

Table 17-1. (continued)

Shape*	Size Code	Bond Hardness Grade Resinoid	Bond Hardness Grade Rubber	Size, in. (mm) Diameter	Size, in. (mm) Length	Availability of Bond and Grit** Resinoid	Availability of Bond and Grit** Rubber	Grit Size
Cylinder	W179	H, J	I–N	0.375 (9.53)	1.250 (31.75)	x	x	16–320
	W183			0.500 (12.70)	0.250 (6.35)	x	x	
Cylinder	W184	G, H, I, J	I–N	0.500 (12.70)	0.375 (9.53)	x	x	16–320
Cylinder	W185	G, H, I, J	I–N	0.500 (12.70)	0.500 (12.70)	x	x	16–320
Cylinder	W186	H, J	I–N	0.500 (12.70)	0.750 (19.05)	x	x	16–320
Cylinder	W187	H, J	I–N	0.500 (12.70)	1.000 (25.40)	x	x	16–320
	W201	J–N	F–H			x	x	24–320
	W201	J–N	F–H			x	x	24–320
	W203	J–N	F–H			x	x	24–320
	W204	J–N	F–H			x	x	24–320
	W205	J–N	F–H			x	x	24–320
	W207	J–N	F–H			x	x	24–320
	W216	J–N	F–H			x	x	24–320
	W217	J–N	F–H			x	x	24–320
	W218	J–N	F–H			x	x	24–320
	W220	J–N	F–H			x	x	24–320
	W221	J–N	F–H			x	x	24–320
	W222	J–N	F–H			x	x	24–320
	W225	J–N	F–H			x	x	24–320
	W226	J–N	F–H			x	x	24–320
	W227	J–N	F–H			x	x	24–320
	W228	J–N	F–H			x	x	24–320
	W230	J–N	F–H			x	x	24–320
	W232	J–N	F–H			x	x	24–320
	W235	J–N	F–H			x	x	24–320
	W236	J–N	F–H			x	x	24–320
	W237	J–N	F–H			x	x	24–320
	W238	J–N	F–H			x	x	24–320
	W242	J–N	F–H			x	x	24–320
Flex disc				4.500–9.000 (114.30–228.60)		x	x	36–80
Miniature straight disc				1.0 (25.4)	0.063 or 0.125 (1.60 or 3.18)		x	80–320
Wheel				1.000–12.000 (25.40–304.80)	0.032–1.000 (0.81–25.40)	x	x	16–120
Sticks								
Round				0.188–0.500 (4.78–12.70)	4.000 (101.60)	x	x	54–320
Square				0.188–0.500 (4.78–12.70)	4.000 (101.60)	x	x	54–320
Dressing stone				2.500 (63.50)	1.500 × 0.500 (38.10 × 12.70) high			

* See Table 13-1 and figures 13-1 through 13-4 for details on shapes referenced in this table.
** An "x" indicates that the tool is available in the indicated bond. A "—" indicates that it is unavailable as a standard. Special shapes are easily made. Epoxy-bonded wheels are available up to 16 in. (406.4 mm) in diameter.

diameter wheels. For both types of bonds, a flexible disc is produced and an epoxy-bonded set of tools is used for grinding and finishing knives and razor blades, cutting tools, and jewelry.

PRODUCT MATERIALS

Table 17-1 lists the various abrasive degrees of materials used in cotton-fiber tools. One manufacturer supplies them in 24-, 36-, 54-, 80-, 120-, 180-, and 320-

grit particle sizes. Other manufacturers may offer different sizes. A look at one of these tools under a microscope shows that it is constructed of laminated layers of nonwoven muslin (cotton fiber). Aluminum oxide (or silicon carbide) grit and resin are blended together with the layers before the wheel is molded and cut into shape. Resin-bonded, mounted wheels cut harder, like a grinding wheel, and the rubber bonding cuts with a softer action. Where the lamination runs parallel to the shank, tool ends are prevented from chipping. Where laminations run perpendicular to the shank, heavy pressure is not recommended. Shank size is 0.125 in. (3.18 mm) in diameter.

USAGE

Cotton tools are used in a way similar to rotary mounted wheels, and require only light to moderate pressure to accomplish the deburring action. They may be operated up to 15,000 rpm because their small size does not represent a safety hazard at high speeds. If a particular tool is not aggressive enough, engineering or management should determine the availability of a more abrasive grit for that size of tool. A tool allowed to dwell at an edge for 5–10 seconds will produce a dished-out area (Figure 17-2); consequently, it is important to continuously move the tool back and forth over an edge.

Cotton tools, which are relatively soft compared to the hard stone-mounted points, are used on stainless steel, aluminum, and some of the more exotic metals. They are found in aerospace, aircraft, shipbuilding, gear, jewelry, tool and die, automotive, food processing, cutlery, and other industries.

Figure 17-3 illustrates the use of an abrasive cotton-fiber tool on a production part where the tool removes a heavy drilling burr left in a stainless-steel part. The tool is passed back and forth over the flat surface until all of the burr is removed. Normally, the tool is used over an edge, but because the tool is soft enough and provides an excellent finish, it will not damage the flat surface.

Figure 17-4 illustrates a common use for abrasive-cotton tools. Their small size enables users to reach into small areas easily. Chapter 18 compares these tools with rubber abrasive-filled tools. In general, however, abrasive-filled cotton tools are a bit more aggressive than abrasive-filled rubber tools. There are no firm rules for their use, but cotton tools will shed fewer particles than rubber tools. In rectangular or round bar form, cotton tools are used on lathes or other rotating machines. In this operation, they are held against the burr-laden edge in the same way one holds an abrasive-filled rubber product.

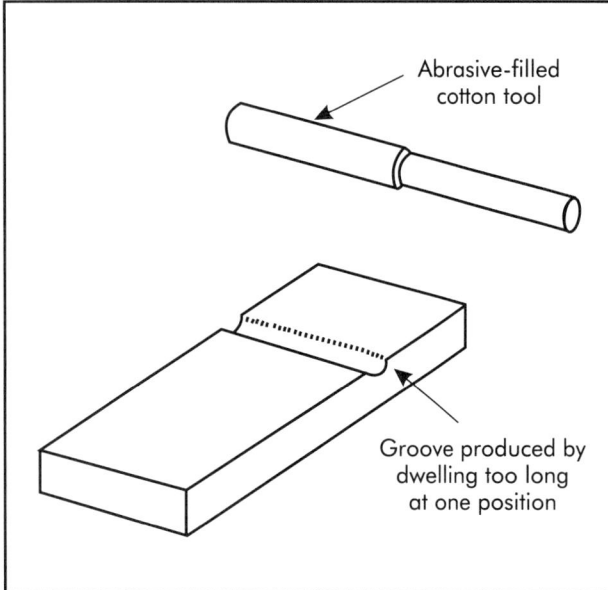

Figure 17-2. Dished-out area caused by dwelling at one spot too long. (Courtesy Honeywell)

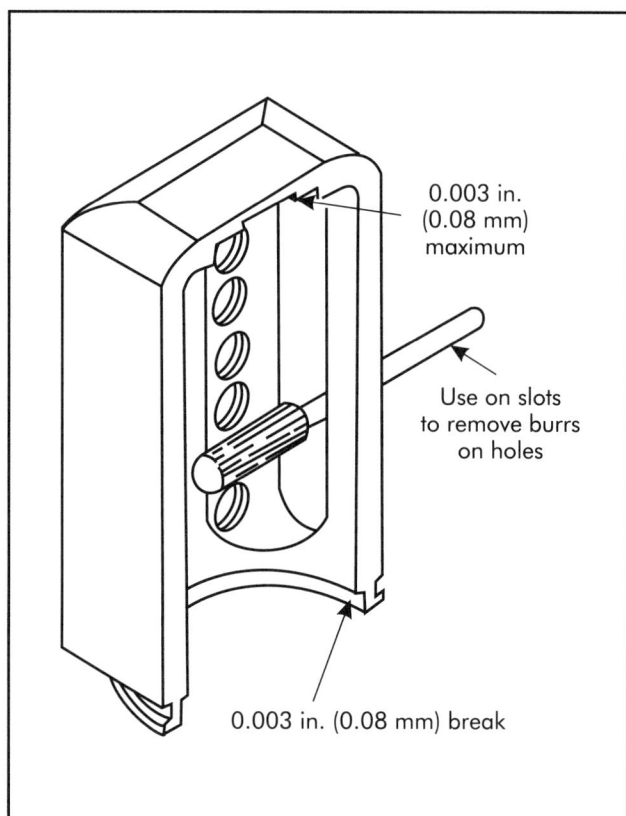

Figure 17-3. Production use of an abrasive cotton-fiber tool to remove drill burrs. (Courtesy Honeywell)

Figure 17-4. Abrasive-cotton tool finishing a gear. (Courtesy Rex-Cut Products, Inc.)

Abrasive Rubber Tools 18

Abrasive rubber tools are used on almost every part deburred at many companies. These tools consist of a rubber wheel, or shape, impregnated with particles of silicon carbide or aluminum oxide. This relatively soft tool does an excellent job removing very small burrs and providing a radius of up to 0.005 in. (0.13 mm). When used correctly, these tools neither damage precision surface finishes nor change the dimensions of parts, except at the edge on which they are used. Figure 18-1 illustrates some common commercial industrial abrasive-filled rubber tools. Figures 18-2 and 18-3 illustrate some of the smaller and gentler tools used in the dental industry.

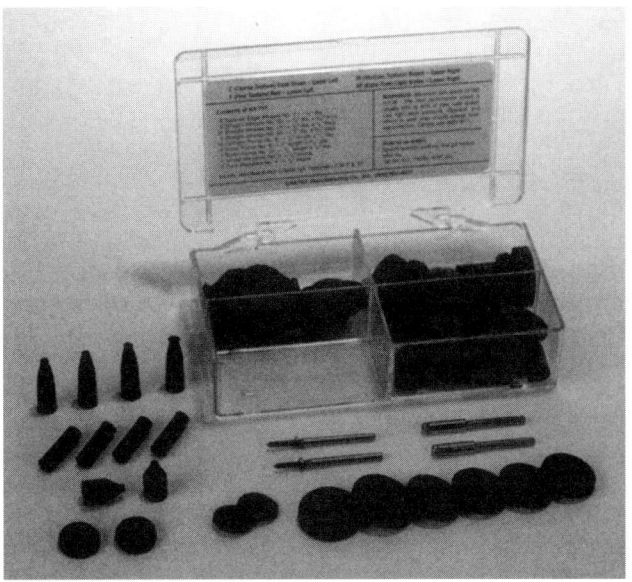

Figure 18-1. Typical abrasive-filled rubber tools. (Courtesy Cratex)

CHARACTERISTICS

Abrasive-filled rubber tools are color-coded by the manufacturer to indicate aggressiveness. In Figure 18-1, for example, the manufacturer uses four color codes for shapes:

- dark green is for coarse or aggressive action;
- brown is for medium action;
- red is for fine action, and
- light green is for extra-fine action.

The products shown in Figure 18-1 are made from oil-resistant chemical rubber, which provides a soft cushion action and allows the abrasive to cut freely and smoothly without gouging or digging into the work surface. The abrasive in this tool is silicon carbide, as in most of these types of products.

Three variables affect aggressiveness: binder, grain size, and grain type. Users may not have their choice of these because of product differences among manufacturers. If they do, however, they should be aware of the technical details described in the following paragraphs.

Binder

The function of the binder is to cushion the abrasive and prevent the tool from becoming loaded or filled with particles from the workpiece. The binder must gradually wear away to constantly present a clean tool surface. The binder largely determines the hardness of the tool.

For proper use, select the abrasive-filled rubber tool with the appropriate hardness for the particular application. Soft action is generally recommended for working on soft metals, polishing, and deburring and finishing. Although soft-action tools can be used on

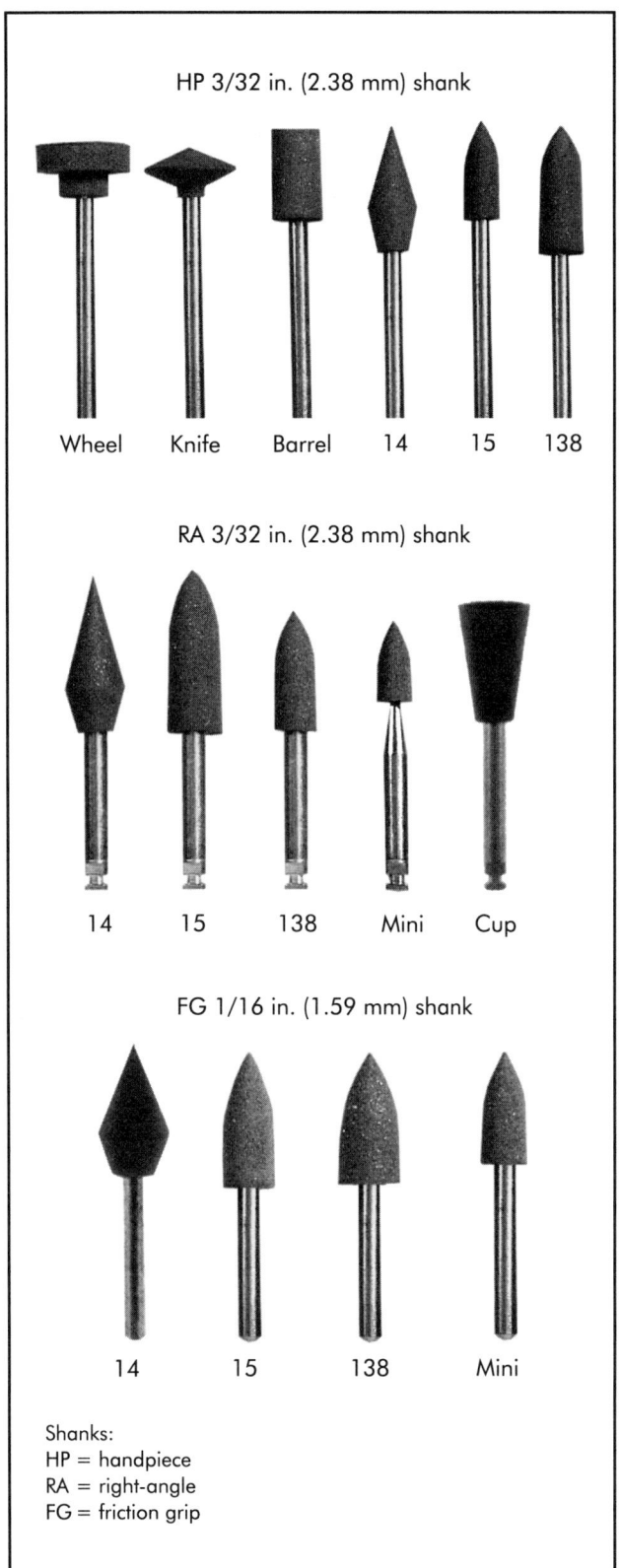

Figure 18-2. Typical dental abrasive-filled rubber tools. (Courtesy Dedeco)

surfaces and the edges of all metals, they are not aggressive cutting tools.

A semi-soft binder, which contains a pumice abrasive and acts primarily for polishing, does not significantly remove metal. A semi-firm binder has longer durability and is used for a combination of resiliency and durability. A firm binder is less friable (it does not fracture as easily) and is widely used for fast deburring action on soft metals. It is also used for removing tool marks and burrs from medium and hard metals. A tough binder is recommended for removing sharp burrs and rough work that would wear other types of binders too quickly. Although not a grinding wheel, the tough binder is much harder than other binders.

Grain Size

The second variable is grain size. All but the semi-soft binders include the following size ranges of abrasive particles:

- 36-grit, extra-coarse action;
- 54-grit, coarse action;
- 70-grit, medium action;
- 90-grit, medium-to-fine action;
- 120-grit, fine action, and
- 180-grit, extra-fine action.

Coarse abrasives remove stubborn burrs and are for rough or semi-rough applications. For general deburring work, finishing, and pre-polishing most metals, 70-grit or finer is preferable. As noted in previous chapters, "finer" action means a larger number. So, 180-grit (roughly 180 particles per 1 in. [7 particles per mm]) is a finer size than 70-grit. The 90-grit products provide a better finish than the coarser ones. They can be used for removing scratch marks from rough-work after tools have been used. For hard metal polishing, use a 120-grit abrasive, which also can be used for copper and other soft metals. Products with a 180-grit grain are used for polishing all metals and any work in which a minimum of stock removal is needed.

Grain Type

Tools can be filled with either aluminum oxide or silicon carbide, but silicon carbide is typically used. Aluminum oxide is not as hard or as sharp as silicon carbide, but it is less brittle. For rubber-bonded tools, the difference is probably insignificant. Silicon-carbide particles are very brittle, and during use have a tendency to break down into a finer grain. This makes them extremely useful for obtaining a high polish, particularly on hard metals. The semi-soft binder, a pink color, is filled with a pumice abrasive. Various

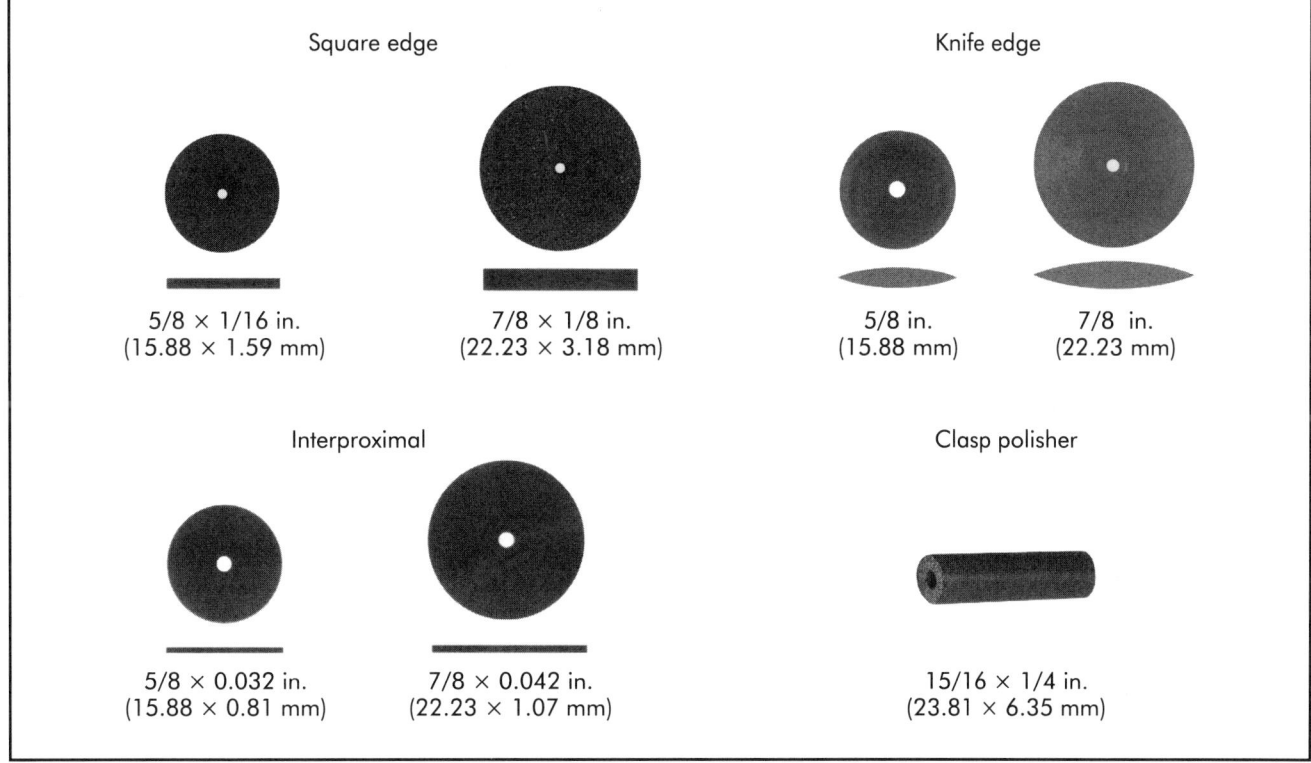

Figure 18-3. Abrasive-filled rubber dental discs. (Courtesy Dedeco)

grain sizes are unavailable in this binder combination.

On many of the tools in which various binders, grain sizes, and material types are available, there is inadequate room to identify its entire composition. For this reason, it is difficult for operators to know a tool's composition. If one of these products performs inadequately, or if a harder or softer or more or less aggressive product is required, ask management or engineering to obtain the appropriate product and some alternate material compositions.

Abrasive-filled dental tools typically are not as aggressive as some of the industrial tools just described. The purpose of abrasive-filled rubber tools in the dental industry is to polish teeth or dentures, so they are not designed to remove heavy amounts of material or large, stubborn burrs. Again, several manufacturers produce these tools and each has its own color code. For the simple dental tools shown in Figure 18-2, the color codes are as follows:

- dark brown—extra coarse, long-lasting action (aluminum oxide);
- black—coarse action (silicon carbide);
- green—medium action (silicon carbide);
- red—fine action (silicon carbide);
- white—very-fine action (aluminum oxide), and
- gray—used for instant high glaze (silicon carbide).

Because of the nature of dental tools, they are significantly softer and smoother. Thus, they provide a better surface finish than the industrial products discussed in this chapter. Some dental tools are filled with soft silicone rubber and are noticeably different. Silicone-rubber products are usually dark gray and have a sheen (which, of course, does not comply with the above color code). Other dental products have a dull surface finish. Usually, the abrasive in the dental products, again, is of very small silicon-carbide particles. In most cases, the particle size is 120–180 or finer grit.

AVAILABLE SIZES

Table 18-1 illustrates the available sizes of frequently used commercial industrial tools. Larger sizes and special designs may be available, but they are not typical. In addition to these and other shapes discussed in this chapter, abrasive rubber tools are available in square, oblong, or round sticks or bars. They are as small as 0.186 in. (4.72 mm) in diameter and as large as 1 in. (25.4 mm) in diameter. They also can be obtained in 10-in. (254-mm) diameter wheels for use in a motorized bench grinder. Because these are soft rubber, the wheels or points can easily be cut or machined into unique configurations to match part contours. Natural wear also provides "form-fit" shapes.

Table 18-2 provides dimensions of commonly used dental abrasive-filled rubber tools. Manufacturers provide a wide variety, so the list is representative, rather than complete. A slightly different size or shape can be created if necessary by holding a file or stone to any of the tools.

SHANK CONFIGURATIONS

Dental tools come in five basic shank sizes or shapes. Also, some dental cups and many of the industrial tools require use of screw-in mandrels. Contra-angle dental shanks consist of a flat-and-groove milled in the end of the shank (see Figure 18-2). In some situations, it is desirable to alter the tool length to reach inside smaller areas. Numerous industrial motors can handle these shanks.

POTENTIAL LIMITATIONS

Abrasive-filled rubber products are simple tools to apply and do not generally cause significant usage problems. However, problems can occur in the following areas:

- Dwelling too long on surfaces will affect dimensions.
- Too aggressive a tool will result in large edge breaks.
- Loose particles can become embedded in soft materials.
- Loose particles can become embedded in machine ways.
- If the mandrel that holds these tools pokes through the rubber tip, it can damage the part.
- These tools may fly apart if used at excessive speeds (some motors run at 400,000 rpm).

Edge finishing requires broad-range capabilities. Rubber tools can perform relatively aggressively or they can be gentle giants. When used gently, many of the tools shown in Figure 18-1 will maintain a 0.003 in. (0.08 mm) maximum edge break. Similarly, dental tools can maintain these small edge breaks. The more aggressive tools will provide edge breaks of up to 0.010 in. (0.25 mm) in stainless steel. Typically, however, that would be an upper limit. Normally, other tools would provide larger edge breaks, then be used to add the smooth radius that makes holes more attractive and functional.

Dwelling too long on a surface or hole with these tools will create changes in dimensions. These changes can be verified easily in first-time usage by holding the tool against the surface for 30–60 seconds. Aggressive bonds and large grit sizes will result in too much stock removal at the edges of precision miniature parts. These tools, however, are convenient and produce satisfactory results on larger parts with less critical edge breaks.

During normal usage, particles do not become impregnated in soft materials. Through carelessness, however, it is possible for particles to fall into chuck jaws or collets. When these are closed onto the workpiece they will mash the loose particles into the workpiece, resulting in a scrap part.

Some supervisors will not allow machinists to use abrasive rubber tools on precision machines because they believe abrasive particles fall into the machine ways and erode them very quickly. Erosion makes the machine less precise and necessitates additional maintenance. To eliminate this problem, protect the ways by covering them with a large white sheet of tissue or cloth. Also, rubber may combine with some coolants to make a sticky glue that can build up on machine ways.

When bullet-shaped tools wear excessively, a sharp point of the mandrel will break through the tool. This action can scratch or gouge the workpiece, although this is not a problem in most uses.

TOOL COMPARISONS

Figures 18-4 through 18-8 illustrate the amount of edge break and size change that can occur from aggressive use of abrasive-filled tools. Figure 18-4, for example, indicates that after 60 seconds of aggressive use, even rubber dental points can produce up to a 0.010-in. (0.25 mm) edge break. Under gentle use, however, these products render an edge break of only 0.002–0.003 in. (0.05–0.08 mm), as indicated in Figure 18-5. Industrial products, if used aggressively, also can produce edge breaks of 0.010–0.015 in. (0.25–0.38 mm) after 60 seconds (Figure 18-6). Even a round dental product can change the diameter required by 0.0015 in. (0.038 mm) after 30 seconds. Again, this illustrates very aggressive use on a stainless-steel workpiece. Typically, one would not use these products on surfaces. They are designed primarily to finish edges and, under normal use (as shown in Figure 18-8), remove only 0.0002 in. (0.005 mm) of material from surfaces. However, aggressive use for longer times can remove up to five times that amount. The surface finish produced by the dental tools typically is about 4–8 μin. (0.1–0.2 μm). Under normal usage, neither large edge breaks nor damage to surfaces will occur.

Table 18-1. Commercially available sizes of abrasive-filled rubber industrial products

Shape	Diameter, in. (mm)	Thickness, in. (mm)
Square-edge wheel	0.5000 (12.700)	0.0625 (1.588)
		0.1250 (3.175)
		0.2500 (6.350)
	0.6250 (15.880)	0.1250 (3.175)
		0.2500 (6.350)
	0.8750 (22.230)	0.1250 (3.175)
		0.1880 (4.775)
		0.2500 (6.350)
	1.0000 (25.400)	0.0938 (2.383)
		0.1250 (3.175)
		0.1880 (4.775)
		0.2500 (6.350)
	1.5000–8.0000 (38.100–203.200)	0.1250–1.5000 (3.175–38.100)
Tapered-edge wheel	0.3750 (9.525)	0.9380 (23.825)
	0.6250 (15.875)	0.9380 (23.825)
	1.0000 (25.400)	0.1250 (3.175)
Cylinder	0.2500 (6.350)	0.5000 (12.700)
	0.2500 (6.350)	0.8750 (22.225)
	0.8750 (22.225)	1.0000 (25.400)
	1.0000 (25.400)	1.5000 (38.100)
Bullet	0.2810 (7.137)	1.0000 (25.400)
	0.3750 (9.525)	1.0000 (25.400)
	0.5000 (12.700)	0.8750 (22.225)
	0.8750 (22.225)	1.7500 (44.450)
Cone	0.3750 (9.525)	0.6250 (15.875)
	0.6250–0.2500 (15.875–6.350)	1.0000 (25.400)
	0.8750–0.2500 (22.225–6.350)	1.2500 (31.750)
	1.0000–0.2500 (25.400–6.350)	1.2500 (31.750)
Narrow cone	0.3750–0.3130 (9.525–7.950)	0.8750 (22.225)
Round-nose cone	0.6250–0.1250 (15.875–3.175)	0.8750 (22.225)
	1.0000–0.5000 (25.400–12.700)	2.0000 (50.800)
Blunt cone	0.8750 (22.225)	1.2500 (31.750)
	1.0000 (25.400)	1.7500 (44.450)
Rods	0.2500 (6.350)	6.0000 (152.400)
	0.3750 (9.525)	6.0000 (152.400)
	0.5000 (12.700)	6.0000 (152.400)
Square blocks	0.5000 (12.700)	6.0000 (152.400)
Parallelogram tablet		2.2500 × 1.1250 × 0.3750 (57.150 × 28.575 × 9.525)

Table 18-2. Characteristics of abrasive-filled rubber dental tools

Shape	Size, in. (mm)	Normal Use	Color	Grit Size	Shank Type, in. (mm)	Shank Size, Shank Type	Overall Length, in. (mm)
Bullet	0.200 (5.80)	Dental	Gray	Ultra fine	0.094 (2.39)	Straight	1.700 (43.18)
Point	0.125 (3.18)	Dental	Green	Fine	0.062 (1.58)	Straight	1.000 (25.40)
Point	0.125 (3.18)	Dental	Green	Fine	0.094 (2.39)	Contra	0.800 (20.32)
Point	0.125 (3.18)	Dental	Brown	Coarse	0.062 (1.58)	Straight	1.000 (25.40)
Cylinder	0.250 (6.35)	Industrial	Brown	Coarse	None	None	1.000 (25.40)
Point	0.125 (3.18)	Dental	Brown	Coarse	0.094 (2.39)	Contra	1.000 (25.40)
Bullet	0.250 (6.35)	Industrial	Red	Fine	None	None	1.000 (25.40)
Cup	0.250 (6.35)	Dental	Brown	Coarse	0.094 (2.39)	Contra	1.100 (27.94)
Cup	0.250 (6.35)	Dental	Green	Fine	0.094 (2.39)	Contra	1.100 (27.94)
Bullet	0.250 (6.35)	Dental	Gray	Ultra fine	0.094 (2.39)	Contra	1.625 (41.28)
Wheel	0.500 (12.70)	Dental	Gray	Ultra fine	0.062 (1.58)	Straight	1.625 (41.28)
Knife-edge wheel	0.500 (12.70)	Dental	Gray	Ultra fine	0.062 (1.58)	Straight	1.625 (41.28)
Knife-edge wheel	0.500 (12.70)	Dental	Gray	Ultra fine	0.094 (2.39)	Contra	1.625 (41.28)
Wheel	0.500 (12.70)	Dental	Gray	Ultra fine	0.094 (2.39)	Contra	1.625 (41.28)

Figure 18-4 indicates that, for the conditions studied, abrasive-filled cotton MX tools produce two to three times the edge cutting of abrasive-filled rubber tools. At the same time, it shows that with the B132 shape, 120-grit MX wheel on large burrs, holding the tool at the edge for only a few seconds can minimize the edge break to 0.002–0.005 in. (0.05–0.13 mm). The data in figures 18-4 through 18-7 are based on typical usage at a precision aerospace plant that employs a few individuals to produce what most people would call ultra-high quality. Performance under other conditions and with other goals in mind (higher productivity and lower quality) would yield other results.

TOOL USAGE

Abrasive-filled bullet tools are used for deburring and radiusing holes of any size and on edges of parts long enough to be traversed over the edges. Soft-lipped dental cups are widely used where a small edge break is desired. The dental cup's thin lip deflects adequately to prevent the extreme pressure that would cause excessive edge break; and, dental cups and points can easily reach into corners. The round tools are used in some type of air motor on a mandrel, and larger products, such as 6-in. (152.4-mm) long bars or rods, are handheld on a rotating part. Dental tools in particular require only a light touch to operate successfully. With very high speeds, heavy loading stalls the motors.

Dental tools are used on aluminum, stainless steel, and most metals. The dental industry also uses them on acrylics and ceramics.

If burrs are large or very sharp, abrasive-filled rubber tools are only used to finish edges. Cutting tools would remove most, or all, of the burr, and rubber tools would provide only the smooth blend or chamfer. Sharp burrs and aggressive hand pressure will slice through many of the rubber tips, and the tip will fly through the air.

Figure 18-9 illustrates some of the handpieces (air motors) used in the dental industry. These will be discussed in more detail in Chapter 23, but note the angle of the head, which provides more comfortable hand positions and may be easier to use on some part geometries. The motors are available in right, straight, and contra angles.

ENVIRONMENTAL, HEALTH, AND SAFETY ISSUES

It is important when using larger abrasive-filled tools to recognize that, even though they are soft, they are essentially grinding wheels and must be operated with the same considerations. For example, tools 0.50 in. (12.7 mm) in diameter or larger should never be

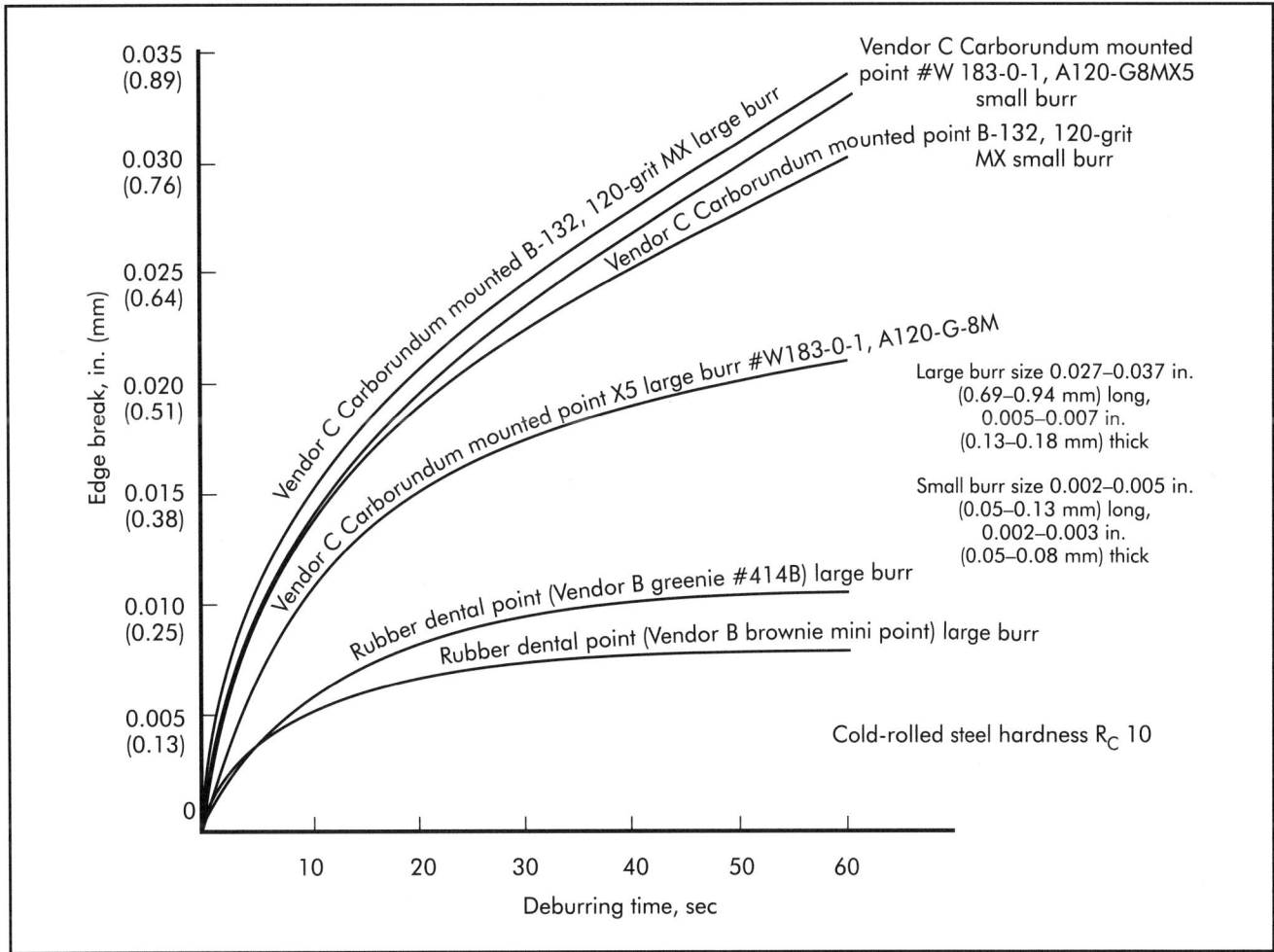

Figure 18-4. Effect of deburring tool, deburring time, and burr size on edge break.

operated at more than 40,000 rpm. On larger sizes, these tools should not be used in industrial air motors that go up to 50,000 rpm. Larger sizes can be safely used on the bench motors at many plants, however, because they rotate at slow speeds. For smaller sizes, such as the 0.125 in (3.18 mm) diameter dental point, high speeds do not represent a problem. Dentists regularly use miniature tools at speeds up to 400,000 rpm.

Any of the motorized tools can result in repetitive motion injury. Since abrasive rubber tools are softer and intended to be used with less force, the tendency to suffer such injuries should be less, but plants should address the potential for injury.

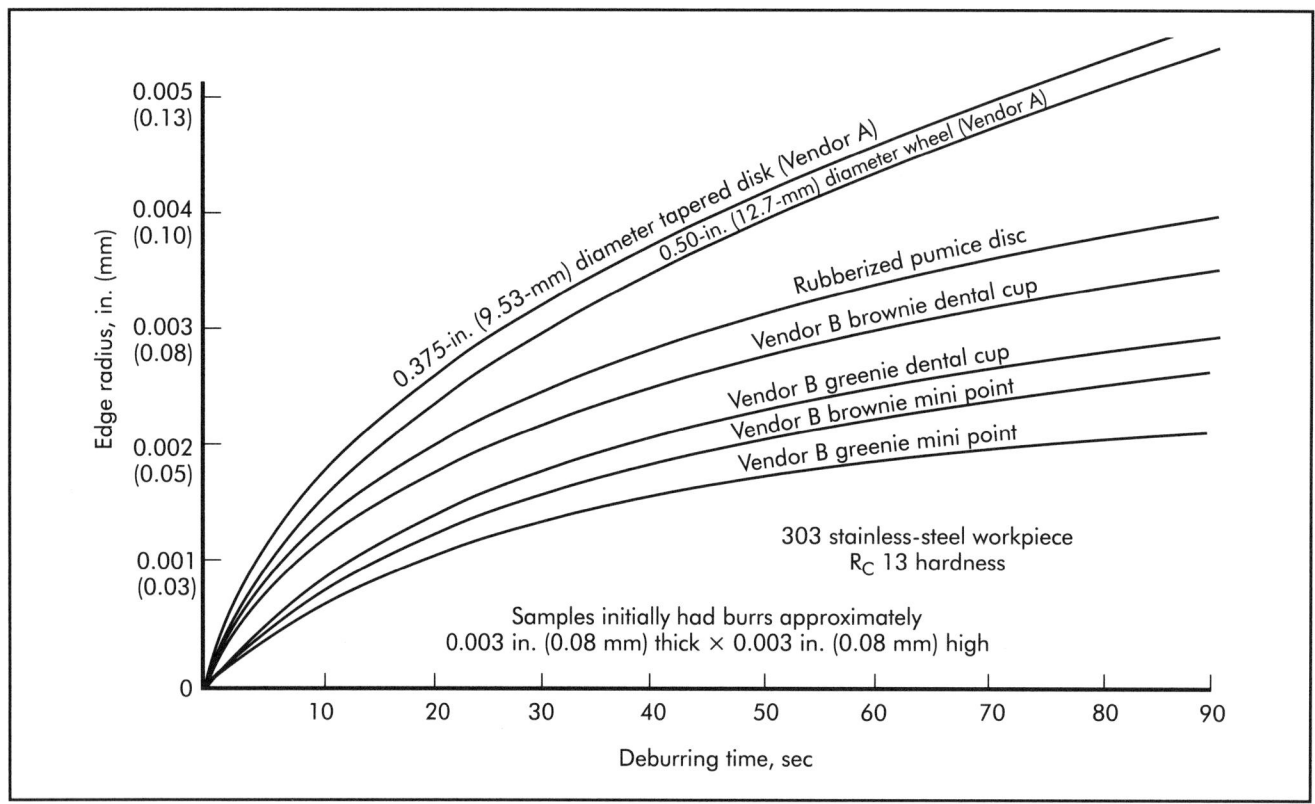

Figure 18-5. Effect of deburring tool and deburring time on edge break with normal use.

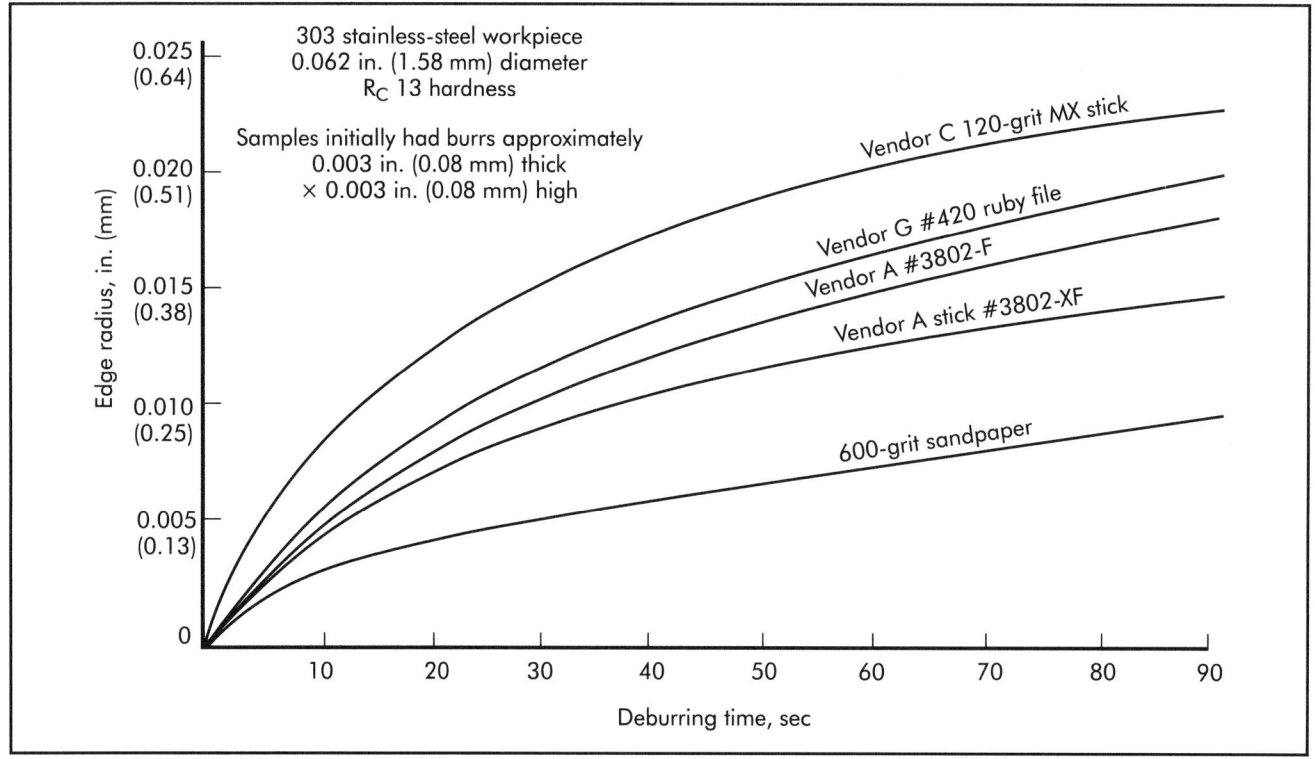

Figure 18-6. Effect of deburring time and tool on edge break with aggressive use.

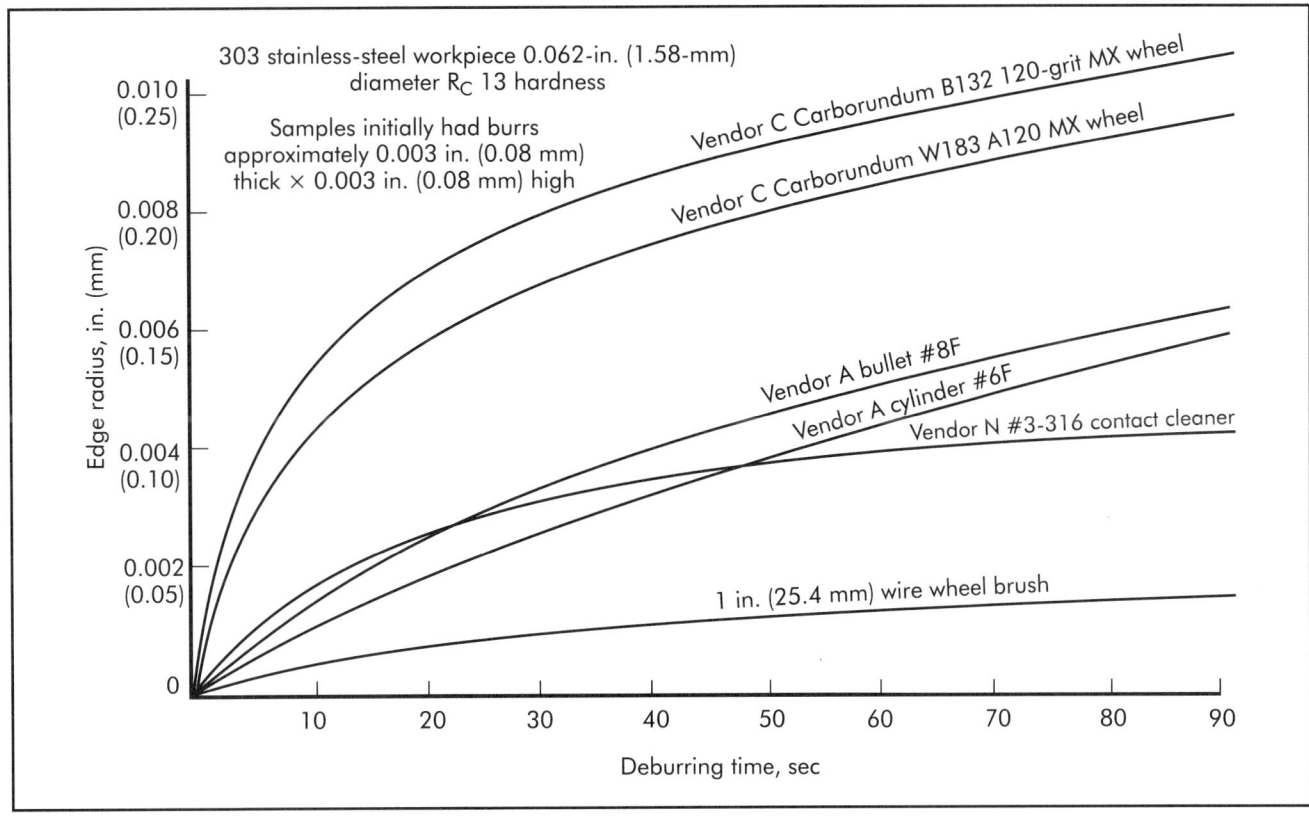

Figure 18-7. Effect of deburring tool and deburring time on edge break with normal use.

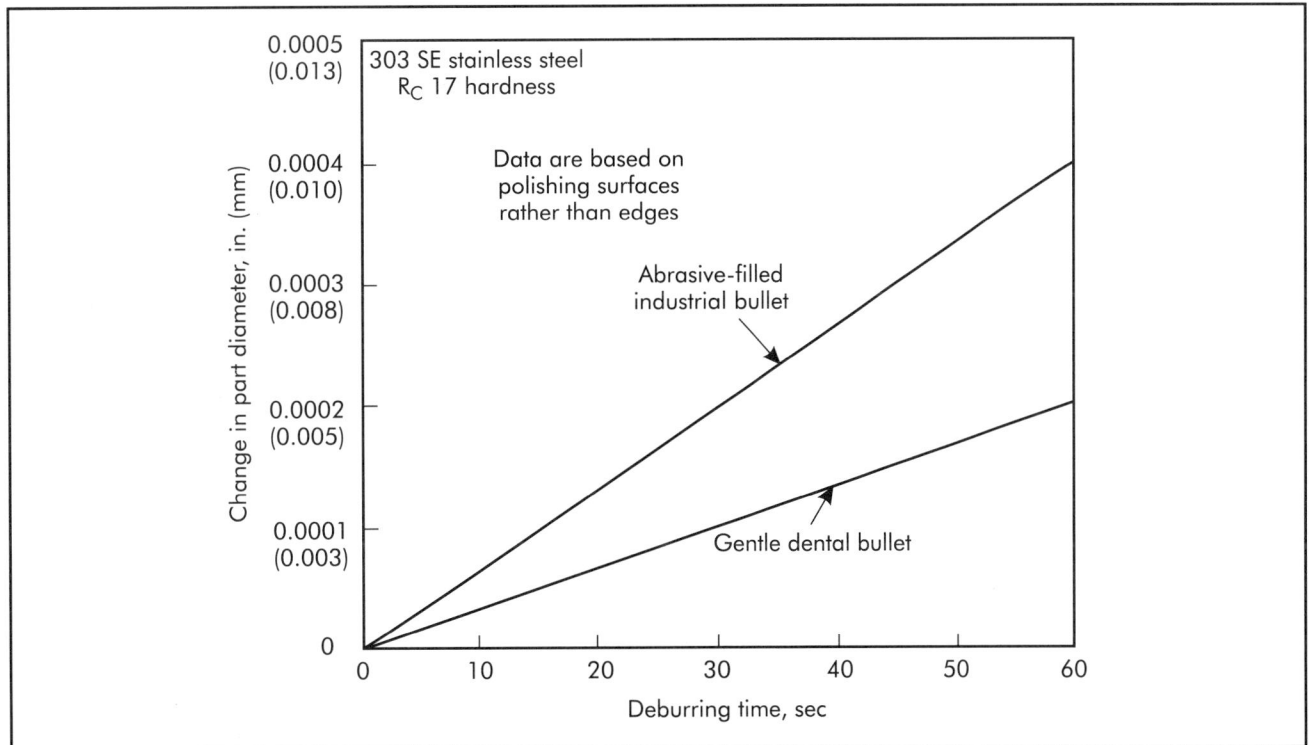

Figure 18-8. Abrasive-filled tools under normal use.

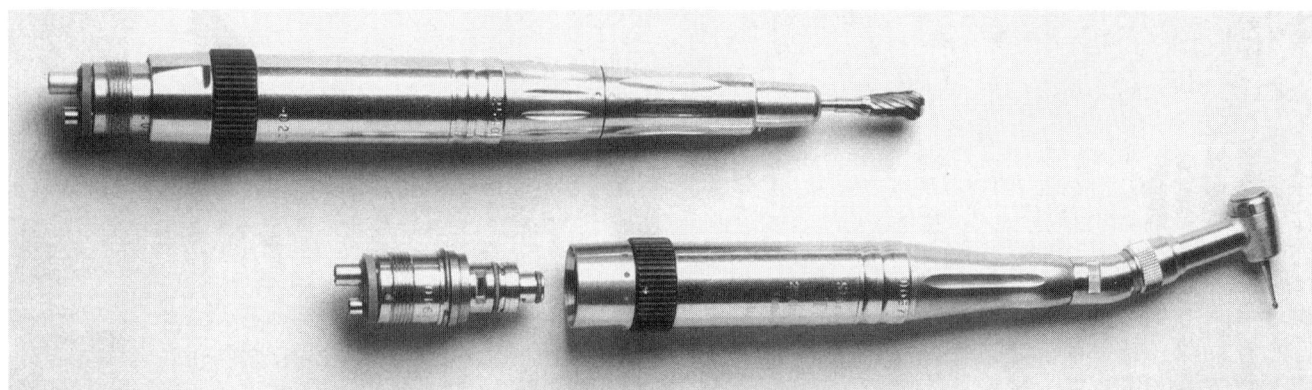

Figure 18-9. Dental handpieces used for deburring. (Courtesy Star Dental)

Bonded Abrasive Tools (Sanding Products) 19

The most widely used deburring tools in the United States are abrasive paper products, often called "sandpaper." That word is not exactly correct, however. Although sandpaper is used to work on wood products, metal parts require something other than a sand abrasive. In literature, abrasive paper products are also referred to as bonded abrasive products or coated abrasives. In this text, the term *abrasive paper products* is used because the majority of products are abrasives bonded to paper or paper-like material.

Bonded abrasive products are found in almost every shop in the world because they are flexible, convenient, inexpensive, and available in a wide variety of shapes and sizes. The variety includes normal shop abrasives, as well as high-precision tape products that produce finishes in the range of 2 μin. (0.05 μm) or better. The products in this family perform equally well for deburring and surface smoothing, and for manual and automated operations. The variety of equipment and material is vast—it is estimated that more than 1,500 product variations exist. Over 35 manufacturers produce equipment, and another 60 companies produce handheld motors and grinders that use these products. The media used is produced or marketed by another 60 companies.

In addition to part variables, process variables include:

- abrasive material and size;
- media configuration type and size;
- backing material and properties;
- operating speed and pressure, and/or
- density (of nonwoven materials).

SHAPES AND SIZES AVAILABLE

Tables 19-1 and 19-2 note the 40 basic shapes of abrasive paper products commercially available and some of the obtainable size ranges. Figures 19-1 and 19-2 illustrate many of these shapes, which are the most commonly manufactured items (although others are available). No single manufacturer produces all the tools defined or material combinations discussed here.

Sheets

Sheets of coated abrasives are produced in two sizes for use in manual operations (by hand or hand block), or on machines with straight-line or orbital action. Precut sheets are also used on a number of drum machines. They come in plain-backed and adhesive-backed configurations. The latter can be cut and attached to any hard backing to provide more aggressive cutting.

Belts

Bonded abrasive belts come with cloth and paper backing and contain aluminum oxide, silicon carbide, or a zirconia abrasive. Some have a waterproof material, which provides a longer life for wet operations. Cork-based materials are used for glass polishing, as well as some metal deburring and chamfering. Manufacturers provide several grades of belt within a specific abrasive and backing type. One source provides eight different backing stiffness grades. Abrasive sizes range from 16–1,200-grit (see Figure 19-3). Belt sizes range from 0.25–100 in. (6.4–2,540 mm) wide and up to 168 in. (4.3 m) long. Nonwoven synthetic material is used to provide a cushioned action and create a product that consistently exposes new sharp grains as the belt material wears away. It is a bit like the skin on a finger; as the old, useless skin peels away, a new layer appears ready to work. Nonwoven synthetic material can be used wet or dry, and its design essentially eliminates the problem of loading up. At least five different end-splicing designs are used to

Table 19-1. Forms of abrasive paper products (nonwoven nylon products are presented in Table 19-2)

Product Shape	Smallest Standard Size Available in.	mm	Largest Standard Size Available in.	mm
Belts	1-1/2 × 60	38.1 × 1,524	52 × 142	1,321 × 3,607
Narrow belts	1/8 × 24	3.175 × 609.6	3/4 × 20-1/2	19.05 × 520.7
Sheets	4-1/2 × 5-1/2	114.3 × 139.7	9 × 11	228.6 × 279.4
Strips	1 × 11	25.4 × 279.4	5-1/2 × 9	139.7 × 228.6
Shop rolls	1 × 1,800	25.4 × 45,720	65 × 3,924	1,651 × 99,670
Flap wheels	0.4 diameter × 0.2 width	10.16 × 5	16 diameter × 4 width	406.4 × 101.6
Cup-shaped flap wheels (See flap discs)	2-3/8 × 1-1/2	60.3 × 38.1	7.2 × 0.62	182.9 × 15.7
Cartridge rolls	1/8 diameter × 3/4 length	3.18 × 19.1	1 diameter × 2 length	25.4 × 50.8
Lapped points	3/16 × 1	4.763 × 25.4	1 × 1-1/2	25.4 × 38.1
Cord/tape	0.012	0.31	0.093	2.36
Cross pads	3/4 × 3/4 × 3/8 thick	19.1 × 19.1 × 9.5	6 × 6 × 2 thick	152.4 × 152.4 × 50.8
Square pads	1 × 1 × 1/4 thick	25.4 × 25.4 × 6.4	4 × 4 × 1/2 thick	101.6 × 101.6 × 12.7
Discs	1/2 diameter	12.7	36 diameter	914.4
Slotted discs	1-1/2	38.1	6	152.4
Slashed discs	6	152.4	8	203.2
Drums/sleeves/bands	1/4 diameter × 1/2 width	6.4 × 12.7	3 diameter × 9 width*	76.2 × 228.6
Brush-backed flap wheels	10 × 1	254 × 25.4	16 × 6	406.4 × 152.4
Spiral rolls/bands	1/2 × 1	12.7 × 25.4	1 × 2	25.4 × 50.8
Overlapping slotted discs	3/4	19.1	6	152.4
Flap discs	4-1/2	114.3	7	177.8
Shaped sleeves (cones, bullet nose) (See also body cones)	0.12 × 0.4	3.1 × 10.2	1-1/2	38.1
Sponge products	4-3/4 × 3-3/4 × 1/2	120.7 × 95.3 × 12.7	4-3/4 × 3-3/4 × 1/2	120.7 × 95.3 × 12.7
Mops (slashed flap wheels)	2 × 1	50.8 × 25.4	10 × 4	254 × 101.6
Nonwoven-backed flap wheels	2-1/2 diameter × 1-1/4 width	63.5 × 31.8	7 diameter × 1-3/4 width	177.8 × 44.5
Raised bump discs	1	25.4	4	101.6
Raised bump drums	0.4 × 0.8	10 × 20	4 × 36	101.6 × 914.4
Raised bump pads	2 × 4	50.8 × 101.6	2 × 4	50.8 × 101.6
Sticks	1-1/2 × 3/4	38.1 × 19.1	2-1/2 × 1/4	63.5 × 6.4
Microfinish film	1	25.4	4	101.6
Fiber discs	2	50.8	9-1/8	231.78
Brush-backed strips	10 × 1	254.0 × 25.4	16 × 6	406.4 × 152.4
Angled flap wheels	4 × 0.8	101.6 × 20.3	4 × 0.8	101.6 × 20.3

Table 19-1. (continued)

Product Shape	Smallest Standard Size Available		Largest Standard Size Available	
	in.	mm	in.	mm
Goblet wheels	4	101.6	4	101.6
Sponge strips	4-3/4 × 3-3/4 × 1/2	120.7 × 95.3 × 12.7	4-3/4 × 3-3/4 × 1/2	120.7 × 95.3 × 12.7
Fingernail files	3/8 × 4	9.53 × 101.60	3/8 × 4	9.53 × 101.60
Body cones	3-3/16 × 1-5/16 × 1-3/16	80.963 × 33.338 × 30.163	5-5/8 × 1-3/4 × 1-1/4	142.88 × 44.45 × 31.75
Abrasive caps	3/8 × 60°	9.53 × 60°	1-1/2 × 60°	38.1 × 60°
Cloth cones	3/4 × 60°	19.05 × 60°	1-1/2 × 60°	38.1 × 60°
Sandscreen	9 × 11	228.6 × 279.4	9 × 11	228.6 × 279.4
Flexible discs	4-1/2	114.3	7	177.8

* Drums 8-in. diameter × 3-in. long (203.2 × 76.2 mm) are also available.

Table 19-2. Forms of bonded abrasives available in nonwoven nylon

Product	Minimum Size Available		Maximum Size Available	
	in.	mm	in.	mm
Belts	1/2 × 18	12.7 × 457.2	6 × 48	152.4 × 1,219.2
Sheets	6 × 9	152.4 × 228.6	9 × 11	228.6 × 279.4
Shop rolls	4 × 360	101.6 × 9,144	6 × 360	152.4 × 9,144
Discs	2	50.8	8	203.2
Drums	3-1/2	88.9	5-3/8	136.53
Flap wheels	4 × 1	101.6 × 25.4	12 × 2	304.8 × 50.8
Paper interleaved flap wheels	4 × 1	101.6 × 25.4	4 × 1	101.6 × 25.4
Hand pads	6 × 9	152.4 × 228.6	6 × 9	152.4 × 228.6
Wheels	4 × 1	101.6 × 25.4	14 × 3	355.6 × 76.2
Side-cutting paper-interleaved flap wheels	4 × 0.60	101.6 × 15.2	4 × 0.60	101.6 × 15.2
Side-cutting flap wheels	4 × 0.60	101.6 × 15.2	4 × 0.60	101.6 × 15.2
Sponge pads	3-5/8 × 5-7/8	92.08 × 149.23	3-5/8 × 5-7/8	92.08 × 149.23
Radial discs	4 × 0.60	101.6 × 15.2	7 × 1-1/2	177.8 × 38.1

ensure the smooth action of any belt design. Belt joints, in most cases, are at least as strong as the coated abrasive product itself, run smoothly, and perform an integral role for the belt. Different machines may require different belt joints to extend belt life or reduce vibrations.

The Coated Abrasive Manufacturers' Institute (CAMI) has endorsed a program of specific belt lengths, as follows:

- for a length range of 12–36 in. (30.5–91.4 cm), belt lengths are made in 3-in. (7.6-cm) length increments;
- for a length range of 42–168 in. (1.1–4.3 m), belt lengths are made in 6-in. (15.2-cm) length increments, and
- for a length range of 180–504 in. (4.6–12.8 m), belt lengths are made in 36-in. (91.4-cm) length increments.

Figure 19-1. Overview of bonded abrasive products. (Courtesy 3M Coated Abrasive Products)

Figure 19-2. Bonded abrasive products. (Courtesy Norton Abrasives)

Discs

Circular abrasive discs are produced in conventional, slotted, slashed assemblies; raised bumps; flap discs; and fiber discs. Conventional discs are round with adhesive backs, snap-on features, twist on/off designs, Velcro® fasteners, and integral threaded or washer and screw attachments. Large adhesive-backed discs are used for "flat lapping," a form of sanding in which the disc is applied to a rotating table and

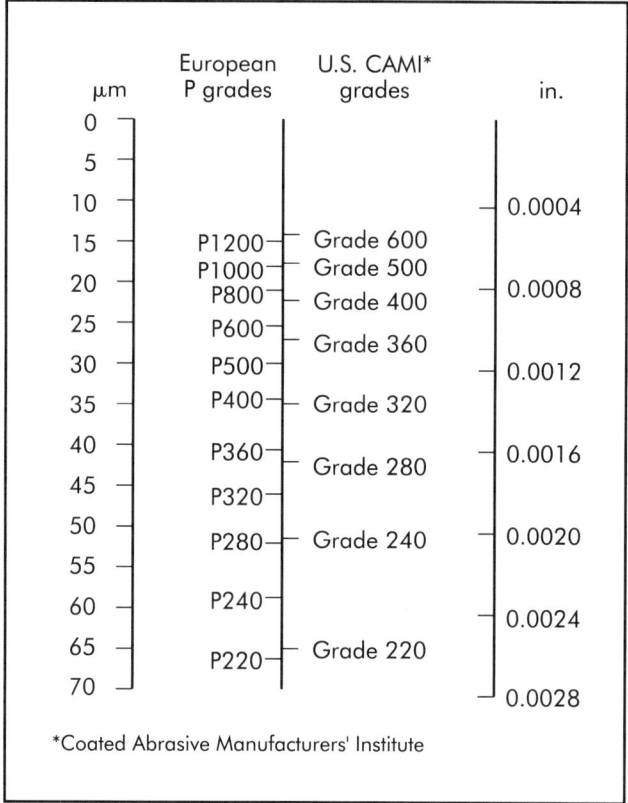

Figure 19-3. Comparative sizes and grades of fine abrasive grains (Gillespie 1999).

parts are placed on the table and held down by rollers or by hand. Discs up to 30 in. (76.2 cm) are standard and available in various abrasive sizes. The disc material can consist of a traditional bonded material, abrasive-coated nonwoven nylon, or other synthetic materials. Nonwoven material provides a softer pad to cushion cutting forces and is generally used for finer work, but aggressive nonwoven materials are also available. Precision finishes may require the use of a scalloped disc with an undulating edge pattern, rather than constant diameter.

One brand of discs can be procured with holes to reduce the heat buildup during heavy grinding and allow operators to better see the surface while it is being finished. Also, it is reported that this disc design provides more airflow to remove debris from the work surface. A few manufacturers produce a center water-feed disc hand pad that enables users to apply a fine spray of water to the operation. Water and other lubricants typically provide a smoother finish.

Slotted Discs

Slotted discs have radial cuts (Figure 19-4) to allow the disc to conform better to hole entrances and

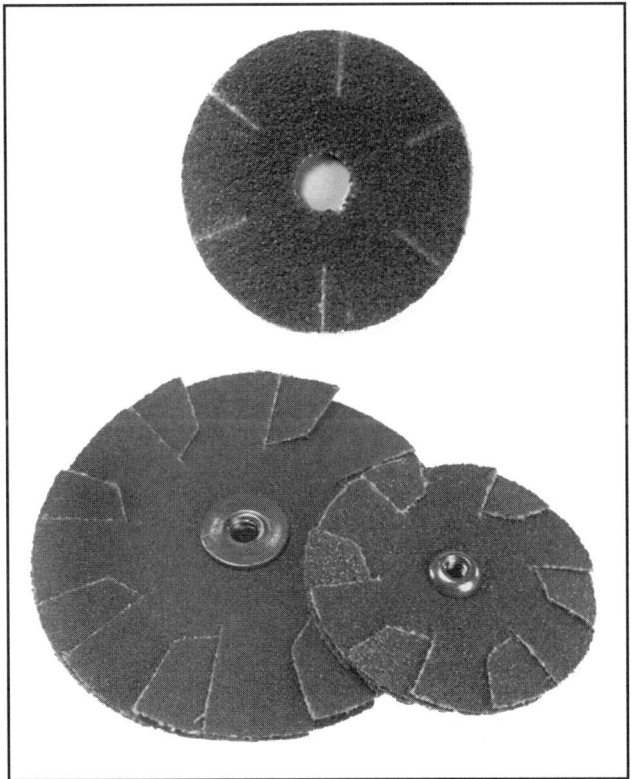

Figure 19-4. Bonded abrasive slotted discs and overlapped slotted discs. (Courtesy Superior Abrasives)

walls. Essentially, the slots allow the paper to fold back over itself, which results in a softer action and prevents some tearing.

Slotted overlapping discs create a heavy cutting action by providing more backing and breaking up the cutting action as sections pass over part edges. They are used widely for deburring holes, polishing bores, and chamfering tube ends. They are made in sets of two or four discs per assembly. The two-disc sets provide cutting action only on the forward motion. The four-disc set allows deburring upon entering and pulling out of a hole.

Fiber Discs

Fiber discs are thin grinding wheels designed to allow some lateral deflection. They are used for heavy stock removal, generally through the use of disc grinders. One company provides a fiber disc cutter to make cutting of any sized disc easy. These are abrasive products; they are not sandpaper-like tools.

Raised Bump Products

Raised bump products are hard, wheel-like fiber discs or cylinders, but they have a raised bump on their surfaces (Figure 19-5). This bump of abrasive provides cutting, while the non-bump surface provides an interrupted cut. One manufacturer refers to these products as island-pattern resin-fiber sanding discs. The minerals on these include aluminum oxide, silicon carbide, alumina zirconia, and boron carbide.

Flap Discs

Like a flap wheel, the flap disc (also called mop disc) consists of a series of abrasive strips laid over each other (Figure 19-6). The layers of strips look like the teeth of a face mill cutter. The disc is rotated so the open edge trails over the surface rather than leading into the cut

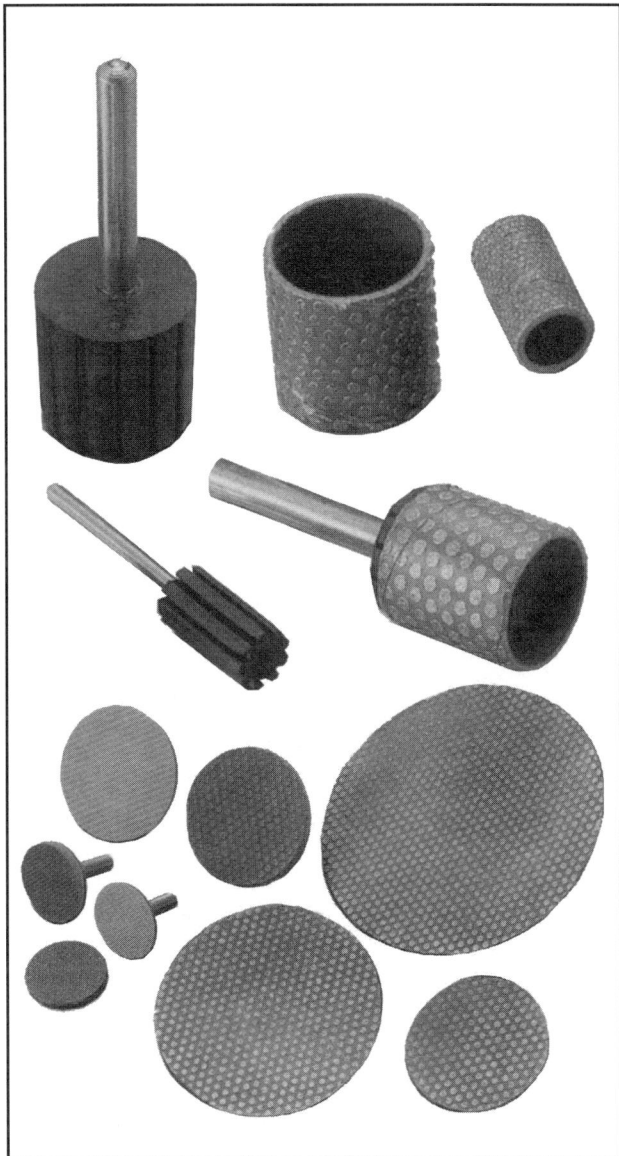

Figure 19-5. Raised bump abrasive cylinders and discs.

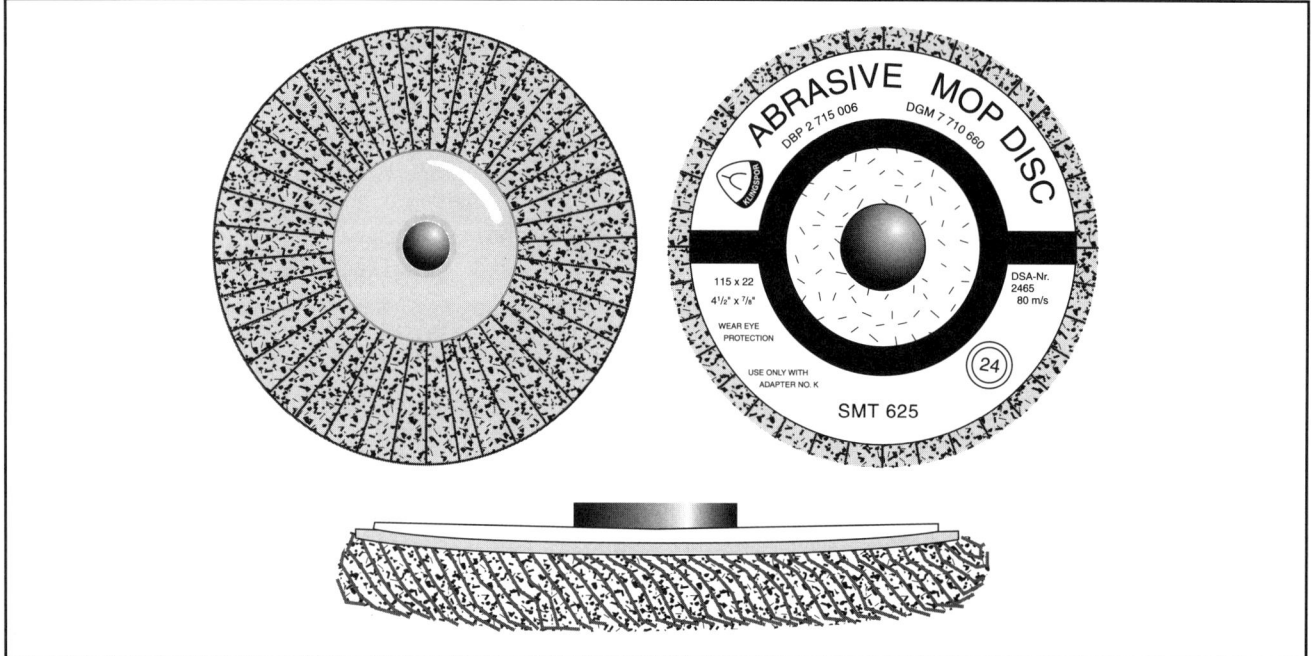

Figure 19-6. Flap discs. (Courtesy Klingspor Abrasives)

similar to a cutter. These products provide a cooler cut, and no backing pad is typically required. They are widely used for stainless steel, steel, aluminum, and most alloys. As the layers wear, they expose the layer below them for additional cutting.

Stars

Stars are six-pointed discs stacked together in layers of two and offset so that a continuous face is visible. When stacks of these sets are put on a mandrel, they act as thin discs, which can reach into fine grooves and deform into holes (see Figure 19-7).

Flap Wheels

Flap wheels are used to reach into cutouts, over steps, and inside tubes, holes, and cylinders (Figure 19-8). Their overlapping design exposes more abrasive as the layers wear. The combination of abrasive paper and abrasive-coated nonwoven nylon provides a softer touch than just the abrasive paper. The same effect is provided by flap wheels that include fiber brush sections among the abrasive paper strips. Coated-abrasive flap wheels deburr and polish with a slapping or flapping action. They are made in a variety of designs, sizes, and materials. The basic design uses strips of abrasive paper bound together at the hub. Since the outer end has more space between the paper strips, a gap opens, which is used to perform the deburring. Materials other than abrasive paper can be placed between the paper strips. Common fill materials include soft brush strips and abrasive-coated nonwoven nylon (see Figure 19-8). To further soften the action, each strip of paper can be slit into multiple thin strips (fingers), which conform better to fine detail. Flap wheels as small as 1 in. (25.4 mm) in diameter by 0.625-in. (15.88-mm) wide to 80-in. (2.0-m) wide are standard.

Mops

Mops are another name for flap wheels. In this chapter, the word *mop* is reserved for a flap wheel with serrated or slashed strips, which make it more flexible than a standard flap wheel. As Figure 19-9 illustrates, when cut to the correct widths, these mops completely fill recesses.

Wheels

A wheel-style product is designed to cut on the outer diameter, just like a grinding wheel. Wheels up to 12 in. (30.5 cm) in diameter are common and can be made by wrapping adhesive-backed abrasive paper into a tight spiral, then cutting it in strips.

One style, often called a unified wheel, involves compressing discs of abrasive-laden nonwoven material and bonding them together with a resin to form a solid. The relatively soft wheel acts as a soft grinding wheel. These wheels are not actually bonded abrasive wheels in the sense that grit is bonded onto the

Figure 19-7. Star discs. (Courtesy Dynabrade)

Figure 19-8. Flap wheel with interleaves of nonwoven abrasive-coated nylon. (Courtesy Superior Abrasives)

surface of the backing material. Rather, they are often sold interchangeably with bonded abrasive products.

Wrapping the nonwoven nylon strip into a tight wheel shape forms convolute wheels. The resins used for these wheels vary considerably, and can provide either very soft or exceptionally hard, yet open, cutting. In addition to deburring metals, unified wheels are used for finishing printed-circuit-board edges and glass edges. Low-density wheels are open-cutting, and do not develop as much heat as the harder, high-density wheels. These wheels can be shaped to match part configurations, and external lubricants can be used, like those that buffing wheels employ. Wheel speeds often range from 3,000–6,500 surface feet per minute (sfm) (914.4–1,981.2 m/min), but manufacturer literature will indicate maximum safe speeds. Wheel-style products are sold with integral or replaceable shafts.

Goblet Wheels

Goblet wheels are almost-hemispherical sanding surfaces designed for reaching into internal corners for deburring and polishing. They are widely used in the food processing industry.

Shop Rolls

Shop rolls are reels of coated abrasives, most commonly 50 yards (45.7 m) in length. They vary in width depending on their usage. They are unending strips of narrow abrasive paper. Rollstock usage varies from precision-slit abrasives used in crankshaft and bearing-polishing operations to wide rolls used on drum machines to handheld operations on lathes or operations that require a shoeshine polish action on fixed parts. Conventional materials and nonwoven nylon materials are available in shop rolls. Some rolls are

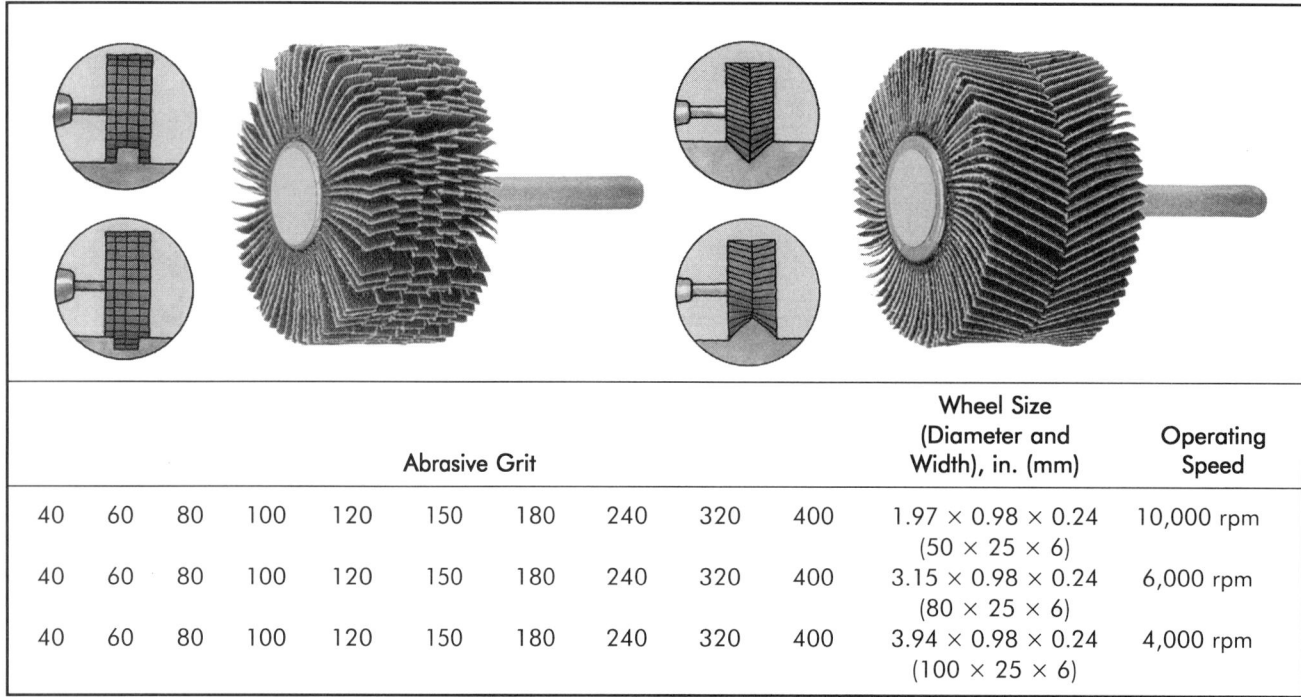

Abrasive Grit										Wheel Size (Diameter and Width), in. (mm)	Operating Speed
40	60	80	100	120	150	180	240	320	400	1.97 × 0.98 × 0.24 (50 × 25 × 6)	10,000 rpm
40	60	80	100	120	150	180	240	320	400	3.15 × 0.98 × 0.24 (80 × 25 × 6)	6,000 rpm
40	60	80	100	120	150	180	240	320	400	3.94 × 0.98 × 0.24 (100 × 25 × 6)	4,000 rpm

Figure 19-9. Slashed flap wheel applications.

the correct size to provide a continuous roll of material that can be cut to length for vibrating sanders.

Cartridge Rolls

Cartridge rolls are small, tightly wound rolls of coated abrasives glued to prevent unwinding. They are press-fitted on mandrels and offer fresh abrasives as an individual winding wears away. They are used as deburring and polishing tools and are commonly available in end-tapered, full-tapered, and regular (untapered simple cylinder) shapes. Cone-shaped coated abrasives are press-fitted to a mandrel for grinding and finishing relatively inaccessible areas, such as curved surfaces and punched or drilled holes. These spiral-wound strips are flexible uniform strips of abrasive wound to a specific diameter for finishing and polishing difficult-to-reach areas, such as grooves, slots, flutes, threads, holes, and other small-diameter openings.

As cartridge rolls wear, they can be unwrapped or merely allowed to wear through to the next layer of abrasive. They are available in sizes as small as 0.125 in. (3.18 mm). Sizes larger than 1 in. (25.4 mm) are commonly called a wheel.

Cartridge rolls come in straight cylinders, tapered rolls (used to reach into corners and small crevices), and cone or lapped points (which come to almost a sharp point for hard-to-reach areas and deep crevices). Figure 19-10 illustrates two shapes. Lapped points are used for scalloped surfaces, venturi tubes, aircraft skins, and automobile parts.

Bands, Drums, and Sleeves

Bands are coated abrasive materials, spiral wound and glued to an inner liner that is similarly wound. They are placed over either a hard rubber or an inflatable holder. The band is held on to the shaft or arbor by expanding the rubber with air pressure or tightening a jam nut to expand the rubber mandrel out radially. Bands deburr and grind areas not reached easily by other abrasive tools. Most sleeves have a single ply of material, but some manufacturers produce a multiple sleeve that may last four times longer than a single one. Diameters range from 0.125–8.000 in. (3.18–203.20 mm), and lengths go up to 72 in. (1.8 m).

Pads and Hand Pads

Hand pads are hand-size strips of material (typically, 6 × 9 in. [152.4 × 228.6 mm]), and are similar to sheets. They are produced in conventional bonded abrasives and abrasive nonwoven-nylon materials, which are often color-coded to indicate composition or aggressiveness.

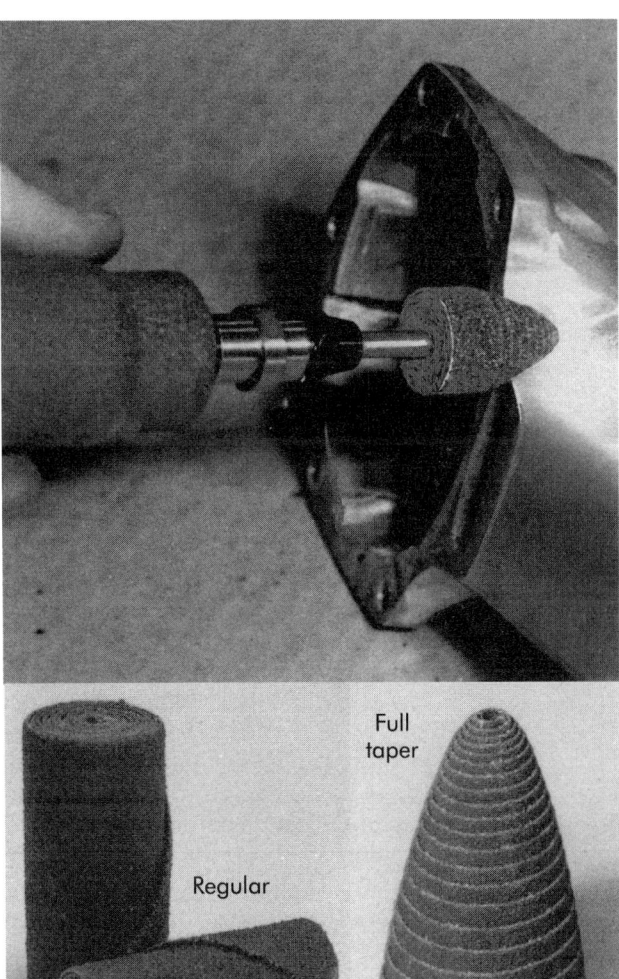

Figure 19-10. Cartridge roll in use. (Courtesy 3M)

Sponge Pads

Sponge pads are made by laminating a sheet of either conventional bonded abrasive or nonwoven nylon material to a cellulose sponge. This will not remove heavy burrs, but will provide fine finishes and conform to the product and the hand.

Abrasive Sponges

A sponge-like material is coated with abrasive grain to provide a soft cutting action. This is different from the sponge pad just mentioned. This product conforms readily to part surfaces and is easy to hold and use by hand. Some sponges are made from polyvinyl alcohol (PVA) to provide a continuous pore structure. These tools have abrasive particles locked into the cell structure that do the polishing and grinding. Hardness and elasticity can be balanced to produce slow, constant wear. As the abrasive wears away, new abrasive is revealed. Sponge materials can be produced in a variety of shapes and porosities for applications from grinding to cleaning semiconductor wafers (*Machine Design* 1995).

Square Pads

Square pads, as the name implies, are a square shape with a hole in the middle to attach a mandrel. The intermittent strokes of the corners of the pad provide a fast action. They are used for forging dies, molds, metal patterns, filets, and corners. These are a heavier duty tool than most of those discussed so far. Typically, they are 0.25–0.50-in. (6.4–12.7-mm) thick and range from 1–4 in. (25.4–101.6 mm) on a side.

Cross Pads

Cross pads (see Figure 19-11) are used to deburr holes. Their shape allows the deburring motor to push them down into the holes to provide effective chamfering and deburring action. Each size will accommodate a wide variety of part hole sizes. Operators can see through cross pads, so the location of the tool is much easier to verify. Typically, they are more aggressive than stars.

Butterfly Tools

Butterfly tools, which have a variety of other names, are created when users place one or more strips of abrasive paper between a clothespin-like mandrel (Figure 19-12). When the mandrel is rotated, the paper spreads and forces itself against the hole walls

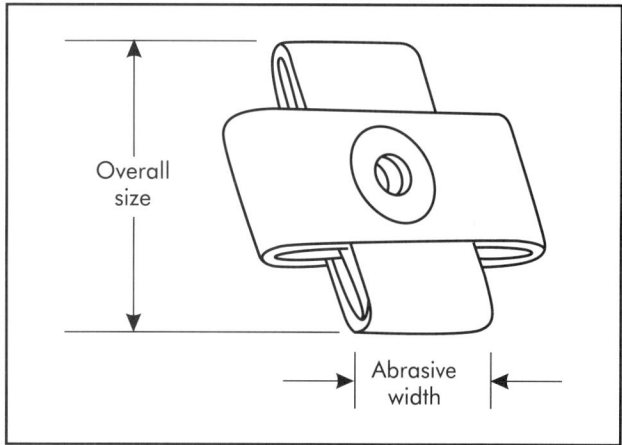

Figure 19-11. Cross-pad configuration. (Courtesy Norton Abrasives)

Mandrels used with abrading pads need not be removed from the power unit to change pads. They can be attached to the driving mechanism of any lathe, drill press, flexible shaft, pneumatic or electric hand drill.

B shape
Where chamfering is required, this abrading pad deburrs and chamfers in one operation.

S shape
The square-shaped abrading pad is designed primarily to remove burrs and polish the bottom of blind holes, as well as side walls. This pad is first inserted into the job and the power is then applied.

BB shape
This abrading pad performs in one operation many of the finishing jobs that used to require removal of the mandrel and resetting of work before completion. This pad can travel entirely through a cylinder, tube, or hole and re-enter from opposite end to chamfer both ends in one continuous operation.

Figure 19-12. Butterfly abrading pads on center mandrel deburr holes and polish bores in one operation. (Courtesy Dayton Abrasive Products)

and edges. Mandrels can be purchased for this purpose, but many users make their own from wood dowels or slit steel rod. Some producers provide a blunt nose series rather than a square end. These tools can be used to reach through a hole and deburr the bottom side. However, cross pads are normally a better choice.

Chamfering Cones

Chamfering cones are small, single thickness cones. They mount to a metal mandrel or throwaway rubber insert, and are designed to provide a 45° chamfer.

Center laps designed for 60° angles offer a variation. While used to lap center holes, they are also deployed for fine deburring. They look like a small dunce cap with adhesive on the backside.

Body Cones

Body cones are made for large parts, and use diameters of 1.00–1.50 in. (25.4–38.1 mm) and cone lengths of 3–5 in. (76–127 mm). They are used on vertical spindle sanders for large, curved surface parts.

Abrasive Cords and Tapes

Cloth cord as small as 0.012 in. (0.31 mm) in diameter is coated with fine abrasives to provide a flexible tool that can reach into small holes and slots. Sizes up to 0.150 in. (3.81 mm) in diameter and tapes up to 0.125-in. (3.18-mm) wide are available. These tools are convenient when nothing else will reach into an area, but the abrasive wears quickly when deburring. Figure 4-8 illustrates abrasive-coated cord used with a shoeshine motion to deburr intersecting holes.

Radial Discs

Radial discs are formed by standing strips of bonded abrasive or nonwoven nylon on edge and radially bonding them to a tough plate. The result is a tool that looks something like a conventional end brush.

Cup-shaped Flap Wheels

Cup-shaped flap wheels are used on the outer diameter, as well as the end or face, of the tool. A normal flap wheel is very stiff on its face, but the open cut of this tool allows the leaves to spread and finish the bottom of the hole.

Diamond Sticks

Handheld diamond sticks are used for sharpening tools and deburring and smoothing edges. These hone-like tools have an open mesh of diamond. Typical diamond particle sizes range from 0.00079–0.00492 in. (20–125 μm). They come in either flat or half-round sticks. Note that these tools may be listed in some catalogs as diamond files.

TOOL MATERIALS

Bonded abrasive products are made with aluminum oxide abrasives, ceramic aluminum oxide, alumina zirconia, silicon carbide, emery, flint, garnet, boron carbide, and even diamond- or tungsten-coated particles. For most situations, the particular material involved is not critical. For most metalworking applications, aluminum oxide or silicon carbide abrasives are used. Silicon-carbide abrasives are most suitable for use on tough metals, such as titanium and steels, which require high-heat treatments, or where extremely sharp cutting edges must remain on the abrasive sandpaper.

Aluminum oxide outperforms other minerals when the ability to resist fracturing is the measured factor. Alumina zirconia breaks down, which provides continuously new cutting edges, and is preferred when heavy grinding pressures are employed. It is not particularly effective on stainless steel, steel, or cast iron. Garnet dulls too rapidly for metalworking. Heat-treated aluminum oxide is available for fast and cool cutting on stainless steel. Silicon carbide is widely used on softer metals, such as aluminum, brass, bronze, magnesium, titanium, rubber, glass, and plastic. Silicon carbide cuts fast under light pressure, and products made with it work well on porcelain, terrazzo, glass, marble, and plastic. Crocus is used for polishing gold and other soft metals. Ceramic alumina has a submicron structure, which allows each grain to break down into more cutting edges. The ceramic material is used for ferrous and nonferrous metals, carbon steel, and exotic alloys.

Abrasive paper products come with five types of backings:

- cloth,
- paper,
- fiber,
- paper and cloth combination, and
- polyester film.

Backings provide the base to which the mineral grain is bonded. Cloth backing is by far the most common. It can be lightweight or tough. (It is strong and generally the most durable.) Lightweight backing provides flexibility for reaching into small areas not possible when using a strong backing. Paper backings come in three weights: lightweight, intermediate, and durable heavy stock.

The abrasive particles are usually bonded with two adhesive layers that lock the mineral to the backing. The "make" coat anchors the mineral to the backing and the "size" coat locks it. The two most common types of bonding agents are glue and resin over glue; but resin over resin and waterproof adhesive are also used.

Open-cut products use fewer grains per surface area to allow room for swarf and chips. Closed-cut products essentially fill the surface area with abrasive grain to maximize the number of cutting points. Soft materials normally require open-cut products to prevent clogging and reduce heating. Open-coat products have 50–70% of the backing surface covered with mineral.

Abrasive papers or abrasive paper products can be obtained in a variety of abrasive sizes, such as a coarse 24-grit—for working on metal products—or a fine 600-grit. Finer and coarser sizes are available; however, they are not commonly used. One of the most frequently used products for precision parts is known as polishing paper or polishing cloth. This particular material has no abrasive size used with it, but it is a very fine-grain abrasive and is suitable for getting finishes close to 8 µin. (0.20 µm).

Abrasive grain size is based on U.S. and international standards. Many domestic coated-abrasive manufacturers make products that adhere to the standards established by the Coated Abrasive Manufacturers' Institute (CAMI). The Federation of European Producers of Abrasives (FEPA) produces a "P" grade of abrasives. Comparisons between the two systems of size grading are shown in Figure 19-3.

One innovation in abrasives is the development of a manmade compound that makes each "grain" the same configuration (Figure 19-13). Each "grain" is a pyramid of exactly the same size (Gagliardi and Duwell 1989; McLean 1997). These precise structures provide an even coating on every belt or surface. Both aluminum oxide and silicon carbide can be made in these pyramids. The pyramids are not a grain, however. Rather, they are clusters of grains, such that as the tip of the pyramid wears, more new (and sharp) small grains appear. As a result, some users have found these materials to last 2–5 times longer than other belts.

Table 19-3 presents some unique bonded abrasive items. Diamond-coated flexible products and lapping film used for crankshafts as well as computer hard drive discs are some of the more unique uses. Abrasive belt cleaners and metal-backed abrasive paper may be useful for some operations.

Abrasive-coated Nonwoven Nylon Products

Abrasive-coated nonwoven nylon products are as widely used for finishing as are bonded paper products. These products consist of clumps of abrasives and resin along the nylon (Figure 19-14). The design of the tools provides a softer action for cutting and polishing than what is available with traditional papers and related products (Figure 19-15). Materials can be made in very aggressive forms. Typically, these products will not gouge a part or cause scratches (if properly used).

Table 19-4 compares nonwoven materials and other products. These are general comparisons—many exceptions can be found. Each user must judge how well the comparisons hold up.

TOOL USE

Figures 19-16 through 19-21 illustrate common uses of bonded abrasive products.

Figure 19-16 illustrates an operator holding a part against a large disc grinder to remove burrs and finish surfaces. These large discs are often overlooked as a deburring tool.

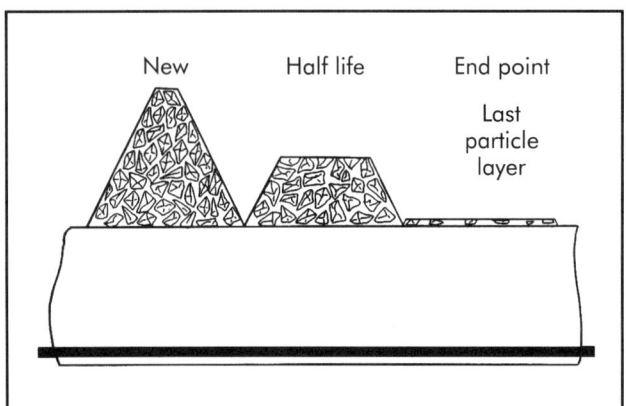

Figure 19-13. Structured abrasive. (Courtesy 3M Coated Abrasive Products)

Table 19-3. Unique bonded abrasive items

Product	Minimum Size Available, in. (mm)		Maximum Size Available, in. (mm)	
Diamond flex-foam backing	3-1/2 × 2-1/8	(88.9 × 53.98)		
Diamond sticks/files	1-1/2 × 3/4	(38.1 × 19.1)	2-1/2 × 1/4	(63.5 × 6.4)
Lapping film	2	(50.8)	4	(101.6)
Diamond-coated metal	3/8 × 4	(9.53 × 101.6)	2 × 3	(50.8 × 76.2)
Honing sticks	N/A			
Abrasive belt cleaners	N/A			
Metal-backed paper	N/A			

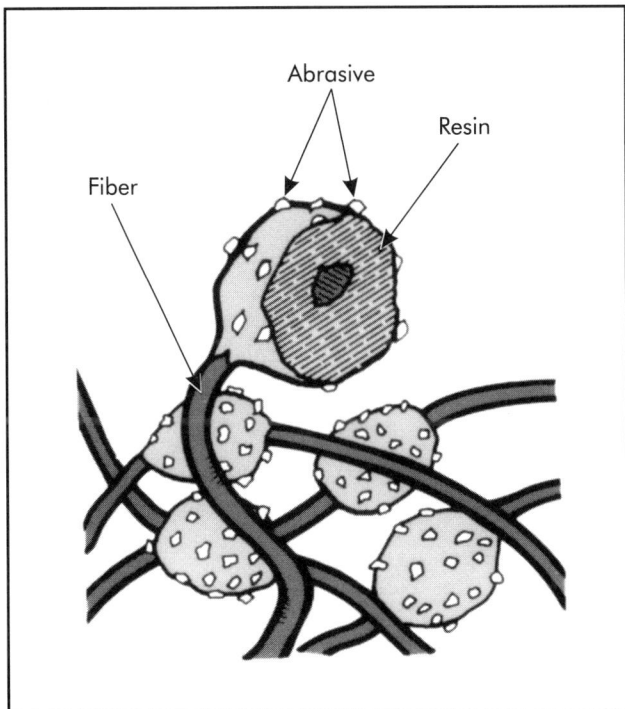

Figure 19-14. Schematic representation of nonwoven abrasive material. (Courtesy 3M Company)

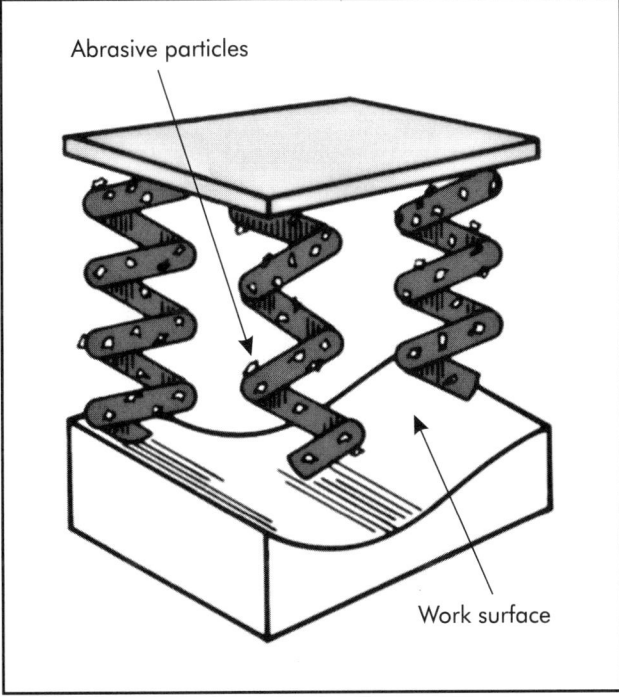

Figure 19-15. Schematic representation of controlled cutting action of nonwoven abrasive material. (Courtesy 3M Company)

Figure 19-9 shows the use of some slashed flap wheels (mops) to deburr grooves and ridges. Figure 19-22 shows the usage of a cup-shaped flap wheel.

Figure 19-23 illustrates typical surface-finish ranges obtained by deburring with various grit sizes. As indicated, a 320 grit or finer abrasive will provide an 8-μin. (0.20-μm) finish when used under normal conditions. Although it is possible to obtain less desirable finishes under normal practice, these values are representative.

There are no magic feed-and-speed combinations used in manual deburring, but some nonwoven nylon products are recommended to be run at 5,500–8,000 sfm (1,676–2,438 m/min). Faster speeds cause a harder cutting action with these materials.

ENVIRONMENTAL, HEALTH, AND SAFETY ISSUES

There are several dangers associated with the use of bonded abrasive tools, including:

- repetitive motion injury;
- parts and abrasive thrown by the energy from motorized tools;
- inhalation of metal dust, and
- fire and explosions from dust.

Repetitive motion injury may be the most common problem. When operators finish same-design parts day after day, particularly when they use heavy motors and exert a large amount of force, they may develop repetitive-motion injury symptoms. Solutions include varying the work content throughout the day, using less force, and taking frequent breaks. Consider using robots for routine work on same-design parts or rotating workers in two-hour shifts to reduce the risk of injury.

A variety of handheld pads are available—some more ergonomically designed than others. They come with a number of features:

- flat surfaces and straps for holding with the palm of the hand;
- foam-filled pads for comfort;
- thick rubber;
- form-fitting palm pieces;
- two-handed, wood-plane-style grips;
- sanding pencils (Figure 19-24), and
- finger pads, which allow users to apply only one finger for pressure.

Some adhesive strips of these products are designed to adhere to the flat side of paint sticks, which provides a hard backing (Figure 19-25). This allows easy use on slots, long edges, and flat surfaces. The same

Table 19-4. Comparisons with nonwoven nylon products (Courtesy 3M)

Competitive Products	Abrasive-coated Nonwoven Nylon Product	Nonwoven Material Advantage
Steel wool	Hand pads/sheets/rolls	• Leaves no metal fines • Leaves no carbon contamination on surface • Leaves no oil contamination • Will not rust
Sandpaper/coated abrasives	Belts/flap brushes/discs/hand pads/sheets/rolls/unitized wheels/wheels	• Consistent finish • Resists loading • No secondary burrs • Uniform radius • Will not undercut workpiece
Wire brushes	Flap brushes/discs/unitized wheels/wheels	• Consistent finish • Complete burr removal • Uniform radius • Safer, no flying wire bristles
Buff and compound	Discs	• Faster • Cleaner
Tampico and bristle	Flap brushes	• Faster • Better radiused edges
Files, burs, scrapers	Flap brushes/unitized wheels/wheels	• Faster • Won't damage workpiece
Setup wheels	Unitized wheels/wheels	• No undercutting • Cleaner • Faster • Less setup time
Sandblasting	Discs/hand pads/sheets/rolls/unitized wheels	• Fewer airborne particles to be contained
Rubber-bonded wheels	Unitized wheels/wheels	• Cleaner finish • Faster • Can level surface

Figure 19-16. Stationary disc grinder deflashing a casting.

Figure 19-17. Flap wheel deburring inside a large housing.

Figure 19-18. Handheld narrow-belt sander deburrs louvers.

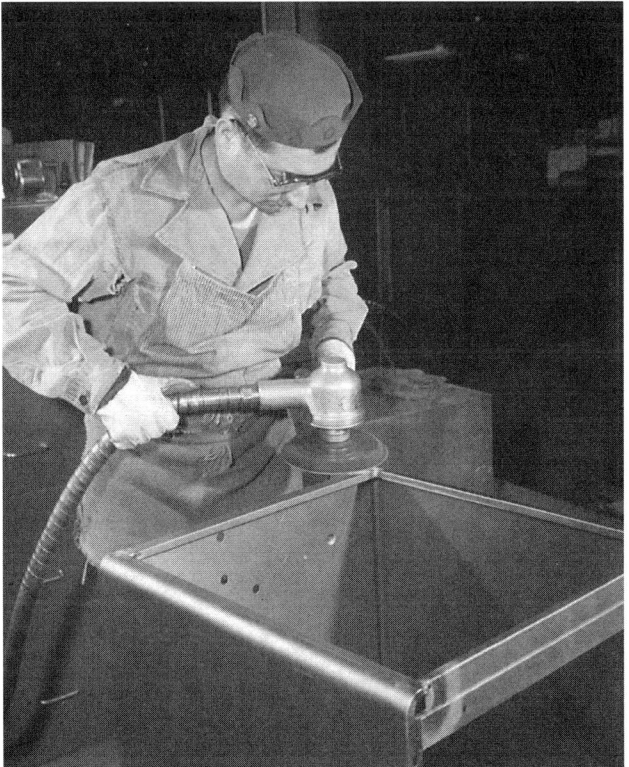

Figure 19-19. Heavy-duty flexible-shaft disc grinder uses fiber discs to deburr and chamfer edges.

paper can be adhered to shim stock or any other special shape to provide the correct contour and backing strength.

Parts must be held firmly to prevent the tool from grabbing and throwing them. Under normal use, this is not a problem, but it can be—a flying part is a deadly missile. Since the tools are rotating at speeds up to 30,000 rpm and they wear while they are being used, both paper and abrasive particles will break free. This is hazardous to the eyes, so eye protection is a requirement.

Nickel is a suspected carcinogen, and it is a principal element in stainless steel. Breathing metal dust containing carcinogens should be controlled for the health of the worker. Beryllium copper, as its name states, contains beryllium. Dust, fumes, or smoke from anything containing beryllium can cause berylliosis, which has no known cure. The dust must be controlled to very tight levels to ensure worker safety. Abrasive dust from papers containing silicon can cause silicosis, also a serious health problem. This dust also must be controlled. Titanium and magnesium dust both ignite easily and can cause serious fires. These materials may need to be sanded wet or used with good dust collectors. Note, however, that dust collectors also can be a source of fires or explosions. Dust from any sanding-like operation generates fine paper particles, fine metal particles, and glue and resin particles. Each of these can be a fire source, and some resins, when burnt, may be an inhalation health hazard.

REFERENCES

Gagliardi, John J. and E. J. Duwell. 1989. "Coated Abrasives Made with Abrasive Grit Clusters for Constant Performance Surface Finishing." Technical paper MR89-139. Dearborn, MI: Society of Manufacturing Engineers.

Gillespie, LaRoux K. 2000. *Guide to Deburring, Deflashing, and Trimming Equipment, and Supplies,* 2nd ed. Kansas City, MO: Deburring Technology International.

———. 1999. *Deburring and Edge Finishing Handbook.* Dearborn, MI: Society of Manufacturing Engineers.

Machine Design. 1995. "Versatile Material Absorbs Cleanly." *Machine Design,* 9, March: 54.

McLean, Ross. 1997. "Structured Abrasives: An Innovative Approach to Finishing and Polishing Operations." In *Proceedings of the 2nd International Machining and Grinding Conference.* Dearborn, MI: Society of Manufacturing Engineers, pp. 455–87.

Figure 19-20. Overlap slotted discs and pyramid discs deburr edges and internal hole diameter.

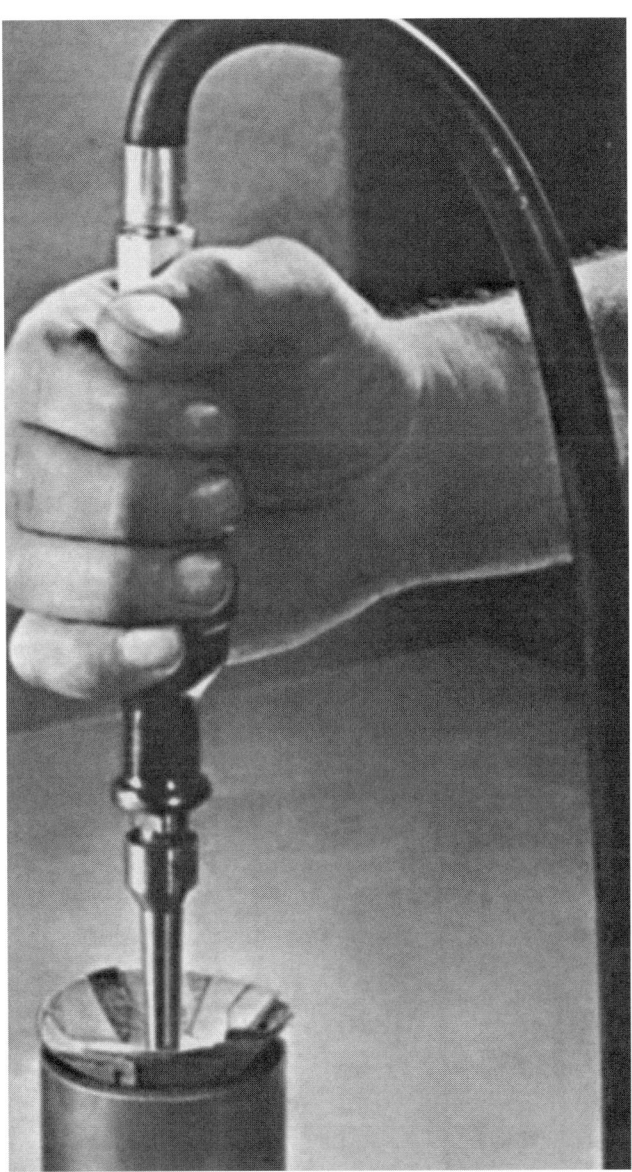

Figure 19-21. Overlap slotted discs deburr tube ends. (Courtesy Dayton Abrasive Products)

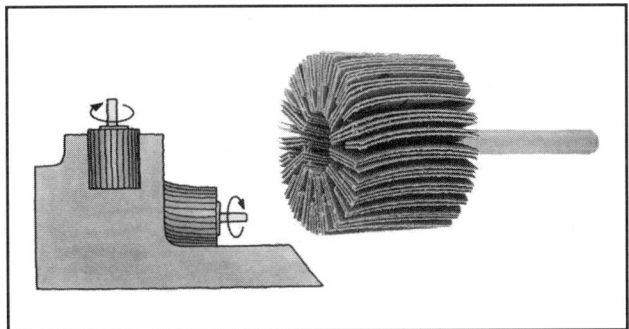

Figure 19-22. Cup-shaped (bottom cutting) flap wheel provides effective cutting against shoulders and down in pockets.

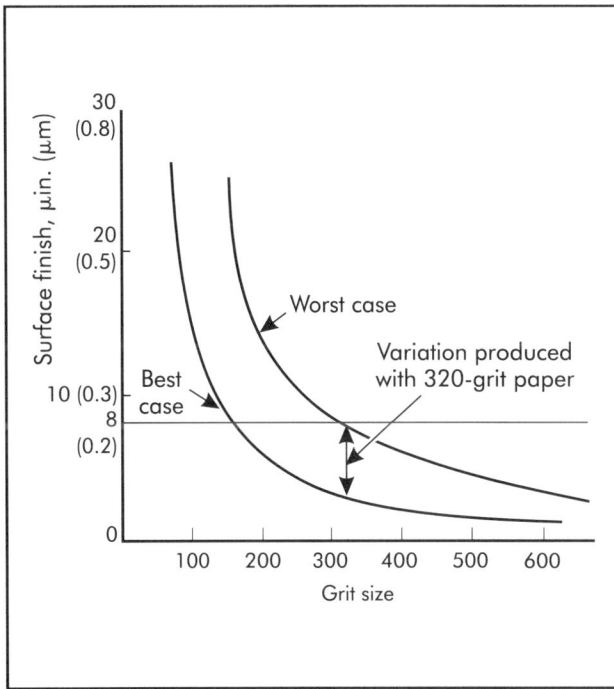

Figure 19-23. Surface finish produced as a function of grit size used.

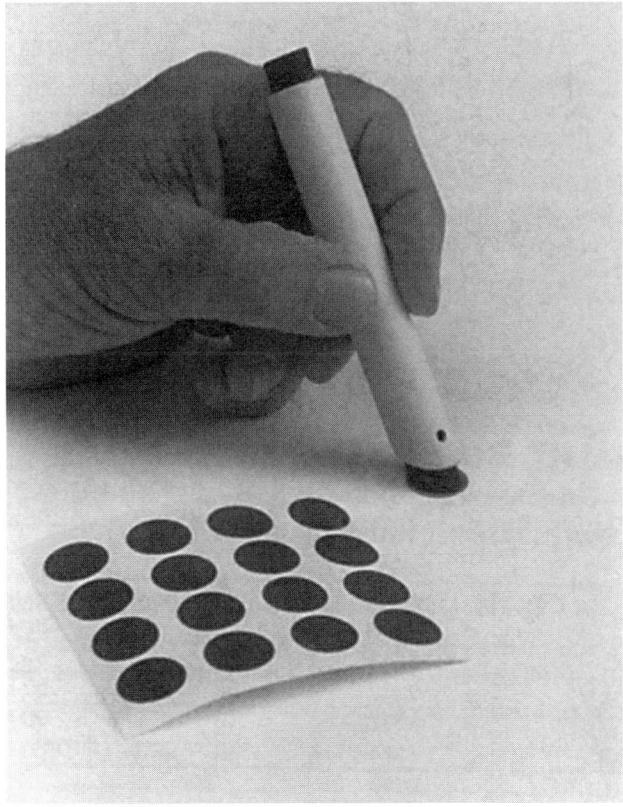

Figure 19-24. Sanding pencil. (Courtesy 3M)

Figure 19-25. Adhesive-backed strips used with paint stirrer for hard backing. (Courtesy Dayton Abrasive Products)

Brushes 20

Power-driven brushes are used in most major metal industries. They deburr, clean, descale, polish, edge blend, and texturize parts. Brush deburring offers many advantages, which is why it is widely used. Brushing is fast, relatively inexpensive (just a few cents per part in some instances), and exceptionally flexible. Brushing can be noncontaminating and is usually an environmentally friendly process. It is readily adaptable to manual equipment and requires little training or floor space. Brushing is often combined with bonded-abrasive deburring and buffing to produce the final part's finishes and edges. Many equipment manufacturers provide brushing motors.

The rotary action of a brush enables the use of a tremendous variety of driving motors and fixtures, which is a major advantage. The industrial usefulness of brushes is broadened by their availability in wide varieties of bristle materials, diameters, and lengths. Brushes are safe and simple to use, which add to their list of advantages.

Variables in brush deburring include:

- brush style;
- brush design within a given style;
- brush material;
- filament length;
- face width;
- externally applied abrasives;
- coolants;
- brush rotational speed;
- brush longitudinal feed rate;
- brush contact area (degree of interference);
- burr size;
- burr location, and
- part material.

With the variety of nontraditional materials available today, lines of distinction are blurred among brushing, sanding, and buffing. A single tool can include components of all three processes.

BRUSH TYPES

Six types of brushes are used for hand deburring, as follows:

- radial or wheel brushes,
- cup and disc brushes,
- end brushes,
- tube brushes,
- nontraditional brushes, and
- miniature brushes.

Each type (see Figure 20-1) varies in design and composition, and can be modified to provide a wide range of use. Each type is produced with different filament materials, lengths, and crimps or knots to produce varying degrees of stiffness. Filament materials are selected based on the intended application. Some filaments are synthetic; others are natural. However, they are collectively referred to as *brush fill material*.

Brushes made of metallic filaments use one of four basic forms: crimped wire, knotted or twisted tuft, straight wire, or multiple-strand crimped. Many types of crimped wire are used in brushes; however, the best crimping occurs when the wire is displaced over multiple planes. Although many crimping techniques produce multiple-plane displacement, the result should be similar to the crimp configuration produced by running a wire through two sets of perpendicular gears. Each set of gears should have a different pitch. The finer of the two gear sets governs the number of crimps/in. (/mm) Crimp wavelength (Figure 20-2), sometimes called frequency, should be uniform over the length of wire. Wire displacement, usually referred to as amplitude, should be held to a tolerance of +0.004 in. (+102 μm).

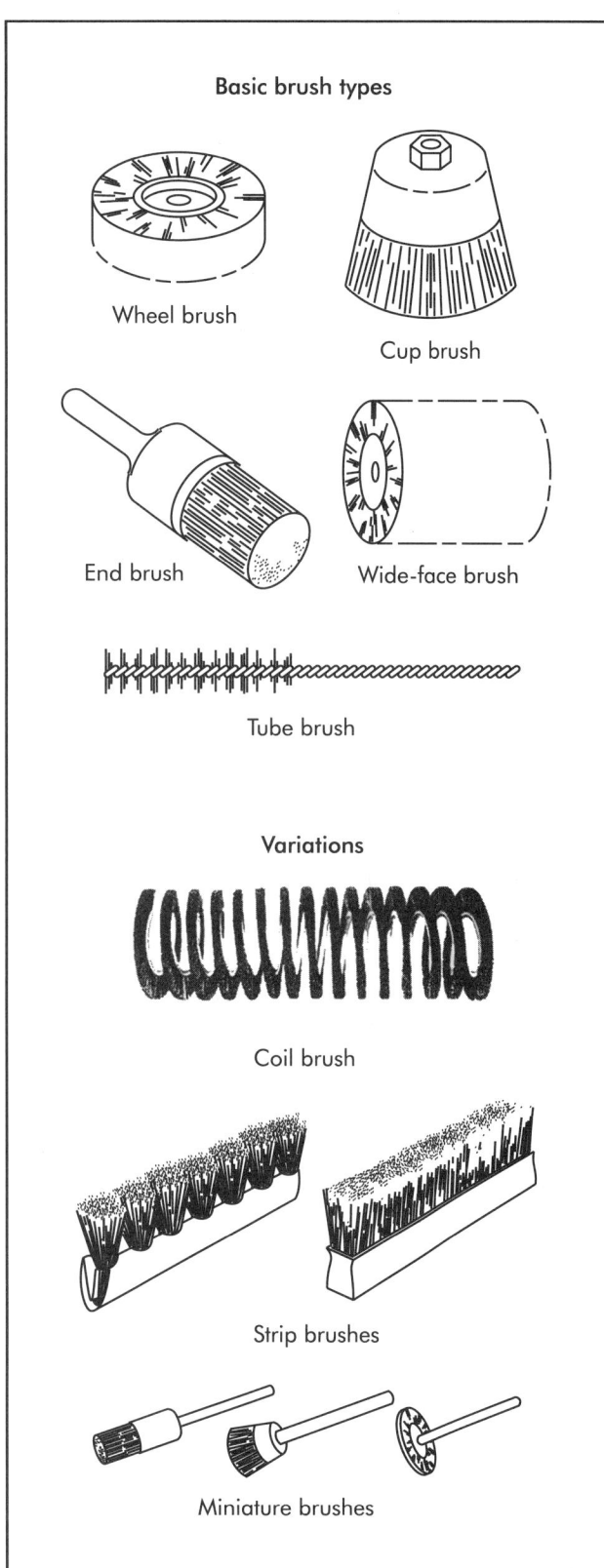

Figure 20-1. Primary types of deburring brushes. (Courtesy Weiler Corporation)

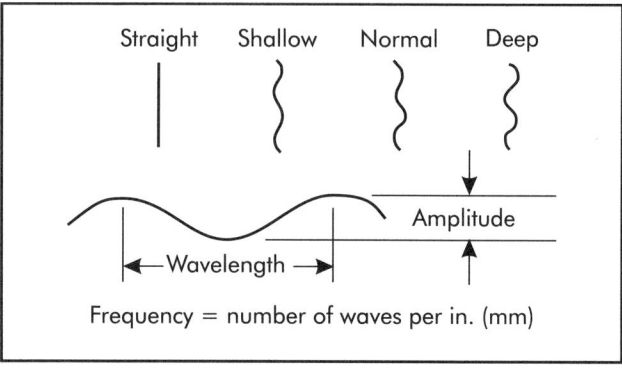

Figure 20-2. Terminology of crimped filaments. (Courtesy Anderson Products, Division of Weiler Corporation)

When making knotted or twisted-tuft filaments, a bundle of straight wires of equal length is formed into hairpin shapes around a retaining member. The bundle is then twisted into a cable with a helix angle of approximately 45°. The standard twist has the tuft twisted for two-thirds of its length; however, in applications requiring maximum impact, such as for extremely severe cleaning or heavy encrustation removal, the entire length of the tuft is twisted (figures 20-3 and 20-4). This is sometimes referred to as cable twist. Multiple rows of twisted tufts, instead of a single row, are sometimes used to provide additional stiffness for the most severe operations.

To fabricate straight wire brushes, rows of holes are drilled in a radial pattern on the periphery of a circular wooden or plastic hub. Small bundles of straight wire are bent into a hairpin shape, pulled into the holes, and locked firmly in place with staples or retaining wire. The protruding ends of the filament wire are relatively unsupported by other wire filaments, so they are highly flexible with minimal filament entanglement. This permits them to penetrate narrow slots and other shapes better than other filament configurations.

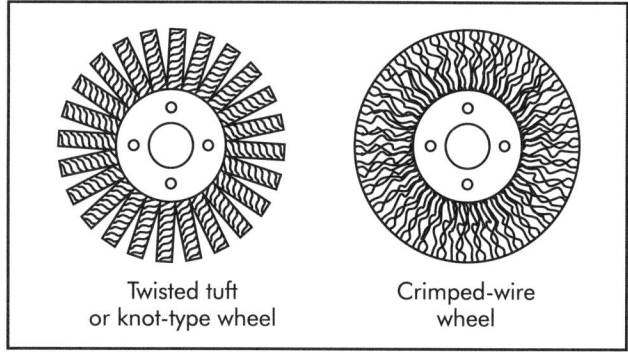

Figure 20-3. Configuration of twisted tuft and crimped-wheel filaments (Gillespie 1999).

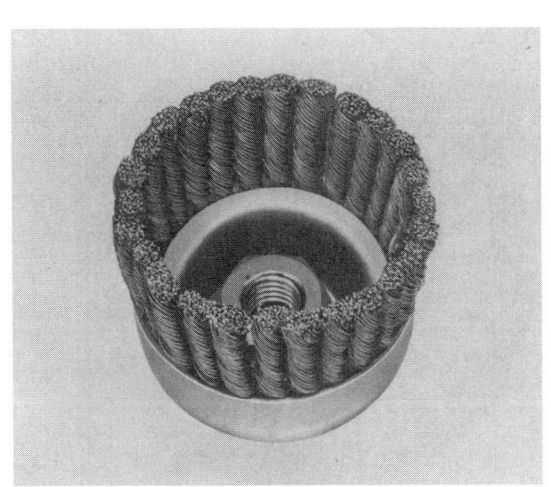

Figure 20-4. Tufted and multiple-row tufted end brushes. (Courtesy Anderson Products, Division of Weiler Corporation)

Multiple-strand crimped filaments are made by stranding together 5–25 straight, fine wires into a steel cord, usually ranging from 0.020–0.050 in. (0.51–1.27 mm) in diameter. Like conventional crimped wire, the cord or cable is then crimped in multiple planes. This type of construction is used in applications requiring the stiffness of coarse wire with the fatigue strength and multiple-point advantages of finer wires (Bateman 1979).

Radial (Wheel) Brushes

Radial (wheel) brushes are generally used for deburring readily available external edges. They are generally placed on fixed machines, such as bench motors; however, small and miniature wheels can be used in handheld motors. Some handheld motors can accommodate larger brushes, but they become too heavy for continuous use and can result in repetitive motion injuries. Table 20-1 provides details about the various sizes of brushes available for each shape or type.

Cup and Disc Brushes

Cup-shaped brushes are used for open surfaces. Typical uses include the tops of cylinder heads and valve bodies with many edges. They have clearance to reach over some low bosses and features, but their main use is to reach down into short cavities to scrub burrs away. Cup brushes have bristles in the same axis as the brush arbor. Bevel brush bristles are at an angle to the arbor. Disc brushes, with bristles parallel to the arbor, are made of abrasive-filled nylon and look more like a floor-cleaning disk than a cup. Bevel brushes are primarily used with right-angle grinders to reach into corners and irregular surfaces. Generally, they are metal bristle tools.

End Brushes

End brushes reach into small, deep pockets and holes to remove burrs and clean counterbores. They come in both tightly bound and flare styles. The flare style allows the bristles to reach sideways into pockets and corners that cannot be reached by a straight up-and-down fiber. Pencil-end brushes, as the name implies, are the size of a pencil and provide an exceptionally compact brush. To protect critical part surfaces, some end brushes can be procured with a rubber coating over the large, nonworking end. Some manufacturers refer to the tightly held end brushes as "banded," which implies that a plastic band is placed around the bristles to minimize bristle spreading and add rigidity.

Some elastomer-coated brushes have a hollow center core to allow use over projecting bosses or studs. Other end brushes contain a solid metal pilot pin to accurately locate the brush around a hole. In this case, the pin slips in the hole and guides the brush. Tufted-end brushes have normal end bristles and radial bristles along a section of the stem to prevent stems from marring hole surfaces.

Tube Brushes

Tube brushes are used almost exclusively for reaching into holes and removing burrs from intersecting holes. They are available with diameters as small as 0.030 in. (0.76 mm) and as large as 6 in. (152.4 mm). Typical tube brushes have helical rows of filaments

Table 20-1. Brush sizes for manual deburring

Brush Style	Metal Bristle		Nylon	
	Minimum Size, in. (mm)	Maximum Size, in. (mm)	Minimum Size, in. (mm)	Maximum Size, in. (mm)
Radial wheel	3.000 (76.20 mm)	15.000 (381.00 mm)	3.000 (76.20 mm)	22.000 (558.80 mm)
Elastomer-coated	2.000 (50.80 mm)	8.000 (203.20 mm)		
Miniature	0.750 (19.05 mm)	2.500 (63.50 mm)	0.750 (19.05 mm)	2.500 (63.50 mm)
Miniature elastomer-coated	1.250 (31.75 mm)	3.000 (76.20 mm)		
Cup	1.750 (44.45 mm)	6.000 (152.40 mm)	2.500 (63.50 mm)	7.000 (177.80 mm)
Elastomer-coated	3.500 (88.90 mm)	6.000 (152.40 mm)		
Bevel				
Miniature	0.563 (14.30 mm)	1.000 (25.40 mm)	0.500 (12.70 mm)	1.000 (25.40 mm)
Disk (for manual use)		2.000 (50.80 mm)	4.000 (101.60 mm)	
End				
Tight	0.500 (12.70 mm)	1.125 (28.58 mm)	0.500 (12.70 mm)	1.000 (25.40 mm)
Elastomer-coated tight	0.500 (12.70 mm)	1.000 (25.40 mm)		
Flare	0.500 (12.70 mm)	1.000 (25.40 mm)	0.500 (12.70 mm)	1.000 (25.40 mm)
Circular-flared	1.000 (25.40 mm)	4.000 (101.60 mm)		
Elastomer-coated circular flare	1.750 (44.45 mm)	2.750 (69.85 mm)		
Miniature	0.188 (4.78 mm)	0.313 (7.95 mm)	0.188 (4.78 mm)	0.313 (7.95 mm)
Pencil	0.188 (4.78 mm)	0.250 (6.35 mm)		
Drill	0.438 (11.13 mm)	0.500 (12.70 mm)		
Tube	1.000 (25.40 mm)	4.000 (101.60 mm)	0.250 (6.35 mm)	4.000 (101.60 mm)
Miniature	0.125 (3.18 mm)	1.500 (38.10 mm)	0.030 (0.76 mm)	0.765 (19.43 mm)
Elastomer-coated miniature	1.250 (31.75 mm)	3.000 (76.20 mm)		
Ball on ends		0.500 (12.70 mm)	24.500 (622.30 mm)	
Egg beater				
Side action	0.250 (6.35 mm)	0.250 (6.35 mm)		
Square	1.063 (27.00 mm)	1.880 (47.75 mm)		
Flare	3.000 (76.20 mm)	8.000 (203.20 mm)		

Brushes are not available in all materials. Tampico and nylon (nonfilled) are available in roughly the same sizes as abrasive-filled nylon radial and miniature brushes.

and range from 0.250–1.000 in. (6.35–25.40 mm) in diameter. Note, for closed-end holes, it is important to procure tube brushes identified as "bottom end–for closed holes," which do not have a long projection of the twisted wire stem at the brush end.

Nontraditional Brushes

Nontraditional brushes for hand deburring include:

- an abrasive ball on bristle ends;
- the metal ball-ended flare brush;
- brush-backed flap wheels;
- elastomer-bonded wire-filled brushes;
- nonwoven abrasive brushes, and
- an eggbeater-style brush.

All these will be discussed later in the chapter.

Miniature Brushes

Miniature brushes are small versions of end, wheel, and cup-type brushes. They are used for cleaning, deburring, and finishing delicate or miniature components. Brush sizes range from 0.156–1.500 in. (3.96–38.10 mm) in diameter. Fill materials include steel, stainless steel, brass, bronze, and nickel-silver wire, as well as various types of animal hair and nylon filaments. The hair-filled types are sometimes used with abrasive media for deburring and polishing small and detailed parts. Miniature brushes are generally mounted on shanks 0.094 or 0.125 in. (2.39 or 3.18 mm) in diameter.

BRUSH FILL MATERIAL

Brush filaments and related materials include tempered and untempered high-tensile steel wire and tempered and untempered stainless-steel wire. A number of nonferrous wires, such as brass, nickel-silver, beryllium copper, Inconel®, and other heat- and corrosion-resistant alloys are also used. Synthetics, such as nylon, polypropylene, and styrene are frequently used in many power brushes. Animal hairs, such as pig bristle and horsehair, and a large number of vegetable fibers, including tampico, habia, bassine, and other combination mixes are frequently used depending on the application. Note that beryllium copper is a cause of chronic beryllium disease, a serious health problem. Fibers from this material may not be a safe choice in the future.

Carbon-steel Wire

In the past, steel wire was probably used more in power brushes than all other fill materials combined, with the tempered, rather than the untempered, variety predominating. The steel wire is a high-quality material similar to that used in springs. It is generally made of a 0.60–0.75% carbon steel with tensile strengths ranging from 300,000–380,000 psi (2,068–2,620 MPa). Untempered steel wires are hard-drawn to their highest tensile strength and usually have lower hardness values and tensile strengths than the tempered wire.

Stainless-steel Wire

Some nuclear applications, and others involving elevated temperatures or corrosive environments, call for stainless-steel wire. Two types are used.

1. Bright-finished type 302 stainless steel hand-drawn to tensile strengths of 300,000–360,000 psi (2,068–2,482 MPa) is usually used in deburring stainless steel and aluminum or other nonferrous materials in an effort to preclude after-rust, which is caused by foreign deposits that occur with the use of carbon steel wires (West 1975).
2. Type 420 tempered stainless steel is used with tensile strengths ranging from 240,000–255,000 psi (1,655–1,758 MPa).

Type 420, however, is only 80% as corrosion resistant as type 302. In spite of this deficiency, it offers distinct advantages in many deburring applications. It is vastly superior to untempered stainless in its cutting capabilities, fatigue resistance, and abrasion resistance. In some instances, it has a life of 7–10 times that of untempered stainless steel, while doing substantially more cutting. It is not recommended, however, for use in nuclear applications or those where it is desirable to deburr with brushes constructed of nonmagnetic stainless steel. Type 420 is magnetic at all times; therefore, type 302 should be used in these applications. Do not make the mistake of checking the wire with a magnet. When drawn to spring temper, as used in brushes, wire from all heats of type 302 can be attracted to a magnet by varying degrees. The degree of the magnetic attraction of a given wire size in the spring tensile range is a function of the degree of cold working that the wire has undergone. Cold work increases tensile strength and, simultaneously, the magnetic permeability of the wire. Only in its fully annealed condition can type 302 stainless steel be considered essentially nonmagnetic; because, in this state, it exhibits permeabilities of less than 1.02 oersteds/gauss.

Tampico

Tampico, a cellular vegetable fiber, is used more widely than any other fiber (nylon is more predominant today than fiber) in deburring brushes. It produces a flexible, resilient brushing wheel. Fibers are used singly or in multiples to make a wide-face deburring brush. Tampico brushes can be treated and cured with a lacquer for improved stiffness and grit retention. This treatment also minimizes shedding, knifing, and weaving, which can occur at high speeds. Treated brushes, therefore, can perform an infinitely wider range of deburring and edge blending than can be accomplished with natural, untreated fiber. The stiffness and tackiness attained by the treatment makes it easy to apply grease stick and abrasive compounds, which are used with these brushes.

One of the main reasons treated tampico brushes are used is to increase the life of the part being brushed. Fatigue failure due to widening microscopic cracks (which result from stress concentrations), accounts for over one-half of all part failures subjected to repeated cyclic stresses. Radii and surface finishes, in proximity to the deburred edge, can almost completely eliminate the scratches, burrs, and sharp corners that cause cracks and progressive fractures.

Nylon

Nylon brushes (without abrasives) are used for gentle finishing. Nylon may be more widely used in finishing plastics than metals, but it will remove very fine metal burrs.

Abrasive Monofilament

Nylon monofilaments with permanently encased aluminum oxide, silicon carbide, or pumice grit are commonly used in deburring and edge-finishing tools. Silicon-carbide grit offers the best combination of hardness, sharpness, and toughness. It has harder and sharper grains than aluminum oxide, is more friable, and is better at finishing metal. Use it for most nylon brushing jobs. Aluminum oxide grit is less likely to fracture and is generally used for finishing soft metals. Because aluminum oxide prevents carbon contamination, the aerospace industry usually prefers it for jet engine blades and other parts.

Additional abrasive particle materials include aluminum silicate, diamond, and boron carbide, but none are standard. Rather, they may be necessary for some exotic part materials. Between 20 and 40% of the abrasive particles, by weight, are uniformly distributed throughout the filament. Grit sizes range from 80–800, with the coarsest grit encapsulated in relatively large-diameter filaments, and the fine sizes in small filaments. This material can be used either wet or dry. Some users believe it provides better action when used along with water, mineral oil, or soluble oil as a coolant. Grease sticks have also been used with this carrier. These filaments are used for finishing the edges of wood, leather, plastics, cork, rubber, ceramics, and glass.

Abrasive elements are sparsely distributed in the nylon; therefore, more nylon than exposed abrasive—which generates heat—rubs the part surface. Smearing (embedding small brush particles into the work material) occurs readily on poor heat conductors, such as stainless steel. This can be eliminated by reducing the friction between the nylon bristle and the part. To do this, spray a lubricant onto the brush surface when the tool is stopped or barely moving. The brush can then accept twice the workload and forces before smearing occurs. Maintain the nylon filaments' temperature below 150° F (66° C) to maximize brush life. Hydrocarbons, oils, and most organic solvents have no lasting effect on the filaments.

As with wire brushes, applying less pressure to abrasive-nylon brushes increases life. Unlike wire, they run at low speeds with enough pressure to force the sides, as well as the tips, of the filaments to cut. Optimum cutting for this material results when the workpiece penetrates, or "mushes," into the brush face.

While the filaments are normally round, they can be made rectangular, oval, or any other shape that can be extruded. Figure 20-5 illustrates the difference in contact area when rectangular cross sections are used. Usually, the monofilament diameter is 0.014–0.050 in. (0.36–1.27 mm).

Figure 20-6 illustrates the commercial sizes of some filaments. Rectangular filaments can be used for more contact and impact with the workpiece in applications such as carbide edge honing, deburring injector nozzle cutouts, and finishing decorative hinges. Round, uncrimped filaments can be used for applications such as deburring slots in cast-magnesium computer cage backs and putting a uniform scratch-finish on aluminum chair bases.

Materials other than nylon can be used for brush filaments, but nylon has several desirable properties. It is resistant to most chemicals; however, strong acidic solutions degrade and embrittle nylon. It is relatively resistant to abrasion. It also "remembers" its original shape, which is significant because brush filaments are constantly being bent. Stiffness of the nylon filaments drops as temperature rises. They are twice as stiff at 32° F (0° C) as they are at 75° F (24° C). Stiffness drops another 50% at 140° F (60° C). Visible smearing occurs when temperatures reach near the melting point (410–482° F [210–250° C]). Temperature effects are minimized by controlling brush speed and pressure and using a coolant, such as water.

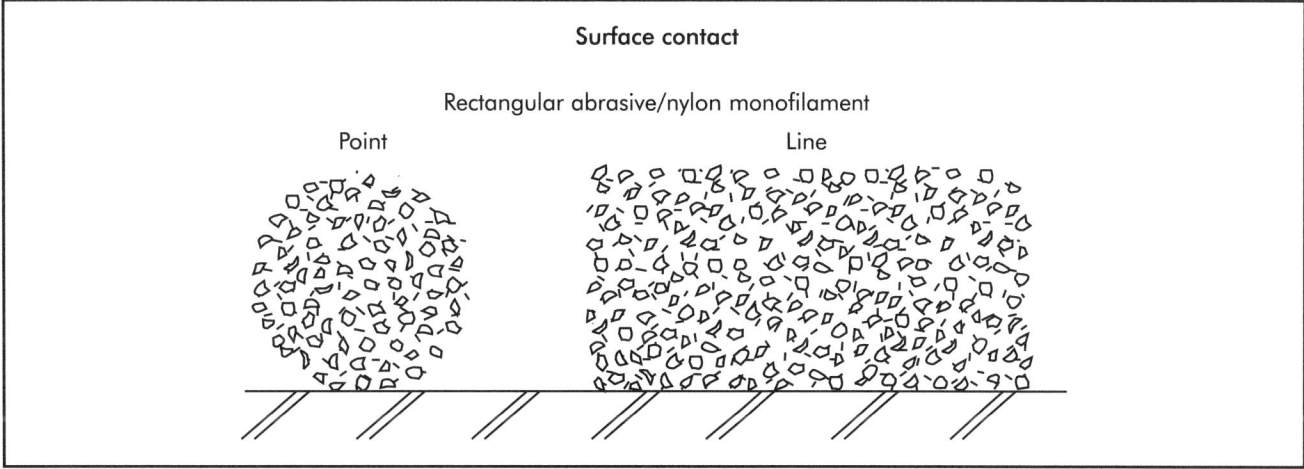

Figure 20-5. Contact area for round nylon filaments versus rectangular nylon filaments (Gasser 1993). (Courtesy Osborn Manufacturing)

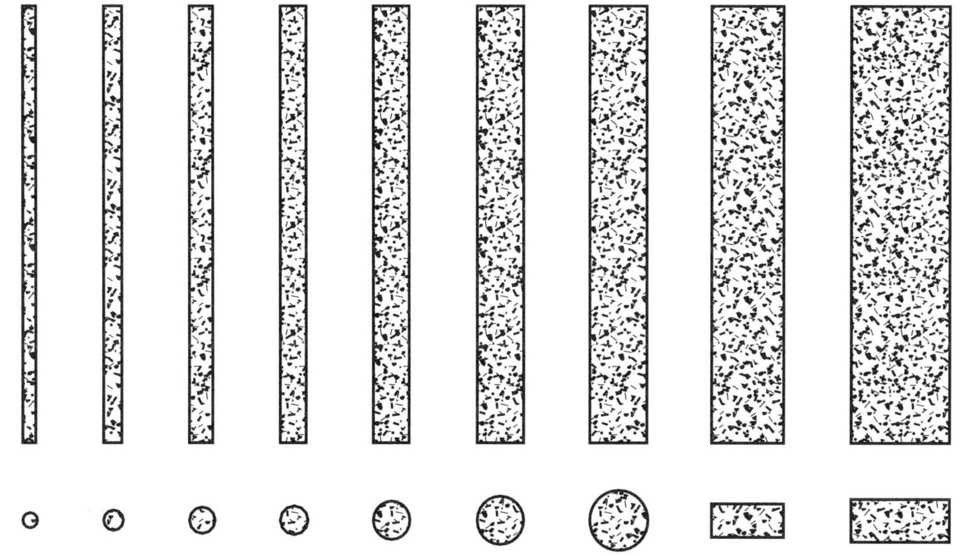

Shape									
Round, rectangle	◉	◉	◉	◉	◉	◉	◉	▭	▭
Size									
mil		8	12	18	22	35	40	50	30 × 70	45 × 90
mm		0.20	0.30	0.46	0.56	0.89	1.02	1.27	0.76 × 1.78	1.14 × 2.29
Grit									
Aluminum silicate	AS								
Aluminum oxide		AO		AO	AO	AO	AO	AO	AO
Silicon carbide			SC	SC	SC	SC	SC	SC	SC
Mesh									
USA	1000	600	500	320	180	120	80	80 120 180	80 120 180
				120				320 600	320 400

Figure 20-6. Abrasive-filled nylon filament shapes and sizes (Gasser 1993). (Courtesy Osborn Manufacturing)

Nonwoven Abrasives

As discussed in Chapter 19, abrasive grains are dispersed uniformly throughout a low-density, open-type, nonwoven web of synthetic fibers and are held there with a binding adhesive (figures 19-14 and 19-15). This produces a cushioned abrasive ideal for fine deburring. Abrasives include aluminum oxide, silicon carbide, flint, and garnet. Although specific grain sizes are used (80–800-grit), these tools are often specified by grades such as coarse, medium, fine, very fine, super fine, and ultra fine. Other nomenclatures are also used.

Nonwoven abrasives are often substituted for flap wheels. They frequently replace setup wheels, steel wool, and abrasive flap wheels. They are waterproof and chemically resistant, can be used wet or dry, are nonloading, and are highly flexible. Nonwoven abrasives efficiently remove light burrs; however, they should not be used on large burrs. The burr removal capability of nonwoven abrasives is substantially enhanced by the availability of a wide variety of densities (degrees of hardness).

Abrasive filament and nonwoven nylon deburring brushes can be used wet or dry. Experience indicates some advantages of wet use when water, mineral oil, or water soluble oil is used as the cooling medium. Grease sticks have also been used efficiently. Generally, these brushes can be used in the same applications as treated tampico brushes. Compounds and slurries are unnecessary, which increases their advantage. Such compounds, however, can be used with abrasive filament brushes to accelerate cutting action. The abrasive filament brushes can also be used in applications for which the tackiness of treated tampico brushes would be objectionable, such as deburring wood, leather, plastics, cork, rubber, ceramics, and glass.

BRUSH AGGRESSIVENESS

Ten factors influence the ability of brushes to remove burrs. For maximum aggressiveness, use a large-diameter brush with closely-packed, short filaments. Knotted or crimped filaments add stiffness, as does a wider wheel. When the diameter of the filaments performs the deburring (rather than the ends) a tough filament surface provides greater abrasion. Typically, it is desirable to cut with the ends of the filaments, although in abrasive-filled nylon brushes, sliding filaments over an edge adds to the radiusing action. Faster velocities increase the effective stiffness of brushes. The following factors affect conventional brush aggressiveness:

1. diameter of filaments;
2. free length of filaments;
3. filament material;
4. filament configuration;
5. density of filaments;
6. texture of filaments;
7. velocity of wheel;
8. width of wheel;
9. type of contact with part, and/or
10. configuration of filament end.

Typically, one thinks of brush filaments as having a uniform diameter since they normally begin that way. After use, however, the brush ends may not be uniform. For example, nylon brush-ends taper with use. Tapered brush ends are less stiff, which means they cut less.

As a basic economic rule, throw away brushes when the filaments become obviously worn. However, there are exceptions. One engineer, for example, tried to get workers who were using miniature abrasive-filled nylon wheels to throw them away when the filaments were obviously worn. But, the workers noted that these dog-eared brushes were ideal for precision, ultra-miniature gears. An investigation revealed that the 0.014-in. (0.36-mm) diameter nylon filaments had tapered to 0.006 in. (152 μm), which allowed them to reach the bottom of gear teeth on 96 diametrical-pitch gears. In addition, because the brushes had become much more flexible, they could maintain a required maximum edge radius of 0.005 in. (127 μm).

Brush flexibility refers to the brush's capability to conform to irregular or contoured surfaces as measured by filament deflection. This quality, which determines resiliency or stiffness, is equated by resistance to filament bending. Flexibility can be varied in a number of ways:

- Change the modulus of the fill material. The modulus differs for different materials. For example, steel, brass, hair, nylon, etc., differ.
- Increase or decrease the diameter of fill material. For example, 0.020 in. (0.51 mm) steel wire is stiffer than 0.005 in. (127 μm) steel wire of the same analysis.
- Increase or decrease the trim length of the fill material. For example, a 0.010-in. (0.25-mm) diameter steel wire filament that is 1 in. (25.4 mm) long is stiffer than a similar 0.010-in. (0.25-mm) diameter steel wire that is 4 in. (101.6 mm) long.
- Increase or decrease the speed of rotation. For example, the faster a brush is rotated, the stiffer it becomes due to the influence of centrifugal force.
- Change the construction of the brush. For example, filaments or wire twisted together (as in a knotted or twisted-tuft brush) are stiffer than

those in a crimped wire brush. In the knot or twisted-tuft construction, the twisted filaments act as a family of filaments; whereas, in a crimped wire brush, they almost act as individual filaments.
- Increase the number of brushing points per square unit of brush face. This changes brush flexibility through wire packing, load division, or both. Flexibility increases as density decreases, so there is less capability to conform results as density increases.
- Encapsulated, or otherwise treated or coated brushes, are less flexible than untreated brushes.

A basic relationship in brush stiffness is the standard formula for beam deflection (Bateman 1975).

$$D = \frac{(WL^3)}{(3EI)} \qquad (20\text{-}1)$$

where:

D = deflection of filament, in. (mm)
W = force causing deflection, lbf (kN)
L = length of filament, in. (mm)
E = modulus of filament, psi (MPa)
I = moment of inertia of the cross section of the filament, in. (mm)

For a round filament, Equation 20-1 becomes:

$$D = \frac{WL^3}{(0.1473Ed^4)} \qquad (20\text{-}2)$$

where:

d = diameter of filament, in. (mm)

Clearly, this equation shows that:

- the stiffness (resistance to bending) of a single round filament, fixed at one end, is proportional to the 4th power of the diameter, and
- the stiffness of a single round filament, fixed at one end, is inversely proportional to the 3rd power of the length of that filament.

In many instances, brushes are made with tufts of filaments rather than single filaments. In the case of a tuft, Equation 20-2 becomes:

$$D = \frac{WL^3}{(0.147Ed^4)} \qquad (20\text{-}3)$$

The area of a round filament is $\pi d^2/4$. Therefore, the number of filaments, N, for a given size tuft is inversely proportional to the 2nd power of the diameter of the filament. Thus, the following conclusions can be drawn for tuft stiffness:

- The stiffness (resistance to bending) of a tuft of round filament is proportional to the 2nd power of the filament diameter.
- The stiffness of a tuft of round filament is inversely proportional to the 3rd power of the length of that tuft.

From these conclusions, the following formula can be written, which relates filament diameter and trim length of one brush to filament diameter and trim length of another brush for equal stiffness:

$$\frac{L_1^3}{D_1^2} = \frac{L_2^3}{d_2^2} \qquad (20\text{-}4)$$

The remaining unknown is computed by substituting the known values in Equation 20-4 and assuming either a new trim length or diameter. You can use the nomograph in Figure 20-7 to graphically solve this equation. As an example, if bristles are 0.5 in. long and 0.050 in. in diameter on an existing brush, and a smaller diameter of 0.030 in. was desired to reach into crevices, the trim length (l_2) of the new brush with equal stiffness should be:

$$\{[(0.5)^3/(0.050)^2](0.030)^2\}^{1/3} = 0.355 \text{ in.}$$

Longer bristles will result in a softer-acting brush, while shorter bristles will act harder than the original brush. Obviously, the illustrated nomograph is for use only when brushes are similarly constructed. It would be impractical to develop a chart that included all varieties of brush constructions. In most brushes, one filament touches another, so the filaments do not function individually, but rather as a family of filaments. Interaction also varies with tuft size, trim length, and the rigidity of the tuft in the retaining member.

Abrasive-filled Nylon Brushes

The effectiveness of abrasive-filled nylon brushes can be significantly altered by simply changing the filament. The following points have been demonstrated by using different filaments in the same tool and using the tool on similar parts.

- A smaller diameter of filament in the brush tool is often more effective. The filament can adjust to the contour of a part more easily, and more surface area can be exposed to the part. For example,

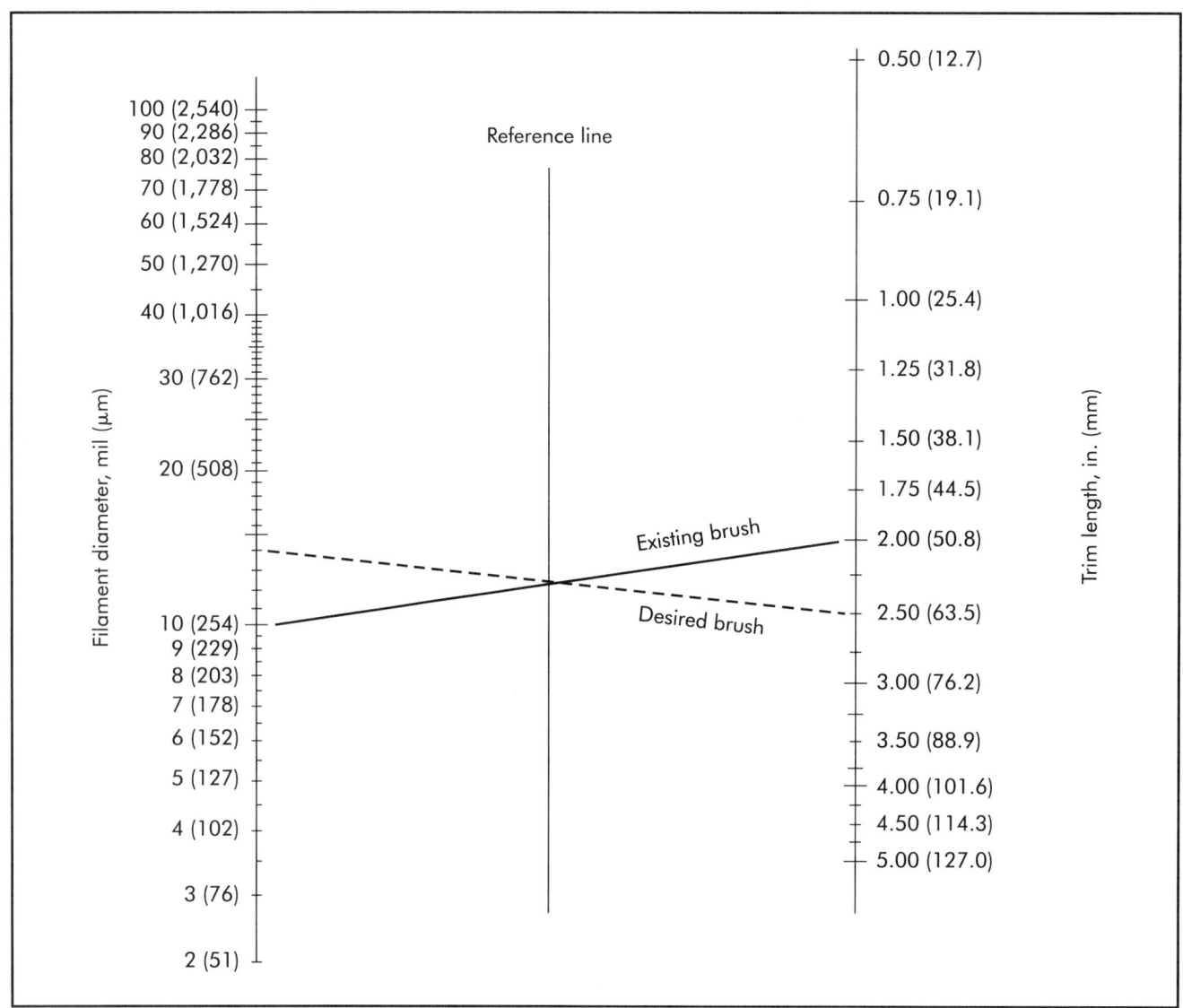

Figure 20-7. Nomograph for relating trim length and filament caliper for equal brush stiffness (Bateman 1979). (Courtesy Anderson Products, Division of Weiler Corporation)

the difference between 50 and 40 mil (1.27 and 1.02 mm) filament is 25% less of the amount of abrasive-filament surface area that can contact the part, given that the tool contains the same volume of filament. A filament that is too large may hit and bounce off the part.
- Small grit can become more effective by offering more cutting surfaces. Using 120-grit versus 80-grit in the same size filament offers about 50% more grit particles on the surface of the filament.
- Increasing grit loading from 30% (which was standard for years) to 40% increases grit particles by one third, offering another level of aggressiveness.

These variations do not include changes that can be made in the way the tool is used, such as altering the tool speed and location, part position, angle of contact, etc.

TOOL WEAR

Tools become worn, and brush tools are no exception. As brush tools wear, their performance characteristics change. Short-trim brushes wear from the tip. As the length of the filament gets shorter, the brush tool becomes more aggressive since the filament becomes shorter and stiffer. The original filament deflection, usually 0.125 in. (3.18 mm) with proper pressure setting, cannot be maintained without increasing pressure. Higher pressure increases stress on the wire filament at the flex point, usually where the filament joins the tool base. This added stress can eventually cause the filaments to break.

Longer trim brushes wear out for a different reason. As with short-trim brushes, the normal wear pattern is a gradual tapering of the filament as the nylon and grit wear. Eventually, the filament diameter narrows and the tips begin to break. Of course, as shown earlier, the stiffness is reduced as the diameter changes. In continuous operations, these tools are changed regularly to maintain a consistent finish effect.

Although flex fatigue can occur, brush construction methods have been developed to better support filaments at the flex point. Filament fatigue, if it occurs, normally results from placing the part too close to the tool core, causing excessive filament flexing. Circulation of coolant at the flex point minimizes heat generation and filament fatigue.

POWER REQUIREMENTS

Three primary factors determine the amount of horsepower (kW) required in any given operation:

- the brushing pressure required to accomplish the work;
- the resistance developed between the work and brush, and
- the brush speed.

Obviously, other influencing factors include brush configuration and filament type and size. Table 20-2 provides a working guide for brushing horsepower (kW) requirements; however, even this guide is based merely on medium brushing action. For more complex conditions, additional research would be necessary to determine the correct horsepower (kW). Note that Table 20-2 shows the horsepower (kW) for a 1-in. (25.4-mm) brush face (1-in. [25.4-mm] wide brush). Power needs for wider brushes are directly proportional to face width.

Effective, low-cost brushing also is influenced by the speed at which the brush rotates. For each particular application, there is an optimum operating speed. Using a higher or lower rotation affects efficiency. Table 20-3 shows general surface speeds for various types of brushing applications using metal brush filaments. In automated and semi-automated applications, brushes should be operated at the highest practical speed with the lightest possible pressure. Wherever practical, equipment should be designed with force or load meters and aligned to measure the load on the drive motor continuously, so the amount of brushing pressure in the particular application is indicated and may be controlled.

In any brushing application using metal filaments, the sharp tips of the brush filaments do the work. Excessive pressure displaces the filaments, causing a wiping, rather than cutting, action (Figure 20-8). Worse, excessive pressure causes excess filament displacement, which overstresses the filament wire and causes premature fatigue failure. Brushing, by its nature, produces forces that compress, flex, and overheat (sometimes) the filaments. When these forces exceed endurance limits, they cause fatigue. Brush failure is almost immediate, and the greater the mechanical stress, the sooner the damage to the brush filaments occurs.

Fatigue failure is possible for any size of wire, but susceptibility varies with wire diameter—finer wires are much more capable than larger-diameter filaments of resisting overstress conditions. In selecting brushes, this factor must be recognized so load tolerances, cycle stress, and other fatigue-inducing characteristics can be compensated for by varying the filament diameter.

Table 20-2. Power ratings for 1-in. (25.4-mm) wide brush

Brush Diameter, in. (mm)	Recommended Motor Size, hp (kW)	Operating Speed, rpm
4 (102)	0.25 (0.19)	3,450
6 (152)	0.50 (0.37)	3,450
8 (203)	0.75 (0.56)	3,450
10 (254)	1.00 (0.75)	1,750
12 (305)	1.00 (0.75)	1,750
15 (381)	1.50 (1.12)	1,750

(Bateman 1975)

Table 20-3. Recommended surface speeds for wire-brushing applications

Application	Recommended Surface Speed	
	sfm	m/s
Burr removal	5,500–7,500	28–38
Scale removal	7,500–10,000	38–51
Weld cleaning	7,200–9,400	36–47
Edge blending	4,700–7,500	24–38
Cleaning (dry)	4,000–5,500	20–28
Cleaning (wet)	1,900–4,000	10–20
Surface polishing	6,400–8,000	32–40
Surface buffing	8,000–10,000	40–51

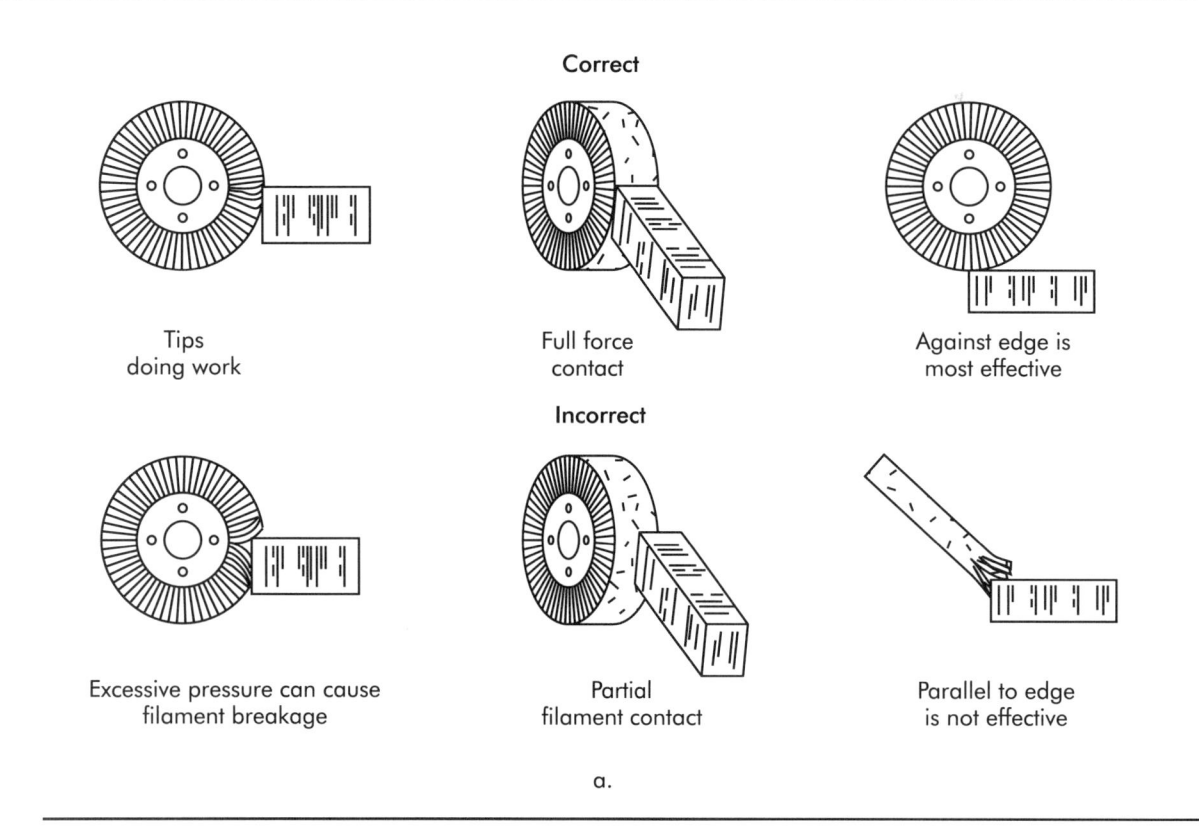

a.

Set tube ℄ on ℄ of brush for I.D. and O.D. brushing

Set tube ℄ above ℄ of brush for I.D. brushing

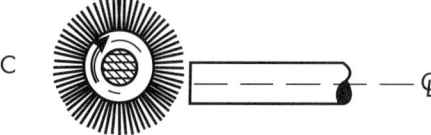

Set tube ℄ below ℄ of brush for O.D. brushing and chamfering

A. If you have a similar burr on the I.D. and the O.D., set the centerline of the tube on the centerline of the brush and you should get equal brushing action on the O.D. and I.D.

B. For an excessive I.D. burr, or if you just want to work the I.D. harder, set the tube centerline above the brush centerline.

C. If you wish to work the O.D. harder, set the tube centerline below the brush centerline.

b.

Figure 20-8. Some correct and incorrect brushing techniques for wire brushes. (Figure 20-8b Courtesy TESGO)

Many operations have attained more efficient brushing by periodically reversing the direction in which the brush rotates. Obviously, filament tips become dull with use so reversing them tends to bring a new, sharp cutting edge to the work. This accelerates cutting capabilities, which reduces the amount of pressure required to do the work and, consequently, operator fatigue.

BRUSH SELECTION

Brush flexibility is a major variable, and the degree of flexibility determines the degree of cutting capability. So, flexibility must be considered in selecting the best brush for the work. The most flexible brushes, regardless of style or configuration, are made of synthetic or natural fibers, such as horsehair, tampico, nylon, etc. These are medium-flexible brushes used in:

- wet and dry cleaning;
- application of oils, including compounds, powders, etc.;
- light finishing, buffing, and polishing, and
- edge blending.

Medium-flexible brushes are used in these applications because of their inherent flexibility and low brushing strength. Generally, they are used without filament treatment or abrasive compounds.

Highly flexible brushes of synthetic or natural fibers can be made slightly stiffer by chemical treatments. This slightly reduces their flexibility and slightly increases the speed of their cutting capabilities. It also permits the use of abrasive compounds for low μin. (μm) surface finishes. These brushes are also used in:

- small burr removal;
- light scale removal;
- surface blending;
- surface finishing for appearance, and
- various cleaning operations.

A variation on this type of brush is the abrasive-filled filaments, where aluminum oxide or silicon carbide is impregnated in a nylon monofilament. This combination still produces a medium-flexible and fast-cutting brush; however, it does not require additional abrasive compounds since the abrasive is already imbedded in the filament. It, too, can be used wet or dry, and the filament diameter can be varied to attain different degrees of stiffness.

If experience indicates that synthetic or natural fiber brushes are not stiff enough for the application (even if they have been treated), fine wire brushes can be used with abrasive compounds. Usually, the wire diameter in this type of application would be 0.006 in. (152 μm) or finer. Fine-wire brushes, in conjunction with an abrasive compound, are used for:

- medium-size burr removal;
- light scale removal;
- edge blending, and
- surface finishing for appearance.

If the brushing application allows a surface finish that exceeds approximately 30 μin. (0.8 μm), another group of brushes is used. The most flexible of this group of brushes are the narrow-face, highly-flexible standard knot type (Figure 20-3). These brushes are used if surface impact action is necessary. They can be used for:

- weld splatter and scale removal;
- general-purpose cleaning;
- surface roughening;
- edge blending;
- burr removal, and
- surface finishing.

To decrease the flexibility of knot brushes, they sometimes are made in what is called *high complement* style. This produces a brush of maximum density with only medium flexibility. These brushes are used for rugged production jobs, especially for automatic or semiautomatic machine application. High-complement knot brushes have relatively short trim length, and the knots are close for increased density. They would generally be used for:

- heavy scale removal;
- heavy burr removal;
- heavy cleaning of rust, paint, rubber, and plastics;
- surface roughening, and
- edge blending.

Regular-crimped wire brushes with a narrow face are medium density and are used where medium-brush flexibility is required. These brushes are used in:

- general-purpose cleaning;
- surface finish for appearance;
- surface roughening;
- edge blending;
- surface refining, and
- heat-scale removal.

These same brushes in a wide face and with a more dense fill would have much faster cutting action and produce fine finishes. Usually, they are used on a general-purpose wheel. Uses for this wider-faced, crimped-wire brush include:

- burr removal;
- general-purpose cleaning;
- surface finishing;

- edge blending;
- heat-scale removal, and
- surface refining.

A high-density, regular-crimped wire brush can be made that yields a super-dense, low-flexibility brush. It can be used during high production, on automatic or semiautomatic brushing equipment for:

- heavy burr removal;
- heavy scale removal;
- edge blending;
- surface finishing;
- general-purpose cleaning, and
- surface refining.

When maximum rigidity of wire filament brushes is required, an elastomer-bonded brush is used. This is a brush with the lowest possible flexibility of filament. It is a rigid-wire wheel covered with a poured plastic, which makes it look somewhat more like a colored grinding wheel than a brush. The elastomer brush has fast action and a controlled (non-flaring) face. This type of brush is used for:

- burr removal;
- edge blending;
- scale removal;
- roughening;
- confined areas, and
- general-purpose cleaning.

Brush flexibility, finishing ability, and brushing action of all brushes can be modified by changing the diameter or density of the wire. Table 20-4 provides commonly used suggestions to improve brushing performance. Obviously, when large areas are to be brushed, the widest possible brush face will provide maximum efficiency.

By applying the basic brush technology and fundamental information provided in this chapter, Table 20-4 can be easily used to select brushes. Note that the table is divided into two parts—Chart 1 and Chart 2. Make a tentative brush selection from Chart 1, which is divided in half based on surface finish requirements after deburring. If requirements are less than 30 μin. (0.76 μm), use the lefthand side of Chart 1. If requirements greater than 30 μin. (0.76 μm) can be tolerated, use the right half. Note that both sides are similarly arranged, with the most flexible brushes listed on the outside. Proportionate relationships exist on the finishing abilities' cutting characteristics. Brush types are listed sequentially and are easily related to the major brush characteristics. Also listed are major brush applications in the order of general usage.

After selecting a brush based on Chart 1, test it. Its performance should be observed and notes taken. Based on those observations and notes, refine the brush specifications or operating conditions, if necessary, by applying the information to Chart 2. Table 20-5 provides some basic wheel diameter and speed information, which is useful for initial tests.

Researchers have generally failed to record relationships between initial burr size, the resulting edge radius, and brushing time. Therefore, literature contains little quantitative information about brush deburring. However, tests have shown that when the intent is to remove a 0.0005-in. (13-μm) thick grinding burr and produce a minimum radius, miniature brushes can easily do this at a feed rate of 1–2 in./min (0.004–0.008 m/s) and speed of 1,800 sfm (9.1 m/s). Table 20-6 illustrates some comparative results. As it shows, nylon brushes maintained significantly smaller edge breaks than stainless brushes with smaller filaments.

Edge breaks as small as 0.003 in. (76 μm) have been maintained on both ferrous and aluminum parts while removing burrs. As in all deburring processes, however, maintaining small edge breaks requires small initial burrs. Smaller burrs allow the use of softer wheels, which can hold finishes as low as 4 μin. (0.102 mm).

Most users will be happy with very large edge breaks, so more-aggressive brushes are usually used.

Because of the need for verifiable comparative data on brushes, many users must perform their own testing. Consequently, many workers rule out potentially suitable products before testing them. For example, one manufacturer typically removed large burrs on miniature parts by using specially designed knives. Parts were then brushed with the nylon brushes described in Table 20-5. This produced edge radii of the desired size and finish. Too much edge breakdown had caused previous efforts with brushes to fail. When a wider, denser brush was obtained with a 0.024-in. (0.61-mm) diameter bristle and a trim length of 0.250 in. (6.40 mm)—instead of 0.375 in. (9.53 mm)—engineers and workers felt that the brush was still too stiff to produce the desired result. The brush remained in a box for several weeks until a worker, who had tired of knife work, decided to try it. The worker's efforts with this brush, which "obviously would not work," produced the desired result and saved the manufacturer several thousand hours of labor time. Brushes are inexpensive, and it may be advantageous to try several before ruling out their effectiveness.

BRUSH DEBURRING EQUIPMENT

At least 15 U.S. manufacturers produce standard or specially designed machines for brush deburring. A number of other manufacturers also produce bench motors, pedestal grinders, and single- and two-spindle

Table 20-4. Recommendations for selecting brushes (Courtesy Anderson, Division of Weiler Corp.)

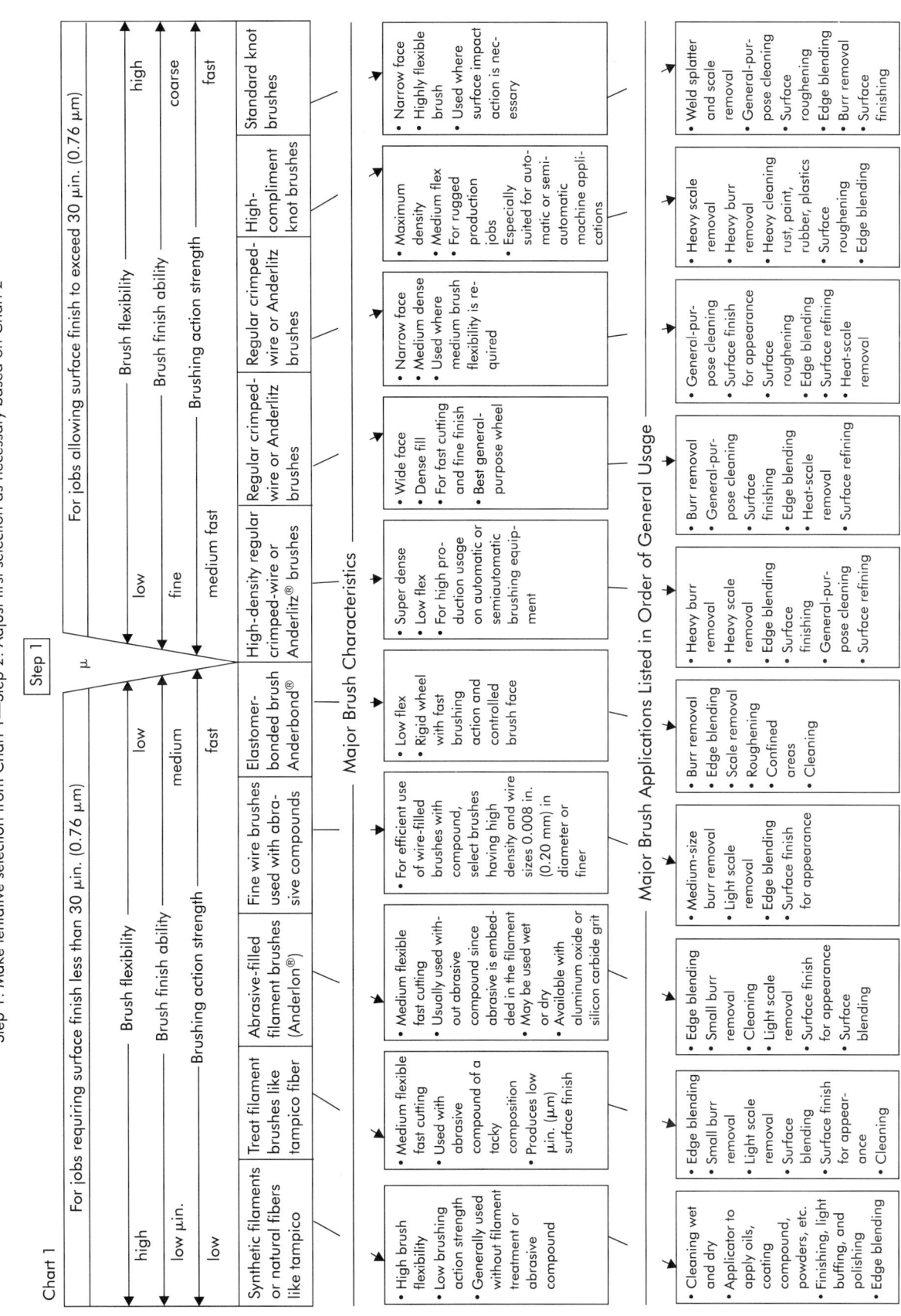

Table 20-4. (continued)

Step 2

After testing brush first selected based upon Chart 1, the following refinements may be made in brush specifications or operating conditions

Chart 2

Observed results

Brush works too slowly	Brush works too fast	Finer or lower μin. (μm) surface required	Surface finish is too smooth and lustrous	Brushing action not sufficiently uniform	Action of brush peens burr to adjacent surface
• Increase brush outside diameter • Increase brush speed • Decrease trim length, thereby increasing the brush fill density • Increase filament diameter • Change brush to heavier duty type	• Reduce trim length • Decrease filament diameter • Reduce brush outside diameter • Reduce brush speed • Select brush types having lower brushing action strength	• Decrease trim length, thereby increasing brush fill density • Decrease filament diameter • Select brush capable of producing lower μin. (μm) surface finishes • Use auxiliary abrasive compounds or oil-base lubricants with brush • Increase brush speed	• Increase trim length • Increase filament diameter • Select a brush type capable of producing coarser finishes • Reduce brush speed	• Increase trim length • Change brush type from sectional to helical construction • Select suitable brushing machinery • Oscillate brush to avoid "tracking"	• Decrease trim length • Select high-density crimped-wire wheel • For ductile metals change from wire-filled brush to nonmetallic-filled brushes

Brush face loads with material being removed	Brush develops excessive heat	Better control of brushing pressure	Brush wire filament ends become dull	Improve brush dynamic balance condition	Brush face wear uneven or grooving
• Increase brush trim length • Increase brush filament diameter • If wet application, apply solution through brush arbor and disperse outward through brush face • Where cleaning solutions are used with brush, modify cleaner for better solubility • Change brush fill specifications	• Use internal cooling of brush and brush arbor • Reduce brushing pressure • Select heavier duty brush • Increase brush trim length • Reduce brush peripheral speed • Use external cooling of brush	• Increase brush trim length • Install meters to measure brushing loads • When part size varies, design pressure relief in part-holding fixture • With automatic brushing equipment, allow brush to float or provide automatic pressure sensing and pressure adjusting	• Reverse rotational direction of brush to automatically recondition • Dress the brush face with an abrasive stone with the brush operating in a direction reverse of normal use • Select a heavier duty brush • Keep brushing heat as low as practical	• When wide-faced brush units are made up of thin sections, statically balance each section and stagger heavy sides • Change from sectional brush assembly to a helically formed brush • Dynamically balance brush arbor and brush assembly • Reduce brush speed • Check entire brush drive train	• If possible, have part contact brush so the full face of brush is on the part • Consider oscillating brush or part • Provide a means for periodically dressing the brush face • Avoid excessive brushing pressure • Change from sectional brush assembly to a helically formed brush part

(Bateman 1975)

Table 20-5. Relationship of surface speed to rotational speed for various brush diameters

Rotational Speed (rpm)	Brush Diameter, in. (mm)					
	4 (102)	6 (152)	8 (203)	10 (254)	12 (305)	15 (381)
	Surface Speed, ft/min (m/min)					
1,000	1,050 (320)	1,575 (480)	2,100 (640)	2,625 (800)	3,150 (960)	3,925 (1,196)
1,500	1,575 (480)	2,350 (716)	3,150 (960)	3,925 (1,196)	4,725 (1,440)	5,900 (1,798)
1,750	1,800 (549)	2,750 (838)	3,650 (1,113)	4,550 (1,387)	5,500 (1,676)	6,800 (2,073)
2,000	2,100 (640)	3,150 (960)	4,200 (1,280)	5,250 (1,600)	6,275 (1,913)	7,850 (2,393)
2,500	2,625 (800)	3,925 (1,196)	5,250 (1,600)	6,550 (1,996)	7,850 (2,393)	9,825 (2,995)
3,000	3,125 (953)	4,725 (1,440)	6,275 (1,913)	7,850 (2,393)	9,425 (2,873)	11,775 (3,589)
3,450	3,600 (1,097)	5,400 (1,646)	7,200 (2,195)	9,000 (2,743)	11,000 (3,353)	13,500 (4,115)
3,750	3,900 (1,189)	5,900 (1,798)	7,800 (2,377)	9,800 (2,987)	11,800 (3,597)	
4,000	4,175 (1,273)	6,275 (1,913)	8,375 (2,553)	10,475 (3,193)		
4,500	4,700 (1,433)	7,075 (2,156)	9,425 (2,873)			
5,000	5,225 (1,593)	7,875 (2,400)				
6,000	6,275 (1,913)	9,425 (2,873)				
8,000	8,375 (2,553)					

Table 20-6. Comparative results of miniature brushes

	Workpiece Material and Brinell Hardness Number (BHN)		
	303Se Stainless Steel 226 BHN	7075-T6 Aluminum 150 BHN	6061-T6 Aluminum 95 BHN
Brush Description	Resulting Edge Radii, in. (mm)		
Stainless-steel brush 0.003 in. (76 μm) filaments	0.0064 (0.163 mm)	0.0174 (0.442 mm)	0.0186 (0.472 mm)
0.014-in. (0.36-mm) diameter nylon filaments with 600-grit silicon carbide	0.0064 (0.163 mm)	0.0070 (0.178 mm)	0.0082 (0.208 mm)
0.018-in. (0.46-mm) diameter nylon filaments with 500-grit aluminum oxide	0.0058 (0.147 mm)	0.0062 (0.158 mm)	0.0066 (0.168 mm)

polishing lathes, also used for brushing (Gillespie 2000). Because simple machines are inexpensive and readily available, it is easy to initiate brush deburring and experiment without affecting production.

Shaft Sizes

Shaft sizes for brushing machines have become fairly standardized through many years of use; however, it is still essential to calculate shaft diameters for special machines and operating conditions. In general, for semiautomatic machines with a shaft length required to provide an 8–12-in. (203.2–304.8-mm) overhang, a 1.25-in. (31.8-mm) diameter shaft is satisfactory. For similar service with a maximum overhang of 20 in. (508 mm), a 2-in. (50.8-mm) diameter shaft is generally suitable.

Deburring on the Cutting Machine

Many companies have overlooked the fact that brushing parts on the machine that produces them can be more economical than using a secondary brushing operation. Even screw machines can have brushing cycles designed into cams and tooling. Putting a live spindle and brush in a lathe turret, for example, can allow one to brush the inside or outside diameter without additional handling. It also has the advantage that if the brush does not remove the burr, the operator knows immediately that sharper tools or different machining techniques are required. In this case, the operator controls both the burr and its removal. This eliminates the unreasonably large burrs found when these responsibilities are separated.

The abundance of numerically controlled (N/C) equipment provides great flexibility in on-machine deburring. N/C milling machines, for example, can machine, deburr, and inspect the part in a single setup. It does not appear, however, that many companies have taken advantage of the potential for fast and thorough brushing offered by their N/C equipment (Gasser 1993; Mahadev 1995; and Scheider 1985, 1989a, 1990).

Figures 20-9 and 20-10 illustrate sturdy, large, abrasive-filled nylon brushes used with computer numerically controlled (CNC) and transfer machines. The tufted-disk brush in Figure 20-9 is shown in the second illustration (Figure 20-10) in a machine toolholder as it would exist in a tool changer. This style of brush is used in the automotive industry for cylinder head deburring and other applications involving sharp and heavy burrs (Figure 20-11).

Figure 20-10. Abrasive-filled nylon-tufted disc brush in CNC machine tool holder. (Mahadev 1995). (Courtesy Weiler Corporation)

Figure 20-9. Tufted abrasive-filled nylon disc brush (Mahadev 1995). (Courtesy Weiler Corporation)

Figure 20-11. Tufted abrasive-filled nylon-disc brush deburring aluminum valve body (Mahadev 1995). (Courtesy Weiler Corporation)

Even small cross-hole deburring brushes of 0.075-in. (1.91-mm) diameter are used in CNC toolholders for removing small burrs in aluminum valve parts after reaming. At 2,000 rpm, a break less than 0.001 in. (25 μm) is produced (Scheider 1990). Abrasive-filled nylon cup brushes have been used in CNC equipment routinely since at least 1988 (Scheider 1989a, b). These 1.50-in. (38.1-mm) diameter cups feed at up to 2.5 ft/min (12.7 m/s) on aluminum parts.

BRUSH DEBURRING APPLICATIONS

The following sections provide some in-depth application issues or examples. This is provided because there is little comparative data in hand-deburring literature, and few useful articles go beyond the basics.

Removing Heavy Burrs

Most burrs with a thick base must be removed by brushes with very substantial cutting action and minimal impact action. These characteristics are found in radial-type brushes with short trim. Because they are short trim and densely filled, they provide maximum cutting action. They can be furnished either in crimped wire or knotted types, with wire size ranges of 0.0118–0.0200 in. (0.300–0.508 mm) most widely used. Knot-type brushes in diameters of 12–15 in. (305–381 mm) are extensively used in the automotive and aircraft industries for removing burrs from gears and spline bores. Deburring occurs immediately after hobbing or spline broaching, just before shaving. These short-trim brushes impart the proper radii, adequately blend gear edges, and remove all remaining minute metal fragments from previous operations.

One distinct advantage of a short-trim brush is its capability of removing the burr from an edge without changing the dimensional characteristics of the two intersecting surfaces. Short-trim radial brushes are used extensively for heavy-duty deburring and edge blending of stampings, metal parts, pipe and tube ends, expanded metal, and flash removal in the rubber industry. They are designed for efficient, fast, low-cost performance on automatic, semiautomatic, and manual machines. In some types of heavy burr removal, it is necessary to impart a circular action, such as that used for deburring drilled holes in condenser tube plates. Knot-type cup brush configurations lend themselves to this type of work.

To preclude side flare and reduce trim length, a bridle or band is used on the brush, which adds a new dimension of stiffness. This not only deburrs the holes in the plates, but also puts a radius around the periphery on a uniform basis, without evidence of hole elongation.

Removing Light Burrs

Light and feather burrs are usually removed with medium- or long-trim brushes. These are available in both knotted and crimped-wire construction. Each type has its own sphere of excellence. The knotted brush, which is usually used singly rather than in multiples, is a high-impact tool. Operating speed radically influences its performance. At low speeds, the brush is extremely flexible and soft-acting. As surface speed increases, it becomes stiffer. When it reaches high speeds, it becomes extremely hard and, consequently, extremely fast-cutting. Normal speeds are 4,500–6,000 sfm (22.9–30.5 m/s), depending on the required stiffness. Wire sizes of 0.0118–0.0230 in. (0.30–0.58 mm) are common. Deburring rubber flash from molded products can be accomplished with knotted, medium- to long-trim brushes by using substantially higher surface speeds. Speeds up to 12,000 sfm (61 m/s) are frequently used. To attain these surface speeds, 12–15-in. (305–381-mm) diameter brushes are used.

Burrs produced by grinding broach teeth are being deburred with 3-in. (76.2-mm) diameter wire wheels having 0.008-in. (0.20-mm) diameter filaments. These are deburred at 17,000 rpm, moving handheld motors from the back to the front of the broach tooth. In this instance, the bristles rotate from the back to the front of the tooth. Excessive radiusing would occur if the cutting edges of the broach teeth were hit with a brush this stiff. In some instances, vapor honing, rather than brushing, is used to remove the burrs (Przyklenk 1985).

Medium- to long-trim crimped-wire brushes are used fairly universally for many types of light deburring operations. Because they are not very stiff, they can follow and conform to the contours of the irregular surfaces being deburred. The flexibility of their filaments enables them to reach relatively inaccessible areas. These wheels can remove feather-like machining and grinding burrs from edges, as well as light flash on rubber and plastic. Typically, smaller-diameter wires—0.006–0.014 in. (0.15–0.36 mm)—are used. Generally, surface speeds are 4,500–6,500 sfm (22.9–33 m/s).

Long-trim crimped-wire brushes of small diameter have inherent softness and flexibility, so they are not considered efficient deburring tools. Light burrs are sometimes removed with brushes of approximately 0.005-in. (0.13-mm) diameter wire to which a deburring compound has been applied. This gives the brush a cutting capability that it would otherwise lack.

Centerless brushing is sometimes used to remove burrs from cylindrical surfaces. The brushes used on centerless grinders normally have diameters of 20 in.

(508 mm) and brush faces of 5–6 in. (127–152 mm). Fill material for these larger radii brushes consists of the following: wire (0.005–0.010 in. [0.13–0.25 mm] in diameter), treated tampico, abrasive impregnated plastic, or nonwoven abrasive web. When the fill is wire, these brushes remove grinding burrs and produce a minimum edge of 0.005 in. (0.13 mm). Finishing work can be performed dry or (preferably) with normal grinding coolant. These fine wire-fill brushes, working with a conventional coolant, remove feather burrs and improve surface finishes such as those on automatic transmission control valves. It is important to remember the following facts about centerless brushes.

- They will not remove metal from a cylindrical surface. Parts, therefore, must be ground to size before brushing.
- The greatest improvement in finish (for unbrushed parts) is in the 25–30-μin. (0.64–0.76-μm) range. A part with a 25-μin. (0.64-μm) finish can be brushed rapidly to deburr and simultaneously obtain a 10–15-μin. (0.25–0.38-μm) finish. A part with a 10-μin. (0.25-μm) finish will have a 1–6-μin. (0.03–0.15-μm) finish after brushing.

Brushes filled with treated tampico fiber or cord material, used in conjunction with grease-based polishing compound, are capable of producing extremely fine finishes in the range of 2–7 μin. (0.05–0.18 μm). Brushes filled with abrasive-impregnated nylon filaments are recommended for fast, thorough burr removal without compound and subsequent part cleaning. This type of brush is often used to deburr perforated steel tubing.

There are at least three basic approaches for applying the brush to the edges of a workpiece, as Figure 20-12 illustrates. When in use, the brushes shown in the middle and lower part of the figure would often be canted or inclined to the edge of the part. One study indicated that a 25° angle of attack is optimal. Other needs, however, may make 45° angles more suitable. Brush oscillation and use of multiple brushes at more than one orientation are also used to ensure complete burr removal. As shown, the location of one centerline to another affects where most of the reaction takes place.

Gear and Sprocket Deburring

High production and the use of automated equipment on primary gear operations cause heavier burr conditions. These burrs result from excessive tool-feed speeds and using tools beyond their most efficient cutting point.

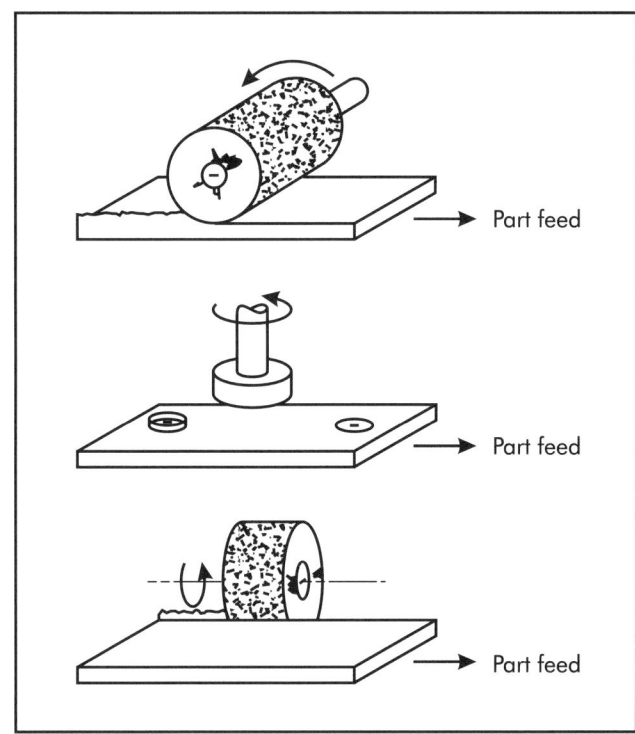

Figure 20-12. Brush arrangements for deburring (Przyklenk 1985).

For a majority of gear deburring jobs, radial, short-trim wire brushes (knot type) are recommended, and gears are brushed in the green (prior to heat treat). Fill wire can be any of several diameters; however, a 14-in. (355.6-mm) outside diameter brush with 0.0118 in. (0.300 mm) special analysis steel wire is most widely used. Although this type of brush develops relatively little impact action, its short trim and dense construction give it a high degree of cutting action. At 6,500 sfm (33 m/s), this brush is extremely fast cutting.

When deburring gears, mechanized operations typically rotate the gear under two brushes operating in different directions to catch the front and back sides of the teeth. This setup looks much like the wheels of a car passing over a large manhole cover. The brushes just kiss the teeth ends like a wheel would pass over the edge of the manhole cover. In this approach, the gear brushes are held perpendicular to the flat-faced gears. In contrast, operators of handheld motors typically traverse each tooth in a down-and-out action.

To brush spline bores, such as those found on driveshaft ends, brushes must be located off center but moved in from the outside of the diameter of the part. In the car-wheel analogy, the wheels would be moved much closer to the middle of the axle so the brush in some fashion falls down into the inner diameter. Typically, operators traverse each cut outward when using a handheld wheel, which may not be the most economical approach.

When helical gears are brushed, sometimes it is necessary to favor the acute side of the gear tooth to develop a generous radius. Such a radius is required because subsequent shaving of the gear produces an otherwise extremely sharp edge. This may require moving the brushes in slightly from the outer diameter of the part (Gillespie 1999).

Another advantage of brushing is that the fillet on the gear tooth can be broken, thus relieving this high-stress-concentration area. This is particularly essential on hardened gears. Many gear manufacturers find fewer failures and greatly reduced replacement problems after brush finishing. Added benefits of this method include easier assembly and safer handling.

Elastomer-bonded wire brushes are used for deburring fine-pitch gears. These brushes have minimum flexibility on the brush face and remove burrs without leaving a secondary roll. Bonded brushes are also effective on powdered metal and aluminum die-cast parts. At maximum efficiency, speeds of these brushes range from 2,600–6,000 sfm (13.2–30.5 m/s). Because they have such a fast cutting action, surface speeds can be lower than those used for conventional wire brushes.

Hardened pump-gear finishing requires treated tampico brushes used with a grease-based deburring compound. The same type of gear positioning mentioned previously, in relation to the brushes, applies to these gears.

End Deburring of Tube, Pipe, and Rod

Power brushes are widely used for deburring the ends of tubes, pipes, solid rods, and tubing of all sizes and shapes. A variety of methods are used, such as simple, offhand operations involving a bench motor or polishing lathe. Figure 20-13 illustrates a means to extend the bristle life of end brushes.

Figure 20-14 illustrates an ingenious method of deburring and cleaning the ends of copper tubing after cutting it to length. Three small-diameter radial brushes filled with crimped steel wire are fixed (they do not revolve individually) on a metal backing plate. A wire tube-type brush is mounted in the center. The tube brush is inserted into the tubing and the plate fixture is rotated by a portable tool. End burrs are removed, and both the inside and outside tubing surfaces are cleaned. Tubes are then ready for soldering or brazing. This same type backing plate and brush assembly can be mounted on a polishing lathe or bench motor to handle long lengths of pipe or tubing. The best results are obtained if a simple angle-iron fixture (trough) is used as a steady rest to present the tubing to the brush cluster in the proper position.

When brushing a recessed area or the inside diameter of a tube, use an end brush in the following manner: insert the brush, start the brush rotating. Turn off the power before removing the brush. This will prevent the brush from flaring out and permit it to fit into the recessed area again.

Insert brush — Apply power — Shut off power — Remove brush

Figure 20-13. Preserving the life of end brushes. (Courtesy Weiler Corporation)

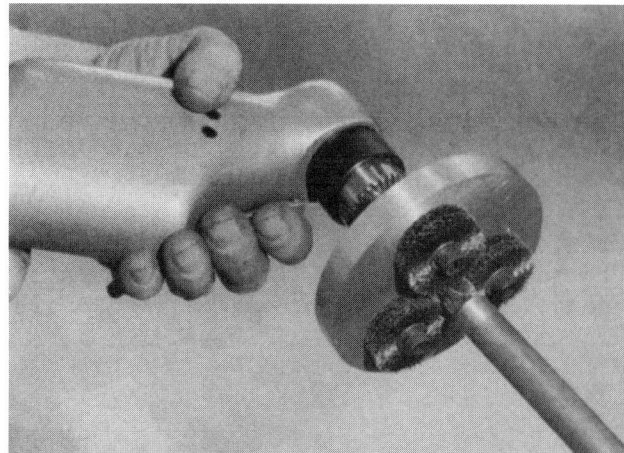

Figure 20-14. Device for deburring tube ends—tube brush is surrounded by three wheel brushes (West 1975). (Courtesy Osborn Manufacturing)

Deburring Internal and Threaded Areas

Small power brushes in a wide selection of sizes, shapes, and fill materials are available for deburring and cleaning recessed areas where limited space is a factor. These brushes are designed for rotary, end, or side action; and they are readily adaptable for use on portable tools, drill presses, bench grinders, lathes, and special equipment (Figure 20-15). Figure 20-16 provides some comparisons on the use of these tools.

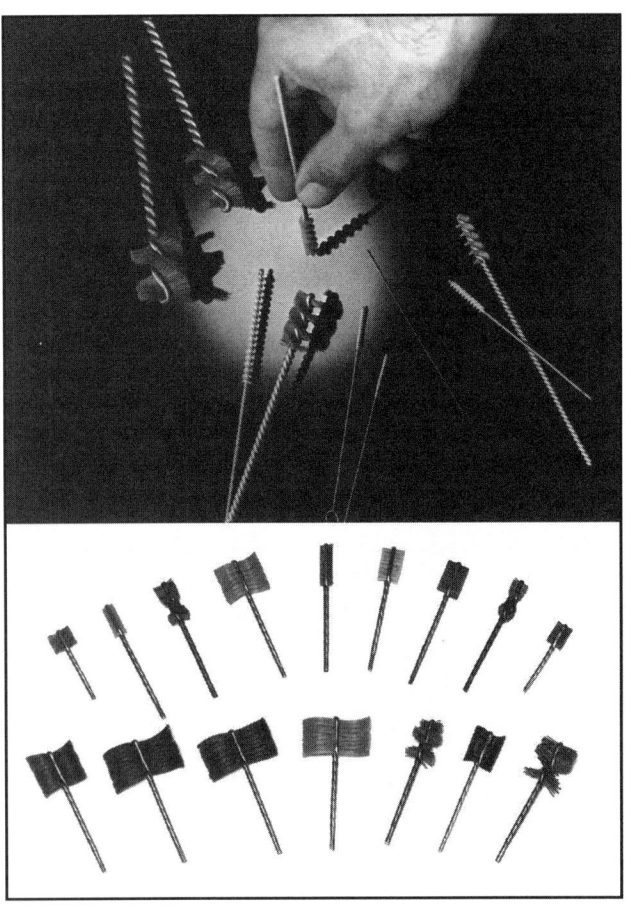

Figure 20-15. An array of cross-hole and other brushes used for deburring holes and edges (Scheider 1989b). (Courtesy Osborn Manufacturing)

Small radial brushes range from 1–3 in. (25.4–76.2 mm) in diameter and are available in wire sizes between 0.0025–0.0140 in. (0.064–0.356 mm). Fill materials include steel and stainless-steel wire, nylon, tampico, abrasive-impregnated nylon filaments, and elastomer-bonded steel wire. A 1-in. (25.4-mm) diameter brush filled with 0.005-in. (0.13-mm) nylon is used to deburr small DC armatures. The brush is driven at 8,000 rpm by a 0.125 hp (0.09 kW) motor and removes burrs and other foreign particles from the narrow commutator slots. "S"-shaped brush filaments, such as those shown in Figure 20-17, maximize contact between the abrasive and surfaces. In some applications, the more traditional open-tube brushes are better at deburring and removing contamination. For more aggressive needs, an elastomer-bonded wire brush mounted on a drill press can clean internal surfaces and deburr intersecting holes in machined castings.

Side-action brushes are effective for internal deburring and removing burrs from threaded areas. Diameters range from 0.156–1.250 in. (3.96–31.75 mm) and are available in a broad range of metallic and nonmetallic styles and fill materials. Brushes with wire-side tufts feather and remove sharp areas from internal threads and chips left from the threading operation. Square-trimmed brushes filled with abrasive-impregnated synthetic webbing clean and deburr threaded and counterbored areas of an aluminum casting in a single operation.

Elastomer-coated tube brushes provide long life and add to the aggressiveness of the tool. For reaching into

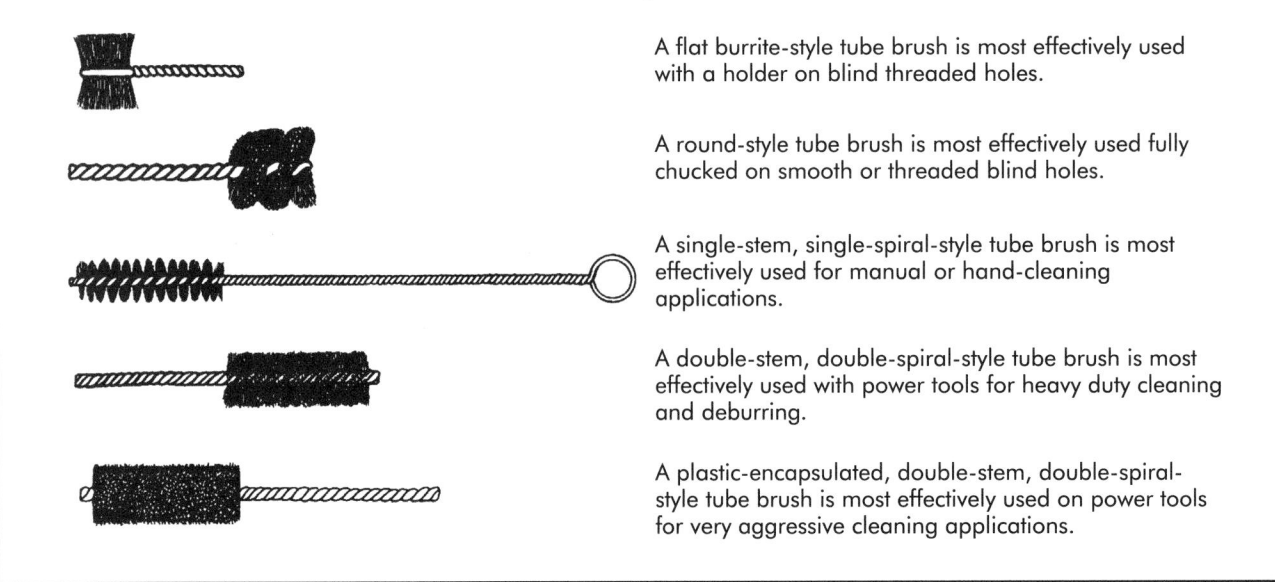

Figure 20-16. Power and hand-tube brush selection tips. (Courtesy Weiler Corporation)

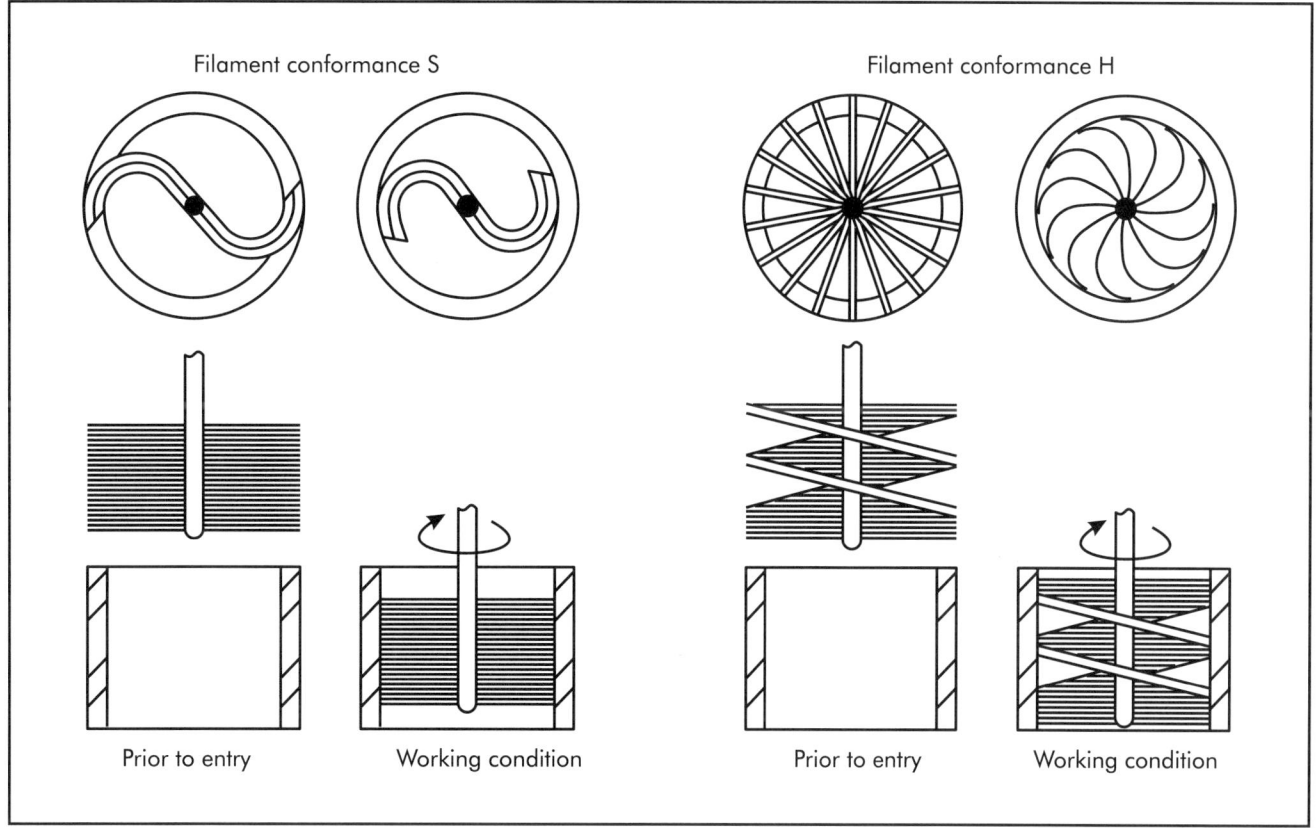

Figure 20-17. Filament bending in holes (Scheider 1991). (Courtesy Osborn Manufacturing)

deep holes, at least one manufacturer provides extra-long extenders. These extenders can add 16 in. (406.4 mm) to the standard tube brush handle (Figure 20-18).

Egg-beater-style brushes are used to finish hydraulic valve holes. Some of the holes have intersecting grooves. These tools, which are made from ragged-edge metal fibers, act as a diamond-coated elastic brush to remove burrs when other tools, such as boring or reaming cutters, only move the burrs into the other hole.

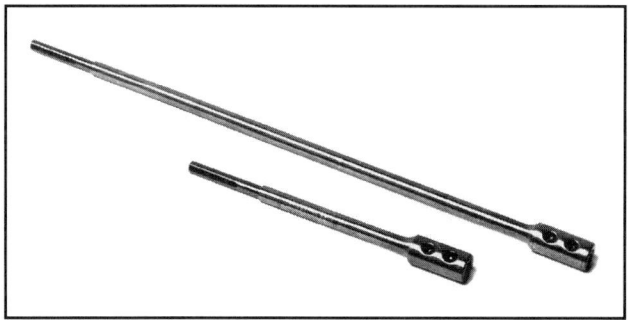

Figure 20-18. Extension holders for tube brushes. (Courtesy Osborn Manufacturing)

Several sources provide additional deburring uses for the brushes discussed in this section (Gasser 1993; Mahadev 1995; Scheider 1989a, b; Scheider and Carmichael 1992).

Deburring Flat Shapes

Edge deburring of steel strips, sheets, or plates is effectively done by using elastomer-bonded steel-wire radial brushes either offhand, with portable tools, or passing them under motor-mounted brush heads. An important characteristic of this brush is its perpetually short trim, with each wire point firmly supported at the brush face. This produces extremely fast cutting action, maximum work efficiency with minimum pressure, and positive control over the area of brush contact.

Many manufacturers have replaced grinding wheels with abrasive-impregnated nylon-filament brushes for finishing and cleaning edges of strip steel (pail stock) prior to welding. Abrasive brushes are quiet in operation (meet Occupational Safety and Health Administration [OSHA] standards), conform to slight distortions in strips, and provide longer service at less cost.

Rounding and blending the edges of cutouts on an aluminum window frame are accomplished efficiently with an interrupted-face radial brush (Figure 20-19).

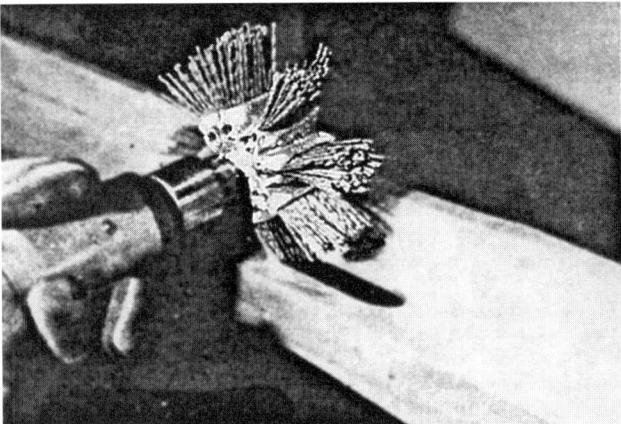

Figure 20-19. Interrupted face radial brush (Scheider 1991). (Courtesy Osborn Manufacturing)

WIRE-FILLED, ELASTOMER-BONDED BRUSHES

Elastomer-bonded, wire-filled brushes have proven to be extremely fast-cutting and provide maximum operator safety. Because the bonding material supports the wire to the very tip, the brush provides a maximum amount of cutting action. By periodically reversing the brush, the operator has a tool that maintains its maximum cutting efficiency throughout its life. Unlike a conventional nonbonded wire-filled brush, there is no loss of wire through fatigue. Bonded wire-filled brushes are made in radial, cup-style, flared-end, and straight-end designs, which permit them to be substituted for conventional wire-filled products. These brushes are effective for many deburring operations because they maintain a uniform face throughout their life and perform at maximum efficiency. Because elastomer wire-filled brushes are so fast-cutting, lower surface speeds can be used. Examples of the types of jobs these brushes can perform include removing burrs from:

- powdered metal components;
- aluminum die castings;
- copper tube ends, and
- metal stampings.

They are also used for edge deburring on steel strips and sheets, and are used extensively for removing burrs from key slots and other hard-to-reach parts. In deburring ductile materials, radii are produced without producing a secondary rolled condition. Usual wire sizes are 0.0118–0.0140 in. (0.300–0.356 mm) in diameter, and appropriate speeds range from 850–6,500 sfm (4.3–33 m/s).

Two factors should be considered when elastomer-bonded brushes are used for deburring. First, periodically dressing the brush face produces a brush life that is 10 times as long as that of conventional nonbonded wheels, thus keeping end-of-service costs for a given deburring operation much lower. Second, because the cutting action is so fast, brush adjustment is important—the part must be held precisely if the brush is to be used efficiently.

The following are three applications for using elastomer-filled brushes.

- They remove burrs from key slots and other hard-to-reach areas. The wire-fill material for these applications is generally 0.012 in. (0.30 mm) in diameter, and the brushes are rotated at 5,500–6,500 sfm (28–33 m/s).
- They remove stamping burrs and produce radii on ductile metals without producing secondary rolled conditions. The wire-fill material is usually 0.008 in. (0.20 mm) in diameter and the brushes rotate at 5,000–7,000 sfm (25.4–35.6 m/s).
- They remove burrs on powdered metal parts, aluminum die castings, and fine-pitch gears. The wire-fill material for these applications is usually 0.006–0.008 in. (0.15–0.20 mm) in diameter and the brushes are rotated at 5,000–6,500 sfm (25.4–33 m/s).

Nonwoven Abrasive Brushes

In many instances, nonwoven abrasives have replaced treated tampico, abrasive monofilament, and fine wire brushes in deburring applications. This has happened because they are available in a variety of densities and grades of hardness, and are filled with a large assortment of grit sizes (in aluminum oxide and silicon carbide). Light deburring can be accomplished with the uncompressed material, but the harder brushes are best for deburring applications. They are used to deburr grinding lines in all manner of metal components, appliances, and parts, including stampings and machined parts.

Other applications for nonwoven abrasive brushes include deburring many types of plastics. Hole deburring (both wet and dry) of printed circuit boards is one of the largest applications for nonwoven abrasives. Even wood items are finished with them. Jobs now assigned to this material include removing the parting lines on jet blades and radiusing gears and machine parts. Generally, the surface speeds used with nonwoven abrasive brushes are 5,000–6,500 sfm (25.4–33 m/s).

TROUBLESHOOTING ISSUES

Some of the potential problems that may be encountered in power brushing, along with suggested corrections, are presented in Table 20-4, Chart 2. Since power-driven brushing tools are expendable items, they must be replaced periodically. Analysis has shown that abrasive wear and fatigue are the two primary causes of rapid brush wear or failure.

Abrasive Wear

To increase the wear life of a brush, a number of steps can be taken. Since the abrasion resistance of a brush is decreased in proportion to the stiffness of the filaments, longer life requires a reduction in the stiffness of the filaments in the brush. This reduction can be made by altering any of the stiffness factors discussed previously, including modulus of elasticity, trim length, and rotational speed.

Brush life also may be increased by simply reducing the load on the brush; that is, by changing the relationship between the brush and the work being brushed. A secondary advantage of load reduction is a decrease in the amount of filament deflection. Wear increases more rapidly than the proportional increase in load when the filaments are wiping instead of cutting.

Some correct and incorrect brushing techniques are illustrated in Figure 20-8, and Table 20-7 illustrates common solutions to abrasive-filled nylon brush problems.

SIDE EFFECTS

Brushing can create a number of side effects that can be beneficial or detrimental, depending on the application. The degree to which some of these items occur depends in part on whether brushing involves entire surfaces or merely edges. Side effects include:

- surface finish changes;
- surface texture changes;
- removal of stress-riser tool marks;
- residual stress changes;
- surface hardening;
- fatigue life changes;
- contamination of part surfaces;
- dusty environment;
- part-size change;
- generation of new burrs, and/or
- color changes.

Brushes can produce finishes as fine as 4 μin. (0.10 μm), which is a distinct advantage for most compa-

Table 20-7. Suggestions for troubleshooting abrasive-filled nylon brushes

Observation	Corrective Measures
Tool not aggressive enough	Use larger grit (mesh size) Use silicon-carbide mineral Use rectangular-shaped filaments Increase density of filaments Use shorter trim length Increase surface speed within range Decrease feed speed—more dwell time
Tool too aggressive	Use smaller grit Use round filaments Decrease filament density Use longer trim length Reduce surface speed Increase feed rate Use aluminum-oxide mineral
Filaments smearing nylon	Use coolant at point of contact Reduce dwell time on workpiece Reduce in-feed pressure on workpiece Reduce surface speed
Surface finish too rough	Use smaller grit (mesh) size Use aluminum-oxide mineral Use smaller filament diameter or shape Apply honing compounds to filaments
Tool life too short	Verify consistency of burrs on incoming parts Reduce surface speed Increase density of filaments Use coolant at point of contact Use rectangular-shaped filaments Increase tool diameter

(Scheider 1991)

nies. Other companies take advantage of brushes because they can leave a decorative texture on surfaces. Turbine and air-frame manufacturers employ brushes in part because they smooth tool marks, which are points of high-stress concentration. Jet-engine manufacturers, for example, have found that half of their part failures can be traced to stress concentrations and progressive fractures, such as scratches, sharp edges, or burrs. Eliminating these stress concentrations greatly extends fatigue life.

Wire-filled brushes can greatly affect residual surface stresses (and, as a result, harden a surface). Research over the past 100 years indicates that surface residual stresses can increase dramatically as a result of chemical, thermal, and mechanical interactions during surface brushing. For example, one researcher noted that hardness near a wire brushed part surface was five times greater than in the center of the part. Others have shown that work hardening, as opposed to brushing, can only double the hardness. If it is assumed that hardness is proportional to yield strength, then residual stresses near the surface would be five times greater than the bulk stresses. Those high stresses occur in a layer only a few thousandths of an inch (mm) thick. Typically, these greatly help increase fatigue life; however, they can decrease the life of thin materials or those whose inner core has high-tensile residual stresses. In addition, part distortion can easily occur on thin, nonsymmetrical parts. These stresses can cause surface textures that resemble an orange peel. This particular condition, however, can be easily seen and corrected by reducing the frictional heat generated by the brush.

If any foreign material comes into contact with a part, it contaminates the part. Thus, when stainless steel is deburred with a carbon steel brush, carbon steel is deposited on the part. This will allow rust spots to occur on a normally rust-proof surface. If brass brushes are used on steel, the part becomes yellow and brass-coated. Nylon, if used at high speeds, melts on parts, leaving black splotches and probably a thin, invisible layer of nylon. Tampico wheels used with grease-based burring compounds leave obvious muddy-like deposits on the part, which are sometimes harder to remove than the burrs. Similarly, diamond lapping compounds can be almost impossible to remove from hard-to-reach recesses. In some aluminum alloys, anodizing will not be complete if foreign particles are embedded in the surface. In addition, if large amounts of aluminum oxide are embedded in steel parts, it can be impossible to satisfactorily perform welding, soldering, and brazing on them. Today, however, abrasive-impregnated nylon or synthetic brushes do not appear to impregnate abrasives into the workpiece in quantities sufficient to cause problems for most industries. In contrast, tumbling or blasting processes produce gross surface contamination. In some situations, brushing can slightly discolor a part, which can either improve or diminish its appearance.

Contamination can be carried from one part to another. For example, if a nylon brush is used on aluminum parts, then used on stainless steel, the stainless steel part will have traces of aluminum. Similarly, nylon brushes, in particular, will pick up oil, grease, or other coatings from one part and deposit them on the next. This can be a source of concern for workers who are hand deburring parts that require high levels of cleanliness.

The use of burring compounds with tampico wheels, metal-filled wheels, buffing wheels, or felt bobs results in a large quantity of particles and clumps of compound thrown into the atmosphere and work area. Because the process is inherently dirty and dusty, and particles are thrown, safety regulations must be followed.

By nature, a brush changes the dimensions of a part each time it is used. After all, brushes are metal removal tools. Generally, changes are small (typically 50 μin. [1.3 μm]), except at edges where burrs are to be removed. However, significant size changes are possible.

Another advantage of using brushes instead of other products is that they can reach into holes *while* the surface is being brushed. This removes some of the burrs frequently left by flat-belt sanding. If, however, an incorrect wire brush or speed is selected, the brush can generate burrs on edges that were initially burr-free.

NONTRADITIONAL BRUSHING

In addition to the common brushes that have been mainstays in the recent and distant past, there are many variations available today. The rules governing success for common brushes are not the same for these nontraditional ones. Nontraditional brushes include:

- miniature flap wheels;
- elastomer-filled radial brushes;
- abrasive-filled, nonwoven synthetic wheels and shapes, and
- abrasive, ball-tipped, nylon-filament tube brushes.

Such configurations do not generally come to mind when the word *brush* is mentioned. These are the types of deburring tools that have tremendous promise for rapid burr removal. In many cases, they can achieve better surface finishes and faster cutting than traditional brushes.

Brush-backed Flap Wheels

Some flap wheels use sections of brushes to provide polishing and cutting actions. It is possible to combine almost any brush material with bonded-abrasive products to provide longer life and different edge finishing.

Wire-filled Elastomer Brushes

Elastomer-bonded, wire-filled brushes are extremely fast-cutting and safe to use. Because bonding material supports the wire to the very tip, the brush provides maximum cutting action. By periodically reversing the brush, optimum cutting efficiency is maintained throughout its life. Compared to conventional nonbonded wire-filled brushes, there is minimal wire loss through fatigue.

Bonded wire-filled brushes are made in radial, cup-style, flared-end, and straight-end designs, and can be used in place of conventional wire-filled products. These brushes maintain a uniform face throughout their life and perform at maximum efficiency, so they are effective for many operations.

Abrasive Ball-ended Bristle Brushes

Ball-type flexible hones have a ball-shaped cluster of fine abrasives at the end of nylon fibers (Figure 20-20). Each abrasive globule consists of built-up laminations of abrasive materials. In some sense, the globules are like pearls coated with layer upon layer of bonded abrasive. Continual wear of these little balls opens new surfaces for fresh cutting. The nylon bristles bend to allow the brushes to reach into large holes and radius cross-holes, as well as hole entrances and exits. As with any tube brush, the brush's diameter should be larger than the hole diameter to ensure continuous scrubbing. In addition to deburring, these tools provide cylinder wall-surface finishes that are much admired by the automotive industry (Miller 1993; Lin 1992). Table 20-8 illustrates recommended materials for cross-hole deburring.

Although little detailed data have been published on the selection of abrasive ball-ended bristle brushes for deburring, some excellent research has been reported describing ball-ended brush applications for honing. In general, ball-end brush honing results claim many similarities to those of abrasive-filled nylon-bristle brush honing (Frazier 1991; Scheider and Carmichael 1992). Surface finishes of 15.75 μin. (0.4 μm) are produced with 120-grit media. Finishes of 3.94 μin. (0.1 μm) are produced with 400-grit media. These grits produced stock removal of 394 μin. (10 μm) on 3.62-in. (92-mm) bores honed for 30 seconds. The overall diameters of the ball-ended brushes used were 2.80 in. (71.1 mm)—14% larger than the hole diameter.

Table 20-8. Recommended abrasive material for deburring cross-holes using ball-ended, nylon-filament brushes

Part Material	Ball Material
Hard steel	120-grit tungsten carbide
Mild steel	80-grit zirconia alumina
Cast iron	120-grit silicon carbide
Stainless steel	120-grit silicon carbide
Aluminum	120-grit aluminum oxide
Chrome plating	120-grit aluminum oxide
Carbide	80-grit boron carbide
Brass/bronze	320-grit aluminum oxide
Nicasil®	120-grit silicon carbide
Nickel	120-grit boron carbide
Boron	120-grit boron carbide

(Miller 1993)

Metal-ball Flare Brushes

Figure 20-21 illustrates a very aggressive brushing tool. It has a cobalt-based hard facing flame, which

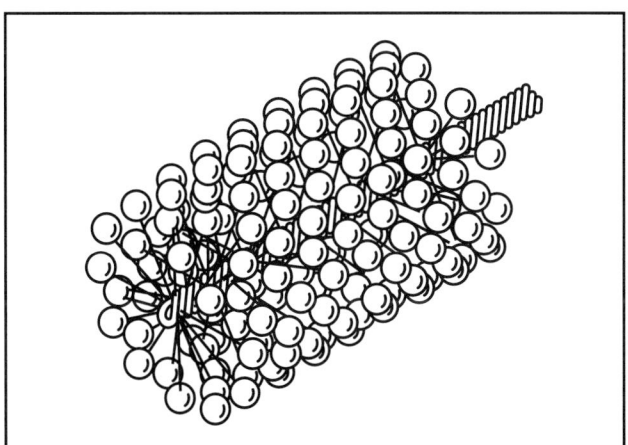

Figure 20-20. Flex-Hone® ball-ended bristle brush (Frazier 1991).

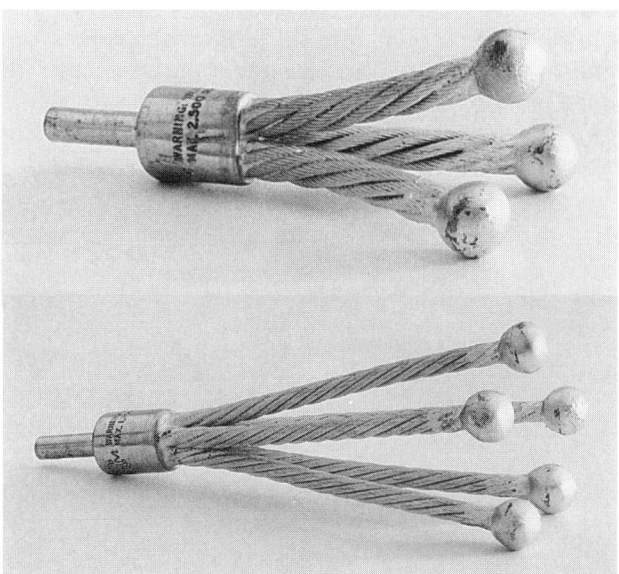

Figure 20-21. Metal ball-ended flare brushes. (Courtesy Brush Research Manufacturing Company)

is coated to the ends of stainless steel aircraft cables. While normally used for removing carbon from ports and pipe, it can break loose flash or burrs at intersecting holes.

Egg-beater Brushes

One manufacturer produces a tool that resembles a kitchen whisk or egg beater (Figure 20-22) (Yinsen et al. 1992). These metal-bladed tools squeeze together when they enter a hole, and expand again when they arrive in an intersecting hole. The surfaces of these tools have either fine sheared teeth or a diamond-plated spring wire. They come in sizes of 0.055–1.575 in. (1.40–40.00 mm) in diameter, and in several lengths. Some have just a few bristles, others have dozens. A nylon band, crimped at either end of the bristles, determines the major diameter of the brush.

Drill Brushes

Figure 20-23 illustrates a drill brush used to remove burrs from the top of the drilled hole. This unique tool works while the hole is being drilled. It cleans up spot faces on rubber-bonded motor mounts and similar parts. Standard drill bits are inserted through the center of the brush and holder, and are locked into place by set screws.

Valve-hole Brush

Figure 20-24 illustrates a valve-hole deburring brush (Yinsen et al. 1992). This is a cylindrical tool used to finish and deburr valve holes that have several grooves cut in the hole. An abrasive is adhered to the top of the nylon wire in the brush, as shown in Figure 20-24. The outer diameter of the brush is 0.078 in. (1.98 mm) larger than the hole. The brush is held by a drill spindle and rotated at 400 rpm, and it com-

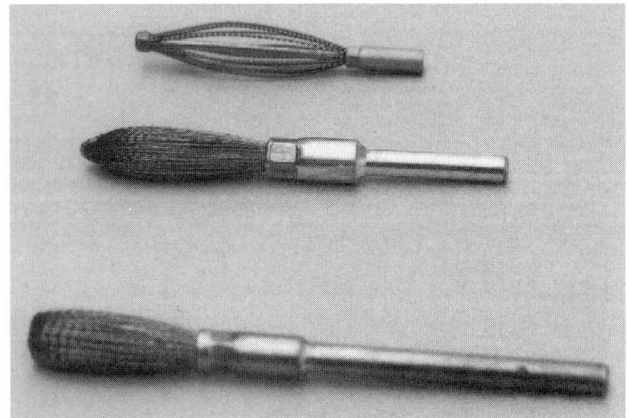

Figure 20-22. Egg beater brushes. (Courtesy Burrtec Company)

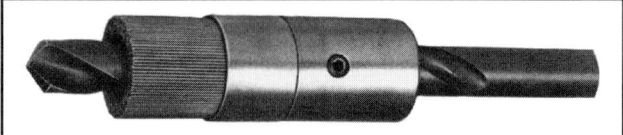

Figure 20-23. Drill brush. (Courtesy Osborn Manufacturing)

Figure 20-24. Valve bore brush (Yinsen et al. 1992).

presses as it enters the hole. Using both vertical motion and rotation, the abrasives touch every edge. In addition to removing hard-to-reach burrs, this tool improves the surface finish from 248 μin. (6 μm) to 126 μin. (3 μm) Ra.

COMBINED PROCESS DEBURRING

As discussed in previous chapters, the most effective solution for many burr problems is to combine more than one process type. For example, combining bonded abrasive and brushing on a single part provides greater capability than a single removal process. These combination approaches are fast and relatively low cost.

ENVIRONMENTAL, HEALTH, AND SAFETY ISSUES

When properly used, power-driven brushes are safe, although they inherently have all the dangerous attributes of any rotating body. Therefore, it is essential that they be mounted, handled, and used carefully, observing all necessary precautions. The operator, as well as others in the work area, must wear safety goggles or full-face shields over safety glasses with side shields and other suitable protective equipment.

The machine on which a brush is being used should not be operated at an excessive speed, and the brush itself should never exceed the manufacturer's maximum safe speed. Before an operation, the machine should be jogged to ensure it is ready for use and the brush fastened securely and concentric with the axis of rotation. After the machine has been started, the brush should be run at the operating speed in a protected enclosure for at least one minute before applying the work. During this time, no one should stand in front of or in line with the brush.

During operation, the work should never be pressed into the face of a wire bristle brush far enough to slow the drive motor. Visual evidence of excessive pressure can be noted when the sides of the filaments are in contact with the work—only the tips should be in contact.

Wire brushes should be stored in an atmosphere that precludes rusting or oxidizing and in a manner that protects the brushes from physical distortion, which might result in brush imbalance.

Excellent safety standard requirements and explanatory information are contained in the American National Standards Institute's (ANSI) Standards B165.1, *Safety Requirements for the Design, Care, and Use of Power-driven Brushing Tools,* and B165.2, *Safety Requirements for the Design, Care, and Use of Power-driven Brushing Tools Constructed with Wood, Plastic, or Composition Hubs or Cores.* These standards should be referred to and fully understood before setting up any new application, and they should be periodically reviewed for present and existing brushing applications. With separate sections for the user, machine tool builder, and brush manufacturer, the standards contain a wealth of engineering and technical information.

Owing to their basic construction, the stems in power-mounted tube brushes are not as strong as the shanks in most other brushes. Therefore, it is very important to avoid any load conditions and brush speeds that can cause excessive stem deflections and destructive bending. A suggested guideline to avoid this unsafe condition is minimizing the overhang of the stem to under 1 in. (25.4 mm) and running the brush at speeds below 2,000 rpm. Increasing the overhang could decrease the safe speed at which the brush can operate. To reach deeper holes, use drill extension rods instead of increasing stem overhang. Before starting the brush, do the following:

- Secure the brush in a chuck, collet, or similar workholding device.
- Ensure a clockwise brush rotation when mounted stem up. (Running the brush counterclockwise could cause the brush to fall apart.)
- Clamp the work in a workholding device and position all guards in place.
- Align the brush with the work so it rotates on its true centerline to prevent stem deflections.
- Guide the brush into the hole before starting the brush rotation.
- Always wear eye protection.

Some work materials present unique hazards. Beryllium is a known carcinogen, and recent research indicates that even the small amounts in beryllium copper can affect workers who breathe or ingest its dust. OSHA and the Department of Energy each have established regulations designed to protect workers from the hazards of this and other materials. The nickel in stainless steel is a suspected carcinogen, and its dust may be harmful if inhaled at levels above documented limits. Although brushing may not produce visible dust in many operations, it may produce enough to create worker health issues. HEPA-filtered vacuum exhausts, if carefully controlled to prevent contamination to those who clean such systems, may provide adequate control. Each plant must perform its own assessment and establish its own controls because general guidelines may be inadequate, particularly in the case of beryllium and its alloys and compounds.

REFERENCES

Bateman, James. 1975. "Capabilities and Limitations of Brush Deburring." Technical Paper MR75-826. Dearborn, MI: Society of Manufacturing Engineers.

———. 1979. "Brush Techniques for Cleaning, Finishing, and Deburring." Technical Paper MR79-757. Dearborn, MI: Society of Manufacturing Engineers.

Frazier, Ronald. 1991. "Deburring Update Utilizing Flexible Radial Wheel Type Processes." Technical Paper MR91-122. Dearborn, MI: Society of Manufacturing Engineers.

Gasser, Joseph. 1993. "Advancements in Flexible Abrasive Finishing Tools." Technical Paper MR93-135. Dearborn, MI: Society of Manufacturing Engineers.

Gillespie, LaRoux K. 1999. *Deburring and Edge Finishing Handbook*. Dearborn, MI: Society of Manufacturing Engineers.

———. 2000. *Guide to Deburring, Deflashing, Trimming Equipment, Supplies and Services* 2nd ed. Kansas City, MO: Deburring Technology International.

Lin, Y. T. 1992. "Honing with Abrasive Brushes." Technical Paper MR92-132. Dearborn, MI: Society of Manufacturing Engineers.

Mahadev, Prasad S. 1995. "Nylon Abrasive Filament (NAF) Brushes—An Alternative to Deburring, Edge Radiusing, and Surface Finishing Problems." Technical Paper MRR95-02. Dearborn, MI: Society of Manufacturing Engineers.

Miller, Michael L. 1993. "Flexible Honing: A Study of Cylinder Wall Microstructure." Technical Paper MR93-146. Dearborn, MI: Society of Manufacturing Engineers.

Przyklenk, Klaus. 1985. "Brushing—New Research Results with an Old Method." Technical Paper MR85-848. Dearborn, MI: Society of Manufacturing Engineers.

Scheider, Alfred F. 1985. "Deburring, Edge Contouring, and Surface Finishing Applications Using Abrasive Brushes and Buffs." Technical Paper MR85-821. Dearborn, MI: Society of Manufacturing Engineers.

———. 1989a. "Emerging Trends in Flexible Abrasive Tools." Technical Paper MR89-113. Dearborn, MI: Society of Manufacturing Engineers.

———. 1989b. "New Technology in Flexible Abrasive Tools for Nontradiitonal Finish Machining of Holes, Bores, and Internal Surfaces." Technical Paper MR89-805. Dearborn, MI: Society of Manufacturing Engineers.

———. 1990. "Advanced Flexible Abrasive Finishing Tools." Technical Paper FC90-153. Dearborn, MI: Society of Manufacturing Engineers.

———. 1991. "New Developments in Flexible Abrasive Finishing Tools, Buffs, and Brushes." Technical Paper MR91-392. Dearborn, MI: Society of Manufacturing Engineers.

Scheider, Alfred F., and Guy Carmichael. 1992. "New Technologies in Abrasive Brush Honing." Technical Paper MR92-135. Dearborn, MI: Society of Manufacturing Engineers.

West, Richard. 1975. "Deburring with Power Brushes." Technical Paper MR75-476. Dearborn, MI: Society of Manufacturing Engineers.

Yinsen, Ye, Guo Hongdi, Huang Yaowen, Lin Feng, and Chen Qili. 1992. "The Technique of Finishing and Deburring Parts for Hydraulic Valves." In *Proceedings of the 2nd International Conference on Precision Surface Finishing and Burr Technology,* pp. 363–67, Dalian, China (September). Kansas City, MO: Deburring Technology International.

BIBLIOGRAPHY

Behringer, J. 1981. "Automated Deburring of Flat Sheet Metal." Technical Paper MR81-387. Dearborn, MI: Society of Manufacturing Engineers.

Cariapa, V., R. J. Stango, S. K. Liang, and A. Prasad. 1989. "Measurement and Analysis of Brushing Tool Performance Characteristics—Part II: Contact Zone Geometry." In *Proceedings of the ASME Production Engineering Division, Symposium on the Mechanics of Deburring and Surface Finishing Processes.* R. J. Stango and P. R. FitzPatrick, eds. PED-vol. 38, pp. 159–172. New York: American Society of Mechanical Engineers.

Duwell, Ernest J., and Ulrich Bloecher. 1983. "Deburring and Surface Conditioning with Brushes Made with Abrasive-loaded Nylon Fibers." Technical Paper MR83-684. Dearborn MI: Society of Manufacturing Engineers.

Flores, Gerhard. 1992. "Mechanical Deburring: Process, Tools, Machines, and Applications." Technical Paper MR95-272. Dearborn, MI: Society of Manufacturing Engineers.

Heinrich, S. M., R. J. Stango, and C. Y. Shia. 1989. "Effect of Workpart Curvature on the Stiffness Properties of Circular Filamentary Brushes." In *Proceedings of the ASME Production Engineering Division, Symposium on the Mechanics of Deburring and Surface Finishing Processes.* R. J. Stango and P. R. FitzPatrick, eds. PED-vol. 38, pp. 27–40. New York: American Society of Mechanical Engineers.

Henderson, J. A., C. Y. Shia, and R. J. Stango. 1993. "Design, Selection, and Machining Performance of Advanced Filamentary Brushing Tools." Technical Paper MR93-325. Dearborn, MI: Society of Manufacturing Engineers.

Keating, Thomas V. 1983. "Today's Applications of Low-density and High-density Nonwoven Nylon Abrasives." Technical Paper MR83-688. Dearborn, MI: Society of Manufacturing Engineers.

Russel, W. N. 1987. "Nylon Abrasive Monofilaments—New Brush Applications." Technical Paper MR87-162. Dearborn, MI: Society of Manufacturing Engineers.

Scheider, Alfred F. 1990a. *Mechanical Deburring and Surface Finishing Technology.* New York: Marcel Dekker.

———. 1990b. "Production Radius Grinding of Carbide and Ceramic Cutting Tool Inserts." Technical Paper MR90-532. Dearborn, MI: Society of Manufacturing Engineers.

———. 1991. "Precision Edge-radius Grinding of Diamond Cutting Tools and Inserts." Technical Paper MR91-195. Dearborn, MI: Society of Manufacturing Engineers.

Shia, C. Y. 1988. "Analysis of Constrained Deformation of a Circular Brush System." Master's thesis. Milwaukee, WI: Marquette University.

Shia, C. Y., R. J. Stango, and S. M. Heinrich. 1989. "Theoretical Analysis of Frictional Effect on Circular Brush Stiffness Properties." Technical Paper MR89-143. Dearborn, MI: Society of Manufacturing Engineers.

Sockman, John. 2001. "Deburring with Nylon Filament Brushes." *Manufacturing Engineering.* June: pp. 70–76.

Stango, R. J. "Damage Assessment of Wire and Nylon Abrasive Filamentary Brushing Tools." Technical Paper MR91-133. Dearborn, MI: Society of Manufacturing Engineers.

Stango, R. J., S. M. Heinrich, and C. Y. Shia. 1989. "Analysis of Constrained Filament Deformation and Stiffness Properties of Brushes." *Journal of Engineering for Industry (Transactions of ASME),* 111 (3): pp. 238–43.

Stango, R. J., V. Cariapa, and J. M. Kanion. 1989. "Experimental Evaluation of Circular Brush Stiffness: Preliminary Results." Technical Paper MR89-144. Dearborn, MI: Society of Manufacturing Engineers.

Stango, R. J., V. Cariapa, A. Prasad, and S. K. Liang. 1989. "Measurement and Analysis of Brushing Tool Performance Characteristics—Part I: Stiffness Response." In *Proceedings of the ASME Production Engineering Division, Symposium on the Mechanics of Deburring and Surface Finishing Processes.* R. J. Stango and P. R. FitzPatrick, eds. PED vol. 38, pp. 143–157. New York: American Society of Mechanical Engineers.

Suzuki, Kiyoshi, Koichi Kitajima, and Tetsutaro Uematsu. 1994. "Methods for Efficient Finish Polishing with Abrasive Wheel Brush." In *Proceedings of the 3rd International Conference on Precision Finishing and Burr Technology,* pp. 161–172, Seoul, Korea (November). (Available from Kansas City, MO: Deburring Technology International.)

Tarrab, K. M. H. 1990. "Dynamic Properties of Circular Brushes and Evaluation of Filamentary Stress: Photographic Analysis." Master's thesis. Milwaukee, WI: Marquette University.

Tynex A. 1996. "The Finishing Touch." Technical Bulletin. Washington, WV: DuPont Filaments.

Watts, James H. 1988. "Abrasive Monofilaments—Critical Factors that Affect Brush Tool Performance." Technical Paper MR88-138. Dearborn, MI: Society of Manufacturing Engineers.

Yera, Harvey J. 1990. "Proper Care and Maintenance of Broach Tools." Technical Paper MS90-361. Dearborn, MI: Society of Manufacturing Engineers.

Ballizing 21

Ballizing, also known as ball broaching, is performed by forcing a ball through a hole to improve its surface finish or change the hole size. While ballizing is normally considered a surface-finishing improvement process, it has been employed for burr removal. It has been used to remove large, drilling rollover and intersecting-hole burrs (Gazan 1977).

In one application, illustrated in Figure 21-1, a hardened steel ball is forced through a part that has oblique cross-holes (Ohshima et al. 1993). The ball, which is larger than the main hole bore, compresses the hole walls as it descends. When it gets to the intersecting holes, the compression breaks off most of the burr, and extreme compressive forces remove most of the plasticity from the hole intersection, making the burr root more brittle. Figure 21-2 illustrates the use of ballizing removing burrs in mismatched holes.

The ballizing process is simple. Place a loose ball on top of a hole, then push the ball through the hole with an arbor press or drill press. In a matter of seconds, the task is complete. In some instances it removes burrs, and in others it produces them. The key to success is to either use it on parts that have relatively brittle burrs or to apply it in applications that have liberal definitions of allowable material at edges (which means that loose burrs must be removed, but firmly adhered material is acceptable).

Ballizing can make significant burrs in ductile materials. In a series of studies in 304L stainless steel, a 0.0625 in. (1.588 mm) ball was forced through holes varying in size from 0.0612–0.0621 in. (1.554–1.577 mm) (Gillespie 1975). The burrs produced at the bottom of the through-hole ranged in thickness from 0.0008–0.0018 in. (20–46 μm) and in length from 0.0025–0.0076 in. (64–193 μm).

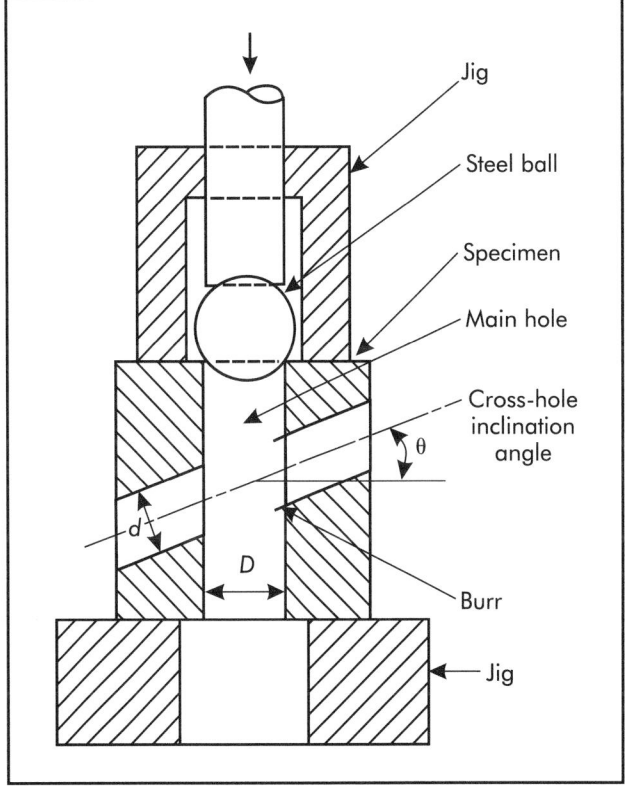

Figure 21-1. Ballizing for deburring (Gillespie 1975). (Courtesy Ohshima, Japan Society for Precision Engineering)

The 300 series stainless steel, including 304L, has high strain-hardening exponents and high ductility, which makes any of these steels a poor choice for the ballizing process. Nevertheless, for long or thick burrs on intersecting holes, the process might make burr removal easier.

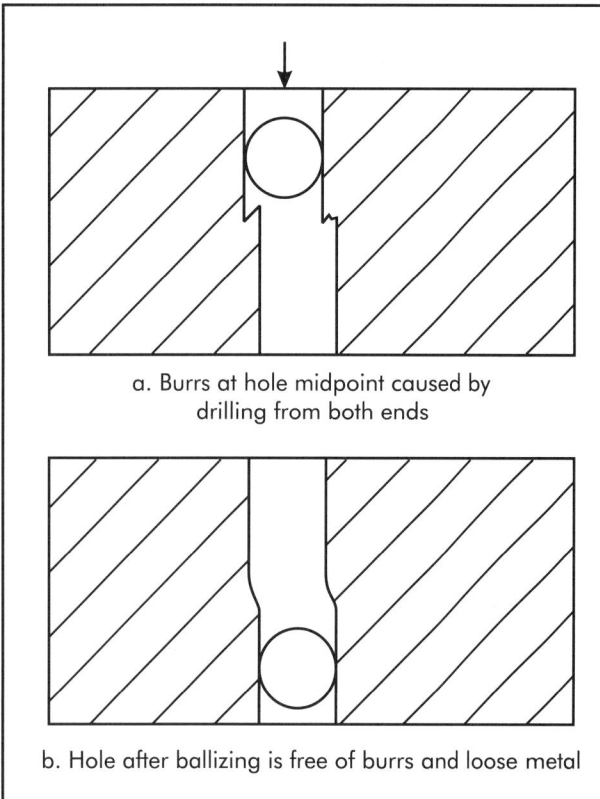

Figure 21-2. Ballizing to remove burrs at mismatched hole intersections (Gillespie 1975).

SIDE EFFECTS

The ballizing process increases the hardness of hole walls. In the previously mentioned stainless-steel study, wall hardness increased from Rockwell C38 to 46. Also, the hole size changes. In the study, typical size changed 0.0008 in. (20 µm). Hole surface finishes can improve from an average of 50 µin. (1.3 µm) Ra to 7 µin. (0.2 µm) Ra in the one-second operation.

In a second study of 304L ballizing, when holes were gundrilled to the point where 0.0003-in. (7.6-µm) thick smearing occurred, ballizing caused minute fissures in the work-hardened walls, which is unacceptable in many aerospace applications.

As noted, the process can make burrs; however, burrs can be minimized by providing a tight-fitting hole in a bottom plate below the workpiece. The plate essentially prevents a burr from forming. Note that burrs or bulged material can form on the top, as well as the bottom, surface. Figure 21-3 illustrates use of a bottom plate with a closely-matched clearance hole.

Ballizing may not be the answer for some applications; however, it is fast, exceptionally low cost, leaves no contamination, and improves hole finish. The lit-

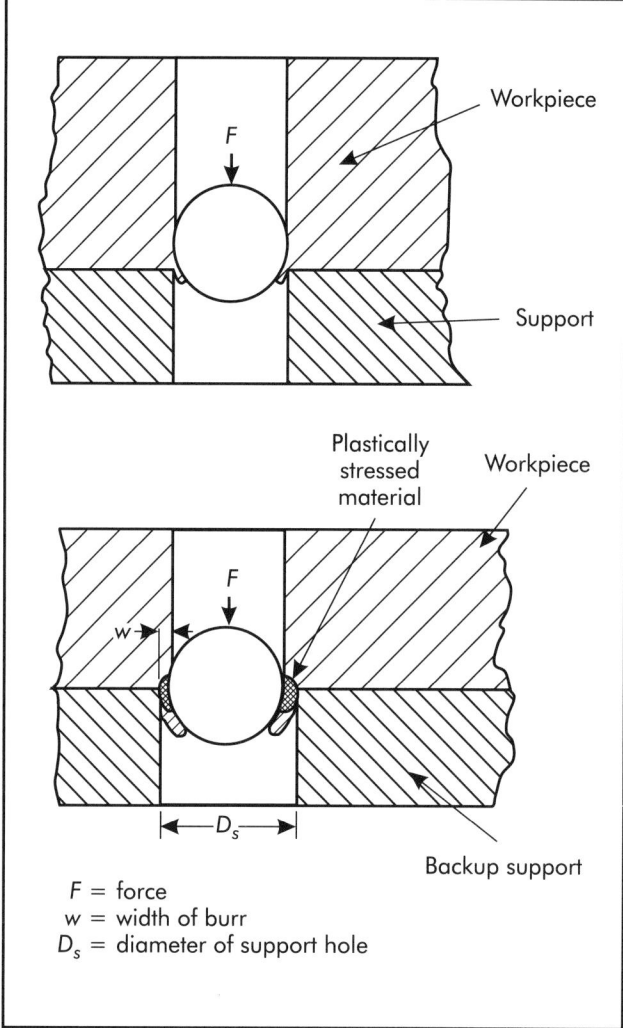

Figure 21-3. Bottom-plate clearance hole minimizes ballizing burr when it is near ball size (Gillespie 1975).

erature offers little guidance in predicting success. Regardless, it is a process that should be investigated, particularly for intersecting holes.

Balls are available in size increments of 0.0001 in. (2.5 µm) and in several hard metals or ceramics (Gillespie 2000). The applications cited are for manual use, but mechanized systems can also be easily built.

ENVIRONMENTAL, HEALTH, AND SAFETY ISSUES

The arbor or drill press can pinch hands or fingers. Care must be exercised in fixturing to make sure the fixture does not fall or that the ball does not squirt during loading. Operators can use honey or high-lubricity lubricants to reduce friction and produce smoother holes.

REFERENCES

Gazan, George. 1977. "How to Make a Good Hole Better." Technical Paper MR77-461. Dearborn, MI: Society of Manufacturing Engineers.

Gillespie, LaRoux K. 1975. "Properties of Burrs Produced by Ball Broaching." Report BDX-613-1084. Kansas City, MO: Bendix Corporation. (Available from NTIS.)

— 2000. *Guide to Deburring, Deflashing, and Trimming Equipment, Supplies and Services*, 2nd ed., p. 59. Kansas City, MO: Deburring Technology International.

Ohshima, Ikuya, Katsuhiro Maekawa, and Ryoji Murata. 1993. "Burr Formation and Deburring in Drilling Cross Holes." *Journal Japan Society of Precision Engineering*, vol. 59, no. 1, pp. 155–160.

Miscellaneous Special Hand Tools 22

This chapter focuses on a variety of often-overlooked tools that support deflashing and hand deburring. Many of these tools are very effective in specific situations. They include:

- side-cutting pliers (nippers);
- hones;
- split-mandrel stones and burrs;
- dental picks;
- buffs;
- midget laps;
- broaches, and
- watchmaker broaches.

TOOLS

The items described in this chapter are simple tools and, with the exception of pliers for deflashing, are used only occasionally for special problems. They are all commercially available but will not be normally found in discussions of deburring. Other publications provide additional details of how these tools are used in the plastics industry, and where to find them (Maroney 1991; Murphy 1999).

Side-cutting Pliers

Side-cutting pliers are common tools for degating and deflashing many plastic parts. Some supply houses for the plastics industry include dozens of designs. While these tools are not applicable to the metal industry, they are important for all molded plastics. Various types of side-cutting pliers include flat face, nipper, angled head, concave/convex, and miniature nipper.

Flat-face Pliers

Flat-face pliers have a flat surface that allows cutting to the very bottom of a plastic sprue (Figure 22-1).

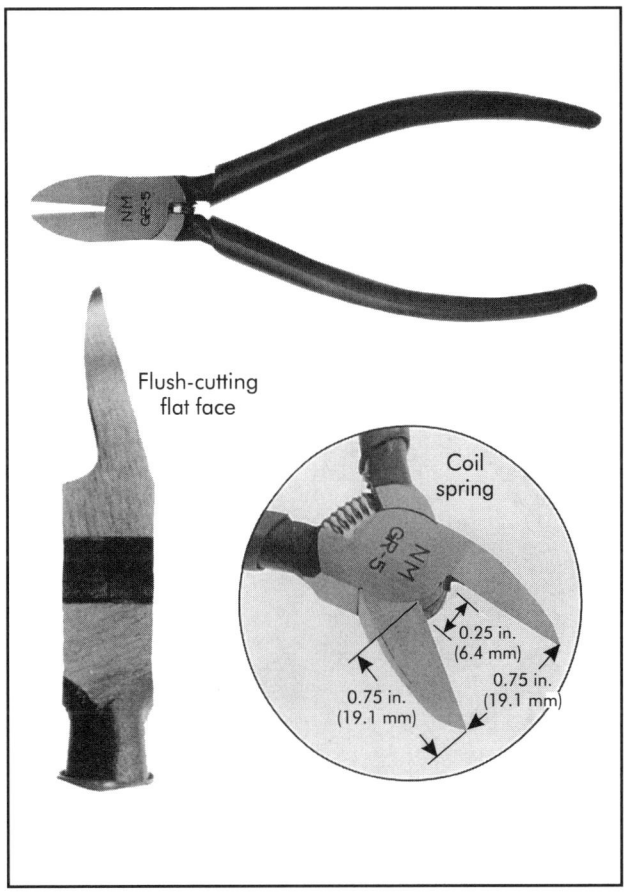

Figure 22-1. Flat-faced pliers provide flush cuts. (Courtesy Nickerson Machinery)

Note that some have an extra-low profile to allow entry into shallow openings. Many have spring-loaded ends to ensure that pliers are fully open for easier cutting.

Hand Deburring: Increasing Shop Productivity

Nipper Pliers

Nipper pliers do not greatly differ from flat-face pliers, but they may have a rounded back to accommodate cutting a sprue's inside diameter (Figure 22-2).

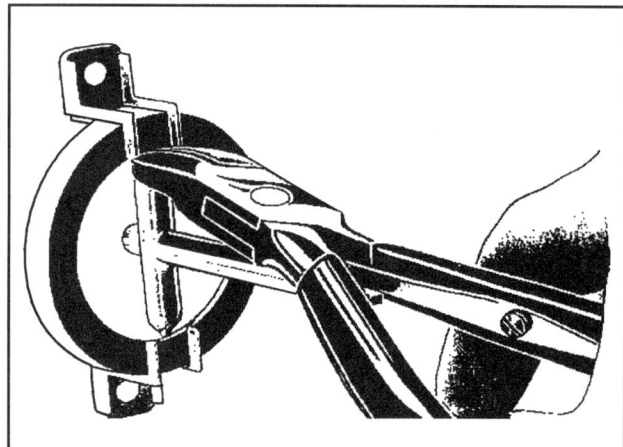

Figure 22-2. Nipper-type pliers have a rounded surface to better fit inside part diameters. (Courtesy Nickerson Machinery)

Angled-head Pliers

Figures 22-3 through 22-5 illustrate angled-head pliers. They come in 45, 85, and 90° angles, and can be flat-backed or rounded.

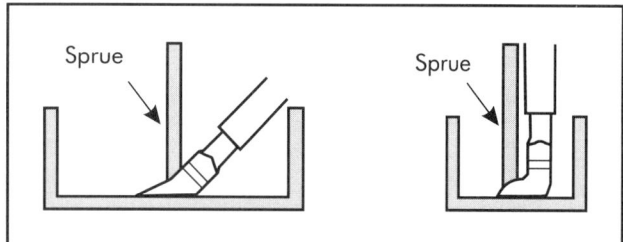

Figure 22-3. Angled-head pliers allow access to the bottom of cavities. (Courtesy Nickerson Machinery)

Concave/Convex Pliers

Concave/convex pliers have a large radius that matches the surface from which the sprue emerges (Figure 22-6).

Miniature Nippers

The miniature nipper is a lightweight tool for small hands (Figure 22-7). This may provide a significant advantage if operators tend to develop carpal tunnel syndrome. The pneumatic nippers are probably even better, since air power provides the cutting force.

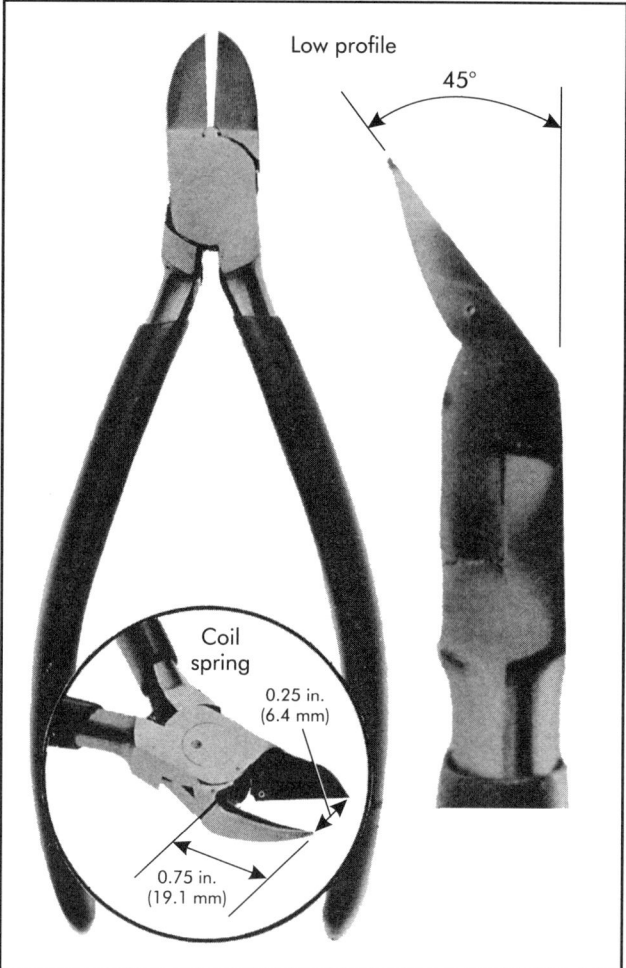

Figure 22-4. Angle-cut pliers with 45° flat back for sprue cutting. (Courtesy Nickerson Machinery)

Availability of Pliers

Pliers are available in a variety of sizes, shapes, and strengths. Ergonomically designed pliers are also obtainable, one of which is shown in Figure 22-8. Use of these specially designed pliers may be an important way in which some companies can reduce injuries.

Air-powered pliers are also available (Figure 22-9). They are used to cut reinforced plastics, nylon, acetal, polycarbonate, fiberglass, and other thermoplastics. They can be purchased in flat-face, 18-, 25-, 45-, and 60°-angle blades. They produce forces ranging from 530–1,650 lbf (2.4–7.3 kN).

Hones

Occasionally, hones are useful for removing small burrs in holes. Brushes with abrasive balls on their

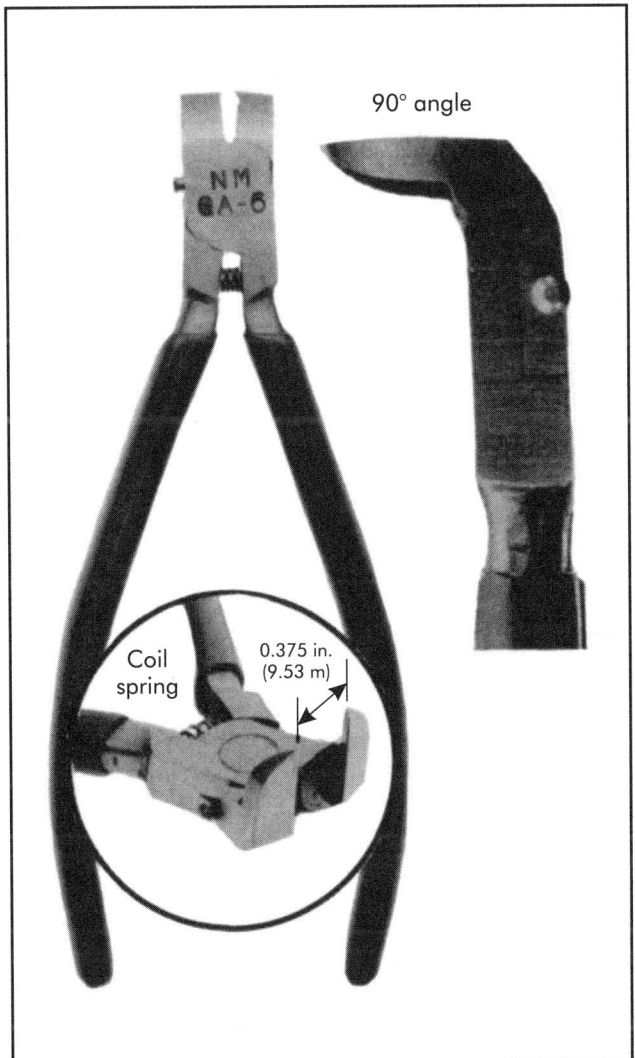

Figure 22-5. Angle-cut pliers with 90° flat back for sprue cutting. (Courtesy Nickerson Machinery)

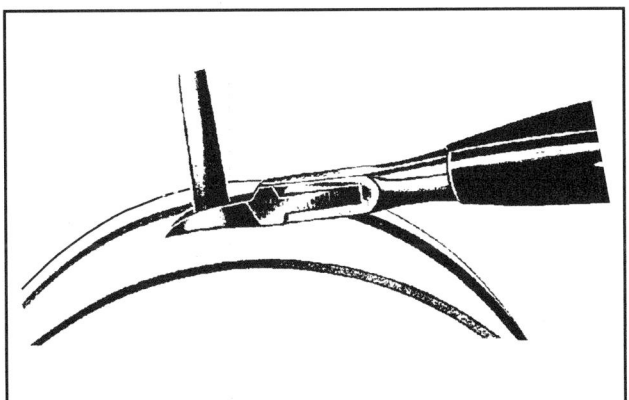

Figure 22-6. Concave/convex pliers for sprue cutting. (Courtesy Nickerson Machinery)

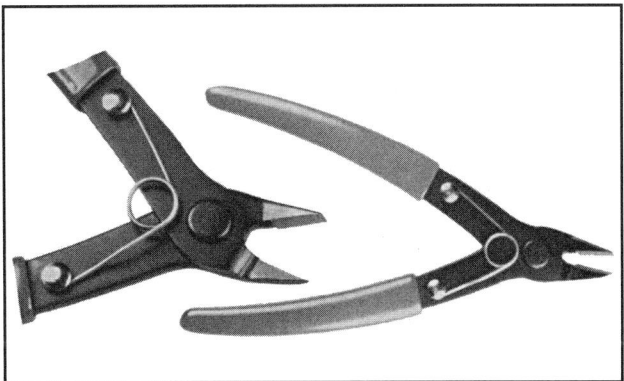

Figure 22-7. Miniature nippers. (Courtesy Nickerson Machinery)

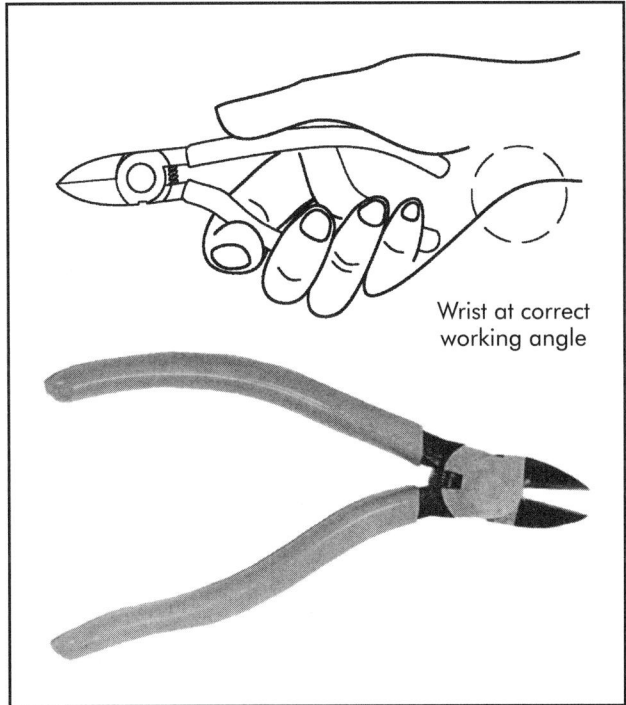

Figure 22-8. Ergonomic pliers. (Courtesy Nickerson Machinery)

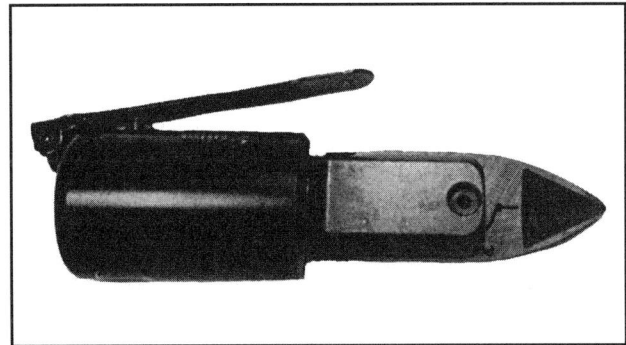

Figure 22-9. Air-powered pliers. (Courtesy Nickerson Machinery)

ends (Figure 20-20) are an example of a hone widely used for deburring. Traditional hones do the same thing. However, they typically do not remove large burrs and do not leave rounded edges. Machines are available to hone edges of carbide and ceramic cutting-tool inserts without chipping them, but these are automated devices (Gillespie 1999).

Spilt-mandrel Hones

Split-mandrel hones have a basic shaft that is split in half to allow the hone to spread apart, creating a diameter that can rub against the entire hole wall.

Laps

Lapping can remove small burrs from flat surfaces, as well as from holes. Flat laps are used on flat-surface parts. They produce improved surface finishes and nearly burr-free edges. Normally, lapping is not a hand operation, although it can be when only a few parts are involved and little material must be removed. Lapping uses fine abrasives (90–800-grit) and a solution called a *vehicle* to carry the abrasive and fine chips from the parts. The finer grit produces finishes of 1–2 μin. (0.03–0.05 μm) (Gillespie 1999).

Hole Laps

Figure 22-10 shows some common, adjustable hole laps. They are available in both through- and blind-hole configurations. Through-hole tools are called *barrel laps* and are available as small as 0.186 in. (4.72 mm). They will only remove tiny burrs in cross-holes.

Midget Laps

Figure 22-11 illustrates midget laps, which remove fine burrs from intersecting holes. They are inappropriate for normal drilling and large burrs. The eye of the tool can be opened to change the lap diameter. Midget laps range in size from 0.032–0.281 in. (0.81–7.14 mm). Each expands to the next larger size and has a slot like the eye of a needle. An accompanying tool expands the slot, which, in turn, increases the diameter. As is the case for the larger laps, these tools will only remove minute burrs. They will not radius edges.

Buffs (Buffing)

Some plants use buffing to deburr and polish at the same time. It is a dirty task, but the hard buffing compound caked onto the face or diameter of the wheel acts like a pliable grinding wheel. It is not generally listed as a deburring process, but it does remove small- to average-size burrs. Because this process both polishes and deburrs, it saves an operation for many users. Several supply houses sell miniature bench units, and others sell stand-alone machines on pedestals. The science of buffing is beyond the purpose of this text. Readers may wish to review some of the current literature (Wick and Veilleux 1982; Murphy 1999).

Dental Picks

Dental picks come in dozens of shapes (Figure 22-12). Their thin and bent configurations offer many

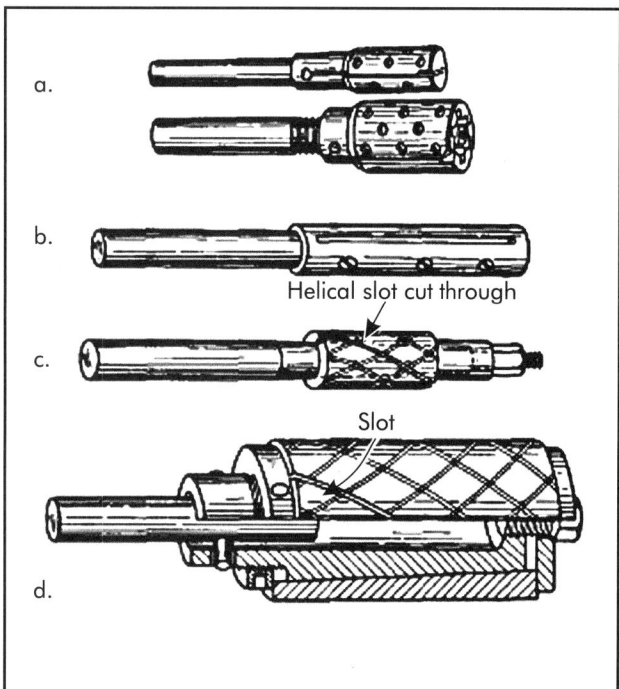

Figure 22-10. Adjustable hole laps (Wick and Veilleux 1982).

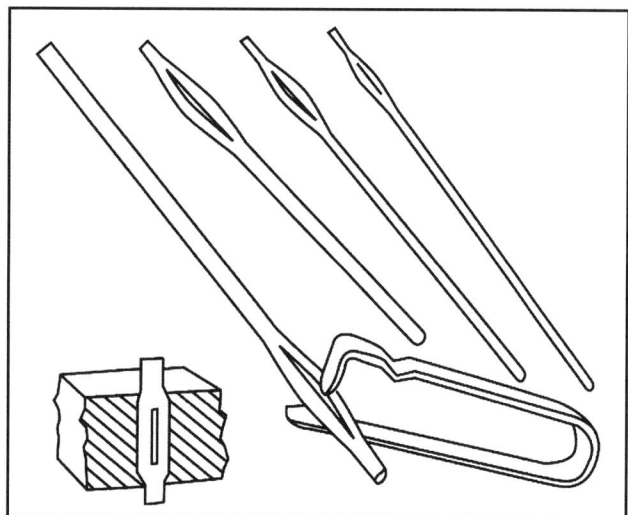

Figure 22-11. Midget laps. (Courtesy Reid Tool Supply)

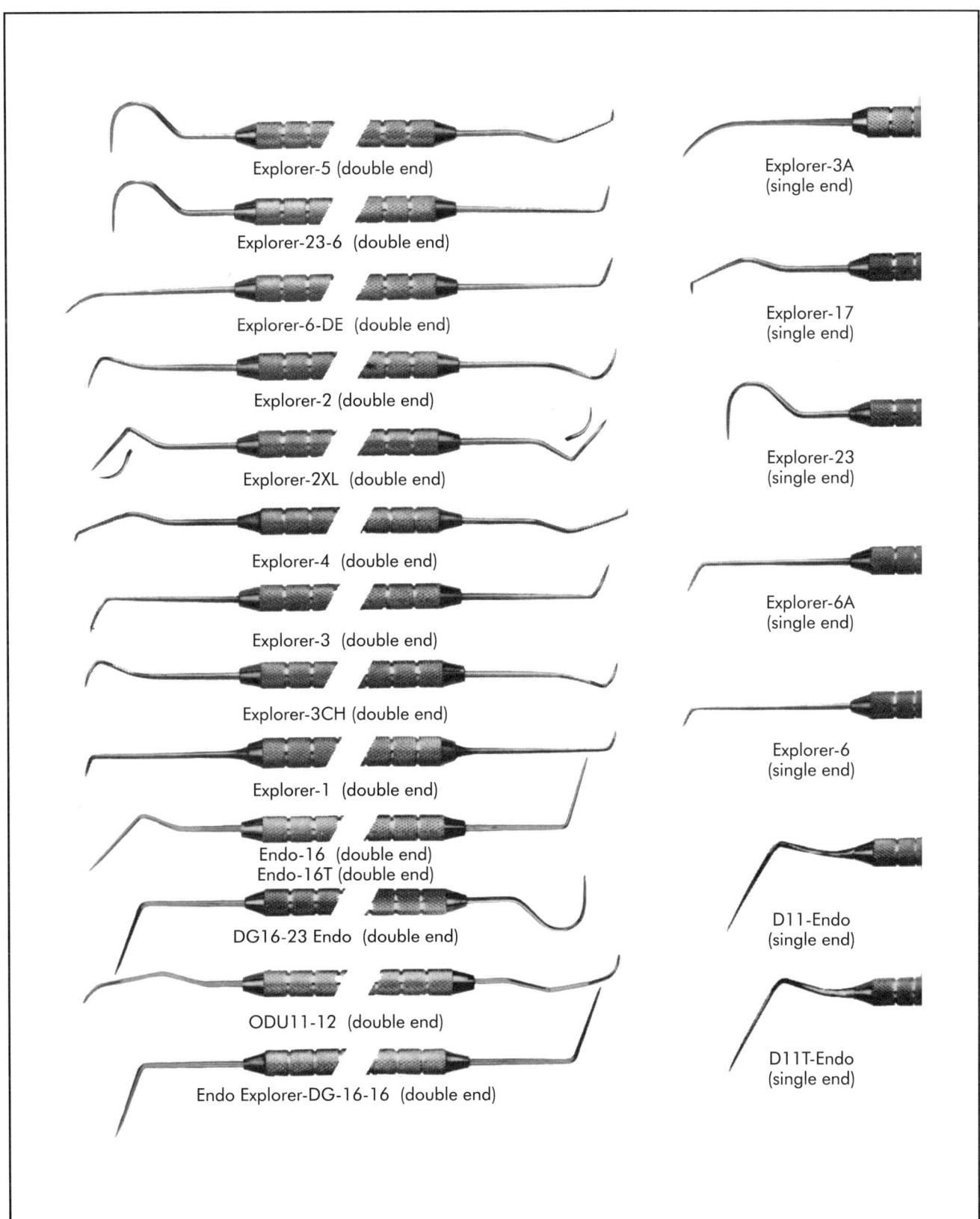

Figure 22-12. Dental picks. (Courtesy Brasseler)

uses when removing small particles and finishing difficult crevices. They do not remove significant material, but are allied tools, required to complete the task. It is possible to cut fine serration on the edges of some dental picks to create an abrading tool.

The dental profession has many names for these tools. Those shown in Figure 22-12 are called "explorers." Users will also find periodontal probes, excavators, chisels, hatchets, hoes, amalgam condensers, carvers, scalers, gingival margin trimmers, gingival cord placement instruments, ball burnishers, periodontal curettes, Gracey curettes, gingivectomy knives, periodontal files, periodontal chisels, elevators, periosteals, root-pick elevators, root picks, bone curettes, and bone files. Shops working with hard-to-reach cavities should have a catalog of some of these tool types (Gillespie 2000).

Broaches

Broaches are used in some situations for deburring blind features and holes. Figures 32-35 and 32-36 in Chapter 32 illustrate such use. Broaches have a sharp leading edge that cuts off burrs. They leave either a sharp edge or an edge with some burr, but much smaller than the original. They can be shaped for any configuration; however, they are limited to a relatively short cutting length. A long, straight edge can be broached, as well as intersecting holes.

Watchmaker Broaches

Watchmaker broaches are made from alloy steel, sharpened and tapered for cutting and smoothing corners, edges, and cavities. The cutting broaches range in size from 0.0123–0.2570 in. (0.312–6.528 mm [Stubbs size F-30]). Size is measured at the largest end of the broach. Cutting broaches are five-sided tools; smoothing broaches are circular. Over 50 broaches are available. Some come with knurled handles and others require a pin vise to hold them. Pivot broaches are used for cutting tubes or other long, extremely small holes. Extra-fine and lens-cutting broaches are also available, as well as part holders specially designed for use with these tools. They are used to clean flash in small holes or cavities, smooth edges of miniature parts, and deburr small pockets.

OTHER TOOLS

Every shop has used unusual tools to help deburr parts, so the preceding list is incomplete. It does, however, provide insight for working on difficult parts.

Diamond-coated, split-point tools compress as they enter the hole, then expand after reaching the cross-hole. In some plants, this is called a coated clothespin tool, because it compresses like an old-fashioned wooden clothespin.

Figure 22-13 illustrates one manufacturer's variety of diamond-coated clothespin tools. As shown, they can be produced with a leading chamfer and a tailing chamfer, rather than ball shapes. Practically any cylindrical shape can be produced with them. One U.S. source produces these split-pin shapes in diameters of 0.059–1.180 in. (1.50–29.97 mm).

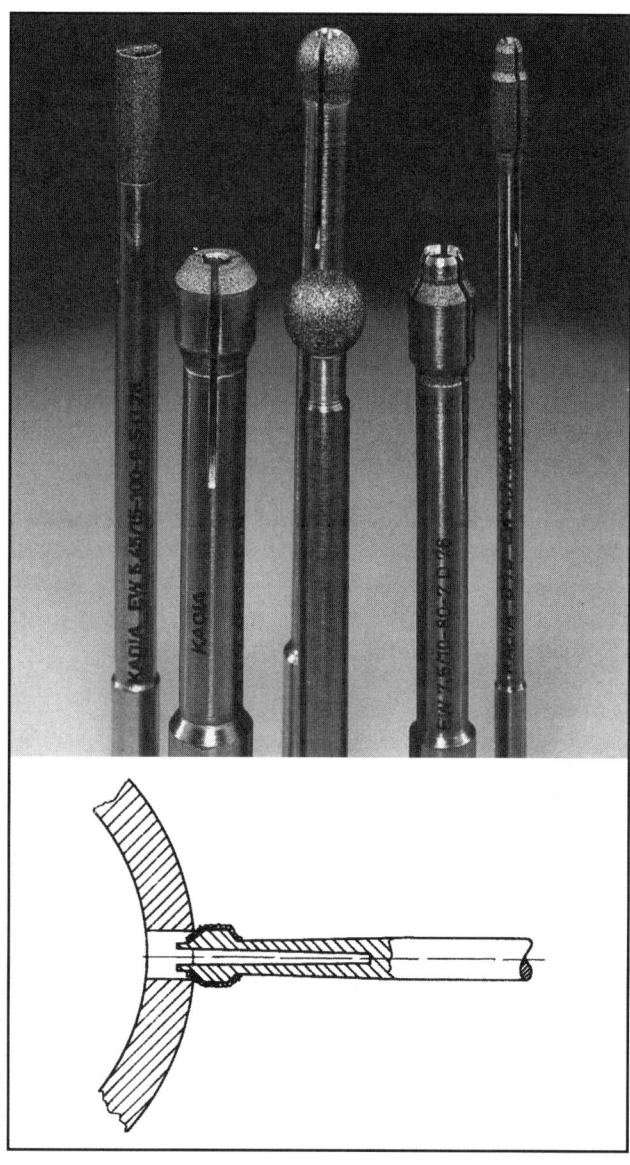

Figure 22-13. Diamond-coated, cross-hole deburring tools. (Courtesy Kadia)

ENVIRONMENTAL, HEALTH, AND SAFETY ISSUES

Daily use of pliers can lead to repetitive motion injuries. When workers use them on hot sprues, they also risk face burns, as well as cuts from sharp, nearby flash. Honing and lapping do not normally create dangerous situations; however, workers' clothing can become wrapped in rotating spindles, causing injuries. Many dental picks have sharp points that can puncture skin. Buffing requires a face shield to protect eyes and face from flying compounds.

REFERENCES

Gillespie, LaRoux K. 1999. *Deburring and Edge Finishing Handbook*. Dearborn, MI: Society of Manufacturing Engineers.

——. 2000. *Guide to Deburring, Deflashing, Trimming Equipment, Supplies, and Services*. Kansas City, MO: Deburring Technology International.

Maroney, Marion L. 1991. *A Guide to Metal and Plastic Finishing*. New York: Industrial Press.

Murphy, Michael. 1999. *Metal Finishing Guidebook and Directory*. New York: Elsevier Science, Inc., pp. 50–65.

Wick, Charles, and Raymond Veilleux, eds. 1982. *Tool and Manufacturing Engineers Handbook*, 4th ed., Vol. 3. *Materials, Finishing, and Coating*. Dearborn, MI: Society of Manufacturing Engineers.

Handheld Motors 23

Hand deburring involves a variety of motorized tools. In simple terms, they can be divided into either handheld units or motors that operators hold parts against (that is, handheld parts). Handheld motors are discussed in this chapter, and bench motors and allied motorized devices in the next chapter.

Nine categories of handheld motors are used for deburring and deflashing. They are industrial motors, dental motors, flexible-shaft motors, angle grinders, portable sanders, reciprocating files and stoneholders, portable drills, portable edger and chamfering machines, chippers, and scalers. Each category includes subcategories (Table 23-1).

Most individuals performing deburring use several of these tools on a daily basis—a bench motor and one or more air motors, for example. The advantage of motorized tools is that they are much faster than hand motions. In many cases, a motorized tool speeds deburring 2–5 times the normal rate. There are some 500 different motorized tools used in the United States for deburring, deflashing, and polishing applications. This chapter deals with those commonly used on metal parts.

TYPES OF HANDHELD MOTORS

Deburring motors can be segmented into typical industrial motors, dental motors, sanders, chippers, scalers and other related tools. These are further divided into air-driven, electric, or cable-driven tools. They can also be divided into finish work, and roughing work, or heavy-duty motors. Handheld motors come in an estimated 300 different variations to accommodate unique work, individual operator differences, and duty cycle differences (that is, continuous or sporadic use).

Industrial Motors

Most users procure rugged industry motors for deburring, deflashing, trimming, and related tasks. They may not realize that dental motors provide a level of skill and fineness not available in industrial motors. Industrial units are by far the most often used hand tools.

Industrial Air Motors

Air motors are the most widely used handheld tool (handpiece) for deburring and come in a wide variety of sizes, shapes, speeds, and powers. They are universal motors, designed to accommodate an unending variety of tools. These motors come in a straight (standard), short handpiece, extended handpiece, and angle tool (Figure 23-1). Some provide power with a squeeze handle, others with a twist ring.

Air die grinder. The most commonly used conventional industrial die grinder air motor accepts a 0.125-in. (3.16-mm) diameter collet and uses air pressures of between 50–80 psi (345–552 kPa)—30–50 ft^3/min (0.85–1.42 m^3/min). These air motors run at 14,000–60,000 rpm, with the higher speeds dedicated to smaller tool diameters. Many, if not most, of the high-speed motors require filtered air containing a small amount of oil. These provide the muscle for effective use of rotary burs, mounted points, small abrasive-filled rubber, cork or cotton products, small sanding devices, and dental stones. Air grinders are handled in one of two ways:

Table 23-1. Summary of handheld motors

Motor Category	Motor Type	Speed Range	Power, hp (kW)
Industrial air motor	Standard die grinder (air)	950–100,000 rpm	0.30 (0.2)
	Angle-head die grinder	15,000–60,000 rpm	0.20 (0.1)
	Angle polisher	2,200–13,200 rpm	0.50–1.60 (0.4–1.2)
	Straight grinder	5,770–17,000 rpm	0.67–3.20 (0.5–2.4)
	Angle grinder (air)	6,000–13,000 rpm	0.67–3.20 (0.5–2.4)
Industrial electric motor		1,000–45,000 rpm	
	Flexible-cable motor	2,000–40,000 rpm	
	Flexible-motor die profiler	0–14,000 strokes/min	0.100 (0.07)
	Angle grinder (electric)	5,000–12,000 rpm	0.50–3.50 (0.4–2.6)
Dental air motor	—	250–430,000 rpm	—
Dental electric motor	—		—
Flexible-shaft motor (electric)	—	0–20,000 rpm	0.10–0.25 (0.07–0.19)
Portable sander	Belt (air)		—
	Belt (electric)	7,000–20,000 rpm	—
	Narrow belt (air)	5,000–20,000 rpm	—
	Narrow belt (electric)	30,000 rpm	—
	Orbital	900–10,000 rpm	—
	Right-angle disc	3,000–20,000 rpm	—
Reciprocating file and stone holder (air)	—	0–3,000 strokes/min	0.07 (0.05)
Portable drill electric/battery	—	0–1,200 rpm	—
Portable edger and chamfering machine	—	2,500–55,000 rpm	0–2.00 (0–1.5)
Chipper and scaler	—	2,500—4,600 blows/min	—

- like a small pencil, in which only light finger pressure is used to "kiss" the edge and remove the material, and
- like a heavy grinder, in which the hand and arms apply considerable pressure to grind off the burrs.

One industrial pencil grinder is available that operates at speeds up to 340,000 rpm. This is far above the normal speeds used by other air grinders. It uses diamond or carbide rotary burs and requires a very light touch (Figure 23-2). Most die grinders do not use coolant or lubricant, but this particular device comes with a spray mist attachment for working on glass, ceramics, precious stones, and other crystalline materials.

Although most die grinders are air-driven, some DC-driven electric handpieces provide the same size of tool and results with less noise (Gillespie 2000). Both air and electric handpieces are available in weights as low as 6 oz (0.17 kg), but heavy-duty systems can be much heavier. The air-powered units are extremely sturdy and permit temporary overloading

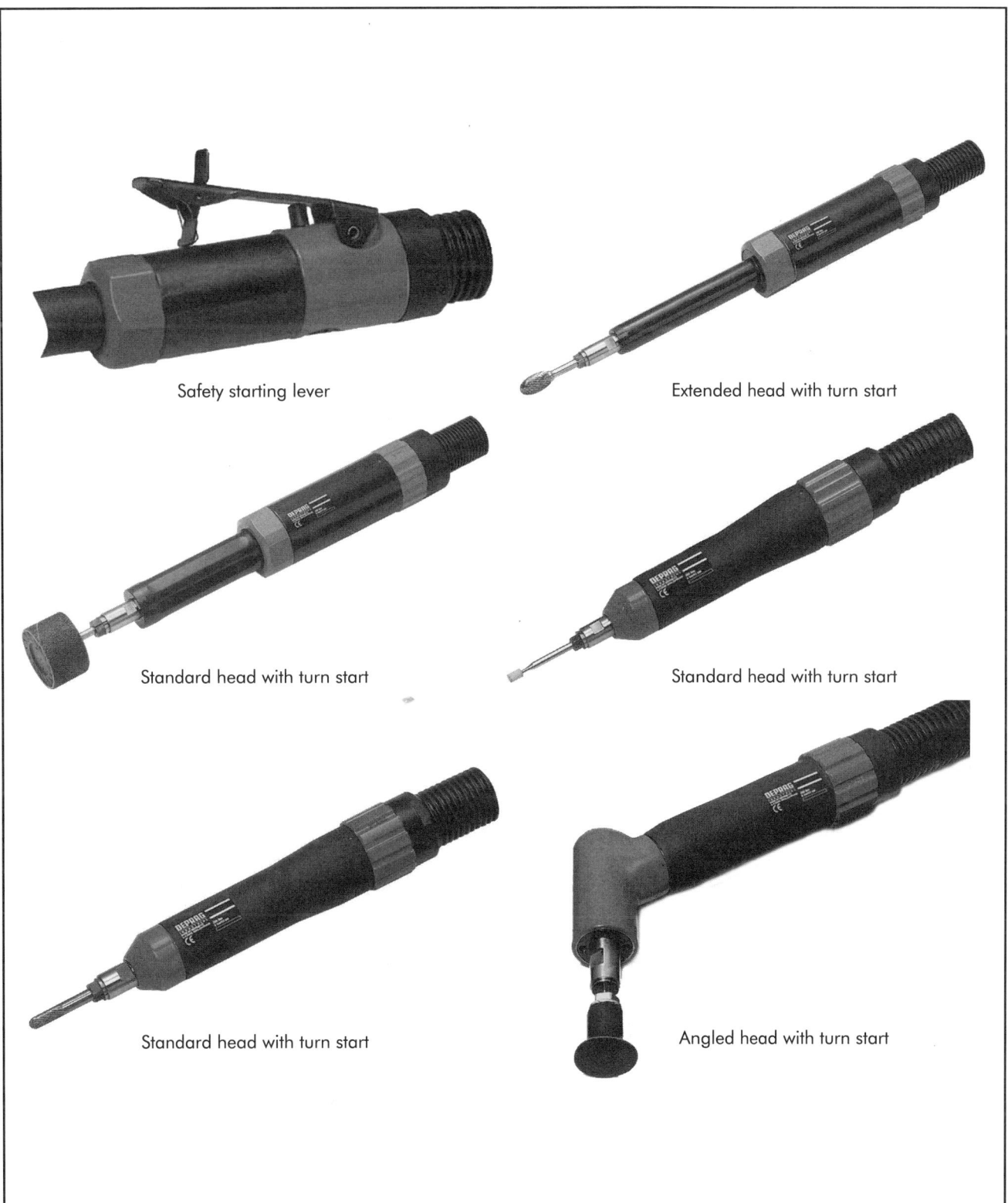

Figure 23-1. Typical die grinders. (Courtesy Deprag)

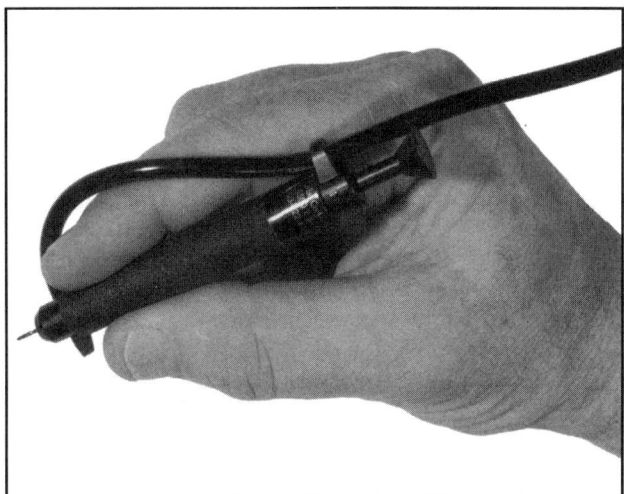

Figure 23-2. High-speed die grinder. (Courtesy Hunter)

without harming the drive motor (Production Engineering Research Association of Great Britain 1956).

Straight grinders. A handheld straight grinder is a heavy-duty grinder that weighs from 2–12 lb (0.91–5.44 kg), as illustrated in Figure 23-3. This type of grinder removes parting lines and heavy flash, grinds large chamfers or reliefs, and attacks the workpiece. Because the rotating tool is larger than rotary burs and mounted points, the rotary speeds are lower than for the die grinders. Straight grinders operate at 1,000–10,000 rpm, provide 1–3 hp (0.8–2.2 kW), and consume up to 2 ft^3/min (0.06 m^3/min) air. The complete assembly can be up to 24-in. (610-mm) long.

Angle grinders. Angle grinders, as referenced in this chapter, are heavy-duty tools with a spindle axis at 90° or other relative angle to the handle axis (Figure 23-4). Typically, they use grinding wheels or abrasive-filled nonwoven nylon products.

Many angle grinders are also called angle die grinders or *disc grinders,* whose disc sizes range from 7–9 in. (177.8–228.6 mm) in diameter. They are available in various industrial configurations to accommodate light, intermittent, and continuous (heavy) work. The heavy-duty units come with integral high-temperature wiring and powerful ventilating fans for efficient cooling. High-cycle electric motors operate at nearly constant rotor speeds and low operating expense.

The backup pad in disc grinding is similar to the contact wheel in belt grinding—it provides support when the part is pushed into the abrasive disc. Pads come in stiff, flexible, and mushroom backup hardness grades. Stiff pads produce flat and smooth surfaces and may have a mildly curved form to reduce gouging. Flexible pads distribute pressure more evenly over the surface and cushion the abrasive action. Mushroom backup pads are used with conformal sanding discs.

Most of these tools allow users to provide extensive pressure on parts for heavy stock removal. They work surfaces as well as edges. The tools illustrated in various figures here have rather rigid abrasive wheels in them, but flexible sanding discs also can be used.

Angle polishers. Angle polishers are much the same as grinders, but are designed for finer finish work. Many equipment manufacturers do not distinguish between polishers and grinders.

Industrial Electric Die Grinder Motors

Electric motors come in many of roughly the same designs and sizes as air die grinders. Generally, the electric motor is heavier since the copper windings are heavy. Electric motors typically run quieter, but they may get hot if used in continuous applications as previously noted.

Dental Air Motors

Dental air motors are one of the most popular deburring tools. They offer a number of significant advantages, including small size and light weight. Also, many styles of heads (which allow a variety of shank sizes and shapes) and numerous cutting tools (see Chapter 10) are available for use with dental air motors.

It is important to remember that the driving pressure of some of these motors should not exceed 40 psi (276 kPa), because high pressures will burn the motor in one minute or less. They have both forward and reverse directions and an adjustable speed control, which provides speeds of 1,000–60,000 rpm (dental motors with speeds up to 400,000 rpm are commercially available).

Shown with a 7.000 × 0.375 in. (177.80 × 9.53 mm) wheel

Figure 23-3. Typical straight grinder. (Courtesy Deprag)

Figure 23-4. Typical angle grinders. (Courtesy Deprag)

Although many head styles and shank sizes are available for use with dental air motors, only three different heads and shank sizes are typically used for deburring (additional heads can be purchased).

- One head will accept a 0.094-in. (2.39-mm) diameter straight-shank tool.
- Another will take a 0.063-in. (1.60-mm) diameter straight-shank tool.
- The third head will take a 0.094 in. (2.39 mm) diameter, latch-type shank tool (see Figure 10-18), which has a grooved, flat end.

By removing the heads, a 0.094-in. (2.38-mm) shank tool can be inserted into the handpiece itself (Figure 23-5). These motors allow horizontal (on axis) tools, as well as right- or slight-angled head orientations. Figure 23-6 illustrates a variation: note the miniature bur in the contra-angle head. Cutters are changed with a simple push-button exchange action. Unlike industrial motors, these can be cleaned in autoclaves. This particular motor will operate at speeds as low as 100 rpm.

Dental motors will not accept many of the available large industrial rotary deburring tools. However, they accept hundreds of dental tools, including rubber dental points, rubber dental cups, rotary files, rotary burs, diamond-coated burs, and rotary dental stones (see chapters 10, 13, and 18). Like conventional industrial air motors, these motors require only light pressure for normal deburring. Heavy pressures tend to break down the bearings and collet bushings.

A few dental air motors include a fiberoptic light near the head, which allows users to see into hard-to-reach areas. These tools will not allow the heavy hand pressure that industrial motors do, but such pressure is not necessarily required to remove burrs. They are inappropriate for heavy burrs. Dental air motors are precision tools used in precision deburring of critical edges. They can be used with foot treadles, on-off switches, or speed adjustments located on the handpiece or controller. Some also provide a water spray directed at the cutter.

Flexible-shaft Electric Motors

Flexible-shaft motors (Figure 23-7) are widely used for deburring small, miniature, and even larger parts. They have variable speeds and allow use of up to 10 different handpieces (Figure 23-8). Some available handpieces have three-jaw chucks, others have collets, and a few have sets of collets that can be inserted in them.

The most commonly available handpieces have 0.094-in. (2.39-mm) diameter straight shanks or 0.094-in. (2.39-mm) diameter shanks with a flat and a groove cut in the end. Flexible-shaft motors, like dental air

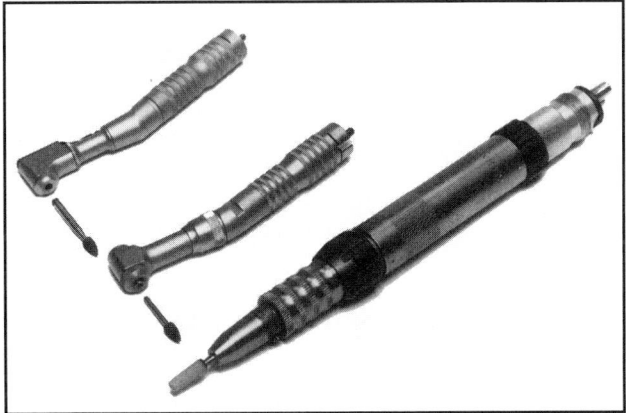

Figure 23-5. Heads and shanks used with dental air motors. (Courtesy Honeywell)

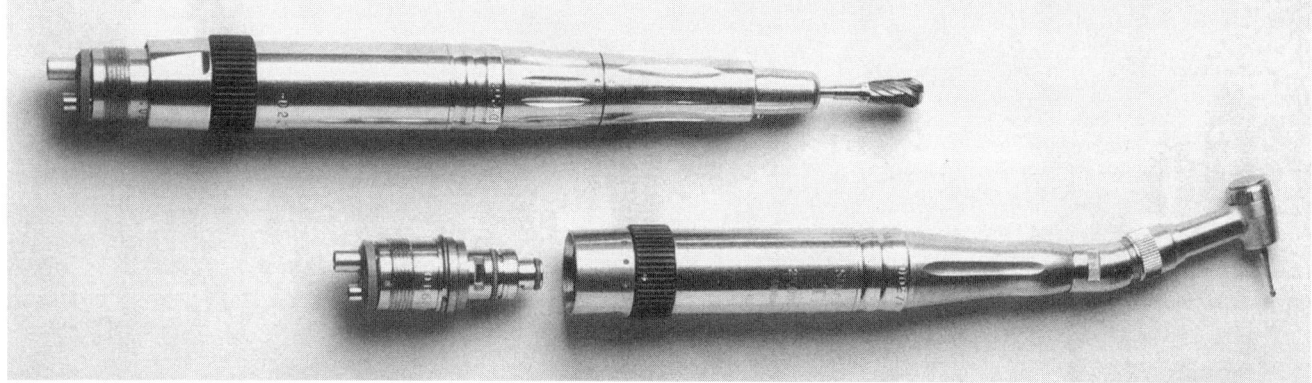

Figure 23-6. Two low-speed dental motors. (Courtesy Star Dental)

Figure 23-7. Flexible-shaft motor and handpiece. (Courtesy Honeywell)

motors, accept all dental tools and many commercial cutting tools. The advantage of these tools is that they are sturdier, in many cases, or more useful in industrial applications than the lighter dental tools.

Flexible-shaft motors can be hung from ceiling hangers or floor stands when desktops are inappropriate. Although most units are relatively low in power, some are available for heavy use, with motors up to 0.25 hp (0.2 kW). Foot controls are standard accessories.

Many flexible-shaft motors include a spring between the handpiece and cable connection, which gives the handpiece extra flexibility (for example, see the long handpieces in Figure 23-8). Most of the handpieces connect to the cable with a simple push-pull ball lock, which works well once users get used to the cable. In some situations, the relatively stiff cable does not flex enough to reach the desired feature, or it puts additional strain on the user's hands or arms. Despite that, tens of thousands of these tools are used in industry each day.

Portable Sanders

Many handheld sanders (Figure 23-9) are used for metal and plastic finishing. At least 100 commercial variations are available at relatively low cost. Their advantage, as with any rotary tool, is that they are much faster than hand motions alone, and tools can be changed from very coarse to ultra fine in a matter of one minute or less. Newly introduced bonded-abrasive materials can be applied immediately, without hardware costs. Commercially available handheld sander types include rotary, orbital or jitterbug, and belt.

Belt and Narrow-belt Sanders

Handheld belt sander belts are available in a variety of widths (Figure 23-10) and can be as narrow as 0.25 in. (6.4 mm). These miniature or narrow-belt sanders enable users to reach into small slots and holes to finish edges. They are also effective when used in something of a crossbow configuration for finishing large tube ends. Belt and narrow-belt sanders come in either air or electric models.

Orbital Sanders

Orbital sanders eliminate swirl marks from finished surfaces and are often preferred for meeting exacting painting surface requirements. They perform at about 10,000 orbits/min and can hold five to six sheets of abrasive paper at a time. Both electric and air-driven units are available. The air-driven units, without the weight of an electric motor, are lighter. However, they must be used near an air supply, which is a disadvantage. Orbital disc grinders are available in addition to the standard rectangular orbital systems.

Right-angle Disc Sanders

Right-angle disc sanders are similar to right-angle grinders. They employ sanding discs rather than rigid

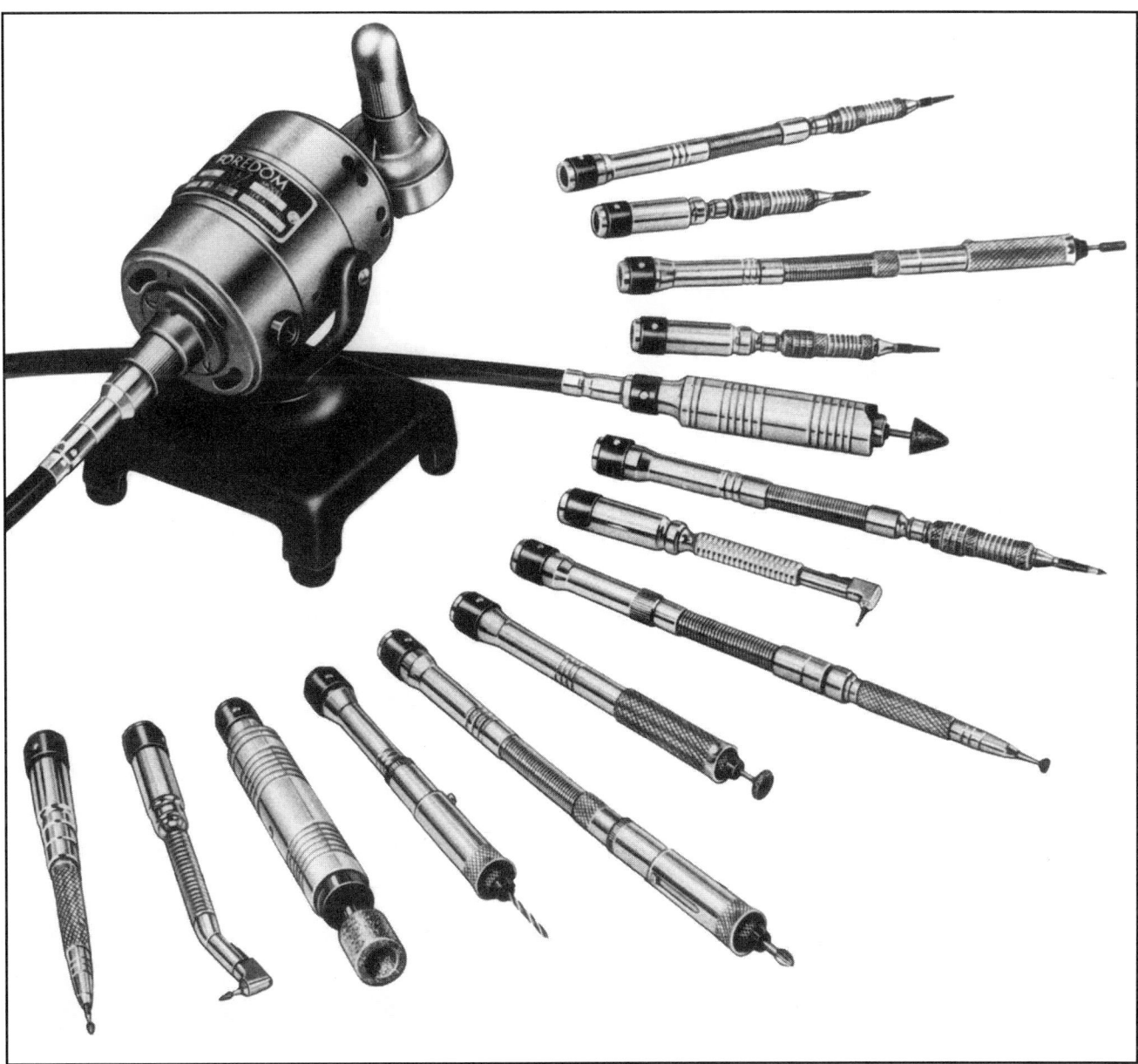

Figure 23-8. Handpieces used with flexible-shaft motor. (Courtesy Foredom)

grinding wheels, and are particularly applicable to deburring holes and large surfaces if swirl patterns can be left on the parts.

Reciprocating Files and Stone Holders (Air)

Figure 23-11 illustrates a reciprocating file or die profiler used to polish rather than deburr. These tools use a variety of miniature files or sanding products to provide a back-and-forth motion in hard-to-reach areas. They are not widely used for miniature parts, although miniature files as small as 0.014 in. (0.36 mm) thick are available.

Reciprocating files are used primarily on dies, molds, sheet metal, and large parts. Their advantage, of course, is that they supply the back-and-forth filing motion so the operator does not fatigue from such work. Files reciprocate at 9,000 rpm and have an adjustable stroke and tilt angle. Smaller units are available (Figure 23-12).

Reciprocating tools can use files, polishing stones, or diamond-coated flat tools (Figure 23-13) with two discrete speed families. A typical reciprocating tool runs at 2,000–10,000 strokes/min. Ultrasonic units vibrate at 20,000–40,000 cycles/sec. Electric or flex-

Figure 23-9. Handheld sanders. (Courtesy Dynabrade)

ible cable devices, as well as air-driven models, are also available.

Portable Drills

The common electric drill is used widely as a deburring and chamfering motor. Figure 23-14 illustrates a ball-ended nylon brush powered by a drill motor.

Portable Edgers and Chamfering Machines

Handheld edgers provide a specific edge chamfer on straight-edged parts (Figure 23-15). Although these units are among the smallest available (they fit in the palm of the hand), they are used for relatively heavy-duty work. These devices chamfer pipe ends, as well as straight, curved, and cut-out areas. Figure 23-16 shows a unit designed for straight-edge work.

Chippers and Scalers

Figure 23-17 illustrates an air-driven chipper used to provide reciprocating action to chisel-type tools. These are used for cleaning metal surfaces covered with slag or other materials. Figure 23-18 shows a needle scaler at work. As illustrated, a series of small "needles" reciprocate to attack different contours.

HANDHELD MOTORS IN USE

Figure 23-19 illustrates a typical workplace system. Two or more motors can be set in place with different tools or different-sized tools to perform the work. Extra heads are available that can reach into specific areas. An electric motor is shown in this illustration, but there may be two to 10 different air-motor lines overhead to allow rapid tool changes.

Figure 23-20 shows how right-angle heads are used to deburr the internal break-through of a hole into another hole. This is an application area in which dental tools would also excel. Extended-nose tools allow users to reach into deep areas, and the various handheld diameters provide a comfortable grip. Note the positions of the hand. A finger control is used for precision work. For other jobs, the fist provides the pressure. For deep holes, the hand provides the pressure; whereas, for intersecting holes, the fingers provide control.

Figures 23-21 and 23-22 illustrate the use of several angle grinders, die grinders, and extended-nose tools. Figure 23-23 shows a variety of the bonded-abrasive tools used on sanders or die grinders.

Other sources provide additional insight into these tools (Farago 1980; Production Engineering Research Association of Great Britain 1956).

ENVIRONMENTAL, HEALTH, AND SAFETY ISSUES

Rotary motors present safety hazards. They rotate, so they can throw things, and workers can injure themselves if they come in contact with the rotary portion of the motor.

On three-jaw chucks, do not forget to remove the chuck key before starting the motor. Even though these chuck keys are very small, they develop tremendous force when thrown from the motor. Remove all chuck keys before starting the motor.

Brushes should not be used in high-speed motors. They will throw metal fibers, and shanks will twist around the chuck. Slower speeds (than offered by these motors) are required for most brushes. In all cases, eyeglasses should be worn around any handheld motorized tool, even if it seems harmless.

If dust is visible in the air, an exhaust system should be used to keep it from the lungs of operators. Beryllium is a known carcinogen, and recent research indicates that even the small amounts in beryllium copper can affect workers who breathe or ingest its dust. OSHA and the Department of Energy each have established regulations designed to protect workers from the hazards of this and other materials. Nickel is a suspected carcinogen (including products that contain nickel, such as stainless steel), and its dust may be harmful if inhaled at levels above documented limits. Although handheld motorized tools may not produce visible dust, they may make enough to create worker health issues.

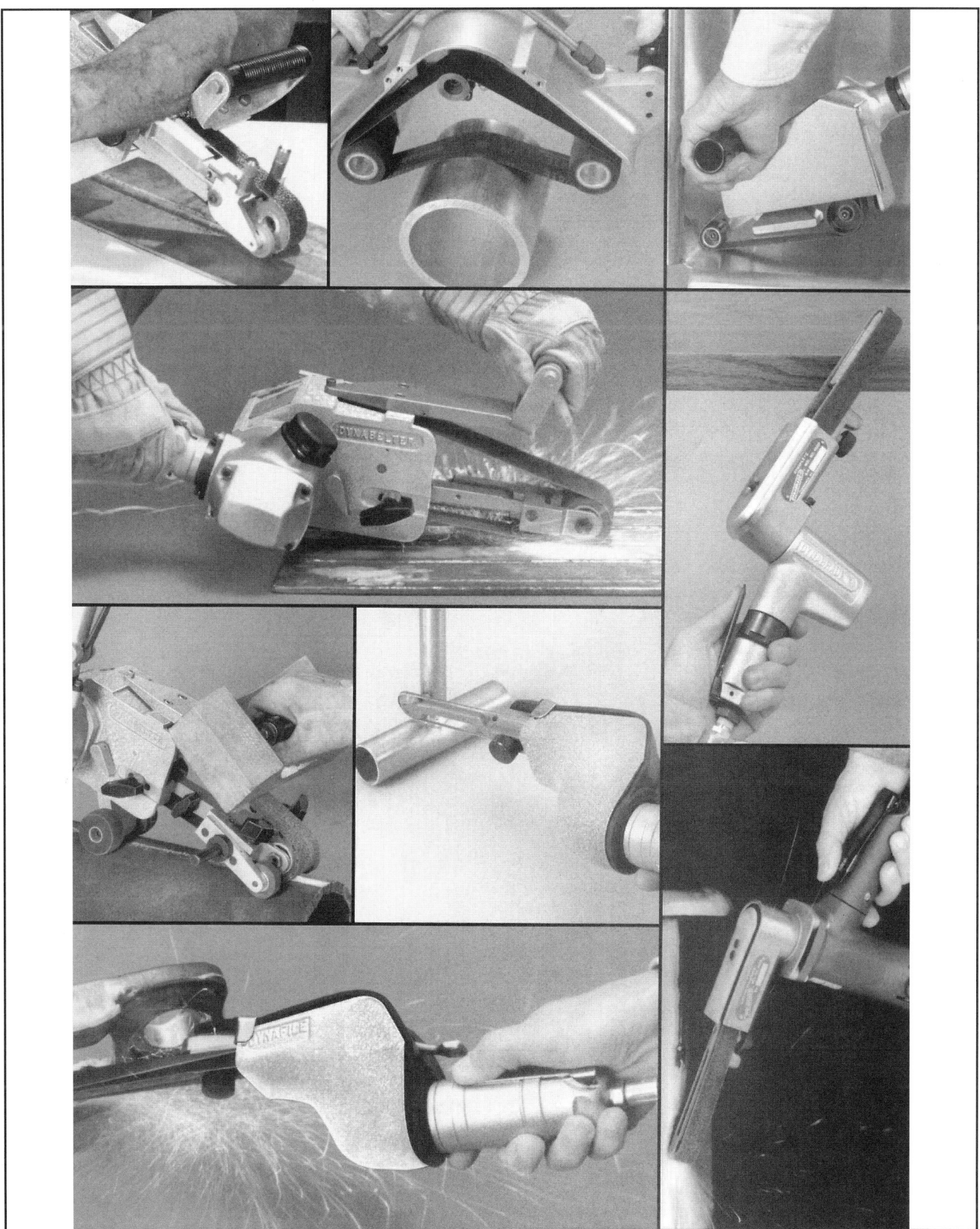

Figure 23-10. Handheld belt sanders in use. (Courtesy Dynabrade)

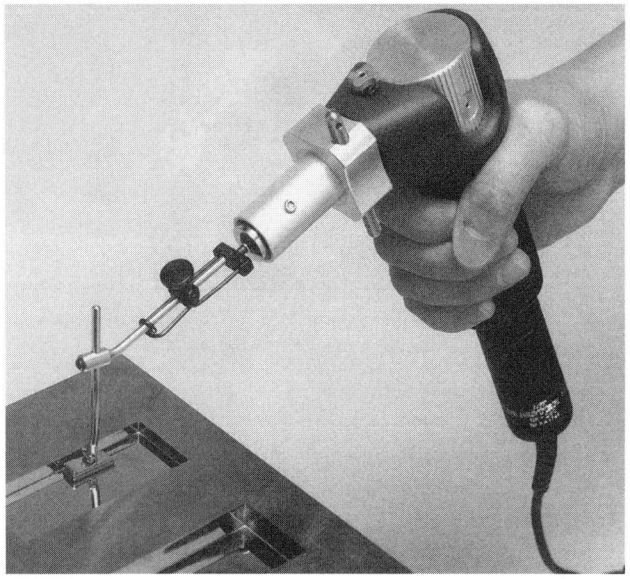

Figure 23-11. Reciprocating file. (Courtesy NSK America)

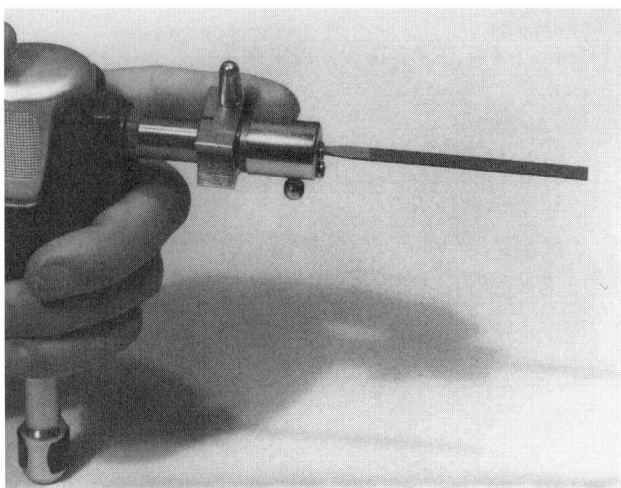

Figure 23-12. Small reciprocating file. (Courtesy Honeywell)

High-efficiency particulate air (HEPA)-filtered vacuum exhausts, if carefully controlled to prevent contamination to those who clean them, may provide adequate controls. Each plant must perform its own assessment and establish its own controls because general guidelines may be inadequate, particularly in the case of beryllium and its alloys and compounds.

REFERENCES

Farago, Francis T. 1980. *Abrasive Methods Engineering.* New York: Industrial Press.

Gillespie, LaRoux K. 2000. *Guide to Deburring, Deflashing, and Trimming Equipment, Supplies, and Services.* Kansas City, MO: Deburring Technology International.

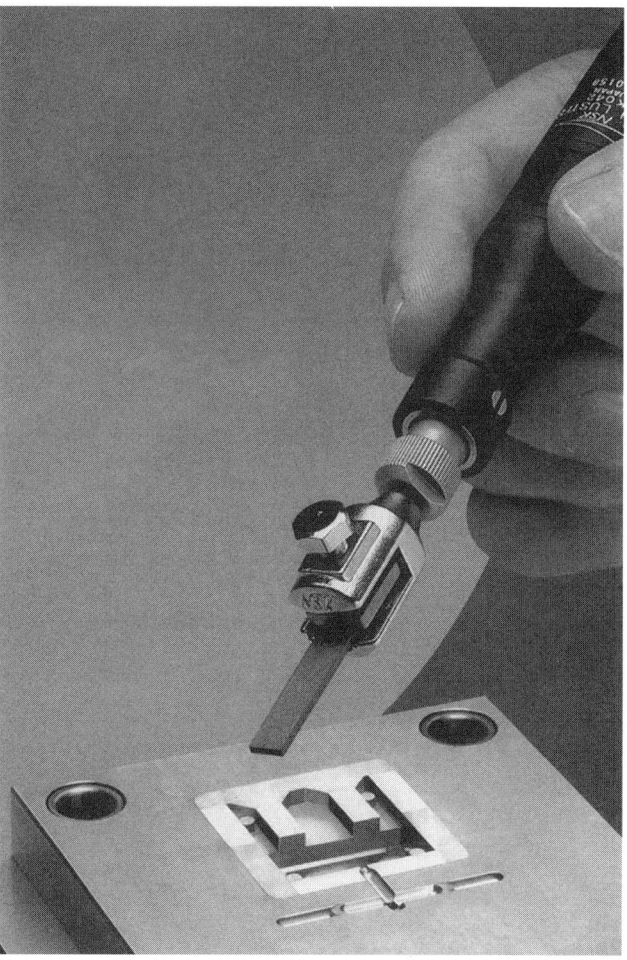

Figure 23-13. Reciprocating stone holder. (Courtesy NSK America)

Figure 23-14. Handheld drill used for deburring intersecting holes and honing bores. (Courtesy Brush Research)

Figure 23-15. Handheld edge-chamfering machine. (Courtesy Nitto Koki USA)

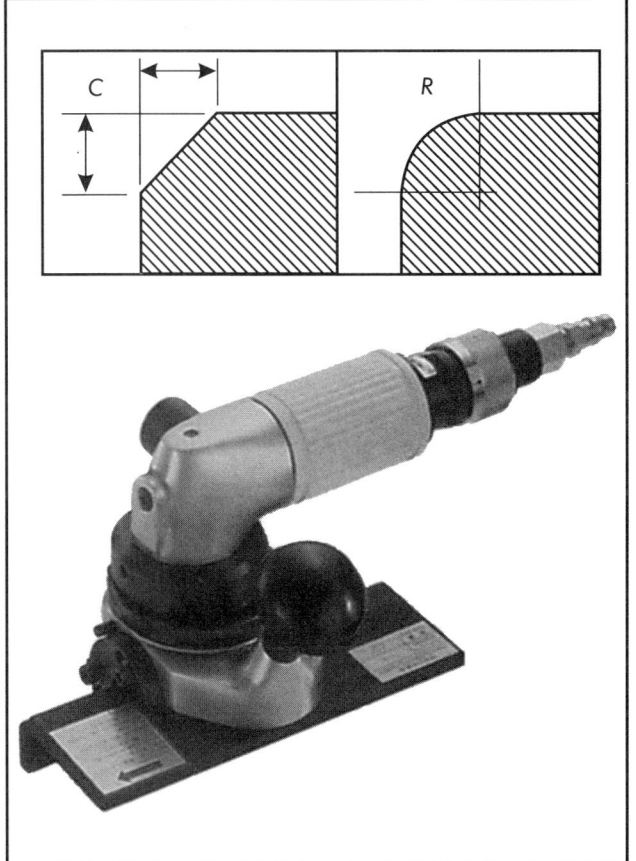

Figure 23-16. Straight-edge beveler. (Courtesy Nitto Koki USA)

Production Engineering Research Association of Great Britain. 1956. *Power Tools: Guide to Deburring of Machined Components: Part IV.* Melton Mowbray, Leicestershire, England: Production Engineering Research Association of Great Britain.

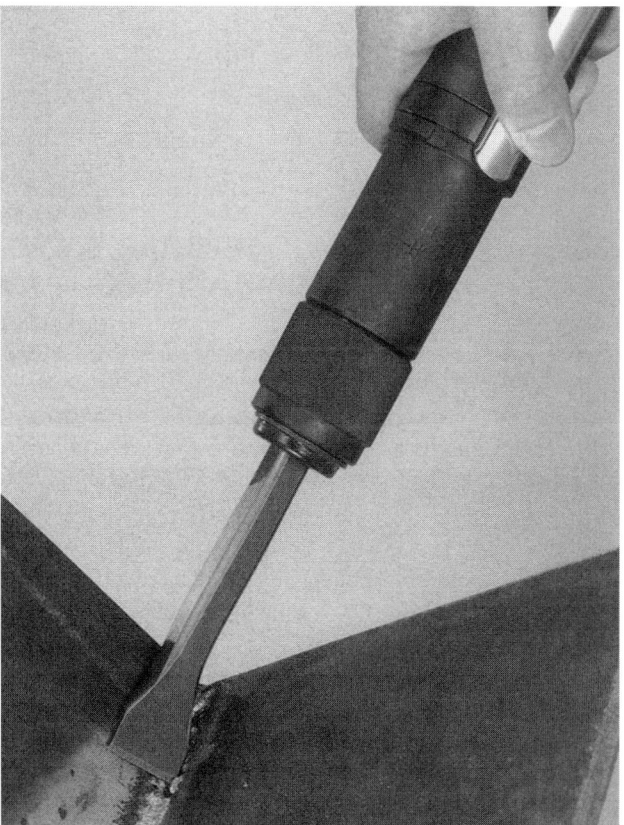

Figure 23-17. Air chipper. (Courtesy Nitto Koki USA)

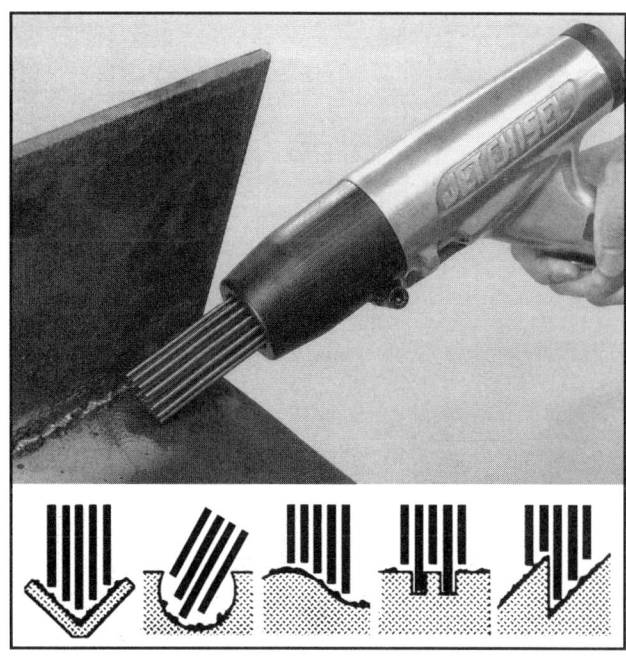

Figure 23-18. Needle scaler. (Courtesy Nitto Koki USA)

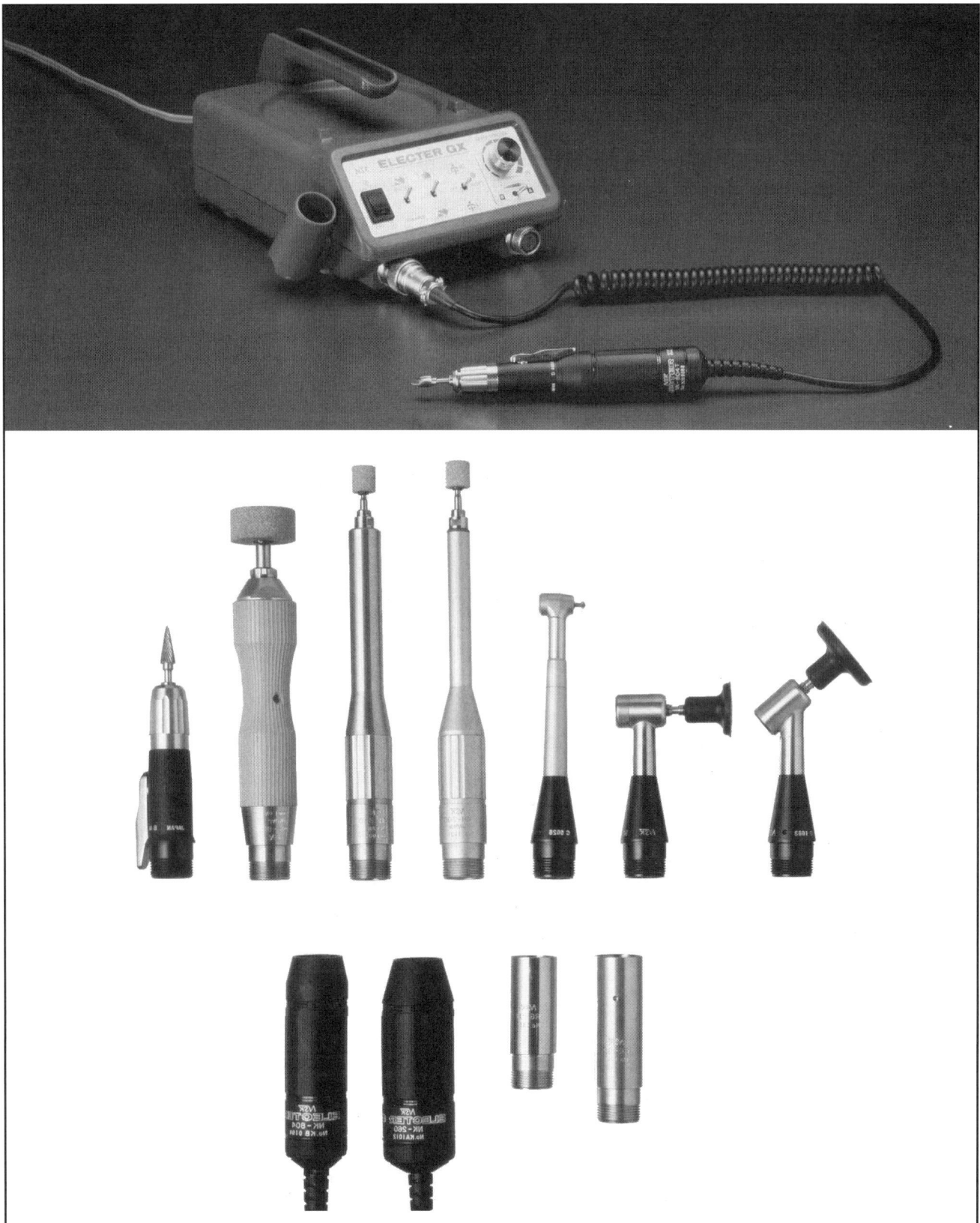

Figure 23-19. Die grinder and handpieces. (Courtesy NSK America)

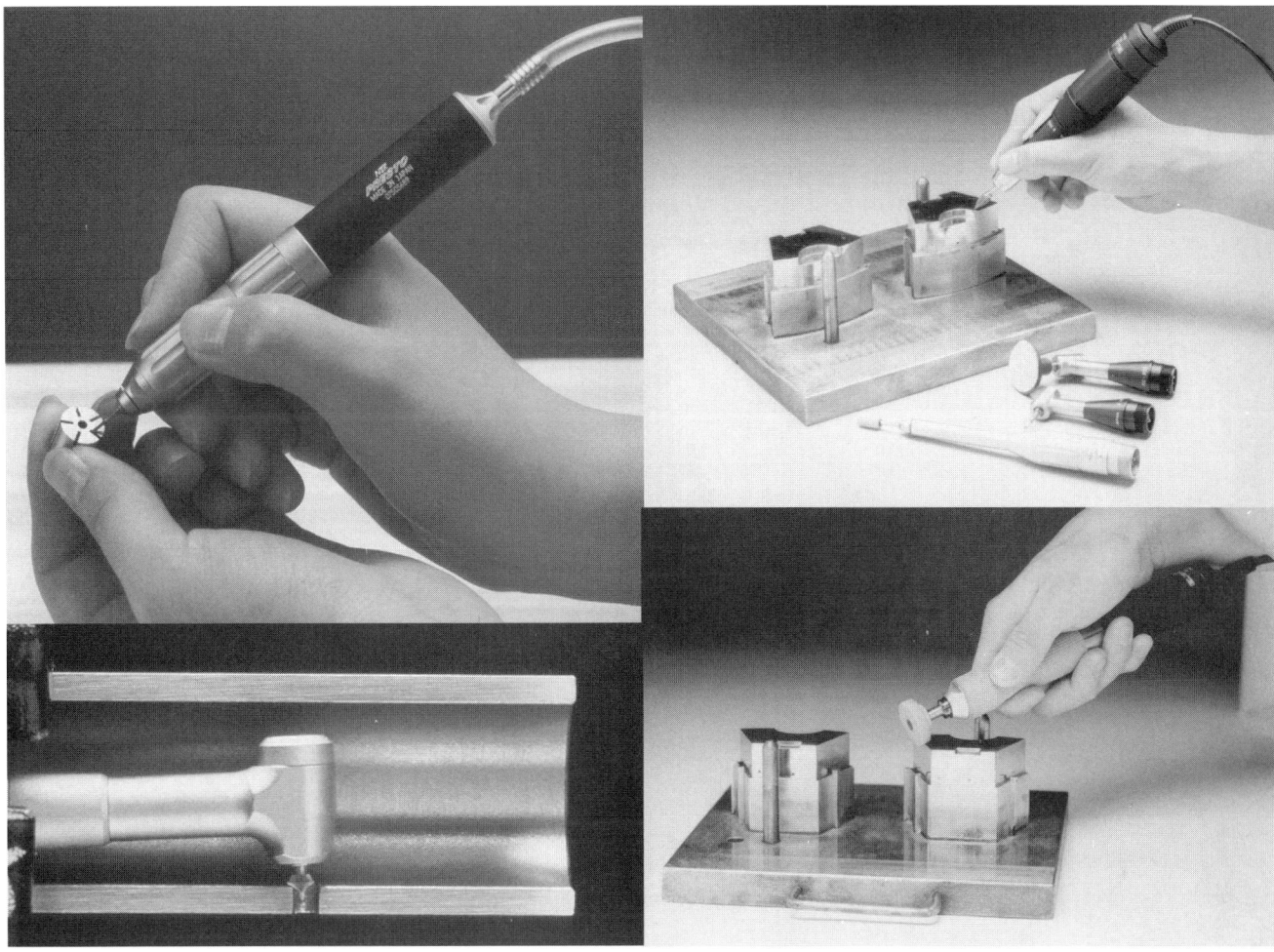

Figure 23-20. Applications for die grinders. (Courtesy NSK America)

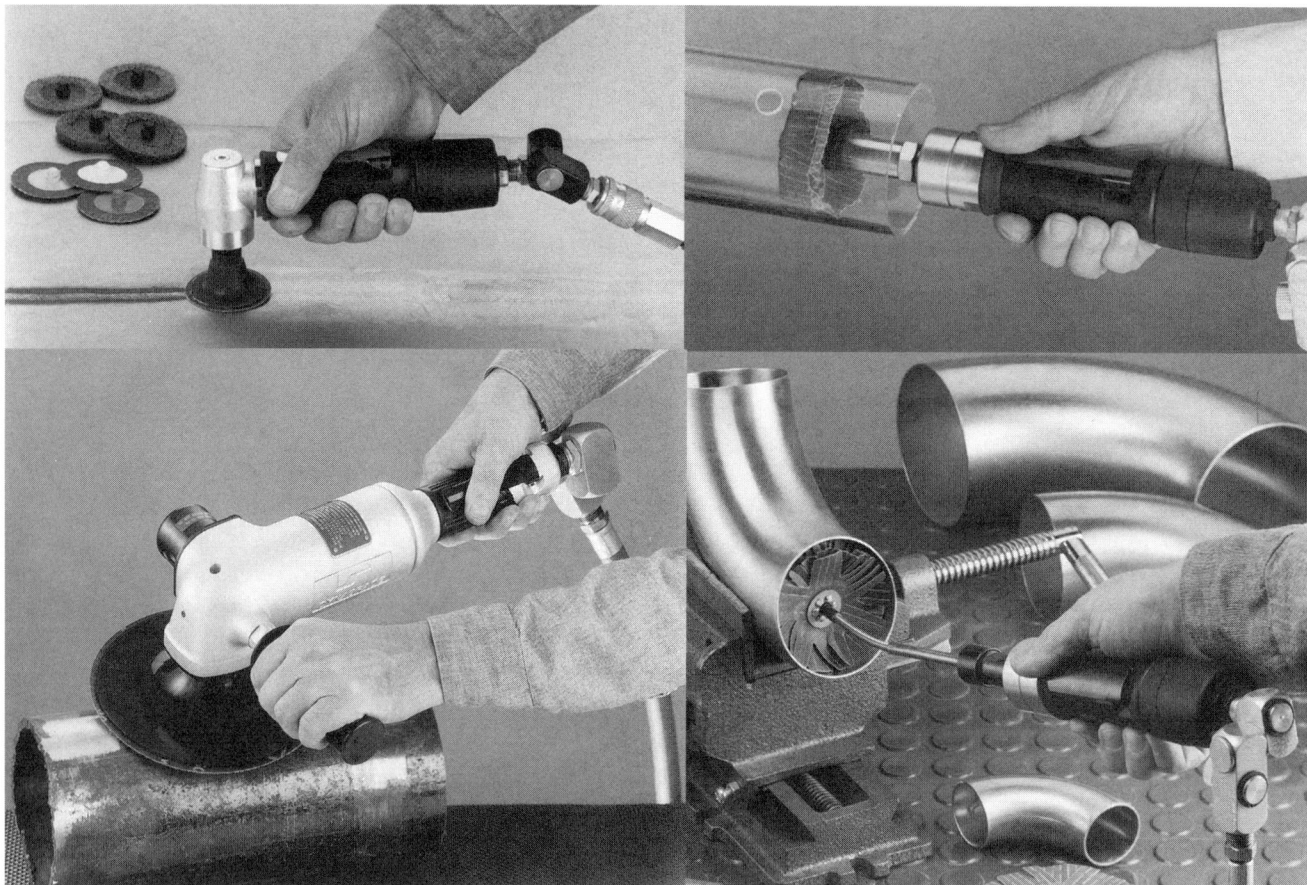

Figure 23-21. Handheld sanders in use. (Courtesy Dynabrade)

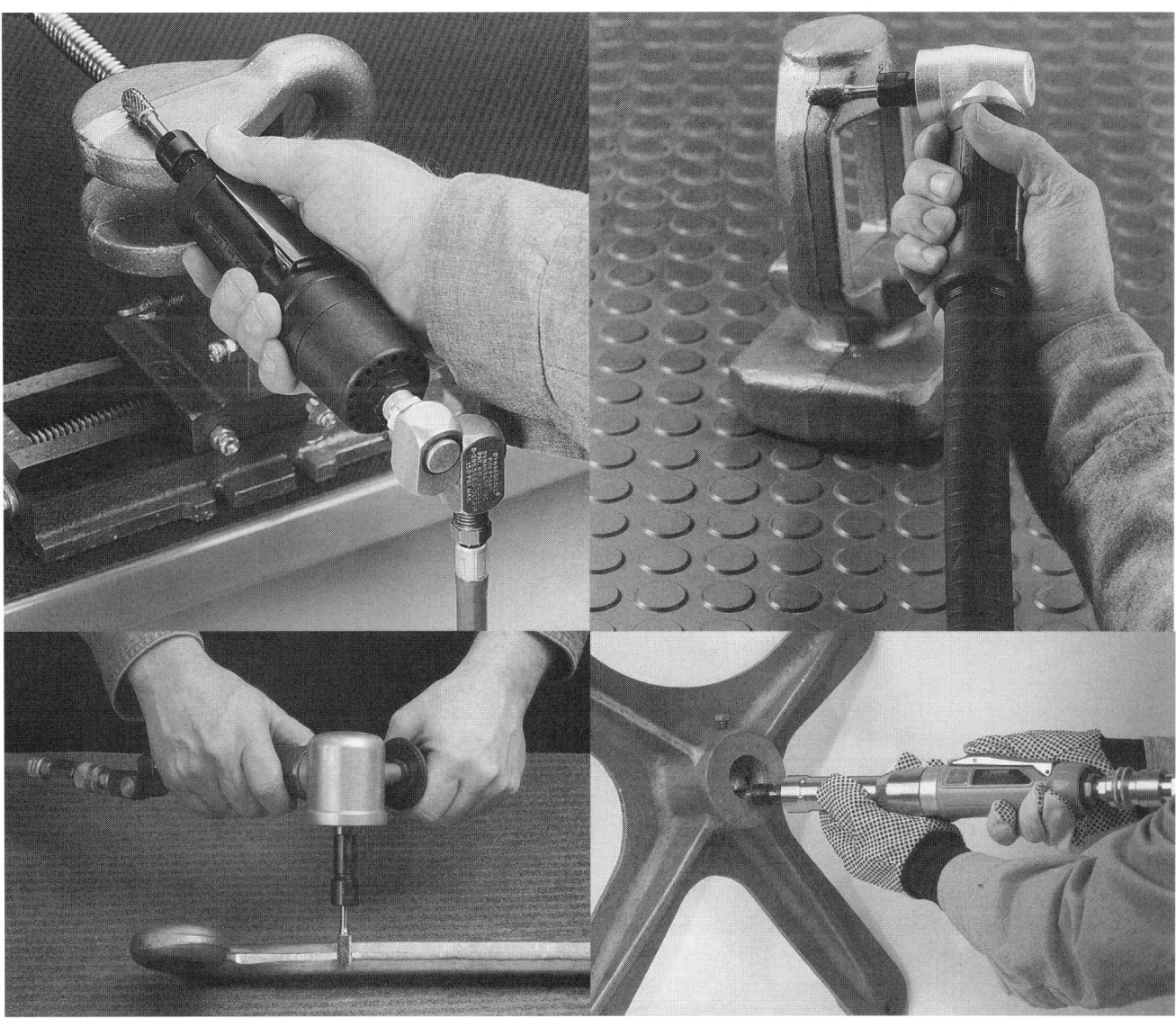

Figure 23-22. Die grinders and angle grinders in use. (Courtesy Dynabrade)

Figure 23-23. Handheld sanding motors. (Courtesy Dynabrade)

Bench and Pedestal Motors 24

Almost every small shop uses some form of bench-top bench motor or pedestal-mounted motor. These are identified by several names and each is used for a specific market segment. With the exception of bench and pedestal grinders, these motors are relatively difficult to find. The most commonly used motors include bench motors, small, second-operation machines, mechanized thread chasers, bench lathes, speed lathes, polishing lathes, alloy grinders, bench grinders, pedestal grinders, and industrial buffers.

BENCH MOTORS

At least three types of bench motors are commonly used in the deburring industry. They include the belt-driven motor with three-jaw chuck, the belt-driven motor with a series of collets, and a direct drive with three-jaw chuck (Figure 24-1).

The advantage of three-jaw chucks is that they easily accept any available shank size. Motors with collets require selecting the proper collet, removing it from the motor, and replacing it with the correct size. Considerable time is required to use two or three different rotary tools when deburring. The collet-type motor, however, has one advantage—it will not normally damage precision parts if the part must be chucked in the spindle.

Available belt-driven motors require the operator to use different pulleys on the motor to obtain different speeds. Although this is not difficult, it slows down some operations. Small pulleys on the tool spindle will result in high speeds. Using the large-diameter pulley on the same spindle will slow down the motor.

Common direct-drive motors operate at speeds of approximately 3,100 rpm. Belt-driven motors have approximately the same maximum speed.

The most commonly used bench motors are relatively heavy units, designed to be placed and left on a bench. Typically, these tools are used to hold small brushes for brushing after the large burrs have been removed. In some cases, they can be used to hold felt bobs or rubber-filled abrasive products. They also may be used to hold a part while it is being deburred. In this type of usage, however, it is important not to tighten the chuck too much because it will damage the part, either by squeezing it out of shape or indenting it and leaving marks. In typical use, the part is held to the brush in the motor and rotated around the part until all of it has been brushed (Figure 24-2).

The bench units in figures 24-1 and 24-2 were made by one plant for their own use, since adequate units of similar size could not be found commercially. Like most of the deburring tools described in this text, they are easy to find when using some form of aide (Gillespie 2000). Notice the large plate beneath the belt-driven units. It provides a mounting platform and adds enough weight to prevent vibrations from causing the motor assembly to "walk" from the table. Two-belt units employ a commercial electric motor and pulley and a machine-tool spindle block to hold the spindle.

The direct-drive system, illustrated in the middle of Figure 24-1, also uses a heavy steel base to prevent walking. Special hand wheels were designed, and the order for the motors had to specify the shaft runout since many motors have far more runout than tools like these can tolerate. These are heavy-duty systems that will last for decades under daily use. Some of the lighter commercial units may not last that long, but they are relatively inexpensive, so if they must be traded out each year it is not a big expense.

Figure 24-3 illustrates a 0.33 hp (0.3 kW) "second-operation" machine. This particular motor is widely

Figure 24-1. Three bench motors. (Courtesy Honeywell)

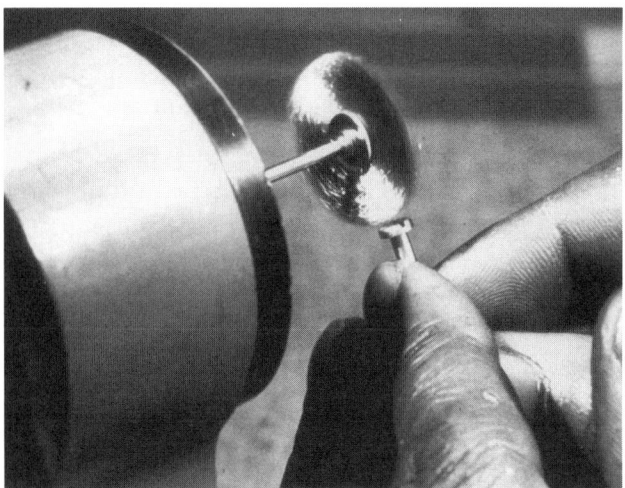

Figure 24-2. Typical bench motor use. (Courtesy Honeywell)

Figure 24-3. Second-operation machine. (Courtesy Somma)

sold for retapping holes and chasing threads. It has a double-disk friction drive that reverses rotation instantly by forward or backward hand pressure on the part. This unit operates at 1,300 rpm. It is sold with a 0.50-in. (12.7-mm) chuck, and its weight (27 lb [12 kg]) eliminates the need for a heavy base plate. It countersinks and deburrs, as well as finishes threads.

Small bench motors are also available (figures 24-4 through 24-6). They may not be designed to handle all the industrial tools operators require, but they satisfy the needs of many plants and do fine work.

A 0.125 hp (0.09 kW) machine is shown in Figure 24-4. Its variable speed control allows it to operate from 1,500–4,000 rpm. This unit also comes with a 36-in. (914.4-mm) long flexible cable and handpiece.

Hand Deburring: Increasing Shop Productivity

Chapter 24: Bench and Pedestal Motors

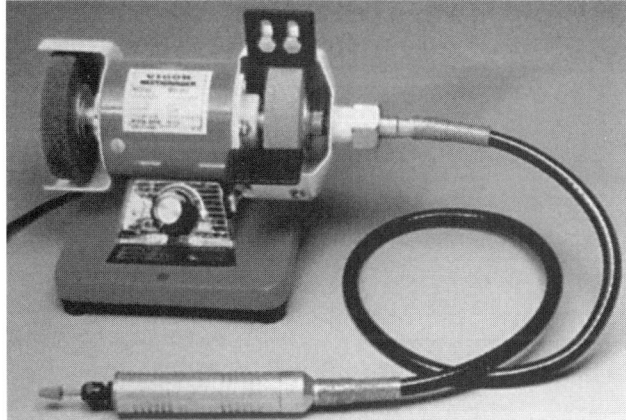

Figure 24-4. Small bench motor. (Courtesy Vigor)

Figure 24-5. Small bench lathe. (Courtesy Foredom)

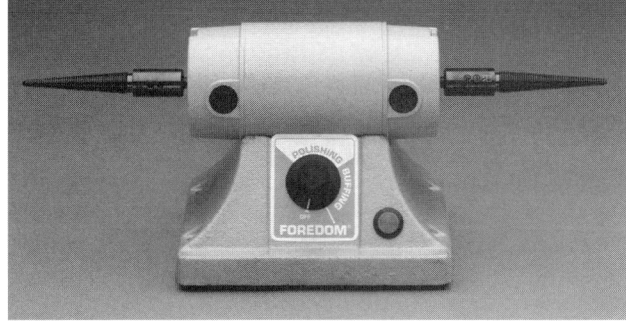

Figure 24-6. Small bench-top polishing lathe. (Courtesy Foredom)

It uses wheels up to 3 in. (76.2 mm) in diameter and 0.75 in. (19.1 mm) wide.

A 0.067 hp (0.05 kW) variable-speed motor (called a bench lathe) is shown in Figure 24-5. It uses wheels up to 2.50 in. (63.5 mm) in diameter. A buff is shown on the left side. Collet holders can be used to hold tools or parts. The motor operates at speeds of up to 17,000 rpm.

Figure 24-6 illustrates a 0.167 hp (0.12 kW) bench-top polishing lathe. It has a speed range of 1,800–7,000 rpm and is designed to buff parts or brush and finish with abrasive rubber tools.

Figure 24-7 illustrates a belt-driven "alloy grinder," which is used to polish small parts or jewelry. It comes with variable speed or two fixed speeds.

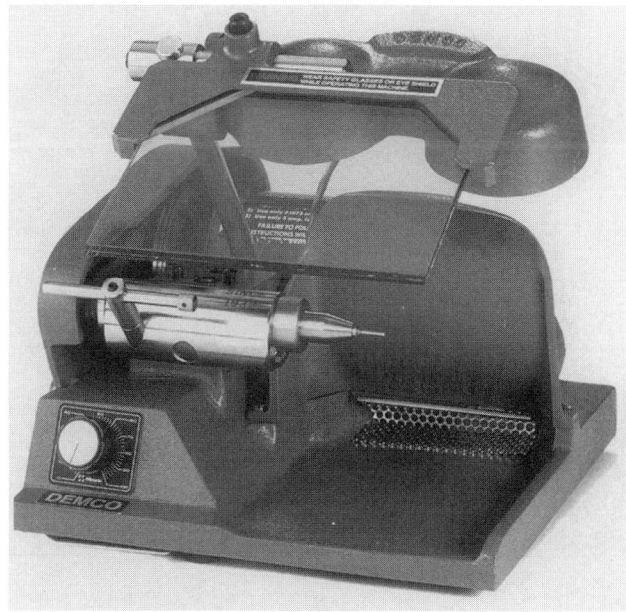

Figure 24-7. Bench-top alloy grinder. (Courtesy Demco Dental Maintenance)

SPEED LATHES

Figure 24-8 illustrates a bench-top speed lathe that operates at speeds of 100–2,500 rpm. It rotates in two directions. Figure 24-9 portrays another variation of a speed lathe with hand-operated locking collet and single- or two-speed options. These are provided with 0.167–1.000 hp (0.12–0.75 kW) drives and weigh up to 140 lb (63.5 kg). A foot-treadle collet release is also available to allow both hands to remain on the part.

BENCH GRINDERS

Almost every plant has one or more commercial bench grinders that hold fixed wheels or brushes (Figure 24-10). They are not designed to hold parts or allow rapid change outs of deburring tools. Bench grinders come in several wheel diameters and most have a protective plastic shield to prevent chips and dust from hitting the worker's face.

Bench grinders typically use 0.33-, 0.50-, or 0.75-hp (0.3-, 0.4-, or 0.6-kW) motors, 6–8-in. (152.4–203.2-

Figure 24-8. Speed lathe. (Courtesy Rovi Products)

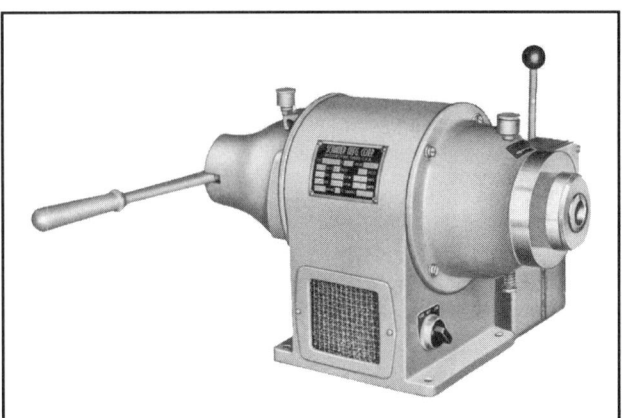

Figure 24-9. Hand-operated speed lathe. (Courtesy D. C. Morrison Co.)

Figure 24-10. Bench grinder. (Courtesy Baldor)

mm) diameter wheels, and have speeds of 1,700–3,400 rpm. Most have fixed speeds, but a few are two-speed systems.

PEDESTAL-MOUNTED MOTORS

Bench and pedestal grinders are similar. The pedestal grinder, as the name implies, sits on a pedestal to enable operators to stand while performing their work (Figure 24-11). They tend to be heavier-duty machines than the bench grinders (motors up to 7.5 hp [6 kW] are available). At least one comes with an integral dust collector.

INDUSTRIAL BUFFERS

Industrial buffers come as either bench-top or pedestal-mounted units (Figure 24-12). They are designed to hold buffing tools and, as such, require lower power

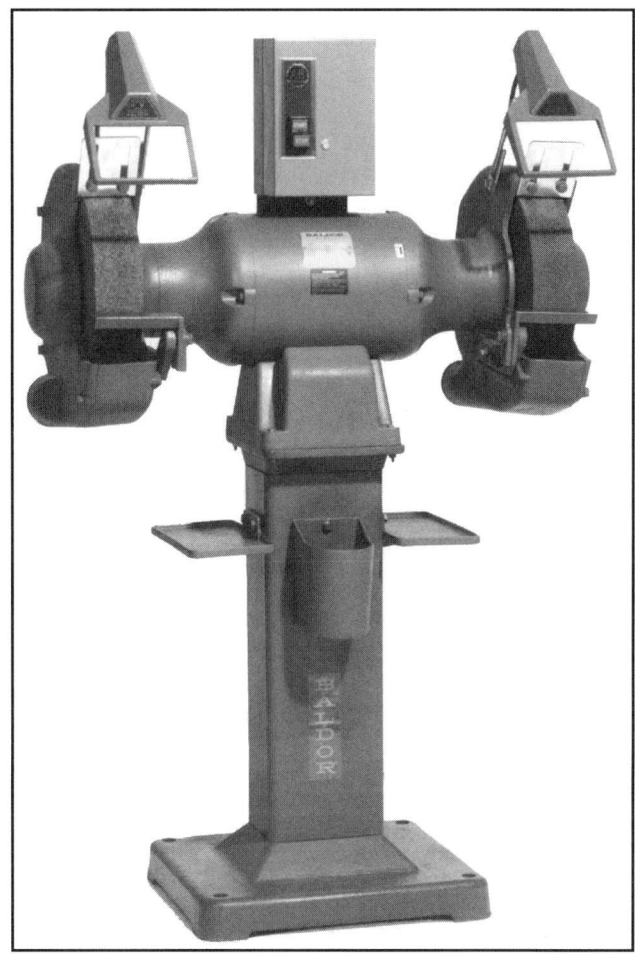

Figure 24-11. Pedestal grinder. (Courtesy Baldor)

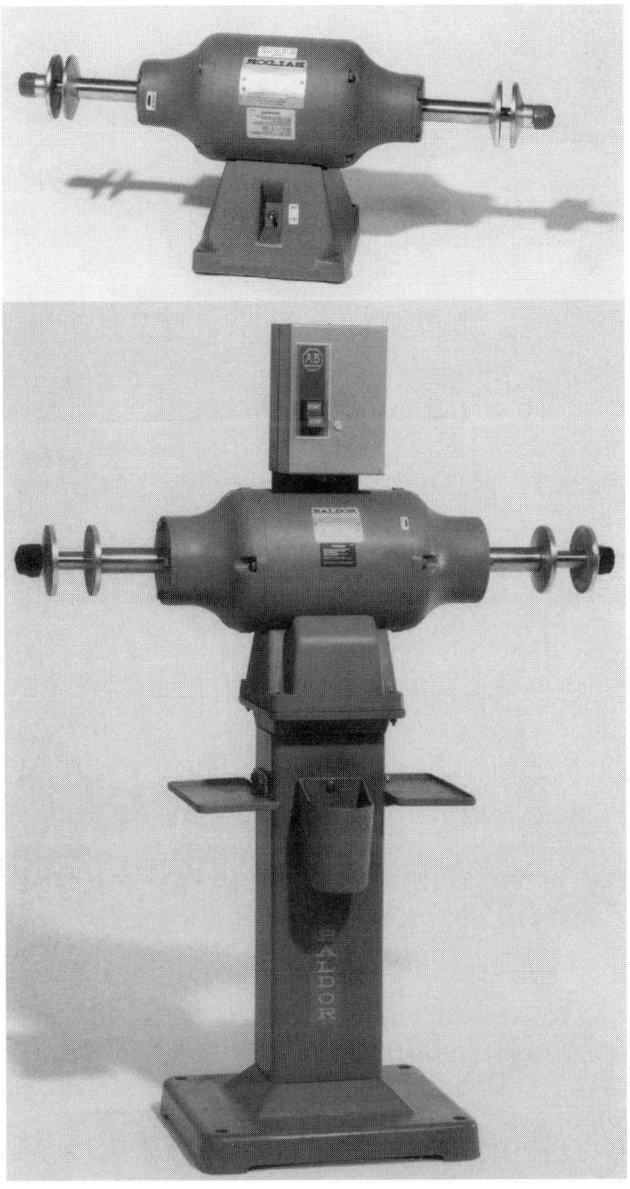

Figure 24-12. Industrial buffers. (Courtesy Baldor)

than bench or pedestal grinders. Typical units begin at 0.25 hp (0.2 kW), but 10-hp (7.5-kW) machines are available. Most are single-speed machines at either 1,800 or 3,600 rpm. They require dust collectors and shrouds before use.

ENVIRONMENTAL, HEALTH, AND SAFETY ISSUES

Rotary motors present safety hazards. They rotate, so they can throw things. Workers can injure themselves if they come into contact with the rotary portion of the motor. Specific problems include the following:

- On three-jaw chucks, do not forget to remove the chuck key before starting the motor. Even though these chuck keys are very small, they develop tremendous force when thrown from a motor. They are safety hazards. Remove all chuck keys before starting the motor.
- As mentioned earlier, bench motors are mainly used for brush deburring. The brushes will grab clothing and wrap it around the tool in the motor. For this reason, do not allow clothing to come in close contact with the end of the chuck or the tool.
- In all cases, eyeglasses should be worn around rotating machinery, even the innocent-looking handheld motorized tools (Production Engineering Research Association of Great Britain 1956).
- All motorized tools should have a cover over the tool area and any exposed edges that can scrape or grab clothing, hands, or fingers. In many instances, covers or shields must be custom made by the user.
- The possibility of fibers flying from metal brushes is a real and ever-present danger.

REFERENCES

Gillespie, LaRoux K. 2000. *Guide to Deburring, Deflashing, and Trimming Equipment, Supplies, and Services.* Kansas City, MO: Deburring Technology International.

Production Engineering Research Association of Great Britain. 1956. *Power Tools: Guide to Deburring of Machined Components: Part IV.* Melton Mowbray, Leicestershire, England: Production Engineering Research Association of Great Britain.

Hand-operated Mechanized Machines 25

Most small shops use one form or another of mechanized machinery—in addition to bench-top motors and pedestal-mounted motors and handpieces—in a manual mode to remove burrs and finish edges. Several varieties are available, and each is used for a specific market segment.

TOOL TYPES

Categories of hand-operated mechanized motors include the following: bench-top chamfering machines, tube-end finishing machines, bar-end finishing machines, sheet-finishing machines, cutoff tip removers, fixed sanding machines, stand-alone polishing lathes, pipe bevelers, degating machines, and trimming presses. Other special purpose machines are also available.

Bench-top Chamfering Machines (Workpiece Edgers)

Two basic approaches are used for general-purpose workpiece edging. In one approach, rectangular workpieces are placed in an angle-iron trough and hand-fed over a small grinding wheel to provide a small, adjustable chamfer (Figure 25-1). This device is used for steel, aluminum, and plastic parts and uses solid carbide cutters. The table is fixed at 45° to the horizontal. Other units have adjustable angles. The table tilts up to allow the cutter to be changed. The 0.75-hp (0.6-kW) motor runs at 10,000 rpm on 110V alternating current (AC). Round collars can be chamfered by placing the collar against an optional stop and rotating it above the cutter.

The second approach uses a piloted chamfering tool extended upward through the flat table of a machine (see Figure 25-2a). Workpieces are moved into contact with the pilot on the cutter and fed across the cutter, using the pilot as a guide or stop. Adjusting the cutter upward or downward (as in a home wood router) controls the chamfer depth. Another variation of this approach is to install a permanent stop (Figure 25-2b) or magnetic device and use a standard chamfering tool. Straight and contoured edges are deburred with such machines.

Figure 25-1. Bench-top machine with angle trough for rectangular parts. (Courtesy Simco Industries)

Figure 25-3 illustrates the use of a ball bearing, instead of a pilot, to provide easy movement inside holes. This machine allows vertical cutter adjustment in 0.002-in. (0.05-mm) increments or finer. The size of the table can be up to 12 × 20 in. (304.8 × 508.0 mm). Both corner radii and chamfers can be milled; and cutter speeds are 13,000–28,000 rpm.

Figure 25-4 illustrates a heavy-duty handheld unit, such as that discussed in Chapter 24. Note the differences between it and the unit in Figure 25-1. Both can apply the same chamfer, but the unit illustrated

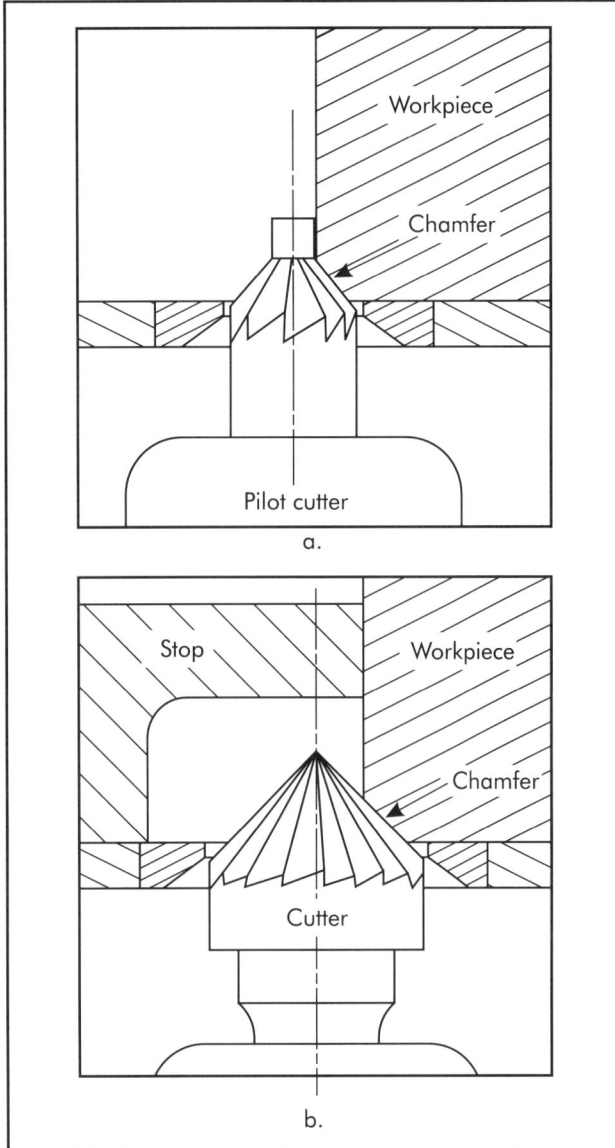

Figure 25-2. Two concepts for locating edges on a general-purpose workpiece edger: (a) piloted chamfering tool, and (b) permanent stop. (Courtesy Senn Company/Reishauer Corporation)

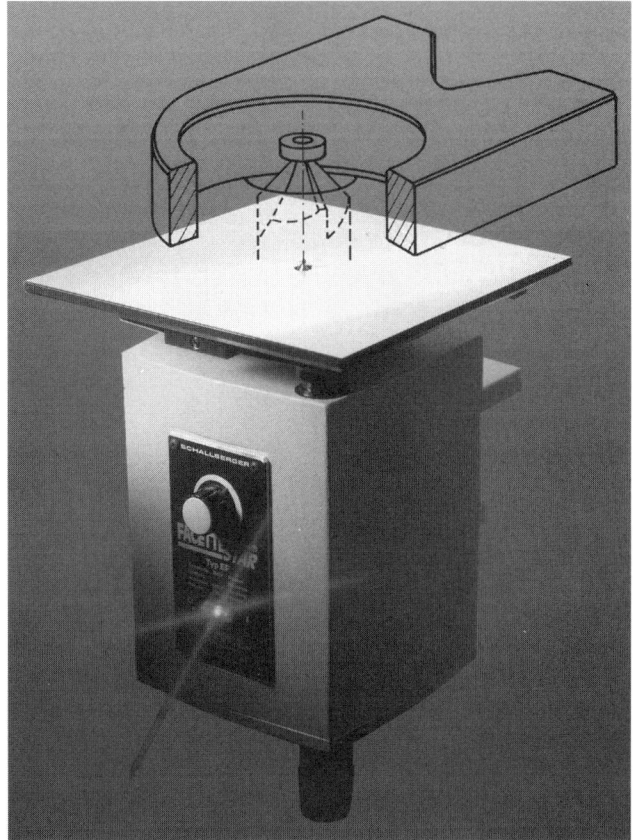

Figure 25-3. Concept of ball bearing on chamfering tool for tracing feature contour. (Courtesy Hansco Enterprises)

in Figure 25-1 enables the operator to take the part to a convenient location. In contrast, the handheld giant in Figure 25-4 allows the chamfering machine to be brought to a heavy workpiece.

Tube-end Finishing Machines

Figure 25-5 illustrates a tube-end finishing machine in use. The operator slides the tube into the machine collet and, if necessary, locks it down and advances a lever to bring the cutter in contact with the tube ends.

Most of these machines can chamfer or deburr both internal and external diameters, and can square-up ends (Spinelli 1984). Some machines are bench-top units, some are stand-alone, and some require a small platform under them to allow proper height for the worker.

Bar-end Finishing Machines

Bar-end finishing machines usually work the same as tube-end finishers. Some lack clamp-down and precision sizing capabilities. The machine illustrated in Figure 25-5 performs both bar- and tube-end finishing. These machines can be made with stiff brushes, rather than cutters, to perform the deburring and edge rounding. They are used, for example, to remove saw-cut burrs. When wire brushes are used, operators may need to wear gloves and use a protective shield on the machine.

Sheet Finishing Machines

The tabletop machines shown in figures 25-1 and 25-2 are used to deburr both rectangular and irregu-

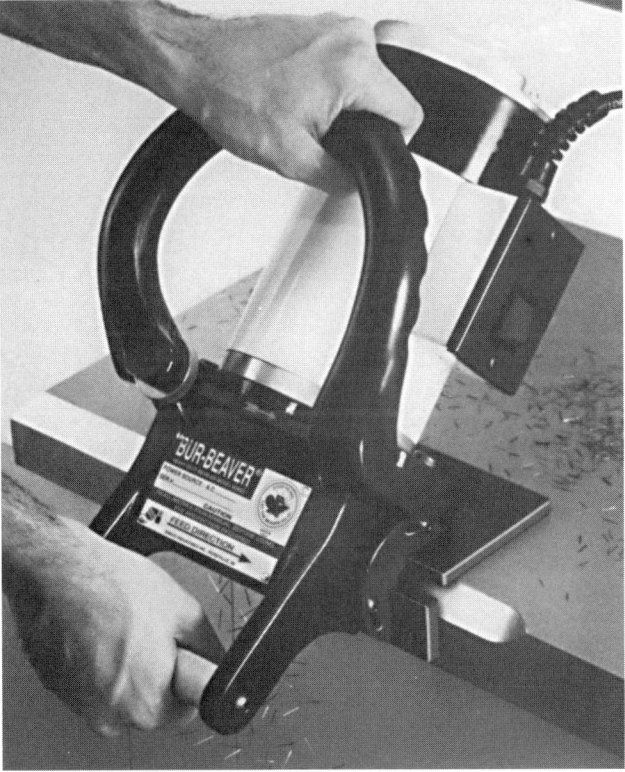

Figure 25-4. Heavy duty hand beveler. (Courtesy Simco Industries)

lar parts with flat surfaces. However, they may be too large to handle sheet stock. Figure 25-6 illustrates a chamfering machine for sheet metal. This device uses two motors with flap wheels to remove the burrs from sheared sheet metal.

Cutoff Tip Removers

One machine is designed to remove the small cutoff tip left when cylindrical parts fall from the lathe. This small nubbin, often only 0.030 in. (0.76 mm) in diameter and the length of the cutoff-tool width, is placed in a rotating head that has a small clipping device in it that "whacks" and shears the cutoff tip (Gillespie 2000).

Fixed Sanding Machines

Sanding machines are produced in both automated and manual styles. Wherever possible, the automated units should be used for flat sheets since they save considerable labor. Every shop seems to have at least one manual unit and, in many instances, several other designs. Common designs include hard-backed belt, slack belt, disc sander, oscillating vertical-spindle sander, radius and internal sander, and combination

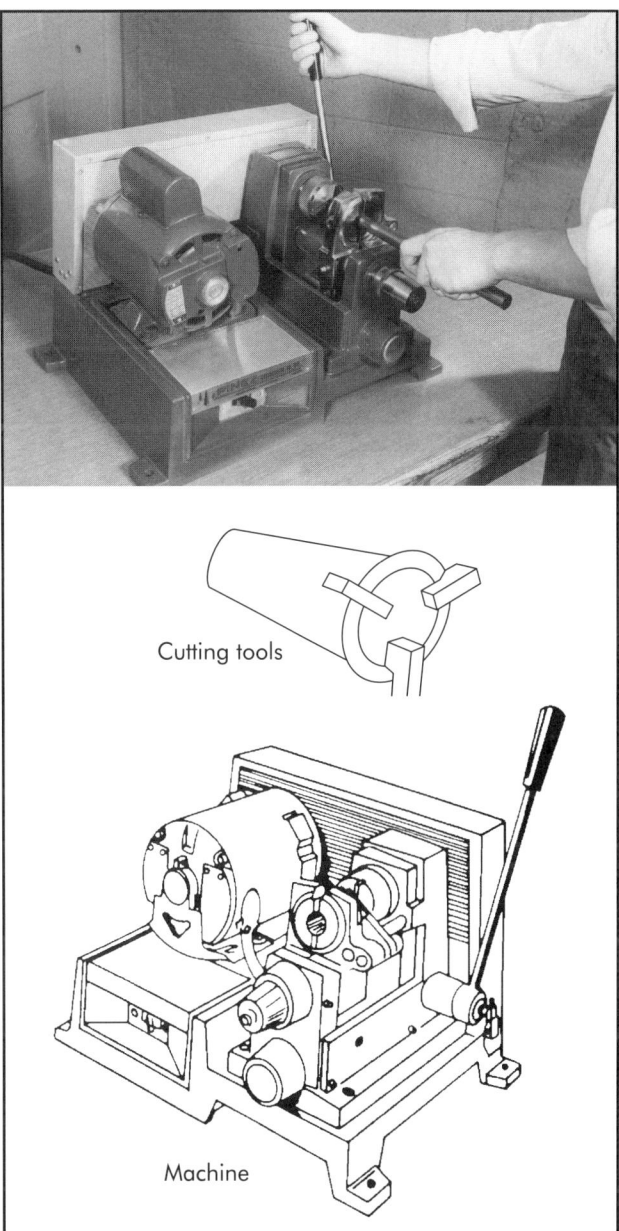

Figure 25-5. Tube-end finishing machine. (Courtesy Simco Industries)

sander. Handheld sanders are discussed in Chapter 23 and in other sources (Farago 1980).

Hard-backed Belt Sanders

Figure 25-7 shows a standard vertical belt sander. These units are particularly useful for straight edges. They come with a small, horizontal metal rest on which workers can place their hands or a part. Generally, they operate at one speed.

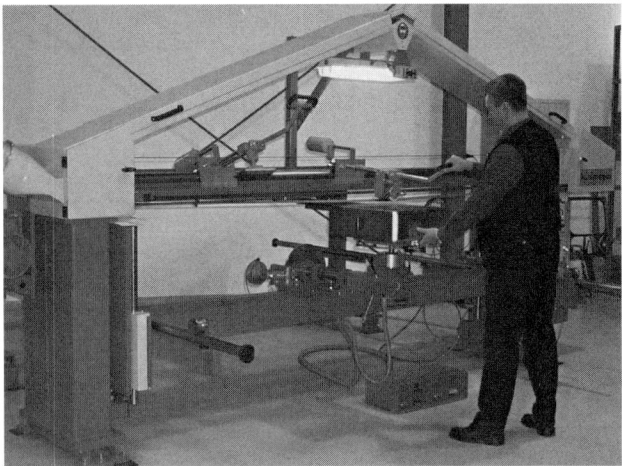

Figure 25-6. With manual assistance, this machine deburrs sheet metal, formed parts, and large metal components and assemblies. (Courtesy CIMID Corporation)

Contact-wheel Belt Sanders

Contact-wheel sanding uses the firmness of a rubber rolling wheel, over which the belt rides, to provide hard sanding.

Slack-belt Sanders

Slack-belt sanders allow the belt to run slack so the part can be held over it for a softer action, like the back-and-forth motion of a shoeshine rag. Some have long, narrow belts to finish intricate features. Figure 25-8 illustrates a sander that has a slack belt with a hard-back roller at the bottom. It has a small radius attachment for intricate cutouts.

Disc Sanders

Several disc sander designs are commonly used in manufacturing applications. Most have a horizontal axis, which puts the full width of the wheel or disc in front of the operator's hands. Some flat-disc sanders are called "lapping machines" and have a vertical axis so the part can be laid on the rotating surface. These can be built in diameters up to 36 in. (914.4 mm). Figure 25-9 illustrates a horizontal axis sander.

Combination Sanders

Some machines provide backup platens, a contact wheel, and slack belt sanding in a single machine, depending on where the operator holds the workpiece. Others combine disc and belt sanding in a single machine (Figure 25-9). These units can be procured as bench-top or full standing units to accommodate small or large parts. Belt and drum sanders provide the convenience of belts and a single-sized drum.

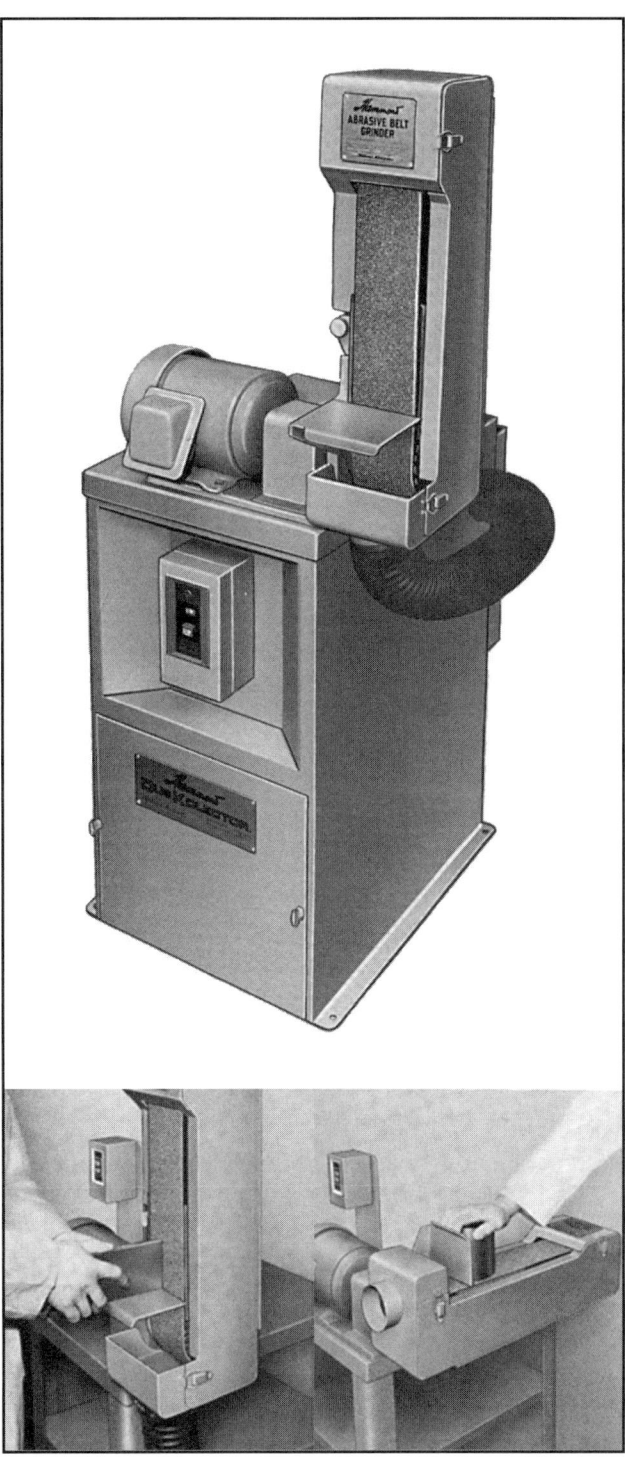

Figure 25-7. Hard-backed belt sander. (Courtesy Hammond Machinery)

Radius and Internal Sanders

Radius and internal sanders are unique and employ a narrow belt adapter on a fixed, stand-up machine to access hard-to-reach areas (see Figure 25-10).

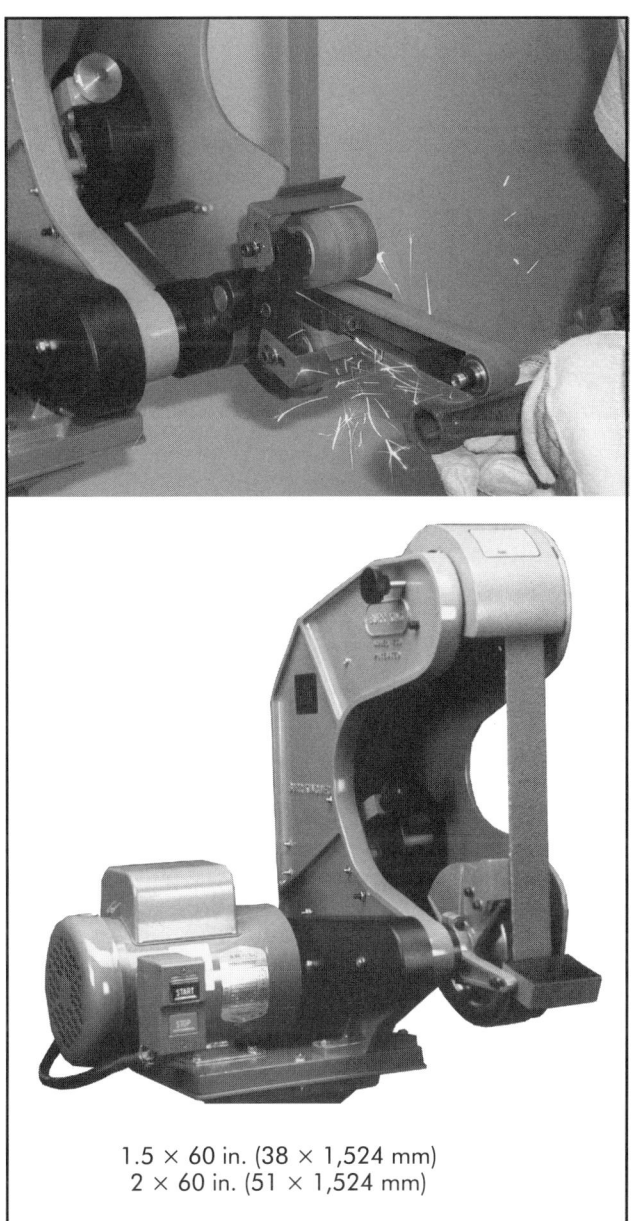

Figure 25-8. Slack-belt sander with contact wheel at bottom. (Courtesy Burr King Manufacturing Co.)

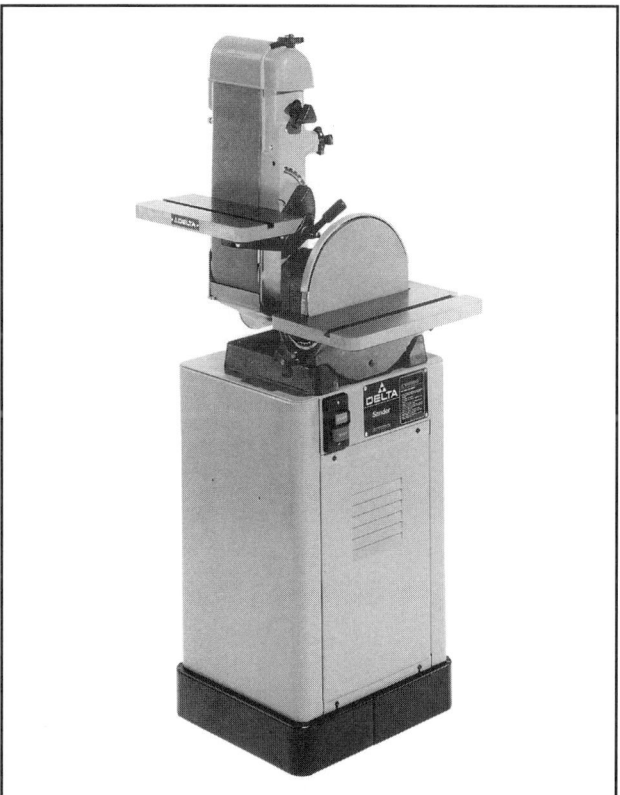

Figure 25-9. Combination belt and disc sander. (Courtesy Delta International)

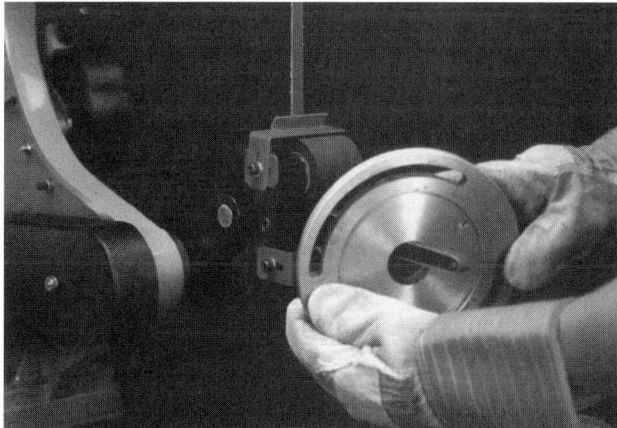

Figure 25-10. Internal- and small-radius wheel sander. (Courtesy Burr King Manufacturing Co.)

Oscillating Vertical-spindle Sander

One sander uses an oscillating vertical spindle to rotate and reciprocate vertically (Figure 25-11). This is designed for cylindrical or small, flat parts.

Sander Usage

When selecting a sander, it is important to consider part size, finished-edge locations, throat depth (the depth at which the part must move into the belt), and throat height. For ergonomic reasons, pedestal height is important. Belt width also affects use. Narrow belts allow access to most part areas; whereas, wide belts allow more sanding to be performed at one time.

To reduce manual deburring costs, flat sheets should be finished on automated sanders. Figure 25-12

Figure 25-11. Oscillating vertical-spindle sander. (Courtesy Delta International)

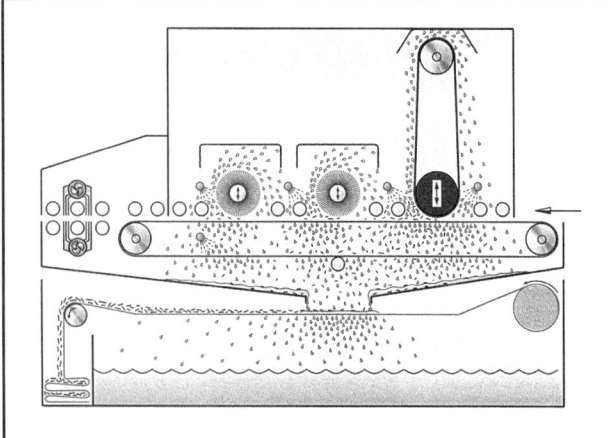

Figure 25-12. Schematic of a flat sander. (Courtesy ASM Machinery Sales)

illustrates the inside working details of one such machine. This machine relies on a sanding belt and two brushes rotating in opposite direction to remove burrs and radius edges. They can sand hundreds of parts per day and provide consistent uniformity.

The following points are other considerations to keep in mind regarding sanders.

- Many of the sanders discussed here are designed to apply a coolant across the sanding surface to reduce dust and improve finish.
- Disc life and the total amount of stock removed are proportional on stainless-steel and hot-rolled steel when surfaces are finished, but burr size

Figure 25-13. Buffing lathe. (Courtesy Munson)

and applied forces can easily tear abrasive papers and reduce predicted life to a fraction of that found in surface finishing.

See Table 25-1 for rpm recommendations that apply to surface sanding with discs.

Table 25-1. Recommended rpm for surface sanding with discs

Material	Disc Diameter	
	7 in. (178 mm)	9 in. (229 mm)
Mild steel	6,000	4,500
Stainless steel	3,600	3,000
Aluminum	6,000	4,500

Stand-alone Polishing Lathes

Many companies use small, manual, second-operation lathes to deburr and finish parts, which include manual chucker and turret lathes. Lathes designed solely for polishing exist, as do many buffing lathes, such as the one shown in Figure 25-13.

Pipe Bevelers

Figure 25-14 illustrates a portable pipe-beveling machine. This is not a handheld unit, despite the term "portable." Rather, the unit is carried to the pipe, which is often in the field and being repaired. A chucking device locks the unit to the pipe (any pipe diameter from 0.50–2.50-in. [12.7–63.5 mm], 3–14 in. [76.2–355.6 mm], and 40 in. [1,016 mm]). Bidirectional, variable-feed cutting heads provide a quick

Figure 25-14. Pipe beveler. (Courtesy D.L. Ricci Corporation)

bevel, flat face (flange facing), or light deburring. Units used to face pressure vessels can finish 110-in. (2.8-m) diameter surfaces and edges. Smaller units finish holes cut into pipe. To obtain a consistent edge, these holes have a three-dimensional contour that must be followed.

Some units can operate under water, in nuclear environments, or in other hazardous conditions. Drive motors can be applied from front faces, back faces, or at angles to the work. While most use a single-point tool on slides as cutters, a plasma flame system is also available. Hydraulic power is used with drive systems up to 20 hp (14.9 kW). Some lighter systems rely on pneumatics and mounted stones. These devices are used for steel, stainless steel, aluminum, and alloy steel pipes and vessels.

Degating Machines

Degating machines remove the gates and risers from plastic molded parts. They can use shears, trimming presses, or cutters.

Trimming Presses

Trimming presses are widely used in the sheet metal industry to shear unwanted metal. They are also used for trimming molded plastic flash from parts.

Miscellaneous Machines

Figure 25-15 shows a small drill press tool (machine) that enables users to reach inside a threaded nut to deburr the cross-hole. Many single-purpose devices can be used to increase manual deburring and edge finishing efficiency by factors of two to five with little investment. Many simple machines can provide savings that many plants often overlook.

ENVIRONMENTAL, HEALTH, AND SAFETY ISSUES

Rotary motors present safety hazards. They rotate, so they can throw things, and workers can injure themselves if they come in contact with the rotary portion of the motor.

On three-jaw chucks, do not forget to remove the chuck key before starting the motor. Even though these chuck keys are very small, they develop tremendous force when thrown from a motor. Remove all chuck keys before starting the motor.

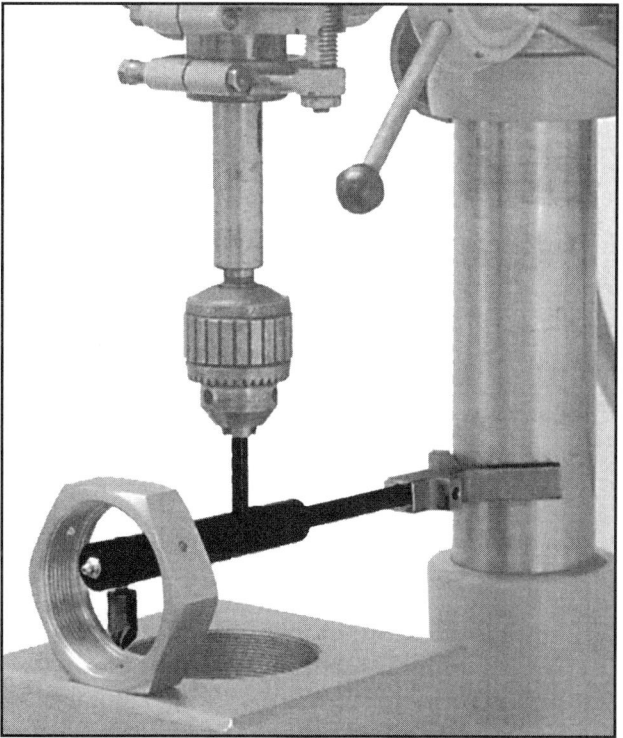

Figure 25-15. Drill press deburring device. (Courtesy Ace Manufacturing Co.)

In all cases, eyeglasses should be worn around rotating machinery, even the innocent-looking handheld motorized tools (Production Engineering Research Association of Great Britain 1956).

All motorized tools should have a cover over the tool area and any exposed edges that would scrape or grab clothing, hands, or fingers. In many instances, covers or shields must be custom-made by the user.

Dust is a major explosion and fire hazard, particularly when paper particles from bonded abrasives pile up with the resins, causing titanium or magnesium dust to cover other materials. When steel sparks at the sander, these other materials may catch fire.

The health effects of carcinogenic dust particles are described in Chapter 19. Other materials may present similar concerns.

REFERENCES

Farago, Francis T. 1980. *Abrasive Methods Engineering.* New York: Industrial Press.

Gillespie, LaRoux K. 2000. *Guide to Deburring, Deflashing, and Trimming Equipment, Supplies and Services.* Kansas City, MO: Deburring Technology International.

Production Engineering Research Association of Great Britain. 1956. *Power Tools: Guide to Deburring of Machined Components: Part IV.* Melton Mowbray, Leicestershire, England: Production Engineering Research Association of Great Britain.

Spinelli, Richard. 1984. "Aircraft Tube-end Finishing." Technical Paper MF84-136. Dearborn, MI: Society of Manufacturing Engineers.

Support Items 26

Deburring and its verification require various types of support devices. Burrs can be measured in over 40 ways by using optical comparators, rubber replica materials, or even via toothpicks. Other support items might include a lighting device to see edges inside motors or magnifying aids for deburring under magnification. This chapter discusses many inspection aids and allied support devices. Specific techniques for using some of these tools to inspect for burrs are discussed in Chapter 34.

SUPPORT DEVICES

Numerous catalogs sell many support devices, including: magnifiers, microscopes, lights, mirrors, sharp-edge measuring devices, chamfer mikes, probes for burrs, toolholders, cryogenic probes, ergonomic devices, protective gloves, pliers and nippers, surface-finish measurers, workbenches, and exhaust systems. These items are discussed in the following paragraphs and summarized in Table 26-1 (Gillespie 2000).

Magnifiers

Burrs may be located in hard-to-see areas; therefore, many shops rely on magnification and lighting aids to detect burrs. Magnification and lighting aids are used at deburring and inspection stations. When manufacturing and inspection departments use the same level and quality of magnification, they are more likely to agree on whether parts are burr-free. To increase inspection's acceptance of parts, manufacturing must be provided with better-quality magnifiers. That may sound strange to some users, but the key is to provide the worker with the tools to adequately do the job, not inspect quality in an inspection operation.

Ring-light Magnifiers

The magnifying ring light is a common tool in manufacturing for evaluating the absence of burrs. Typically, it is mounted on a telescoping or flexible arm that allows it to be moved across a broad portion of the workbench (Figure 26-1). It can be moved upward or downward to bring it into focus. It is large enough (5–8-in. [127.0–203.2-mm] diameter lens) so that almost anyone can view the area in question. It is a rugged tool that lasts years under normal usage. Magnifying ring lights are available in magnifications of 1–4×. Most incorporate a fluorescent light around the lens to provide a cool, diffuse, shadow-free light over the work area. Lighting power varies from 20–60 W (20–60 J). At least one manufacturer provides a glare-free bulb to reduce eyestrain.

An 8-in. (203.2-mm) diameter lens provides almost 2.5 times the viewing area as that of a 5-in. (127-mm) diameter lens and makes inspection easier. However, it also requires a larger portion of the bench area. These tools are generally mounted with a flexible arm, but more solidly built units that stand on a bench top are available for smaller parts and where greater rigidity is desired. Some are equipped with integral carbon-activated fume absorbers or fans. Most of these units are mounted on the back of the workbench; however, some can be mounted on a vertical rail beside the bench.

Pocket Scopes

Small, handheld monoculars or magnifiers (also called tube microscopes) are used to peer into close quarters (Figure 26-2). These are produced in at least 24 styles and some include an integral light. They are

Table 26-1. Deburring and edge finishing support tooling

Tooling Style	Categories
Magnifiers	4× magnifiers
	Pocket scopes
	Optical comparators
	Video systems
	Optical profilers
	Borescopes
	Loupes
	Eyeglass magnifiers
	Hand magnifiers
	Headbands
	Flexible-arm magnifiers
Microscopes	Monocular
	Stereo
	Full-screen systems
	Surgical operating
	Portable microscopes
Lights	Microscope illuminator
	Fluorescent
	Halogen
	Fiberoptic
	Otoscope
	Long stem
	Headband
	Deep hole
	Flexible
	Articulated arm, incandescent
Mirrors	
Sharp-edge measuring devices	
Chamfer mikes	
Probes for burrs	
Toolholders	Pin vises
	Stone holders
	Hand chucks
Ergonomic devices	Flex-arm holders
	Chairs
Cryogenic probes	
Protective gloves	
Pliers and nippers	
Surface-finish measurers	
Workbenches	Shop workbenches
	Down-draft workbenches
Exhaust systems	Exhaust hoods
	Vacuum systems
	Air-cleaning systems

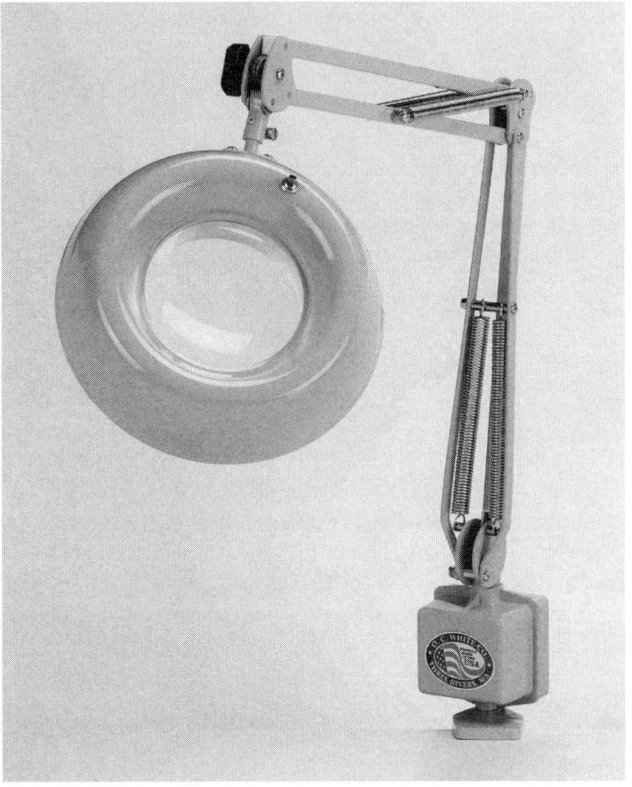

Figure 26-1. Lighted magnifier, 5 in. (127 mm) in diameter. (Courtesy O.C. White)

convenient but not as widely used as some other devices. They include magnifications up to 40×. Their name comes from their size—they will fit in either pants or shirt pockets. Some pocket scopes not only magnify but also have calibrated reticules, such as shown in Figure 26-3, to measure burrs and chamfers.

Video Systems

Video systems project an image onto a small screen to allow workers to look straight ahead to see the surface on which they are working (Figure 26-4). Some have stereo-like viewing and some allow video or still photographs to be taken for historical records. Other video systems include measuring capability, but this feature is not usually applicable to burr measurement. Video systems can be used for training and better viewing. However, depth of field (DOF) and perception are not as good as those obtained with a good stereomicroscope. As technology improves, these systems will become more viable for deburring applications.

Optical Profilers

Optical profilers are also known as *laser surface-finish measuring systems*. Laser systems today can

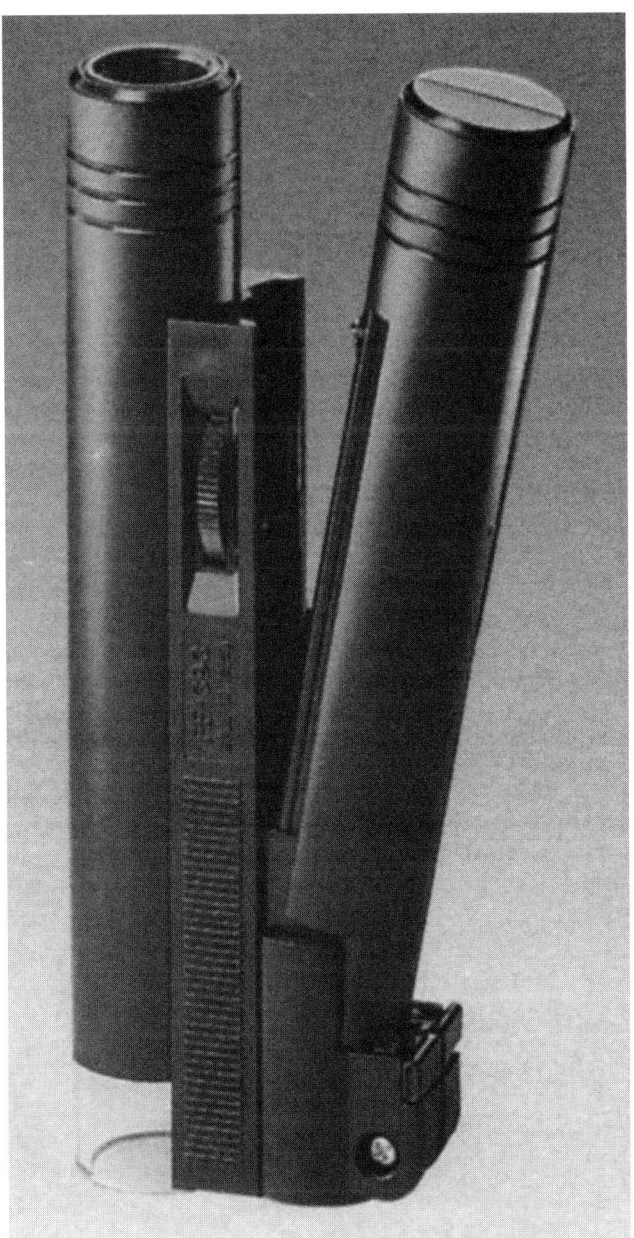

Figure 26-2. Pocket scope. (Courtesy Eschenbach)

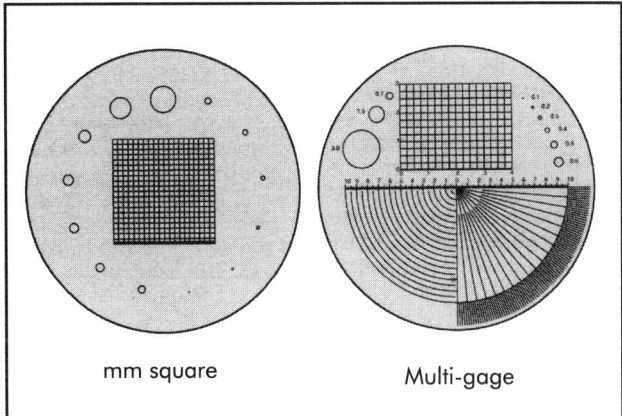

Figure 26-3. Reticles used with microscopes and pocket scopes. (Courtesy Edmund Scientific)

rameters. The analysis packages allow several options: 2D, 3D solid plots, contour plots, and histograms. Data analysis can be enhanced, data can be trimmed and edited, and subsections can be selected. Colors identify specific features. Their design limits use to relatively flat surfaces.

Borescopes

Figure 26-6 illustrates a small borescope used to inspect interior walls and intersections of small holes. Borescopes are produced in sizes down to 0.055 in. (1.40 mm) in diameter, and special research devices can be smaller. Their cylindrical view requires training to interpret, but they are widely used in many industries, including automotive, aerospace, medical, and in pressure vessels to survey surfaces otherwise impossible to see.

Borescopes enable users to look forward, sideways, and backwards; and they can be attached to video and still cameras to record the views. They can provide either black-and-white or color images. They are normally used with a stiff outer jacket to keep the breakable fiberoptic from hitting part surfaces. They are also available with a "driveable" end that can articulate in five directions so users can see edges better. Although normal scopes are short (about 6 in. [152.4 mm]), they are available in flexible snakes up to 71-in. (1.8-m) long.

Borescopes have a specific direction of view (Figure 26-7). To investigate a large cavity, several probe ends may be required. For a single intersection, only one or two ends would be necessary. As shown in Figure 26-8, borescopes have an exceptionally large depth of field—often from infinity to 1 in. (25.4 mm) or less. The closer an object is to the lens, the greater the magnification. To calculate magnification, the distance of

detect and provide 3D images of the topography of a flat surface and edge (Figure 26-5). These are usually too expensive for normal deburring use; however, they can detect burrs as small as 0.000000004 in. (0.1 nm) tall, and some even as tall as 0.020 μin. (500 μm).

Typically, optical profilers are used in semiconductor and scientific investigations. They would only be used for exceptionally high-quality edges or for burr-related studies. They provide printouts of burr height and width, as well as many other surface finish pa-

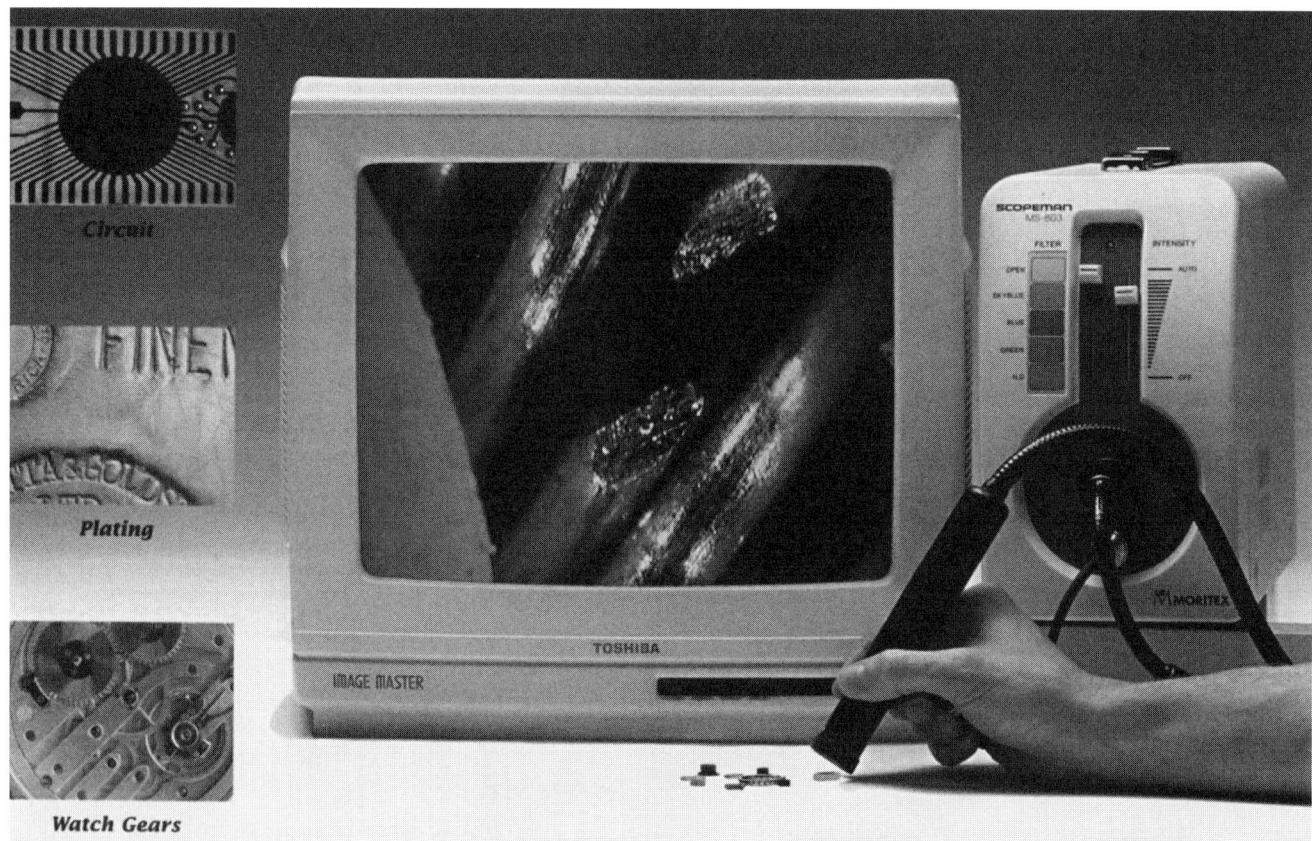

Figure 26-4. Video system for part and edge inspection. (Courtesy Moritex)

Figure 26-5. Optical profiler with output display. (Courtesy WYKO)

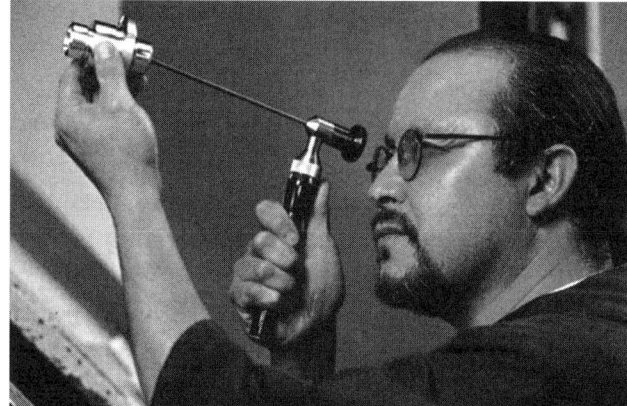

Figure 26-6. Small borescope in use. (Courtesy Gradient Lens Corp.)

the subject from the lens must be known. In contrast, microscopes and loupes have a limited depth of field. To see around a hole intersection, the barrel of the device is rotated by hand. Eyepieces on some of these devices can also rotate. Typical weights are around

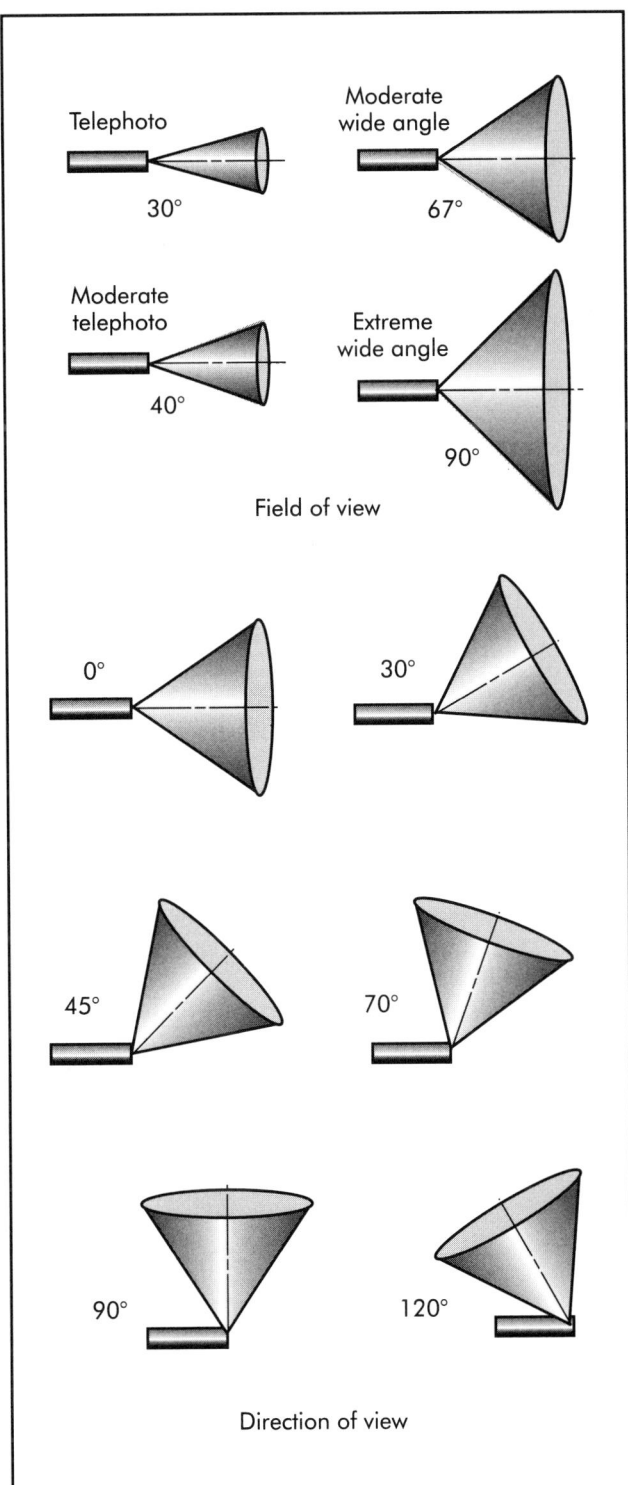

Figure 26-7. *Borescope views. (Courtesy Gradient Lens Corp.)*

2–4 lb (0.9–1.8 kg). AA-size batteries usually power the lights.

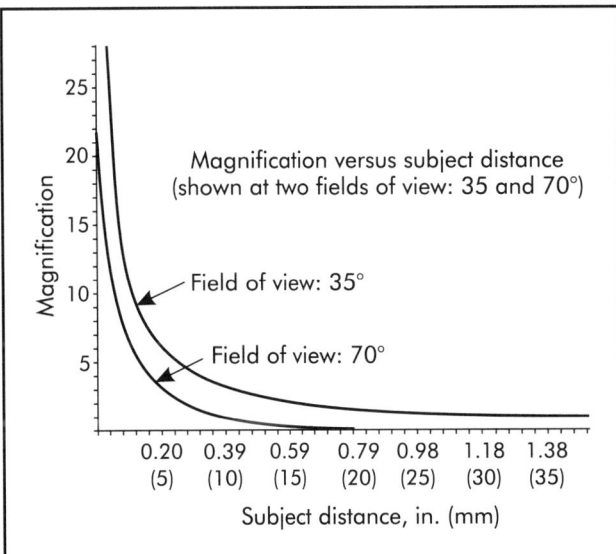

Figure 26-8. *Borescope magnification as a function of distance to subject. (Courtesy Gradient Lens Corp.)*

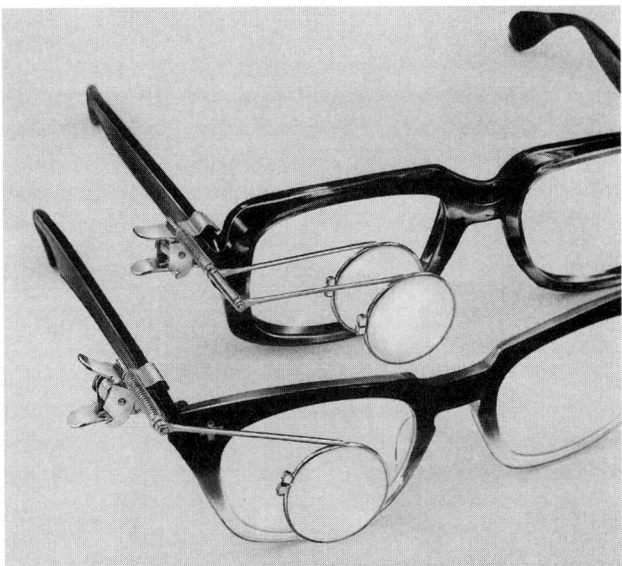

Figure 26-9. *Spectacle loupes. (Courtesy Gesswein)*

Loupes

The old watchmaker of several decades ago wore a loupe over his glasses (Figure 26-9) to closely view his work. Those same little aids are used today for deburring and inspection, and can be found in many tool catalogs. They can be clipped to a hat or eyeglasses. Lenses can either swing out from eyeglasses (eyeglass loupe), or the small and solid handheld piece can be temporarily held in the eye by squinting (jeweler's loupe). Loupes are available in powers of 2.5–18×, and in a stereo style.

Loupes that rest on the surface of the part are also available. These handheld or hand-sized units employ a wide variety of etched scales to enable measurement without moving parts (Figure 26-10). Some are called *graphic arts comparators* or *contact reticules*. The latter are designed to rest on a flat surface, and most have a screw design that enables vertical focusing (Figure 26-11). At least one design also incorporates a small fluorescent light. Over 24 different designs exist for these tools. They come in powers up to 200× and have integral light sources and vertical adjustment. The latter tools are useful for flat parts; however, they are not ideal for complex 3D parts.

Eyeglass Magnifiers

Designed for microsurgery, these eyeglass magnifiers (Figure 26-12) provide users with a stereoscopic view over a depth of 1.50 in. (38.1 mm). Clip-on monocular styles are available. Figure 26-13 illustrates an eyeglass monocular used with a finger ring.

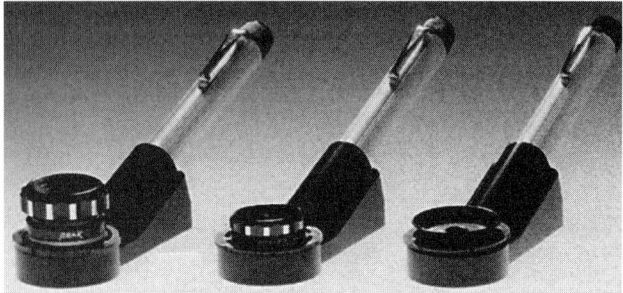

Figure 26-10. Contact scale-measuring loupes. (Courtesy Edmund Scientific)

Figure 26-11. Graphic arts comparator. (Courtesy Edmund Scientific)

Figure 26-12. Stereo eyeglass magnifier. (Courtesy Edmund Scientific)

Hand Magnifiers

Handheld magnifiers are used for detecting burrs and edge quality (Figure 26-14). They come in many sizes and shapes, and provide low-cost viewing of dark and hard-to-see areas. Fixed and folding designs are available. Folding styles are small enough to be carried in pockets and have magnification powers of 4–20×. Magnifications are fixed, so they must be ordered with the desired magnification.

Flexible-arm Magnifiers

Handheld magnifiers can be attached to a flexible arm for more permanent, hands-free placement (Figure 26-15). They come in many sizes, but the most commonly used have 4–5-in. (101.6–127.0-mm) round lenses. Their heavy base makes them stable devices for most work areas.

Headband Magnifiers

Headband magnifiers allow workers to flip down the magnifier when needed, then move it back on the head, out of the way, when it is not needed (Figure 26-16). They come with up to six different powers of magnification, and one unit has an additional loupe to further increase the magnification. These devices have the advantage of a long focal distance (4–20 in. [101.6–508.0 mm]), which allows peering into deep areas.

Optical Comparators

Generally, optical comparators are used to measure the contours of relatively large surfaces. Their design, however, enables users to quickly and easily detect the presence of burrs and, if present, measure them as well as chamfers. Most shops have one of these general-purpose machines.

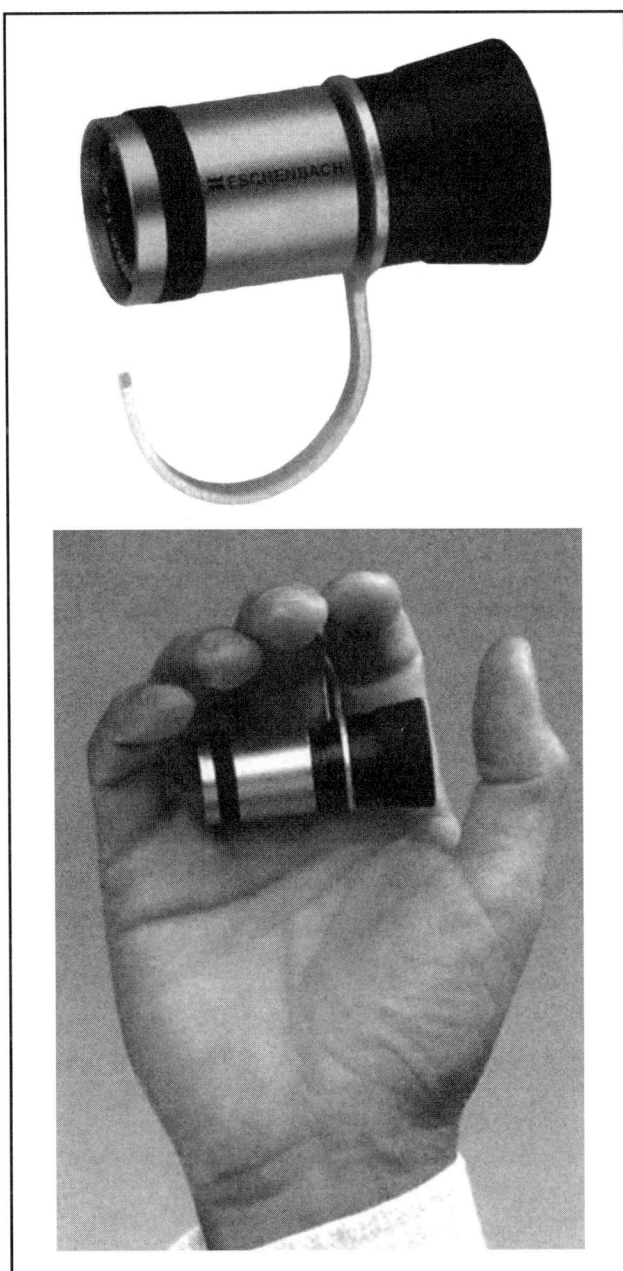

Figure 26-13. Finger-ring monocular. (Courtesy Eschenbach)

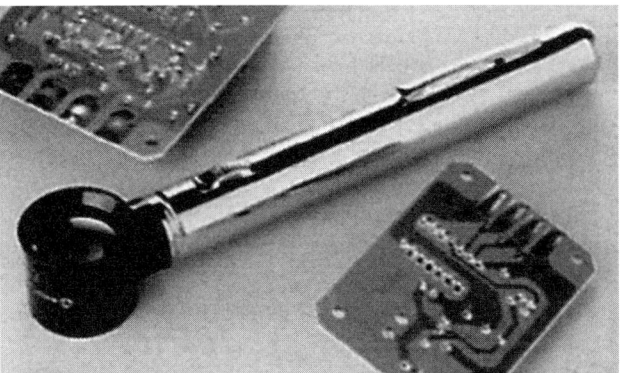

Figure 26-14. Lighted hand magnifier. (Courtesy Edmund Scientific)

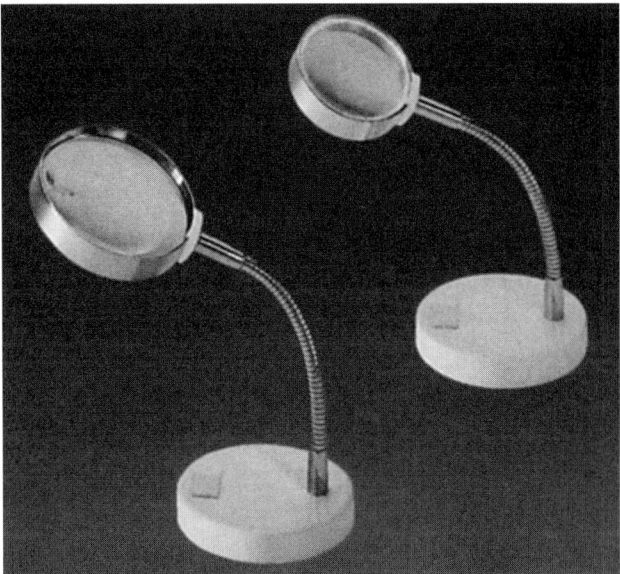

Figure 26-15. Gooseneck magnifier. (Courtesy Eschenbach)

Microscopes

Microscopes are used to view the part while it is being deburred and to verify that features are burr-free after the work has been performed. Most microscope-related comments in this chapter apply to both applications. Clearly, one advantage of microscopes is that they allow hands-free operation. Typically, the part is held in one hand while the deburring knife (or another tool) is held in the other. Monocular and stereo are the two basic microscope variations, but there are major utilitarian distinctions among stereo systems. Specifically, microscopes used in medical operating rooms have superior optics and lighting, which make them particularly useful in microscopic deburring.

A portable, digital video microscope can store 500 images. The lens can be changed to accommodate 10–1,000× magnification. One system claims to have up to 20 times the depth of field of conventional microscopes. Portability allows the microscope to be carried to giant pieces, as well as small parts.

Two-million pixel, charge-coupled device cameras with associated computers and flat-display screens

Figure 26-16. Headband magnifier. (Courtesy Gesswein)

enable users to send an image immediately to other parts of the plant or across the world for immediate decision-making. One system allows split-screen viewing, so a standard can be directly compared to the part in question. Computer electronics and software allow real-time edge enhancement for finding flaws in metal or glass surfaces. They also allow measurement of burrs or scratches.

Monocular Microscopes

Monocular microscopes have a single tube, as the name implies, which allows one eye to view the surface. A toolmaker's microscope is one example of a monocular microscope. These microscopes provide *XY* table measurement capability that other monocular systems lack. Other monocular microscopes include those used in high-school biology classes to view cell structure and minute life forms. For normal shop use, a magnification of 20× is appropriate. These tools are convenient and inexpensive. They are not as useful, however, for parts that require depth-perception viewing. Stereomicroscopes provide that ability, which is often critical for precision deburring.

Stereomicroscopes

Stereomicroscopes are the most commonly used form of microscope in situations where operators must work for long periods. Stereo vision provides greater detection ability of burrs, dirt, or other features that could cause a part to be rejected. They come with either fixed or continuously adjustable magnification. Several common microscopes have four settings that cover a 6–40× range. Others have a zoom lens that provides continuous, variable magnification (Figure 26-17). As mentioned previously, any power above 20× greatly increases the time required to validate the absence of burrs. Zoom magnification is preferable, but it may double the price of fixed-setting microscopes.

Magnification. When using a microscope to view miniature parts and hard-to-see features, the question of magnification arises: "What magnification should I use?" or "What magnification level is best?" These questions are not easily answered. For most uses, 7–10× magnification is easiest on the eye and, generally, adequate for most deburring. Many individuals can see better at 10× than 30× magnification.

Although eyepiece magnification is clearly marked, the value shown does not reflect the actual magnification of a part. To illustrate, there are three lenses,

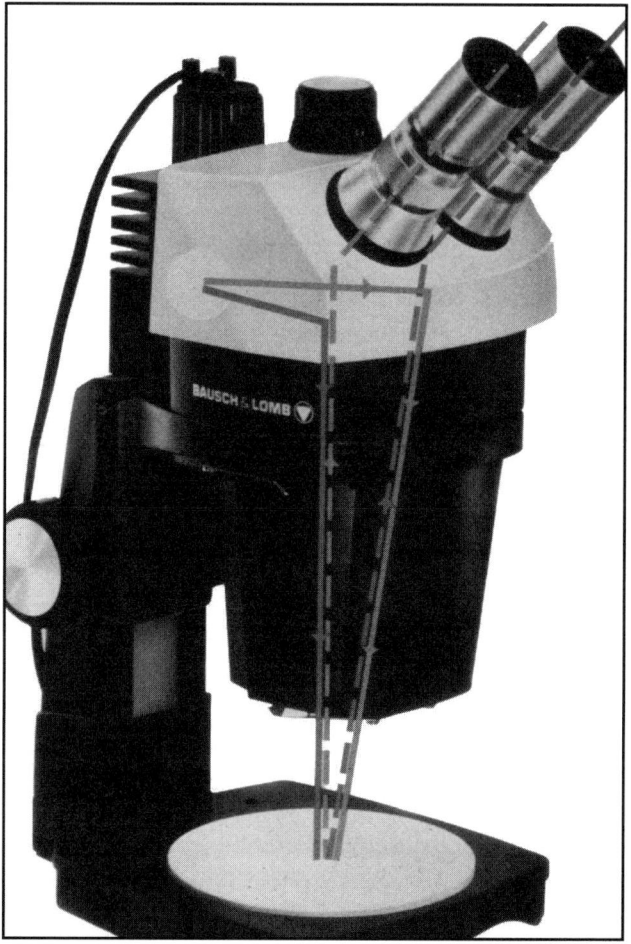

Figure 26-17. Stereo-zoom microscope. (Courtesy Bausch & Lomb)

basically, in a microscope: an eyepiece lens, a lens in the center of the microscope body, and an auxiliary lens at the bottom of the power body. Total magnification is the level of each of these multiplied by itself. The equation is:

$$T_m = E_m \times B_m \times A_m \qquad (26\text{-}1)$$

where:

T_m = total magnification
E_m = eyepiece magnification
B_m = body magnification
A_m = auxiliary magnification

The following examples illustrate the equation:

- Example 1: Eyepiece magnification is normally 10×. Assume the zoom setting is 0.7× on the zoom magnification knob at the right of the microscope, and an auxiliary lens is not screwed into the bottom of the power body. In this case, the calculation is 10 × 0.7 × 1.0 = 7× total magnification.
- Example 2: Assume that for a 15× eyepiece, the zoom magnification knob is set at 2×, and an auxiliary lens of 1.5× is attached to the bottom of the power body. Total magnification is 15 × 2 × 1.5 = 45×. In other words, the part features viewed through this lens system would be 45× their actual size.

While most eyepieces available on microscopes are 10× magnifications, 20× magnification eyepieces are available and other magnification levels can be purchased if necessary. Typically, an auxiliary lens is not used because the highest magnification level required on regular parts is 30×. Generally, more than 20× would not be used. Since microscopes with 10× eyepieces and zoom levels of 7–3× will accommodate magnification up to 30×, there is no need to use an auxiliary lens.

A problem with high magnification is that the more magnification used, the smaller the depth of field. Just like a camera or telescope at very high magnification levels, only a small area is actually within view or focus at that level. Regardless of the microscope used, high magnification means only a small area of the part can be seen. A smaller lateral area (field of vision), as well as a smaller vertical area (depth of field), are in view (Figure 26-18). Resolution is the smallest feature that can be resolved. At 20× magnifications, if the worker uses proper lighting and is properly trained, burrs only 0.0002 in. (5 μm) tall can be detected.

Working height is the amount of open area beneath the power body available for placing hands, parts, and

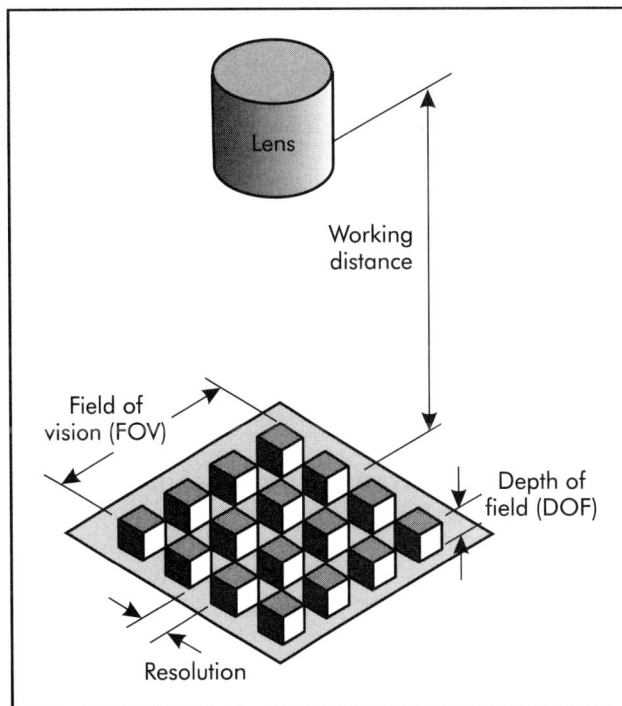

Figure 26-18. Microscope vision variables. (Courtesy Edmund Scientific)

tools. Generally, this is fixed by the microscope design. If necessary, the working height can be increased with an extension sleeve below the microscope body.

Some of the commonly used microscopes have a Nicholas illuminator, which can be placed in a hole at the back of the microscope stand—a location that provides adequate light positioning for normal deburring. For situations in which it is unacceptable in these positions, it is placed in a stand in front of the microscope. A cool fluorescent light that moves with the power body can be attached to the microscope. The advantage of this setup is that, in most cases, the light will permanently illuminate the area being deburred. A disadvantage of lights mounted off the microscope is that when the microscope body is adjusted upward, the light is still focused at the lower position.

Microscope reticles, such as those shown in Figure 26-3, can be inserted within the microscope eyepiece. Figure 26-19, for example, illustrates reticles inscribed with small radii. These radii indicate the actual amount of edge break to put on parts. Note, however, that since the microscope power is adjustable, the proper magnification must be set on the microscope for the reticle to represent a specific radii. For example, the scales available at one company have small arcs representing 0.003-, 0.005-, 0.007-, and 0.009-in. (0.08-, 0.13-, 0.18-, and 0.23-mm) radii when a microscope is

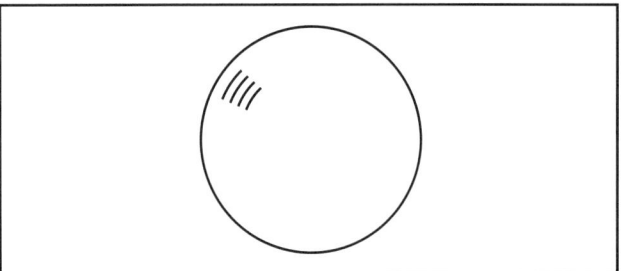

Figure 26-19. Edge radii on reticles measure edge break.

set at 30× magnification. Other scales are commercially available or can be designed to suit specific applications.

Full-screen stereomicroscope. Figure 26-20 illustrates a full-screen microscope. The larger picture makes it easier to view and, reportedly, reduces tired eyes, sore necks, and back pain that can result from using conventional microscopes. Video systems also can be attached to this style of device. Operators can wear their normal eyeglasses. Reportedly, the viewing screen provides 64× larger images than conventional microscopes provide.

Portable Microscopes

When a part is too large to take to the microscope, it is possible to take the microscope to the part (Figure 26-21). Foot pedals provide hands-free zooming and focusing operations.

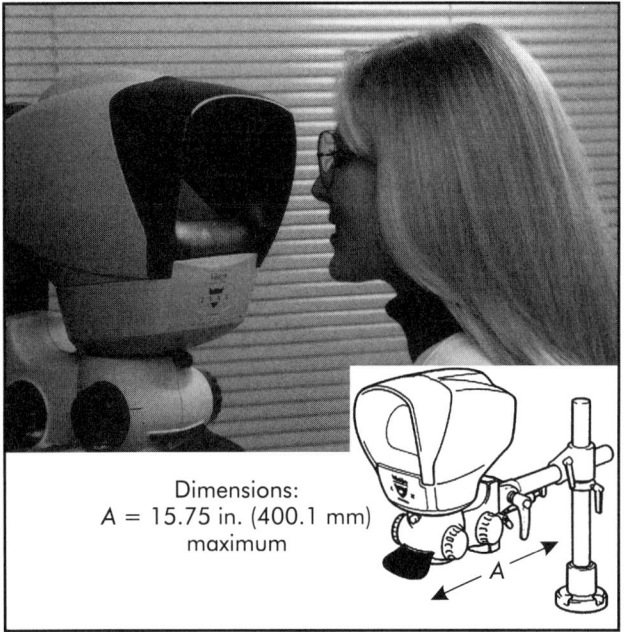

Figure 26-20. Full-screen microscope. (Courtesy Vision Engineering)

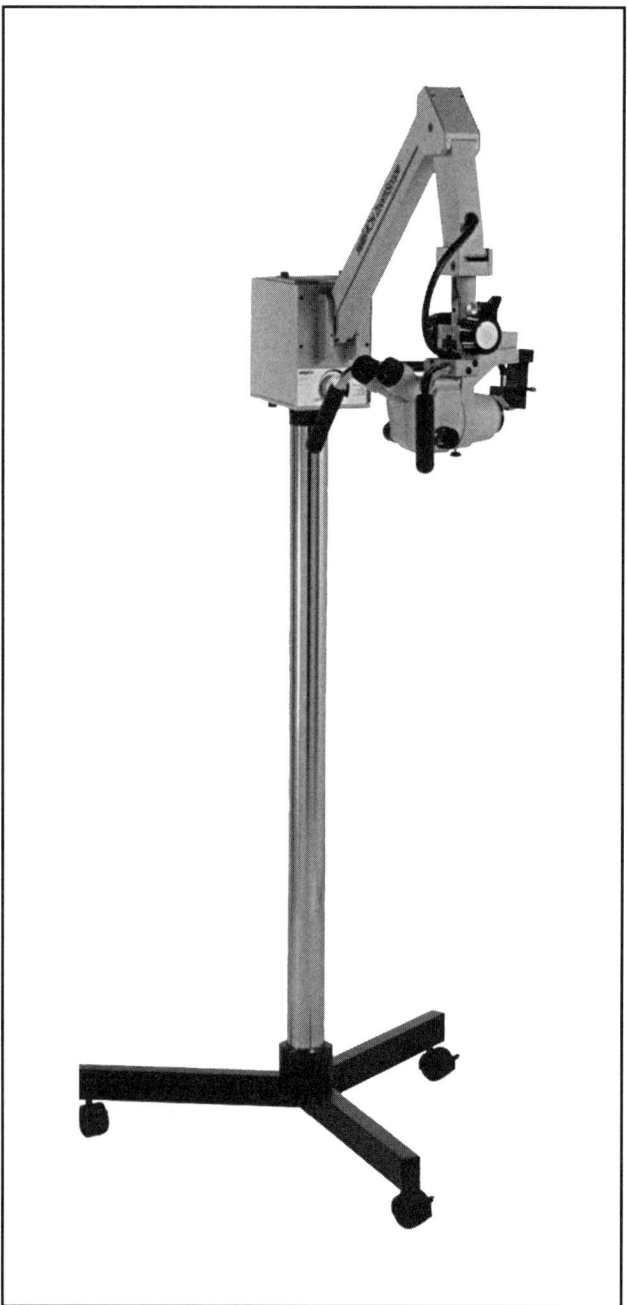

Figure 26-21. Portable microscope. (Courtesy Wallach Surgical Devices)

Teaching Microscopes

Figure 26-22 illustrates one of several arrangements possible for two-person, same-object viewing. This allows both student and teacher, or worker and observer, to view exactly the same object. This is a major advantage if users are being taught to recognize easily overlooked, minute problem areas.

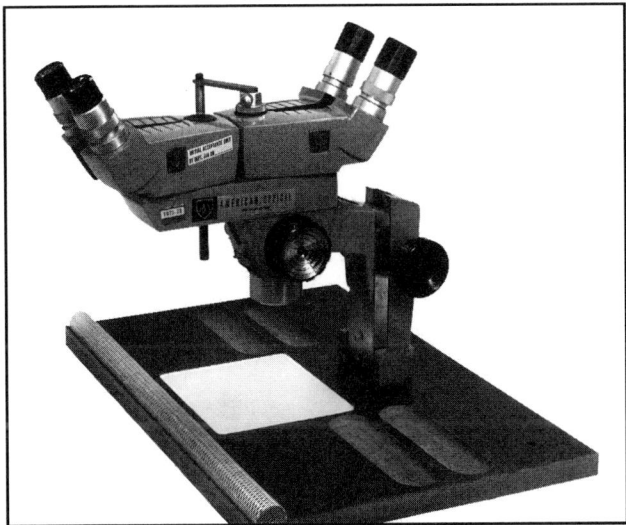

Figure 26-22. Teaching microscope. (Courtesy Honeywell)

Surgical Operating Microscopes

Figures 26-17 and 26-23 illustrate microscopes that have light passing through the lens system. This lighting is known as *coaxial illumination*. Coaxial, in this case, implies that it is coincident with the viewing light rays.

Advantages

The microscopes discussed in this chapter offer three principal advantages.

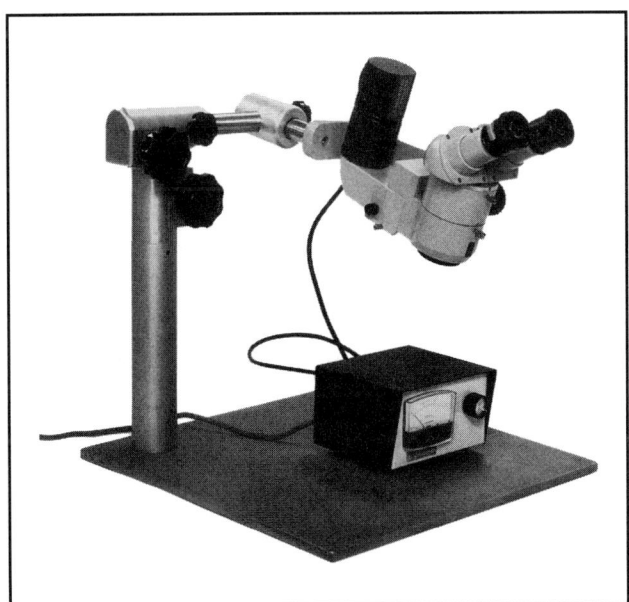

Figure 26-23. Surgeon's operating microscope. (Courtesy Leica Microsystems Ltd.)

- The light is always shining on the part and requires no adjustment while the part or power body is moved.
- The working focal length of the microscope (that is, the distance from the auxiliary lens to the point in focus) can be easily adjusted. For example, with a 3.94 in. (100.1 mm) objective or auxiliary lens, the focus is 4 in. (101.6 mm) below it. A 15.75 in. (400.1 mm) objective lens extends the focus to 16 in. (406.4 mm) from the bottom of the power body. Lenses are easily interchangeable and readily available.
- Some microscopes employ more expensive optical systems. In many cases, users feel that parts are in clearer view, or more easily viewed, than they are with less expensive microscopes. It should be noted, however, that even the microscopes shown in figures 26-17 and 26-23 can have adapters added to them that provide coaxial illumination and permit changes in the focal length of the microscope.

Note that in Figure 26-23, the microscope is extended some distance from the vertical post. This is an advantage in that very large parts may be deburred under these microscopes. It is also a disadvantage because the arms magnify table vibrations. Thus, for proper utilization of any microscope extended on an arm or boom, elimination of vibrations is essential (at least while the part is being viewed).

When lights are used with microscopes, it is important to use only the required amounts of power or light. When some microscope lights are turned to their maximum power, the microscope bulbs will only last a few hours. In one case, the microscope light may only last 40 hours at full power. When turned to its lowest level, which is normally adequate for general viewing, it would last up to 500 hours.

Care and Cleaning

Every four to six months, under heavy use, microscopes should be checked and, if necessary, repaired. Over time, the optics become dirty and operators cannot see as well with this tool. Unfortunately, only a few services offer cleaning and repair. Traveling cleaning services are available, however. A company's staff members go to the plant and not only clean and repair the microscopes, but also provide training and problem identification tips.

To extend the time between cleaning and repairing, always use the accompanying plastic dust cover when the microscope is not in use. Eyepieces should always be kept in the microscope to prevent dust from collecting within the eyepiece tube. When cleaning the microscope, do not attempt to disassemble the power body. The lens system within the body was carefully cleaned and aligned at the factory.

Dust on the eyepiece lens is seen as specks that rotate with the eyepiece when it is turned. If the viewing field does not appear clear, carefully inspect the lower lens of the power body. Subtle loss of contrast and definition due to dust or a slight smear on the lenses can be avoided with routine inspection and cleaning. If an optical surface becomes badly coated with dust or dirt, blow it off with a syringe or dust it with a camels-hair brush before attempting to wipe the surface clean.

Optical surfaces should be cleaned with a lint-free, soft, linen cloth, lens paper, or a cotton-tip swab moistened with distilled water or alcohol. It is important to avoid excessive moistening—the cloth, lens tissue, or cotton-tip should be just moistened with the solvent, rather than wet enough for the solvent to run onto the lens, which could loosen the cement on interior surfaces. Immediately wipe the surface dry, using a circular motion, before allowing it to air dry.

Glass surfaces should never be touched by bare fingers because this will leave a greasy smear and, frequently, corrosive perspiration. Do not clean optical parts unless it is necessary. Operators in one plant noted that one worker's body salt and acid were so corrosive that his fingerprints would etch glass.

Lights

Microscope Illuminators

Microscope illuminators come in many configurations. Some provide a broad path to highlight a wide part area. The Nicholas illuminator employs narrow beam, 600-ft (182.9-m) lights specifically designed for use on or near microscopes (Figure 26-24). Most of the lights have a variable transformer to adjust intensity. Many have adjustable necks to put the light where it is needed. Figure 26-25 shows flexible-arm lights used with a stereomicroscope.

Fluorescent Lights

Fluorescent ring lights fit below the lenses of most microscopes (Figure 26-26) to provide a diffuse light at the point of use. Fluorescent lighting provides the advantage of "cool" light, which reduces fatigue, eliminates temperature effects on microscopes, and prevents operator burns.

Halogen Lights

Halogen lamps provide high-intensity, long-life lighting for microscope work. Flexible gooseneck arms provide the adjustment necessary for universal work.

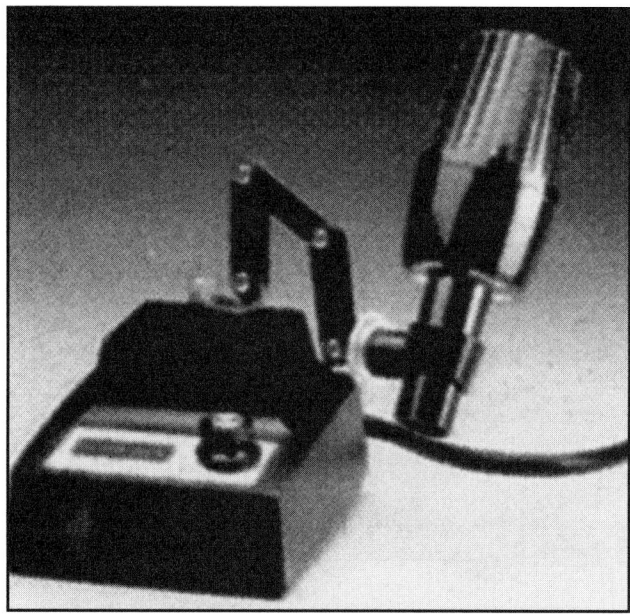

Figure 26-24. Nicholas illuminator for microscopes. (Courtesy Edmund Scientific)

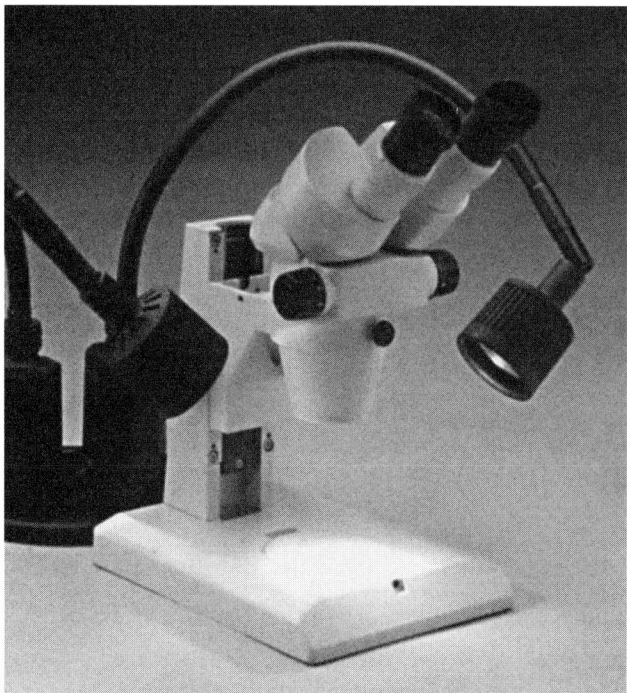

Figure 26-25. Dual halogen lamps used for microscope lighting. (Courtesy Sunnex)

Fiberoptic Lights

Fiberoptic bundles carry light from a source to many hard-to-reach areas (Figure 26-27). They can provide light to ring lights as well.

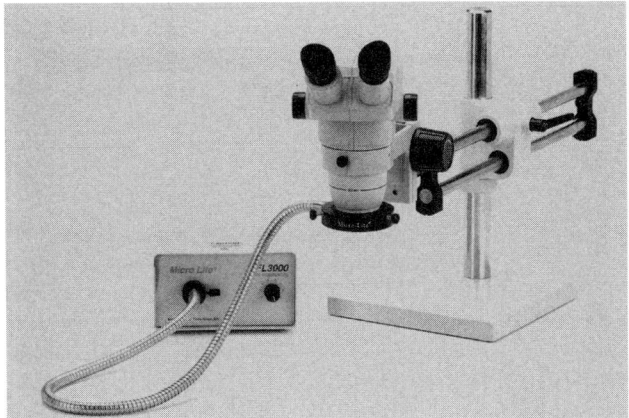

Figure 26-26. Ring light around microscope. (Courtesy Micro-Lite)

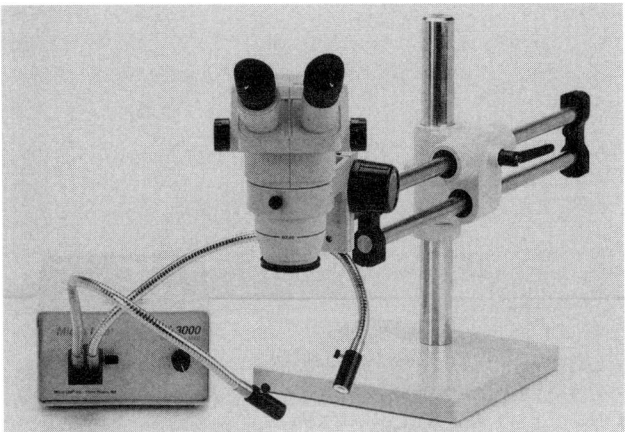

Figure 26-27. Fiberoptic illumination. (Courtesy Micro-Lite)

Otoscope

Doctors use otoscopes to peer into ears (Figure 26-28). The devices are small and provide convenient concentrations of light.

Long-stem Lights

Figures 26-24 and 26-25 show some long-stem flexible-arm lights used for deburring and machine-tool lighting. They come in a multitude of mounting arrangements and can be found in several lengths, but users will have to search many catalogs to find extra-length units. Figure 26-29 shows a long-stem light at a drill press and a machine-mounted magnifying glass above the part. Figure 26-30 illustrates three lamps with attached magnifying aids.

Headband Lights

Headband magnifiers often carry a light source to illuminate hard-to-reach areas (Figure 26-16).

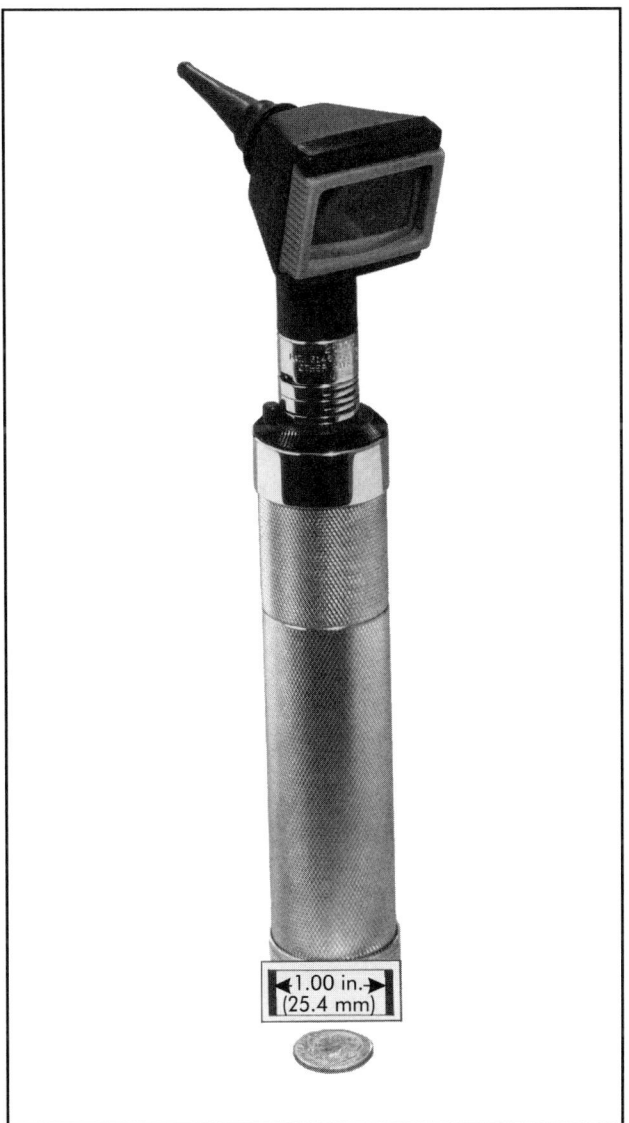

Figure 26-28. Otoscope. (Courtesy Honeywell)

Deep-hole Lighting

Some fiberoptic tools are designed specifically for deep-hole applications. Figure 26-31, for example, illustrates a small-diameter light that can be bent to any angle. Some of these also come with miniature mirror attachments. The unit shown in Figure 26–31 is 10-in. (254-mm) long.

Articulated-arm Incandescent Light

Articulated-arm incandescent lights are widely used in industry for many inspection purposes. Despite their low cost and flexible positioning, however, the heat from incandescent lights can burn operators and make the work area uncomfortable.

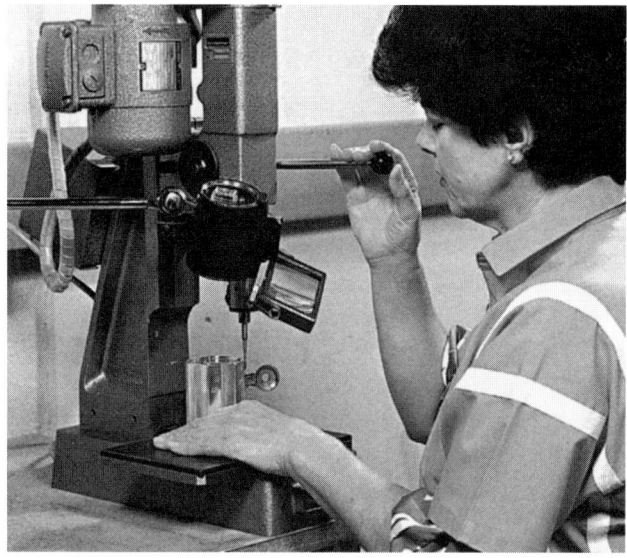

Figure 26-29. Long-stem light and magnifier at drill press. (Courtesy Waldman Lighting)

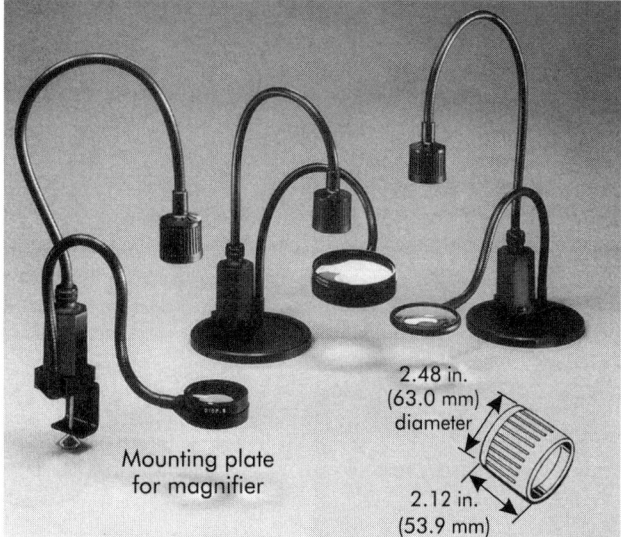

Figure 26-30. Gooseneck lamps with magnifiers. (Courtesy Sunnex)

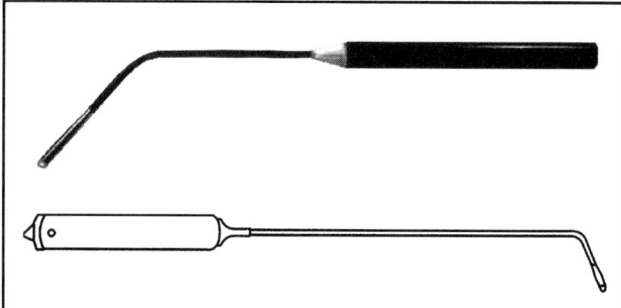

Figure 26-31. Small-diameter light can bend to reach remote areas. (Courtesy Reid Tool Supply)

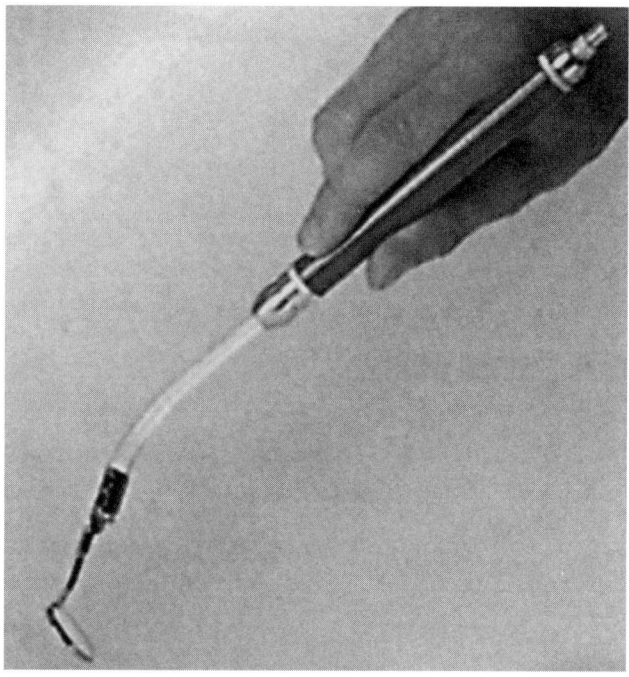

Figure 26-32. Deep-hole light and mirror. (Courtesy Reynolds Machine & Tool)

provide objective measurements of sharpness (Gillespie 2000). It does not measure burr size—it tells users that the part edges will cut hands or fingers.

Chamfer Mikes

Chamfer mikes measure the depth or diameter of hole- and rod-end chamfers (Figure 26-33). These are quick and easy to use and validate that burrs have been removed. They come in analog dials or digital units.

Ball-end Probes

Several plants use a ball-end probe to check for burrs (Figure 26-34). The stainless-steel ball prevents surface scratching. These tools can be found in a few

Mirrors

Small mirrors are used to see into crevices. Dental mirrors are the most widely used. One of the industrial variations has an integral light to facilitate viewing (Figure 26-32). The smallest simple dental mirror is about 0.375 in. (9.53 mm) in diameter.

Sharp-edge Measuring Devices

Underwriters Laboratories developed a handheld sharp-edge tester many years ago that is still used to

Figure 26-33. Inside- and outside-diameter chamfer micrometers. (Courtesy Reynolds Machine & Tool)

office supply stores, as well as some dental supply houses (Gillespie 2000).

Toolholders

Pin Vises

Pin vises are the most commonly used support device for holding small deburring tools. Several designs are available, including cylindrical (figures 26-35a and b), dog nose (Figure 26-36), and shortened dog nose (Figure 26-37). Cylindrical pin vises accommodate tool diameters up to 0.125 in. (3.18 mm), while the dognose design accommodates items up to 0.25 in. (6.4 mm) in diameter. Extended-length pin vises (extension tap wrenches) are available up to 10-in. (254-mm) long. At one time, over 25 different designs of these tools were produced. They also may be found under the following names: hand vises, adapter chucks, graver holders, and pin chucks.

Hand Chucks

Miniature, three-jaw chucks accommodate tools from 0.039–0.109 in. (1.00–2.77 mm) in diameter (Figure 26-38). Larger sizes are also used.

Stone Holders

Handheld stones are often chucked in special round holders. Some have slim, pencil-like designs; whereas others are specifically designed for a bolder hand (Figure 26-39). Most are plastic, although brass is also used in one design.

Surface-finish-measuring machines

Deburring and surface finishing are closely allied fields. Frequently, users must demonstrate that deburring did not mar a finish, or that it provided the necessary finish. Surface profilometers (finish-measuring machines) provide this information.

Cryogenic Probes

A fine blast of cryogenic temperature may be useful for deflashing some plastics. A handheld unit exists that employs 25 different solid tips (Figure 26-40). A small handheld nozzle is also reportedly available, although not shown here, to shoot cryogenic materials at plastic edges. The solid tips would only be useful on very thin plastics. A spray of cryogen would enable use on more typical plastic parts. Note that cryogens are hazardous materials since they can freeze skin quickly (one of their major applications). Too much cryogenic fluid can crack many plastics.

Workbenches

Workbenches are taken for granted, but several considerations must be made for effective use, including: ergonomic height, adequate number of safe electrical outlets, drawer space (and ease of use when filled with dozens or hundreds of tools), adequate bench lighting, and dust-collecting capability.

Shop Workbenches

Standard workbenches are equipped with many drawers, shelves, and holding features to allow workers ready access to tools. The number of tools stored in a workbench may surprise management, since several thousands of dollars' worth of tools are unseen—and perhaps unused. Management tours of the deburring area and questioning workbench adequacy could help reduce some unwarranted costs.

Exhaust Systems

Exhaust systems include devices that suck dust or those that push it away. These are critical for areas that generate many particles or some particles of hazardous materials. Workers with allergies may also need some of these devices.

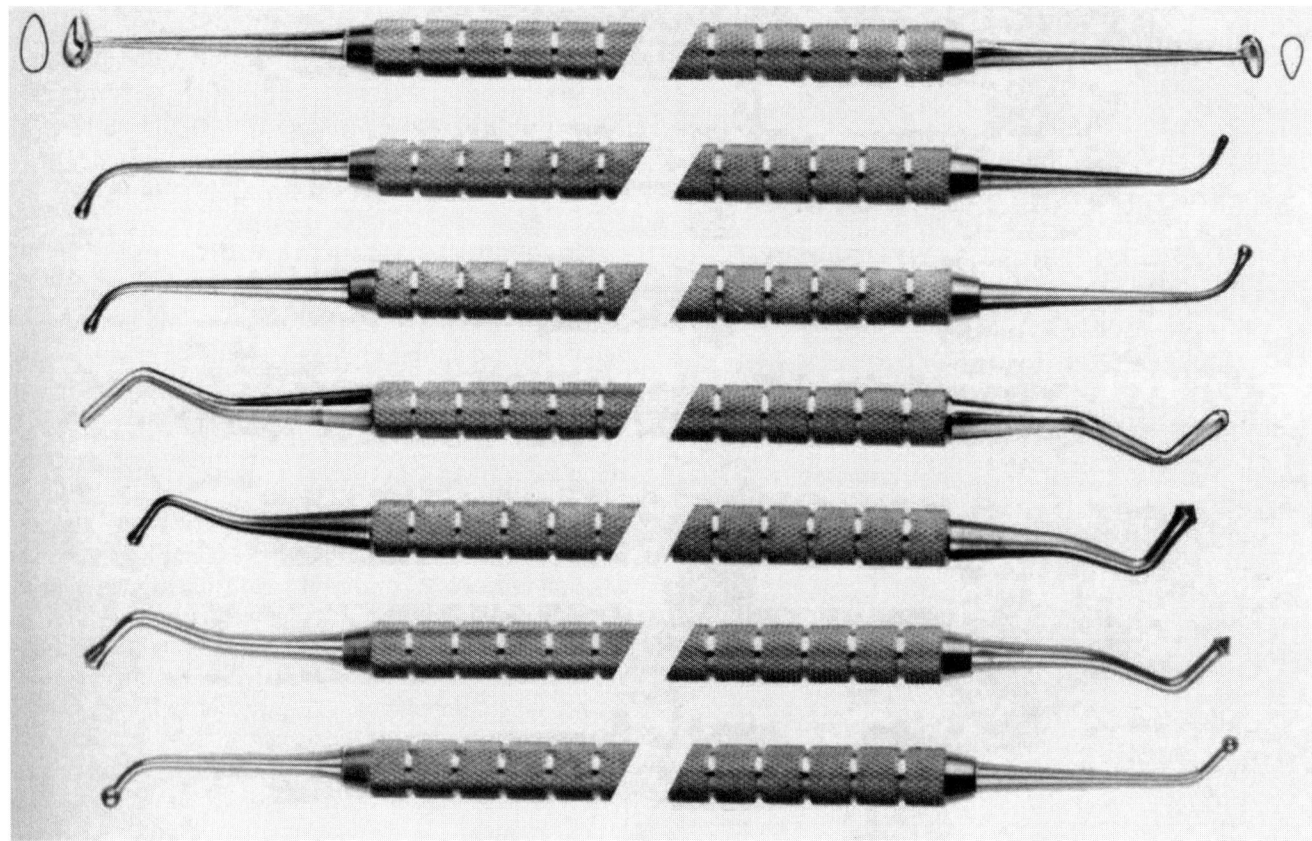

Figure 26-34. Ball-end probes to detect burrs. (Courtesy Brasseler)

Exhaust Hoods

Bench-top exhaust hoods are commercially available to remove dust and debris from the work area. Figure 26-41 shows one such small collector used with a bench motor.

Down-draft Workbenches

Down-draft benches push dust away from the work area; whereas, dust collection devices suck up dust. Clean benches force air outward to ensure no dust remains on the product.

Portable Clean Bench

A portable clean bench provides a table or plastic bubble with gloves that encapsulate the product while it is being worked on. Clean air is provided to ensure the product is maintained in a clean-room environment while being deburred.

Vacuum Systems

Several portable and fixed vacuum systems remove dust from the work area. Some come with high-efficiency particulate air (HEPA) filters. Many of these devices are noisy and not well-liked by workers.

ENVIRONMENTAL, HEALTH, AND SAFETY ISSUES

Support tools are either the source of many safety or ergonomic problems or the solution to such problems. Improper lighting, improper chair adjustment, and burns from lights are a few examples of the issues regarding support tools. The following paragraphs provide more in-depth examples.

Protective Gloves

Some industries manufacture parts with such sharp edges their products must be handled with gloves. Many varieties of protective gloves are available. In addition, one company produces surgical-grade gauze coated with natural rubber that sticks to itself. Users wrap this material around key fingers to protect them from cuts (Figure 26-42). This material is used where gloves are awkward or otherwise unsafe to use. In addition to protecting fingers from sharp edges, this

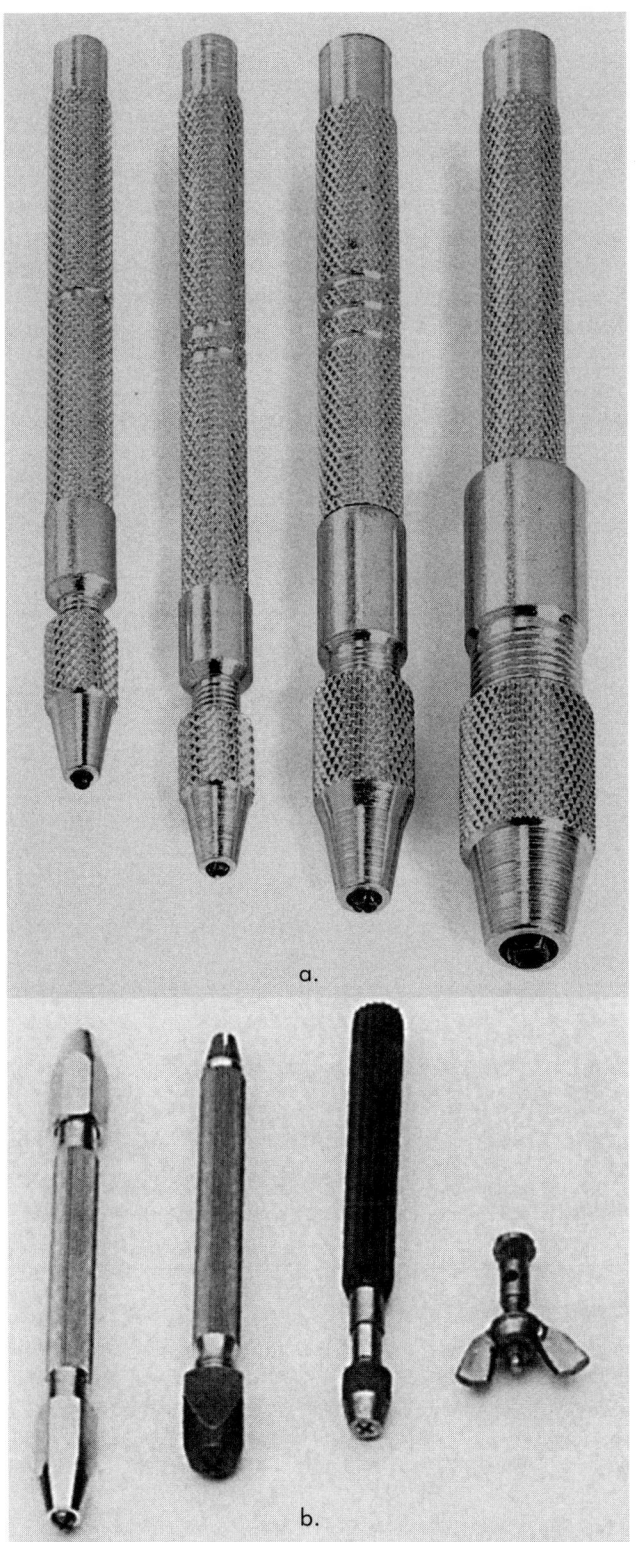

Figure 26-35. (a) Cylindrical pin vises and (b) small pin vises. (Courtesy Gesswein)

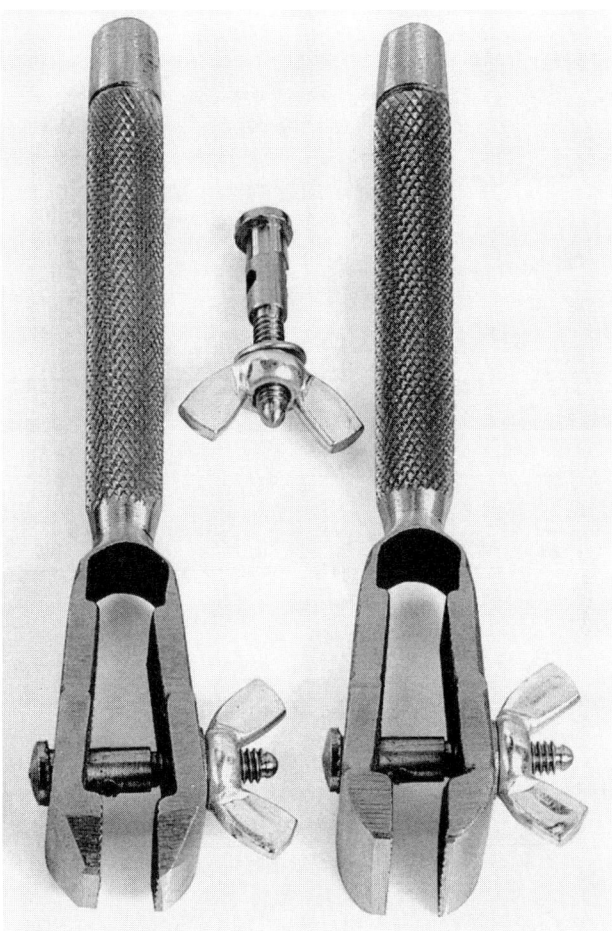

Figure 26-36. Dog-nose pin vises. (Courtesy Gesswein)

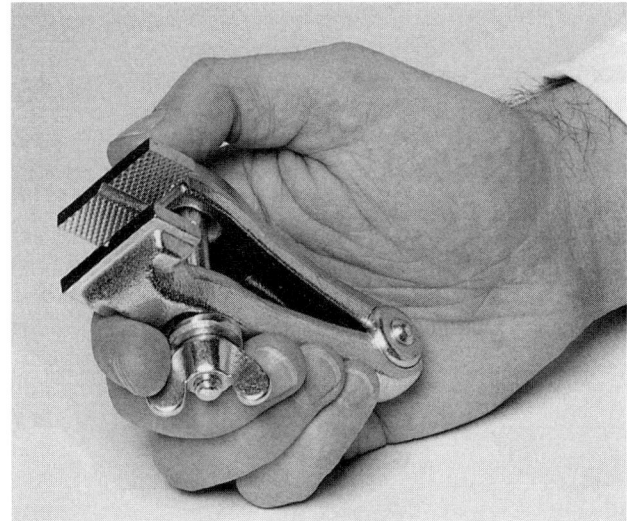

Figure 26-37. Short-dog-nose pin vise. (Courtesy Gesswein)

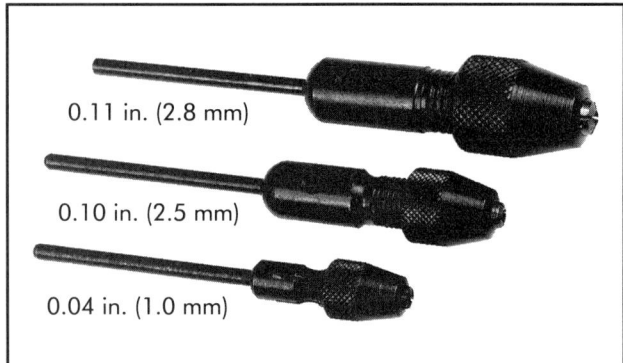

Figure 26-38. Small hand chucks. (Courtesy Gesswein)

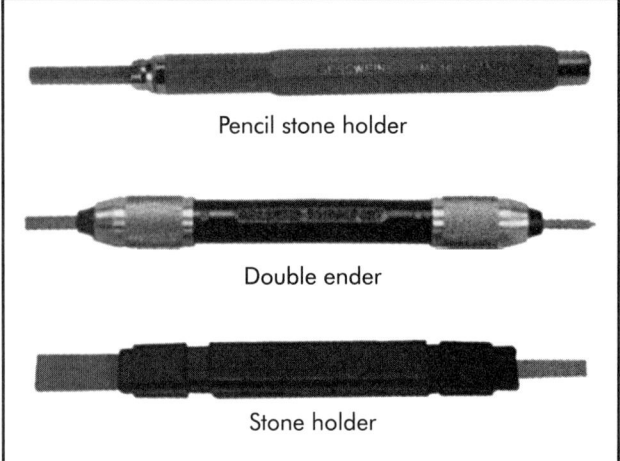

Figure 26-39. Stone holders. (Courtesy Gesswein)

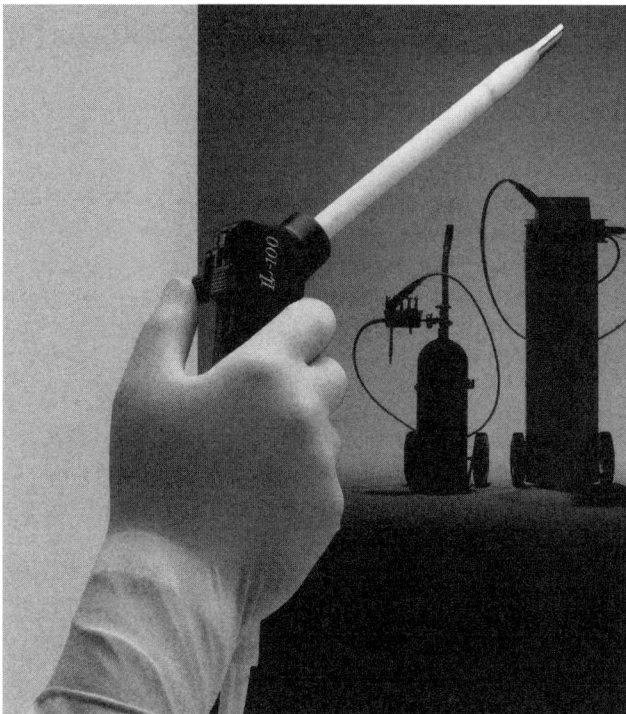

Figure 26-40. Cryoprobe. (Courtesy Wallach Surgical Devices)

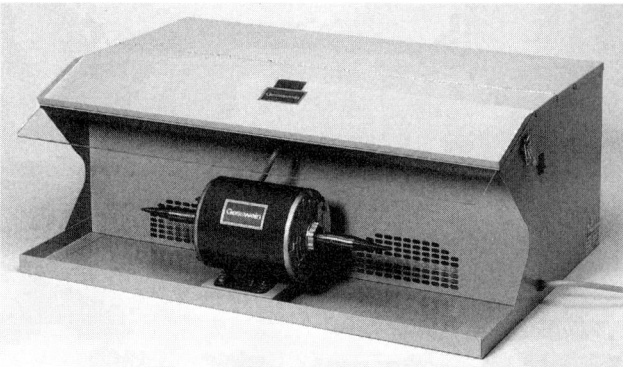

Figure 26-41. Bench-top dust collector. (Courtesy Gesswein)

material reduces burn potential when hot materials are involved (Gillespie 2000).

Chairs

Chapters throughout this book have discussed ergonomic features and considerations with regard to deburring tools. For general ergonomic support, however, chairs and benches also should be considered. Today, chairs are available with vertical and horizontal adjustment controls that can help prevent some repetitive injuries. They provide great support for hands, arms, and backs; and many provide adjustments to accommodate different body sizes and leg lengths. Some are sold solely for their ergonomic characteristics.

Flexible-positioning Arm

Figure 26-43 illustrates an ergonomic arm that can help in some deburring operations, such as hole chamfering.

Heat Considerations

Heat is an enemy of high production rates and quality. Improving heat control in a facility is beyond the scope of this work; however, most ergonomic texts provide several considerations on this topic. Chapter 31 provides some additional thoughts on this problem.

Other Issues

In all cases, eyeglasses should be worn around rotating machinery, even innocent-looking handheld motorized tools. At some sites, protective eyeglasses are not required for working under microscopes. Be-

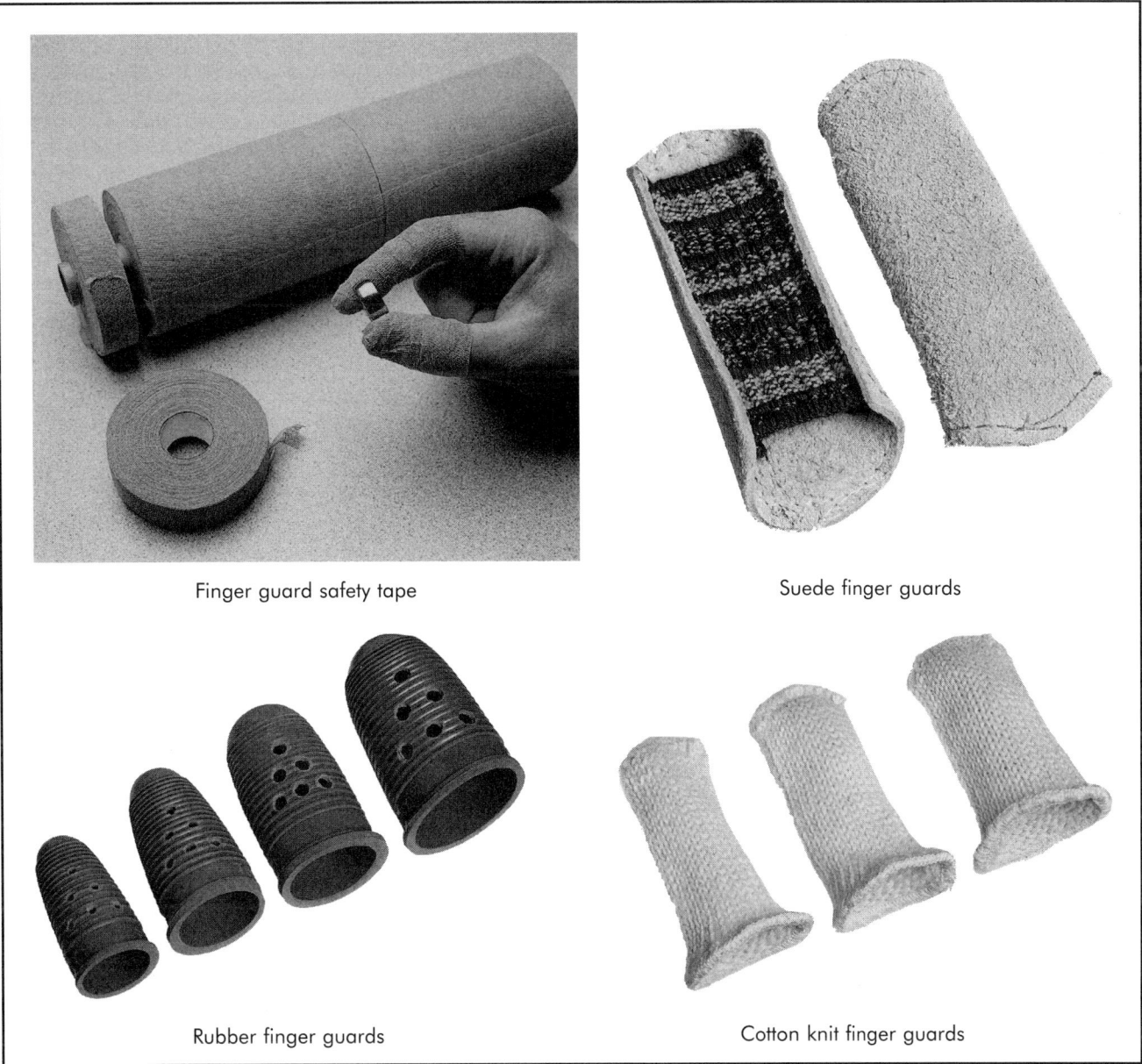

Figure 26-42. Finger guards. (Courtesy Gesswein)

cause rotating tools are still used, and the face is not always against the microscope eyepieces, some risk is involved if protective glasses are not worn.

Dust is a major explosion and fire hazard, particularly when paper particles from bonded abrasives pile up with the resins. When titanium or magnesium dust covers other materials, and steel sparks at a sander, these other materials may catch fire. The health effects of carcinogenic metal dusts are described in Chapter 19. Other materials may present similar concerns, which is why a discussion of vacuum and exhaust systems is included in this chapter.

Air guns are commonly used for blowing dust and other deburring debris from parts. However, they are widely used as part of the deburring and cleaning cycle. They introduce the danger of blowing debris into a worker's face or lungs. Safety-supply catalogs offer several kinds of air guns that provide some form of protective design to prevent facial injuries.

Good posture is important for personal comfort when using a microscope. The worker's back should be straight, and he or she should look through the eyepieces without bending the neck. It is also important to work at a height that enables looking through

Figure 26-43. Flexible arm for ergonomic use. (Courtesy Flexarm)

the microscope and deburring parts without bending. At first, neck and back fatigue may occur, because the microscope may be unfamiliar, or the worker did not properly adjust it to suit his or her specific needs. With proper setup, however, fatigue will quickly go away. Tall individuals can use riser blocks to adjust the microscope. Also, the microscope base can be tipped at an angle for more comfortable viewing. Wider bases provide greater stability to hold microscopes. In addition, some bases provide a number of convenient pockets for temporarily storing tools or miniature parts.

REFERENCE

Gillespie, LaRoux K. 2000. *Guide to Deburring, Deflashing and Trimming Equipment, Supplies, and Services.* Kansas City, MO: Deburring Technology International.

Tools that Beat Down Edges 27

The word "deburring" conjures images of cutting or grinding off burrs. For many applications, however, the real need is to ensure that part edges are smoothed to the extent that they do not cut. If a smooth, noncutting edge is the only requirement, beating the burrs down to a smooth surface accomplishes the goal just as well as cutting them off.

TOOL TYPES

In the 1970s and 1980s, two tools were produced to hand deburr by beating down burrs. Both could be used in a drill press or motor to deburr thin sheet metal, plastic strip, or tube ends.

Rotary Metal and Plastic Edge Deburring Tool

The first deburring tool was developed to peen hole surfaces to harden and smooth them. U.S. Patent 3,934,443 describes a peening tool used for hole walls (Keen 1975). As illustrated in Figure 27-1, the tool consists of two pancake-like ends that have a cylindrical groove between them. The tool is rotated in a drill chuck and the operator passes the groove over the edge to be deburred. A ball-bearing race is within the groove. This allows the balls to move radially outward when one of four pins or other balls push against them (Figure 27-2). The outer raceway is held against the edge to be deburred in such a way that it has almost zero velocity. The inner cylinder, however, turns due to the drill rotation. This turning rotates the inner cylinder, which forces the four inner short rods (or balls) against the outer balls. When this happens, the outer balls push against the edge in a hammering, or peening, action (McMaster Carr 1983). The width of the groove is such that two rows of these balls can be used, allowing both edges of the sheet to be deburred at the same time. This particular tool allows holes larger than 2.25 in. (57.2 mm) to be deburred, as well as outer edges.

Drill Press Vibrating Hammer

Figure 27-3 illustrates a vibrating hammer (also known as a sheet-metal edge deburrer) used on a drill press to deburr unhardened sheet metal. It accommodates irregular edges and corners and is normally

Figure 27-1. Rotary ball-peening tool (McMaster Carr 1983).

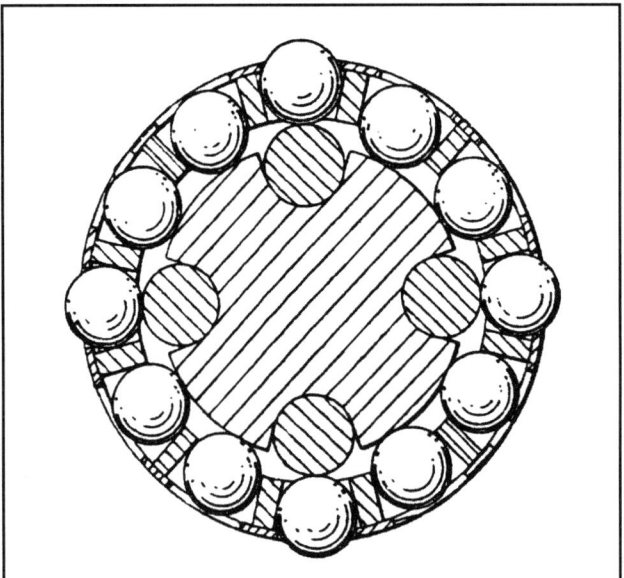

Figure 27-2. Principle of ball-peening tool (Keen 1975).

used in a drill press at 1,800 rpm. Drill motors of 0.25-in. (6.4-mm) capacity also can be used with these tools, and some have been used in lathes. As the drill spindle turns, a cam inside the tool changes the rotary motion to a vertical motion. A V-notch tool at the end is forced up and down over the edge of the sheet metal. These tools come with six peening blades that can be reground and stoned to meet unique needs (McMaster Carr 1983). This tool is called a "sheet metal edge deburrer."

Reciprocating Hammers

Today, many variations of vibrating hammers exist that do not require a reciprocating hammer. The hammer itself provides the required vibratory action. Users just add their own V-notched replaceable blade to hammer off burrs. Hammering burrs away is not as sophisticated as cutting or grinding them, and it probably is not as fast a removal method as grinding. However, it may offer advantages to industries and economies in which labor is cheap. Generally, it will neither break an edge more than allowed nor gouge other part surfaces.

ENVIRONMENTAL, HEALTH, AND SAFETY ISSUES

Environmental, health, and safety issues associated with vibrating tools have not been documented. Certainly, noise is a factor; and repetitive pulsing can damage arms, wrists, and hands. These tools move burrs; however, they do so with less force than some

Figure 27-3. Drill press vibrating hammer (McMaster Carr 1983).

of the other tools discussed in previous chapters, so there should be fewer flying projectiles and parts thrown from the tool. Keep in mind that hammering the burrs may still leave sharp projections, which could cut the hands of subsequent operators or users.

REFERENCES

Keen, D. P. 1975. U.S. Patent and Trademark Office, Patent number 3,934,443. Redondo Beach, CA.

McMaster Carr. 1983 Catalog, No. 83, p. 1,219. Chicago, IL: McMaster Carr.

Hot-wire Tools for Thermoplastic Parts 28

Heat treatments and cold cutting processes can be used to deburr and deflash thermoplastic parts. Most of the cold efforts are based on cryogenic tumbling processes rather than hand operations because cryogenic temperatures cause severe burns. As mentioned in Chapter 26, cryogenic probes are used in medical applications; and, with proper safety considerations, they can also be used in unique manufacturing situations.

THE HOT-TOOL CONCEPT

The hot-tool concept is simple. A tool is shaped like the desired edge finish, then passed along the edge to produce the shape. Figure 7-14 illustrates several such knives. These tools can be made from bent wire or machined steel. Even soldering-iron tools can be used if the temperatures produced match the needs of the plastic. It seems that there is only a single company in the U.S. specializing in this tooling (Gillespie 2000).

Thermocutters are handheld tools that typically use an integral transformer of 0.2–0.3 hp (120–250 W) and 110 V. They are used more for cutting than edge finishing, but the requirements are the same. They deflash, finish tube ends, and remove sprues from rubber and plastics. The handheld unit is attached by a power cord to the controller. Tool weight is 1–2 lb (0.5–0.9 kg) and blades can reach 220–1,400° F (104–760° C), depending on which one is used. They are designed for continuous use; and, they work on foam as well as sheet or other plastic configurations. The blades are changed with four screws. Figure 28-1 illustrates one thermocutter design.

Chamfers can be produced with any tool held at an angle to the edge. Radii must be produced with a tool that mirrors the image of the radii desired. Large ra-

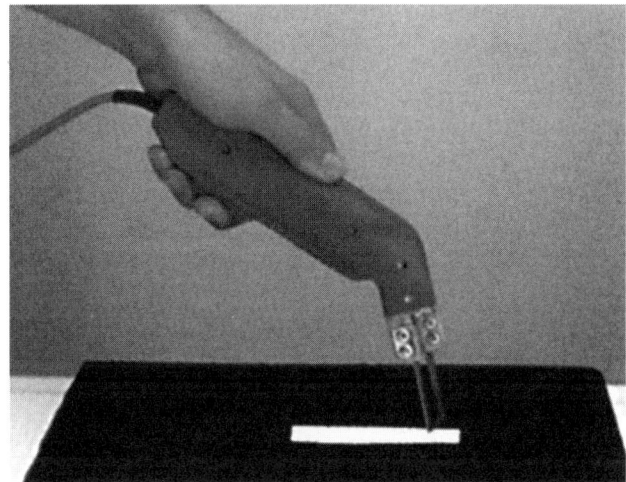

Figure 28-1. A handheld thermocutter in use. (Courtesy Abbeon Cal)

dii can be produced, but it considerably slows the edge-cutting rate. Figure 28-2 illustrates a double-edged blade for cutting. The deburring blade is ground on one side only. It allows anti-static trimming of excess plastic and sprues with tension cracks. The illustrated blade is 0.031-in. (0.79-mm) thick.

SUITABLE PLASTIC MATERIALS

Any thermoplastic material can be deflashed with hot tools, but some tools are more effective than others. Unfortunately, no known compilation of suitable material exists to help the user. Studies on polyacetal, polycarbonate, and phenol have been conducted (Kanda and Shimada 1966). One study used 0.016- and 0.020-in. (0.40- and 0.51-mm) thick flash, temperatures of

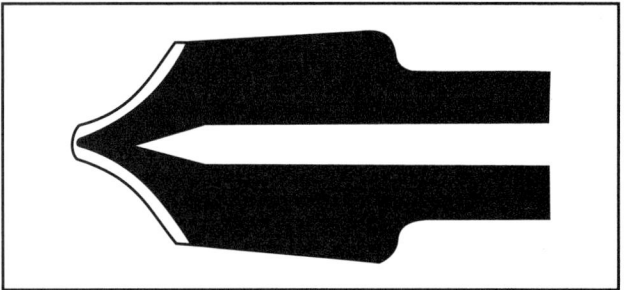

Figure 28-2. Handheld thermocutter. (Courtesy Abbeon Cal)

302–752° F (150–400° C), and feed rates of 1.97–11.80 in./min (50–300 mm/min). A secondary flash occurred, which increased with feed rate and temperature (Figure 28-3). Figure 28-4 illustrates the effect of feed rate and temperature on the surface finish when a robot fed the tool.

One plant had to deflash a 0.010-in. (0.25-mm) glass-filled polyetheretherketone (PEEK) slot. To accomplish this, the plant designed a small rectangular plug that mated to the slot, pushed it into the slot, and heated it to melt the edge material back into the basic material. Although some glass fibers still projected, many were recaptured in the matrix. In this instance, the engineer designed the tool with fillet radii to generate the radii on the finished part. PEEK can withstand temperatures up to 230° F (110° C) before softening.

ENVIRONMENTAL, HEALTH, AND SAFETY ISSUES

Hot tools can burn flesh; therefore, flesh burns are a prevalent issue with this type of tool. The electrical cord leading to the tool also carries electrical safety concerns. It must be protected from the hot knife, and it must not cause a tripping hazard. Part edges heated by these tools also are safety hazards for a short time. A less obvious issue is that hot tools generate fumes that can irritate breathing passages and, for some plastics, could generate health problems. Review the material safety data sheet (MSDS) for each material before using hot tools on any plastic or rubber material.

When deflashing, gloves may be necessary to prevent injury from the thin, sharp flash, as well as from the hot tool. (Figure 28-1 illustrates use with bare hands.) Flash will cut rapidly, and some plastic parts have many flash areas, so flash may be hard to avoid. The tool design shown in Figure 28-1 allows heavy hand and arm pressure, but any repetitive hand motion may require special consideration to prevent repetitive motion injury.

REFERENCES

Gillespie, LaRoux K. 2000. *Guide to Deburring, Deflashing and Trimming Equipment, Services, and Supplies.* Kansas City, MO: Deburring Technology International.

Kanda, Yuichi and Hideyuki Shimada. 1966. "Micro Deburring System for Engineering Plastics." In *Proceedings of the 4th International Conference on Precision Surface Finishing and Burr Technology,* Bad Nauheim, Germany. Kansas City, MO: Deburring Technology International, pp. 92–100.

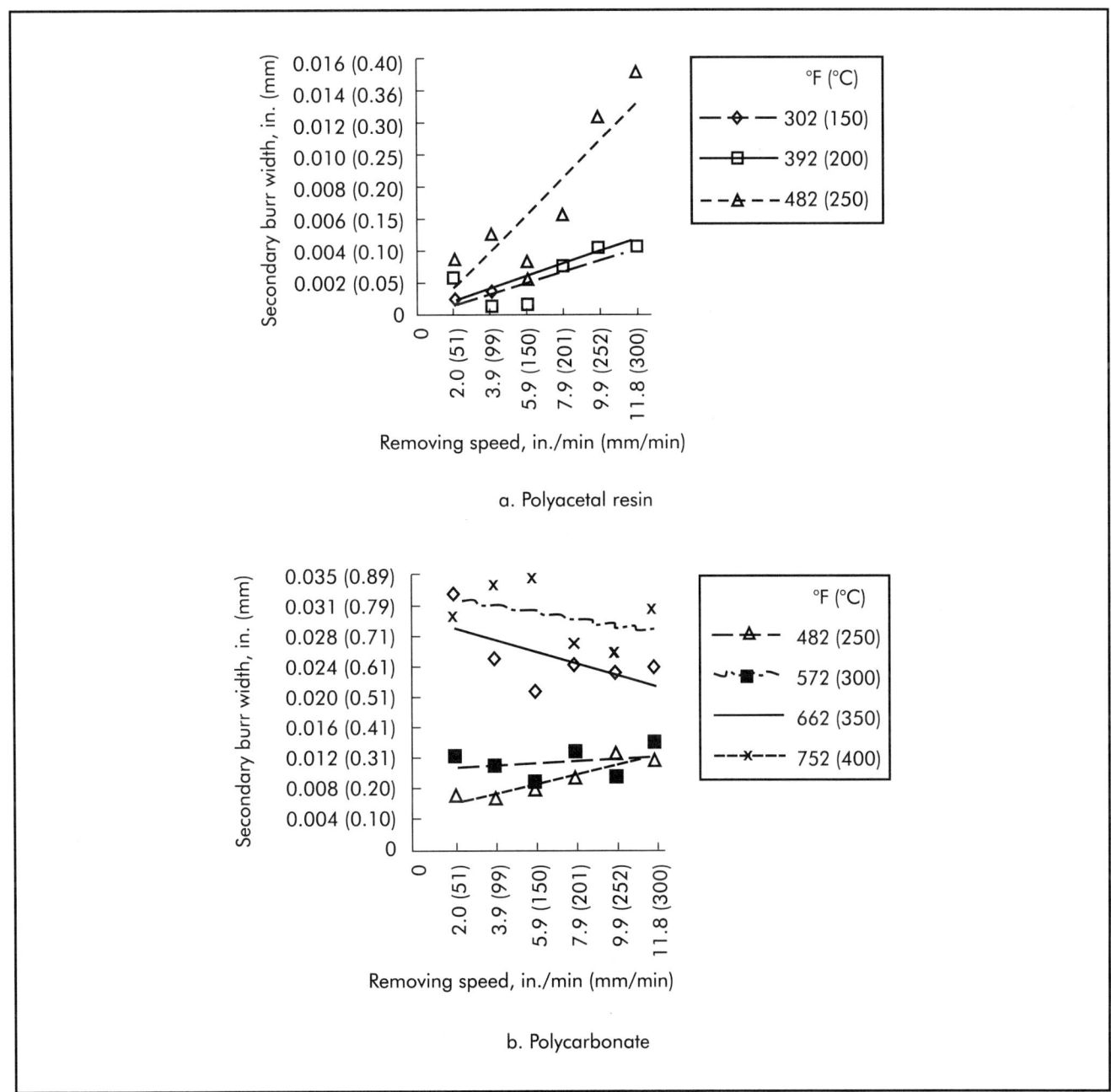

Figure 28-3. Width of secondary flash from hot tool. (Courtesy Worldwide Burr Technology Committee)

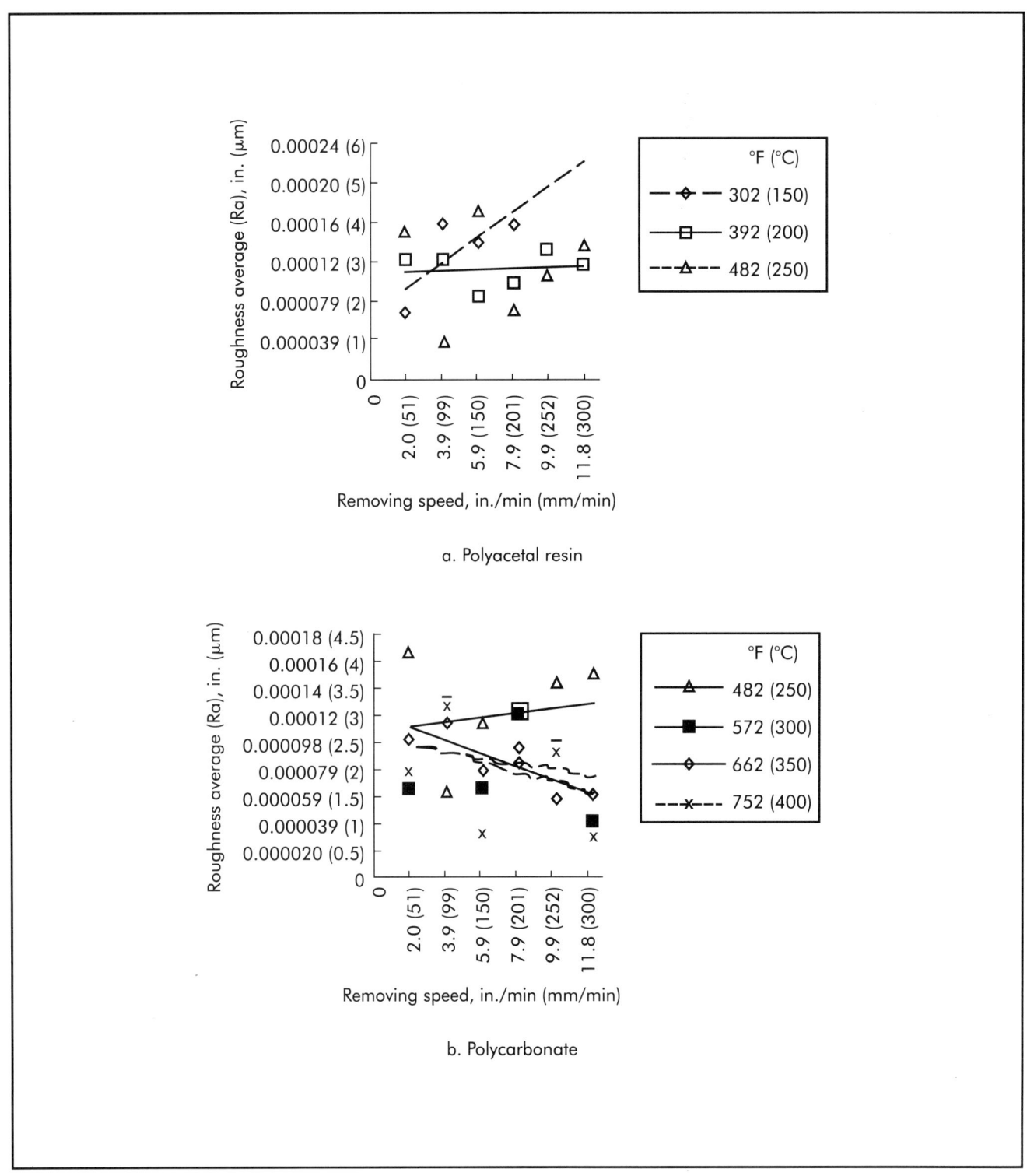

Figure 28-4. Effect of feed rate on surface finish. (Courtesy Worldwide Burr Technology Committee)

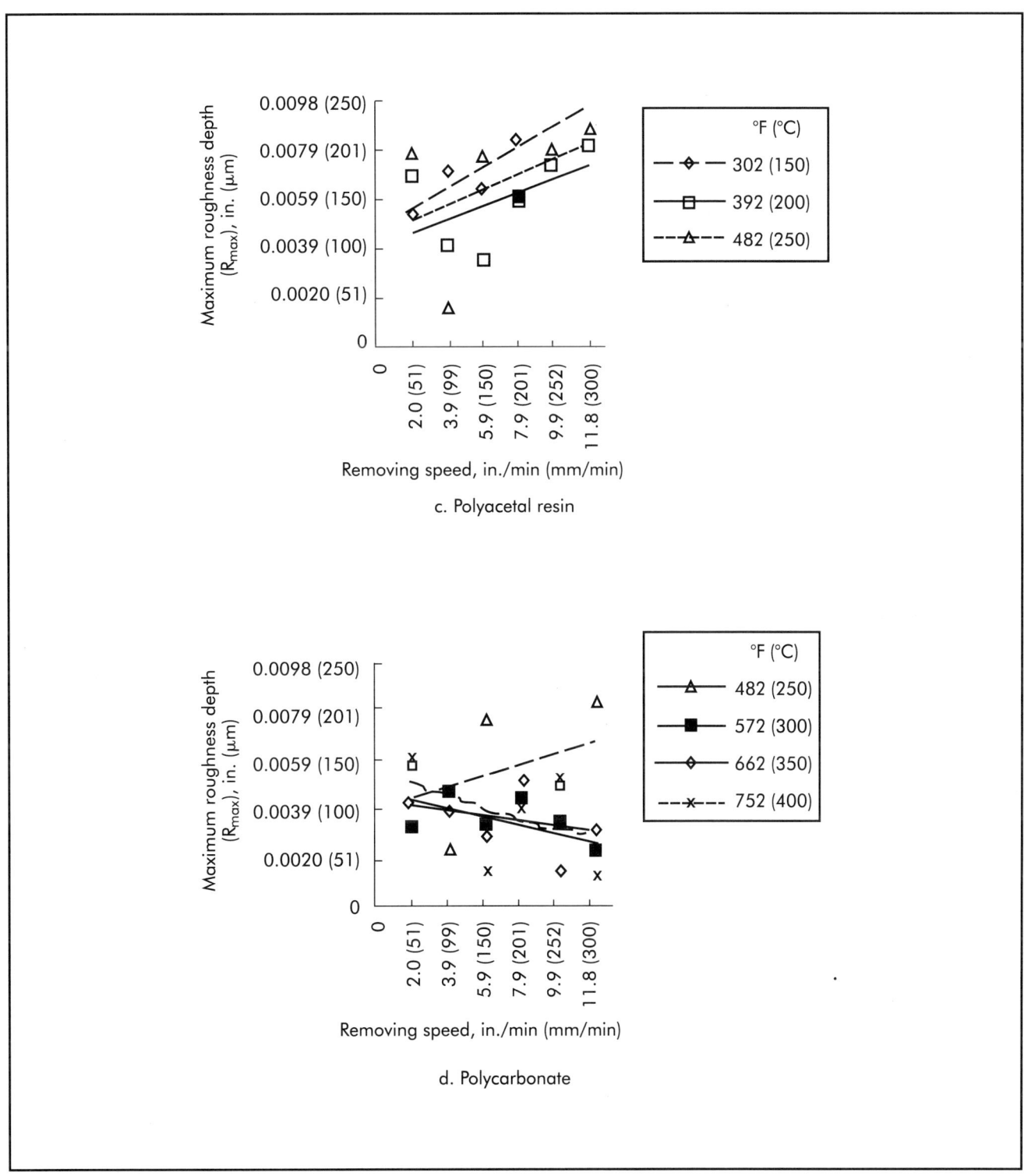

Figure 28-4. (continued)

Mechanization and Automation 29

So far, this book has been dedicated to identifying new or better means to manually deburr and deflash parts. Hand-operated, mechanized machines discussed in this chapter are exceptionally versatile in some instances; in others, they are limited but fast in what they do. The machines in Chapter 25 are the first consideration companies should turn to if current techniques are not cost-effective and the many other suggestions in this book are not applicable. If, after assessing ideas presented in other chapters of this book, a plant still needs to make mechanized or automated improvements, this chapter provides useful information.

Buying the equipment described here involves tens of thousands of dollars and requires supporting technical staffs (in many instances), shop space, and other utilities. These machines are the correct answer to many needs, but many will not be cost-effective in job-shop operations with constantly changing part configurations.

The term *mechanization,* as used in this chapter, means to add mechanized equipment to assist the operator. Operator assistance, however, is still required to load, unload, push against the deburring tool, or adjust machines. *Automation,* in contrast, means that an operator is not required for any part of the deburring effort. For more in-depth insight into mechanized deburring processes, please refer to this book's companion work, the *Deburring and Edge Finishing Handbook* (Gillespie 1999).

MECHANIZATION TO REPLACE MANUAL DEBURRING

Of the 109 deburring processes listed in Table 5-4, only a few have the ability to provide true deburring to meet normal industry demands. Several provide some level of acceptable deburring in nonprecision applications. Most mechanized processes deburr only a portion of a part. That portion, however, may be either the most time-consuming or the most likely to be rejected if not deburred by that mechanized process. The processes most employed to mechanize deburring include:

- vibratory finishing;
- centrifugal barrel finishing;
- roll-flow finishing;
- spindle finishing;
- abrasive-jet deburring;
- electrochemical deburring (ECD);
- abrasive-flow deburring;
- brush deburring;
- bonded-abrasive deburring (sanding), and
- thermal-energy method (TEM) deburring.

For thermoplastic materials, variations using cryogenic temperatures are also used. The reference list at the end of this chapter provides resources on each of these processes. One book provides detailed summaries of equipment and supplies for these processes (Gillespie 2000a), while others provide more operational discussion. Another provides a comprehensive listing of all known publications on deburring and deflashing (Gillespie and Repnikova 2001). A third book, *Cost Guide for Automatic Finishing Processes,* provides an example of cost comparisons for a wide variety of applications (Rhoades 1981). Also, Chapter 5 of this book provides cost estimate formulas and insight.

Each mechanized process has side effects. Two books provide extensive discussions of side effects and current advancements and practices, which include changes to part color, texture, and size; small nicks and dings; corrosion resistance; application of smut-like deposits; streaking; etching; residual stress changes; bent flanges; oxide coatings; and changes to part corners as well as edges (Gillespie 1978, 1999).

Purchasing Issues

When equipment must be purchased, it is generally procured to meet the removal needs of specific burr properties. Most mechanized processes are affected by variations in burr size. Bigger burrs are not as easily removed in a fixed-cycle operation. When burrs grow, or a different material becomes the part standard, the equipment may not be able to accommodate the change. Equipment should be bought to meet expected manufacturing variations rather than to accommodate the needs of today's burrs.

Usually, the most cost-effective approach to deburring is to combine mechanization with manual deburring. It is quicker and easier to implement than a solution that demands total burr removal with a single process. For example, electrochemical deburring (ECD) can deburr one or 100 holes in 15 seconds. However, there also may be a need to clean threads and reach intricate passageways to ensure absolutely untouched surface contours. In this situation, ECD will not normally meet the need without time-consuming and expensive tooling. It is more cost-effective to allow people to deburr hard-to-reach areas and let ECD take care of the holes. This method also enables a quick response and minimizes the engineering cost requirements of total mechanization. Many similar solutions should be considered. Each requires a capital equipment investment, unlike manual deburring.

Vibratory Finishing

Vibratory finishing is one of the most cost-effective processes used by industry. For external edges, it allows thousands of parts to be deburred unattended in a 1/2–3-hour batch in a continuous operation. Many variables can be controlled to allow the completion of surface finish, edge radiusing, cleaning, polishing, protective coating, drying, and other allied processes on the same equipment or finishing line. Supplies are available throughout the world, and there are many authorities to help resolve problems. Miniature parts are adaptable, and wing spars 40-ft (12.2-m) long have also been finished on this type of equipment. Almost any material can be finished, from gold to tungsten, plastics, and ceramics.

Costs for vibratory finishing vary depending on the quantity of parts run, size of parts, finishing conditions required, and other issues, but costs in the United States are typically cents per part. Costs are lower in less industrialized countries. Large parts, or parts requiring special handling, might cost $1–2 each. Giant wing spars would cost much more, of course, because special machines are required. There is one text available that provides an extensive discussion on calculating economics for this process, as well as many variations that accelerate production (Gillespie 1999).

Figure 29-1 illustrates a common vibratory bowl machine. The parts are dumped into the bowl and, after vibrating for a specific amount of time in the abrasive media, emerge on top of a coarse screen and fall out the end of the chute. The media falls through the screen holes and back into the machine. Parts are normally processed wet, which allows abrasive compounds to accelerate finishing action and solutions to clean the parts of obvious particles. Parts can be run dry, if needed; and, if rust is a concern, they can be oiled.

Generally, internal features are not fully deburred, although there are many tricks of the trade that, for specific parts, enable these areas to be finished. Surfaces can be improved to as fine as 8 μin. (0.2 μm) in this type equipment, although 16–32-μin. (0.4–0.8-μm) finishes are more common. Manual deburring can be used to finish features not fully deburred by the mechanized process. Burr properties and part tolerances affect cycle time, with larger burrs and radii requiring more time.

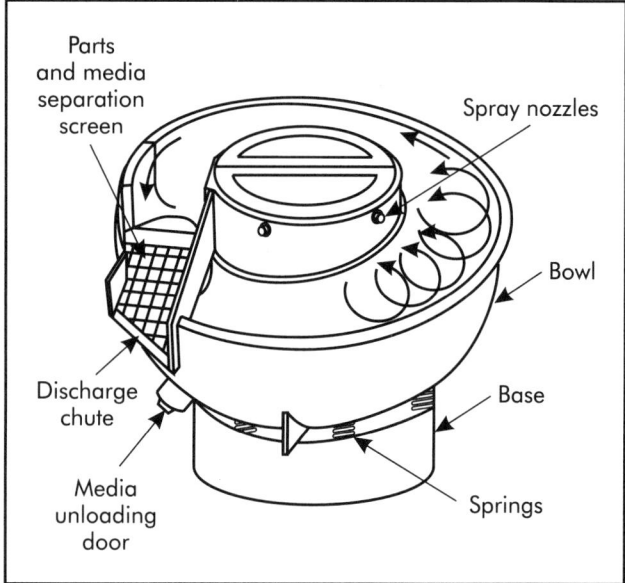

Figure 29-1. Automated, continuous-feed vibratory finishing machine with screen for separating parts and media, and conveyor and chutes for returning media to the machine. (Courtesy Gyromatic Manufacturing Company)

Common side effects of vibratory finishing include changes to part color, texture, and size; small nicks and dings; residual stress changes; bent flanges; and changes to part corners and edges. Operator skill-level requirements are low, although higher knowledge helps improve the process. Environmental, health, and safety issues are relatively few, although operators are continually working in wet and noisy conditions.

Centrifugal Barrel Finishing

Centrifugal barrel finishing shares the same general issues and benefits as vibratory finishing. The major difference is that centrifugal barrel finishing uses closed containers, so continuous finishing is impossible. When big machines are used, batches can be relatively large. This process cuts finishing time to approximately one-third that of vibratory finishing; and, if parts require less finishing time, deburring times can be cut even more. The same media and compounds are used, and the same side effects are possible. Centrifugal barrel finishing does induce compressive stresses, which generally benefit part strength and fatigue.

Roll-flow Deburring

Roll-flow deburring is a faster version of vibratory finishing. While it is a batch operation, its open top and faster cycle make it a popular process for many applications. Costs, operations, benefits, and side effects are comparable to vibratory finishing. Regular maintenance may be required to keep the thin opening below the spinning bowl from clogging. If properly considered, however, this is a minor issue.

Spindle Finishing

Cylindrical parts are good candidates for spindle finishing. In principle, the process operates like the electric-powered egg beater or mixing bowl. A part is attached to the spindle with a collet, or similar device, then rotated in a bowl of small abrasive media and particles to remove rough edges (Figure 29-2). A 2–5-minute cycle is required to finish a part, and many parts can be ganged on the spindle to finish 5–10 parts in the same 2–5-minute cycle. During the cycle, operators are busy loading and unloading parts from holders.

Side effects and basic operation for spindle finishing are roughly similar to vibratory finishing, but the variety of parts that can be deburred is not as broad as with vibratory finishing. Automotive parts are a key product line for this process.

Abrasive-jet Deburring

Abrasive-jet deburring operations are widely used to remove flash from zinc die castings and many plastic parts. The same process is used to peen metal

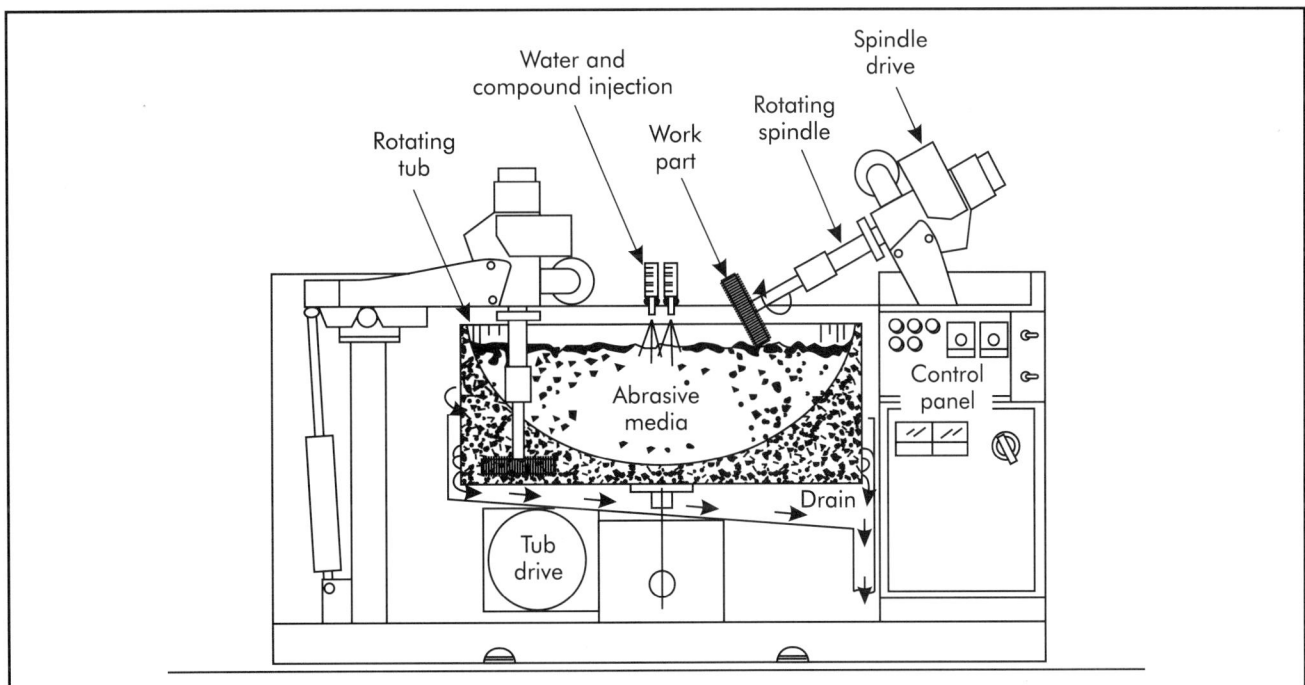

Figure 29-2. Two-spindle, spindle-finishing machine. (Courtesy Almco Industrial Finishing Systems)

surfaces to provide compressive stress and remove dirt and oxides while removing thin burrs or flash.

Cryogenic deflashing is a variation of the abrasive-jet deburring process. It removes burrs in the direct abrasive stream. It dulls edges, but may not remove all burr traces. Deburring ability is a function of burr size and material properties, as well as blasting parameters.

Abrasive-jet deburring is one of the most versatile processes, and almost all shops have a blasting machine. Side effects include distortion of thin sections, abrasive particles left inside small passageways, compressive stresses, and obvious surface appearance changes, with small indentations visible in some instances. The abrasive lodged in surfaces may inhibit electroplating and welding.

Electrochemical Deburring

Electrochemical deburring (ECD) is a fast operation that affects only the edges that are designed to be deburred. It is tolerant to burr thickness variations; however, long burrs must be removed before using this process, or the tooling will short out on the burr, which will damage either the tooling or the part. In this instance, "long burrs" refers to burrs that are long enough to bend and touch electrodes that are 0.020–0.030 in. (0.51–0.76 mm) from the part edge.

Two ECD solutions are commonly used. Salt solutions may allow stray etching at a distance from the edge, and some carbon-like deposits may have to be scrubbed off with brushes. Stainless steel is an ideal material for this process, but it works equally well with any steel or aluminum. The single requirement is that the part must conduct electricity. Titanium is not a good part material for salt-based solutions. A glycol-based solution reportedly provides better cutting for many parts and prevents stray etching. Both solutions should be reviewed before purchasing equipment.

Multiple parts can be deburred in the same cycle, but it increases tooling costs and requires quick loading and unloading by the operator. A tool must be designed for each part with an electrode positioned at each edge to be deburred (Figure 29-3). Tools are relatively simple and are commonly made from brass, Teflon®, and Teflon tubing.

Low-voltage electricity is used in ECD and provides for safer operation than household voltages. Solutions may cause skin dryness or irritation. Fumes should be removed by a vent system, since nickel is a suspected carcinogen, and nickel is a key ingredient of stainless steel. Hydrogen is liberated, so it must also be removed to prevent risk of explosion. Wastes may have to be disposed of according to environmental

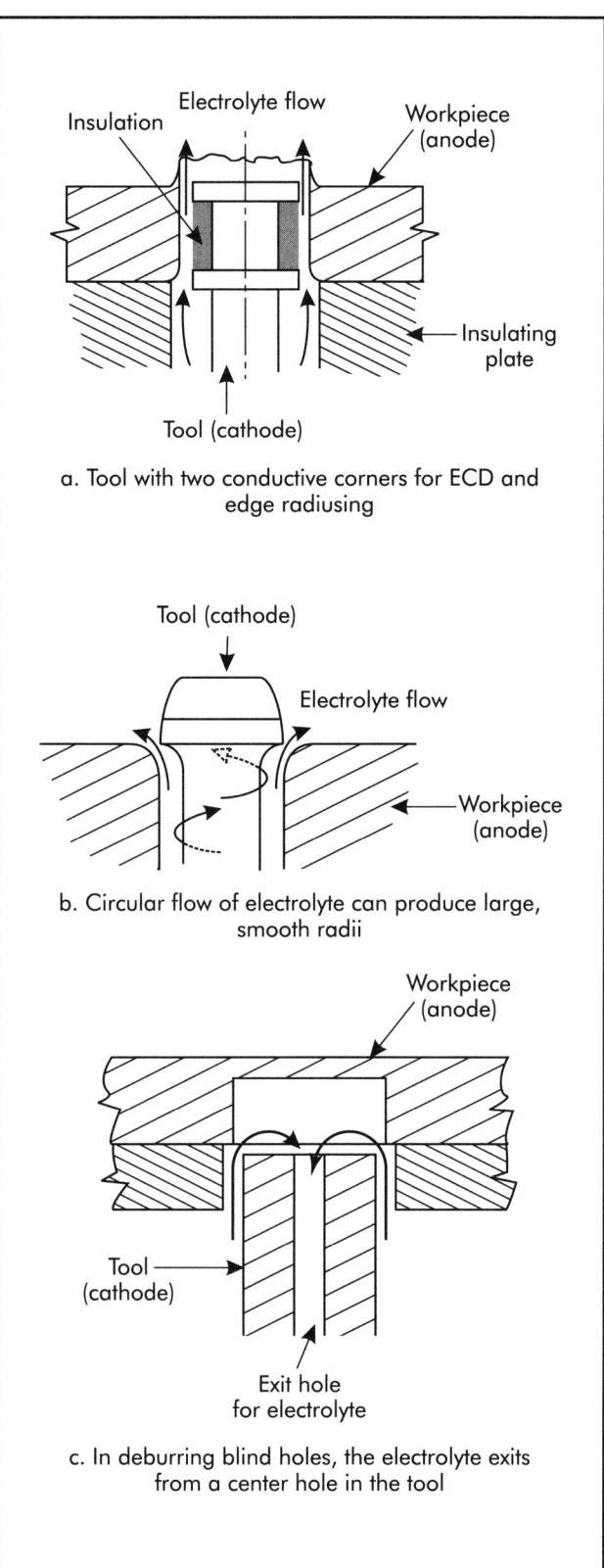

Figure 29-3. Tooling concepts for electrochemically deburring holes. (Courtesy Dynetics)

waste guidelines. Since the operator is working around solutions all day, slipping and falling is a concern.

Abrasive-flow Deburring

Abrasive-flow deburring uses a putty-like material filled with abrasive to rub burrs from edges. The machine looks like two hydraulic presses pushing a viscous media back and forth from one cylinder to another (Figure 29-4). The media passes through all part openings that the tooling allows. Just a few passes remove burrs and provide a smooth edge radius. Typical cycle times are 1–2 minutes, but part loading and media cleanup from the part require additional time. The process is simple, quick, and requires little operator training.

Abrasive-flow deburring machines produce surface finishes as fine as 2 μin. (0.05 μm) if needed and, in most instances, significantly improve the finish of the original part interior. This deburring process can be used on external features, but it is most applicable to intersecting holes and holes in general. The machine has polished turbine blades and similar parts. The media lasts for hundreds or thousands of parts and is easily replenished.

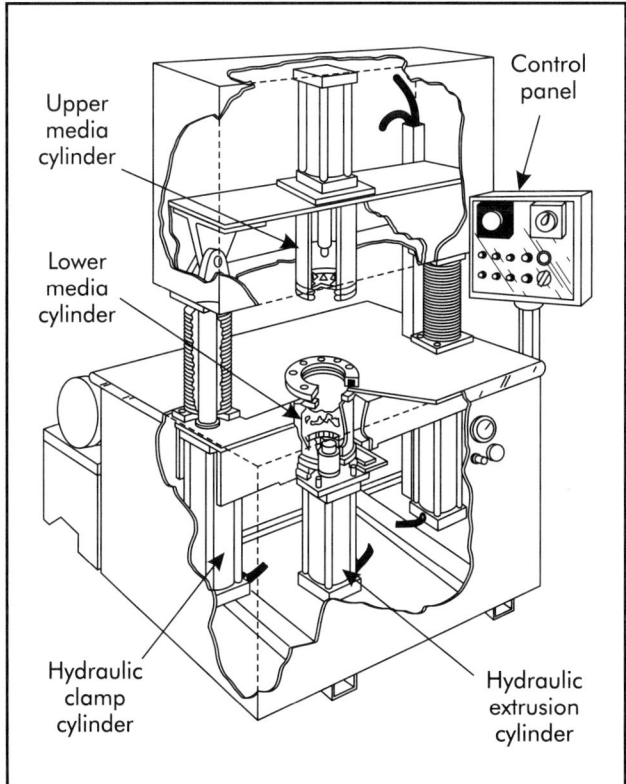

Figure 29-4. Cutaway of high-production machine for abrasive-flow deburring. (Courtesy Extrude Hone Corporation)

The hydraulic cylinders present normal press-pinching concerns, so dual palm buttons are provided to ensure that hands are away from the rams. Tooling is generally simple, often just a short length of large-diameter piping to contain the media. The media is blown from the part interior with normal shop air pressure, then the parts are rinsed in a solution to remove media, abrasive, and any remaining oils. The abrasive tends to migrate from the work area on shoes, clothes, and arms, as well as on parts, so effort should be made to capture the abrasive in an area that will not harm precision surfaces or machines.

Bonded Abrasive Deburring

Sanding equipment employs conveyor belts to draw parts under the sanding heads (Figure 29-5). With soft rollers behind the belt, the belt can reach into large holes and cavities to not only remove burrs, but also apply a small edge break. Abrasive belts can provide fast rough finishes or exceptionally fine finishes. A wide range of abrasive grit sizes and types provide many variations.

As with some of the other processes discussed, it is important to consider the bonded abrasive process, along with manual deburring, for quickly removing the majority of burrs and accommodating what the machine cannot do. Dust from metals and bonded abrasive material must be removed for health, fire, and safety reasons.

Brush Deburring

Brushing is still one of the least expensive processes for external edges. Semiautomated machines look like sanding machines, and several machines combine brushing with sanding belts to finish edges. The latter approach provides much more aggressive and complete deburring. Several vendors provide standard machines, and special machines can be produced from standard elements relatively inexpensively. Deburring time on these machines may be as short as 1 minute per part, and there are few associated environmental, health, or safety issues, as long as protective guards are in place to protect workers from hand and arm pinches and thrown parts or brush fibers. Care also must be taken that the machine does not grab clothing.

Thermal Energy Method Deburring

The thermal energy method (TEM) is advertised as the fastest deburring process because burr removal is done in a fraction of a second. In practice, a load of parts is loosely placed in an open container, which is rotated under a hydraulic ram that closes the container

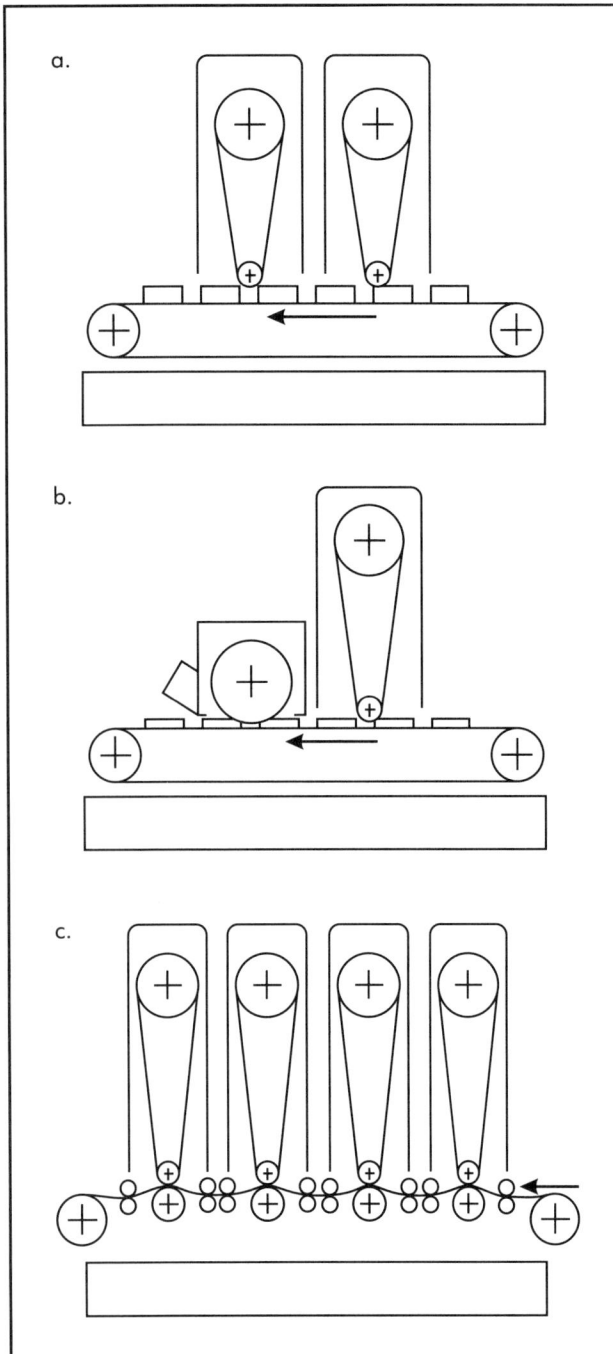

Figure 29-5. Multiple-head, conveyorized abrasive-belt machines for one-pass edge finishing.

produces an oxide or other surface coating that generally must be removed in a solution or by brushing in a separate operation. Many parts can be deburred in a single cycle, and part orientation is not critical for most parts.

Even though TEM deburring uses combustible gases and surfaces must be sealed to prevent the ignition from escaping, the process is widely used in mass-production facilities around the world. The process requires up-front facilities engineering to meet safety codes, but day-to-day operation is straightforward (Schäfer 1975; Thilow et al. 1992).

Deburring on the Machine that Made the Part

Today, many companies use on-machine deburring. Multi-axis machining centers can handle stiff, abrasive-filled brushes that deburr and radius edges. One company provides a robotic-compliant tool that fits in the tool changer, just like any other cutting tool (Gillespie 2000a). This pneumatically operated spindle follows the edge of the part, regardless of whether the edge is exactly where the tool is positioned. The tool allows rapid chamfering without extra machines, providing a relatively consistent deburring operation on any machine. Another machine provides high-pressure water jets in the machine enclosure for removing thin burrs immediately after or during machining.

AUTOMATION TO REPLACE MANUAL DEBURRING

Any deburring process can be automated. In practice, however, only a few are widely automated, including:

- vibratory finishing;
- centrifugal barrel finishing;
- roll-flow finishing;
- abrasive-jet deburring;
- electrochemical deburring (ECD);
- abrasive-flow deburring;
- brush deburring;
- bonded-abrasive deburring (sanding);
- thermal energy method (TEM) deburring;
- cryogenic abrasive-jet deburring;
- electropolish deburring, and
- robotic deburring.

Fully automated lines are more expensive than stand-alone machines that an operator loads and unloads. Large-volume companies that make the same product for years can afford and need to use such capabilities.

Vibratory lines are attached to continuous conveyors in stamping facilities and could be in machining

in a sealed chamber (Figure 29-6). Then, combustible gases are introduced and ignited. The instantaneous flash-front heats the burrs to a temperature that causes them to ignite. The burrs ignite because they have a thin cross section and can reach ignition temperature much faster than the part. The flash-front

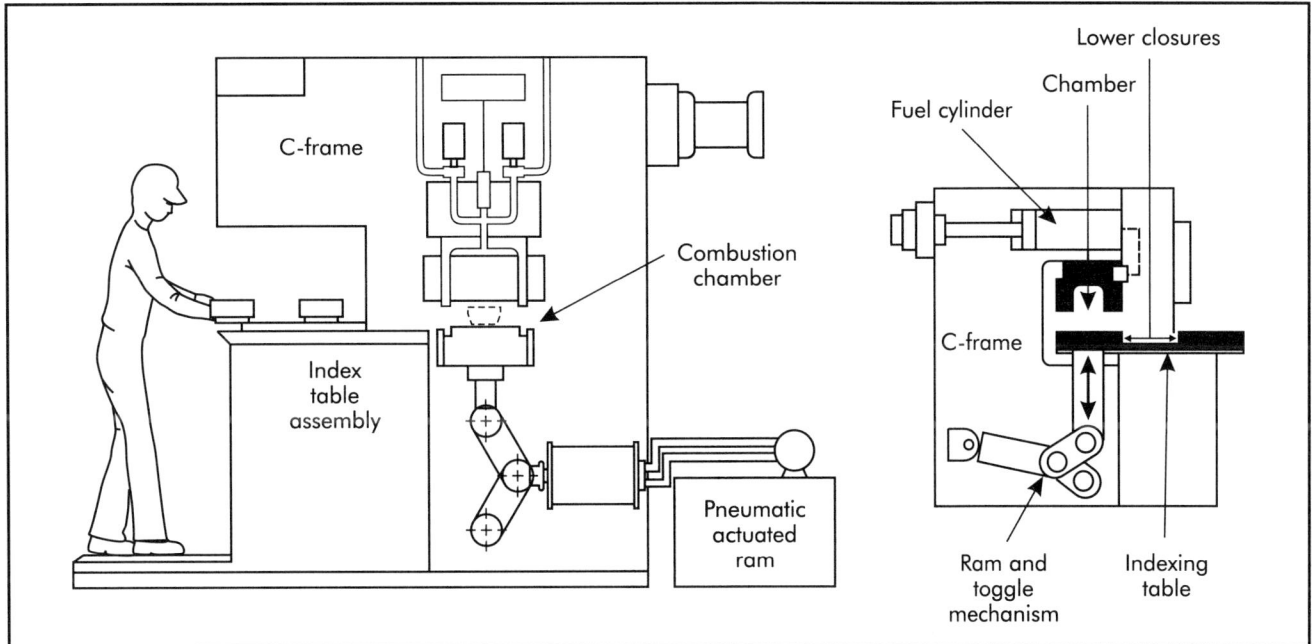

Figure 29-6. TEM machine with C-frame. (Courtesy SurfTran)

facilities as well. They can run continuously without supervision. While centrifugal barrel finishing generally uses closed drums to finish parts, fully automated units are also available (Figure 29-7).

Roll-flow and spindle-finishing machines could be automated, but generally are not. Blasting machines have been automated for decades for metal, plastic, and rubber products. Even relatively delicate products, such as semiconductors, have been blasted to remove flash from the carrier. Plastic and rubber deflashing often relies on a cryogenic blast of gas to cool the flash to a brittle point so that it easily breaks free. Selecting the correct temperature and time requires some expertise and the process cannot be used on all plastics, since many will craze or crack before the flash comes off.

Generally, ECD had not been considered a process that can be fully automated. That changed in the late 1990s, when a company introduced a continuous line to electrochemically deburr a specific part. The process, however, is still not easy to automate.

A version of the abrasive-flow deburring process (Figure 29-8) can be automated with the addition of a part-loading device.

Electropolish deburring is often automated. It produces highly reflective surface finishes while removing small burrs. Hypodermic needles have been deburred in continuous lines on this equipment. Electropolishing removes stock from the part surfaces,

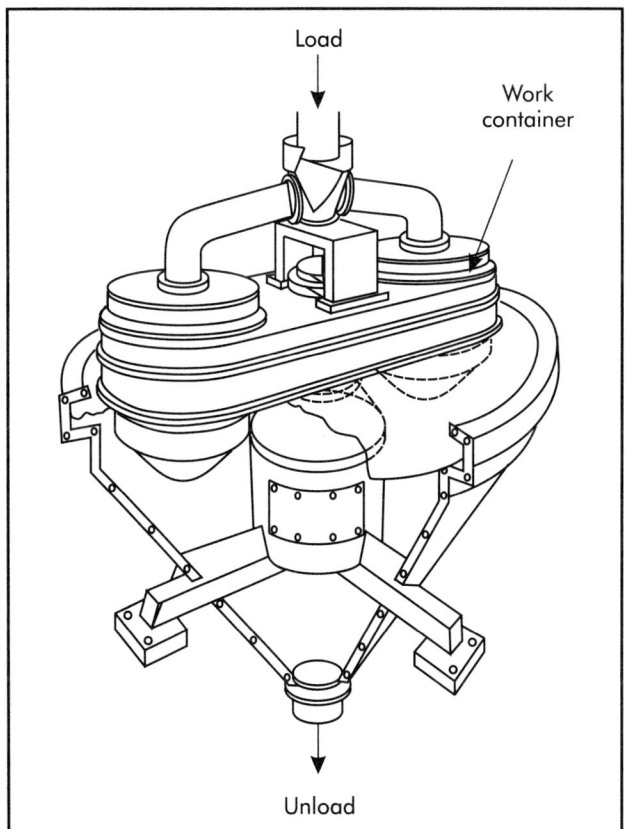

Figure 29-7. Automatic centrifugal barrel finishing machine. (Courtesy Harper Division of the Charles. G. Allen Co.)

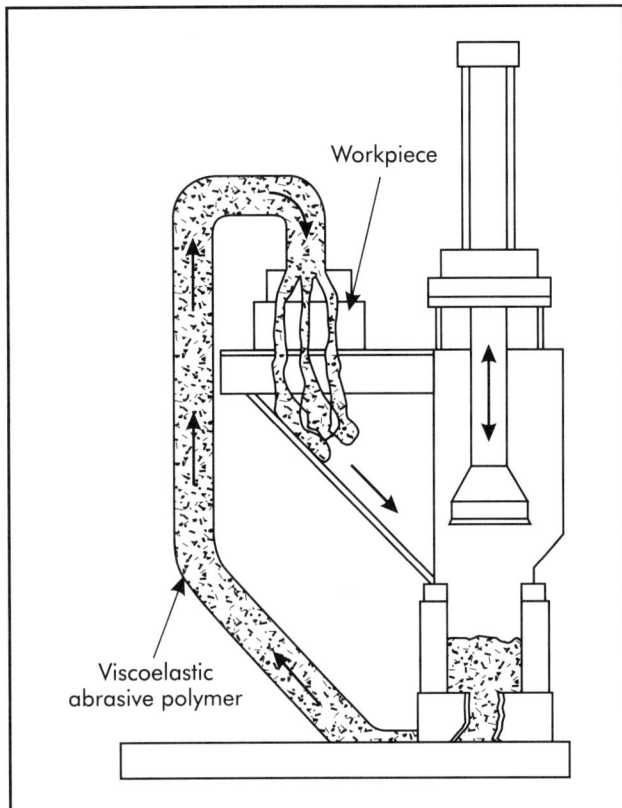

Figure 29-8. Schematic of one-way flow, abrasive-flow machine. (Courtesy Extrude Hone Corporation)

as well as edges, so it is not widely used for machined parts. Most electroplating shops have an electropolishing capability since the anode and cathode can be switched to remove metal from the part instead of removing metal from the electrode to plate onto the part. Tens of thousands of parts can be deburred in a day using this equipment. Figure 29-9 illustrates several equipment types.

Environmental, health, and safety considerations are the same for electropolish deburring as for any electroplating operation. They include hydrogen gas buildup, toxic fumes, hot solutions, electrical currents, and wet floors. Waste disposal can be an issue for the same reasons that it is for plating operations.

Robotic deburring encompasses many mechanical finishing processes—brushing, cutting, sanding, blasting, etc. (Gillespie 1999). It is estimated that there are 1,500 or more robots in the world performing deburring, deflashing, or allied operations (Gillespie 2000b). Considerable time is required to set up robots for new applications, and using them requires maintenance and engineering staff support, which many other deburring processes do not require. Robots are used for aircraft engine parts, expensive automotive parts, and deflashing hot plastic-molded parts. Numerous studies of robot usage have been reported, but few written reports of actual use or economics are available (Gillespie and Repnikova 2001).

Robotic deburring can prevent ergonomic injuries. The equipment is versatile, so it can be reapplied to other parts and other tasks. Robot safety is a field in itself. A number of safety approaches are in use to ensure that people do not inadvertently walk into a swinging robot arm and that they are out of the way if the robot suddenly flails its arms or parts.

REFERENCES

Gillespie, LaRoux K. 1978. *Advances in Deburring*. Dearborn, MI: Society of Manufacturing Engineers.

———. 1987. *Robotic Deburring Handbook*. Dearborn, MI: Society of Manufacturing Engineers.

———. 1999. *Deburring and Edge Finishing Handbook*. Dearborn, MI: Society of Manufacturing Engineers.

———. 2000a. *Guide to Deburring, Deflashing, and Trimming Equipment, Services and Supplies*. Kansas City, MO: Deburring Technology International, Inc.

———. 2000b. "State of the Art in Deburring and Edge Finishing in the U.S. in 2000." September. Kansas City, MO: Deburring Technology International.

Gillespie, LaRoux K., and Elena Repnikova. 2001. *Deburring: A 70 Year Bibliography*. Kansas City, MO: Deburring Technology International.

Rhoades, Lawrence J. 1981. *Cost Guide for Automatic Finishing Processes*. Dearborn, MI: Society of Manufacturing Engineers.

Schäfer, F. 1975. *Entgraten*, (in German). Mainz, Germany: Krausskopf-Verlag.

Thilow, Alfred P. and Klaus Berger, Helmut Prüller, Klaus Przyklenk, Freidrich Schäfer, Siegfried Pießlinger-Schweiger. 1992. *Entgraten-technik Entwicklungsstand und Problemlosungen*, (in German). Stuttgart, Germany: Expert Verlag.

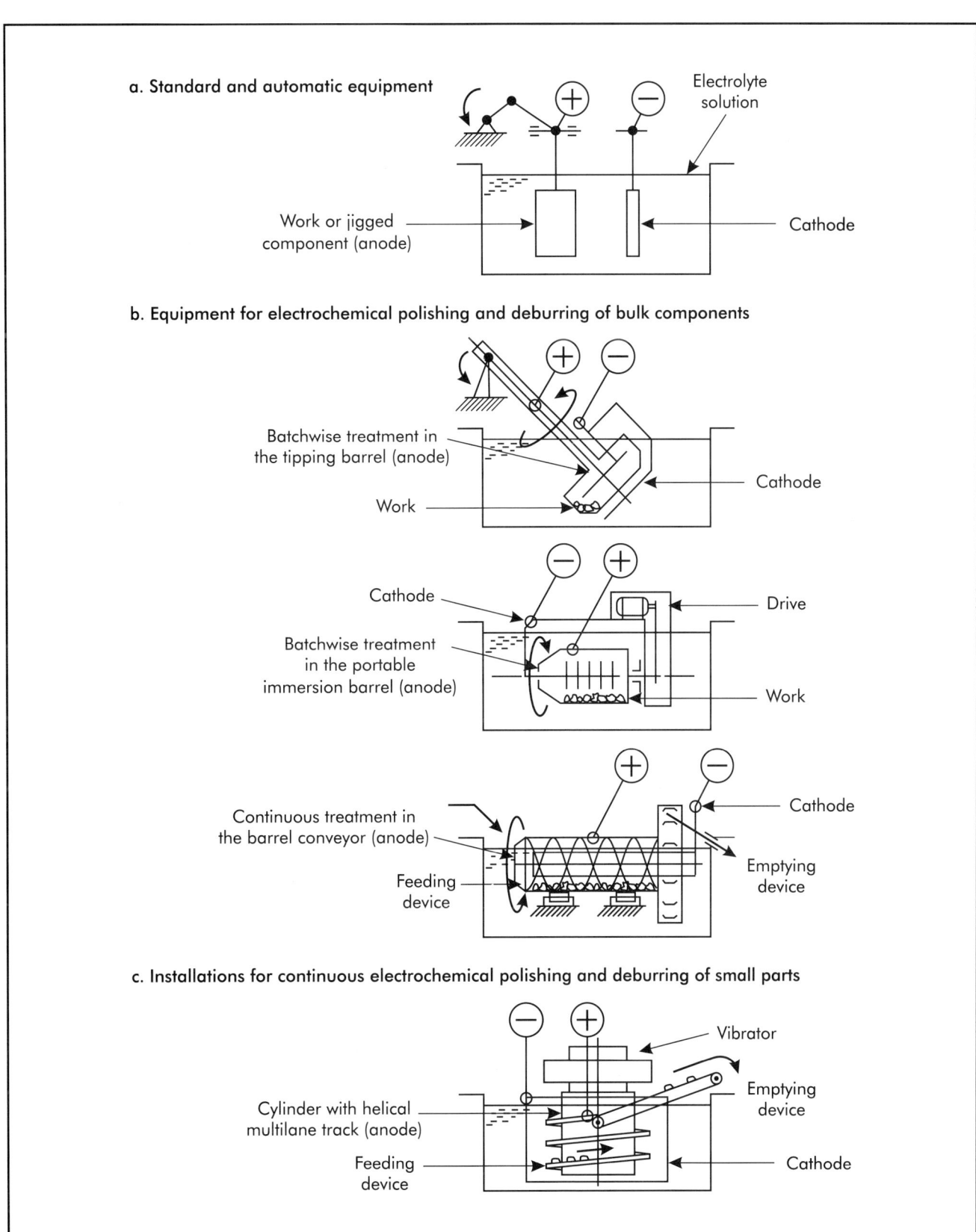

Figure 29-9. Electropolish deburring units for high-volume production of small parts. (Courtesy Polygrat Corporation)

Environmental, Health, and Safety Issues 30

Environmental, health, and safety (EHS) issues have been discussed in almost every chapter of this book. This chapter provides an overview that may make it easier to discern which approaches offer reduced risks. Manufacturers of deburring tools should have specific EHS insight that can help tailor local preventive programs. With some 2,000 manufacturers and suppliers, there is a wealth of knowledge that can be used (Gillespie 2000).

Table 30-1 cites the most common EHS issues, but others can exist because of part size, part design, materials used, or production rate. Table entries are based on the assumption that workers have received adequate training and communicate clearly with management and plant staff. Training programs must be designed with this aspect of operations in mind (Gillespie 1979). As Table 30-1 indicates, principle environmental concerns include pollution and dust. Major health concerns include dust, fumes, and carcinogens. Safety concerns include cuts and burns, hearing and eye impairment or loss, dangers associated with flying objects, and ergonomics issues.

EHS issues are a major concern; yet, it is important to remember that literally tens of thousands of workers safely perform their jobs every day using manual deburring and edge-finishing techniques. Proper training, safety auditing, and diligent reviews of accidents can ensure safe operations. It is also important to remember that mechanized deburring presents unique EHS issues. In many instances, the potential for permanent injury is larger for mechanized operations than for well-controlled manual operations (Gillespie 1999).

KNIVES

Knives are inherently dangerous because they are designed to cut. In normal use, deburring knives do not present a safety problem. Any time a knife is used in a fist position, however, there is an ever-present risk of cutting oneself or nearby workers. For this reason, knives should only be used in the fist position for special circumstances. Knives should never be used for anything except their intended purpose, and they should not be used by workers who have not received training in their effective and safe use. Knife cuts are more likely to occur if the worker is required to apply heavy force, such as when the knife blade is dull or the burr is thicker than normal. The key is to use sharp tools and minimize burr thickness. Chapter 7 describes many of the hazards associated with knives.

Thin scalpel and hobby knives are particularly sharp and dangerous. Swivel-blade knives are generally a safer design than scalpel-style knives. Swivel tools allow the operator to make a simple rotary motion to complete the deburring. Such tools do not require a fixed wrist position, which might irritate repetitive motion injuries. Surgical and hobby knives are widely used to remove fine flash from molded rubber and plastic parts. They are not well-suited for metal parts or burrs.

Germ-related hazards also exist in the workplace and can be introduced via a knife cut. A part carries the germs of every person who has handled it since it was last cleaned. At the end of the day, most deburring workers' hands are blackened and germ-laden from handling dirty parts and metal. No study has been performed, to this author's knowledge, to evaluate risk potential, but it is a concern that companies should consider. To reduce the possibility of contamination, workers should wash their hands frequently or wear gloves.

Some people are more sensitive than others to common materials. For example, if beryllium copper falls into a wound on some workers, the wound, reportedly,

Table 30-1. Common deburring and deflashing environmental, health, and safety issues

Process/Product Entity	Environmental Issue	Health Issue	Safety Issue
Knives			Cut hands Cut arms
Scrapers		Heavy forces from scraper use may cause repetitive motion injury	Cut hands Cut arms
Motors			Clothing caught in powered cutter may pull worker into machine Use speeds designated for shank size to prevent tool from bending and being thrown from chuck
Countersinks/chamfer tools			
Hand driven			
Motor driven			Thrown tool Cut hands from tool in live spindle
Rotary burs			
Hand driven			
Motor driven			Thrown tool Cut hands from tool in live spindle
Reamers		Hand reaming can cause ergonomic injuries because of the heavy torque required	Reamer grabs, forcing part edges into hand
Files		Continuous use of files can cause repetitive motion injuries, because of the forces and angles of arm used	Slip while thrusting forward and cut arm File tangs are sharp and will puncture the torso Filing on live spindle lathes is dangerous, because of potential to slip or lunge into machine and across sharp edges on part
Mounted points		Breathing metal dust or stone dust Dust from beryllium alloys can cause chronic beryllium disease (known as CBD or berylliosis)	Shattering stone particles lodge in eye
Hand stones		Repetitive motion injury is possible with heavy stoning	
Abrasive wood tools			
Abrasive cork tools			When used to finish glass edges, any slipping can cause severe glass cuts
Abrasive cotton tools			
Abrasive rubber tools			
Handheld			
Motor driven			Small particles in eye Punctures from pointed mandrel

Table 30-1. (continued)

Process/Product Entity	Environmental Issue	Health Issue	Safety Issue
Bonded abrasive tools	Metal or plastic dust may affect surrounding plants or people	Breathing dust from paper, glue, and metal can affect lungs Repetitive motion injury may be highest for these tools, since they require contact force and may need heavy motors	Sanding heat burns hands Abrasive chafes hands Abrasive particles in eye Parts thrown by belts or disks cut or injure body parts Dust is both a fire and explosion hazard and must be prevented
Brushes			Heat from rubbing can burn hands Wire bristles can chafe hands Abrasive particles or filaments can harm eyes Parts thrown by brush may cut or injure body parts
Ballizing			Pushing device can provide pinch point to injure hands or fingers Inadequate ball-part alignment can cause sudden flinging of item at worker
Special tools			
Handheld motors		Ergonomic damage to hands, wrists or arms Hearing damage from motor or cutting noise Workers may be allergic to some metals Dust from rotary tools can affect lungs	Cut body parts with tools in motor Body struck by tools thrown by motor
Bench motors		Dust from rotary tools can affect lungs	Body parts cut by tools in motor Body struck by tools thrown by motor
Hand-operated mechanized machines		Ergonomic damage to hands, wrists or arms Hearing damage from motor or cutting noise Dust from rotary tools can affect lungs	Body parts cut by tools in motor Body struck by tools thrown by motor
Support items		Microscope use may affect eyes temporarily or long-term Improper adjustment of seating and microscopes may cause back or leg pain	
Tools that beat down edges		Ergonomic damage to hands, wrists, or arms Hearing damage from motor or cutting noise	
Hot-wire tools		Fumes affect lungs Fumes may be carcinogenic	Hot tool or part burns worker Hot material impacts eye
Product issues		Germs from handling parts Metal dust or by-products from handling parts may affect lungs	Protecting workers from sharp edges when they must deburr many edges on a single large part may require special temporary cut protection until the majority of the part is finished
Training issues			Untrained workers are at high risk for safety incidents Include EHS issues early in training

may not heal well. Worse, if left under the skin, particles of beryllium, beryllium compounds, or alloys may make a person sensitive to beryllium. This could later result in beryllious, a serious health hazard. Some personnel have an allergic reaction to nickel or the nickel in stainless steel. Others suffer skin or blood allergies when such materials are introduced. Pregnant women may be more susceptible than other workers to metal allergies.

Extendable-length tools may help access hard-to-reach areas and prevent arm cuts and ergonomic damage. Most tools can accommodate an extender, if needed, and some have more ergonomic designs than others. Also, knurled handles may be easier to hold. Scrapers, in general, reduce the risk of cuts to hands or arms. Figure 7-15 illustrates some utility knives that are commonly used for plastics deflashing and trimming. These tools are generally unsuitable for metal burrs. Note that some have rounded leading edges to reduce the risk of damaging the product and preventing puncture-type wounds from pointed ends. Hook-shaped knives, such as those shown in Figure 7-16, are used in limited applications where the edge is drawn toward the operator. Generally, they are not used for metals but can be used for plastics and rubber. Ceramic-blade knives may not be as sharp as metal-blade knives although their life may be longer.

Before using a deburring knife, the user should investigate its edge under a microscope. It is too easy to grind deburring knives in a way that leaves edges ragged or chipped. In some cases, these tools are reground 1,000 at a time; thus, it is easy for workers to overlook those that are improperly ground. Improperly ground tools do not last. If a deburring knife has burrs, it will not successfully deburr parts and may cause injuries. Knife storage is also important. A styrofoam block (see Figure 30-1) safely stores sharp knife blades, yet leaves them readily accessible.

Relationship Between Knife and Part

When deburring with knives under a microscope, the worker should not hold the part down on the microscope base or table. It is important for both part quality and personal safety to hold the part above a solid surface. If necessary, wrists or arms may rest on the tabletop. When using the table to support the part, a sudden lunge made by the hand that is holding the knife will gouge the part or cut the other hand. By holding the part in the air, the grasping hand easily and quickly responds to sudden motions by the knife-wielding hand. Abrupt changes in burr sizes or location can easily cause minute lunges.

The simple, self-adhesive bandage materials described in Chapter 26 are designed to prevent cuts

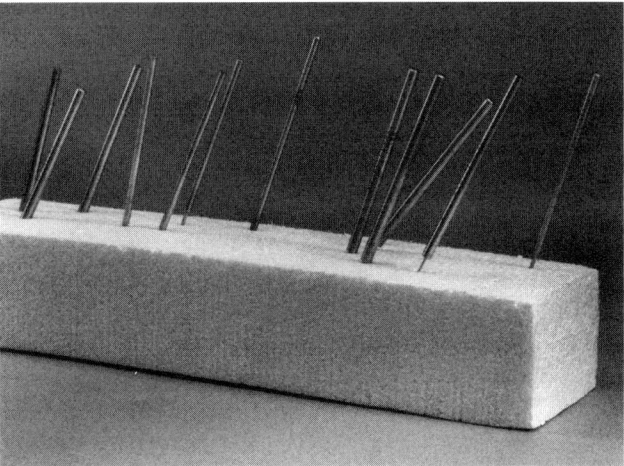

Figure 30-1. Styrofoam block to store knife blades helps prevent operator cuts. (Courtesy Honeywell)

when using knives. In general, these are relatively unobtrusive (far less than they appear to be) when compared to gloves, and should be considered for those doing knife work.

Techniques for better knife use include:

- selecting the right tool for the burr size, part geometry, and part material;
- moving dangerous tools away from the body when heavy force is required;
- not forcing the tool to remove excessive burrs;
- knowing the edge condition required by the specification;
- having an overview of all plant deburring tools;
- knowing which tools to use for the job that caused the accident (and that prompted the safety course);
- never using dull tools, and
- wearing appropriate safety equipment (such as safety glasses, steel-toe shoes, etc.).

FILES

File tangs, if inadequately covered, are particularly dangerous. Similarly, the use of files to deburr edges of parts rotating on lathes is a dangerous practice. Operators need to be standing in safe positions to prevent falling into edges or slipping on wet surfaces.

MOTORIZED OPERATIONS

Motorized operations often throw tools in the chuck or grab and hurl parts. Bonded-abrasive products in these motors also generate heavy dust that can be a fire safety concern, as well as a long-term health issue. Many sources provide video instruction on the safe use of grinding wheels and abrasives in the foundry industry. For foundry applications, the principle contact for safety instruction is the Non-Ferrous Founders' Society (Park Ridge, Illinois).

INJURY STATISTICS

The economic effect of industrial injuries is an area of some speculation, but it is estimated that industrial injuries cost U.S. industry tens of millions of dollars per year in lawsuits alone. Bureau of Labor Statistics (BLS) data indicates that occupational injuries and illnesses occurred at a rate of 8.9 per 100 full-time workers in 1991. By 1999, it had fallen to 6.3 per 100. Dust caused 2.2 injury cases per 100 workers in 1999, while respiratory illness had a rate of 16.5 per 100. Physical injuries occurred at a rate of 15 per 100 workers in 1999, and repeated trauma occurred 247 times per 100 workers in the same year (Bureau of Labor Statistics 1996). Table 30-2 provides some insight into accident types.

Operators, fabricators, and laborers in manufacturing sustained 418,200 injuries in 1994 that caused days away from work, representing 72% of all manufacturing-related injuries. Thirty-one thousand (5%) manufacturing injuries resulting in time away from work involved eyes; 177,300 (30%) involved hands, fingers, and wrists. In 1994, 38,600 injuries in manufacturing operations involved tools, instruments, and equipment, representing about 6.6% of all manufacturing lost-time injuries (Bureau of Labor Statistics 1996).

The cited Department of Labor report notes that some conditions, such as long-term, latent illnesses caused by exposure to carcinogens, are difficult to relate to the workplace and are not adequately recognized and reported. These long-term illnesses are believed to be understated in the survey.

TRAINING FOR SAFETY

A training program for deburring workers has not been cited in the literature; however, at least one has been developed. It is an interactive program entitled "Is Hand Deburring Inherently Unsafe?" and provides a 10-year history of plant accidents. It addresses the following questions:

- Why were there not more accidents?
- What kind of health and safety concerns does hand deburring cause?
- What can we do differently?

Plant Example

Safety became an issue for one particular plant when an untrained worker slipped and ran his arm

Table 30-2. The manner in which workers sustained injuries and illnesses resulting in days away from work, 1992–1994

Event or Worker's Exposure*	Distribution, 1994 Cases, %	Cases per 10,000 Workers		
		1992	1993	1994
Total	100	305	286	277
Bodily reaction and exertion	44	135	128	121
Overexertion	27	86	81	76
Bodily reaction (slip, twist, etc.)	11	34	32	30
Repetitive motion	4	12	12	12
Contact with objects and equipment	27	83	78	75
Struck by object	13	39	37	36
Struck against object	7	22	21	20
Caught in or compressed by equipment or objects	4	13	12	12
Falls	18	49	47	49
Fall on same level	12	31	31	33
Fall to lower level	5	15	14	14
Exposure to harmful substances or environments	5	15	14	14
Transportation incidents	4	9	9	10
Assaults and violent acts	1	4	3	3
Assaults by person(s)	1	3	3	3

* Total and broad event or exposure categories may not include data for classifications in addition to those shown separately. Note: Because of rounding and classifications not shown, percentages may not add to 100 (Bureau of Labor Statistics 1996).

down a long piece of sharp sheet metal. His sliced arm required 16 stitches. During the accident review, it became clear that the worker had not been trained in proper deburring-tool use and selection. It also became apparent that no one in the plant had been instructed formally for at least 10 years on safety issues for this type of work.

In most situations, hand deburring is not inherently unsafe. As proof, the instructor noted that for 30 years the same plant employed a blind worker to deburr parts and there were no safety events for that person. This worker had to take his time and study parts with his hands. To prevent cutting his hands, he knew he had to move them carefully and slowly over edges. Like other operators, he used power tools. He asked for help and insight when he needed it. He worked at a consistent pace and did not take unnecessary risks. The company appreciated his dedication and reliability. Safety was always a concern for management; but, in fact, it was never a problem for this blind worker.

The accident report also pointed out that 100,000 parts per month were passing through the deburring department without causing safety problems. The difference between now and then was that parts had been miniature and required little force and delicate motions. The accident had occurred in a large-part department staffed with untrained, new workers. The size of the part added to accident probability, and the deburring approach used by the operator added to the problem.

The nature of the part adds to problems. (Many long, sharp edges are more likely to cause cuts than many small holes that must be deburred.) The operator's deburring style (impulsive, unheeding of the advice of others) also added to the problem. In some situations, the pace to meet expected quotas also increased the potential for safety problems. Last, a lack of emphasis on safety in some plants increased the likelihood of accidents. In other words, by not making safety a prominent plant issue, accident occurrence may be higher than necessary. People respond to emphasis.

The training course previously discussed identifies several EHS issues: heavy parts; long, sharp edges; and use of heavy, handheld motors. It also identifies numerous movements and situations that can cause injuries (see Table 30-2) and discusses proper knife-deburring techniques, as well as the use of other tools. It is important to instruct workers that they can hurt themselves in many different ways. For example:

- you can slide your hand across sharp edges (cut);
- you can cut yourself with the burr tool;
- you can drop a heavy part on yourself (cut, bruise) or drop a burr knife on yourself (cut);
- brushes can throw parts (hit, cut, bruise);
- brushes can pull a worker into the brush (abrasions, bruises);
- the rotary sander can pass over a hand (abrasions, cuts);
- your hand can be thrown into a hard object (bruises, strains);
- you can slip on sanding debris left on the floor and fall (bruises, cuts, broken bones, eye damage);
- hot parts from a sander can cause burns;
- hot cleaner causes burns;
- large parts can be pushed into the stomach (bruises, internal damage, cuts);
- you can slip on greasy parts or the floor (bruises, cuts, broken bones, eye damage);
- when you push down you can tip a part and/or hand tools into the other hand (bruises, cuts);
- ergonomic problems can develop (hand, back, eyes);
- dust or chips can get in the eye;
- breathing dust can irritate allergies or cause lung or throat damage;
- you can be pulled into a lathe during on-machine deburring (cuts, abrasions, loss of hand or arm, death);
- your hand can be pulled into a sanding belt conveyor (cuts, abrasions, loss of hand or arm, death);
- dust in the air can cause an explosion or fire, and/or
- some metal dusts on equipment or in flues/ducts can cause fires.

In over 12 million hours of deburring, this plant had only three known burr-related injuries. It is an excellent record; however, with such numbers, one must wonder if all injuries were reported.

The course previously discussed became the official training course for workers in the deburring category. In all safety courses such as this, it is essential that management is made aware of any perceived or known health or safety issues. Holding such a course allows that feedback to occur.

REFERENCES

Bureau of Labor Statistics (BLS). 1996. "Characteristics of Injuries and Illnesses Resulting in Absences from Work, 1994." USDL 96–163, 1. Washington, DC: U.S. Department of Labor, Bureau of Labor Statistics.

Gillespie, LaRoux K. 1979. "A Training Manual for Precision Hand Deburring." Pt. 1. Report BDX-613-2245 (November), 69–90. Kansas City, MO: Bendix Corporation. (Available from NTIS.)

———. 1999. *Deburring and Edge Finishing Handbook.* Dearborn, MI: Society of Manufacturing Engineers.

———. 2000. *Guide to Deburring, Deflashing, and Trimming Equipment, Services and Supplies.* Kansas City, MO: Deburring Technology International.

Ergonomics 31

Ergonomics issues are a major concern and cost to most companies. There is more benefit to a safe ergonomic environment than just preventing injuries. Cost-effective ergonomics can result in the following benefits:

- Productivity can be increased when ergonomic issues are removed.
- Nonproductive time can be reduced when ergonomic issues are removed.
- Legal costs can be reduced when ergonomic issues are clearly prevented or removed.
- Medical and insurance costs can be reduced when ergonomic issues are removed.
- Social costs can be reduced when ergonomic issues are removed.

The first four elements can be quantified. Although the fifth element may not be quantifiable, it can be a very real cost.

ERGONOMIC DATA

Injuries are costly to companies and workers. Companies must pay for workers' lost time, and those with high injury rates pay higher unemployment rates than those with low injury rates. Workers, in turn, often receive smaller paychecks and suffer pain of incalculable cost. They may suffer from hidden, lifelong illnesses related to work environments or accidents.

Industry lacks data for ergonomic injuries relating specifically to deburring and edge finishing workers. Data prepared by the Occupational Health and Safety Administration (OSHA) and the Bureau of Labor Statistics (BLS) do not identify specific occupations; however, key areas for which some information exists include:

- lifting heavy objects,
- carpal tunnel syndrome, and
- repetitive motion injuries.

A 1994 BLS report indicates that carpal tunnel injuries represented about 3% of manufacturing injuries. It caused 17,600 cases of time away from work in manufacturing plants, with an average of 30 days away from work. The same data indicate that repetitive motion, such as grasping tools, represented 4% of recorded cases. It caused 49,300 cases of time away from the job, with a median of 18 days away from work. An estimated one in 1,000 industrial workers sustain this type of injury. See Table 30-2 for additional insight on this subject (Bureau of Labor Statistics 1996).

COST OF POOR ERGONOMICS

There are immediate and apparent costs of an ergonomic injury (Alexander 1986). These include the:

- cost of the injury, including medical treatment, lost work hours, and any make-up wages paid to supplement disability income;
- cost of long-term disability paid to an employee who is no longer able to work;
- cost of workers who are no longer able to work at a 100% level on a job, such as the inability to meet the pace or perform all tasks;
- cost of damaged equipment;
- cost of damaged or lost products;
- cost of time to get production operating again after an accident or injury, and
- cost of time for review meetings and accident investigations and reports.

IMPROVING PRODUCTIVITY

Preventing ergonomic injuries can improve operator output. The operator will not have to stop and rest or slow down because of pain. Work slowdown is often hard to detect until it has gone on for some time.

Reducing Nonproductive Time

Resolving ergonomic issues can reduce absenteeism caused by sore or painful arms, fingers, or hands. Ergonomic problems can also cause high labor turnover. Often, this plant cost is ignored, although it takes many hours of administrative effort and recruiting costs to replace workers. Lack of qualified staff, in turn, causes plant shortages, which result in late deliveries or at least extra costs to accommodate the absence of people. The cost of untimely delivery is not easy to calculate; but, if lost orders occur, the impact is greater than ever accounted for by traditional cost accounting.

The time a worker spends away from the job site to seek medical attention takes hours away from productive work. Completing administrative reports on accidents and delays reduces administrative and managerial effectiveness and bottom-line profit.

Heat stress may cause health issues. If heat forces workers to reduce output while they recover their strength or "get their wind," productivity is reduced. Ergonomic solutions may involve cooling a room, increasing air flow, changing the work pace, or changing the nature of the work so that heat is not a factor. Cold temperatures, vibration, poor lighting, noise, and related issues also can reduce operator efficiency and, potentially, cause ergonomic injuries or illnesses.

Tasks that require high endurance place a load on the body that also can reduce efficiency. Reducing the need for physical effort, such as adding a sling that carries most of the load of a handheld motor, will provide increased output at almost no detectable cost. Reducing strength requirements allows more operators to perform the work and increases output.

An analysis of errors can improve productivity in some instances. For example, one book provides an excellent study of inspection errors that affected output and consequent plant costs. It involved many of the issues discussed here, including work area layout, lighting, and training. Ergonomic or human factors in plant studies are initiated for:

- analysis of accidents;
- analysis of hot, heavy jobs;
- analysis of potential heat-stress areas, and
- investigation of operational errors (Alexander 1986).

LEGAL LIABILITIES

Today, both real and perceived injuries too commonly result in lawsuits. These can be much more expensive than medical and insurance costs and, clearly, add to image problems. Violations of state and federal regulations may add more costs for evaluations, fines, and legal defense.

MEDICAL AND INSURANCE COSTS

Medical payments for ergonomic injuries, whether paid directly by companies or by insurance premiums, cost money. Each worker's compensation claim increases plant cost to the state agency that accommodates such claims. It is a direct cost that can be seen and tracked each year.

SOCIAL COSTS

The cost of reputation for workers who are in pain is real, but not easy to calculate. The social responsibility of companies to protect and care for their employees is a cost for which there is no dollar value. Some plants are known for their excellence in worker support, and they attract more quality workers because of it. Better workers and a happier workforce bring in more work.

RED FLAGS

The following situations signal potential ergonomic issues.

- Are there requests for transfers to other jobs, even jobs that pay less?
- Do people regularly refuse promotion to a particular job?
- Do complaints and comments come from an identifiable section of the work population (for example, females, older workers, short-statured individuals)?
- What is the absenteeism rate on a particular job relative to other jobs in the plant? In the promotional sequence? In the same pay rate? In nearby work areas?

Questions and information relative to strength and endurance include the following:

- What is the injury rate of one job relative to other jobs and the rest of the plant?
- Are the injuries of the same type (for example, lower back or upper arm)?
- Has the injury rate changed recently, or is there a trend in the data that shows increasing injuries?
- Has the work population changed?

- What do recent accident investigations reveal? Do they show trends that verify and corroborate verbal information?
- Is there a history of visits to the medical department that shows information similar to accident data?
- Do complaints come from an identifiable section of the work population (for example, females, older workers, short-statured individuals)?

Questions relative to human error include the following:

- Has the error rate changed?
- Do errors come from an identifiable group (for example, a specific age group, newer workers, or workers who wear glasses)?
- Are the jobs or job-related tasks difficult to learn?

Anthropometric (effect of human body dimensions) questions include the following:

- Do shorter workers have more problems than taller workers?
- Is there a performance difference associated with heavier people?
- Does a larger size, longer reach, or greater stature make a difference?

PREVENTING ERGONOMIC INJURIES

No known publication is dedicated to improving ergonomic issues in metal finishing. The basic principals, however, are publicized, and the techniques for solving such problems are categorized in several books. Tables 31-1 through 31-3 provide insight into the causes of and some solutions for preventing egonomic injuries. Figure 31-1 provides some simple solutions that relate to deburring tools.

CUMULATIVE TRAUMA DISORDERS

Cumulative trauma disorders (CTDs) are just that—injuries that result from repeated use of the hand or another body part. They are normally seen as carpal tunnel syndrome, tendonitis, bursitis, and tenosynovitis. People are not equally susceptible to these injuries; those working side-by-side will experience different responses to the same job stress. Nonjob activities contribute to cumulative disorders. Some disorders, in fact, are better known by their nonwork names, such as tennis elbow and trigger finger.

Causes

Because of a reasonably low incidence of CTDs (compared to total exposure), there are no absolute causes. Contributors, however, include large forces, repeated and severe bending, and low temperatures.

Large forces can contribute to injuries to delicate hand and wrist structures. Forces from scraping paint or using the palm of the hand to pound things will injure palm and wrist areas.

Extreme wrist movements are hazardous, as are highly repetitive movements. Forces on the wrist at the limits of flexion and extension are severe. Highly repetitive movements also harm hand and wrist joints. When extreme joint motions are combined with high repetitions and large forces, problems are sure to follow.

Low working temperatures are also believed to be a contributor, so working in a cold or wet environment contributes to CTD problems. These environments severely reduce blood flow, causing fatigue and inhibiting healing of any injuries.

For elbow and shoulder joints, extreme motions and repetitive movements are believed to be a strong contributing source of CTD. Extreme motions can come from fully extended elbows or from working with the arms overhead. Highly repetitive movements are required for a variety of tasks, mostly in assembly, where actions are frequently repeated.

Correction and Prevention

CTD injuries often heal with rest. Some injuries may require surgery. The more severe the case, the less likely the person will later be able to perform the troublesome job. A correction to a diagnosed CTD problem will be more effective when there are corresponding changes at the work site or to the job. The same types of improvements will also help prevent new injuries.

Carpal Tunnel Syndrome

Figure 31-2 illustrates the cross-section of a wrist. Most of the muscles that move the fingers are in the forearm. The fingers are connected to these muscles by tendons that run through narrow channels of the wrist bones. More specifically, the flexor tendons run through the carpal (the Latin word for wrist) tunnel. Synovial membranes lubricate tendon movements. Finger movements with a flexed or extended wrist cause the tendons to be displaced past and against the adjacent walls of the carpal tunnel (much like a belt sliding on a pulley). Contact between the tendons and adjacent surfaces irritates the synovial membranes, causing synovitis. This thickens the synovial membrane, which in turn compresses the median nerve (which also passes through the carpal tunnel). This compression is called carpal tunnel syndrome

Table 31-1. Human factors engineering checklist (Courtesy Prentice-Hall)

A positive response to any of the questions indicates a need for careful human factors engineering evaluation of the work situation.

Considerations	Questions	Check Box
General indicators of the need for an engineering evaluation of human factors	Is absenteeism on this task too high? Is turnover on this task too high? Is production efficiency on this task too low? Do employees complain frequently about this task? Is personnel assignment on this task limited by age, sex, or body size? Is training time for this task too long? Is product quality too low? Have there been too many accidents? Have there been too many visits to seek medical attention? Does this task result in too much material waste? Is there excessive equipment damage on this job? Does the worker make frequent mistakes? Is the operator frequently away from his or her workplace? Is this workplace used for more than one shift per day?	
Indicators of the need for workplace redesign	Does the work surface appear to be too high or too low for many operators? Do workers frequently sit on the front edge of their chair? Must the worker assume an unnatural or stretched position to see dials, gages, or parts of the work unit and to reach controls, materials, or parts of the work unit? Is the worker required to operate foot pedals while standing? Does the operation of foot pedals or knee switches prevent the worker from assuming a natural, comfortable posture? Are foot pedals too small to allow foot-position changes? Is a footrest necessary? Do workers frequently attempt to modify their work chair by adding cushions or pads? Are workers required to hold up their arms or hands without the assistance of armrests? Are dials and equipment controls difficult to operate or poorly labeled? Are the design and layout of equipment a hindrance to cleaning and maintenance activities? Does the workplace appear unnecessarily cluttered? Is the worker required to use a nonadjustable chair? Can the worker be relieved of static holding by providing clamps or support for the work units?	
Indicators of the need for task redesign	Is the worker required to lift and carry too much weight? Is the worker required to push or pull carts, boxes, rolls of material, etc., that involve large breakaway forces to get started? Is the worker required to push or pull carts and hand trucks up or down ramps and inclines? Does the task require the worker to apply pushing, pulling, lifting, or lowering forces while the body is bent, twisted, or stretched? Is the work pace rapid and not under the operator's control? Does the worker's heart rate exceed 120 beats/min during task performance? Do workers complain that fatigue allowances are insufficient? Does the task require that one motion pattern be repetitively performed at a high frequency? Does the task require the frequent use or manipulation of hand tools? Does the task require both hands and feet to continually operate controls or manipulate the work unit? Is the worker required to maintain the same posture, either sitting or standing, all the time?	

Table 31-1. (continued)

Considerations	Questions	Check Box
	Is the worker required to mentally keep track of a changing work situation, particularly as it concerns the status of several machines? Is the rate at which the worker must process information likely to exceed his or her capability? Does the worker have insufficient time to sense and respond to information signals that occur simultaneously from different machines?	
Indicators of the need for special consideration of the working environment	Does process noise interfere with the reception of speech or auditory signals? Is process noise of an irritating nature, so that it interferes with the worker's attention to his or her task? Is process noise loud enough to cause hearing loss? Do the work tasks contain significant visual components, thus necessitating careful attention to lighting? Do the worker's eyes have to move periodically from dark to light areas? Are there any direct or reflected glare sources in the work area? Do lights shine on moving machinery in such a manner as to produce stroboscopic effects or distracting flashes? Does task background coloration interfere with the color codes or knobs, handles, or displays? Is the air temperature uncomfortably hot or cold? Is the relative humidity uncomfortably high? Are radiant heat sources located near an operator's workstation? Is the worker exposed to rapid thermal or visual environment changes? Do hand tools or process equipment vibrate the worker's hands, arms, or body? Does process dust settle on displays, making them difficult to see?	
Indicators of the need for restricted employee selection	Is process equipment designed so that only tall persons can reach it? Is process equipment designed so that tall persons must bend too much while performing their task? Is process equipment so close together that large persons cannot easily and safely attend it? Is the task's thermal environment such that only individuals in good physical condition and who are acclimatized to work in the heat can safely perform the task? Does the task require normal color vision? Does the task require heavy physical exertion, so that only persons in good physical condition can perform it? Are any aspects of the task unnecessarily awkward or difficult to perform if the worker becomes pregnant? Is the personal space under the worktable so restricted that only small persons can properly sit at the table and work? Does the task require normal hearing ability? Does the task require rapid decision-making and manual response to machine signals? Does the task require unusual eye-hand coordination?	
Indicators of the need for more effective supervisory control	Do workers appear to be uninformed of proper work postures? Are chairs properly adjusted? Do employees continue to work during scheduled breaks? Do employees appear to be uninformed of lifting, carrying, lowering, pushing, and pulling techniques? Do employees often fail to wear personal protective equipment? Do employees appear to be uninformed of the effects of noise and heat stress on their health? Do employees fail to drink sufficient liquid when they work in high-heat areas?	

(Alexander and Pulat 1985)

Table 31-2. Human factors opportunities checklist (Courtesy Prentice-Hall)

Issue	Questions	Check Box
Are there jobs in the work area that require a long training period or are difficult to learn?	Is the job mentally demanding? Is a high level of skill or experience required?	
Are there jobs with a high error rate or a high error consequence (process control)?	Are the errors from inattention? Are the errors from insufficient training or job knowledge? Is sufficient process control information available (inspection)? Is a high degree of concentration required? Is lighting sufficient for the task? Are the errors from insufficient training or job knowledge (packaging, labeling)? Is the product code or labeling information clear and easy to understand?	
Are there jobs that require a high level of motivation, alertness, or concentration?	Is automation a possibility, or is human control required? Is the consequence of an error serious?	
Are there jobs that are repetitive or boring?	Is the job machine paced? Does the job allow social interaction while working? Can the job be automated?	
Are there jobs that have a high accident or injury rate (especially hand and musculoskeletal injuries)?	Are the injuries similar in nature? Are there tasks that require significant strength? Do job tasks require awkward or uncomfortable body positions or postures? Is repeated lifting or lowering required? Are materials handled or lifted below the knees, above the chest, or away from the body?	
Are there jobs with a high absence, turnover, or job-transfer request rate?	Are there frequent absences from the work area for nonwork reasons? Is there a high rate of medical visits? Are real medical problems reported, or just minor complaints? Is the situation a general one, or does it occur mainly when certain types of work activity are scheduled? Is the problem age- or sex-related? Is the work difficult, tiring, or boring? Is the work area hot? Are worker complaints general, or related to specific problems?	
Are there jobs to which employees refuse promotion?	Are refusals age- or sex-related? Is the job difficult, or does it require above-average strength? Is employee perception of the job accurate?	
Are there jobs that workers complain a lot about?	Is the work area hot? Is the work area dirty or dusty? Is the work considered too hard?	
Are there jobs that involve a lot of manual material handling?	Is manual material handling a major part of the job, or a small part of it? Is mechanization possible? What is the frequency of handling? What is the length of the handling activity? What size, shape, and weight of material is involved? What lifting heights are involved? Can the material be held close to the body? Is there a high accident, injury, or medical visit rate for this job? Are mechanical assists required or needed?	
Are there jobs that require awkward postures or body positions?	Is the worker stooped much of the time? Does the work require body limbs to be held in the same position for long periods? What is the length of time and frequency of the activity? Are heavy tools used overhead?	

Table 31-2. (continued)

Issue	Questions	Check Box
Are there jobs that females or older workers find difficult to perform?	Is the work hot or dirty? Is the job physically demanding? Is physical strength required? Does the job appear to require an employee with certain physical characteristics?	
Are there jobs where environmental conditions, such as heat or noise, affect production or worker productivity?	Do workers complain about job conditions in general, or are there specific complaints? Is the problem one of comfort, or do work conditions present a health concern?	

(Alexander and Pulat 1985)

(Kvålseth 1983). Dequervain's disease is a kind of tenosynovitis associated with the thumb.

The following are solutions for reducing stress:

- reduce repetition (for example, use the left and right hand alternately, or alternate tasks among workers);
- reduce the force required (for example, sharpen knife blades or use machine instead of human power), and/or
- exert force with a more favorable wrist orientation like the handshake position instead of a contorted position, or by using a specialized tool, such as a redesigned handle (Tichauer and Gage 1977; Armstrong et al. 1982).

HAND TOOL DESIGN

There are 10 principles for the design and use of hand tools (Nemeth 1985). These principles help prevent injuries and enhance the performance and quality of workmanship. The principles discuss tool design and tool use, along with workstation and job designs. This is consistent with the systems-type approach followed in this book. Principles and brief comments are provided as follows.

- Maintain straight wrists. Bent wrists encourage carpal tunnel syndrome. Any wrist deviation is further aggravated by repetitive motions or large forces. Tools should be held and used in a neutral position.
- Avoid static muscle loading. Work should be done with the arm and shoulder in a normal position to avoid excessive fatigue. This is especially true when tool weights are large or the tool is used for extended periods of time. Counterbalancing tools is a common solution (Figure 31-3).
- Avoid stress concentrations over the soft tissue of the hand. Pressure on these tissues can obstruct blood flow and nerve function. Figure 31-4 shows a tool that concentrates force on the hand. Such tools should be avoided. Figure 22-8 illustrated an ergonomic pliers design to prevent such forces. Figure 31-5 shows a simple pneumatic-powered nipper that has high cutting forces but requires little hand force.
- Reduce grip-force requirements. Grip forces can pressure the hands or result in tool slippage. Of special note is the distribution of force on the hand, which results in the same problems mentioned in avoiding stress concentrations.
- Maintain optimal grip span. The optimal power grip with the fingers, palm, and thumb should span 2.50–3.50 in. (63.5–88.9 mm). For circular tool handles, such as screwdrivers, the optimum power grip is 1.25–2.00 in. (31.8–50.8 mm). For fingertip use, the optimum precision grip is 0.30–0.60 in. (7.6–15.2 mm).
- Avoid sharp edges, pinch points, and awkward movements. Sharp edges cause blisters and pressure points. Pinch points can make a tool almost unusable. Awkward movements are easily found through observation or use. The movement required to open a tool is a common problem, especially when used repetitively.
- Avoid repetitive finger-trigger actions. Using a single finger to operate a trigger is harmful, especially with frequent use. Use of the thumb is preferred, since the thumb muscles are in the hand, not in the forearm (this avoids carpal tunnel syndrome). Other triggering mechanisms, such as proximity or pressure switches, are possible. Note, however, that the thumb can lock up and cause similar pain, but from different sources.
- Protect hands from heat and cold. Like soldering irons and other heat-generating tools, tools with motors can produce and transfer heat. Cold comes most often from air-powered tools with an exhaust near the hand grips.

Table 31-3. Considerations for human factors improvements (Courtesy Prentice-Hall)

Cues	Possible Causes and Solutions
High vision demands (color, fine discrimination, glare, contrast)	Change lighting; provide specialized lighting; test vision; use optical scanners
Twisting, rotating, or unstable body postures while lifting	Change workplace layout; automate; change weight of lifts
Lack of maintenance accessibility	Plan for maintenance and service work; provide access to equipment; make accessibility part of the equipment design
High data load	Change information rates; use better coding systems; look at and balance workload requirements; automate
Use of cathode ray tube (CRT) screens	Remove glare; design workplace to provide adjustability and remove constrained postures, work pace, and schedule; vary tasks
Make-shift workplaces	Provide a well-designed workplace; create adjustability for a variety of workers
Interference with verbal communication	Lower noise level; change mode of communication to visual or tactile; reduce need to communicate
Stretching to reach	Lower the items being reached for; raise the person's level; select physically capable people; lower shelves; workplace redesign
Constrained or rigid postures	Change job to provide for or require other postures; avoid overuse of equipment that requires a fixed point of contact between human and equipment, such as eye scopes and CRT screens
Workplace not adjustable	Provide adjustability for workplace (chair, table, footrest); vary task requirements
Excessive weight handled	Reduce weight; select physically capable people; use handling aids or equipment; automate
Cannot work double shifts	Reduce workload if job is excessively difficult; provide suitable work/recovery regimen
High job turndown rate, requests for transfers	Redesign for excessive workload or changing work population; improve reputation of job
Multiple-shift operation (concerns for shift work system, use of equipment by different people, communications, etc.)	Provide for adjustability; check operating methods for consistency; train; provide suitable shift schedule
High error rate	Improve coding systems; change sensory input methods; change pace; train
Population stereotypes; incorrect for control	Change to correct stereotype; color code; purchase ergonomically designed equipment
Legibility of labels and signs	Change size of letters, type styles, color of letters, signs, and background; improve viewing distance from label or sign
Sore hands and wrists	Change tool design and/or task requirements; reduce forces required; reduce multiple motions
Static muscle loading	Provide rest for load; use alignment pins; use quick connects; reduce weight; slide material instead of lifting
Different manufacturers' equipment causes lack of conformity	Place nonstandard machines in different areas; color code different machines; purchase standard equipment; use ergonomic criteria in purchasing decisions
High injuries	Reduce task overload; select physically capable people; improve workplace design
Long training time	Poor methods; incorrect stereotypes; redesign complex tasks

(Alexander and Pulat 1985)

Hand Deburring: Increasing Shop Productivity

Chapter 31: Ergonomics

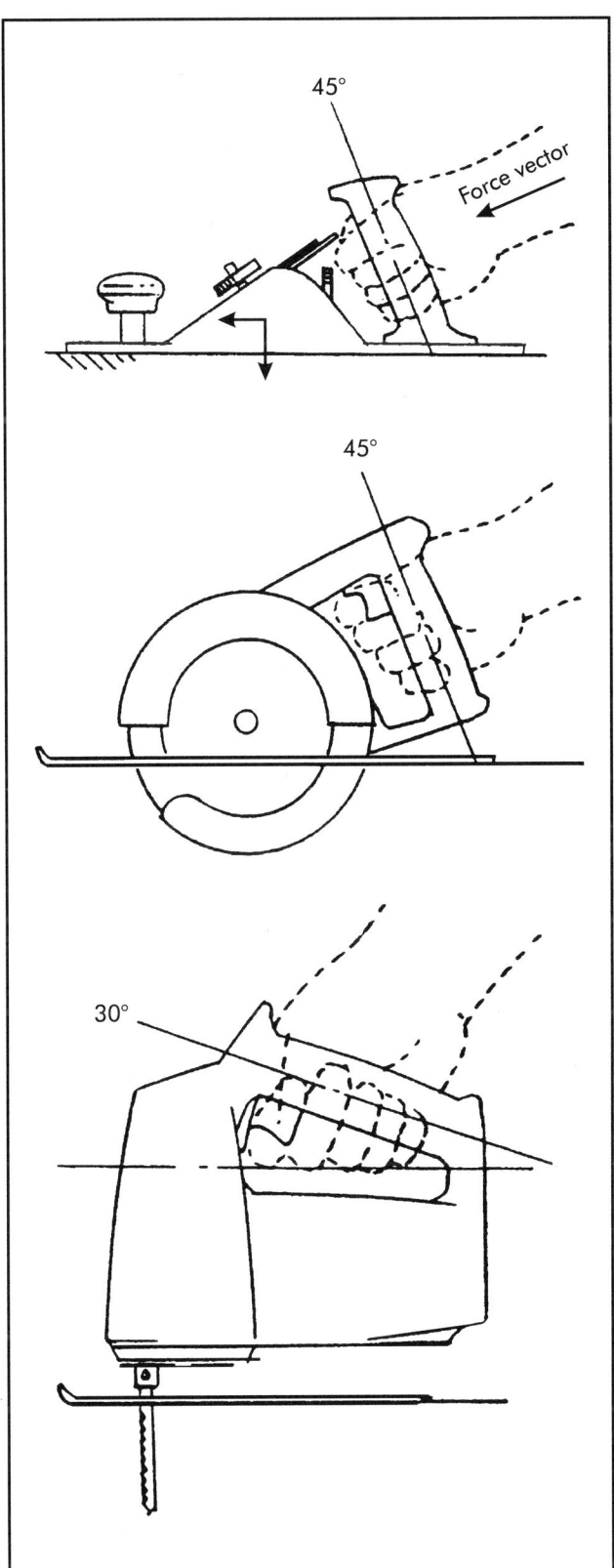

Figure 31-1. Handle angles for typical power tools (Woodson et al. 1981). (Courtesy McGraw-Hill)

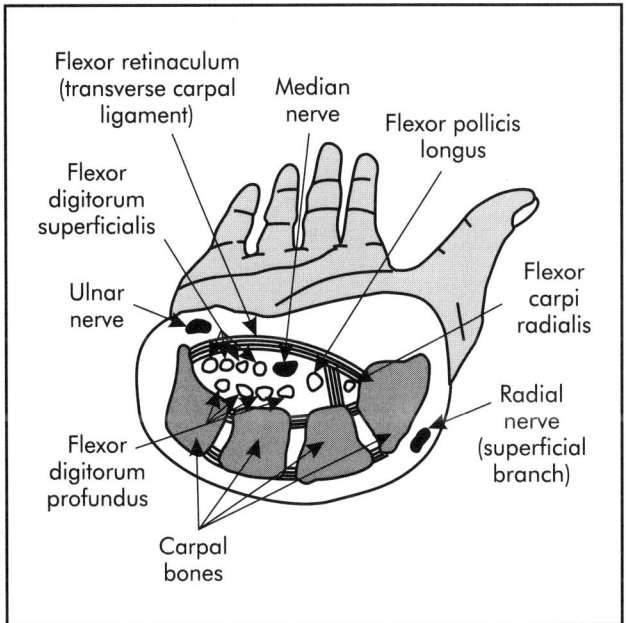

Figure 31-2. The median nerve and tendons pass through the carpal (wrist) tunnel (Nemeth 1985). (Courtesy Prentice-Hall)

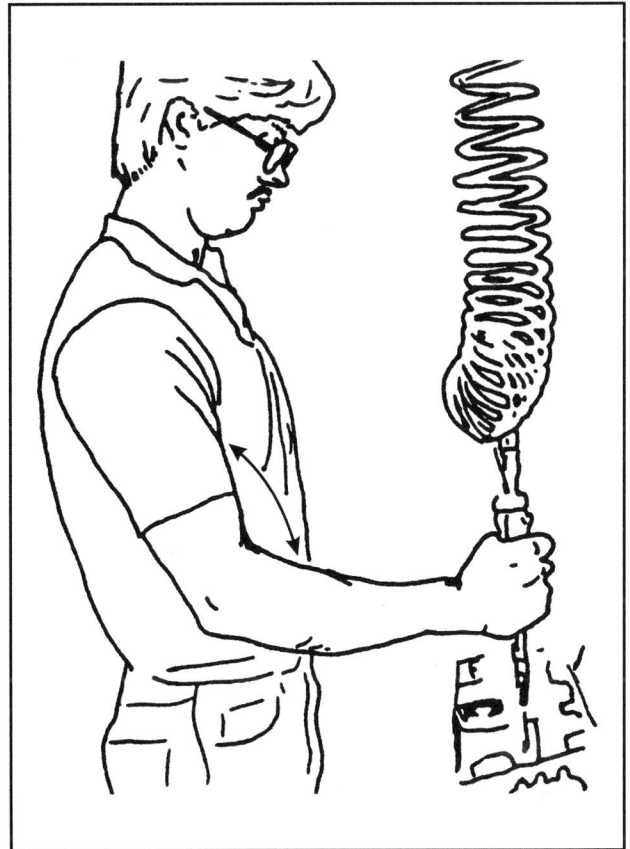

Figure 31-3. Counterbalanced tools avoid static loading (Alexander and Pulat 1985). (Courtesy Prentice-Hall)

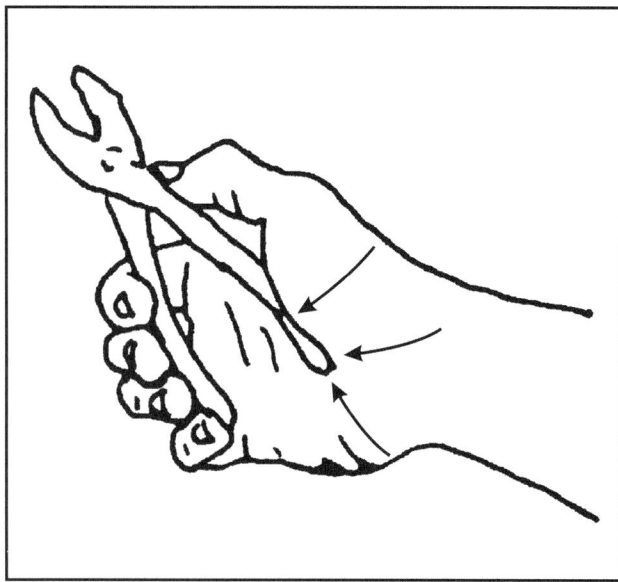

Figure 31-4. Avoid stress concentrations in tool handles (Alexander and Pulat 1985). (Courtesy Prentice-Hall)

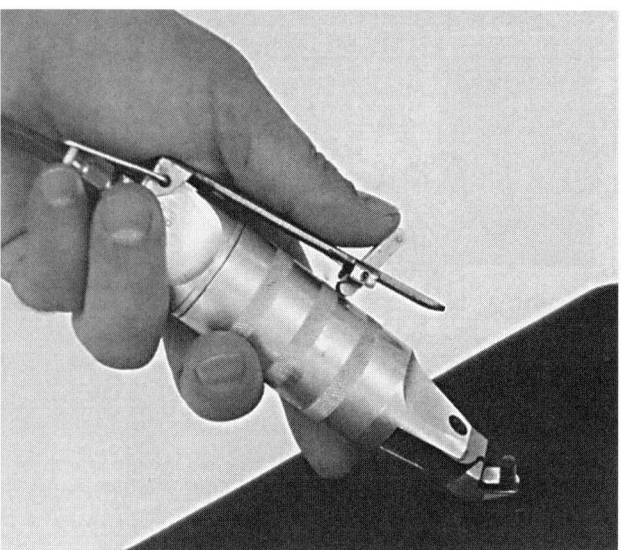

Figure 31-5. Ergonomic cutter replaces manual squeeze with pneumatic shear. (Courtesy Abbeon Cal)

- Avoid excessive vibration. Vibrations cause Reynaud's syndrome, or dead fingers. The best solution is damping or isolating the vibration. Rotating jobs is a possible way to limit exposure.
- Use gloves that fit. Gloves that are too big or thick reduce tool control because they reduce an individual's strength and dexterity. The most economical remedies are using gloves that are the correct weight for the job and stocking gloves in several sizes. Figure 31-1 illustrates several key points for hand tools that can be conceptually expanded to apply to deburring tools (Woodson et al. 1981).

BACK PAIN

Some deburring and deflashing tasks result in undue back pain. There are several simple questions to ask to determine if analysis is needed for this aspect of ergonomics (Table 31-4).

HIGH STRESS

For hot, hard jobs, leaders in ergonomics suggest maintaining a heart rate below those recommended in Table 31-5. Although this is not an absolute measure, it provides a direct, simple means of measuring job stress. Before researching in this area or making changes based on heart rate, employers are advised to check with local medical authorities to ascertain if other factors also should be considered in high-stress jobs.

Preventing excessive stress can take four approaches:

- Reduce the workload peaks (space the workload).
- Reduce the workload total for the shift (have the workers perform other kinds of needed work to reduce the total load).
- Change the ratio of the muscle groups used that caused the peak load (fingers, arms, wrists, torso, legs all may have a role in deburring).
- Use a more appropriate work/recovery cycle (more frequent or longer breaks).

Preventing heat stress can be accomplished by (note that clothing, age, gender, fitness, acclimatization, humidity, wind, airflow, and other factors affect heat stress):

- reducing the amount of heat generated,
- reducing the temperature of the work area, and
- changing the rate of heat generated.

ILLUMINATION

Little research exists for lighting levels in deburring. In one situation, half the workers in a facility performing precision microscopic deburring demanded a very bright room to ensure good quality. The other half was quite vocal about its need of darkness to maximize the effectiveness of microscope lights. Dim rooms with low light and no supporting lights can be a source of eyestrain for some workers. Work-area lighting requirements to ensure the removal of minute burrs only 0.0001 in. (0.025 mm) tall on some parts are up to 500 fc (5,382 lx); whereas, deflashing large castings may only require 30 fc (323 lx).

Table 31-4. Back pain checklist (Courtesy Prentice-Hall)

Issues	Considerations	Check Box
Does your back hurt as you work or at the end of the day?		
Have other people at work had back problems?		
Does your job entail	lifting from high spots? lifting from low spots? pulling or pushing loads? leaning over excessively? twisting or reaching motions? working with a bent neck or spine? standing still too long?	
Is the height of your workbench or the equipment you are operating correct for the work you are doing? Is the chair or stool, if you have one, correct for what you are doing?		
Are there ways to change your job to reduce the pressures on your back?		

(Alexander 1986)

Table 31-5. Maximal heart rate averages published by 10 American and European investigators

Age, Years	Target Heart Rate (50–75%) Zone, Beats per Minute	Average Maximum Heart Rate (100%), Beats per Minute
20	100–150	200
25	98–146	195
30	95–142	190
35	93–138	185
40	90–135	180
45	88–131	175
50	85–127	170
55	83–123	165
60	80–120	160
65	78–116	155
70	75–113	150

(Reproduced with permission of American Heart Association World Wide Web site [www.americanheart.org]. Copyright © 2000, American Heart Association)

Chapter 34 describes several lighting issues associated with burr inspection. In general, however, the following considerations need to be made:

- prevent glare;
- use contrast where it helps; remove it when it detracts, and
- work in shadowless areas.

These objectives can be attained in several ways, including:

- shielding or relocating light sources;
- changing the task to require less fine visual perception;
- changing the work pace so that eyes quickly adjust to lighting changes. Eyes will not readily adjust to the speed of the assembly line if the lighting is too bright, too low, or constantly changing, and
- changing the amount of light that reaches the area.

NOISE

Noise can have three detrimental effects: hearing loss; interference with communications; and annoyance, which result in lower performance (Alexander 1986).

Permanent hearing loss results from higher sound pressures (decibel levels) during high-noise exposure. The frequency of the noise is important, since high-frequency sound is more harmful than low-frequency. The duration of exposure contributes to hearing loss. These factors have been incorporated into a set of limits of permissible noise exposures for industrial jobs. The standards are time-based and weighted for intensity (decibel level) and frequency (the weighting factor for dBA). Table 31-6 provides the standard.

Since susceptibility varies among individuals, there is some controversy over the standard. The classical trade-off for ergonomic limits is present here—the percentage of people who sustain damage at different

Table 31-6. Noise exposure limits

Time Duration per Day (hr)	Sound Level (dBA)
8	90
6	92
4	95
3	97
2	100
1.5	102
1	105
0.75	107
0.50	110
0.25	115 max

(OSHA 1971)

exposures is weighed against the overall costs to industry associated with lower exposure levels. The levels are a compromise for both.

The decibel scale is logarithmic, with significant increases in the intensity of the noise only slightly changing the decibel reading at higher levels. Some representative values of the A-weighted frequency, dBA, are shown in Figure 31-6. Table 31-7 illustrates the impact of noise on communications.

Intermittent noise causes less hearing loss than continuous noise because hearing mechanisms have a chance to recover before each new exposure. Similarly, impulsive noise has a different (usually reduced) effect on hearing loss at the same dBA level. Limits are available for both types of noise, but they are complicated to apply and exceed the scope of this book. Several resources for additional reading are provided in the bibliography section (Harris 1957; Kryter, 1970).

SUMMARY

This chapter has highlighted major ergonomic issues that can affect deburring operations. Many detailed texts are available that can provide more in-depth guidance. (See the bibliography at the end of this chapter.)

In manual deburring, people are the most important asset. They deserve the best care a company can provide. With care and concern for workers, deburring and deflashing can be performed daily with sustained output without ergonomic issues arising.

The principles are simple, but it can be hard to spot the beginning of a problem. Listening to workers and sponsoring ergonomic awareness training can help. Faculty and students of industrial engineering departments of major engineering schools can make suggestions. Trained professionals can help and are found

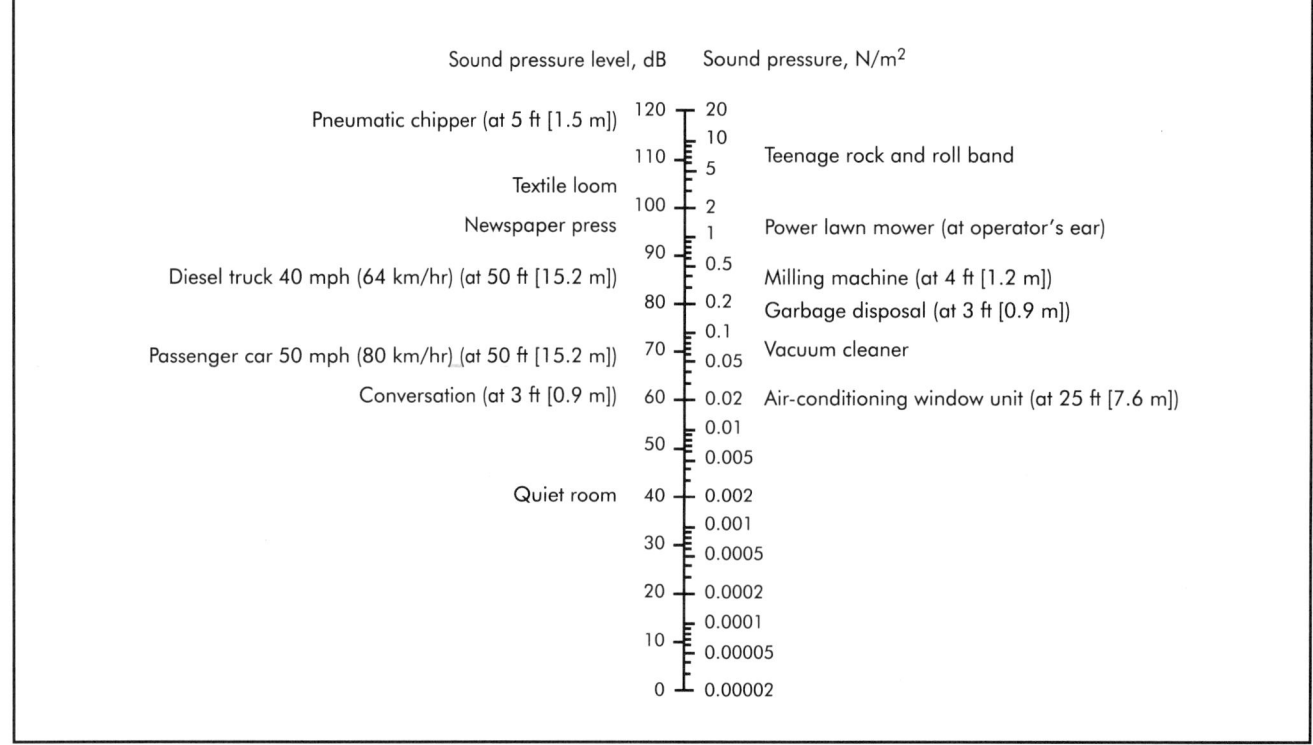

Figure 31-6. Sound levels (Anonymous 1973).

Table 31-7. Noise level (dB) versus ease of communication

Noise Level (dB)	Ease of Communication
30–40	Communication in normal voice satisfactory
40–50	Communication satisfactory in normal voice at 3–6 ft (0.9–1.8m); raised voice from 6–12 ft (1.8–3.7 m); telephone use satisfactory to slightly difficult
50–60	Communication satisfactory in normal voice at 1–2 ft (0.3–0.6 m); raised voice from 3–6 ft (0.9–1.8 m); telephone use slightly difficult
60–70	Communication with raised voice satisfactory at 1–2 ft (0.3–0.6 m); slightly difficult from 3–6 ft (0.9–1.8 m); telephone use difficult; earplugs and/or earmuffs can be worn with no adverse effect on communications
70–80	Communication slightly difficult with raised voice at 1–2 ft (0.3–0.6 m); slightly difficult with shouting from 3–6 ft (0.9–1.8 m); telephone use very difficult; earplugs and/or earmuffs can be worn with no adverse effect on communications
80–85	Communication slightly difficult with shouting from 1–2 ft (0.3–0.6 m); telephone use unsatisfactory; earplugs and/or earmuffs can be worn with no adverse effect on communications

(Alexander 1986)

in major cities and companies. Suppliers to large companies have access to many professionals. OSHA inspectors can highlight areas and approaches for improvement.

REFERENCES

Alexander, David C. 1986. *The Practice and Management of Industrial Ergonomics.* Englewood Cliffs, NJ: Prentice-Hall.

Alexander, David C. and Babur Mustafa Pulat, eds. 1985. *Industrial Ergonomics: A Practioner's Guide.* Norcross, GA: Institute of Industrial Engineers, Industrial Engineering and Management Press, pp. 57–63, 191.

Anonymous. 1973. "The Industrial Environment—Its Evaluation and Control." Washington, D.C.: U.S. Government Printing Office, p. 301.

Armstrong, T., J. A. Foulke, B. S. Joseph, and S. A. Goldstein. 1982. "Investigation of Cumulative Trauma Disorders in a Poultry Processing Plant." *American Industrial Hygiene Association Journal,* 43 (2), pp. 103–116.

Bureau of Labor Statistics (BLS). 1996. "Characteristics of Injuries and Illnesses Resulting in Absences from Work, 1994." USDL 96-163, 1. Washington, D.C.: U.S. Department of Labor, Bureau of Labor Statistics.

Harris, C.M. 1957. *Handbook of Noise Control.* New York: McGraw-Hill.

Kryter, K.D. 1970. *The Effect of Noise on Man.* New York: Academic Press, Inc.

Kvålseth, Tarald O. 1983. *Ergonomics of Workstation Design.* Boston, MA: Butterworths, pp. 50–63.

Nemeth, Susan E. 1985. "Hand Tool Design." In *Industrial Ergonomics: A Practitioner's Guide.* David C. Alexander and B. Mustafa Pulat, eds. Norcross, GA: Institute of Industrial Engineers, Industrial Engineering and Management Press.

OSHA. 1971. "Occupational Safety and Health Act." *Federal Register* 36:105 (May 29). Section 1910.95. Washington, D.C.: Occupational Safety and Health Administration (OSHA).

Tichauer, E. and H. Gage. 1977. "Ergonomic Principles Basic to Hand Tool Design." *American Industrial Hygiene Association Journal,* 38 (11), pp. 622–634.

Woodson, Wesley E., Barry Tillman, and Peggy Tillman. 1981. *Human Factors Design Handbook.* New York: McGraw-Hill, p. 665.

BIBLIOGRAPHY
Books

Burke, Mike. 1992. *Applied Ergonomics Handbook.* Boca Raton, FL: Lewis Publishers.

Corlett, E. N. and T. S. Clark. 1995. *The Ergonomics of Workspaces and Machines,* 2nd ed. Bristol, PA: Taylor & Francis.

Galer, Ian. 1987. *Ergonomics Handbook.* Boston, MA: Butterworths.

Monk, Tim and Simon Folkard. 1992. *Making Shiftwork Tolerable.* Bristol, PA: Taylor & Francis.

Osborne, David J. 1987. *Ergonomics at Work.* New York: John Wiley & Sons.

Pulat, Babur Mustafa and David C. Alexander. 1992. *Industrial Ergonomics: Case Studies.* New York: McGraw-Hill.

Putz-Anderson, Vern. 1988. *Cumulative Trauma Disorders: A Manual for Musculoskeletal Diseases of the Upper Limbs.* Bristol, PA: Taylor & Francis.

Rodgers, Suzanne H. 1986. *Ergonomic Design for People at Work,* Vol. 2 (Ergonomics Group, Health and Environmental Laboratories, Eastman Kodak Co.). New York: Van Nostrand Reinhold.

Periodicals and Journals

Applied Ergonomics

Ergonomics

Human Factors Journal

Industrial Engineering

Getting More from Your Operation 32

Each chapter of this book has provided insight to tools and techniques for improving manual deburring. This chapter wraps up that effort by providing an overview of management considerations, approaches that improve deburring effectiveness, data that helps define capabilities and limitations, and a number of practical hints.

REDUCING DEBURRING COSTS

Table 32-1 defines the approaches that were required to reduce hand-deburring costs by 50% at one facility. It is typical of the kinds of issues that, when improved, can reduce deburring costs at many plants. Clearly, the savings are easy to implement and do not require significant analysis—they are simply the result of paying more attention to deburring. These savings, when multiplied by 62 deburring workers, 173.5 hours per worker per month for 12 months, result in a savings of 77,450 hours per year, or in increased capacity of that amount.

Another improvement review found that the following factors influenced low net efficiency:

- in-process (at machine) deburring was not performed completely (so final deburring workers had to do more);
- considerable deburring and polishing was cosmetic—it was not driven by specifications (inspectors and deburring staff trying to be too good);
- microscope quality was provided on noncritical parts;
- burr size was excessive (machinists were not controlling size), and
- second- and third-shift workers were not performing at the same levels as first shift (motivation, tool availability, lack of effective guidance).

Another company's study found that the issues detailed in Table 32-2 were instrumental in its extensive burr-related rejects. As the table indicates, 72% of rejects were attributed to the following factors:

- untrained deburring workers;
- difference of opinion about part requirements;
- incomplete definition of a burr, and
- incomplete work instructions.

Tables 32-3 and 32-4 provide a relative measure of the impact many of these issues can have on the inefficiency of some deburring operations. These are actual estimates from one plant. The issues are common to many plants, and the tables' format allows readers to input their own estimates to determine how much better their operations could be by solving some of these issues. Table 32-5 contains a list of techniques for improving the process of manual deburring.

PROBLEM IDENTIFICATION

An amazing aspect of any management task is that it is easy to focus on the wrong problems. What are the real problems in deburring? Each company has its own, and problems vary in time and among departments, production shifts, and supervisors. Common problems, however, have been identified in Tables 32-1 and 32-2. Deburring problems, generally, originate from technical requirements, communications, and attitudes.

Solutions

Many solutions are obvious once the problems have been stated. The Pareto principle still applies: concentrating on the correct 15–20% of the solutions will eliminate 80% of the problems. Many solutions involve cost, at least initially. In many cases, formal research

Table 32-1. Improvement opportunities for deburring

Item Number	Time Saved (%)	Estimate of Approach
1	2	Production supervision encourages deburring workers to follow deburring procedures specified in existing work instructions.
2	6	Methods and process engineering specify deburring procedures expected to be followed in work instructions.
3	8	Product and process engineering identify and eliminate the finishing performed that is only cosmetic.
4	5	Production supervisor and machinists follow manufacturing sequence specified in work instructions.
5	5	Increase the use of centrifugal barrel and vibratory deburring on those parts now deburred entirely by hand.
6	6	Eliminate unnecessary in-process deburring specified in work instructions.
7	10	Production supervision continues efforts to increase production workers' personal efficiencies (distinct from engineering improvements).
8	10	Production supervision educates or reassigns production workers who have personal efficiencies less than 33%.
9	5	Utilize electrochemical deburring to its optimum.
10	3	Utilize electropolish deburring to its optimum.
	60 (total)*	

* Total time saved: 27% production supervision related; 33% engineering related.

is necessary, and, in most instances, communication or training is required. A variety of solutions might be necessary as indicated by the following study by one company:

- The use of TV (video) systems to reduce eye fatigue for deburr operators who must continually use magnification has been explored. There does not appear, at this time, to be any TV system on the market with adequate resolution for precision deburring purposes.
- The use of microscopes, both in practice and in comparing long-focal-length microscopes with the more conventional stereo-zoom microscopes, has been reviewed. The long-focal-length microscopes with coaxial lighting offer a number of advantages on parts with hard-to-reach recesses. In addition, high-quality optics provide distinctly better resolution than the standard optics now in use for deburring.
- Conventional microscope lighting will not allow a user to see chips and burrs in deep threaded holes (specifically 4–40 Unified National Coarse [UNC] or smaller). If coaxial lighting is not available, fiberoptic lighting is necessary.
- Use several newly available hand tools for general part applications, as well as for specific parts. These include miniature dental tools and motors; miniature pin vises; special-design knives; commercial, miniature deburring brushes; rubberized abrasive products; probes for detecting burrs; miniature, fast-cutting stones; fast-cutting, muslin-filled wheels; contact burnishers; and watchmaker tools. Dental tools have proven extremely beneficial on many small parts. More than 100 new tools have been evaluated.
- All parts must be cleaned before they are scrutinized to ensure that all burrs have been removed. Recent studies have shown that for scrubbing action or removal of particles, water-based solutions are something in the order of 10 times better than the old trichlorotrifluoroethane or trichloroethylene in ultrasonic cleaners.
- The edge-break capabilities of a number of commercially used deburring tools has been measured. More specifically, the type of edge radius

Table 32-2. Reasons for extensive burr-related rejects

Item Number	Reason	Accounts for __% of Rejects (est.)	Cure	Cure Improves Net Efficiency by __%
1	Deburring workers are untrained or unqualified	30	Provide the training required and require evidence of capability.	20
2	Required small edge breaks cannot be measured.	3	Develop inspection tool (short cylinders with known small chamfers radii) to measure small edge break (see Chapters 26 and 32).	5
3	Deburr worker cannot see burr until pointed out to him or her.	1	Do not use these individuals for final deburring.	
4	There are differences of opinion over part requirements.	15	Provide definitive instructions.	5
5	Work instructions are incomplete.	15	Update work instructions to include more detail on which edges, illustrations, and specific deburr techniques or use videos.	5
6	Inspectors are becoming more stringent or observant.	4	Coordinate requirement review with inspectors, engineering, and customer.	4
7	There is a difference of opinion as to what is and is not a burr.	12	Publish definitive guidelines (see Chapter 3).	5
8	New supervisors do not know burr requirements.	1	Train the supervisors.	
9	Centrifugal and vibratory equipment leaves some burrs.	4	As part of the mechanized deburr operation, scrutinize every lot and extend run time, if needed (validate that more time does not affect part dimensions, finishes, and so forth).	
10	Deburr workers are not verifying that they have finished all edges on each part.	12	Provide checklist or visual aid, or both, for each part, and add burr scrutinize operation by deburr workers—not machinists.	4
11	Parts are shipped into inspection without having been deburred.	1	Supervisor, operators, and inspection review route sheet sign-off before beginning inspection.	
12	Inspection gages or check fixtures generate burrs after final deburr has been completed.	1	Scrutinize parts for burrs after rework and gaging that occurs after final deburr. Resubmit parts to deburr operation to deburr burrs from gages.	
13	Parts that were purchased or made in other departments are rejected for burrs in the next department.	1	Provide closer review of purchased product procedures.	
14	Other (engineering, equipment, misc.)	5		

Table 32-3. Additional common efficiency issues in deburring

Issue	Solution	Reduces Net Efficiency by __%
There is no definition of what "final deburr" implies.	Provide specific instructions on edges; on illustrations, where needed; and use specific finishing approaches. Train operators on the difference of need and intent of "in-process," "rough," and "final deburr."	0.5
Many deburr tools are unavailable when needed.	Increase the minimum and maximum store requirement. Order larger quantities. Determine why there are not enough tools.	0.5
Deburr tools are of poor quality.	Have tool inspection monitor quality. Go to different vendors. Define real needs on tool drawings.	7
Deburr tools are inappropriate.	Train operators. Order correct tools. Assign engineer to the operation.	12.5
Workers do not know if machinist or deburr operator is supposed to remove burrs.	Provide specific work instructions that detail expectations. Make two or three kinds of deburr operations (rough, in-process, final).	
There is not a written guide to in-house practice.	Provide a guide, even if not highly sophisticated—write it down.	0.3
There is no visual aid for identifying available deburring tools.	Provide display box of tools and tool identification. Make a web page showing all tools.	
There is no accountability of which parts were deburred by which operator.	Develop sign-off sheet by part or lot.	5
Operators are not reading written work instructions.	Make sure operators can read. Require work instructions to be at workplace.	5
Operators are not following written work instructions.		5
Operators are striving for microscope quality on noncritical parts.		1
Lighting is unavailable to look into deep holes.	Provide lighting (see Chapters 33 and 34).	5
Workers are not striving to meet rate.		2.5
Workers are not striving to find better approaches (feedback, rather than motivation).		
Excessive polishing (cosmetic) is performed.		
Machining operations are performed out of sequence.		
Scrap parts are carried along with good parts, without identification (deburred without realizing they are scrap).		1
Added deburr costs result from machining rework.		5

Table 32-4. Common burr problems on miniature parts

Problem	Solution
Burrs on internal thread crests	Brush, ream, or tap to remove, then clean.
Burrs on external thread crests	Brush, sand, or tumble to remove, then clean.
Lead-in/lead-out thread damaged	Cut out flexible portion of lead-in/lead-out thread.
Raised metal around tapped hole	Countersink prior to tapping. After tapping, use countersink, abrasive-paper discs, or mounted points.
Burrs in bottom of blind tapped holes	Examine all blind holes carefully; cut out burrs with knives, bottoming taps, or sharp probes. Use vacuum probe to remove particles.
Minute burrs on gear teeth	Carefully examine at 10–20× magnification before submitting as complete.
Dirt on parts hides burrs	Clean until no dirt particles are evident.
Burr rolled over on face of screw machine parts	Examine with shadow-casting lighting at 10–20× magnification.
Burrs beat flush against tumbling diameters	Notify engineer to eliminate or change operation.
Too much edge break	Change deburring method; use edge break standards.
Missed half of all features on entire lot	Scrutinize parts prior to submitting to next operation.
Missed one or more parts completely in a lot	Scrutinize all parts 100% prior to submitting to next operation.
Varying level of quality in a lot	Allow only one person to deburr all parts within a lot to ensure consistency. (Remove the problem of allowing several persons to deburr a portion of parts in a lot.)
Small, secondary burrs remaining on part	Polish all edges with Cratex® or brushes to eliminate burrs or sharp edges produced by knives. Scrutinize at 10–20× magnification prior to submitting to next operation.
Parts have indentations or markings	Use soft-jaw collets (brass). Put tape on chuck jaws. Get special design holders.
Burrs or sharp edges remain in hard-to-reach areas	Ask engineer to provide tools for difficult areas.

that will result, under normal usage, from a wide variety of hand deburring tools, has been shown.
- Repeatability can be obtained from hand deburring sample parts, by both experienced and inexperienced personnel.
- A guide to hand deburring has been prepared and is being used to show production workers, engineers, and supervisors the types of equipment available for hand deburring (Gillespie 1975).
- An illustrated guide to 240 hand tools used for deburring at this plant has been prepared.
- Equipment needs have been previewed and some miniature barrel-tumbling units; a lap for flat parts; an electrochemical deburring machine; a high-power, miniature, ultrasonic cleaner; and a high-pressure miniature blaster have been purchased. Each of these purchases was the result, in part or totally, of evaluating machine capabilities and pinpointing potential problem areas.
- Special studies of contamination on ceramic parts from tumbling operations have been made, and suitable processes to accommodate these types of parts were developed.
- The capabilities of deburring processes for removing electrical discharge machining (EDM) recast metal and weld balls from part surfaces have been reviewed. This particular study will involve some additional effort. Basically, high-pressure miniature jets of abrasive remove this material but may cause unallowable secondary contamination.
- Techniques for specific groups of parts have been developed (for example, specific sequences of operations for deburring miniature flat gears). As a result, procedures for threaded holes have been proposed.
- Some edge-break samples were created that can be used in conjunction with the microscope to evaluate how large edge breaks exist on any given part. These samples are held next to the

Table 32-5. Techniques for improving manual deburring

- Design the product to accommodate burrs.
- Design tooling to accommodate burrs.
- Practice burr prevention.
- Design tooling to minimize burrs.
- Design tooling to put burrs in easiest-to-remove location.
- Select more appropriate cutting tools.
- Control burr size.
- Provide formal, burr-related training.
- Improve deburring processes.
- Improve in-plant standards.
- Initiate or improve accountability for deburring costs and quality.
- Subcontract deburring operations.
- Improve communication.
- Provide a central source of burr-related assistance.
- Make one individual the authority on what will be done.
- Train workers.
- Assimilate industry knowledge.
- Define in-plant problems.
- Purchase more appropriate equipment.
- Combine deburring with a machining operation.
- Perform deburring during another operation's idle time.
- Combine deburring with inspection or packaging.
- Subdivide deburring and add it to machining operations.
- Use correct machine settings.
- Use power tools.
- Add chamfers to fixtures to reduce positioning time.
- Require operators to deburr more than one part at a time.
- Use a combination of deburring processes to more effectively deburr parts.
- Define level of deburring quality required.
- Decide if all edges have to be burr-free.
- Identify where a burr-free condition is required.
- Define what is an acceptable (burr-free) edge condition.
- Define how a burr-free condition is measured.
- Define how an edge-break condition is measured.
- Know why the specified edge break or edge radii is required.
- Know the thickness and height of each burr on the part.
- Know within what size ranges burr characteristics are consistent.
- Decide if the burr has to be removed.
- Decide if the burr has to be removed now.
- Decide if the burr will cause an electrical short circuit, jam a mechanism, cause interference fits, be a safety hazard, cause misalignment, cause unacceptable stress concentrations, accelerate wear beyond allowable limits, or detract from product appearance.
- Know if the burr is accessible.
- Decide if the sequence of operations can be changed to facilitate deburring.
- Decide if the direction of cut can be changed.
- Know which tool feeds generate the smallest burrs.
- Define proper tool-change intervals.

part under the microscope to provide a visual comparison. They do not actually measure the part; rather, they provide a relative comparison for workers (for further discussion on this topic, refer to Chapter 34).
- Microscope eyepieces were evaluated and are being purchased—manufactured to requirements—with radial arcs of specific sizes. This will allow the microscope to be used as a comparator for measuring edge breaks.
- The possibility of electropolishing miniature gears has been explored.

This list is much longer and more detailed than necessary for many companies, and conclusions may be diametrically opposite that of other companies because specific problems and economics may be totally different.

In many cases, solutions are based on consideration of the deburring requirements in the initial stages of manufacturing planning. This requires careful implementation of different machining approaches, since even simple cutter substitution or alteration can double deburring time. When the problem involves a lack of understanding of basic process economics, other information sources may prove beneficial (Gillespie 1999; Rhoades 1981). Normally, however, solutions involve a detailed evaluation of techniques and economics applicable to a specific department or plant. For industry or multi-plant solutions, the needs are larger. The following list, for example, summarizes some of the needs for industry in general:

- papers defining and explaining which hand tools and cutters work best in which situations;
- definitions of typical situations in which handwork is least expensive when performed: (a) before vibratory finishing (abrasive flow, electrochemical deburring [ECD], etc.); (b) after vibratory finishing (abrasive flow, ECD, etc.);
- industrywide time standards—based on burr size, location, or workpiece material—for hand deburring;
- development of improved mechanized processes using microminiature sanders or microminiature laps;
- national centers for operator education;
- major articles defining cost savings incurred by careful investigation of hand operations;
- definition of how burr size affects hand deburring time;
- research of human factors on how to improve hand operations (particularly those involving microscopes), and
- design compendium for edge breaks as a function of past use and a statement of why each condition was chosen.

COORDINATING PEOPLE

People who perform the work (human factors) are one of the most important aspects of deburring. Nothing, however, is published on this subject; yet, it can be more important than all the mechanistic facts described on the previous pages. This aspect of deburring has several facets, including mechanical skills, personal efficiency, social stratification, motivation, coordination, and acknowledgement of expertise.

Although few studies have been performed, industrial engineers have shown that a proper environment can improve worker output, quality, and morale for almost any task. Appropriate noise and lighting levels, comfortable temperatures and workstations, and appropriately arranged worktables are all notable elements in effective planning for maximum output. The following questions summarize considerations specifically applicable to human factors in deburring:

- Do workers have convenient tools for holding parts?
- Do workers have convenient tool-handle sizes that minimize hand fatigue?
- Are deburring tools all designed for right-handed individuals (and you have left-handed workers)?
- Is there adequate work space on the workbench?
- Is a technique available for segregating finished from unfinished parts?
- Is part storage and retrieval at the workbench convenient?
- Does lighting at the workbench generate too much heat?
- Are deburring tools long enough to hold conveniently?
- Are knurled tools advantageous for holding, or are they more fatiguing?
- If microscopes are used, do they have appropriate working focal lengths and appropriate magnification specified to minimize eye strain while providing the required quality?
- Does deburring generate unacceptable levels of dust, noise, or heat?

The preceding list provides leading questions. Responses, however, will vary among companies. For example, parts may be conveniently stored in small vials to prevent interoperation damage. Removing and replacing parts in these vials can be easy or difficult, depending on the clearance between the vial and part.

It can be time-consuming and frustrating, all of which works to the detriment of efficiency and worker morale.

As noted in Chapter 31, a survey at one facility where microscopic deburring was done revealed that 50% of the workers in this area wanted normal lighting, while 50% felt that less light reduced glare in the microscopes. Fifty percent wanted individual cells, while the remainder wanted the companionship found in a "bull-room" facility. Fifty percent wanted normal room temperatures, and 50% wanted a cooler than normal room. In true random fashion, the people who constituted the 50% changed with each query, such that it was impossible to isolate a group possessing common environmental desires.

A common philosophy for microscope usage is that higher power results in better part quality. That may be true in some instances; however, for complex, three-dimensional parts, some individuals have observed that quality starts to drop above 16× magnification, and deburring time increases exponentially because little work area is within the viewable field, and the depth of field is so shallow. Some companies, however, still require deburring at 100× magnification. It may be more appropriate, in such cases, to deburr at low power, then verify at a higher power. This allows the worker to touch up the part as required at a higher magnification.

Operator Mistakes

Now, let us consider a human-factor analysis of deburring. In this instance, we will concentrate on mistakes made in precision hand-deburring operations. For the first example, consider a miniature gear with 16 teeth that requires burr-free edges under 10× magnification and edge breaks that do not exceed 0.003 in. (76 µm). Because each tooth consists of 10 arc or line segments (Figure 32-1), each gear will have 160 line segments. If each part requires 15 minutes to deburr, then 640 arc or line segments are being deburred every hour (10 segments per tooth × 16 teeth × 4 parts per hour). In a 7.5-hour workday, 4,800 segments are deburred. If it is assumed that, on any line segment, an operator has the opportunity to make five errors, then in a single day that operator has 24,000 opportunities for error. Typical error rates are one in 1,000 to one in 10,000. In one day, a single person could reasonably be expected to make up to 24 errors on these gears. For a group of workers, it is reasonable to expect that up to 240 unacceptably deburred parts will enter inspection under these conditions.

In some precision operations, a burr left on a single segment is enough to cause failure in a critical unit.

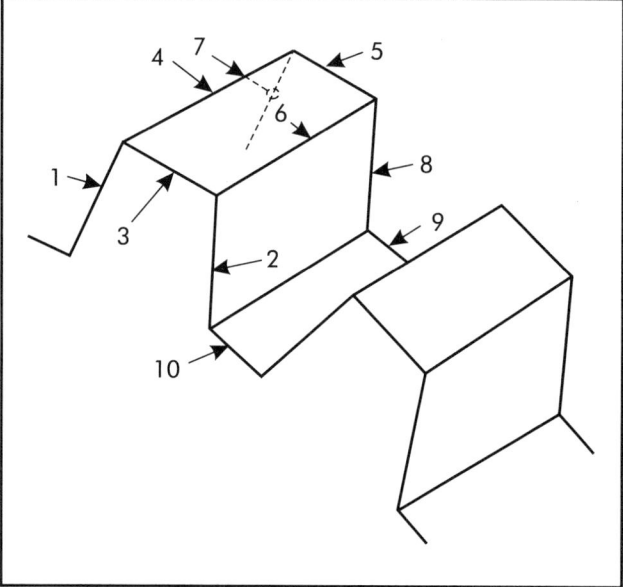

Figure 32-1. Line segments on a gear tooth.

This may not be a major problem on an automobile speedometer, but if an aircraft altimeter fails because of a burr, the consequences are more severe. The following are a number of factors that could constitute an error.

- The operator could forget an edge entirely.
- The part could rotate one tooth in the operator's hand and be missed.
- The knife used to remove the burr may generate two smaller, yet visible, burrs.
- The operator may forget to remove secondary burrs.
- An unfinished part may be inadvertently placed in a box of finished parts.
- Dirt may mask a small burr.
- The operator's angle of view of the part may be rotated causing loss of full vision.
- Someone could distract the operator while he or she is finishing one edge.
- The edge break can be excessive.

Many other factors can cause errors or result in sending unfinished or unacceptable parts to inspection or the next assembly. Consider the errors for an entire factory. Assume, in this instance, that each individual has 20,000 opportunities for error each day, and the error rate is one in 1,000. The result is 20 errors for each workday. If the plant has 1,000 workers who work 250 days per year, some 5 billion error opportunities exist each year, and some 5 million errors occur each year. The point is that human error is inevitable where large numbers of people do large numbers of tasks.

In addition, researchers have noted that error is as likely to occur to one worker as another. There is seldom more than a factor of three difference. Most errors by trained people are sporadic. If human error is the problem, it can be reduced. To reduce human error and its effects, use the following techniques:

- reduce the number of opportunities for error;
- reduce the influence of contributing factors;
- improve human tolerance limits;
- increase the detectability and correctability of likely errors;
- reduce the consequences of likely errors;
- improve feedback to the worker;
- improve analysis and prediction techniques;
- improve attitudes and understanding of error, and
- improve personnel selection and training.

The objective approach to reducing the effects of human error (actual or potential) are:

- understanding human error;
- measuring human error;
- identifying the causes of human error;
- deciding which errors are important;
- determining the frequency of human errors;
- determining recovery factors;
- determining the likelihood of adverse consequences;
- determining the cost of consequences;
- determining the cost of fixing the problem, and
- correcting the work situation.

Communication is a human-factor approach, so it is appropriate to consider written communications. It is important to keep written instructions simple. A National Institute of Education report has shown that 12 million people over the age of 14 cannot read as well as a typical fourth-grade student.

The keys to preparing job instructions (aids) that minimize misunderstanding include:

- listing major tasks or subdivisions of a job;
- writing step-by-step procedures for each task;
- for each step, determining the needed information;
- deciding what information is needed in a job aid;
- selecting a job-relevant format;
- preparing a rough draft, and
- conducting a small-scale tryout.

Two more basic principles are worth noting:

- If the written procedures are difficult to read or locate, or are inconvenient, they will seldom be used.
- The burden of written communication is on the writer, not the reader.

Writing effective deburring instructions or procedures can be difficult. The following is a checklist for writing such instructions.

1. Are written instructions provided to those who perform deburring?
2. Are areas requiring particular care clearly identified?
3. Are specific requirements identified?
4. Are any special tools identified?
5. Is a specific sequence of steps necessary?
6. Is the sequence identified?
7. Are the steps marked to indicate that they must be performed in the sequence presented?
8. Are instructions sufficiently simple to ensure use?
9. Does the procedure provide for verification and sign-off of actions?
10. Are means provided to verify conformance?
11. Do procedures provide instructions for reasonable contingencies?
12. Was a task analysis performed to develop the instructions in the procedure? If yes, was the task analysis performed by someone trained to do task analyses?
13. Was information pertaining to the instructions obtained from persons who use the procedure?
14. Is information pertaining to procedural instructions routinely sought from persons for whom the procedures are intended?
15. Is there a written specification describing the format and content of procedures?
16. Was the procedure tested before it was released for use? (Mark "yes" only if the entire procedure was tested.) If yes, which of the following test methods were used: a) walk-through of the procedure was performed by the writer (or other expert); b) walk-through of the procedure was performed by a user observed by the writer (or other expert); or c) a hands-on test of the procedure was performed by an expert.
17. Was the procedure reviewed for technical adequacy before it was released for use?
18. Does this procedure refer to other documents? If yes, does the procedure refer to one of the following: a) entire documents; b) specific sections within the documents; or c) specific pages or paragraphs (or both) within the documents?
19. Are personnel required to use this procedure for performing the activities it describes?
20. Does the procedure require that users perform computations?

The checklist applies to any work instructions. It has been adapted, in part, by one company, which used the following basic criteria for written deburring work instructions:

1. The instructions shall indicate which edges are to be deburred.
   - For in-process or final deburring operations, an illustration, if needed, should be provided. Such an operation should not reference an earlier machining operation illustration. (These machining sheets may not be available when the deburring is being done.)
   - Deburring done during machining also should be guided by instructions on which edges to deburr.
2. The work instructions shall indicate how much edge break is allowed or required. Since 0.010-in. (0.25-mm) maximum edge break is a plantwide workmanship standard, it need not be specified. All exceptions to 0.010 in. (0.25 mm) maximum break must be noted.
3. Whenever doubt exists as to which edge or detail is to be deburred, an illustration should be provided. (Three shifts of people will use the illustration for guidance. Without an illustration, it may be difficult to distinguish which receives the specified edge break when several features have similar size or shape.)
4. When a part has critical surfaces (close tolerances, fine finishes, callouts for no scratches, etc.) that could be damaged in a deburring operation, they should be indicated in the appropriate deburring instructions.
5. For final deburring operations in which some edges have already been adequately deburred (supposedly), but some have not, indicate the edges that should be deburred by adding words such as: ". . . and verify that the remainder of the edges are burr-free." Edge breaks other than 0.010 in. (0.25 mm) that have previously been done should again be shown, but as reference values. For example, "deburr EDM'ed slot complete per illustration, and verify that the remainder of edges are burr-free at up to 10× magnification."
6. As a general policy, it is unnecessary to list all the deburring tools used. One major reason for not doing so is that each deburring operator has different capabilities, so some individuals may find some tools more useful than others. For example, a few tools are designed for right-handed users. This is an obvious disadvantage for left-handed individuals. There are three exceptions to this general policy:
   - The first exception concerns threaded holes. All threaded holes should have a brush called out in the deburring tool requirements list. This particular requirement is the only way to ensure that miniature threads are reasonably particle- and burr-free. There is only one brush that will adequately deburr any given thread size.
   - The second exception concerns the use of microscopes. When the drawing or the required workmanship standards dictate inspection under magnification, it needs to be indicated to the deburring operators. Currently, this is being done by adding the words "microscope up to 20×" in the requirements list in the deburring operation. A few parts that have specific callouts on the drawing also indicate in the text of the deburring operation to "deburr at 7–10×" or related magnifications.
   - The third exception concerns specific parts for which there is an obvious deburring problem, meaning that something often is overlooked, quality often is not attained, or a special tool is required. These conditions will be handled as necessary so that everyone is aware of the special tools or requirements.
7. To ensure quality or productivity on some parts, special deburring instructions will be added in the text of the operation rather than just in an illustration. Currently, methods engineering is adding this information, with the process engineer's concurrence, after it has completed a time study.
8. The tools required to meet the indicated rate or quality, or both, will be specified in the text as well as the requirements list. The sequence of tools used is typically also specified in the text.
9. The purpose of the information in the text is to add emphasis that this is the required way to finish the parts. If a faster way is conceived, it should be evaluated by methods and process engineering.
10. When any of several tools could be used with equal efficiency, it is unnecessary to specify the tool.
11. Instructions for all deburring operations (in-process and final) should be on a separate page. The practice of listing instructions for more

than one operation per page requires deburring workers or supervisors to search the department or computer to find a sheet that may be in use at a machine.

12. The following is an example of the preceding guidelines:

Operation 130 P/N1023076-301

Deburr per illustration using 320-grit, or finer, sandpaper for edges around the large, flat area. Brush the contoured areas with abrasive-filled nylon brush. A burr knife should only be needed for heavy burrs.

Tools required:
- nylon brush 11223344,
- motor arbor 22114433, and
- 320-grit sandpaper.

Figure XX illustrates the use of specific notes for specific edges.

SOCIAL ASPECTS OF DEBURRING

Most plants have some degree of social stratification of workers. Frequently, other workers view 5-axis computer numerical control (CNC) machine operators, tool and die makers, or setup personnel as elite groups. Machinists who use manual machines may be considered in the middle structure, and trainees may be considered to be in the lowest strata. In many instances, individuals have noted that deburring workers in their plant are in the lowest structure. Deburring is considered a demeaning task, requiring no skill, no knowledge, and presenting no challenge. That may be the case in many plants, but deburring need not carry such a stigma.

Individuals who have studied motivation have demonstrated that on-the-job self-esteem and pride in work can result in dramatic improvements in quality and production rates. Unfortunately, the importance and difficulty of some deburring jobs goes unnoticed and unacknowledged. There are few rewards for some workers who must take someone else's "work of art" and add the final touch of perfection that makes it usable.

When those performing deburring and their fellow workers perceive deburring as a challenging task, morale increases. Everyone appreciates recognition for good work and acknowledgement of a difficult job. Deburring can be the most difficult task in a plant, so deburring management should convey the importance, challenge, and opportunities of deburring.

The question, "What images do deburring and deburring workers project in my plant?" may help identify many motivational limitations. What image do these terms project in the minds of machinists, inspectors, supervisors, engineers, salaried personnel, and plant managers? These images may be the most important of the management areas previously discussed. The list of performance-shaping factors in the appendix of this chapter may help define some of the characteristics of deburring tasks that can be improved for greater productivity and morale.

Worker Stress

Some workers suffer from high levels of psychological stress, which may be caused by the work environment or outside influences. However, all stresses influence deburring performance. Solutions to stress are not offered—the problem is raised for awareness purposes only.

Management can help cause and control stress. For example, during poor economic times, low-seniority staff worry about layoffs. If they cannot speak English, they worry about job longevity. They may worry that their skill levels are not good enough to maintain the job. Some may face family pressure to find a better-paying or more meaningful job, even though they are good at and like what they do. They may face overt derogatory comments from other workers. Clear, open, and caring discussions with staff in team environments improve these concerns, whether real or imaginary. Increased productivity many times is the result.

Supervision

The coordination of deburring personnel is another area in which nothing has been published. Consider the case in which manufacturing supervisors have total responsibility for part fabrication (that is, control over every step and operation). This is a typical arrangement in most companies. It provides total accountability and control in one individual.

The disadvantage is that few machining supervisors have deburring expertise or believe they have the time to devote to major deburring improvements. Because machining operations require the majority of plant hours, deburring emphasis can suffer. The advantage of this structure is that if machinists allow tools to become too dull, and the burrs are too big to remove efficiently, the supervisor possesses the necessary control to correct the situation. He or she also has the responsibility of overall unit efficiency.

When deburring is controlled by a finishing specialist, all finishing operations are coordinated and a body of expertise can be established. On the other hand, Supervisor A, for example, may try to maximize his efficiency, even if it reduces that of Supervisor B. The net result can be an increase in part cost. These tradeoffs are difficult to document and, frequently, many parts have heavy burrs before the problem is identified. Cellular manufacturing arrangements resolve these issues better because everyone is more an integral part of the result.

Besides coordination issues, it is essential to plan the efforts of the second and/or third shifts. There is a syndrome that exists across the country. It can be stated briefly as: "Well, second shift did it again. All the parts are damaged." Many of the problems exist because people do not communicate effectively with those on other shifts. People forget that night shifts do not have the benefit of immediate guidance available from the first-shift professional staff; individuals who do most of the inspection; individuals who traditionally assemble the parts; or those who can order new tools or answer questions about new products. Traditionally, these shifts use individuals who have less experience than their day-shift counterparts.

For example, a first-shift individual wrote a note to a second-shift supervisor to complete a certain group of parts. The note was brief, stating "complete." The second-shift supervisor saw the note and was very pleased to see that the parts were "complete." Since deburring was "complete," he could submit the parts to inspection and meet schedules. However, each supervisor interpreted "complete" differently, and these parts, which still had burrs, were submitted to inspection with the results expected: they were rejected.

EQUIPMENT ALLOCATION

One goal of most plants is to reduce deburring costs. For many individuals, the most obvious means of accomplishing this goal is to purchase mechanized equipment. This is an approach that all companies should consider, but the purchase of such equipment may not be the best approach for some companies. As previously discussed, the first need is to define requirements and existing problems or restraints. When equipment acquisition is appropriate, the following questions should be answered:

- Does the need warrant new equipment?
- What are the economic paybacks?
- Is an adequate, single piece of equipment available?
- Would two different mechanized deburring processes be preferable to a single process?
- Does the cost of expendables exceed labor savings?
- Will the equipment perform well on typical or worst-case burrs?

- Does the process require a professional level of expertise that the plant does not have?
- Will suppliers provide after-sale technical backup?
- What side effects will the new process cause?
- If there is a downturn in orders, is the company still obligated to pay for unused equipment?

Yearly studies indicate that 6% of the capital equipment dollars spent by U.S. industries is dedicated to finishing equipment. Unfortunately, industry does not have the guidelines that consumers would like to see for identifying the most appropriate or economic process or machine.

In many instances, it is important that the use of two mechanized processes can result in more favorable returns on investment than a single process. For example, when burrs become very large, most processes will not remove them without altering the part. If a machine is used that mechanically cuts off the burr, then brushes the edge, almost any size burr can be removed repeatedly. In this case, cutting is one process and brushing the second. The user must determine which combination of processes will provide the desired final result. In some instances, the combination of machines provides desired or required side effects, such as clean parts, aesthetic surfaces, less friction, or better electrical properties.

Deburring is an aspect of plant growth that is easy to overlook. This is particularly true of hand deburring. If machining output is doubled, hand deburring, similarly, must be doubled, or another process must be selected. Increasing hand deburring operations increases the costs for:

- more workbenches and air and electrical hookups;
- purchase of more air and electric motors;
- purchase of more lighting for deburring benches;
- purchase of more hand tools;
- increased number of expendable supplies, and/or
- increased cleaning needs.

At least one company has a standard practice of budgeting for or allocating a specific amount of floor space and equipment to deburring. The amounts are based on the number of machinists the plant expects to hire. For small commercial parts, the ratio may be one new deburring individual to six machinists. For a precision department, the ratio may be 1:3. Immediate financial requirements may be $2,000 for each deburring individual, and supplies may be budgeted for a $6,000 increase each year. Every machinist may mean an increase of 70 ft^2 (6.5 m^2) of bench and cabinet space for deburring. If only mechanized deburring is required, the above allocation guidelines may decrease by a factor of 10. Expendables would increase significantly, however. Additional insight into management treatment of estimates, which is applicable to deburring as well as other fields, can be found in another source (Rowe 1977).

DATA-DRIVEN SOLUTIONS

What are the technical capabilities of a hand deburring staff? How small of a burr can workers see? How big of a burr can workers remove? How does tool #1 compare in capability and speed with tool #2? How different are workers A and B? Will 320-grit sandpaper damage 8-μin. (0.2-μm) finish surfaces? If this same abrasive paper is held against precision diameters, will it change the diameter by more than 0.00001 in. (0.25 μm) in 10 seconds? Will it remove a 0.005-in. thick × 0.005 in. tall (127 × 127 μm) burr in stainless steel? If it does, how long will it take? Are abrasive-filled rubber tools faster than sanding? How do they affect size, finish, and burrs? These are the types of questions that must be answered for long-term cost reductions in hand deburring. They require experimentation, time and motion studies, the purchase of new products, and implementation.

Even when a new product works more effectively than the product it replaced, it may not be accepted by those who use other tools. Figure 32-2 illustrates the measurements of how repeatable one operator was using deburring tools, then compares this person with others.

Figure 32-3 provides measured results of the ability of experienced and new deburring staff when they were required to remove burrs from a stainless-steel part while not chamfering or radiusing the edge more than 0.003 in. (76 μm). These results, taken from the training case history described in Chapter 33, illustrate that new workers could maintain the 0.003 in. (76 μm) requirement 71% of the time. For the same

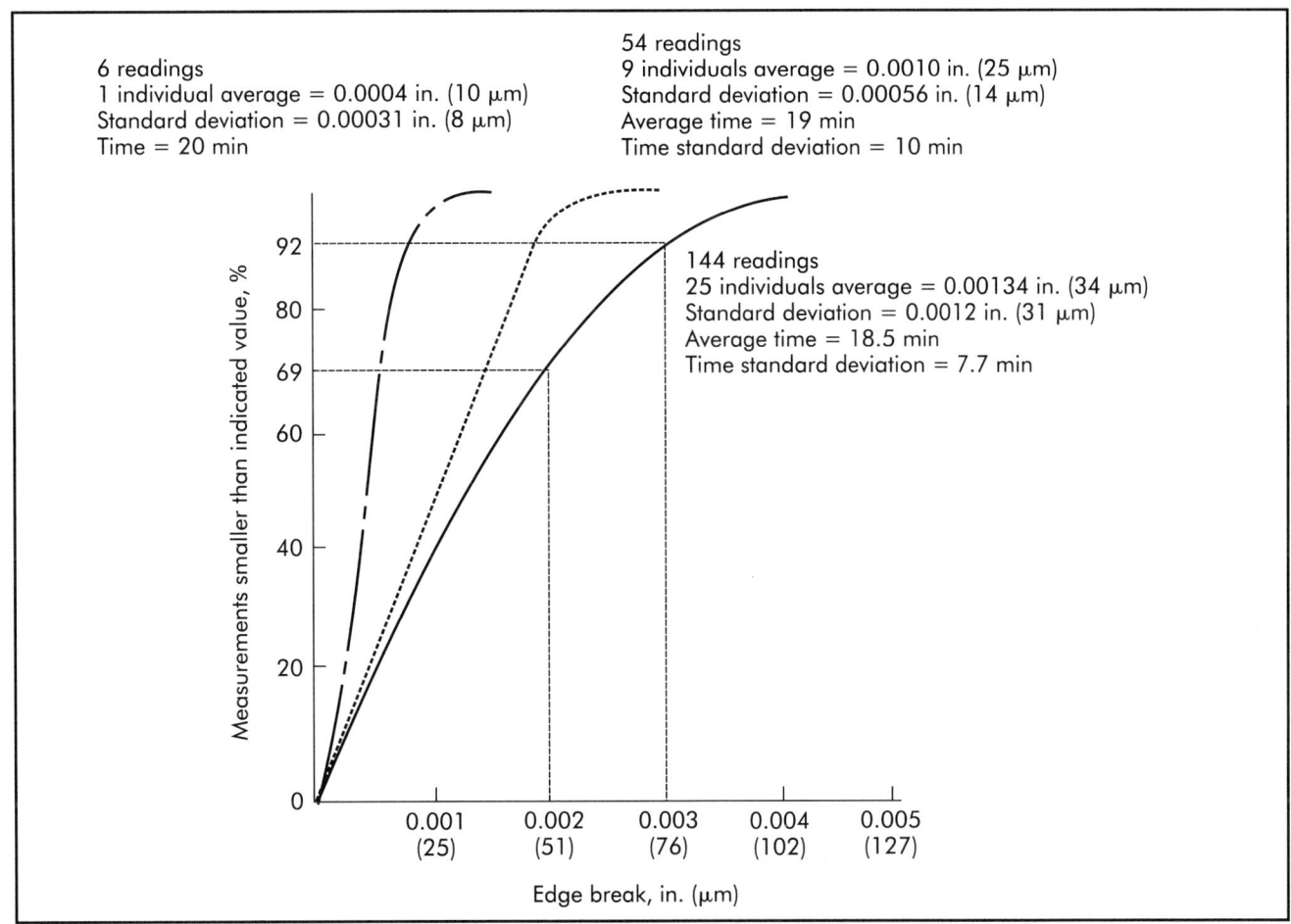

Figure 32-2. Percentage of edges having an edge break smaller than a specified value when experienced workers tried to maintain small breaks (Gillespie 1978b).

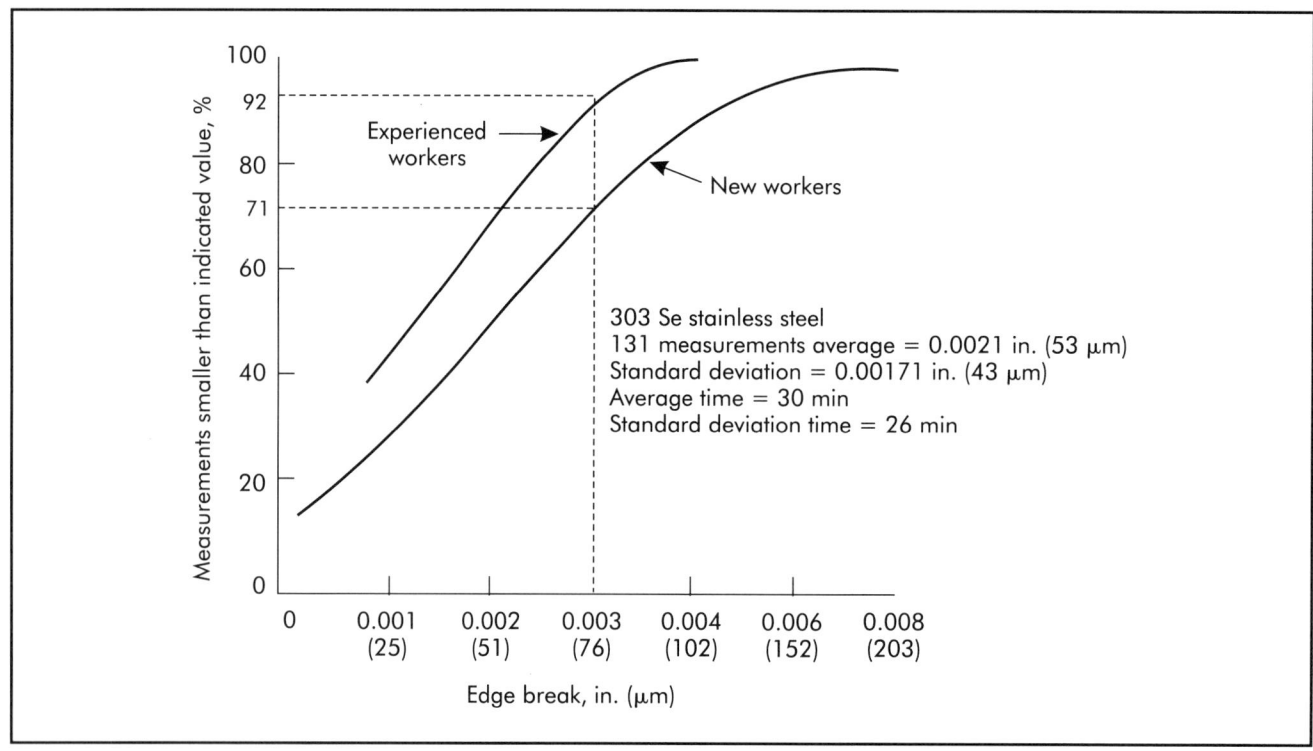

Figure 32-3. Effect of work experience when maintaining small edge breaks. (Courtesy Machine and Tool Blue Book)

parts, 92% of the experienced workers were able to maintain this minute edge break while removing the burr entirely.

The author knows of only one study that provides data on the following topics: repeatability of burr removal or edge radiusing; the effect of burr properties, product size, and geometry on deburring time; and the burr-removal capability of various deburring tools. The importance of a comprehensive evaluation of hand deburring, without actual in-plant observation on each part, becomes obvious when one considers that an estimated 10,000 deburring tools are available. As discussed later, operator technique can create an equally inspiring number of variables in the process.

Tests of Deburring Repeatability

The following paragraphs describe seven tests performed using techniques and tools commonly used on precision, miniature metal components. Some tests explored final edge condition (after the burr was removed) to determine the smallest repeatable edge break maintainable on parts. Other tests explored the effect of product geometry and burr properties on deburring time. More specifically, the product geometry tests analyzed how the accessibility of burrs affected deburring time.

In one study, experienced deburring personnel were asked to deburr 0.50 in. (12.7 mm) cubes (Figure 32-4a). The workpiece material was 303 Se stainless steel, and the burrs produced by an end mill were up to 0.0098-in. (249-μm) long and 0.0037-in. (94-μm) thick (which is normal for this material and shop). The goal on these parts was 0.005-in. (127-μm) maximum break. Extensive measuring indicated that 95% of the edges met the 0.005-in. (127-μm) maximum break. Fifty percent of the edges had radii of 0.002 in. (51 μm) or less.

In the initial test of deburring repeatability, four production workers, whose principal job duties were burr removal, were each given six specimens to deburr. Each worker was told that the final edge break, or radius, must be maintained at 0.005 in. (127 μm) or less. A statistical analysis of the data indicated that the average radius produced was 0.0022 in. (56 μm), and the standard deviation of the data was 0.0015 in. (38 μm). There was no difference in the results produced by the four operators. Statistically, for these values, 95% of the radii should be 0.0052 in. (132 μm) or less. An examination of the test measurements revealed that 95% of the edge radii were, in fact, 0.005 in. (127 μm) or less. From these results and production experience, it is clear that trained operators can maintain a 0.005-in. (127-μm) maximum edge break while removing burrs. Statistically, less than 50% of

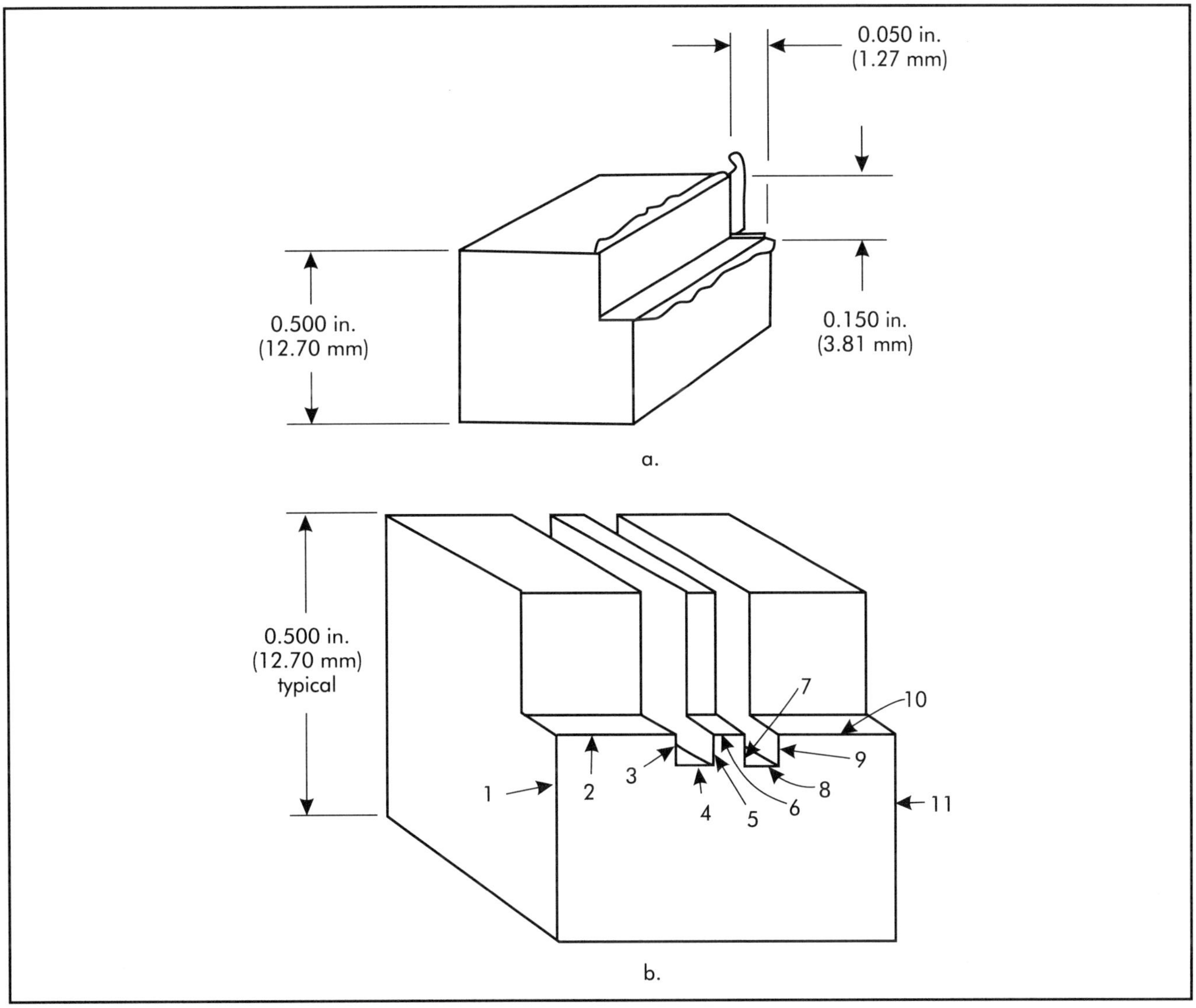

Figure 32-4. Specimen geometry used to assess edge-break repeatability on small parts (Gillespie 1978b).

the edges produced under these conditions would have radii of 0.002 in. (51 μm) or less.

Edge radii were measured on the top and bottom lip of the milled cutout. These measurements were made by sectioning the specimen, mounting it in a metallurgical mount, polishing, and, then, inspecting with a Zeiss inspection machine. In several cases, the edge radius produced was not of uniform blend; but all visible burrs and sharp edges were removed. Figure 32-5 shows some typical edge conditions.

In the second study, burrs had to be removed from thin slots as well as from exposed edges (Figure 32-4b). In this test, one individual deburred 10 specimens each of four workpiece materials. A 0.005-in. (127-μm) maximum edge radius was the goal of this study. Table 32-6 summarizes the results.

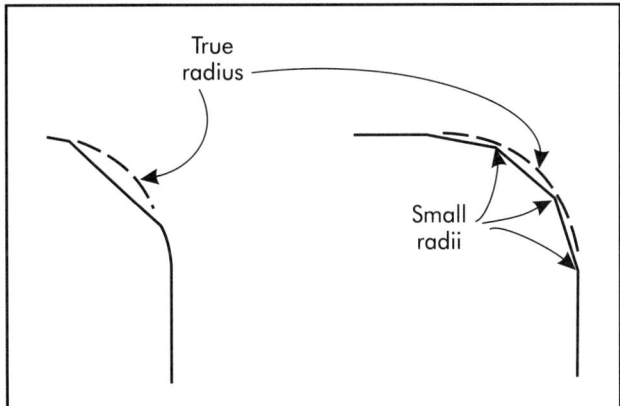

Figure 32-5. Common edge conditions produced by hand deburring (Gillespie 1975).

Table 32-6. Results of hand deburring test (see Figure 32-4b)

Test Parameters	Workpiece Material		
	6061-T6 Aluminum	17-4 PH Stainless Steel (H-900 Condition)	303 Se Stainless Steel
Average time required to remove burrs (min)	2.51	3.08	4.66
Results, in. (μm)			
Edge No. 1			
Average radius ($\bar{r}$)	0.0031 (78)	0.0025 (64)	0.0021 (53)
Standard deviation (σ)	0.0007 (18)	0.0006 (15)	0.0008 (20)
($\bar{r}$) + 2σ	0.0045 (114)	0.0037 (94)	0.0037 (94)
Edge No. 2			
Average radius ($\bar{r}$)	0.0038 (97)	0.0032 (81)	0.0042 (107)
Standard deviation (σ)	0.0007 (18)	0.0004 (10)	0.0012 (31)
($\bar{r}$) + 2σ	0.0052 (132)	0.0040 (102)	0.0066 (168)
Edge No. 3			
Average radius ($\bar{r}$)	0.0026 (66)	0.0016 (41)	0.0025 (64)
Standard deviation (σ)	0.0009 (23)	0.0010 (25)	0.0007 (18)
($\bar{r}$) + 2σ	0.0044 (112)	0.0036 (91)	0.0039 (99)
Edge No. 4			
Average radius ($\bar{r}$)	0.0026 (66)	0.0029 (74)	0.0023 (58)
Standard deviation (σ)	0.0009 (23)	0.0007 (18)	0.0008 (20)
($\bar{r}$) + 2σ	0.0044 (112)	0.0043 (109)	0.0039 (99)

The actual edge break produced was varied, depending on which edge was involved. Edge #2, for example, consistently had breaks exceeding 0.005 in. (127 μm). In reality, the average radius on most edges was 0.002–0.004 in. (51–102 μm). Part requirements at this plant, however, called for a less-than-maximum break at every edge point. Typically, if the average break was 0.004 in. (102 μm), half the breaks varied 0.004–0.0066 in. (102–168 μm). This is not a large variation, but it is large enough that every part in this study would have been rejected for excessive edge break.

As shown in Table 32-6, average ($\bar{r}$) values for edge radii typically ranged from 0.0025–0.0042 in. (64–107 μm). For 17-4 PH stainless steel, 95% of the edges ($\bar{r}$ + 2σ) produced under these conditions had radii of 0.0043 in. (109 μm) or smaller. While generally this is true of the other metals, on some edges, the allowable edge radii would have to be 0.0071 in. (180 μm) to accommodate 95% of the radii on that edge. While the required 0.005-in. (127-μm) maximum radii was maintained on most edges, 4.5% of the measured edges exceeded the requirement.

The graph in Figure 32-6 shows that 3% of the edges in this test had a break of 0.002 in. (51 μm) or less; 44% of the edge had a radius of 0.004 in. (102 μm) or less; and 75% had a radius of 0.005 in. (127 μm). Thus, 25% of the edges had an edge break greater than the specification. This fairly large range of results also indicates some of the difficulty in maintaining edge breaks of 0.003 in. (76 μm) or less on toleranced edges, such as 0.002–0.005 in. (51–127 μm).

The hand deburring in the study tended to be consistent in that the standard deviation typically was 0.0007 in. (18 μm). Table 32-6 shows that the 303 Se stainless-steel specimen required almost twice the time to deburr as the aluminum specimen. The burrs in this test, which were produced by slitting saws and an end mill, were quite small, typically 0.0008-in. (20-μm) thick and 0.0039-in. (99-μm) long.

A statistical analysis of the data indicates that the time required to remove these burrs was not proportional to their length or the final radius; neither was the final edge radius a function of burr length. There was insufficient data to provide a relationship between

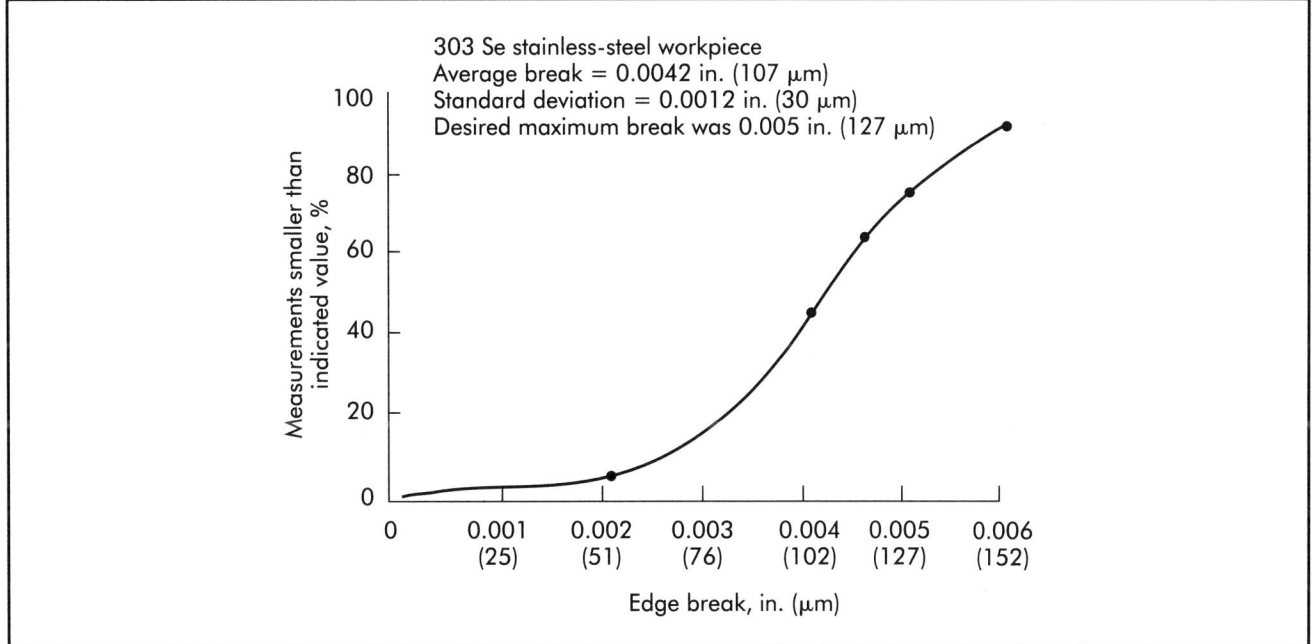

Figure 32-6. Percentage of edge #2 having an edge break smaller than a specified value (Gillespie 1978b).

deburring time and burr thickness. Since knives were used to cut off burrs, it is logical that deburring time is a function of burr thickness. The time required for deburring appears to be proportional to the strain-hardening exponent of the workpiece (Figure 32-7).

At this point, a question arises: "Are requirements, such as 0.003 or 0.005 in. maximum (76 or 127 μm), really needed on precision miniature parts?" In some cases, they are not, but many times there is a real need for small breaks. Sometimes, a small break is required to provide an adequate bearing surface for small-diameter journals (Figure 32-8). If a larger radius existed on the example shown, the hole diameter would quickly enlarge and allow the shaft to change its precision location. Sometimes, the available space for press-fitting pins into holes is so limited that a

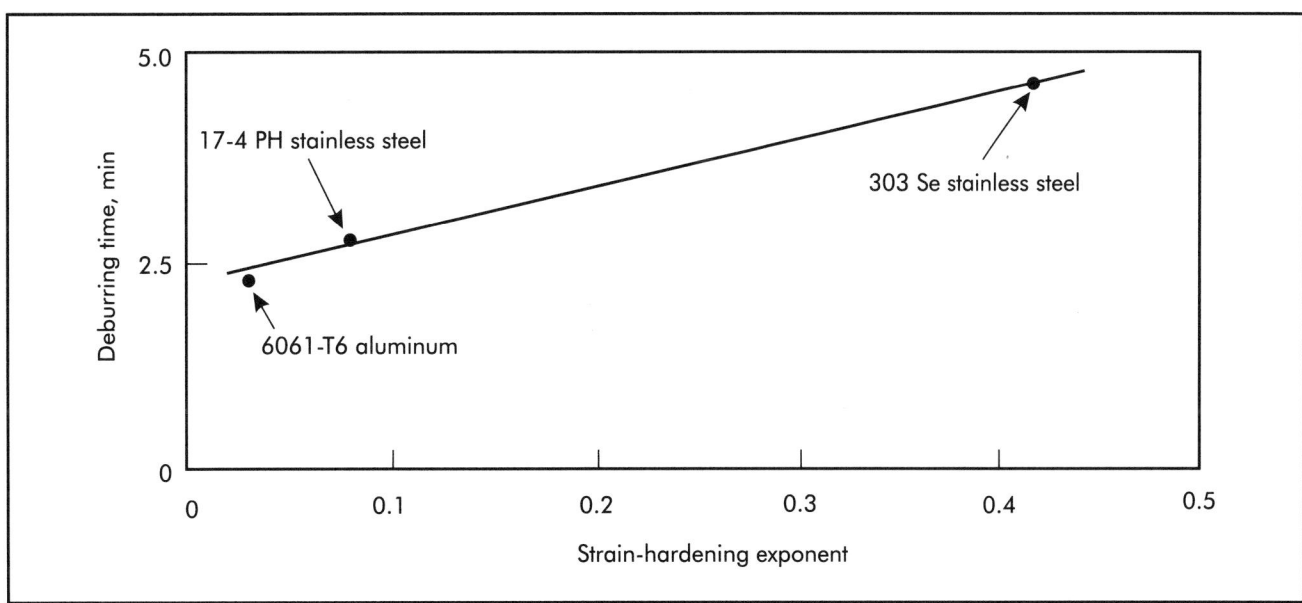

Figure 32-7. Relationship between deburring time and strain-hardening exponent (Gillespie 1975).

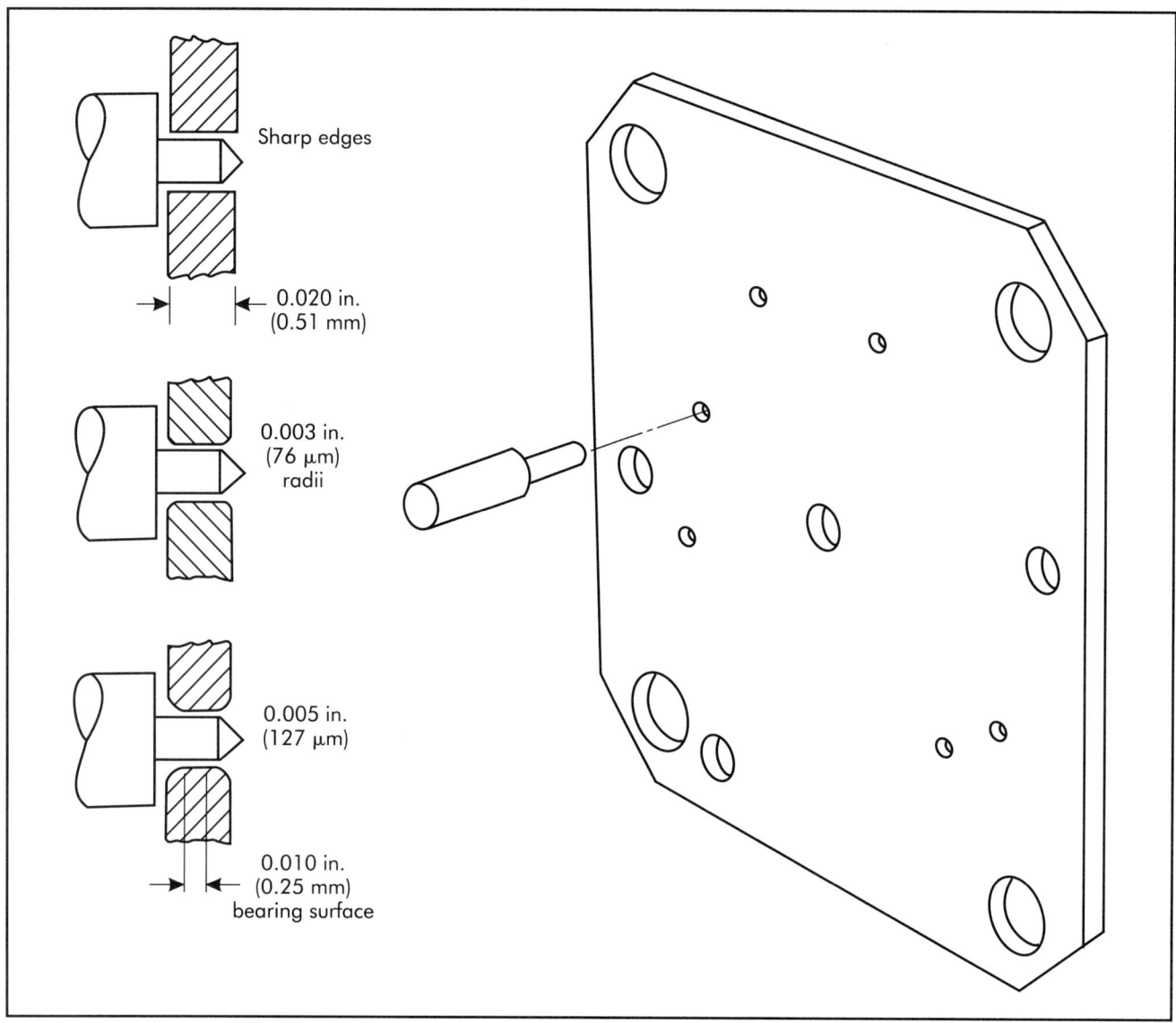

Figure 32-8. Large edge breaks reduce load-bearing width (Gillespie 1978b).

large chamfer on the hole or pin would not provide enough area to hold the pin in the hole (Figure 32-9). With edge break tolerances like this, it is important to challenge the requirement.

For the third test, an operator was asked to produce an edge radius no larger than 0.002 in. (51 μm). In this case, only three materials were studied, but radii were measured on 11 edges (Figure 32-4b). More than one measurement was made on some edges, since the radius is not always constant along edges. A total of 330 edges were measured.

On stainless steel samples, only 18% of the edges actually had a 0.002 in. (51 μm) or smaller break across the entire edge. Figure 32-10 illustrates typical results of this study. Although a large percentage of edges had some portions with 0.002 in. (51 μm) break or less, it is significant that, of so many edges on a part, not a single part had all edges within the required 0.002 in. (51 μm) break. Essentially, one would have had to allow 0.004 in. (102 μm) to accept 95% of the parts.

Average edge radii values were calculated for each of the 11 workpiece edges. As shown in Table 32-7, on the 17-4 PH stainless-steel workpieces, only six of the 11 edges had an average radius of 0.002 in. (51 μm) or less. The objective of this test was to deburr to the extent that all edges (not just 50%) had radii of 0.002 in. (51 μm) or less. On only two of the 11 edges would 95% ($\bar{r} + 2\sigma$) of the radii be 0.002 in. (51 μm) or less.

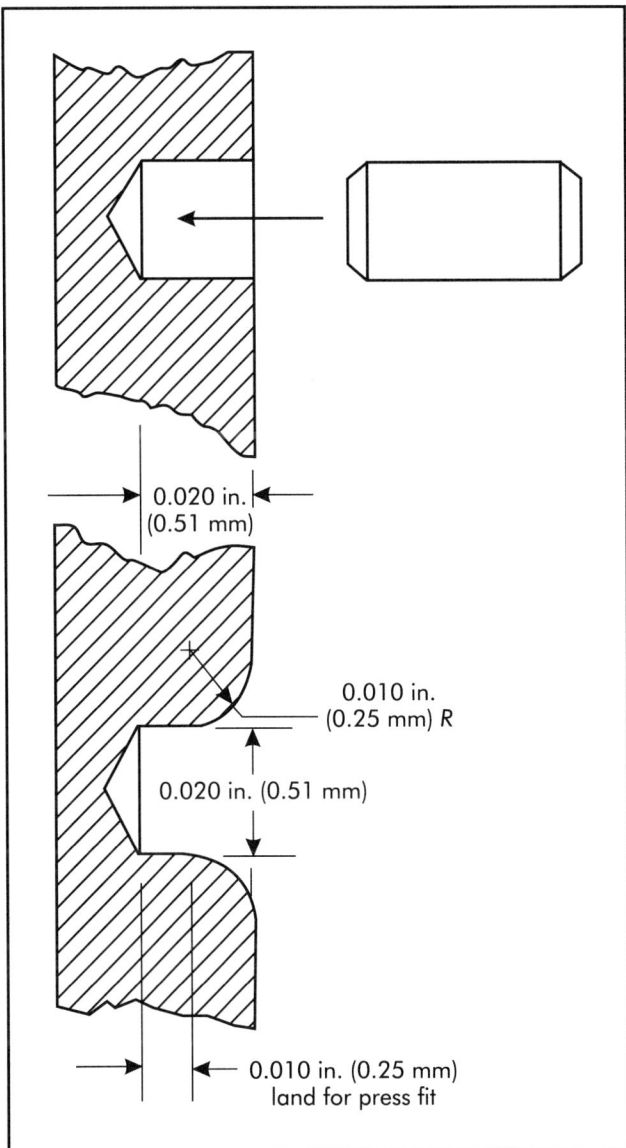

Figure 32-9. Large edge break affects press fit pin-holding strength (Gillespie 1978b).

Table 32-8 provides some edge radii numerical values. If a typical radius (r) for any one edge had to be selected, it would be 0.0021 in. (51 μm) in 17-4 PH stainless steel. Ninety-five percent of the parts ($\bar{r} + 2\sigma$) measured at that edge would tend to have a radius of 0.0030 in. (76 μm) or less. If it were a concern that all 11 edges on 95% of the parts had a radius equal to or less than $\bar{r}$, then $\bar{r}$ would have to be 0.0029 in. (74 μm) for 17-4 PH stainless steel.

Although it may be true that some individuals can produce better results than indicated in figures 32-6 and 32-10, this work was performed by several well-qualified people. In subsequent tests, individuals with only two months of deburring training were told they were incapable of maintaining 0.003-in. (76-μm) maximum breaks, but were to attempt to do so using any technique desired. They were given 303 Se stainless-steel parts (illustrated in Figure 32-11) with normal to small burrs (0.003-in. [76-μm] thick or less). In this study, 24 of 25 people maintained edge breaks of less than 0.0025 in. (64 μm) on all edges; and at least 25% of the individuals maintained edge breaks of less than 0.001 in. (25 μm).

In analyzing the results, several items stand out. These "rookies" were told they were incapable of meeting the requirement. As a result, they took great care in deburring the parts. In a normal situation, they would be trying to meet a production rate of close to 2 min/part. In this study, they had as much time as they wanted—and most took 20 min/part to meet the requirement. Second, they knew their work was going to be measured and everyone would know whether their performance was "good" or "bad." Third, to meet the requirement, most of them realized it would be impossible to use knives or brushes. Most resorted to 600-grit abrasive paper, sanded parallel to edges, and used comparators to check their work. Some used the backside of abrasive paper to limit abrasive action. Finally, the part edges they used were easily accessible and had only 25–50% of the total edge length as had been used in the initial studies. As a result, the difficulty of removing burrs was much less than the specimen shown in Figure 32-4. They did far better than most might have expected, but they used nonstandard methods and took up to 10 times longer than individuals using conventional approaches.

Figure 32-2 illustrates the ability of experienced hand deburring workers, each using his or her own approach to produce a burr-free edge with the minimum possible edge break. The group of 25 maintained 0.003-in. (76-μm) breaks, or less, on 92% of the edges of the simple parts shown in Figure 32-11. Unfortunately, of the 50 samples measured, a small burr remained on 12 parts. Thus, of the samples from the most experienced workers, inspection would have found roughly 25% unacceptable. Although the burrs left were minute and found only on one small portion of an edge, they were present, which could have caused problems in the mechanism of an assembled part.

Of the parts produced by the nine individuals generating the smallest edge breaks and leaving no burrs, essentially 100% of the edges had 0.003-in. (76-μm) breaks or less. The most skillful individual observed in this study had 97.7% of his edges 0.001 in. (25 μm) or smaller.

One possible reason for the presence of burrs on the samples is that some workers prejudged the time normally expected for their work and, therefore, allowed

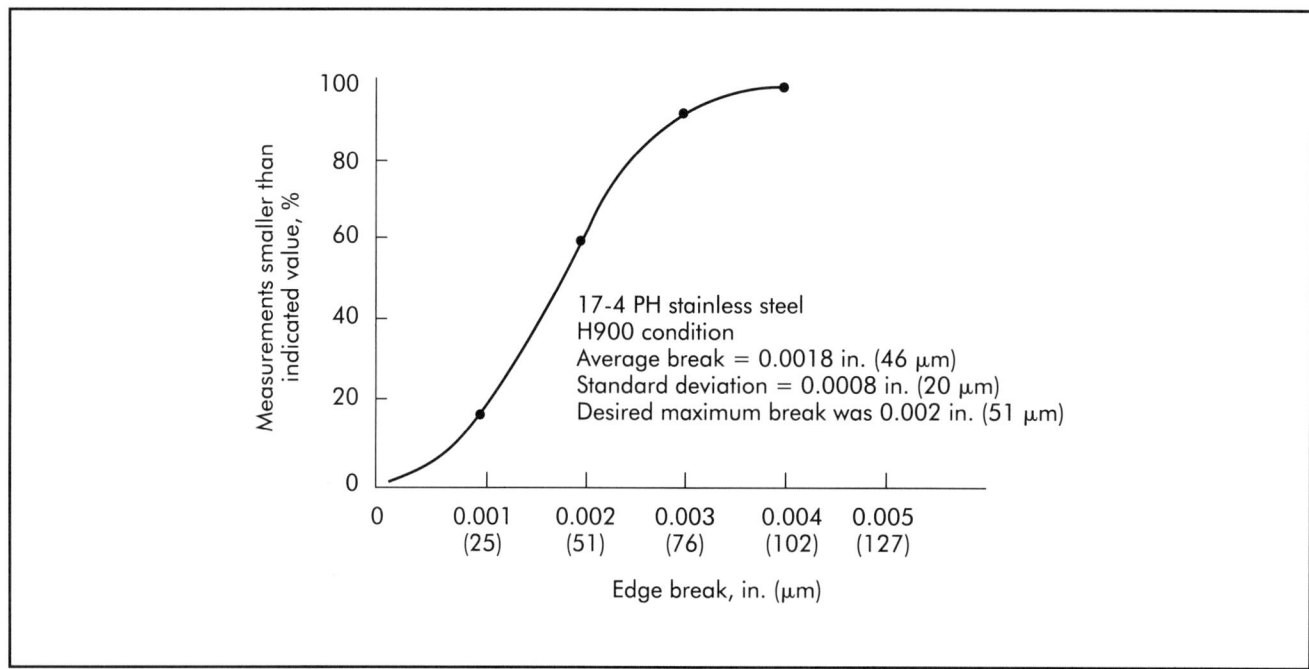

Figure 32-10. Percent of edges having an edge break smaller than a specified value for test three (Gillespie 1978b).

Table 32-7. Hand-deburring results for third study (see Figure 32-4b)

Workpiece Material	Edges with Average Radius ($\bar{r}$) of 0.002 in. (51 μm) or Less		Edges for which ($\bar{r}$) + 2σ would be 0.002 in. (51 μm) or Less	
	Number	%	Number	%
Aluminum	7	63	3	27
17-4 PH stainless steel	6	55	2	18

Table 32-8. Analysis of edge radii from test three (see Figure 32-10)

	Workpiece Material	
	Aluminum	17-4 PH Stainless Steel
	Results	
Average deburr time (min)	2.39	1.89
	Typical Values of Radii, in. (μm)	
$\bar{r}$	0.0017 (43)	0.0021 (53)
σ	0.0004 (10)	0.0004 (10)
$\bar{r}$ + 2σ	0.0025 (64)	0.0029 (74)
	Mean and Standard Deviation for All Edges, in. (μm)	
$\bar{r}$	0.0017 (43)	0.0018 (46)
σ	0.0008 (20)	0.0008 (20)
$\bar{r}$ + 2σ	0.0034 (86)	0.0036 (91)

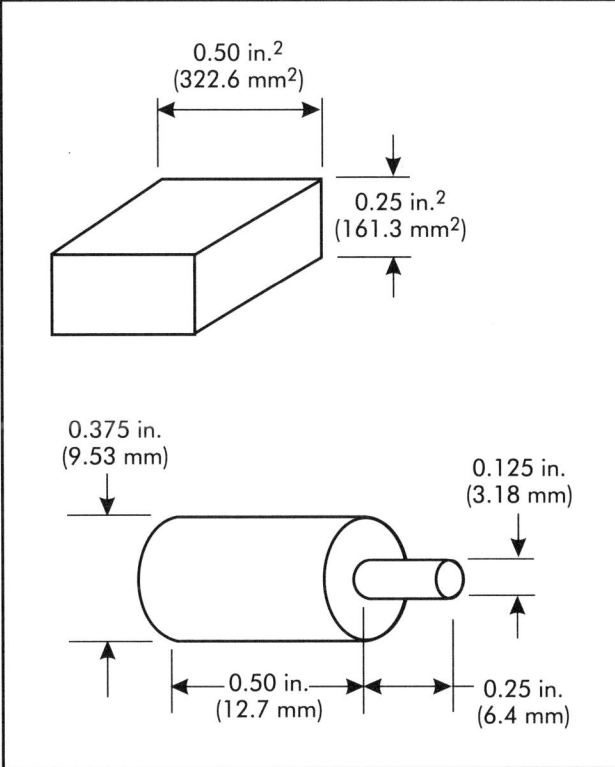

Figure 32-11. Test specimen used to determine smallest edge break individuals could maintain (Gillespie 1978b).

expected production rates to override the goal of optimum quality. Although this is conjecture, each individual knew his or her time and capability were going to be listed by name. These individuals used an average of 10 min/part, which was half the time used by inexperienced workers.

The following implications result from this and the two tests previously discussed, and it has been found that these results generally apply to production parts as well.

- Maintaining 0.002-in. (51-μm) maximum edge radius by existing hand-deburring techniques (knives, abrasive paper, and brushes) is not usually feasible without 50% fallout. It is possible, however, to produce an average radius of 0.002 in. (51 μm). While the majority of edges on a part may have radii of 0.002 in. (51 μm), or less, a significant portion of the total length of edges will have greater radii.
- There is little difference in quality in trying to maintain a 0.005-in. (127-μm) or 0.002-in. (51-μm) maximum edge radius.
- Increasing deburring complexity and the number of edges to be deburred increases the variability in results.

There was little difference in the quality produced by experienced operators in this study; however, time differences were marked. In production situations, additional notable differences exist in part quality.

Repeatability is a function of burr size and location. These variations affect some operators more than others. It is not always possible to guarantee that the most highly capable individuals are assigned to deburr workpieces requiring small edge breaks during production. The time required to deburr specimens will vary considerably among operators.

Effects of Burr Size and Part Geometry on Deburring Time

In the second series of tests, an attempt was made to determine how burr accessibility and size affected deburring time. Four brief tests were performed. As discussed in the previous section, deburring time is a function of workpiece material properties. Short, wide burrs occur in some materials, whereas long, thin burrs occur in others. For any given material, the conditions that produced the burr greatly affected burr size and, therefore, deburring time.

The accessibility of the burr also influenced the time required for deburring, and the quality of the resulting edge. For example, if one had to deburr the intersection of two holes, one would reasonably expect the time required to deburr the intersection to be a function of the hole sizes (Figure 32-12). As will be shown, actual deburring times are a function of accessibility, but they do not follow the regular pattern shown in Figure 32-12.

Effects of Burr Size

For the first study, a round, washer-like part (Figure 32-13) was deburred. Some parts had drilling burrs and some had reaming burrs. These parts were made from 303 Se and 17-4 PH stainless steel, 1018 steel, and 6061-T6 aluminum. Typical properties of the burrs on these specimens are shown in Table 32-9. Burr thickness and length were measured on the side where the drill entered the workpiece, as well as the exit side.

Figure 32-14 illustrates how burr thickness influences removal time. Washer-like parts with 0.005-in. (127-μm) thick burrs took four times longer to remove than burrs 0.0015-in. (38-μm) thick. Obviously, thinner burrs require considerably less deburring time.

Five specimens were deburred with each of the eight material/machining combinations. Table 32-10 shows the time required and indicates that the typically thinner reaming burrs required less time for deburring than did the drilling burrs. The three-step procedure used to deburr these specimens follows:

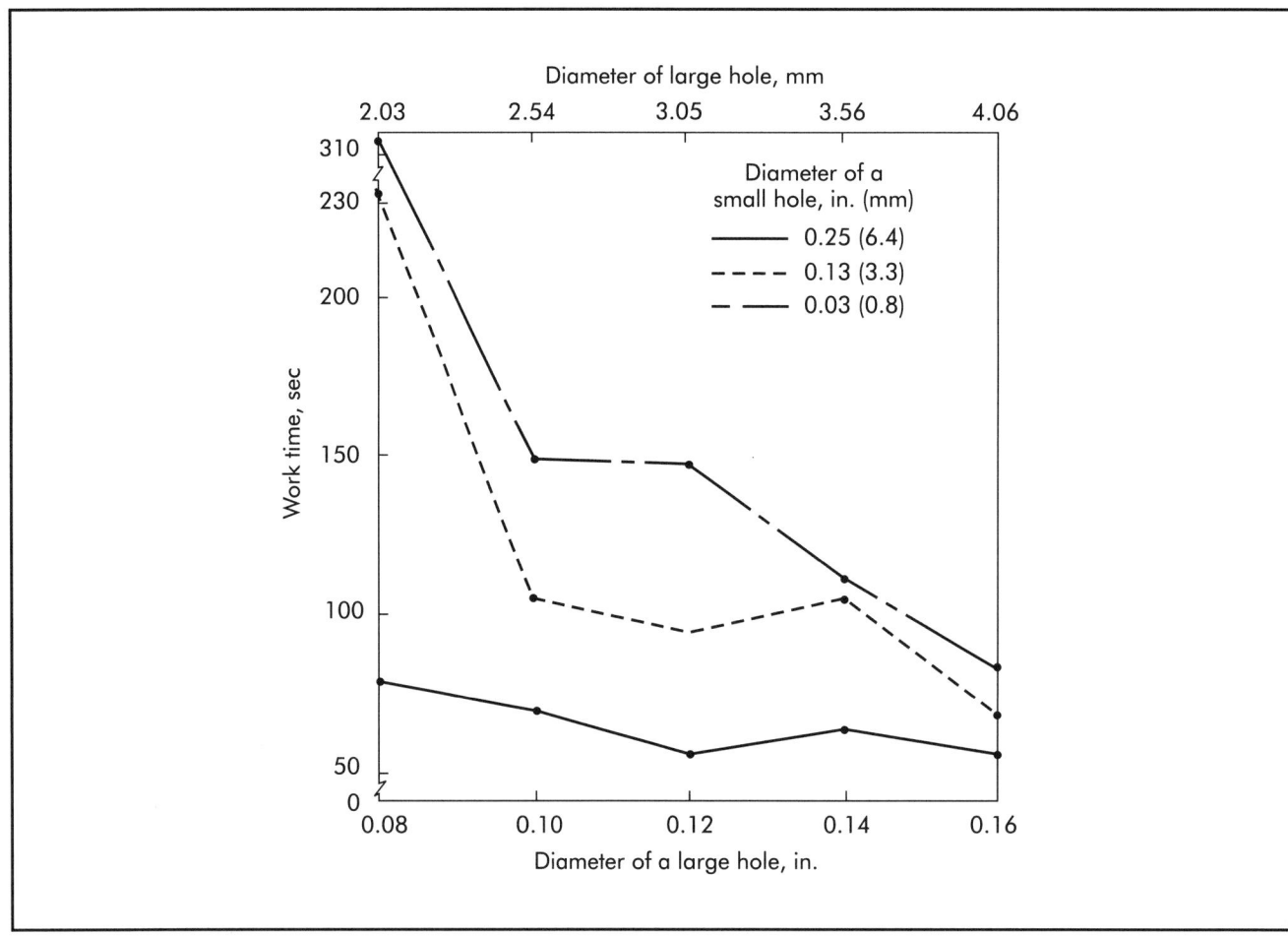

Figure 32-12. Conceptual model of burr removal time as a function of hole size (Gillespie 1975).

Table 32-9. Typical drilling and reaming burr properties

Workpiece Material	Entrance Burr Length, in. (μm)	Thickness, in. (μm)	Exit Burr Length, in. (μm)	Thickness, in. (μm)
Drilled Hole				
303 Se stainless steel	0.0033 (84)	0.0022 (56)	0.0066 (168)	0.0020 (51)
17-4 PH stainless steel	0.0007 (18)	0.0009 (23)	0.0088 (224)	0.0036 (91)
6061-T6 Aluminum	0.0026 (66)	0.0034 (86)	0.0065 (165)	0.0042 (107)
1018 Steel	0.0044 (112)	0.0043 (109)	0.0096 (244)	0.0042 (107)
Drilled and Reamed Holes				
303 Se stainless steel	0.0024 (61)	0.0013 (33)	0.0068 (173)	0.0025 (64)
17-4 PH stainless steel	0.0044 (112)	0.0031 (79)	0.0409 (1,039)	0.0051 (130)
6061-T6 Aluminum	0.0040 (102)	0.0034 (86)	0.0097 (246)	0.0030 (76)
1018 Steel	0.0026 (66)	0.0017 (43)	0.0056 (142)	0.0017 (43)

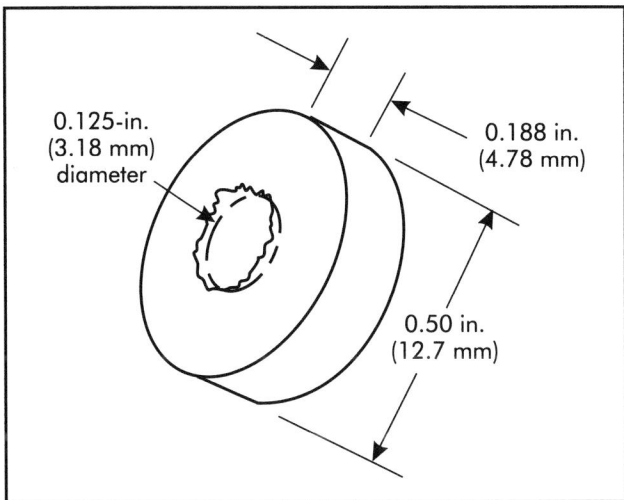

Figure 32-13. Specimen used for analysis of deburring time (Gillespie 1975).

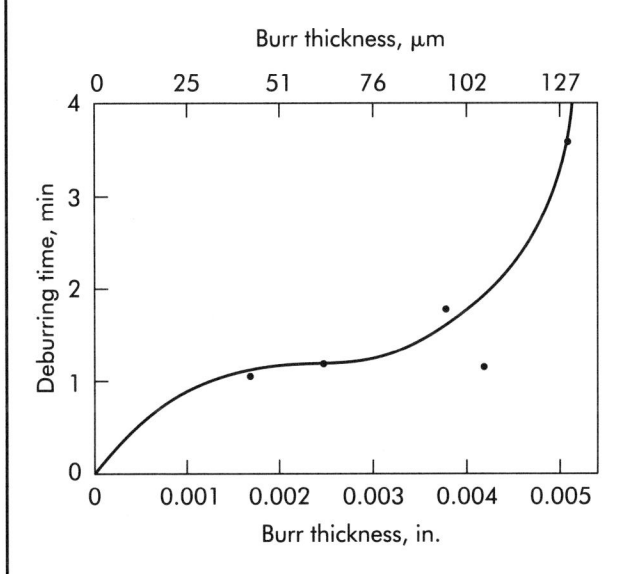

Figure 32-14. Effect of burr thickness on deburring time (Gillespie 1999).

Table 32-10. Time required to deburr washer-like parts

Material	Time Required, min	
	Drill Only	Drill and Ream
303 Se stainless steel	1.464	1.115
17-4 PH stainless steel	1.800	3.545
6061-T6 Aluminum	1.190	0.973
1018 Steel	1.276	1.105

1. rub both sides of the part on 600-grit paper to remove raised metal;
2. cut off the burr with a burr knife on both sides, then
3. rub both sides of the part on 600-grit paper.

A very large burr occurred on the drill and reamed specimen of 17-4 PH stainless steel because rapid drill wear generated high temperatures, causing an extruded burr. Reaming did not remove all of the burr. The 17-4 PH drill specimen, in this test, was made before the drill became dull and, therefore, did not exhibit the extruded burr. Nippers were used to remove the majority of the extruded burr; then the previously described procedure was followed.

As shown in Figure 32-14, which is a plot of the data in Table 32-8, once burr thickness increases beyond 0.004 in. (102 μm), deburring time increases sharply. While only a few points were available from which to construct Figure 32-14, the basic shape and the 0.004 in. (102 μm) breaking point are consistent with conditions observed in production. Although the graph is useful to illustrate the significance of burr size, using the data predicted by the equation describing this curve can provide some misleading estimates of production deburring time on other parts.

$$T = \frac{[\ln(A_1 - b_t) - \ln A_2]^{1/c}}{A_3} \quad (32\text{-}1)$$

where:

T = time required to manually remove burr, min
A_1, A_2, A_3 = constants
b_t = burr thickness, in. (μm)
c = constant

There is a point, undefined as yet, at which each deburring worker will change his or her approach to deburring. For example, nippers were used to cut off the major portion of the 0.005-in. (127-μm) thick burr. Another worker might have used a file, then followed this procedure with sandpaper. Still another might have used a file, a reamer, then sandpaper. In short, because of the wide variety of tools and techniques available, actual times required for deburring will not follow a smooth curve, as shown in Figure 32-14, which ignores the effect of burr length and the time spent on deburring the smaller entrance burr. Equation 32-2 provides a more realistic estimate than Equation 32-1 of deburring time or difficulty.

$$T = C_1(h_e f_e + h_x f_x)\sigma_\mu \quad (32\text{-}2)$$

where:

T = time, min
h_e = height of entrance burr, in. (μm)
h_x = height of exit burr
f_e = Equation 32-1, evaluated for entrance burr thickness, in. (mm)
f_x = Equation 32-1, evaluated for exit burr thickness, in. (μm)
σ_μ = tensile strength of the workpiece material
C_1 = constant

Even this equation is inaccurate for large values of burr length because a burr 0.50-in. (13-mm) long and 0.0005-in. (13-μm) thick does not require 50,000% more deburring time than a 0.001-in. (25-μm) long burr of the same thickness. Since the majority of burrs produced in metal vary from 0.001–0.005-in. (25–127-μm) thick, and 0.001–0.015-in. (25–381-μm) high, Equation 32-2 offers a reasonable and logical baseline for estimating deburring time. Its use, however, requires knowledge of the constants in both equations 32-1 and 32-2.

Effect of Hole Size

In the second test, three different hole sizes were produced in flat plates of four different materials. The diameters of the hole sizes used were 0.1250, 0.0625, and 0.03125 in. (3.175, 1.588, and 0.794 mm). The operator was instructed to maintain a 0.002-in. (51-μm) edge break around both sides of each hole. (Throughout this handbook, it is assumed that the edges after deburring must be free of any material extending past the two 90° surfaces that form the edge. In other words, both sharp fragments and bulging material must be removed.)

Smaller holes frequently require more time to deburr than larger holes (Figure 32-15). More time is spent scrutinizing under a microscope the deburring results of small holes than larger holes. In addition, most deburring tools are designed for 0.125-in. (3.18-mm) diameter holes, rather than smaller sizes. Ten percent of the allowed time is spent looking for burrs on simple parts rather than deburring. On small gears, 50% of the rated time may be spent just looking for minute burr fragments.

Figure 32-15 shows that holes smaller than 0.125 in. (3.18 mm) require more time to deburr than do 0.125-in. (3.18-mm) diameter burrs. The effect of hole size, however, does not follow a smooth curve, as shown in Figure 32-12, partly because it is not possible to produce the same burr properties with three different drill sizes (see Table 32-11). The major reason, however, is that when accessibility of the burr restricts use of one deburring approach, a different approach is tried. While this introduces a non-uniform test condition, it does reflect actual shop practice.

In this test, the 0.12-in. (3.1-mm) diameter holes were deburred by the following three-step procedure:

1. wiping both sides of the hole with 400-grit emery paper;
2. cutting off the burr with a burr knife, then
3. removing slivers produced by the knife by polishing with a 1.00-in. (25.4-mm) diameter motor-driven nylon wheel.

In the case of the two smaller holes, a reamer was used in place of a burr knife. For practical purposes, the thickness of the exit burr was the same for all these holes (Table 32-11). As a result, this variable is not influencing the trends shown in Figure 32-15.

It is interesting to note that 10% of the time spent deburring the samples was actually devoted to scrutinizing the holes to make sure the burrs were removed. On more complex and critical parts, that portion of time for scrutinizing increases. On some miniature parts, the operator spends more time checking for a burr-free condition than actually removing the burrs.

Effect of Burr Location

In the third test, the intersections of four 0.062-in. (1.57-mm) diameter holes with a larger cross-hole were deburred. The cross-hole diameter varied from 0.062–0.500 in. (1.57–12.70 mm) (Figure 32-16). This particular geometry reflects one of the most difficult hand-deburring situations—essentially, there is no room to maneuver the deburring tools.

Figure 32-16 illustrates a part that has four small holes breaking into a larger one. The time required to deburr the intersections depends on the diameter of the large cross-hole. When it was 0.50 in. (12.7 mm) in diameter, a 6061-T6 aluminum part (Figure 32-17) required less than three minutes to deburr. When the horizontal hole was 0.062 in. (1.57 mm), deburring took 11 minutes. Unfortunately, with a very small cross-hole, providing a nice, smooth blend at hole intersections by hand deburring is impossible. As a result, more time on such difficult features is required and deburring quality worsens. Processes (such as abrasive-flow and electrochemical deburring) can, however, provide smooth blending on any of these hard-to-reach intersections.

It proved impossible to deburr the intersection of holes when the cross-hole was only 0.062 in. (1.57 mm) in diameter. As expected, the larger the cross-hole the shorter the time required for deburring (Figure 32-17). As the cross-hole becomes smaller, the effectiveness of a deburring knife diminishes. Essentially,

Table 32-11. Properties of burrs produced on three different hole sizes

Workpiece Material	Hole Diameter, in. (μm)	Entrance Burr, in. (μm)		Exit Burr, in. (μm)	
		Hole Length	Hole Thickness	Hole Length	Hole Thickness
6061-T6 Aluminum	0.03125 (794)	0.0026 (66)	0.0016 (41)	0.0054 (137)	0.0029 (74)
		0.0022 (56)	0.0012 (31)	0.0051 (130)	0.0023 (58)
	0.0625 (1,588)	0.0007 (18)	0.0006 (15)	0.0081 (206)	0.0014 (36)
		0.0010 (25)	0.0009 (23)	0.0102 (259)	0.0017 (43)
	0.125 (3,175)	0	0	0.0064 (163)	0.0010 (25)
		0	0	0.0096 (244)	0.0012 (31)
303 Se stainless steel	0.03125 (794)	0.0008 (20)	0.0005 (13)	0.0034 (86)	0.0019 (48)
		0.0015 (38)	0.0010 (25)	0.0022 (56)	0.0014 (36)
	0.0625 (1,588)	0	0	0.0050 (127)	0.0014 (36)
		0.0007 (18)	0.0015 (38)	0.0059 (150)	0.0020 (51)
	0.125 (3,175)	0.0035 (89)	0.0007 (18)	0.0037 (94)	0.0014 (36)
		0	0	0.0072 (183)	0.0021 (53)
17-4 PH stainless steel	0.03125 (794)	0	0	0.0026 (66)	0.0017 (43)
		0	0	0.0019 (48)	0.0020 (51)
	0.0625 (1,588)	0.0015 (38)	0.0005 (13)	0.0028 (71)	0.0012 (31)
		0	0	0.0028 (71)	0.0012 (31)
	0.125 (3,175)	0.0021 (53)	0.0009 (23)	0.0018 (46)	0.0012 (31)
		0.0003 (8)	0.0009 (23)	0.0025 (64)	0.0011 (28)
1018 Steel	0.03125 (794)	0.0017 (43)	0.0004 (10)	0.0036 (91)	0.0011 (28)
		0.0016 (41)	0.0007 (18)	0.0044 (112)	0.0010 (25)
	0.0625 (1,588)	0.0005 (13)	0.0002 (5)	0.0004 (10)	0.0011 (28)
		0	0	0	0
	0.125 (3,175)	0	0	0	0
		0	0	0.0043 (109)	0.0008 (20)

Note: Holes were produced by high-speed steel (HSS) drills (at 1,200 rpm and 1.20 in./min [0.5 mm/sec]). A water-soluble coolant was used on all holes. No drill produced more than 21 holes.

reamers must be depended on to remove the burrs. By alternately reaming the large hole, then the small holes, and then repeating this sequence, it is possible to remove the large, flexible portion of the burr. This approach, however, will not guarantee that a small burr will not be left on hole intersections. Neither will it radius the intersecting edges. Thus, for manual deburring, the quality of deburring decreases as the

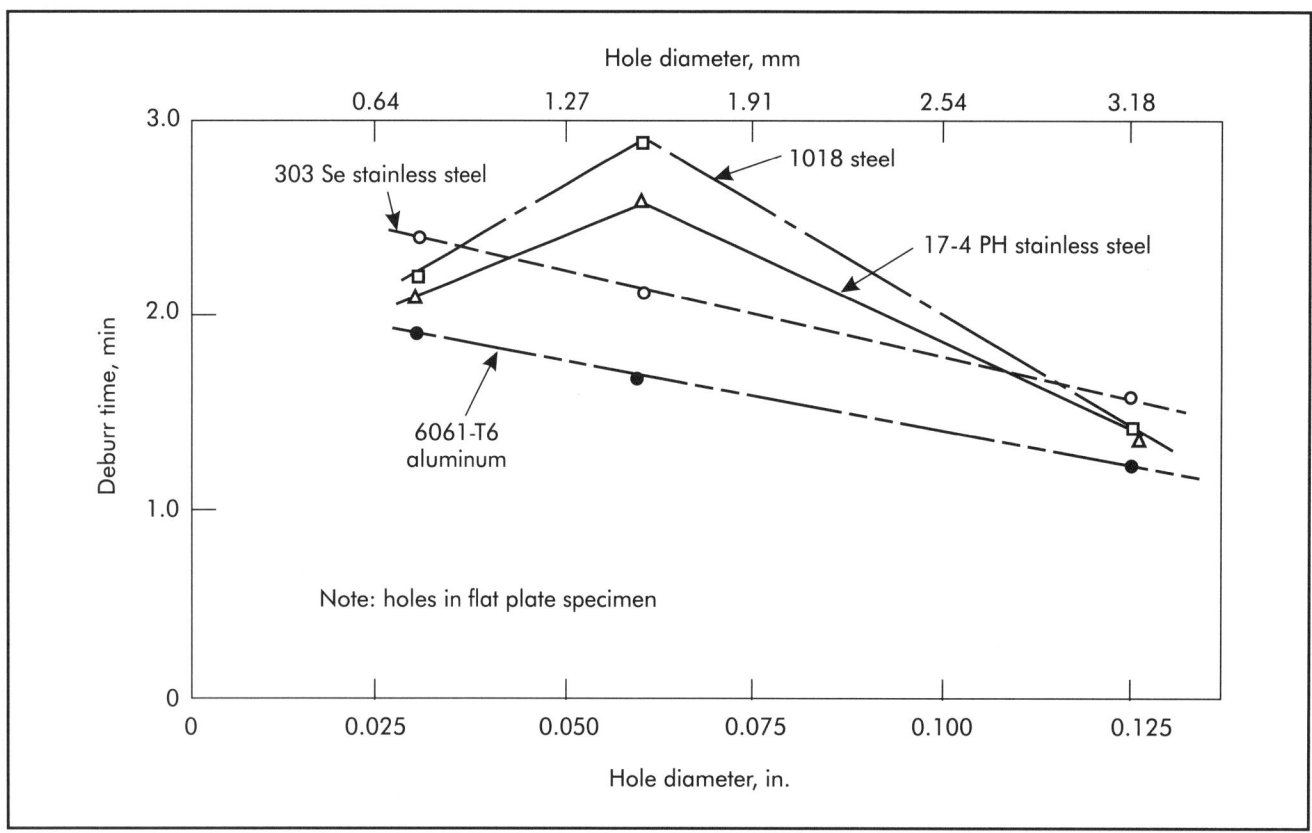

Figure 32-15. Effect of hole size on deburring time (Gillespie 1978b).

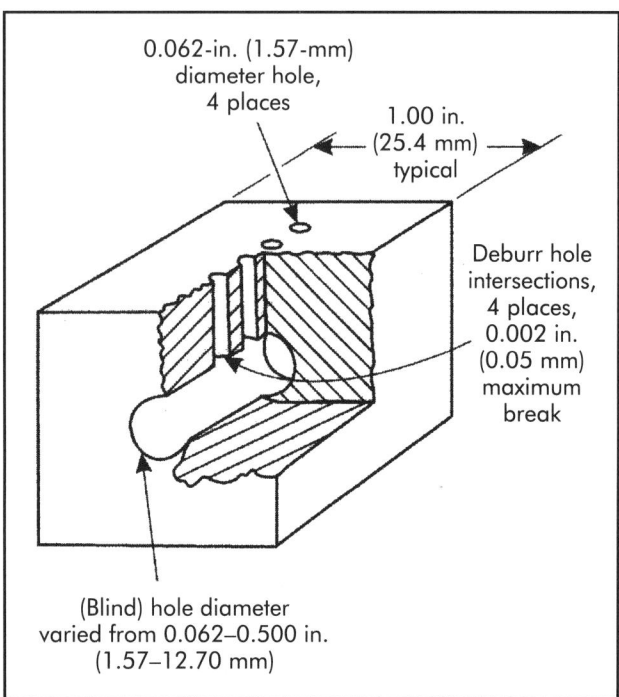

Figure 32-16. Geometry of intersecting hole specimen (Gillespie 1975).

effective working area decreases. No attempt was made, however, to measure the edge break on the intersecting holes. It should be noted that measuring small edge radii becomes extremely difficult for small features and three-dimensional intersections.

Figure 32-18 illustrates another test sample. Figure 32-19 illustrates the time required for deburring slotted parts. As slot width decreases, deburring times rise exponentially when hand deburring is required. Half the workpieces had 0.032-in. (0.81-mm) wide slots, while the other half had 0.25-in. (6.4-mm) wide slots (Figure 32-18). The time required to deburr the four slots manually on both sets of workpieces was recorded and divided by the total edge length deburred. As illustrated in Figure 32-19, the 0.032-in. (0.81-mm) wide slots required more time to deburr per inch (mm) of edge than did the wider slot. It appears that below 0.032 in. (0.81 mm), deburring time would increase sharply.

As previously indicated, deburring techniques vary as a function of part, geometry, feature size, and workpiece material. In this study, a file was used to remove the majority of the burr from the stainless-steel workpieces. Because of the restricted access, it could not be used on the narrow slots. In 1018 steel

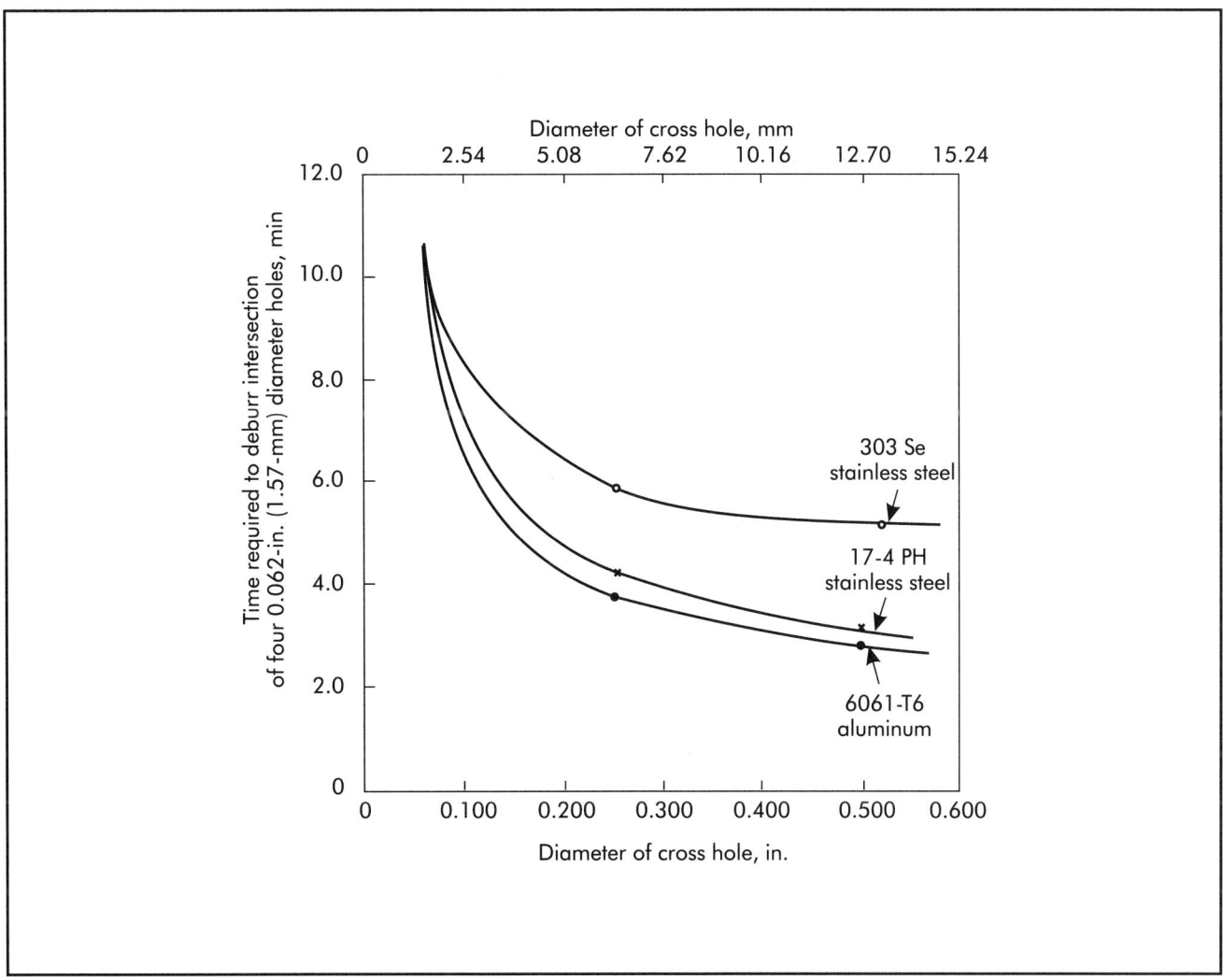

Figure 32-17. Effect of cross-hole diameter on deburring time (Gillespie 1975).

and aluminum, the burr was thin enough that a burr knife was the most effective tool on slots of both widths. Burr properties were not recorded.

Interestingly, it is easier to remove a burr from the flat side than the L-side of parts of the same size. The addition of the ledge can increase deburring time by 50% on this particular workpiece. Since the location of the large cutter exit burr is a function of setup or fixturing, deburring time can be influenced by both the manufacturing engineer and the machine operator.

Deburring Tool Capabilities

The limits that each hand deburring tool and technique are capable of have not been defined. For example, these questions might be asked: "Can a Cratex® bullet remove a burr and produce no more than an 0.001-in. (25-μm) edge break? If it can, how long does it require, and what technique must be used to achieve it?" Although extensive detailed information is unavailable, some recent results provide background information.

Figure 32-20 shows results obtained by a variety of tools. For example, one individual, using only a miniature file, was able to remove burrs and leave an edge break of only 0.001 in. (25 μm). Another person could not produce a break smaller than 0.002 in. (51 μm). By aggressive action, a 0.010-in. (0.25-mm) break can be produced with miniature files.

A deburring knife or a 1.00-in. (25.4-mm) diameter, abrasive-impregnated nylon brush also can maintain a 0.001-in. (25-μm) break, if the user is skillful and careful (Figure 32-20). Typically, knives produce an 0.003–0.005-in. (76–127-μm) break. Some useful guidance on the aggressiveness of various tools is available

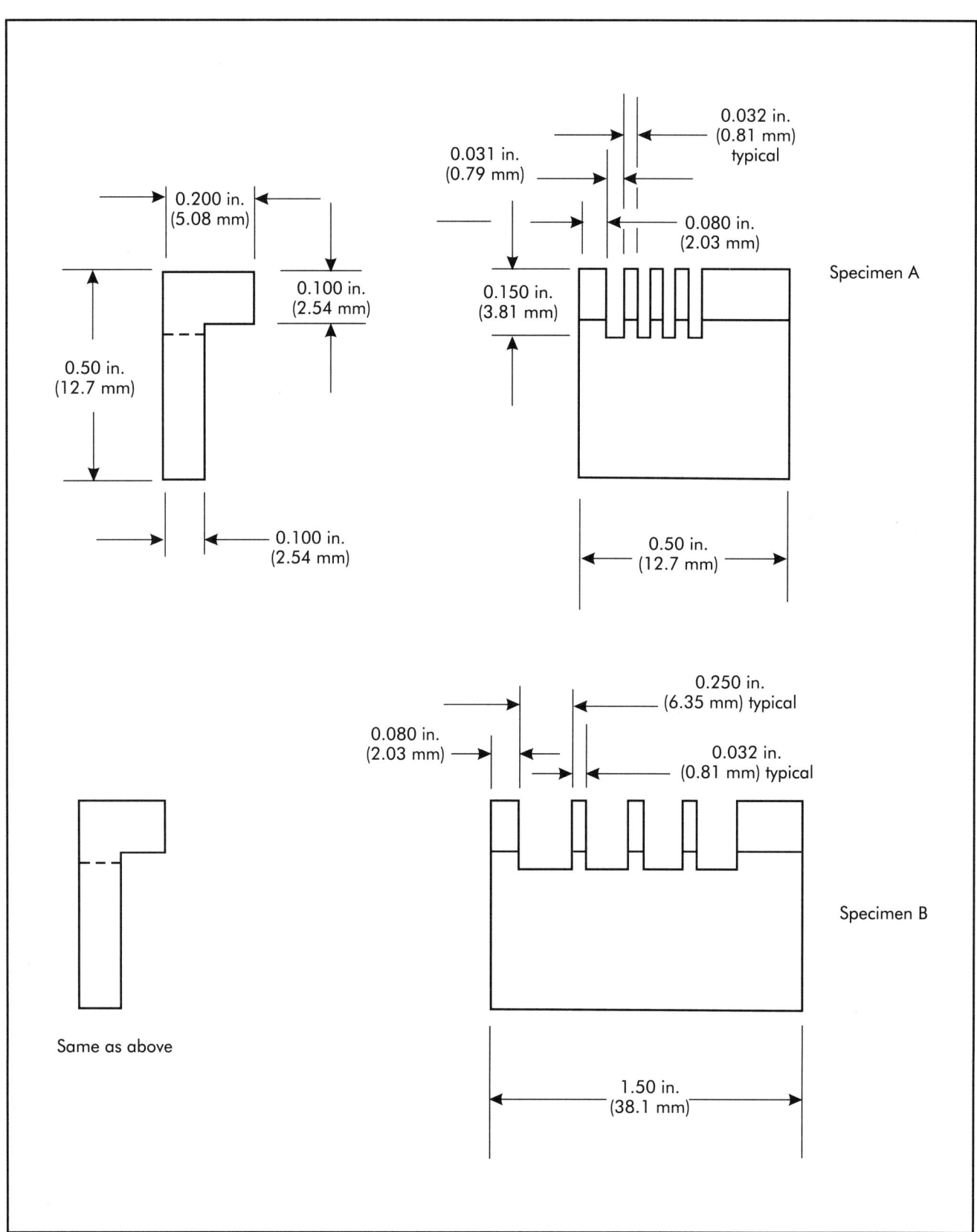

Figure 32-18. Milled slot specimens (Gillespie 1975).

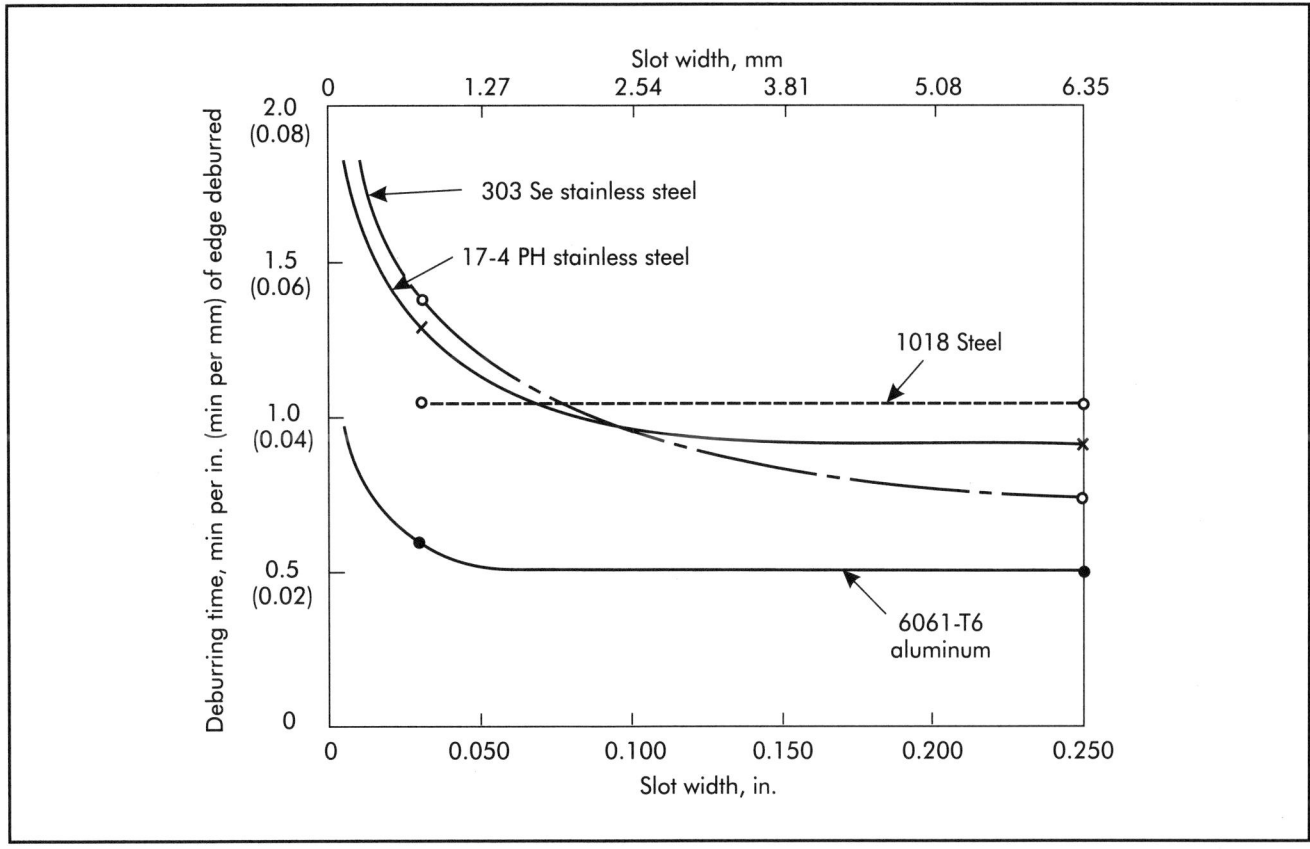

Figure 32-19. Effect of slot width on deburring time (Gillespie 1975).

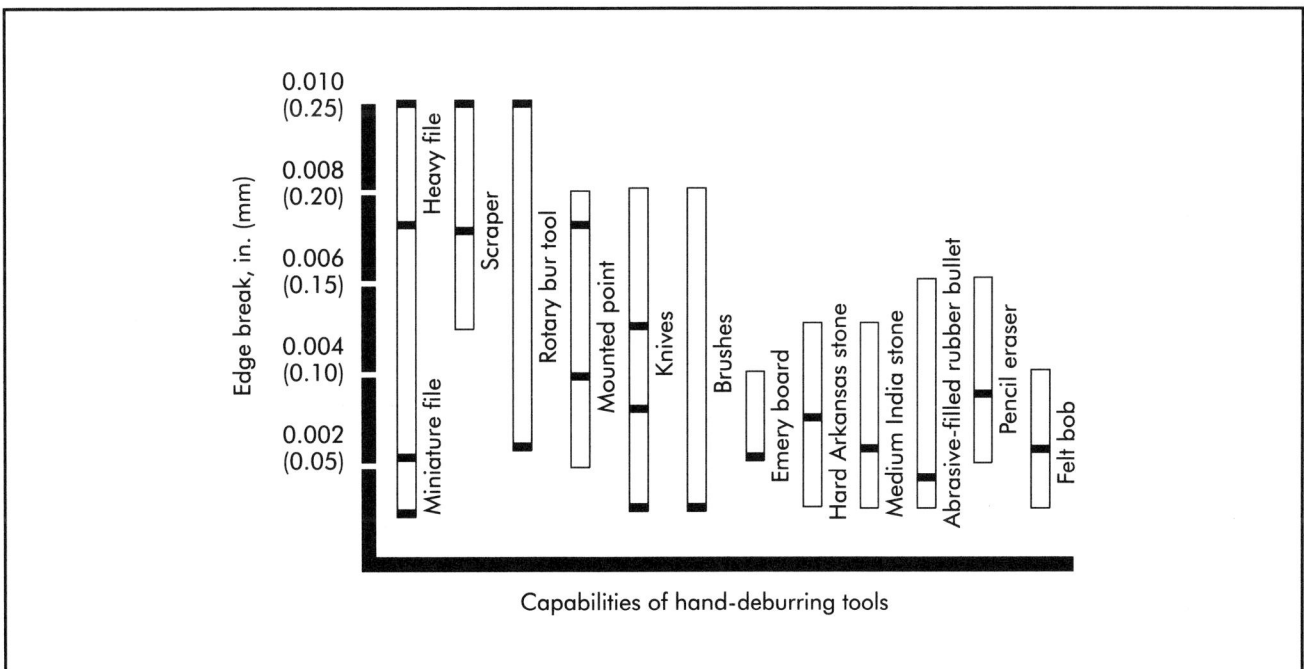

Figure 32-20. Basic edge-break capabilities using specific tools to remove 0.003 in. (76 μm) thick × 0.003-in. (76-μm) high burrs from 303 Se stainless steel (horizontal bars indicate measured results by one operator) (Gillespie 1999).

from this study. However, every individual will generate somewhat different results, even under identical conditions. Part geometry and burr size also influence results.

Figure 32-21 illustrates the effect of grit size on surface finish when various abrasive papers are used. The finish is subject to considerable variation. Worn abrasive paper and wet sanding typically produce the best finishes. Figure 32-22 illustrates the effect of some deburring tools on part dimensions.

General Observations

The extensive review of manual-deburring capabilities was presented to demonstrate the need for data-driven solutions. The plant that performed the studies has facts to convince management that mechanization is needed or will not work. It has facts to estimate deburring costs on the next design. This plant knows which questions to ask. It is an industry leader. The data indicate that the workers at this plant are exceptionally capable of meeting requirements. The tests did not take long to perform. They provided a means of measuring worker quality, especially of those who had been recently trained. The same samples could also provide information on production rate. The plant had facts.

As mentioned earlier, 10% of the time spent in deburring the test samples was actually devoted to determining if all the burr was removed. On precision miniature production parts, this scrutinizing time can exceed 60% of the total hand-deburring time. While 10% does not seem excessive, it occurs on every single part deburred. At this rate, on 200 parts requiring three minutes each to hand deburr, one hour would be devoted to just checking edge quality. That scrutinizing time alone is longer than that used for some mechanized processes deburring 200 parts.

High-speed movies of workers hand deburring production parts indicate that some motions of workers are more habitual than essential. For example, operators develop a certain rhythm for polishing an edge with sandpaper. For many edges, the natural wrist motion is to make three passes over an edge; however, one pass would be sufficient for many edges. These habitual actions are beneficial on some workpieces, but merely time-consuming on others.

Effective Utilization of Hand Deburring

The comments made in this chapter have assumed that deburring the part was done by hand and to precision tolerances. In most instances, it is more efficient to use hand deburring as only a part of the

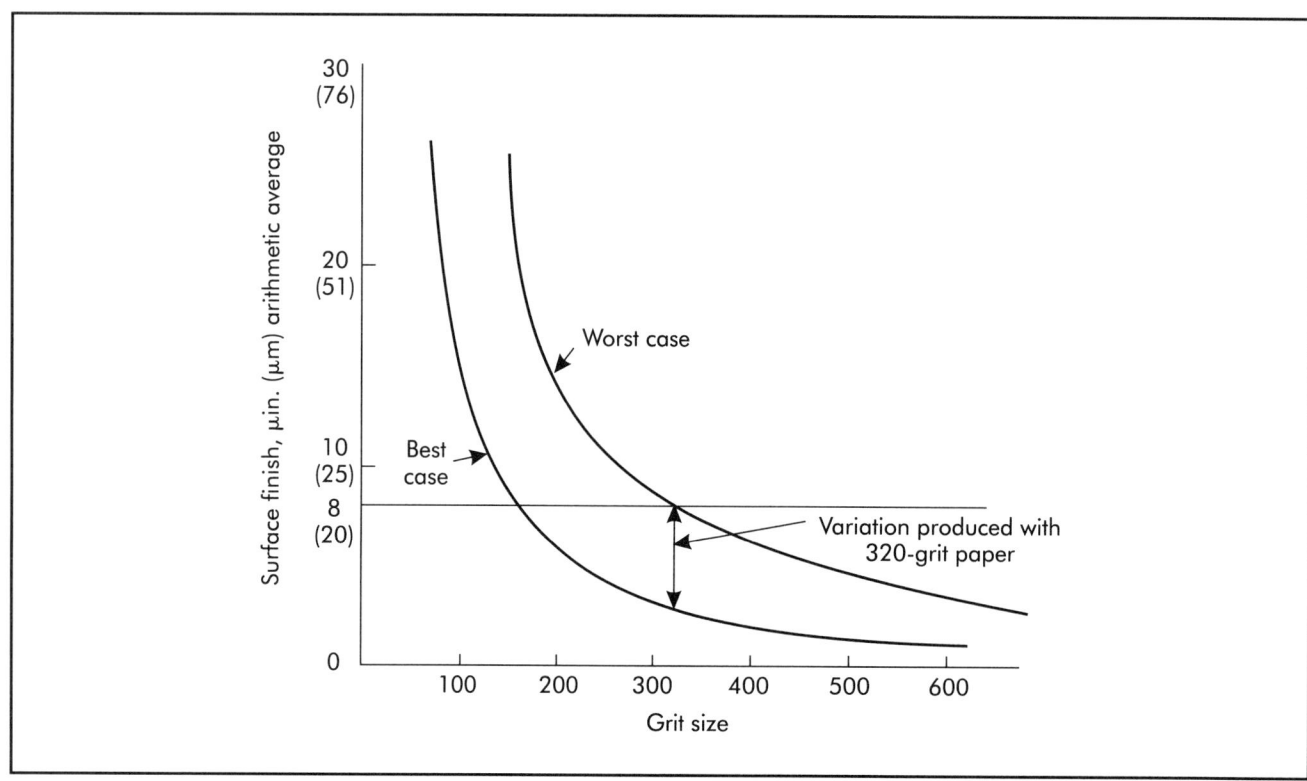

Figure 32-21. Surface finish produced as a function of grit size used (Gillespie 1978b).

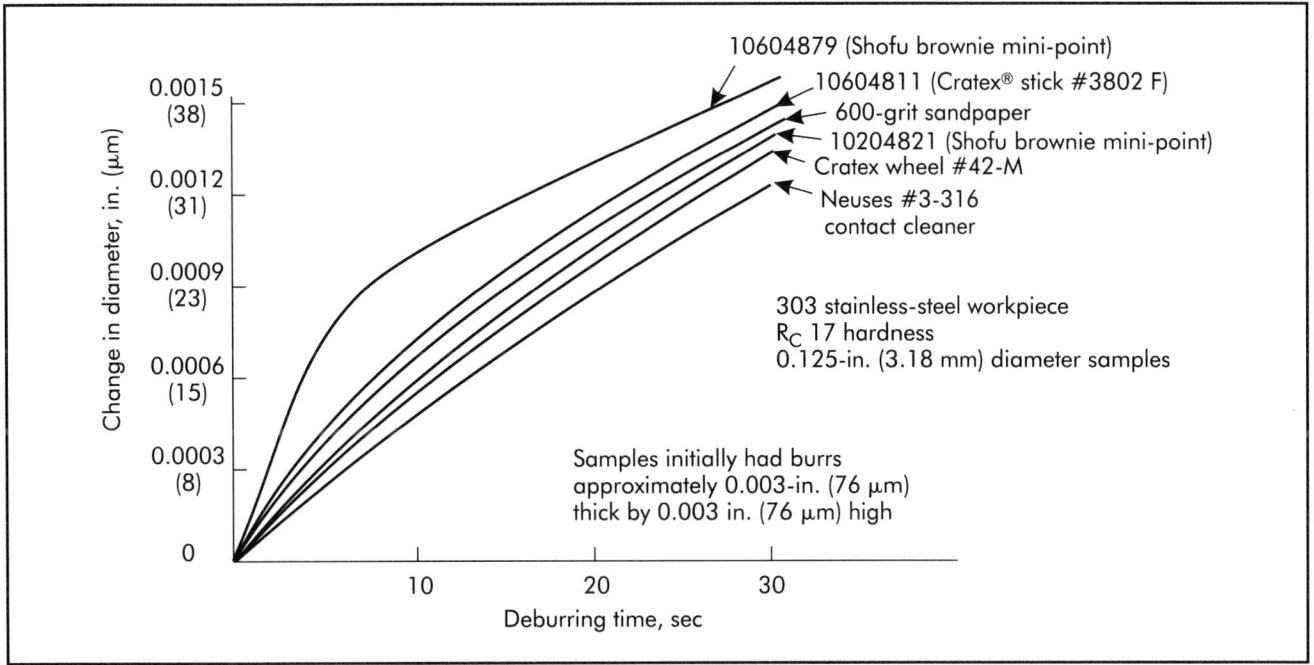

Figure 32-22. Effect of several deburring tools on part size (Gillespie 1978b).

deburring sequence. The basic approaches to utilizing hand deburring are to:

- hand deburr the entire part to print requirements;
- hand deburr only certain edges on the part to print requirements to allow the part to locate in a fixture, and deburr edges not reached by mechanized processes, and
- hand deburr to remove only the big or loose portion of the burrs.

In the second and third approaches to hand deburring, a faster or mechanized deburring process is used to produce the final edge quality required on a part. Most deburring processes can maintain radii requirements of 0.002-in. (51-μm) maximum, if the burr to be removed is not large.

Generally, the majority of time spent in hand deburring precision miniature parts is not in removing the large burr; rather, it is in ensuring that all the burr is removed and edge radii requirements are met. In many applications, the essential requirement is to remove the loose or flexible portion of the burr, not the entire burr. Often, hand deburring can be the most efficient method of doing this. Subsequent deburring by a mechanized process smoothes the edges and removes the small, often flexible burr produced when the larger burr is cut off.

The most difficult aspect of the third approach is defining for the worker how much of the burr must be removed. On some precision miniature parts, which are deburred and radiused by centrifugal barrel tumbling, for example, it is essential that the burr left by hand deburring not be larger than 0.002-in. (51-μm) high and 0.002-in. (51-μm) thick. On miniature parts, it is difficult and time-consuming to measure such features. More significantly, however, no information has been published that describes the amount of burr that can be removed by a given process while maintaining specified workpiece tolerances, surface finishes, and edge radii. Thus, at this time, it is impossible to provide absolute, quantitative information for the person doing hand deburring to determine when he or she has removed enough of the burr.

When hand deburring is required, it is economically essential that as much of it as possible be performed after the part is completely machined. Deburring between machining operations can unnecessarily increase cost for two reasons:

- some of these deburred edges may be subsequently machined away, and this represents wasted work, and
- the individual doing the final deburring typically will spend some time scrutinizing edges that were already deburred and scrutinized.

Other methods and factors also represent wasted work, including the following.

- On edges that are easy to reach, it may be possible to maintain smaller edge radii. Increasing

the number of edges to be deburred and the complexity of the deburring increases the variability of results.
- Deburring time appears to be a function of burr thickness, which, in turn, is a function of the strain-hardening exponent. No relationship was observed between burr height and deburring time.
- Deburring techniques vary among workers and as a function of burr accessibility. This makes it difficult to predict deburring time. Openings around burr-laden edges less than 0.060 in. (1.52 mm) in diameter or width increase deburring time significantly.
- Although hand deburring is an economic necessity on many short-run precision miniature parts, it is possible to minimize the time spent and maintain 0.002-in. (51-μm) maximum edge radii by utilizing both hand and mechanized deburring.
- The development of improved hand deburring results and times requires a knowledge of the tools available, the situations in which they work best, and some data on the results produced.

Operational Cost Data

A major aerospace company investigated improvements in sanding and deburring titanium material. The approach used is well-designed to help other plants find more economical supplies for their own application. The objective of this study was to establish an optimum abrasive finishing tool and a deburring and sanding process that was cost-effective for titanium material. The current shop was set up for aluminum manufacturing with finishing tools that ranged from 12,000–30,000 rpm.

Finishing air motors and sanding products were evaluated under various controlled test conditions that met the 2,000 sfm (10.2 m/sec) maximum requirement found in the internal manufacturing specification for this material. Five air-motor finishing tools were evaluated, ranging from 2,000–9,000 rpm, and five different 2-in. (50.8-mm) diameter abrasive-disc compositions/constructions were also studied. The abrasive disc speed had to be below 3,820 rpm to meet the 2,000 sfm (10.2 m/sec) requirement. A single sanding parameter sample was collected to indicate air-motor and abrasive-product performance. The removal rate was measured in material weight loss in grams, and surface finish was measured with a profilometer in μin., Ra. The initial controlled tests were conducted by robots to provide exact repeatability, but later tests were performed by hand operators.

Test phase one evaluated air motors. Test phase two used one air motor to evaluate five abrasives and two process parameters. Test phase three evaluated firmer abrasive backing. Test phase four evaluated air motors and abrasives with the operators who would use them in production. All test specimens were titanium 6A1-4V with a sanding area of 0.5625 × 5.5000 in. (14.288 × 139.700 mm). Test-monitoring variables included titanium material weight loss in grams and three surface-finish measurements per sample. A digital tachometer was used to monitor rpm under load conditions. Table 32-12 summarizes the test parameters and equipment used.

Air-motor Performance

The first test phase investigated five finishing tools (air motors) from two manufacturers. They included straight and right-angle heads. The intent was to document motor rpm range under load conditions that would limit rpm for the 2-in. (50.8-mm) diameter abrasive discs tested. Four different load conditions were used. Five finishing tools and four sanding parameters, under test conditions, are shown in Table 32-13. Note that the manufacturer's rated rpm differs significantly from the tested condition, with the exception of one finishing tool. Results indicate that a 6,000-rpm finishing motor best meets the 3,820-rpm maximum requirement specified in the process specification. Clearly, manufacturers' data may not reflect real operating conditions for air-motor speeds under either no-load or load conditions. Users must test for themselves to meet the requirements. To meet the 2,000 sfm (10.2 m/sec) requirement, abrasive disc rpm should not exceed 3,820.

Sanding-disc Performance

Table 32-14 (a–d) shows an abrasive-product rating versus the sanding parameter during a five-minute sanding time period. The lower overall rating number indicates a higher-performing abrasive product for material removal and lower Ra surface-finish number. Abrasive products that have two or three numbers grouped together indicate similar performance. In Table 32-14a, disc A was rated "1" in all but one of the four test setups, and "1" was the highest rating (first place). Note that it was far superior to the other products for removal rate when compared to the composite rating in the last row.

Product B, however, provided the best surface finish results (Table 32-14b). Surface roughness was only 21–39 Ra, while product A averaged in the 40+ μin. (1+ μm) range. The robotically positioned air motors positioned the discs at an angle to the surface (either

Table 32-12. Test parameter variables

Parameter	Variable	
Abrasive Type	Alumina Zirconia	Ceramic Aluminum Oxide
Abrasive size	80 grit	120 grit
Applied force	5 lbf (22 N)	9 lbf (40 N)
Applied sanding angle	4°	10°
Abrasive-backing firmness	Hard (phase two tests)	Extra hard (phase three tests)
Abrasive configuration and size	2 in. (50.8 mm) disc	—
Sanding oscillation	25 oscillations/min	—

Table 32-13. Air-motor speed under test conditions

Tool Code	Manufacturer Rated rpm	No Load rpm	Delta rpm (%)	5 lbf (22 N)/ 4°, rpm	9 lbf (40 N)/ 4°, rpm	5 lbf (22 N)/ 10°, rpm	9 lbf (40 N)/ 10°, rpm
1	2,000	2,025	1	1,845	1,725	1,710	1,660
2	3,200	2,696	16	2,530	2,459	2,436	2,249
3	5,000	3,200	36	2,682	2,281	2,277	1,900
4	6,000	4,480	25	4,090	3,704	3,670	3,265
5	9,000	7,199	20	5,430	4,258	4,172	3,580

Abrasive-disc backing rating: hard
Abrasive-grit size: 120
Abrasive type: code "Disc A"

4 or 10°) being sanded and applied either 5 or 9 lbf (22 or 40 N) sanding force to the surface. In Table 32-14c, which used 120-grit abrasive, discs A and E were best for the rate of removal. Table 32-14d indicates disc B again was somewhat better than A.

Although there is no information about the manufacturers of the discs, the data provide useful information, including a simple test matrix and key variables for performing similar tests in other plants. Removal rate ranges are defined. The impact of sanding force and approach angle are shown. Not surprisingly, heavier forces remove more material, but do not necessarily worsen surface roughness. The data also allows readers to perform experimental designs (analysis of variance) to search for more subtle clues to performance.

Test phase three investigated three abrasives selected by best removal rate, best surface finish, and an abrasive with a middle-of-the-pack performance for removal rate and surface finish. An extra-hard abrasive backing was used. Abrasive product A was selected for its removal rate, B for its surface finish, and D for its average rating for removal rate and surface finish. Test parameters followed the same procedures as described for Table 32-14 a–d. Hard backing does not make a significant difference in performance.

Tables 32-15 and 32-16 describe the cost comparisons of the data from Table 32-14.

Shop Operator Ratings

Table 32-17 indicates abrasive-product ranking per air-motor finishing tool, air-motor ranking, and shop acceptance of combination abrasive product and motors as indicated in bold-italic typeface. For this test, three finishing tools and five abrasives were evaluated by a minimum of two shop personnel from each of three shifts. Participants sanded mismatched and rough surfaces to determine performance.

As Table 32-17 shows, the operators preferred disc C, which had the lowest rating when the robot performed the tests. Product E was not a top performer when robots performed the tests, but human operators rated it as second most desirable. The air motor picked by the operators did correspond to those used by the robots.

Cost Analysis

A cost analysis summary, based on 3.5 oz (100 g) of titanium removed, is shown in tables 32-15 and 32-16. The cost analysis utilized the 6,000-rpm finishing tool and hard-abrasive backing. Calculations are based

Table 32-14a. Removal rating for 80-grit abrasive with hard backing

Abrasive Product Code No.	5 lbf (22 N)/4°		5 lbf (22 N)/10°		9 lbf (40 N)/4°		9 lbf (40 N)/10°		Overall Ranking No.
	Rating	Stock Loss oz (grams)	Rating	Stock Loss oz (grams)	Rating	Stock Loss oz (grams)	Rating	Stock Loss oz (grams)	
A	1	0.108 (3.06)	2	0.145 (4.11)	1	0.186 (5.27)	1	0.224 (6.35)	5
B	5	0.033 (0.94)	5	0.029 (0.82)	5	0.047 (1.33)	5	0.063 (1.79)	20
C	4	0.055 (1.56)	4	0.065 (1.84)	3–4	0.105 (2.98)	4	0.101 (2.86)	15.5
D	2	0.064 (1.81)	1	0.154 (4.37)	3–4	0.103 (2.92)	3	0.131 (3.71)	9.5
E	3	0.058 (1.64)	3	0.132 (3.74)	2	0.152 (4.31)	2	0.202 (5.73)	10

Table 32-14b. Surface finish rating with 80-grit hard-backing discs

Abrasive Product Code No.	Surface Finish								Overall Ranking No.
	5 lbf (22 N)/4°		5 lbf (22 N)/10°		9 lbf (40 N)/4°		9 lbf (40 N)/10°		
	Rating	Ra Average	Rating	Ra Average	Rating	Ra Average	Rating	Ra Average	
A	4	61	2	41	2–3	50	2	48	10.5
B	1	39	1	27	1	21	1	30	4
C	3	50	4–5	62	4	65	4–5	69	16
D	2	47	4–5	63	2–3	52	4–5	78	13.5
E	5	77	3	52	4–5	71	3	73	16

Table 32-14c. Removal rating with 120-grit abrasive hard backing

Abrasive Product Code No.	5 lbf (22 N)/4°		5 lbf (22 N)/10°		9 lbf (40 N)/4°		9 lbf (40 N)/10°		Overall Ranking No.
	Rating	Stock Loss oz (grams)	Rating	Stock Loss oz (grams)	Rating	Stock Loss oz (grams)	Rating	Stock Loss oz (grams)	
A	2–3	0.074 (2.10)	1–2	0.108 (3.06)	1	0.133 (3.77)	1	0.181 (5.13)	6.5
B	5	0.022 (0.62)	5	0.028 (0.79)	5	0.036 (1.02)	5	0.043 (1.22)	20
C	4	0.057 (1.62)	4	0.061 (1.73)	4	0.086 (2.44)	3	0.109 (3.09)	15
D	2–3	0.075 (2.13)	3	0.099 (2.81)	2–3	0.119 (3.37)	4	0.099 (2.81)	12
E	1	0.086 (2.44)	1–2	0.108 (3.06)	2–3	0.120 (3.40)	2	0.147 (4.17)	7

Table 32-14d. Surface finish rating with 120-grit hard backing

Abrasive Product Code No.	Surface Finish								Overall Ranking No.
	5 lbf (22 N)/4°		5 lbf (22 N)/10°		9 lbf (40 N)/4°		9 lbf (40 N)/10°		
	Rating	Ra Average	Rating	Ra Average	Rating	Ra Average	Rating	Ra Average	
A	3–4	45	1	23	2	35	2	33	8.5
B	1	36	2	29	1	15	1	22	5
C	5	49	4	46	4–5	50	5	64	18.5
D	2	40	3	39	4–5	50	3	39	12.5
E	3–4	45	5	52	3	37	4	50	15.5

Removal rate: 1 = Most material removed
Surface finish: 1 = Lowest Ra number (best finish)

Table 32-15. Abrasive cost summary with hard backing material

Abrasive Product Code No.	Sanding Parameters								Cost Indicator ($)	Cost Increase ($)
	5 lbf (22 N)/4°		5 lbf (22 N)/10°		9 lbf (40 N)/4°		9 lbf (40 N)/10°			
	80 Grit ($)	120 Grit ($)	80 Grit ($)	120 Grit ($)	80 Grit ($)	120 Grit ($)	80 Grit ($)	120 Grit ($)		
A	153	221	113	153	90	124	74	91	1,019	0
B	484	730	620	582	343	452	251	370	3,832	276
C	289	273	246	262	153	186	158	147	1,714	68
D	250	212	103	158	158	136	120	158	1,295	27
E	280	187	121	148	105	132	82	110	1,165	14

Table 32-16. Abrasive cost summary with extra-hard backing material

Abrasive Product Code No.	Sanding Parameters								Cost Indicator ($)	Cost Increase ($)
	5 lbf (22 N)/4°		5 lbf (22 N)/10°		9 lbf (40 N)/4°		9 lbf (40 N)/10°			
	80 Grit ($)	120 Grit ($)	80 Grit ($)	120 Grit ($)	80 Grit ($)	120 Grit ($)	80 Grit ($)	120 Grit ($)		
A	113	164	119	215	79	119	74	119	1,002	0
B	463	512	365	430	343	337	278	359	3,087	201
D	256	229	338	169	147	130	125	103	1,497	49

Table 32-17. Shop evaluation of abrasives and finishing motors

Abrasive Product Code No.	Air Motor Code–(rpm)–(Ranking)		
	5–(9,000)–(3)	4–(6,000)–(1)	3–(5,000)–(2)
A	4	4	4
B	5	5	5
C	*1*	*1*	*1*
D	3	2	3
E	2	2	2

Note: Shop acceptance of combination abrasive product and motors is indicated in bold-italic typeface.

on an abrasive-disc change frequency of five minutes. Test results indicate material removal for all abrasives, with the exception of one, which no longer removed material after four minutes. Results indicate that product A has the lowest cost when abrasive and user costs are considered.

Summary

The tests indicate that finishing-motor rpm varied considerably. The higher horsepower (0.5–1.0 hp [0.37–0.75 kW]) finishing motors have minimum rpm variation under load. Air-motor rpm was influenced by air supply and line size. A 6,000-rpm finishing tool, ±500 rpm, best met the requirements.

Performance data. Any shop can quickly note which operators produce more good parts in a given time. This efficiency data may be an important consideration in improving hand deburring. One plant, using methods engineering approaches, measured the efficiency of deburring workers and found that they varied from 45% of normal to 150% for one operator. Knowing the efficiency of each person may allow plant leaders to assign specific workers to the type of deburring at which they are most efficient. Putting the least efficient worker on the most exacting, detailed work does not make much sense, unless other personnel factors are involved.

TYPICAL PROBLEM AREAS

Experience has shown that several problems frequently occur during hand deburring. Some of these issues have been covered previously, so the following discussion only centers on a few of the major issues.

Burrs on Holes

Burrs only 0.0001-in. (3-μm) high are often undetected on holes 0.020 in. (0.51 mm) in diameter. In many cases, because of the deburring techniques or manufacturing processes used, these burrs are extremely small and uniform. At a glance of the surfaces or tips that part at an angle, the burrs may be so uniform that they appear to have been removed. This is not necessarily the case. For holes produced with tolerances of +0.0002 in. (5 μm), the removal of all such projections is imperative. To determine whether burrs exist, it is necessary to probe the hole physically with a small tool before using the size wires. Verification that edges are burr-free must be done before gages are used, and not only when a problem is anticipated.

Threads

Chapter 4 discussed several issues relating to burrs on threads. One of the principle ways to improve deburring on such features is to define exactly what the customer or plant expects on threaded parts. One plant described visual thread-quality expectations and practice, just for burr-related criteria, in a 100-page study (Gillespie 1981b).

Plants may want to look at which point chamfering threaded holes occurs. In some instances, chamfering is done after drilling and before tapping. In others, it is done after tapping to remove the swelled material. In others, it is part of the deburring operation. It is much more efficient to chamfer in the mechanized operations than in the deburring operation. Two issues for improvement are:

- wherever possible, holes intersecting threads should be finished with mechanized processes, and
- parts that are purchased as "burr-free" often are not.

Figure 32-23 illustrates two intersecting threaded holes. These present difficult burrs and major challenges. Electrochemical deburring (ECD), however, removes the burrs and produces smooth blends in a matter of 15 seconds in stainless steel. Figure 32-24 illustrates the intersection shown in Figure 32-23 after ECD. One or 50 holes can be deburred at the same time. Note the crests on the small, cold-rolled threads. They also can be removed by ECD, if desired.

Figures 32-25 and 32-26 illustrate a "burr-free," purchased miniature screw. Clearly, "burr-free" means different things to different customers. When such parts are ordered, it is important to understand what "burr-free" means to the supplier and how it is verified. These "burr-free" parts were deburred, eventually, under magnification at a rate of one every 15 minutes.

Figure 32-27 illustrates another purchased item. In this instance, it is a tap used to make miniature holes. The sides of the tap were ground to provide

Figure 32-23. Typical intersection of two threads. (Courtesy Honeywell)

Figure 32-24. Thread intersections after electrochemical deburring, brushing, and cleaning. (Courtesy Honeywell)

Figure 32-25. Threads and head-slot burr on standard purchased screw. (Courtesy Honeywell)

Figure 32-26. Magnified view of commercial "burr-free" screw head. (Courtesy Honeywell)

Figure 32-27. Magnified view of miniature tap. (Courtesy Honeywell)

relief. In doing so, burrs filled the threads of the precision tap, and the tap had more burrs than the finished holes. Inspect the tools used to see what edge conditions the plant is actually purchasing.

The chips left in the bottom of blind threaded holes are not a problem for most operations, but they can be when the screw bottoms out on the raised chip (designers may not have more space to allow a deeper hole). The remaining chip can result from retracting the drill too rapidly. It is estimated that if the drill were allowed to dwell 3–5 seconds longer at the bottom of the hole, deburring time required to remove the incomplete chip might be lowered by several hours per year. Also, if designers allowed more hole depth, the chip would not be an issue.

Gears

It is possible on many of the 96-diametral pitch gears to find very minute burrs rolled into the teeth by sandpaper, knives, or brushes. At 20–30× magnification, burrs as small as 0.0002 in. (5 μm) can be seen.

Each miniature gear should be scrutinized at these magnifications to ensure that no burr exists. In some cases, the use of gear inspection equipment may be desirable to pinpoint burrs on parts.

Dirt and Burrs

On gears and threads, dirt or particles can hide small burrs. Sometimes, it is just a case of confused identity: what looks like dirt is actually an attached particle. Before parts are submitted to inspection, they should be thoroughly cleaned, then scrutinized. Similarly, for the most effective deburring, parts should be clean, and they should be recleaned frequently, as required, during deburring. In many cases, an ultrasonic rinse in warm, soapy water is required. However, some magnetic parts—specifically, solenoid components—cannot be immersed in water.

Screw Machine Parts

On pinion-type parts that rub on part shoulders, it is critical that no burrs exist on the shoulders because they will affect the operating torque of the next assembly. In the past, it has been easy to miss the minute burrs that get bent out of the way and onto these surfaces (Figure 32-28). These burrs are frequently best detected using long-focal-length microscopes and lighting that casts harsh shadows.

Burrs Produced by Deburring

Any time a cutting tool is used to remove burrs, it produces burrs. Chamfering tools, for example, typically leave two small burrs while removing one large burr. For miniature precision parts, even these miniature burrs should be removed. On some parts, a sharp chamfer will act as a cutting tool when a mating pin is pressed. A chip or burr produced by an assembly process, which is highly undesirable (Figure 32-29), is the result.

Burrs Left in Hard-to-reach Areas

A few shapes exist in which burrs are almost impossible to reach. There is a natural tendency to leave them, because they are difficult to remove. These conditions should be brought to the attention of the process engineer, so that adequate tooling can be provided to produce a burr-free part.

Systematic Deburring Techniques

Many burr rejects result from a failure to develop a systematic approach to handling each part. For example, some features on a part may not be deburred because they were inadvertently overlooked. The practice of allowing two different people to deburr parts from the same lot causes similar differences in quality. The solutions to these situations are a consistent approach on every part and an in-depth review of finished part quality. For example, place all parts not totally deburred on the left side of the bench. Place finished parts in a container on the right side. After deburring, make a final check on part quality using higher magnification than used for deburring, if necessary.

On some parts, it may be convenient to make a checklist, then mark each feature when it is com-

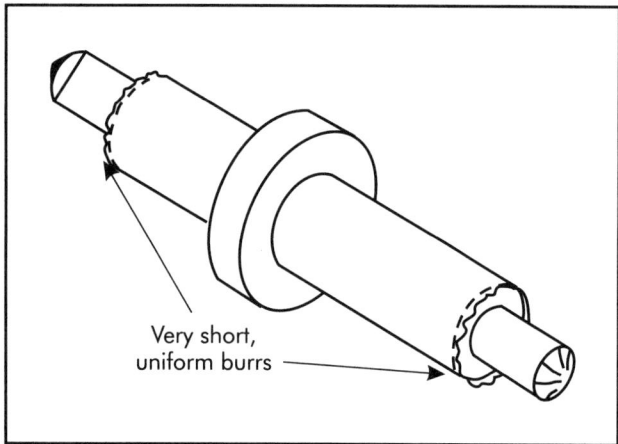

Figure 32-28. Burrs on shoulders of turned part (Gillespie 1978b). (Courtesy Honeywell)

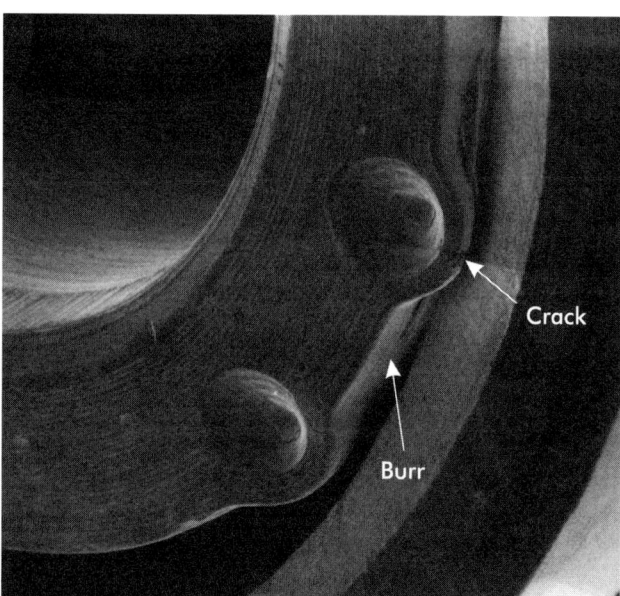

Figure 32-29. Chip or burr produced by press fitting (Gillespie 1978b). (Courtesy Honeywell)

pleted. On some parts, all holes may be deburred first, followed by other edges. Each part should be treated in the same sequence as all other parts. To facilitate deburring, the following questions should be asked.

- Are traveler instructions clear and specific?
- Are special tools required?
- Are special tools supplied, or are they specified on the traveler?
- Would a special brush, tool, or air motor help?
- Are the microscopes being used adequate for the task?
- Is the appropriate lighting available?
- Are all parts in a lot being deburred by the same individual?
- Are any burrs excessive?
- Can burrs be placed in a location from where they can be more easily removed?
- Can the part be totally or partially deburred by machine faster or easier?
- Is a systematic approach used to deburr the parts?
- Have all parts been cleaned prior to final verification of burr-free conditions?

Other Problem Areas

On many parts, workers tend to produce the smallest possible break on all edges. Although this tendency helps to ensure that excessive breaks never occur, it also greatly increases deburring time and, in fact, increases the probability that some burrs remain on parts. The most economical and intended practice is to use techniques that meet the requirements without unnecessarily "creeping up on the burr."

Some purchased parts, such as screws, bearings, and retaining rings, arrive at a plant with larger burrs than those allowed on most parts. Larger burrs are allowable on some parts because these items are purchased to military specifications rather than normal plant standards. These, usually, are less critical parts than those made in the production department. Although an obvious difference in quality expectations exists on some of these parts, the majority of purchased parts are expected to meet the same quality levels as those produced in house. Discussions about burrs prior to order placement may prevent some of these differences.

PRACTICAL HINTS

The following paragraphs discuss many practical hints on how to cut deburring costs by influencing product design. A number of sources in the References section provide further examples.

Figure 32-30 illustrates a cut-off tip on the end of a screw-machine part. The cut-off tip does not have to

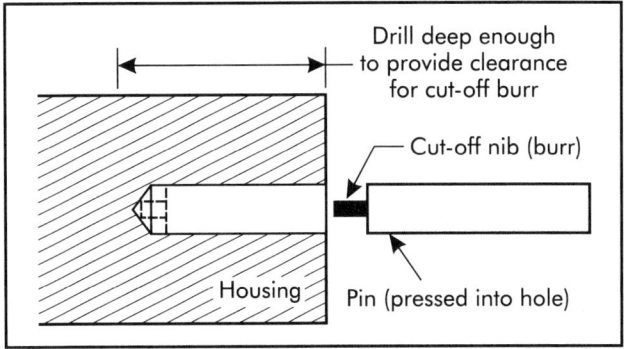

Figure 32-30. Design of part allows cut-off burr to remain on part (Gillespie 1999).

be removed if it is captured in a blind hole by drilling the hole deeper than normal.

Fixturing the part for deburring can save handling time in some operations. Normally, the fixtures will be very simple. They will position the part to most rapidly remove the burrs or mask certain edges so the processes used for other edges do not damage some special or precision edges.

When communication between inspection and manufacturing personnel is a significant source of problems, using a roving inspector may provide the immediate feedback needed to reduce rejects and, at the same time, identify who, where, when, and why defects are occurring. Frequently, the problem of burrs left on parts is the result of one or two individuals consistently failing to provide the needed quality level. When the inspector wanders the department, checking work as it is done, these kinds of problems are quickly identified and, together, the inspector and operator can resolve the issues. Roving inspectors are necessary on all shifts when this approach is used to ensure that needed awareness and training are provided to all who require it.

In most instances, deburring personnel and inspectors should use identical acceptance approaches, which means someone has to review each individual's efforts and note any differences. Team meetings will reveal why the underlying differences exist and provide clues to additional documentation and training needs.

Verify that inspection gages are not creating burrs. Surprisingly, many check fixtures or gages are designed closely—the designers forgetting that sharp edges necessary for location can also cut into edges and raise burrs or scratch parts. In some plants, there is a reluctance for the deburring worker to go to inspection and ask where the burrs are located. The workers would just review the entire part rather than the single area containing burrs. This wastes time and still may not help in locating burrs.

Supervisors often have little experience with deburring approaches. They can benefit from training as much as operators and inspectors. Companies need to develop such training, make it permanent, and ensure that everyone, including engineers, attend the training.

Work instruction notes such as "Microscope—up to 20×" tell manufacturing that inspection will use 20× magnification, but manufacturing can use any power it wants to deburr parts. Some staff members do not need to use that magnification since their techniques are applicable, regardless of magnification. Telling workers that they must use 20× slows output for some and may reduce the quality output of others. Define the acceptance criteria, but allow workers to use the magnification that best fits their abilities and needs (within time-standard requirements).

Cost reduction awards to personnel for deburring improvements are significant motivators. Imagine the cost savings if every employee provided 10 cost suggestions a year. If these save any money, they would make the company more competitive over a two-, three-, or 10-year period. Encourage ideas and reward staff for their involvement.

When parts are rejected in assembly for burrs, verify that the requirements there are the same as for the component parts. In some instances, the assembly operations have higher inspection requirements than for the piece-parts.

Standardizing the deburring approaches and individual deburring steps can reduce costs significantly. One plant saved 32,000 hours using a detailed methods-engineering analysis and standardization on just four part numbers that had high production. Other companies, similarly, have shown that a methods-engineering study of how deburring is actually being performed can reduce costs by 30%.

Do not machine parts out of sequence unless it is clear that it will not result in other penalties. Frequently, this results in bigger burrs in some areas, which costs additional time and money. Also, some of these burrs may not be detected since they are located in nontraditional areas and may not be where the next deburring operation is instructed to deburr.

There are many good deburring equipment choices to balance manual deburring needs in job shops. These will not eliminate manual deburring, but they will complement the goals of minimum total costs and cycle time reduction. For low cost, adaptability, and ease of use, it would be common for a plant to have equipment types for precision deburring, including:

- flat sanding machines;
- vibratory deburring;
- centrifugal barrel tumbling;
- miniature blaster;
- electrochemical deburring, and/or
- electropolish deburring.

In some instances, thermal-energy-method (TEM) and abrasive-flow deburring are also good choices. For shops that produce a single type of product, other equipment may be more suitable.

Once parts are burr-free, it is essential to package them in such a way that no part-on-part damage can occur. If this is overlooked, inspection will reject either all or parts of a lot for workmanship issues that generally are attributed to deburring workers.

When parts are rejected for burrs, make sure the individual who did the deburring has the task of deburring the parts correctly. Passing these parts to other workers who deburred their parts correctly is like a slap in the face and tells workers that the good workers will have to do their share of work, as well as someone else's.

When drill sizes and tolerances allow, use a combination of drill and countersink tools. These are available in solid tool forms, indexable tools, or adjustable-position tools. The countersink portion provides deburring with no extra handling and only a second or two of additional on-machine time.

APPENDIX

The following list of performance-shaping factors and tips may help the reader define some of the characteristics of deburring tasks. Managers and operators can implement and enhance these items for greater productivity and morale.

Quit Scratching the Part

When miniature parts are held in pin vises, steel collets, or vise jaws, it is very easy to leave indentations on them. Many of these marks are unallowable. To prevent such occurrences, workers can use brass jaws or collets, put masking tape over the jaws, or carefully control the amount of closure pressure applied.

Sanding Changes Part Size

Many small screw-machine parts have tolerances of ±0.0002 in. (±5 μm). When abrasive paper is used to remove the burrs from edges, these short diameters can easily be tapered by 0.0003 in. (8 μm) in only 2–3 seconds with 600-grit paper. Test and measure before committing to a deburring process.

Form-milling Burr

Figure 32-31 illustrates a 0.50-in. (12.7-mm) diameter ratchet wheel of 303 Se stainless steel, which was

Figure 32-31. Ratchet wheel has large burrs (Gillespie 1977).

produced on a prototype basis by form milling the teeth one at a time. A burr 0.004–0.005-in. (102–127-μm) thick occurred during milling. If unrolled, the burrs would be close to 0.125-in. (3.18-mm) long. Material toughness and burr thickness make this burr almost impossible to remove by hand.

The normal approach to producing the part would be to turn the diameter, face one end, and produce the hole. Then, in the next operation, face the part to thickness, mill the ratchet teeth, machine off the heavy burr, and deburr all edges. In this particular case, the manufacturing engineer recognized, before the part was made, that a heavy burr would occur from this operation. As a result, the engineer put the thickness-sizing operation after the form-milling operation. This eliminated the machining operations that would have been necessary for removing the heavy burr, which resulted in a substantial reduction in total fabrication time.

The thickness-sizing operation was a grinding process which, while slower than turning, allowed several parts to run at the same time. The grinding burr then was removed by vibratory deburring.

Since the part shown in Figure 32-31 was produced, it has been found that glass-bead peening will effectively remove the burrs in two minutes if some surface roughness is allowable. This is somewhat surprising, since the ductility of 303 Se stainless steel generally prohibits heavy burr removal on precision parts (blasting works better on thin and non-ductile materials). Apparently, the extreme deformation caused by the form-milling cutter work hardens the burr so much that the burr has a very low ductility. This does not happen often in this series of stainless steel. This example highlights one important aspect already mentioned—effective, economical deburring requires careful planning in the initial selection of manufacturing processes.

Figure 32-32 illustrates a similar example. In this instance, a helical gear is being machined off, then chamfered on a stand-alone, semiautomatic deburring machine. In the previous example, few parts were being made in a year's schedule. For production rates of such gears or ratchets, the simple gear-deburring machines greatly reduce labor and improve quality. When such gear-deburring machines are unavailable, use a mill or other cutter to cut off the heavy burr and an abrasive-filled cotton or fiber wheel to quickly chamfer the part manually.

Aluminum Spur Gear

Figure 32-33 shows a spur gear with a 3-in. (76.2-mm) outside diameter and a 0.80-in. (20.3-mm) hub

Figure 32-32. Semi-automatic tools simplify hand deburring.

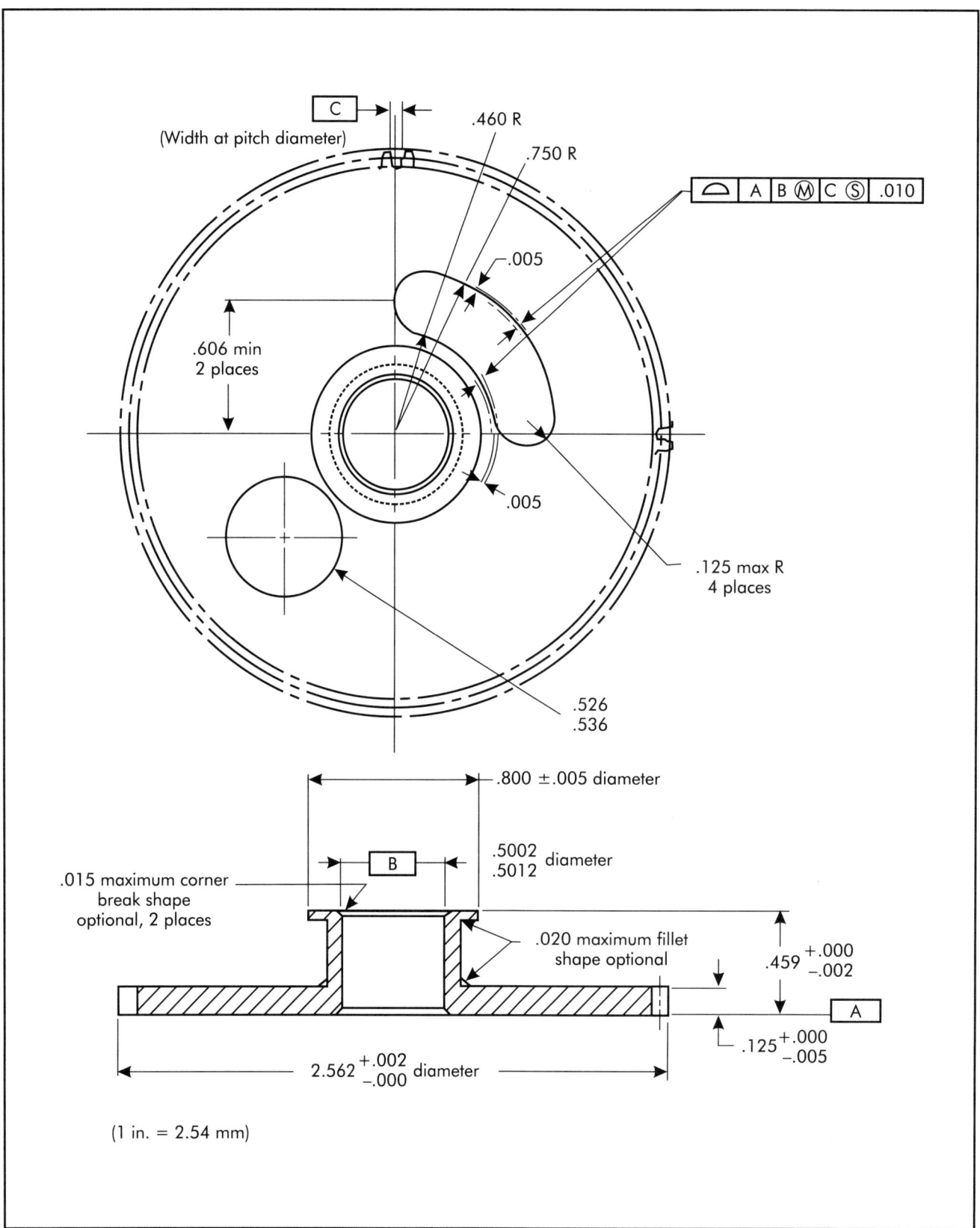

Figure 32-33. Aluminum spur gear (Gillespie 1977).

in the center. It is an 80-tooth, 32-diametral pitch, American Gear Manufacturers Association (AGMA) class 8B gear. This part is made from 6061-T6 aluminum, and the teeth are hobbed. Hobbing produces a relatively large ductile burr on the hub side of the part.

Because of the hub, it is impossible to lay the burr-laden side on sandpaper to sand off the heavy burr. As a result, the production operators deburred this part by hand, tooth by tooth, as they did most other parts for which tooling was not provided. An observant engineer, however, noted that a sanding block with a slot in it (Figure 32-34) would allow heavy burrs to be removed and that brushing would remove the small sanding burrs. The net savings was almost 30 minutes per assembly, with no loss in quality. This almost trivial example highlights three more aspects of deburring that are often overlooked:

- Special-design tools can greatly reduce deburring costs.
- The use of two or more techniques may be required.
- It is easy to get in a rut of doing things the way they have always been done.

The sanding block, inexpensive as it is, is a design tool. Someone conceived it, made it, and used it. While the production operator knows what he or she needs to do a good job, there may not be an interested contact to see that the operator's ideas are translated into tools, especially in large companies. In many large companies, hand-deburring operators also have little or no machining or tooling knowledge; therefore, they do not readily innovate new approaches. Similarly, it has been rare in the past to find operators who have any idea of which tools can readily be purchased, or of the real economic tradeoffs between hand-deburring tool costs and deburring time. Unfortunately, there is a dearth of literature on practical tooling aids.

In the preceding example, the sanding block tool worked well, yet it had limits. The same concept was unsuccessful when applied to stainless-steel parts one-tenth the size. The reason for the difference is that the large part had easily removable small burrs and a large surface area requiring sandpaper. The stainless-steel burrs are much harder, even though they are the same size. This wore out the sandpaper much faster in the vicinity of the sanding block slot, and new sandpaper had to be applied for every part. In addition, the small size of the stainless-steel part made holding and sliding it difficult—the burrs tended to dig in and tip the part as it was being sanded.

The spur gear example used two techniques for removing burrs: one for heavy burrs and another for small burrs. For precision deburring, this multiple approach is almost a necessity for maintaining tolerances.

In any company, the manufacturing engineer is a very busy individual who has to solve a variety of immediate problems. It is easy, in such cases, to fall into the habit of following familiar, well-known routines in deburring processes. While normal practice solves the immediate problem of using a total manufacturing process, often it is not the cheapest approach. Fortunately, the information now being published and presented at seminars throughout the world is providing the education and insight required to break free of the rut of past practice.

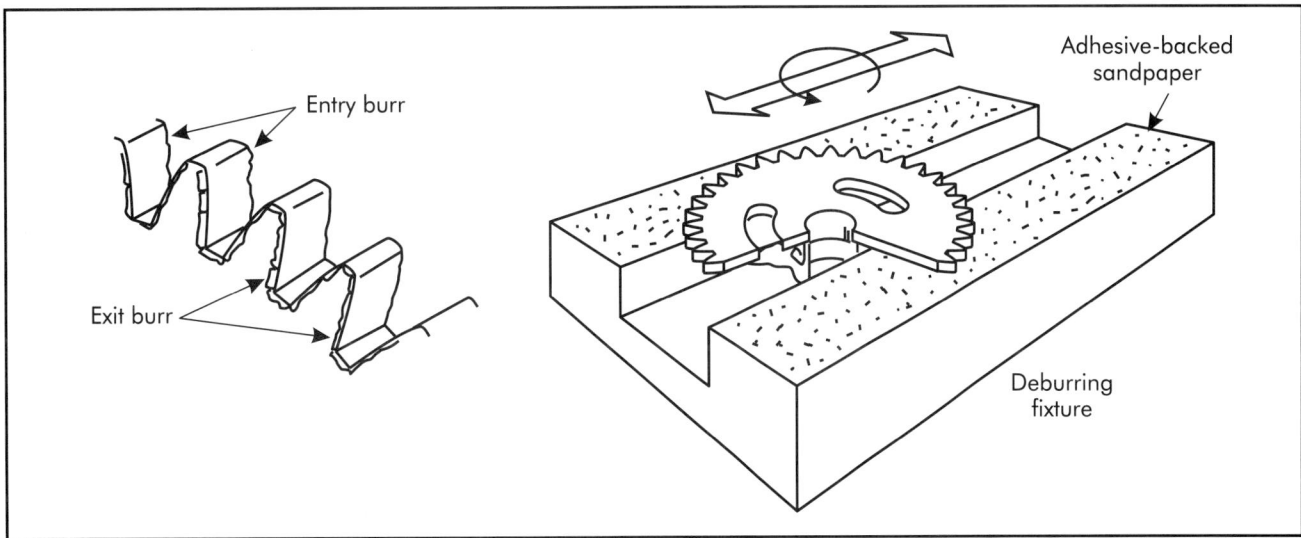

Figure 32-34. Sanding block allows quick removal of burrs on teeth (Gillespie 1999).

Punch It Instead of Deburring It

Figure 32-35 shows a simple tool used to remove burrs from special, drilled Allen-head screws. While tools of this type may produce burrs on the part, the secondary burrs are small enough to be easily removed by faster, more readily controlled processes.

Sequence of Machining Operations

Figure 32-36 shows the approach for an intersecting-hole problem. A judicious choice of the machining operation sequence can help minimize burr size and removal problems. For example, when small holes intersect with a large hole, the large hole, generally, should be drilled first. This permits use of a reamer or broach to shear, in a single operation, the burrs resulting from drilling the small holes. In addition, burrs produced by the large tool tend to be larger than those produced by the small tool.

If the opposite drilling sequence were used, the deburring tool would have to be placed in each of the small holes, instead of only in the large hole. This sequence has an advantage only if burrs are allowed in the small holes but not in the large hole.

The same principles apply to all machining processes. Depending on the part geometry and the processes involved, some combination of processes (without regard to speed, feed, and tool-geometry effects) minimizes the size of the burrs remaining on the part.

The machining sequence can be planned to place burrs in locations that are easy to attack to remove burrs. As shown in Figure 32-37, feeding the cutter in one direction causes the burr to form over the shaft. In this position, the burr is difficult to remove uniformly by any deburring process. If the direction of the cut is reversed, the burr is placed in a position where it can be readily removed. This determination of the burr location must be made before tooling is ordered, however, to provide appropriate support for the workpiece. Any changes made at a later time may involve additional tool design and fabrication expenses.

Paperwork Sign-off

In one plant, operators would deburr most of the parts and, during a break or before leaving the area,

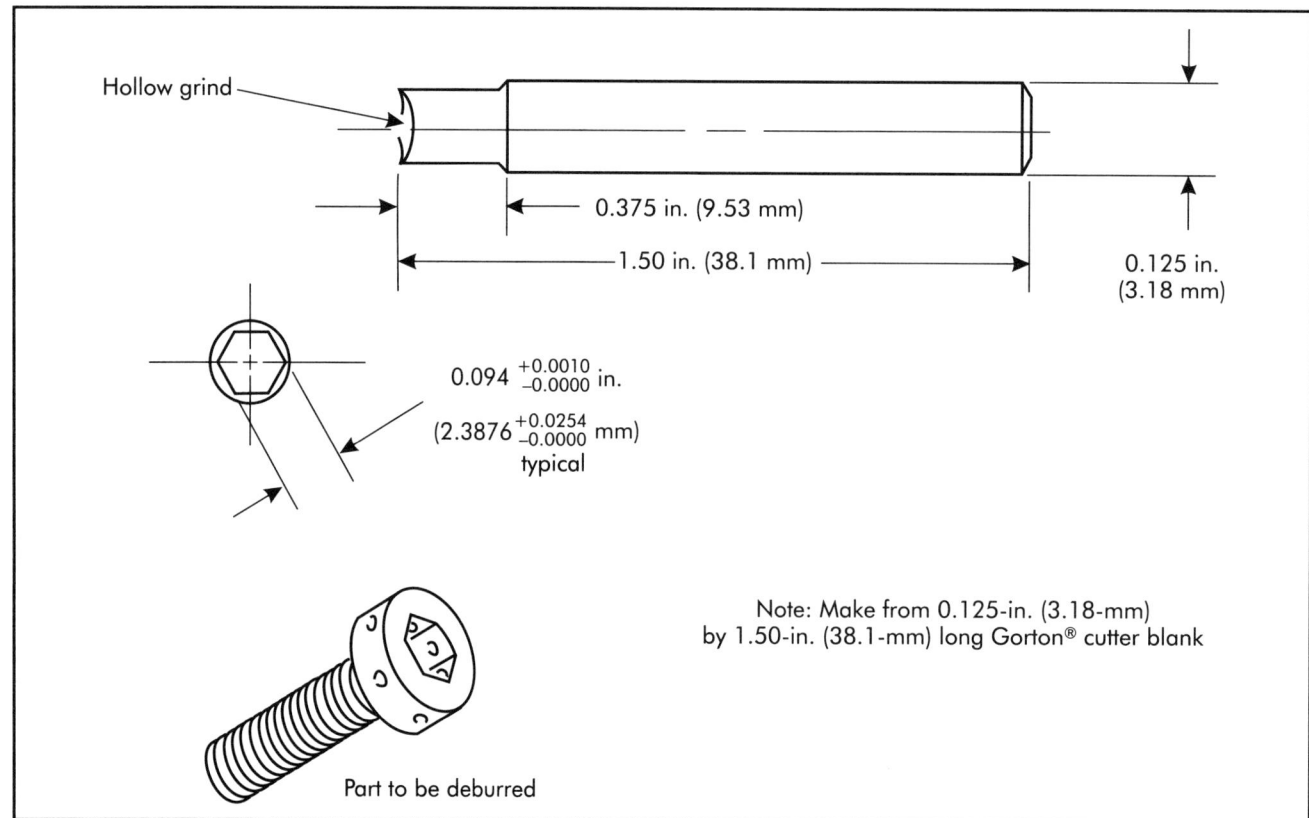

Figure 32-35. Broaching tool used for deburring.

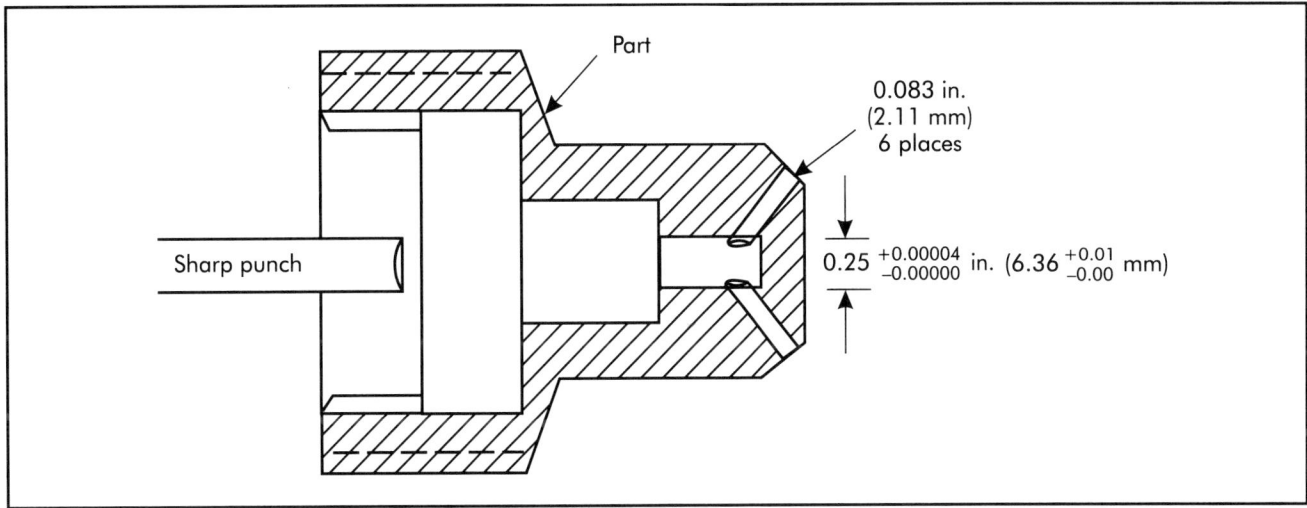

Figure 32-36. Punch designed for deburring intersecting holes.

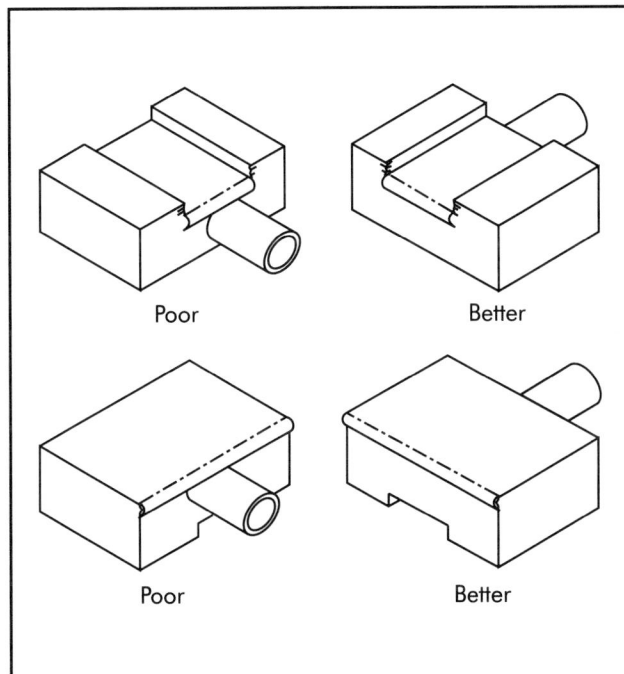

Figure 32-37. Location of rollover exit burr affects deburring costs (Gillespie 1999).

they would mark on the paperwork that the job had been completed knowing that they would be back to finish the parts the next day, later that day, or in a few minutes. Either they forgot, or someone else would pick up the lot of parts marked "complete," and the parts would move to inspection, shipping, or the next operation. Parts should not be signed-off of a final deburring operation until they are completely burr-free or the supervisor has been notified of the burrs.

Knock it Off

Figure 32-38 shows a cut-off tip left on a part when it fell during the cut-off operation. A subsequent operation to face-off this 17-4 PH stainless-steel part broke the facing tool. When placed on a square block, a sharp tap with a brass mallet fractured the cut-off tip in a fraction of a second. In some instances, the fractured area will lie below the machined surface; and, in some materials, a small hub will be left above the surface. For this example, a 0.02-in. (0.5-mm) projection was left. Material choice and part size determine the resulting tip size and removal difficulty.

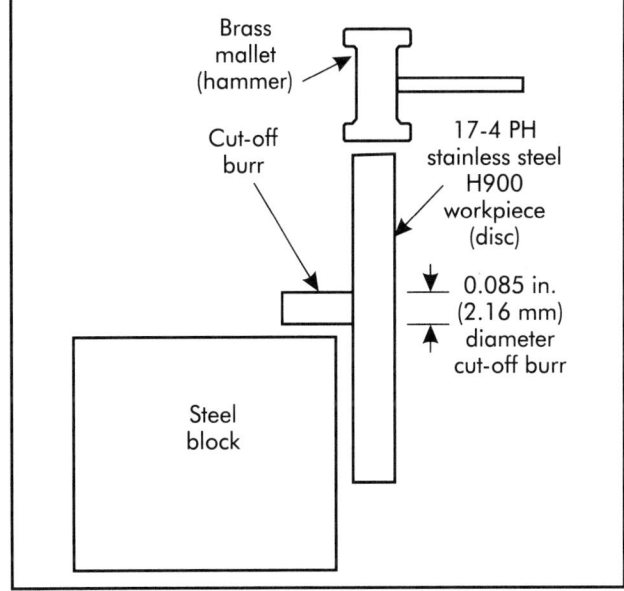

Figure 32-38. A light hammer tap breaks off cut-off burr.

Make a Mirror

One plant noted that when workers were required to look in hard-to-reach areas, they could not see the area they needed to deburr. One worker modified a worn needle file by grinding a flat on it, chrome-plating it, then polishing it. This tool was used as a mirror for viewing the bottom and sides of threaded holes. Many shapes of Swiss precision files allow almost any shape to be viewed with such an approach.

Push it Down

Cutter-wheel-type tools used to sever tubing often leave a small, raised lip around the cut end. The lip must be removed before the tubing can be inserted into a close-fitting hole. A manual cutter was needed to cut tubing of different diameters and leave a smooth edge. The solution was to replace the opposing smooth rollers that, ordinarily, are used in tube cutters with appropriately grooved rollers (Figure 32-39).

The sharp-edged cutting wheel, shown in the upper part of Figure 32-39, raises a lip on the periphery of the tubing because the cutting action forces the tubing wall material to the sides of the cut. When cutting begins, the small lip is cleared by the roller groove because the groove is wider than the initial cut. But as the cut becomes wider and the lip height increases, the edges of the groove in the roller bear down on the lip and prevent it from rising. Just before the tubing is severed, the width of the cut equals the width of the roller groove (note the lower part of Figure 32-39), and the smoothing is complete.

Use a Drill Bushing

Figure 32-40 illustrates a drill bushing altered to provide a cutting edge. The purpose was to remove a burr from four little tabs milled around a counterbore. The pocket was originally 1 in. (25.4 mm) in diameter. The burr made by the milling cutter when it passed through the raised hub was difficult to remove. By grinding teeth on the bushing, and guiding the bushing with a locating pin, the heavy burr was removed in seconds.

Tape it On

A plant was faced with the task of deburring several hundred thin discs of copper. Their small size made them difficult to pick up and hold—180,000 would fit into a #10 thimble (Figure 32-41). They had been punched with a dull die or undersized punch, and a large amount of bent stock remained (Figure 32-42). It was not a burr, it was bent stock. These were 0.001–0.002-in. (25–51-μm) thick and had a 0.024-in. (0.61-mm) diameter hole. The thinness and the bent metal made it impossible to deburr by normal means. Some had been centrifugal barrel finished twice, with no apparent improvement.

The operator who chose to work on the parts exemplified ingenuity and teamwork. Engineering had no good ideas, but she did. She picked up and viewed each one under 20× magnification to determine which side had the burr. She then placed the opposite side on the adhesive side of a piece of masking tape. She

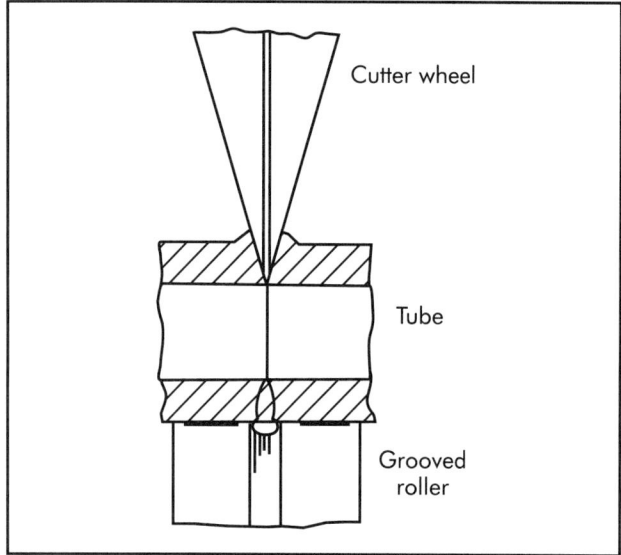

Figure 32-39. Grooved roller pushes burr out of the way. (Courtesy Machine and Tool Blue Book)

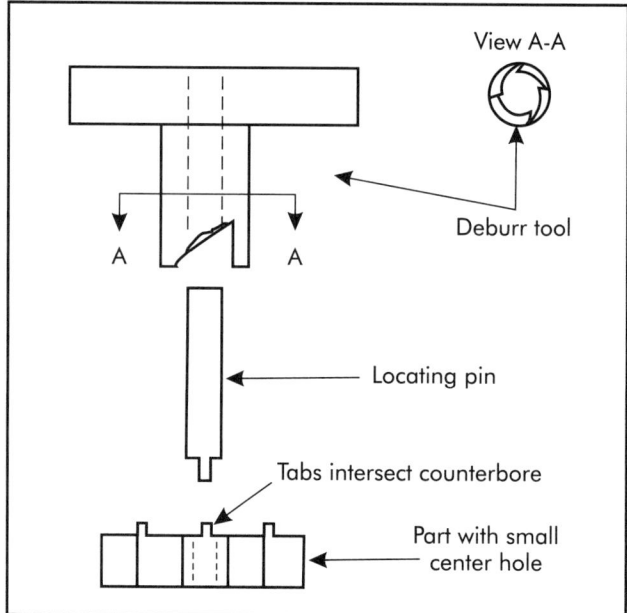

Figure 32-40. Modified drill bushing removes burrs.

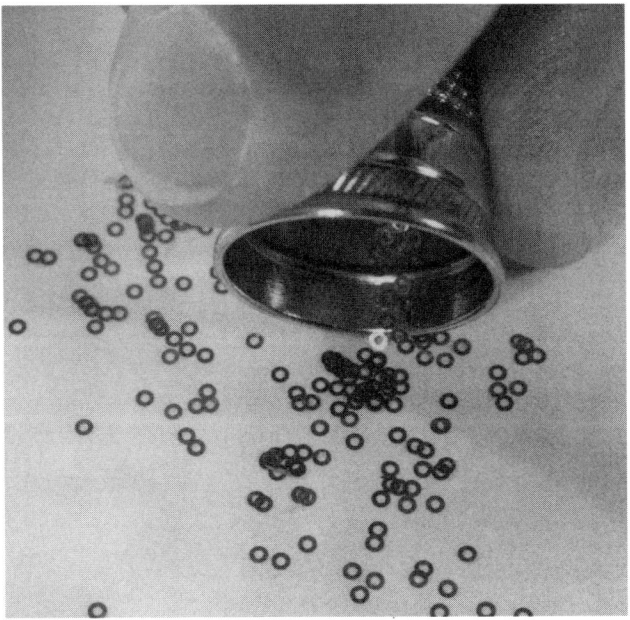

Figure 32-41. Miniature washers and thimble. (Courtesy Honeywell)

Figure 32-42. Bent material around internal hole. (Courtesy Honeywell)

repeated this until she had 75–100 loaded on the tape. Then, she formed a ring of the tape around her finger, almost like a thimble (Figure 32-43). Next, she sanded these parts on 500–600-grit abrasive paper to remove the burrs. After sanding the parts, she re-

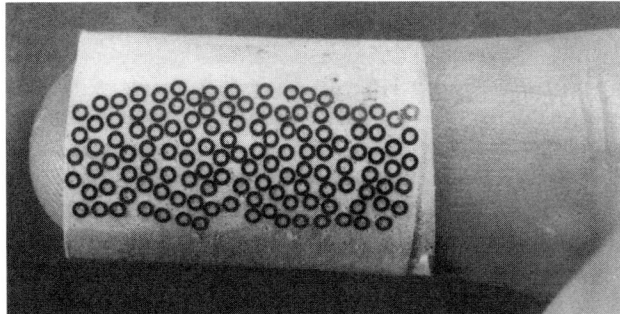

Figure 32-43. Washers on masking tape. (Courtesy Honeywell)

moved them with tweezers and placed them in a solvent to remove the adhesive left on the parts. It allowed her to do 400 parts per day when the parts had waited months for a solution. She solved a production bottleneck.

Watchmaker Broaches

When all else seems to have failed, and there are still burrs in small pockets, consider using watchmaker's cutting and smoothing broaches. These come in a small set of five-sided, tapered broaches (also called reamers). Smoothing broaches are round and tapered. A dozen of these broaches are sold in a small plastic tube with a hand pin vise to drive them. Lens-cutting broaches, pivot broaches, and extra-fine cutting broaches add to the selection of these tools.

Make a Mini-file

Deburring narrow slots is difficult when the slots are 0.014-in. (0.36-mm) wide. One company used two different approaches. In the first approach, the company electropolished a miniature file until it fit the slot. Electropolishing leaves most of the tooth sharpness as it is polished down. A second worker on a larger slot took shim stock and adhered adhesive-backed abrasive paper to both sides of the shim stock. The shim stock provided great stiffness, and the abrasive size could be easily adjusted. Adhesive had to be removed from the parts, though, because minute amounts were squeezed from the side of the shim sandwich by pressure. A solvent readily removed the material. Shim stock is available in thickness down to 0.0005 in. (0.013 mm). In Chapter 19, it is noted that diamond-plated metal foil can be obtained, also, in thicknesses down to 0.0005 in. (0.013 mm).

Toothpicks and Writing Paper

Toothpicks and writing paper have worked on gold parts that are susceptible to damage.

Rocks

In using loose-abrasive tumbling processes, such as vibratory deburring and barrel tumbling, an operator eventually faces the problem of removing undersized media (rocks) from holes and slots. In some instances, it will cause operators to switch entirely to hand deburring. In others, the removal of rocks is a task for which the hand-deburring operator is responsible. In either instance, there are simple improvements. While this is not deburring, it is similar and time-consuming. In removing these rocks (rock picking), or to minimize this problem, it may be necessary to follow some of the following approaches:

- Use different-size media.
- Use soft media, which breaks easily.
- Use limestone, which is consumed by acetic acid.
- Use ultrasonic cleaners to loosen rocks.
- Use a water-flushing dental device to loosen rocks.
- Use ultrasonic plating probes to loosen rocks.
- Use sewing needles as media.
- Plug holes prior to tumbling.
- Do not use tumbling processes.

Make a Chamfer Tool

Figure 32-44 illustrates a small chamfer tool made to center on a hole. A small blade, adjusted with a set screw on the opposite side, allows it to be adjusted for different chamfer sizes. The shoulder of the tool limits its depth.

Use the Correct Drill Point

Figure 32-45 illustrates an intersecting hole and the before-and-after drill-point geometry used to drill the cross-hole. The fishtail design made a very small burr, compared to the one produced conventionally. Brushing removed the remaining burr.

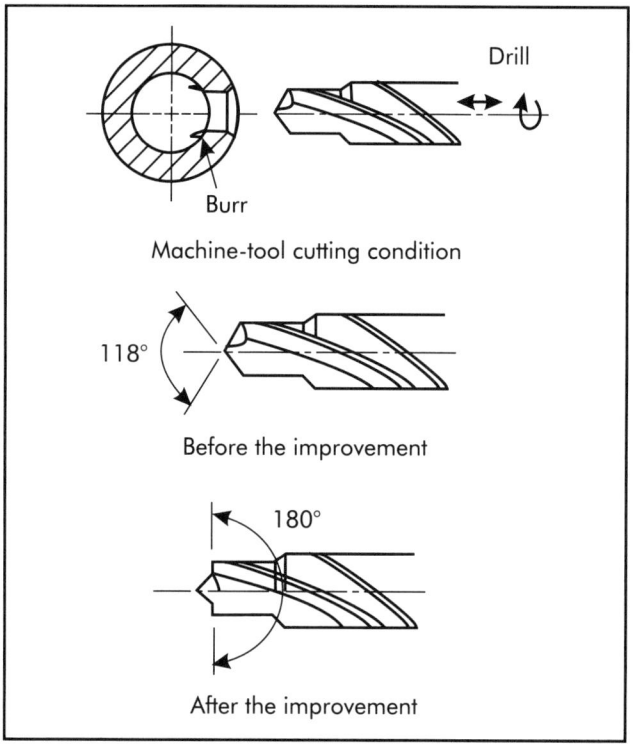

Figure 32-45. Fishtail drill point prevents large drill burr. (Courtesy Koya Takazawa)

Deburr with the Drill

The chamfering collar around the drill shown in Figure 32-46 eliminates the need for further deburring.

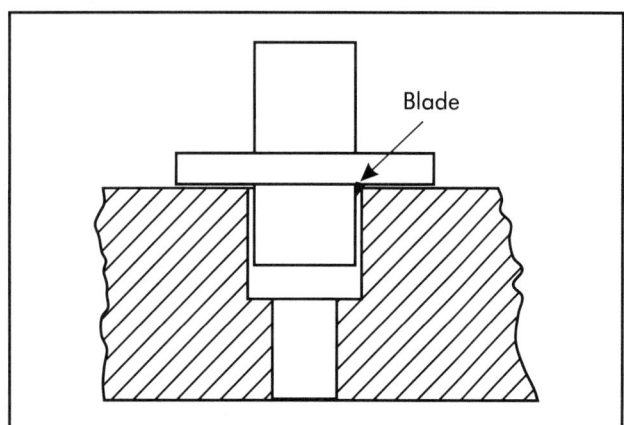

Figure 32-44. Piloted tool with adjustable blade.

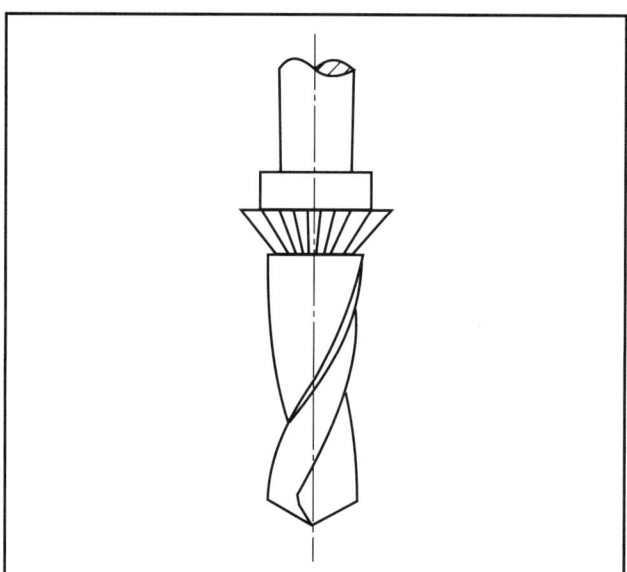

Figure 32-46. Chamfer tool around drill deburrs while drilling. (Courtesy IPA)

Machine Inside Out

Figure 32-47 illustrates an approach to prevent the large cut-off burr when tubing is cut off and the part is allowed to fall. The inner cutter cuts half way into the wall, then allows the outer cutter to finish. There still may be a burr, but it is axial rather than on the diameter.

Machine it Off

When using CNC machines for parts, take advantage of the routines or canned cycles already in the machine to chamfer or radius each edge on the machine (Figure 32-48).

Use the Backside of Tools

Figure 32-49 shows an inverted hart bur that deburrs with both top and bottom edges. Bur balls, and other tools, allow deburring on blind sides of holes.

Five-minute Tool Saves Hours of Time

Figure 32-50 illustrates a paint scraper that has a V-notch ground into it. This tool deburrs and chamfers both sides of sheet or plate, or even pipe. If a kit having V-notch tools is not readily available, a scraper can be modified.

SUMMARY

Many sources in the reference and bibliography sections include several pages that show how to prevent and minimize burrs. One source in particular (Schäfer 1975) provides the most information on this subject.

Remember one last, proven fact. Hand-deburring time, dollars, and frustration are saved by planning ahead and understanding how to balance machining and deburring issues. The recent applicability of computers to mold heating and cooling allows industrious plants to use innovative heating, cooling, and venting techniques with carefully machined and polished mold surfaces to eliminate visible flash (*Machine Design* 2001).

ENVIRONMENTAL, HEALTH, AND SAFETY ISSUES

Most of the health and safety issues have been described in previous chapters. However, it is important again to note that dust and fine particles from materials containing more than 0.1% beryllium are considered carcinogenic and must be treated carefully to prevent inhalation. This health insight may not address all the concerns about this material, and there

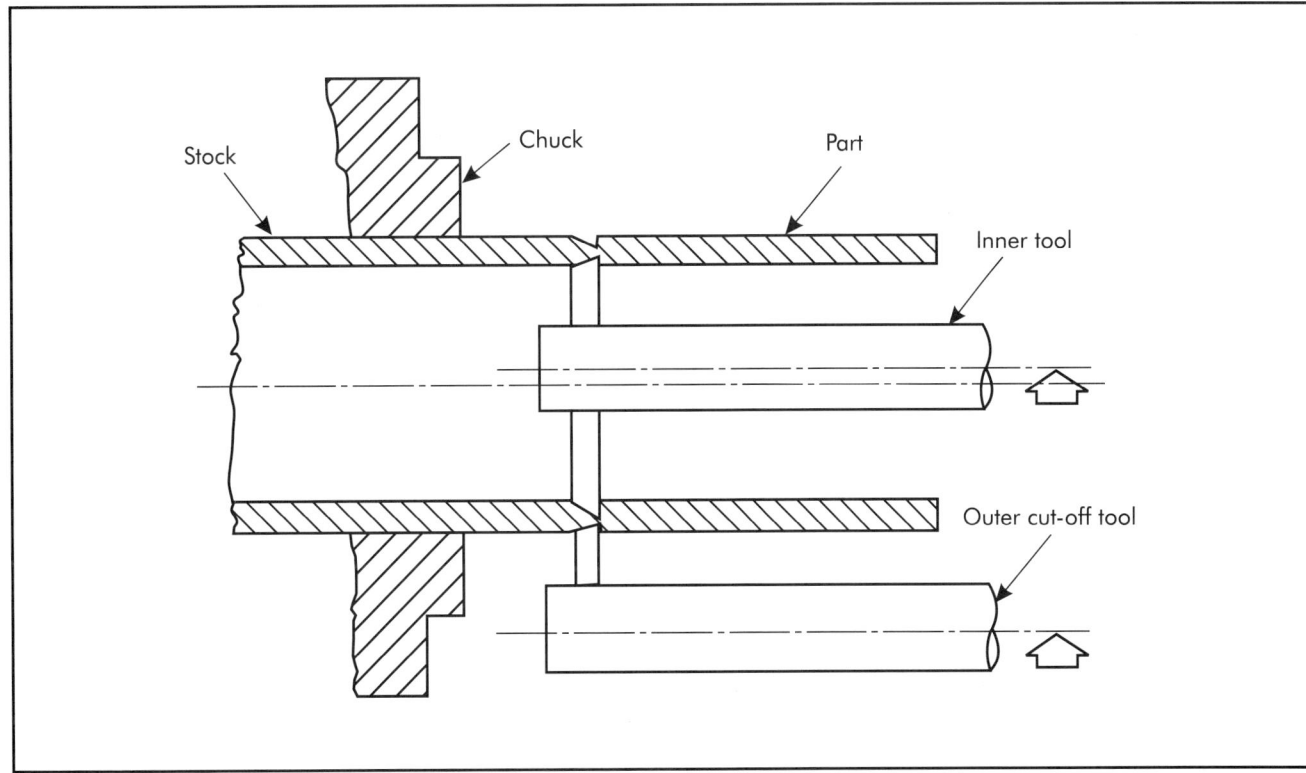

Figure 32-47. Chamfer from inside to prevent large cut-off burr. (Courtesy IPA)

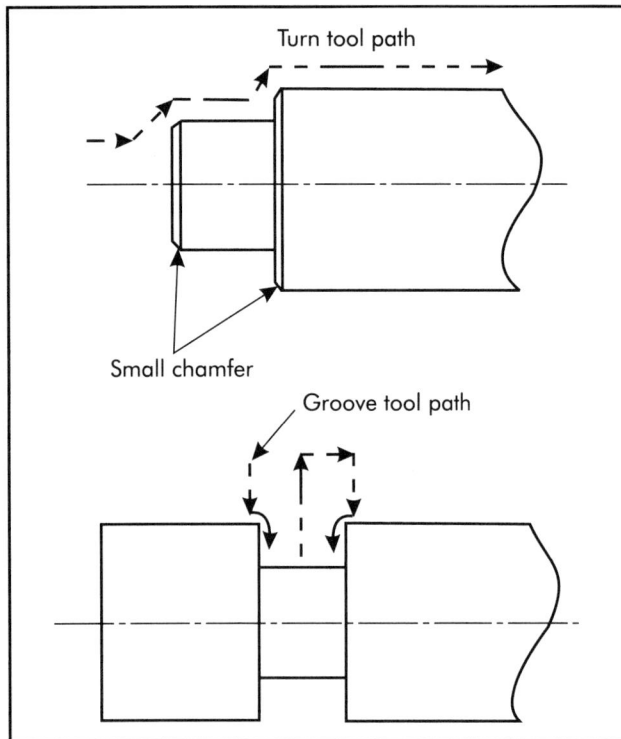

Figure 32-48. Chamfer on the lathe.

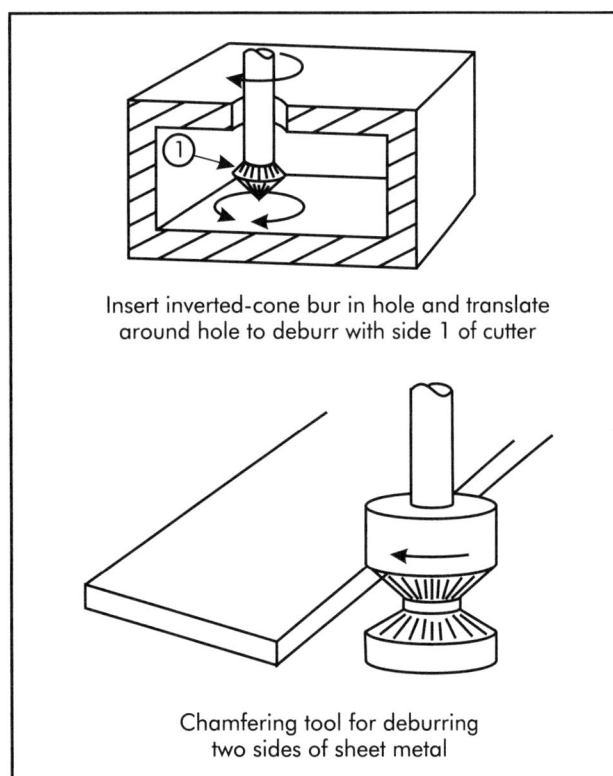

Figure 32-49. Special tools quickly cut costs (Wick and Veilleux 1982).

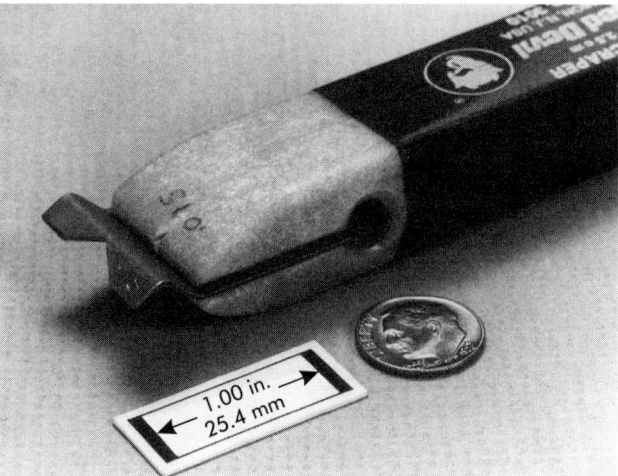

Figure 32-50. V-notched paint scraper reduces deburring time.

are other materials that also must be handled safely. Titanium, magnesium, and other materials, for example, can catch fire when small chips become hot.

REFERENCES

Gillespie, LaRoux K. 1975. "Hand Deburring of Precision Miniature Parts." Report BDX-613-1443. Kansas City, MO: Bendix Corporation. (Available from NTIS.)

——. 1977. "Deburring Case Histories." SME Technical Paper MRR-7706. Dearborn, MI: Society of Manufacturing Engineers.

——. 1978. "Hand Deburring Guide." Report BDX-613-2089 (September). Kansas City, MO: Bendix Corporation. (Available from NTIS.)

——. 1981. "Visual Thread Quality for Precision Miniature Mechanisms." Report BDX-613-2600 (April). Kansas City, MO: Bendix Corporation. (Available from NTIS.)

——. 1999. *Deburring and Edge Finishing Handbook.* Dearborn, MI: Society of Manufacturing Engineers.

Machine Design. 2001. "Soft Grip Overmold Takes the Abuse." *Machine Design* (May 3): p. 107.

Rhoades, Lawrence. 1981. *Cost Guide to Automatic Finishing Processes.* Dearborn, MI: Society of Manufacturing Engineers.

Rowe, W. D. 1977. "Law of the Estimated Fact." *Research/Development* (December): pp. 54, 56.

Schäfer, Frederick. 1975. Entgratgerechtes konstruieren und fertigen. In *Proceedings Fachtagung Entgraten 1975, Institut für Produktionstechnik und Automatisierung der Fraunhofer-Gesellschaft:* pp. 2–17. Stuttgart, Germany: University of Stuttgart.

Wick, Charles and Raymond Veilleux, eds. 1982. *Tool and Manufacturing Engineers Handbook*, 4th ed., vol. 3. *Mate-*

rials, Finishing, and Coating. Dearborn, MI: Society of Manufacturing Engineers.

BIBLIOGRAPHY

Chodakauskas, Stanislaus R. 1998. "Titanium Deburring Process Improvements." In *Proceedings of the Fifth International Deburring and Surface Conditioning Conference,* San Francisco, CA (September). (Available from Deburring Technology International.)

Gillespie, LaRoux K. 1976. *Deburring Capabilities and Limitations.* Dearborn, MI: Society of Manufacturing Engineers.

——. 1978. *Advances in Deburring.* Dearborn, MI: Society of Manufacturing Engineers.

——. 1981. *Deburring Technology for Improved Manufacturing.* Dearborn, MI: Society of Manufacturing Engineers.

Goodemote, DeWitt. 1980. "Custom Deburring Tool." *Production*, July, p. 101.

Society of Plastics Engineers (SPE). 1999. *Mold Finishing and Polishing Manual* (#1083). Brookfield, CT: Society of Plastics Engineers.

Training for Manual Deburring 33

It is amazing that industry provides such poor instruction to deburring personnel. Despite the availability of photographs, sketches, coded sample parts, videos, guides to available supplies (Gillespie 2000), and written standards, much of industry has ignored effective instruction. This is a topic that has been discussed (Gillespie and Bolinger 1979), and at least one set of training manuals is available (Gillespie 1979; Gillespie 1980 a, b, c).

While one typically thinks of training as a cost, it can bring excellence; which, in turn, brings in new business and enhances a company's reputation. For example, one company regularly used the deburring area as the first place to bring admirals, generals, and congressional representatives because 100 workers finishing parts under microscopes conveyed an image of unmatched excellence in quality. It set an image of unsurpassed quality that had immeasurable economic benefit. The company's work became known throughout precision industry, and it was the place to go to obtain quality.

TRAINING ELEMENTS

Consider that you own a plant. How many workers have received technical training on deburring or deburring economics? As the reader of this chapter how were you trained? Would you be more effective if you had some training? Would plant costs be lower? What did it cost your company because there was no training? Table 33-1 provides a quick means for readers to determine the issues of training that may justify costs.

Training needs include:

- identification of the types of available deburring tools and those already in-house;
- training on how to use each type of tool, and many hours of practice;
- training on where and when to use each type of tool;
- training on how to select and use different grades, grits, and sizes of a specific type of tool;
- identification of plant edge-quality requirements, simple deburring economics, and employer expectations;
- training on surface finishes and how deburring can affect finish;
- training on simple cleaning and its impact on seeing burrs;
- identification of critical drawing needs;
- training in ergometric and safety considerations;
- training on how to inspect for burrs;
- training on how to use microscopes (if used in production), and
- introduction to systematic manual deburring for each part in a lot.

TRAINING GOALS

Training usually begins with a clear statement of the expected outcome, such as: "Produce an employee who can immediately provide expected deburring quality at 75% of the production rate of an experienced worker." Another goal might be to "produce an employee who can deburr Part #1 successfully, immediately."

Critical deburring on complex and expensive parts might require 6–9 months of training. That is expensive, but so are parts worth between $5,000–10,000 each, particularly when "excessive deburring" causes an entire lot to be scrapped. Two sources (Gillespie 1979; Gillespie and Dulin 1979) describe an extensive

Table 33-1. Training costs

Cost Element of ___ Weeks of Training	Cost ($)	Cost of Untrained Workers	Cost ($)
Developing training course		Scrap costs	
Trainer, direct labor costs		Rework costs	
Trainee, direct labor cost		Cost of schedule misses	
Trainer and trainee, overhead costs		Warranty costs	
Parts for training		Worker injury costs	
Tools for training		Insurance increase costs	
Equipment for training			
Lost income from trainer unavailability			
Facility costs			

program for training personnel to deburr under 15–30× magnification. This is not required for many plants, but the sources illustrate the nature of a planned training program.

Training goals need to be accompanied by skill assessments to determine if training is successful and if personnel skills match needs. Unfortunately, there are no known assessments publicly available. Note that pre-employment assessments to determine the candidate's ability to learn the skills are as important as performance assessments.

PEOPLE FOR DEBURRING

There is a perception that, when microscopic deburring is required, women make better workers because they are perceived to be more patient and caring than men. That women are better deburring workers, however, is not proven by experience. Gender, race, or age does not seem to affect performance on even the most tedious deburring tasks. Visual acuity, however, affects microscopic work. If workers must work under 30× magnification, eye and neck strain may rule out many potential workers, but no known test exists to screen these kinds of potential problems. Experience indicates that workers can deburr at 7–10× for 8 hours without undue strain. At 20–30×, however, production rates fall greatly, and workers will only be able to work in spurts under such magnification. When heavy parts or tools are involved, relative physical strength makes a difference; but, with ergonomically designed aids, even strength does not have to be a limiting factor.

Communication skills are essential. There are many examples indicating that defects result from language differences among workers or supervisors. Individuals who successfully communicate with others possess good "people skills." Not all management personnel have this skill, and some workers lack the skills or patience to teach their new compatriots. Unfortunately, no known agency exists that teaches deburring or trains deburring teachers. Job-shop deburring companies may be able to teach others, but it is important that the shop use the same general techniques and tools that new workers will be expected to use.

PARTS FOR TRAINING

When parts are inexpensive, there is no need to provide special parts for training purposes. When the parts are expensive, a plan for capturing reject and scrap parts is required, as well as hands-on deburring practice. Implementation demands careful thought to minimize expenses and define the nature and features of burrs to be deburred.

IDENTIFYING TOOLS

This book provides an excellent starting point for identifying available deburring tools. Chapters 7–28 provide a starting place for an educated workforce. Company catalogs (Gillespie 2000) provide additional detail on thousands of deburring and edge-finishing tools. A large company may need to establish a Web-based tool catalog just for its deburring effort. Why? It is not uncommon for a large company to own 300 different tools, and the tools are difficult to locate if it has not identified them by number or crib location. Hunting for numbers for replacement orders is a waste of time. Digital cameras provide a quick means of establishing a plant-oriented, illustrated guide to deburring tools for both training and daily use.

HANDS-ON TRAINING EXPERIENCES

Training involves two distinct aspects: the theory or lecture aspect of understanding why a tool is used, and the hands-on skill portion, which is more important for immediate use. For long-term improvements, both must be taught. As indicated previously, hands-on training involves three aspects:

- practice using specific deburring tools;
- practice selecting the best approach (or tool) for a specific part feature, and
- practice selecting the correct grit size, abrasive-grain type, or variation of a specific tool type.

Different tools are required for finishing holes and sheet metal edges. Threaded features take yet different tools. Plastic parts need yet another set of tools. None of this is obvious until training has been provided to workers. This is the heart of the training. This is the part of the training where workers try different sandpaper products to see how grit size and sanding time affect part size and surface finish, as well as burr removal.

The use of burr knives requires care and experience to produce quality parts. One company initially performed such training with manufacturing supervisors grading edge quality. The approach was not as successful as having a production inspector accept or reject part edges. The inspector was much more critical, because he knew the real requirements.

QUALITY REQUIREMENTS

Every plant has its own edge-quality practice or requirements. Each may differ from anything mentioned in this book—and, quite often, a particular plant's requirements may be markedly different from the requirements of the shop down the street. This is where the worker sees, firsthand, the expected edge quality. In some instances, it may be markedly different from written edge-quality statements. Some shops are more critical than their written requirements specify; some are less.

DEBURRING ECONOMICS

To train workers in deburring economics, provide them with a simple set of examples that will help them understand the economics of their choices, or the choices of others who dictate work approaches. At this point, also stress job significance in relationship to final part salability. For example, emphasize the importance of worker contribution in the manufacture of an important missile part, medical device, or aesthetic shape. Depending on the maturity of the work group, this also can be the place in the program to stress that workers can make cost improvements by finding more effective deburring methods or recommending new approaches or tools.

PLANT EXPECTATIONS FOR DEBURRING WORKERS

Workers are expected to provide a full day of work and advise management when bad parts are found and current approaches are not working. They may be required to produce a specific number of parts per shift and train new workers, as well as perform many other functions. They may be expected to clean, as well as deburr, even though work instructions only say to "deburr." They may be inspectors as well. This is the place to clearly delineate expectations.

One plant lists the following expectations for its workers. These are in addition to general requirements that apply to all employees.

- Read and understand the work instructions.
- Remove all burrs to the extent that they meet the requirements of the master workmanship specification or any other explicitly defined specifications.
- Leave no sharp edges, secondary burrs (burrs made by the deburring process), or humps of material at edges.
- Maintain the required maximum or minimum edge break.
- Do not mark, mar, scratch, distort, or change part dimensions.
- Select the best method for the particular part and the individual performing the deburring.
- Verify the work.
- Identify burr conditions that, if not corrected, would result in scrap, rejects, or other problems.
- Identify to the manufacturing supervisor specific needs for deburring tools.
- Identify any problem, or potential problem, before working on the part.
- Suggest better machining and deburring approaches.
- Complete the appropriate paperwork when the job is completed.

Inspectors are responsible for verifying that parts are burr-free and meet edge-break requirements defined in the inspection traveler. Following standard procedure, this verification of burr-free conditions is to be performed before dimensional inspections are made. To facilitate any rework, the inspector is expected to indicate by adhesive-backed arrows on the part, or by words on the rejection slip, the general location of any burrs found on the part.

Surface Finishes

If a worker has never measured surface finish and does not know how much a scratch or sandpaper will affect a finish, he or she will eventually use something that causes parts to be scrapped. An understanding of surface finish and how the shop handles that issue is essential.

Clean Parts to See Burrs

Small burrs can only be seen when part edges are clean. Deburring personnel need to know how parts should be cleaned so they can adequately verify that their work is free of objectionable edge material.

Critical Drawing Needs— How They are Communicated

Work instructions vary widely, and some workers do not understand the instructions they are given. Figures 3-6 and 3-7, from the chapter on edge requirements, and Figure 33-1, illustrate three simple ways to inform operators of edge requirements. Figure 3-6 requires knowledge of how much 0.001 in. (0.03 mm) represents. More specifically, many workers have no idea how to translate the number into a radius of that magnitude. How big is 0.001 in. (0.03 mm)? That is why a tooling aid, such as a small pin that has a radius of 0.001 in. (0.03 mm), or any other specific size, is important. The worker merely holds the reference tooling aid against the edge that must be verified to see if they are about the same size.

Inspecting for Burrs

Chapter 34 provides great detail on inspection approaches for burrs. A few of these approaches can be used by any shop. Finding burrs, measuring burrs, and recognizing when too much material has been removed are training topics that must be considered.

How to Use a Microscope

Microscopes are more difficult to use correctly than many workers realize. If microscopes are required, some simple training is necessary to show workers how to set them up, keep them clean, and recognize when they are inadequate for viewing items that require detection. Nothing is obvious to the new worker. Chapter 26 provides some discussion on this topic.

Deburring System

When parts have many edges, it is essential for operators to use a systematic approach to ensure that every edge is totally deburred. This may mean starting at the left side and working around the part until every edge is finished. It may mean doing all holes, then all contoured edges, then all splines, etc. A systematic approach must be taught.

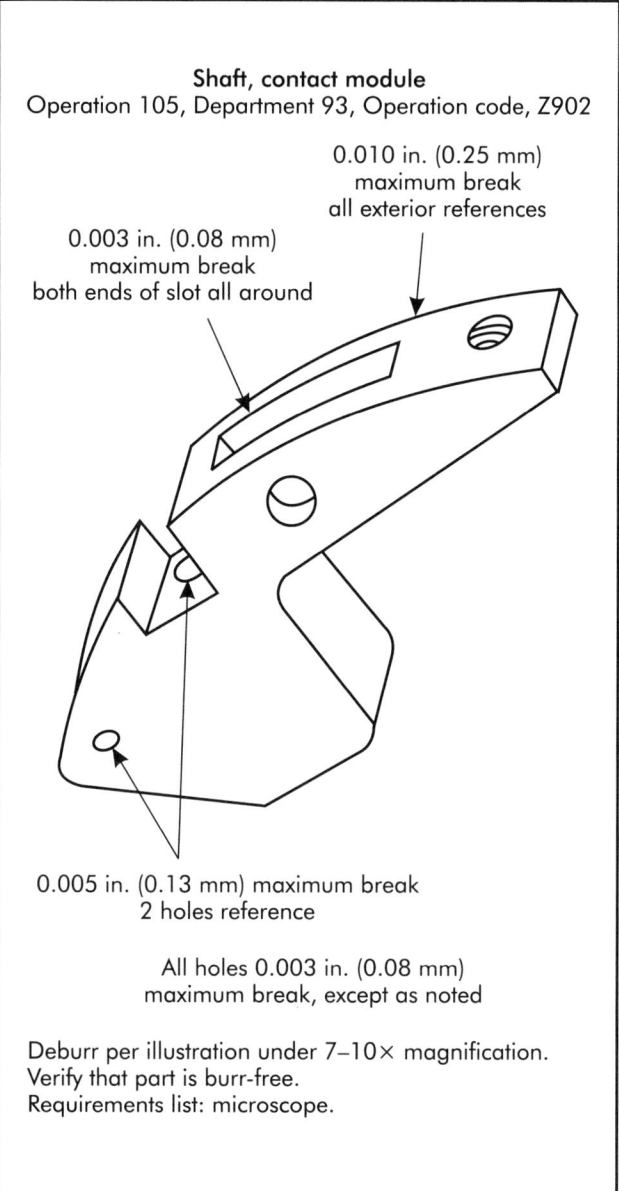

Figure 33-1. Simple communication aid to identify edge requirements.

Ergonomics

Operators must be coached and watched to prevent ergonomic injuries. They should know tool locations relative to operator position; proper microscope placement on the bench; and proper chair height and support. Workbench edges need to be dulled. Sitting straight in chairs is important. Microscopes can be tilted forward or placed closer to edges, but operators may not think of this; so, it should be covered in a section of the training program.

CASE HISTORY

This section describes the training program developed by one large company. It was designed for the production of sophisticated, precision miniature parts requiring high levels of precision and care. The format of the training program and considerations involved are described in the following sections.

Introduction

Because of the company's need to develop expertise in a rather large number (100) of individuals, a training program was developed to produce individuals capable of precision deburring. Deburring requirements were established with the goal of burr-free edges, with edges inspected at up to 25× magnification, and to provide edges that, in many cases, had no more than 0.003-in. (0.08-mm) break, radius, or chamfer. The program addressed the needs of miniature precision metal parts required in small quantities on a regular schedule. In addition to preparing and presenting this training program, formal documentation of the program and training materials also was prepared.

Nature of Deburring Needs

The training program was initiated at a manufacturing facility with more than 2,100 machine tool or workload centers. Approximately 60 distinct departments comprised a self-contained facility for the manufacture of miniature electromechanical components; large housings; plastic parts; integrated circuits; electrical, mechanical, and electromechanical assemblies; and a variety of related components. The diversity of products, wide variance in lot sizes, state-of-the-art tolerances, and rigid schedules necessitated unique considerations.

The specific departments requiring training were those manufacturing miniature mechanical and electromechanical components. The following list defines some of the production area's characteristics:

- More than 350 three-dimensional parts were manufactured each year on 99 different machines, with at least 120,000 pieces inspected each year.
- In addition to the 350 parts, approximately 900 screw-machine miniature parts were fabricated, many of which required some amount of hand deburring (although the majority of the edges were finished by centrifugal barrel tumbling). Deburring was performed on at least two shifts, and in some cases, all three shifts.
- Many of the parts involved were so small they could be hidden by a worker's fingernail.

Table 33-2 defines the basic manufacturing requirements for some of these parts. Their minute size was difficult for many to comprehend: In isolated instances, as many as 6,000–12,000 parts could be placed in a single thimble. Fortunately, however, most parts involved pins, shafts, or gears that are 0.02 in. (0.5 mm) in diameter or larger. Although many parts had quality requirements less rigid than those shown in Table 33-2, training was essential for employees who performed the deburring. It was also essential for employees to have equipment capable of producing the finishing operations required for those tolerances. Because of operator movement among departments, even workers on large parts might be required to deburr to more exacting standards. The company's product line included:

- large turned parts,
- small turned parts,
- gears,
- flat parts,
- box-shaped parts,
- machined, asymmetrical castings and forgings, and
- other configurations that are not easily categorized.

In general, the parts were produced by conventional processes using precision machines, high-quality tools and, in many cases, slower feeds than those used for commercial parts. Because of the high quality demanded and the need to maintain rigid schedules, many parts could not be successfully deburred by commercial mechanized deburring processes. Many parts, however, did undergo such processes as centrifugal barrel tumbling, vibratory deburring, electrochemical deburring and, in a few cases, electropolish deburring if tolerances, surface finishes, and burr size were compatible with these processes.

A 10-year study of the mechanized processes had been conducted to define their limitations and capabilities on these small parts. One problem was that conventional pin vises could easily damage miniature parts because the closing pressures necessary to hold parts indented the precision surfaces. Using an international technical library, the project leaders had read every article published in every language on the field of deburring, so they were knowledgeable of all aspects of deburring.

The plant's large, sophisticated job shop normally produced small quantities of parts—as small as 50 parts per setup. In many cases, the time required to develop familiarity with one part and the techniques for dealing with it were lost, since that part may not have appeared again in the queue of parts for three to

Table 33-2. Basic characteristics of electromechanical products

Part Characteristic	Value
Size	12,000 parts/in.^{3} (732 parts/cm^3)*
Diameter	≤0.280 in. (7.11 mm); 47% of total
	≤0.500 in. (12.70 mm); 63% of total
	≤1.000 in. (25.40 mm); 77% of total
	≤4.000 in. (102 mm); 99% of total
Quality	
Shaft diameter	±0.0001 in. (±2.54 µm)
Surface finish	8–32 µin. (0.2–0.8 µm)
Edges near sharp	0.003-in. (76-µm) maximum radius
Material	Stainless-steel 300 series and precipitation-hardened series; also 6061-T6 and 7075-T6 aluminum
Production quantities	25 each month, 300 every three months, and 2,000 every 3 years
Machine/finishing operations	Average 10 operations per part; maximum 50 per part
Set-up time	22% of total shop hours
Typical sequence	Bar stock—point bar end—grind bar diameter—automatic screw machine—lathe (second lathe)—mill—mill—drill—deburr—inspect—passivate—inspect

*Size of smallest part; most are significantly larger.

six months, sometimes even a year or two. This time lapse, combined with the unique requirements of the parts, demanded a wide range of skills of employees performing the deburring. Some parts required no magnification; others required 7, 10, 20, 30 and, in some isolated instances, up to 50× magnification to verify that edge quality was satisfactory.

Training Program Expectations

The training program was designed so that individuals who knew absolutely nothing about deburring or machining would, within approximately three months, be able to finish most of the parts that required deburring. Each individual also was expected to know the inherent limitations and capabilities of the tools used.

Usually, the instructions provided to employees who perform production deburring are merely written instructions and sketches illustrating part edge-break requirements and specific areas of concern. Written instructions for production parts do not usually specify which deburring tools to use. Figure 33-1 is representative of the production instructions given.

Workers were expected to have a general knowledge of the in-house inspection standards, and standards for a few specific parts. They were expected to be able to deburr any part given to them under the magnification indicated on the paperwork, and to use microscopes to view the work they had deburred. In addition, employees were expected to know the differences among various workpiece materials and to request special design tools where appropriate. As a result of these expectations, detailed training was necessary.

Training Format

Figure 33-2 illustrates the time sequence of the material presented in the three-month training program. The figure also shows an additional three months in which workers performed production deburring before they entered the specific departments in which they would work. The implementation of the program varied somewhat among specific classes—the first two groups each consisted of 25 people, and subsequent groups contained approximately 15 people.

As indicated in Figure 33-2, training began with an orientation. During the first week, workers were introduced to general plant procedures, safety practices, benefits, and worker expectations, and they were given an overview of plant manufacturing requirements and facilities. Individuals also performed some rough or finish deburring for at least two hours per day (sometimes for six hours per day).

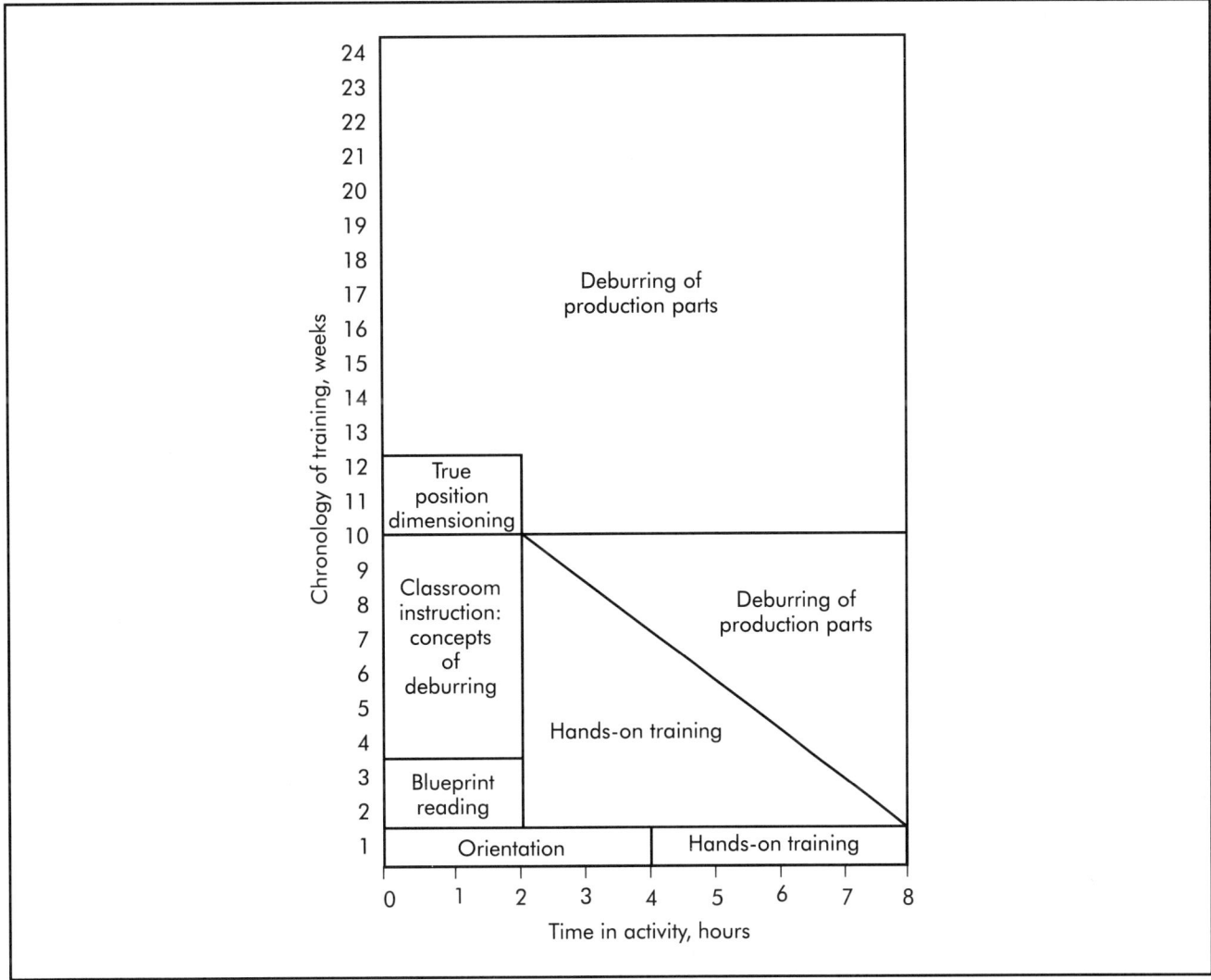

Figure 33-2. Time sequence of training program (Gillespie and Bolinger 1979).

During the second and third weeks, individuals received two hours per day of instruction in blueprint reading. During the fourth through 10th weeks, concepts of deburring were presented for 32 days. During the 10th through 12th weeks, true position dimensioning was presented. Whenever possible, daily formal classroom instruction was limited to a two-hour session. During orientation, however, individuals were in a classroom environment for at least four hours per day.

Classroom and hands-on training involved several teachers. Each group was supervised by an experienced supervisor and a setup person or similar experienced individual. These two individuals monitored progress and conducted some classroom discussions. Additional classroom instruction was provided by the technical training staff or engineers who had as many as 10 years of engineering experience in deburring techniques and product needs.

Table 33-3 lists subjects covered in the 32-day training section. Some 50 aspects of deburring and a wide variety of deburring equipment were presented. Extensive discussions of deburring tools and approaches for specific parts, inspection requirements, cleaning requirements, and capabilities of mechanized processes also took place. A detailed list of the material presented covered nine single-spaced pages. The following list describes the material presented in a single two-hour session. The session topic, for this example, is brush deburring:

- types of brushes available;
- sizes of brushes available;
- brush filament materials;

Table 33-3. Subjects covered in training

Day	Subject
Day 1	Economic impact of burrs Technical impact of burrs Schedule impact of burrs
Day 2	How to look for burrs
Day 3	The 16 basic categories of deburring tools The 54 subcategories of deburring tools
Days 4 and 5	How to use deburring knives
Day 6	How to use motorized tools
Day 7	The use and care of microscopes
Day 8	Basic deburring needs
Day 9	Review of decimal numbers
Day 10	Deburring holes Deburring threaded features
Day 11	Deburring turned parts Deburring flat parts
Day 12	Inspection standards for burrs
Day 13	Standards for surface finish
Day 14	Written deburring instructions
Day 15	Research on hand deburring Assembly requirements
Day 16	Brush deburring
Day 17	Files for deburring
Day 18	Rotary burs and countersink tools
Day 19	Mounted stones
Day 20	Scrapers and related tools Hand stones
Day 21	Abrasive-filled rubber tools
Day 22	Abrasive-filled cotton tools
Day 23	Abrasive paper products
Day 24	Felt bobs and lapping compounds Mandrels and holding devices
Day 25	Miscellaneous tools
Day 26	Personal techniques for ensuring quality Cleaning parts
Day 27	Deburring gears
Day 28	Deburring plastic parts
Day 29	Barrel tumbling Vibratory deburring Centrifugal barrel tumbling
Day 30	Abrasive blast deburring Electropolish deburring Electrochemical deburring
Day 31	Time studies and methods engineering Deburring large parts
Day 32	The burr and the machining process Finding the best deburring method

- variables in brush aggressiveness;
- side-effects of brush deburring;
- comparisons of brush capabilities;
- safety precautions in the use of brushes, and
- specific brush-usage approaches.

At the conclusion of each session, such as the brush deburring lesson, each individual was required to perform an experiment using some of the materials or concepts covered in the lesson. For example, three individuals might be required to use 1-in. (25.4-mm) diameter abrasive-filled nylon bristle brushes. Three other individuals might be required to use stainless-steel brushes. Yet another group might be required to use all brushes on identical parts. Each individual or group reported their respective observations at the next session. Individuals were frequently required to completely deburr a part using only the tool that was covered in that lesson. This requirement provided some experience using specific tools that might not otherwise have been selected by those who preferred to use other tools.

In addition to lectures, case histories and some role playing were used as teaching aids. Students received their own 100-page guide to plant deburring requirements, which described not only plant requirements but also the capabilities of many of the tools used at the plant. It did not, however, indicate how to use specific tools. Participants received another book that illustrated and defined the 250 most commonly used deburring tools at the plant.

A comprehensive test was given every fifth day. The test covered material presented that week, as well as significant points from previous weeks' instruction. These tests were not only indicators of comprehension levels, but also weak points in the presentations. The tests stimulated extensive note-taking, which further encouraged studying and learning.

Training Difficulties

Many training regimens encounter hardships when attempting to delve into specialty areas, a couple of examples of which are listed in the following sections.

Precision Miniature Parts

Several problems were encountered when initiating a new training program for deburring precision miniature parts, including:

- having enough parts or samples available for learning and experimentation;
- showing people how small some edge breaks (such as as 0.003 in. [0.08 mm]) really are, and how difficult they are to maintain;

- defining how to select the best tools for deburring;
- showing people how best to use deburring knives, and
- defining the types of problems typically faced on the floor.

For the class of 15–25 people in training for several weeks, several hundred parts were needed for practice. Many more of the small parts, which have only a small amount of edge length, were needed. Large parts provided the necessary number of inches per part, but, frequently, were not representative of the problems fully-trained workers faced on the job. Showing individuals how to maintain the 0.003-in. (0.08-mm) edge breaks was a time-consuming and difficult task. The instructor had to measure these edge breaks to convince people that their breaks were above the allowable tolerance. Measuring such small edge breaks, however, was difficult for many shapes and sizes of parts.

In spite of classroom instruction on the capabilities and limitations of many of these tools, students still had difficulty assimilating which of the 250 tools should be used for specific applications. Each part was different. Each part had different burr sizes. Lectures alone could not provide all necessary knowledge. Training had to include experienced teachers demonstrating some of the best approaches in a one-on-one situation. Finally, people learned from experience, which came with practice.

Knives

Using deburring knives is an art in itself. This teaching, too, had to be done on a one-on-one basis. Learning, however, was accelerated by using videotapes. Every person in the class was photographed while using a knife. The photographs helped the instructor point out to each person exactly what he or she was doing, both right and wrong. For successful and efficient training, videotape is an indispensable tool. Its capability for immediate feedback is one of its most useful assets.

Usually, there are two or three problem areas in deburring, such as the deburring of gears, flat parts, or holes. In a small job-shop, every part presents its own problem. It is helpful to classify the various types of problems one might expect in deburring. One approach for listing problems is to use a group technology classification, or coding, system (Gallagher and Knight 1973). These sorting, or grouping, techniques do not have to involve computers or be very sophisticated. They do require, however, classifying the type and features of each part. This classification allows effective teaching of deburring specific features to begin quickly.

Assume there is a variety of parts. Some parts have hole diameters of 0.250 in. (6.35 mm), 0.125 in. (3.18 mm), 0.063 in. (1.60 mm), and 0.020 in. (0.51 mm). Each of these holes may require different treatments. For example, to deburr a hole 0.50 in. (12.7 mm) in diameter, different tools are used than for a 0.020-in. (0.51-mm) hole. The size of the features alone may necessitate the use of different tools. On a large hole, a commercial abrasive-filled rubber product or scraper may be best. On a miniature hole, dental or miniature abrasive-filled rubber products may be necessary. Similarly, features can be classified as through-holes and blind holes, threaded externally or internally. A variety of similar classifications can be used.

Training Facilities

Since the program was new, a facility had to be established and equipped with deburring tools. Many new deburring products suitable for use with precision miniature parts had been located commercially before training began. These new products were introduced into production deburring through the newly-trained employees. Motors and other permanent devices were also provided. With the exception of dental air motors, each individual had direct access to all equipment listed.

Training Aids

Teaching deburring of such minute precision parts requires special visual aids. For example, it is difficult to explain to a class of 25 people the impact of minute burrs when the parts can barely be seen. The only suitable way of photographing many of these miniature parts, or their burr-laden edges, is to use scanning electron microscopes (SEMs) with magnifications of 100–200×. Scanning electron microscopes provide tremendous depth of field, as well as necessary clarity of all portions of three-dimensional parts. As an example of the unique nature of this plant to inspect the quality of one set of parts, 30,000× magnification was used. This was not typical, but the holes were only 0.000005 in. (0.13 μm) in diameter. These holes were not deburred manually.

In addition to showing actual parts and photographs of parts, effective training required the use of 35-mm slides, movies, and overhead transparencies. These aids, however, were often still inadequate for showing individuals how to handle each tool and the motions involved in working with delicate, minute parts. In these situations, one-on-one training or videotapes were used to show close-up action. Videotapes have additional advantages—they can be shown at any time and repeated many times, if necessary, to provide essential information.

One of the problems in providing adequate visual aids for hand-deburring tools is that there are at least 10,000 different products commercially available that are and can be used for deburring. With this many products, it is difficult to gather enough visual aids to show not only what is available in-plant, but also what is available commercially when specific problems arise. There is no simple solution to this problem, but having a great number of catalogs on hand is helpful when questions arise. At least 500 pounds of catalog material on deburring tools is now available.

In addition to videotapes, transparencies, 35-mm slides, and actual tools and parts, employees in training need formal descriptions of capabilities and limitations. Insofar as possible, the requirements for deburring within an individual company should be written. This documentation is usually difficult to compile because of the large number of individuals and products involved. Yet, many companies have documented their deburring requirements, some in more detail than others. The magnitude of the problem is illustrated by the plant being discussed—it took more than 100 pages for this company to document and illustrate common approaches, problems, tools, and inspection requirements.

Who Should Receive Training?

After many years of trying to minimize deburring problems and costs, it was clear that those performing deburring as a full-time occupation were not the only people who needed training. General machinists and machine operators also received some training in deburring. Many deburring problems arose from improper machining practices, which were controlled by machinists. Training could help make them more effective on their machines; and, if they were required to provide final deburring, they would be able to meet quality requirements.

Supervisors, assemblers, engineers, and many others in any plant are often uninformed of the complex and not-well-defined area of hand deburring. Some 600 people at the plant being discussed received some degree of deburring training over a three-year period. Some deburring problems were first observed by those assembling the parts. For this reason, many new hires in assembly received this training. At this plant, assembly was performed in clean-room environments.

Although the training was primarily designed for those actually performing deburring full-time, it was gratifying to see the impact of training on engineers who specified machining and blanking operations. Engineers may have received one or two additional sessions highlighting the design factors that could influence burr size or burr removal difficulty. After 32 sessions of seeing, hearing, feeling, and becoming aware of deburring problems, these non-deburring individuals had a better appreciation of and concern for the employees performing deburring and for the problems burrs presented.

Trainee Attitudes

One of the interesting observations made during this particular endeavor was the positive attitude shown by all those involved in the training. In many places in the world, deburring is looked on as a mundane and simple task. However, on sophisticated parts, where precision tolerances exist on miniature features, deburring can be one of the most exacting tasks in the manufacture of the entire part. Deburring, after all, is a skill-dependent operation. After weeks of training, it becomes apparent to many people receiving the training that they, indeed, are being trained for a very difficult task that many others in the world may be incapable of performing. This realization brings about a special pride in workmanship. This pride in workmanship is not automatic. It results from a combination of many factors, including the comments of the individuals working around and with the students.

Another aspect of attitude, which has been observed time and again in many other situations, is that individuals who are new to a job or task are willing to accept and try many things that "seasoned" workers would not. In this training, every effort was made to use the most modern, up-to-date, and fast equipment. Dental motors, dental stones, burs, and abrasive-filled rubber products were obtained for the students because these tools are very effective on many miniature parts. In addition, high-resolution surgical microscopes, special fiberoptic lights, and a variety of other hand motors and bench motors were purchased. Much of this equipment was entirely different than that used on the production floor. These tools were introduced both in training classes and on the production floor among experienced workers.

Many experienced workers, however, even though they were encouraged to use the new tools, did not feel that their efforts would be successful and did not wish to damage any parts in an attempt to use the new items. Those in training, however, accepted the fact that those tools would be acceptable, spent time practicing their use, and used them in their daily routines for sample and production parts. As the trainees entered the main work stream, other workers noticed the effectiveness of these tools and began to request them.

There is something about two people using two different approaches to a task that piques the interest of at least one of the two. When a younger coworker is using a precision instrument or motor and producing good results without causing damage to the parts, a seasoned worker may realize that perhaps there are alternatives and better ways to approach a problem. The individuals who graduated from deburring training helped bring to the production floor new techniques that otherwise might not have been accepted as readily.

Results

The training program was initially implemented for general machinists. Six months later, the program was significantly expanded for new deburring recruits. For each class, training was slightly different. The results have been impressive for those individuals who attended all 32 training sessions. After six months of training, their productivity as a group is not significantly different from a similar-sized group that had been performing the same work for 10–20 years. Although some burrs are missed by people new to deburring, many of the burrs are so minute that they are difficult for anyone to see. Without special lighting, even at 30× magnification, some burrs are almost impossible to see. Some are only 0.0001-in. (2.5-μm) thick, but they are still unallowable.

Observation of parts rejected by inspection and comments made by fellow workers and supervisors indicate that the training program has worked. There are individual differences but, as a group, the trainees have shown that they are interested. They have shown that they can maintain the desired production rates (or close to them), and they recognize the general quality levels required on many parts. In addition, many of these people are more receptive to bringing problem situations to the attention of their supervisors for immediate solution. In the past, many individuals solved problems themselves, as best they could; but, the next person who worked on the same part had to resolve the problem because nothing was documented from the earlier solution. Twenty years after this training program began, the former students were the plant teachers.

The project leaders studied deburring intently for 10 years and learned something new practically every day. There really is no end to what is needed to be known and can be learned about deburring. There appears to be no end to questions about how to determine which is the most effective tool for specific situations. As a result, continued research is in progress and will probably continue for many years. The cost of not using the most efficient approaches, of not being aware of available tooling, and of not knowing what others are doing in the same field, is just too expensive to ignore.

This effort has shown that individuals with no industrial training (fresh from high school) can be trained to do precision deburring on extremely miniature, complex parts. The training can be accomplished in a reasonable amount of time, and the individuals can become productive in a short period of time. All this, however, takes a commitment. It takes a commitment on the part of supervision to provide individuals knowledgeable about deburring for guidance and supervision. It requires a commitment to find the best tools and equipment to do the job. It requires the commitment of a daily teacher working with these individuals, solving the problems, and obtaining the proper tools. Efficient hand deburring can be taught and learned.

REFERENCES

Gallagher, C. C. and W. A. Knight. 1973. *Group Technology*. London, England: Butterworths.

Gillespie, LaRoux K. 1979. "Training Manual for Precision Hand Deburring, Part 1." Report BDX-613-2245 (November). Kansas City, MO: Bendix Corporation. (Available from NTIS.)

———. 1980a. "Training Manual for Precision Hand Deburring, Part 2." Report BDX-613-2534 (November). Kansas City, MO: Bendix Corporation. (Available from NTIS.)

———. 1980b. "Training Manual for Precision Hand Deburring, Part 3." Report BDX-613-2572 (November). Kansas City, MO: Bendix Corporation. (Available from NTIS.)

———. 1980c. "Training Manual for Precision Hand Deburring, Part 4." Report BDX-613-2582 (November). Kansas City, MO: Bendix Corporation. (Available from NTIS.)

———. 1999. *Deburring and Edge Finishing Handbook*. Dearborn, MI: Society of Manufacturing Engineers, p. 32.

———. 2000. *Guide to Deburring, Deflashing, and Trimming Equipment, Services, and Supplies*. Kansas City, MO: Deburring Technology International, Inc.

Gillespie, LaRoux K. and J.C. Bolinger. 1979. "Training for the Deburring of Precision Miniature Parts." Technical Paper MR79-501. Dearborn, MI: Society of Manufacturing Engineers.

Gillespie, LaRoux K. and F. E. Dulin. 1979. "Training Course Description for Deburring." Report BDX-613-2203 (April). Kansas City, MO: Bendix Corporation. (Available from NTIS.)

Inspecting for Burrs and Sharp Edges 34

Every shop seems to have a slightly different way of looking for burrs, differing expectations of what burrs are, and degrees of acceptability. Each inspection technique yields different results. While no single way is best, there are ways to be fast, exceptionally accurate, quantitative, and provide a pass-fail conclusion. This chapter provides an in-depth look at the many aspects of burr inspection. The terminology is consistent with "An Integrated International Standard for Burrs and Edge Conditions" (Gillespie 1998).

There are six elements of burr inspection or measuring systems, including:

- standard for defining pass or fail;
- measurement or detection method;
- consistency in employing the method and standard;
- training of inspection workers;
- equipment calibration;
- knowing the statistic of most significance, and
- documentation of results.

These elements are discussed throughout this chapter.

Whichever technique is used, it is important to base it on a written standard, employ it consistently, perform it with an understanding of what the results imply, employ it with calibrated tools, and document the results. These are standard quality issues; however, they may be overlooked in something as "simple" as burr measurement.

Just to make life challenging, users could find that making a single measurement of a burr on a part will not allow repeatable results or establish normal trends. Establishing a burr's size requires making more than one measurement per feature.

BURR AND EDGE STANDARDS

As discussed in Chapter 3, some companies have written standards for burr inspection, but many do not. The Worldwide Burr Technology Committee (WBTC) publishes a series of standards that companies can use, including callouts for drawings, definitions, and explanations.

In its simplest form, a burr standard defines the size, location, and functionality of edge requirements. Sheet-metal operations, for example, often allow burrs on parts that are smaller than 0.005 in. (0.13 mm), or 10% of sheet thickness for thin-gage materials. That is the standard. A more sophisticated burr standard calls for burr-free edges under 30× magnification over the entire part.

The standard also should refer to the principle attribute that actually causes the problem. Is the problem caused by sharpness or raised metal? Is the problem burr removal or any projection? Burr thickness is often the real issue, but workers measure burr height because it is easier, even if height is unrelated to the problem.

INSPECTION APPROACHES

There are more than 40 techniques for measuring or detecting burrs, as shown in Table 34-1. They can be categorized as:

- tactile,
- visual,
- destructive,
- replica,
- dimensional,

Table 34-1. Summary of inspection approaches for edges and burrs

Code	Process	Reference (See Reference List at the End of Chapter 34)
A	Visual (unaided)	
B	Toothpick probe	
C	#2 pencil point	
D	Fingernail	
E	4× glass	
F	Microscope	
G	Borescope	
H	Metallurgical mount (destructive)	Gillespie 1974
J	Height gage and indicator	Zaima, Yoshio, and Akiyasu 1975
K	Profilometer	Przyklenk and Schlatter 1987; Riis 1979
L	3D laser profilometer	Ko, Lee, and Jun 1988
M	Optical comparator	Bell and Kearsley 1963
N	Toolmaker's microscope	Przyklenk and Schlatter 1987; Zaima, Yoshio, and Akiyasu 1975
P	Light section microscope	Przyklenk and Schlatter 1987
Q	Diode scan	Przyklenk and Schlatter 1987
R	Taper section mount	Przyklenk and Schlatter 1987
S	UL sharpness gage	Bogue 1977; Sorrels and Berger 1974; Underwriters Laboratories 1998
T	Capacitance gage	Jones and Furness 1997
U	Water path	Gillespie 1981
V	Laser line-vision system	
W	Inductive measurement	
X	Handheld micrometer	
Y	Pipe cleaner	
Z	Toothpick shaving	
AA	Fingernail shaving	
AB	Thumb and forefinger	
AC	Plug gage for burr height	
ACstep	Designed gage	*American Machinist* 1973
AD	Plug gage for burr presence	
AE	Picks and probes	
AF	Sight pipes	
AG	Otoscopes	
AH	Stacked part gage	Hamill 1950
AI	Functional testing	
AJ	Replica measurements	
AK	Handheld measuring microscope	Przyklenk and Schlatter 1987; Pahl 1963
AL	Special design optical burr comparator	Pahl 1963, Seidenberg 1967
AM	Burr shadow	Przyklenk and Schlatter 1987; Pahl 1963; Von Obering 1955
AN	Scanning electron microscope	Ko, Lee, and Jun 1988
AO	Electrical contact micrometer	Przyklenk and Schlatter 1987
AP	Leaf-spring gage	Przyklenk and Schlatter 1987
AQ	Force feedback	Elbestawi et al.; Stango 1997
AR	Acoustic	Dornfeld 1984, 1991
AS	Bench-type dial gage	Hamill 1950
AT	Paper clip	
AU	Coordinate measuring machine (CMM)	
AV	Blunt-end form coder	Ko, Lee, and Jun 1988
AW	Laser confocal method	Ko, Lee, and Jun 1988
AX	Blur circle technique	Ko, Lee, and Jun 1988

- mechanical or hard-gaged,
- structured lighting,
- functionally tested,
- capacitive,
- inductive,
- acoustic, and/or
- force feedback.

No studies about measuring abilities or repeatability of these techniques have been unearthed, yet some rule-of-thumb data is mentioned in this chapter when known. Techniques have been evaluated, but they have not been reported in the literature. Some standard gages that were once used commercially may no longer be available, although the concepts employed are simple. Typically, visual and tactile approaches are used by industry, while each of the other categories described are used either for research or by a specific industry. The sharp-edge tester is used by a wide cross-section of industry.

Tactile Methods

Tactile methods tell operators or inspectors if a burr or smooth edge is present. They are cheap, relatively accurate given some experience, and every worker can relate with little training.

Thumb and Forefinger

The most common inspection method for burrs is a combination of tactile and visual. On hand-size or larger parts, workers typically look first for obvious burrs and sharpness, then run either a thumb or forefinger across the edges in question. The quick visual approach provides limited assurance that the fingers will not be cut while inspecting the part. It also provides the quickest means of verifying burr-free conditions. Even coarse, weathered hands can detect burrs 0.002-in. (51-μm) tall. By using the thumb and forefinger together in a scissoring action across an edge, workers can quickly tell if a burr is folded over and if the edge has some radius. Tactile methods cannot determine burr thickness. In general, fingers cannot gage burr height.

Fingernail

Most workers can detect burrs only 0.002-in. (51-μm) tall with a fingernail. Experienced workers with sensitive fingers can detect burrs as small as 0.005-in. (127-μm) tall; and, a few individuals can detect burrs as small as 0.0002-in. (5-μm) tall with their nails. One company uses thin shim stock held flat on a table to show workers how fine a step they can detect with their fingernails (Figure 34-1). This approach provides

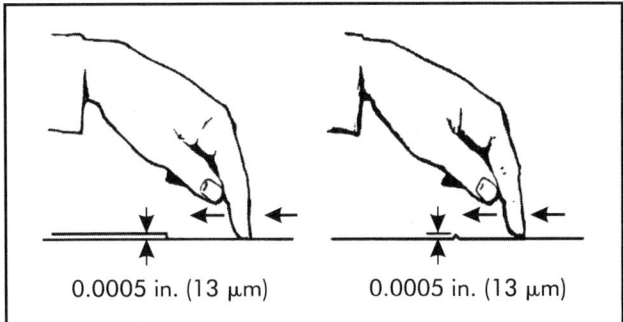

Figure 34-1. Detecting burrs and steps (Gillespie 1979).

one of the few calibration techniques and screening approaches for fingernail inspections.

Toothpick

Wooden toothpicks work for one company. Small obstructions can be detected by dragging this low-cost tool across an edge. Using this technique, even relatively smooth, raised material can be detected by trained operators. Most users rely on round toothpicks because apparently they are stiffer. The pressure required is very low, so strength is not an issue. Some precision shops cannot use this approach, however, because many toothpicks are filled with a white wax that contaminates precision assemblies. Toothpicks are not quite as sensitive as fingernails.

Paper Clip

Paper clips are used by at least one company to probe for burrs. It is inadvisable to use them on precision parts and fine finishes, since the sharp edge of the paper clip (which has a burr on it) will scratch part surfaces. In fact, this burr may be the best example of a common burr.

Dental Pick or Probe

Thin picks and probes (Figure 34-2), and even fine wires, allow access to hard-to-reach areas. These leave a fine scratch on the part surface. Most industrial parts are not affected by such scratches, but some precision parts are rejected for cosmetic scratches. Thus, these tools are banned from departments that prohibit any visual track from tools. Rounded or ball-ended tools also have been used. Some office supply stores carry ball-ended scribes used for stencil machines. Their detection capability is roughly comparable to toothpicks. Electronic probes also can be used. Size wires (gage wires) can be used successfully, particularly in small gears (120-diametral pitch, for example) to probe through the gear to the back edge of

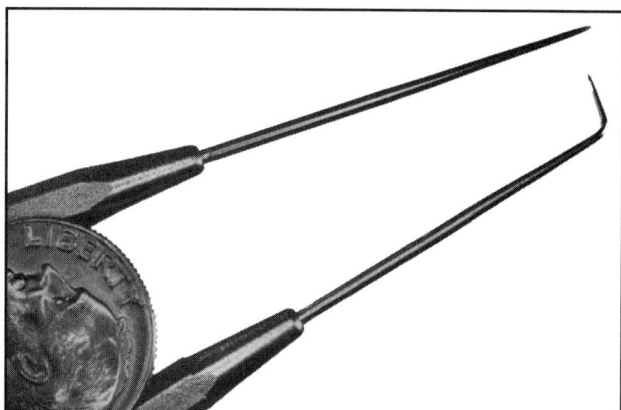

Figure 34-2. Dental picks used for probing (Gillespie 1979).

gear teeth. If the end of the wire stops or drags, a burr is present.

Pipe Cleaner

In some cases, a pipe cleaner is dragged over an edge because any rough edges or burrs will tear pieces of cotton loose. These pieces of fuzz and lint become evident and indicate the edge is not uniformly smooth, which may indicate a burr is present.

No. 2 Pencil Point

The techniques described in the preceding paragraphs indicate the presence of a burr at an edge. Some users poke at an edge or burr with a sharpened No. 2 pencil point. If the burr is firmly adhered, the point breaks before any burr material is torn loose. If the burr moves without breaking the point, the burr is considered loose and must be removed. This, reportedly, was once used to accept automatic transmission edge quality.

Plug Gage

For a few parts, gages have been designed to determine the presence of burrs, or burrs of excessive size. (These will be discussed later.) An individual can use a go-plug gage to determine if a burr exists at either the entrance or exit side of a hole. If the go-gage fails to enter a hole, it indicates the hole is undersized or a burr is thrown over into the hole interfering with the plug gage (see Figure 34-3). Similarly, if the plug gage goes through the hole but not out of the bottom, the hole is slightly tapered, undersized, or a burr is thrown over in the hole interfering with the gage. Plug gages do not absolutely indicate that a burr is present, but they are another weapon deployed in the war against burrs.

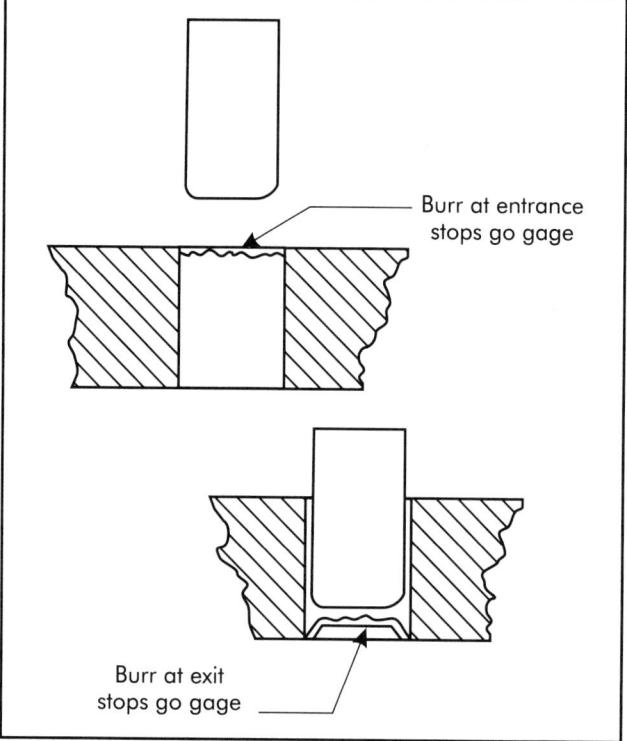

Figure 34-3. Go-gage stopped because of burr at hole entrance or exit (Gillespie 1979).

There are at least two problems associated with gages in checking for burrs. The first problem is that burrs will wear or scratch precision gages. This is not normally condoned by those responsible for monitoring gage size and quality. The second problem is that, in some cases, the burrs may lock the gage in the part. Forcing the gage, either through the hole or back out, will not only scratch the gage but may also scratch the part.

Tactile Probing Problems

Probing, or the tactile approach, to search for burrs has three major problems:

- Probing can easily scratch precision parts.
- Probing the entire edge of a part is time-consuming because many parts have many inches of edging.
- Probing is not foolproof, but it can be used. It works only if done in a direction 90° to the burr.

Fingernail Scrapings

In at least one case, an individual using a fingernail to probe over edges found that scrapings from the fingernail were frequently left on the edge. This was assumed to be the result of a sharp edge or burr, so edges were rejected for being too sharp. Actually, it is possible to easily verify that fingernails drawn over any edge can leave fingernail scrapings. It is not necessarily the presence of a burr that leaves these particles on the part.

Visual Methods

The visual inspection techniques listed here are used only to detect the presence of burrs. Actual burr measurements are described in conjunction with some of the equipment discussed later. Basically, users rely on one of the following approaches:

- naked eye;
- 4× magnifying glass;
- microscope;
- otoscope;
- sight pipe (light pipe), and/or
- borescope.

Naked Eye

The human eye can detect large burrs with relative ease. Smaller burrs can be detected by comparing them with another part feature. The eye is particularly good at detecting differences, rather than measuring actual size. No studies exist on the types and sizes of burrs the eye can readily detect. A typical inspector would be able to detect a burr that is at least 0.010-in. (0.25-mm) tall on a hand-sized part. The same inspector, however, may not be able to detect the same size burr on a 16-ft (4.9-m) diameter bull gear.

4× Magnifying Glass

Many inspectors use a round, 4× magnifying glass on an articulated arm to allow movement over the part, as illustrated in Figure 34-4. There are many suppliers for this type of product (Gillespie 2000). A ring light often surrounds the glass to provide adequate illumination. Inspectors may use eyeglasses if they normally wear them. Visual acuity varies significantly among individuals, so absolute burr-free conditions cannot be ensured. Some plants do not monitor visual acuity and have found, too late, that the inspector can no longer see fine burrs that were once obvious.

Microscope

Microscopes are used by different types of shops and in different situations: some shops may make miniature precision parts, others require exceptional edge quality (for example, deburring suture needles).

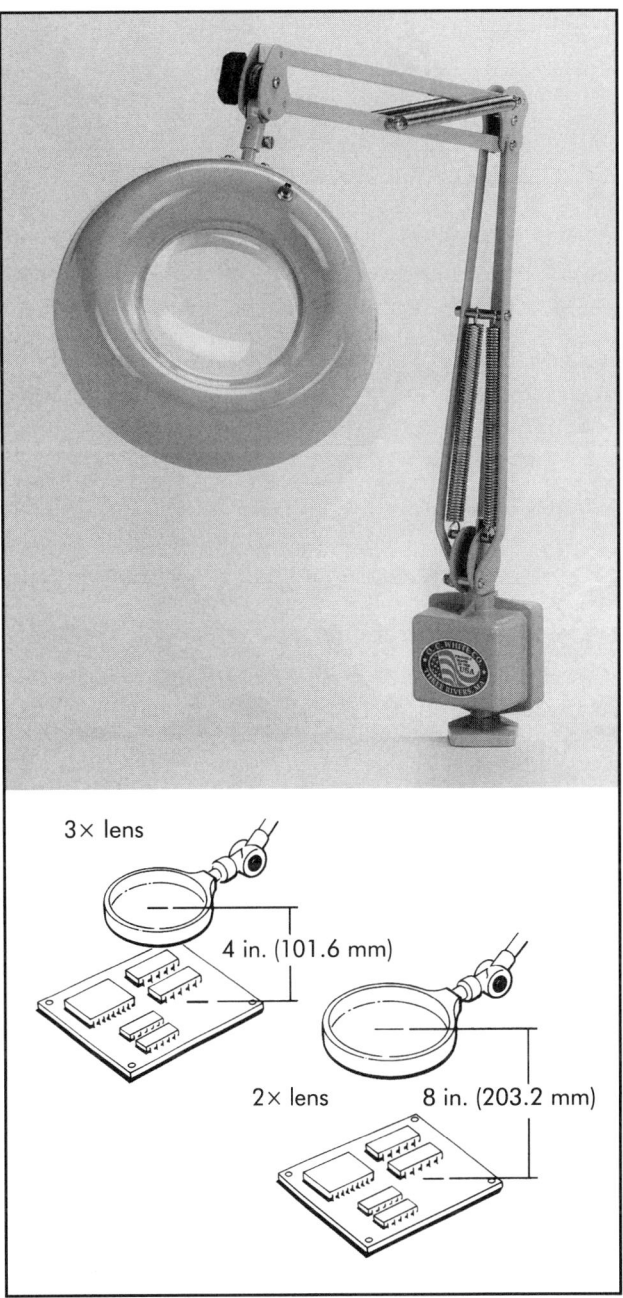

Figure 34-4. 4× magnifying glass for inspecting burrs. (Courtesy O. C. White and Sunnex)

Some may experience perpetual rejects because microscope quality is not dictated by specifications.

Stereozoom microscopes are required for most applications. Some of them have internal, coaxial through-the-lens lighting. This follows the same path as the eye, allows lighting to reach great depths, and illuminates internal cavities (Figure 34-5). One company uses the same scopes as those in operating rooms

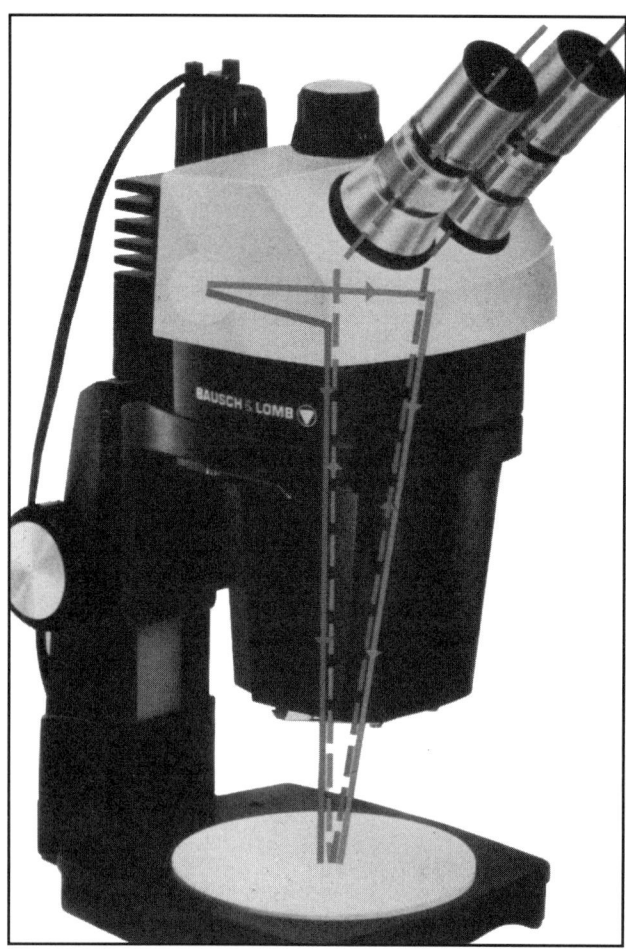

Figure 34-5. Coaxial illumination for a microscope (Gillespie 1979).

Figure 34-6. Stereozoom illuminator mounted on the back ledge of the microscope. Two positions are provided to allow working at different focal lengths (Gillespie 1979).

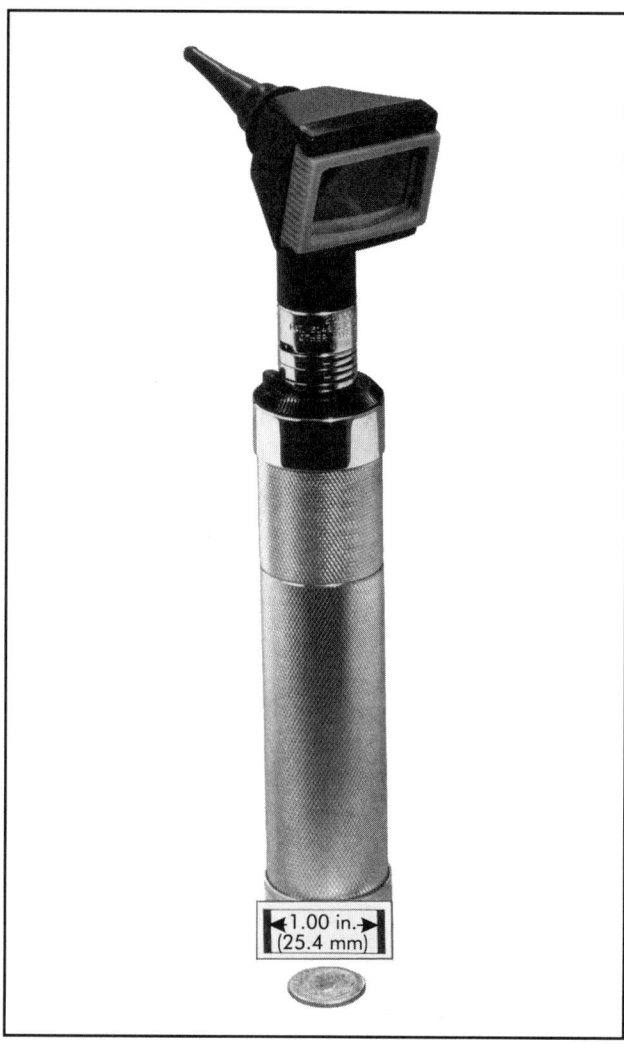

Figure 34-7. Otoscope used for burr detection in hard-to-see cavities (Gillespie 1979). (Courtesy Honeywell)

used by surgeons because the quality of optics is much greater than other systems. The price difference between conventional stereozoom microscopes and those used by surgeons ranges from $1,000–10,000.

Monocular microscopes are not a good choice for high-quality results. Stereozoom offers better appreciation of hard-to-visualize items, such as small burrs. Note that many stereozoom microscopes offer only external lighting and are less desirable for internal inspection (Figure 34-6).

Otoscope

Otoscopes are handheld flashlights used to see inside ears (Figure 34-7). These light sources include a small magnifying lens and concentrate the light in a very small beam. On hand-size and larger parts, this combination of lighting and magnification can adequately determine whether burrs have been successfully removed. Otoscopes are commonly used in situations where holes 0.25 in. (6.4 mm) or larger must be inspected.

Sight Pipe

Sight pipes are handheld plastic cylinders with a 45° end. They are used to inspect walls (Figure 34-8) or bottoms of holes as small as 0.25 in. (6.4 mm). If necessary, a small flashlight can be used to provide additional illumination. Inspectors insert this tool in a hole and look down at the top. A well-lighted, full-size image of the sidewall can be seen.

The sight pipe is made of crystal-clear acrylic plastic, which has good optical and light-piping characteristics. The enlarged head on the sight pipe collects natural light and pipes it internally, illuminating the area to be viewed. The deep groove around the head makes it easy to use supplemental lighting such as a penlight if needed.

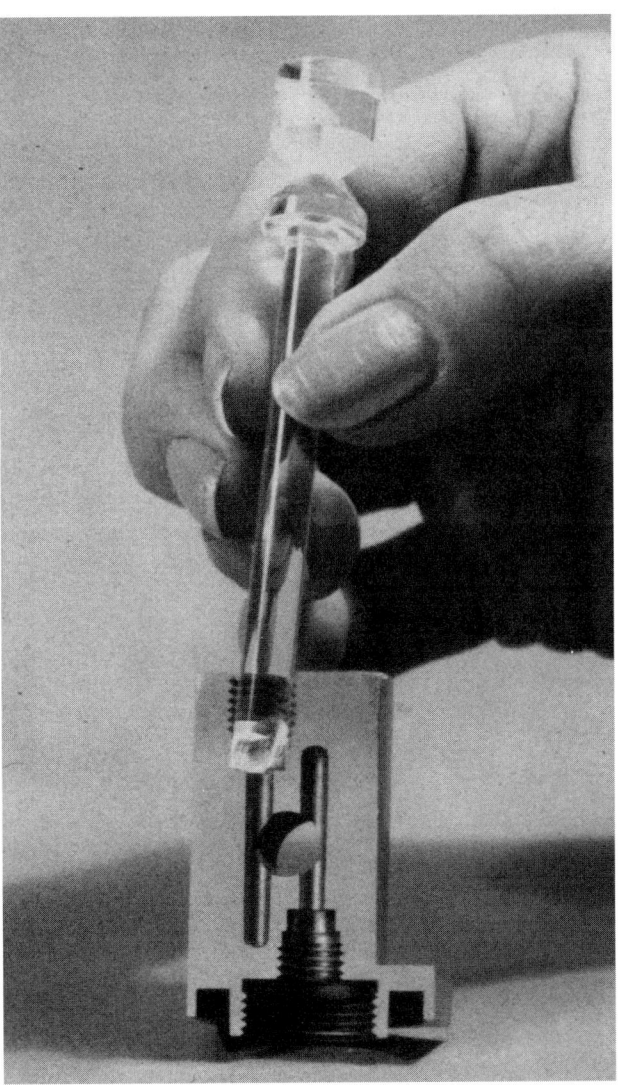

Figure 34-8. Sight pipe for inspecting holes. (Courtesy Honeywell)

Sight pipes can be used to inspect for a number of conditions, including: surface finish of bores; sidewalls of cavities; burrs at intersections of small holes into larger bores; internal areas of castings for porosity; burrs on internal O-ring or snap-ring grooves; and burrs on thread crests.

Borescope

Borescopes are miniature magnifying tubes used to inspect the bottoms or sides of holes or cavities as small as 0.062 in. (1.58 mm) in diameter. Unlike the sight pipe, this tool contains an integral lighting system (Figure 34-9). Several fields of view are possible by using different end pieces. Borescope images are circular, and some practice is required to interpret them. Borescope inspection must be done by operators who are qualified to detect burrs and assess the associated edge conditions. Acceptance may be made by a borescope optic eyepiece or videoscreen, which can provide 100× magnification and be videotaped for permanent documentation.

Lighting for Visual Inspection

Many burrs are difficult to see, particularly for those unaccustomed to looking for minute ones. The ability to detect burrs improves with experience. For those lacking experience, there are a few approaches that may help. Basically, they fall into the following three categories:

- looking in the best direction;
- using optimum lighting, and
- looking at clean parts.

Best direction. Many burrs are almost impossible to see, even when looking at the correct edge but in the wrong direction. For example, burrs viewed from the bottom of a hole might not be visible. By looking through the hole to the bottom surface, many of these burrs can be located (Figure 34-10). To verify that all burrs have been removed, look at the top surface first, then look through the hole to see if any burrs are thrown over into the hole. Next, turn the part over and look at the new topside. Then, look down through the hole again to verify that the hole is completely burr-free.

Looking through the hole to the backside is only one aspect of finding the best direction. To see the burr correctly on most parts it is important to tip the part to an angle in your line of vision (Figure 34-11). It is difficult to define the best angle because each part has so many different geometries. This ability comes with experience, but in general, do not look straight down on a part. Instead, look down at a 30–60° angle to any edge.

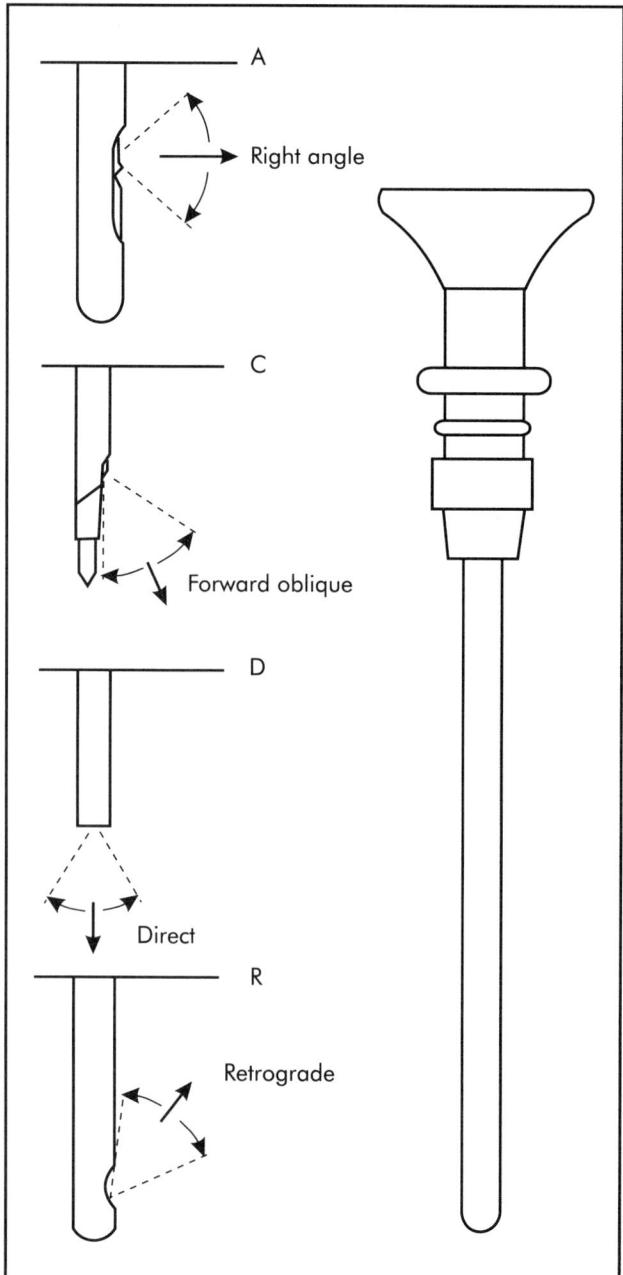

Figure 34-9. Borescope for inspecting small hole intersections.

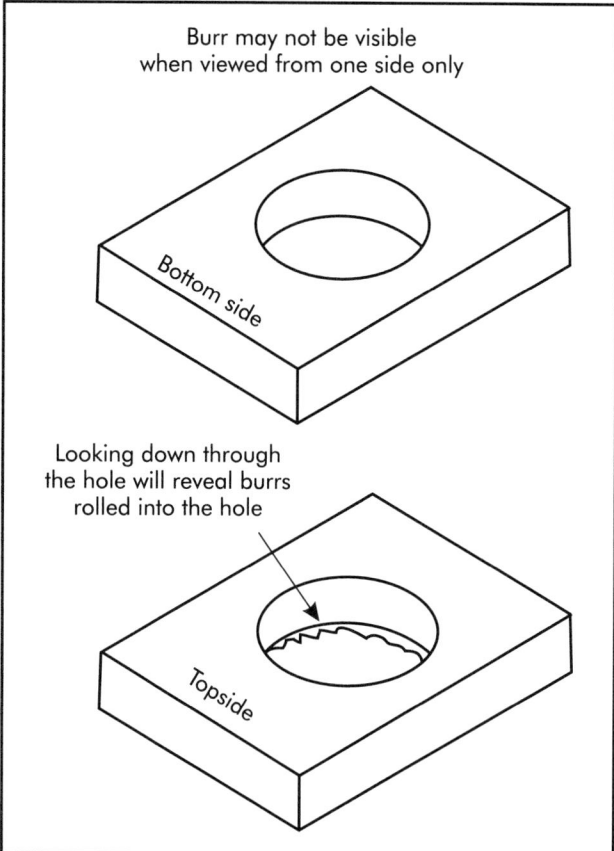

Figure 34-10. Finding burrs on opposite side of hole by looking through holes. (Courtesy Honeywell)

Figure 34-11. Tip the part at an angle to look for burrs. (Courtesy Honeywell)

It is important to have the lighting at the correct angle. The lighting chosen and the angle at which it strikes the part will make a major difference in whether a burr can be seen. One difficult task is to determine how to hold the light in relation to the part to see the burr adequately. Figure 34-12 illustrates the proper method.

Optimum lighting. There are 11 basic lighting techniques for use with microscopes that offer certain advantages in deburring or inspecting for burrs:

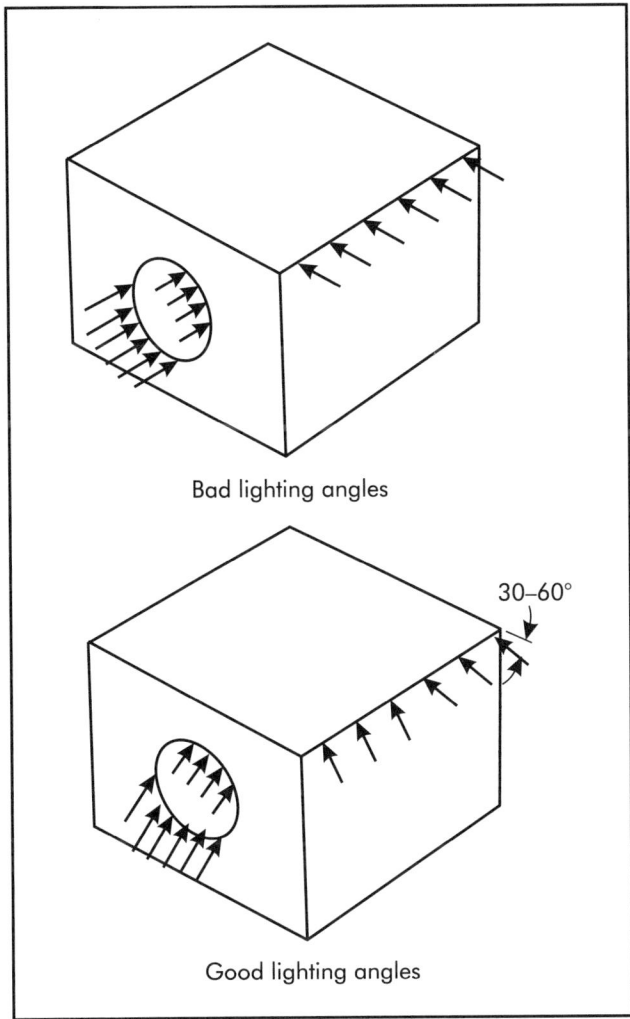

Figure 34-12. Representation of lighting angles for burr inspection. (Courtesy Honeywell)

- tensor lights off the microscope;
- ring lights (fluorescent tube around the lens);
- Nicholas illuminators mounted on the microscope frame;
- Nicholas illuminators off the microscope;
- fluorescent illuminators off the microscope;
- general-purpose illuminators off the microscope;
- coaxial illuminators (through the microscope);
- backlighting;
- polarizing accessories;
- color filters, and
- fiberoptic lights.

Tensor light. Tensor lights (Figure 34-13) provide high-intensity illumination that usually results in easier viewing. They create strong shadows, which is an advantage when inspecting for burrs because burrs cast a visible shadow. This type of light is recommended

Figure 34-13. Tensor lights used for deburring. (Courtesy Honeywell)

when inspecting for minute burrs. Typically, one light is inadequate—two opposing lights are used during deburring and inspection. The disadvantages of this type of light is that it consumes needed space in the work area and it can burn hands.

Ring light. Fluorescent ring lights (Figure 34-14a) provide uniform lighting across the entire workpiece. Uniform lighting results in less eyestrain and provides a shadowless light, which is adequate for many deburring situations. Shadowless light, however, is inadequate for minute burrs or those in hard-to-see locations. When mounted around a microscope lens, these lights reduce the allowable working height. Burr tools, particularly knives, can easily break the ring light.

An off-microscope fluorescent illuminator may also be purchased (Figure 34-14b). A flexible arm may even be attached to the microscope body.

Nicholas illuminator. Nicholas illuminators (Figure 34-15) provide a directional, concentrated beam of relatively high-intensity light. On some lights, a single power setting is available; others offer two or three levels of light intensity. Nicholas illuminators can be placed at the side of the microscope or inserted at the back of most microscope stands to provide nearly vertical lighting. Many microscope stands have two positions on which lights can be mounted (Figure 34-6). On some microscopes, the illuminator can be mounted on a flexible arm at the side of the microscope (Figure 34-16).

General-purpose illuminator. General-purpose lights, which look similar to desk lamps, are available to concentrate an intense spot of light on a relatively small field. Some of these lights can be focused. Also, many have an adjustable level of intensity. They may be mounted on or off the microscope with the addition of an arm. On some general-purpose illumina-

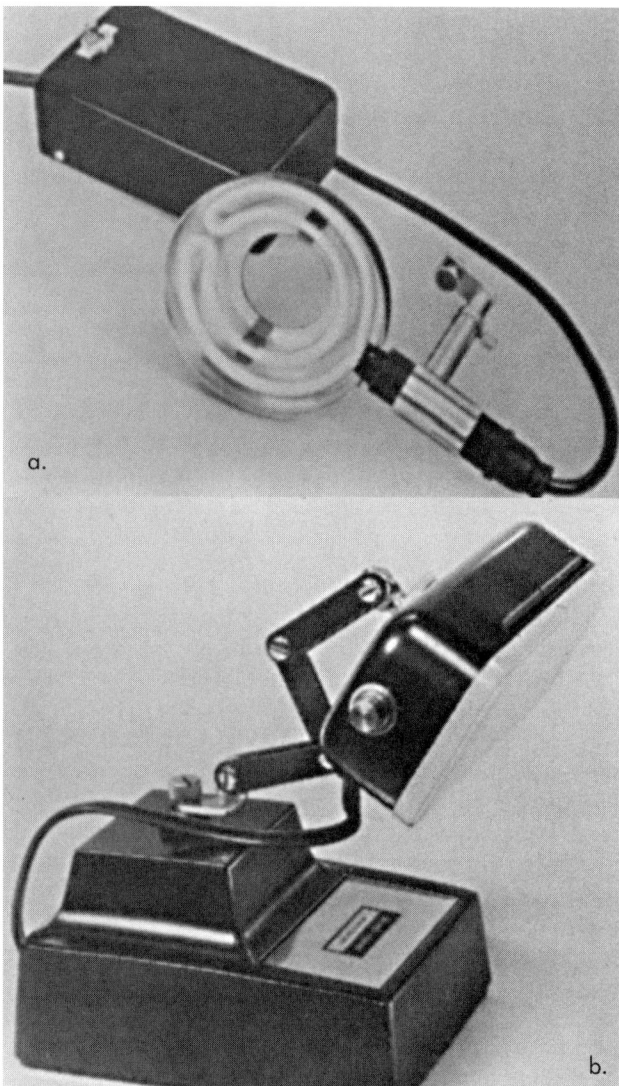

Figure 34-14. Fluorescent illuminators. (Courtesy Honeywell)

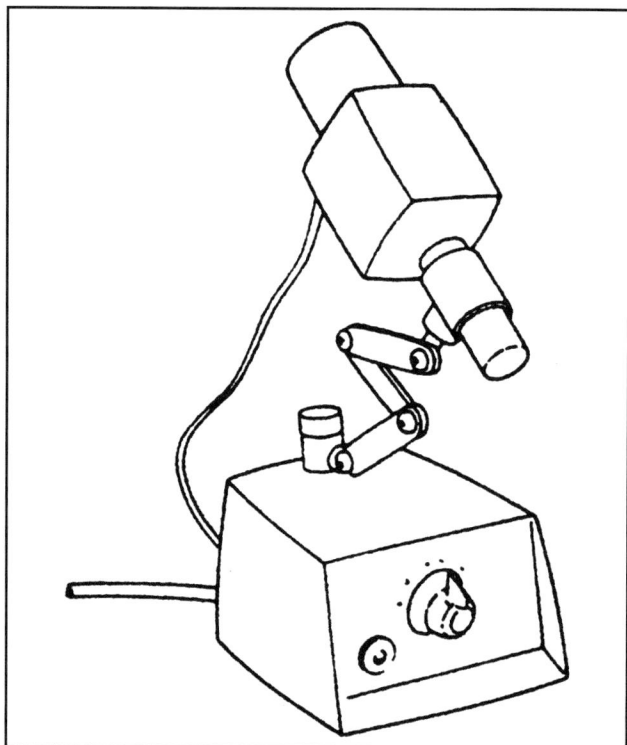

Figure 34-15. Nicholas illuminator mounted on a variable transformer.

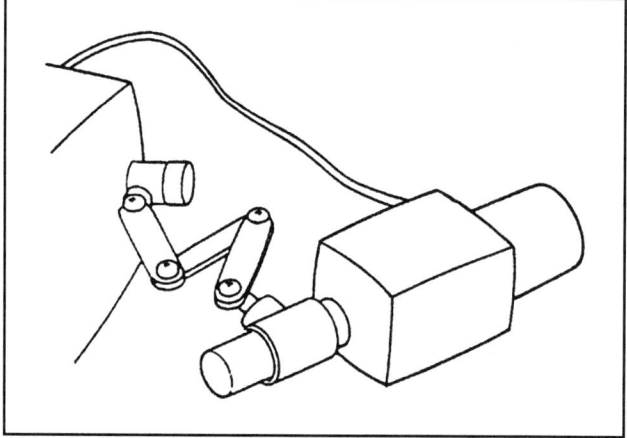

Figure 34-16. Illuminator mounted on the side of the power body.

tors it is possible to adjust the diameter of the light beam.

Coaxial illuminator. Coaxial illuminators are best when it is necessary to look into relatively deep holes or at flat, reflecting surfaces. These devices direct all light down the line of sight (Figure 34-5). Because the light beams straight down onto the workpiece it eliminates shadows unless the part is tipped at an angle to the line of sight. There are some advantages to coaxial illuminators: the light is always shining directly on the part, regardless of where the part is situated; and the light is never near hands or tools—it is completely away from the working area. Coaxial illuminator attachments can be purchased for most microscopes.

Backlighting. Many modern microscopes can provide light from the base of the microscope, upward from beneath a part, and up through the lens system. Backlighting is not normally used, but there are instances (for example, on some through-holes) where it is advantageous (Figure 34-17).

Polarizing accessories. In some instances, glare from lighting makes it almost impossible to see part

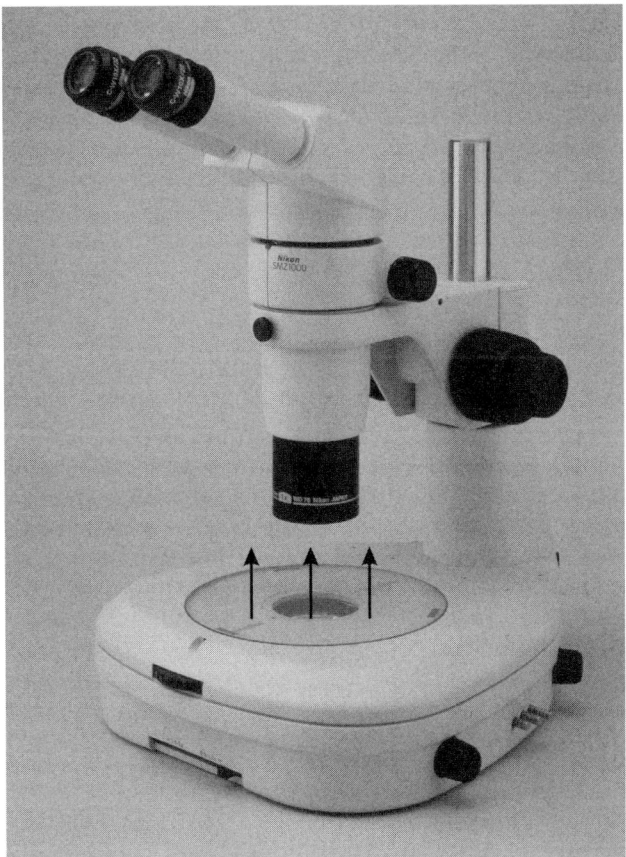

Figure 34-17. Backlighting. (Courtesy Nikon)

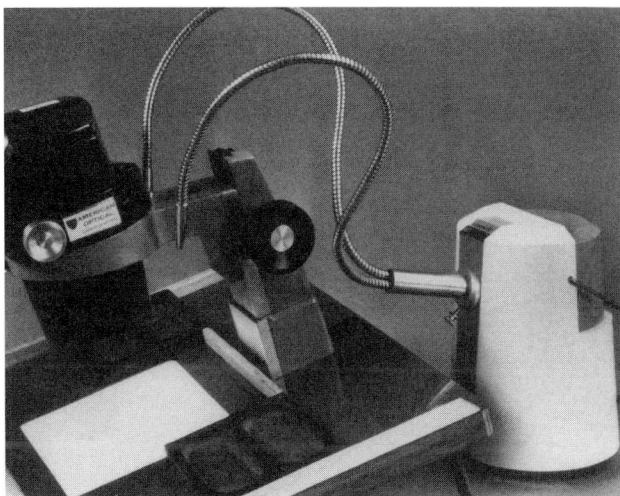

Figure 34-18. Fiberoptic light used with a microscope. (Courtesy Honeywell)

features that require deburring. In these instances, it is possible to use polarizing accessories on the lens to reduce glare. Polarizing accessories are similar to polarizing sunglasses. While eliminating glare, polarization makes it possible to see areas that must be deburred.

Color filters. In other instances, the color of the light makes it possible to see the burr. This may not be an important factor on steel parts, but it is on white, dark, or odd-colored parts. White ceramic or copper-colored parts may require colored filters. Some microscopes come with a built-in system to allow use of green or blue filters in addition to the normal white light.

Fiberoptic light. Fiberoptic lights can concentrate high-intensity light on a very small area. Fiberoptics enable the light to be maneuvered in almost any direction. Figure 34-18 illustrates a fiberoptic light that is commonly used with a microscope. These lights frequently provide so much light that their intensity must be adjusted to the lowest setting to be useful.

When working with microscope lights, it is important to use no more light than necessary to see the burr (this also extends the life of the bulb). One widely used fiberoptic light will last some 500 hours at the low setting. At the high setting, it will last only 40 hours. One advantage of fiberoptic lights is that they permit looking into deep holes and hard-to-reach areas, which are inaccessible using conventional lighting.

While fiberoptic lights are widely used, they may be less convenient than coaxial lights. Whenever the part is moved or tipped at an angle, the fiberoptic light must be readjusted. In contrast, the coaxial light is always in the correct location.

Problems with lights. Whenever light sources are placed around working tools, bulb breakage and burned hands are constant safety hazards that must be considered. In some cases, these light sources generate enough heat to make hands perspire to the extent that they cannot be successfully used for deburring. A major problem with most light sources is that they cannot be quickly focused. If the part is stationary this does not represent a problem, but in most cases the part is not stationary. Rather, it is held above tables and microscope platforms. A constant up-and-down motion brings the part in and out of the focus of the light beam. Also, as previously mentioned, it is easy to receive too much glare from some light sources.

Importance of Clean Parts for Visual Burr Inspection

One of the most frequently overlooked aspects of locating burrs is the necessity of having a clean part to determine whether burrs are present. Often, in the haste to move parts to the next operation, parts are deburred as best as possible, cleaned, then submitted without final scrutiny for burrs. On small features (miniature gears, for example), it is impossible to tell if a particle at an edge is dirt or a burr unless the part is clean. It is absolutely essential that parts are clean when a final review for burrs is made.

Measuring Microscopes

There are at least three variations of measuring microscopes. Most of the time, routine deburring measurements are not required. To improve deburring, a method for obtaining actual burr sizes is needed. Microscopes are one of the most inexpensive approaches. They are widely available in job shops for many applications.

Reticules for Microscopes

Reticules, which enable measurements to be taken, may be purchased for microscope eyepieces. Fine scales are etched or plated onto the glass eyepiece and allow measurements in inch or millimeter unit (Figure 34-19).

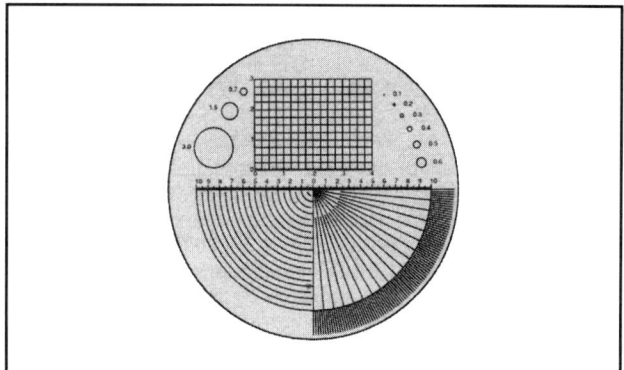

Figure 34-19. Reticule used on microscope for measuring.

Handheld Microscopes

Figure 34-20 illustrates a commercial, handheld microscope used for measuring the height of a burr. These do not work well for many parts because the part geometry has too many projections to allow convenient positioning. However, these microscopes are good for use on sheet-metal parts.

Toolmaker's Microscope

The measuring scales on toolmaker's microscopes make them useful for measuring burr thickness and height, provided the burrs are not bent. Height is measured by a vertical scale using dimensional differences when focused at the top and base of the burr (Figure 34-21).

Structured Lighting Methods

When a special beam is directed over an edge it is called structured lighting. Several measuring systems employ some form of structured light for measuring edges or small features. Optical comparators are a simple form of structured lighting; laser systems are one of the more complex forms.

Optical Comparators

The simplest structured lighting system involves parallel light rays in optical comparators. If the part can be placed so that the base and top, or thickness, of the burr can be seen, it can be measured. For many parts, the burrs go entirely around the part, so it is impossible to determine a reference plane. However, after some edge deburring it may be possible to measure the burr. Comparators are used for specific parts and types of burrs. For very small burrs, high magnification is required. Standard comparators use 20×, 40×, 50×, and 100× magnification.

Dedicated Burr-measuring Comparators

Figure 34-22 upper shows a small, simple optical projector on a base used to measure burr profiles. In this instance, the part is positioned on a staging fixture to place it at the proper height relative to the focal point of the comparator.

Figure 34-22 lower shows another variation, which is a microscopic device with the workpiece clamped on a rotating magnetic clamping fixture for round parts. This enables the round slug to be rotated under the microscope to measure burrs around the contour. The user aligns the burr with a mark on the left microscope by moving the bottom platform either left or right. The right-hand microscope is focused on a scale on the bottom platform. By recording the before and after dimension, burr size is measured. There are two sources of light—one at the top, and one at the bottom. By tipping the magnetic spindle, users are able to see and measure—to the root of the burr—the change in burr size.

Burr Shadow Length

The simplest way to measure burr height is by measuring the length of its shadow to calculate its height using simple trigonometry (figures 34-23 and 34-24). This device was first built to measure burrs on flat, sheet-metal stampings.

Light-section Measurement

Several structured lighting approaches are used to measure edge conditions. Figure 34-25 illustrates a

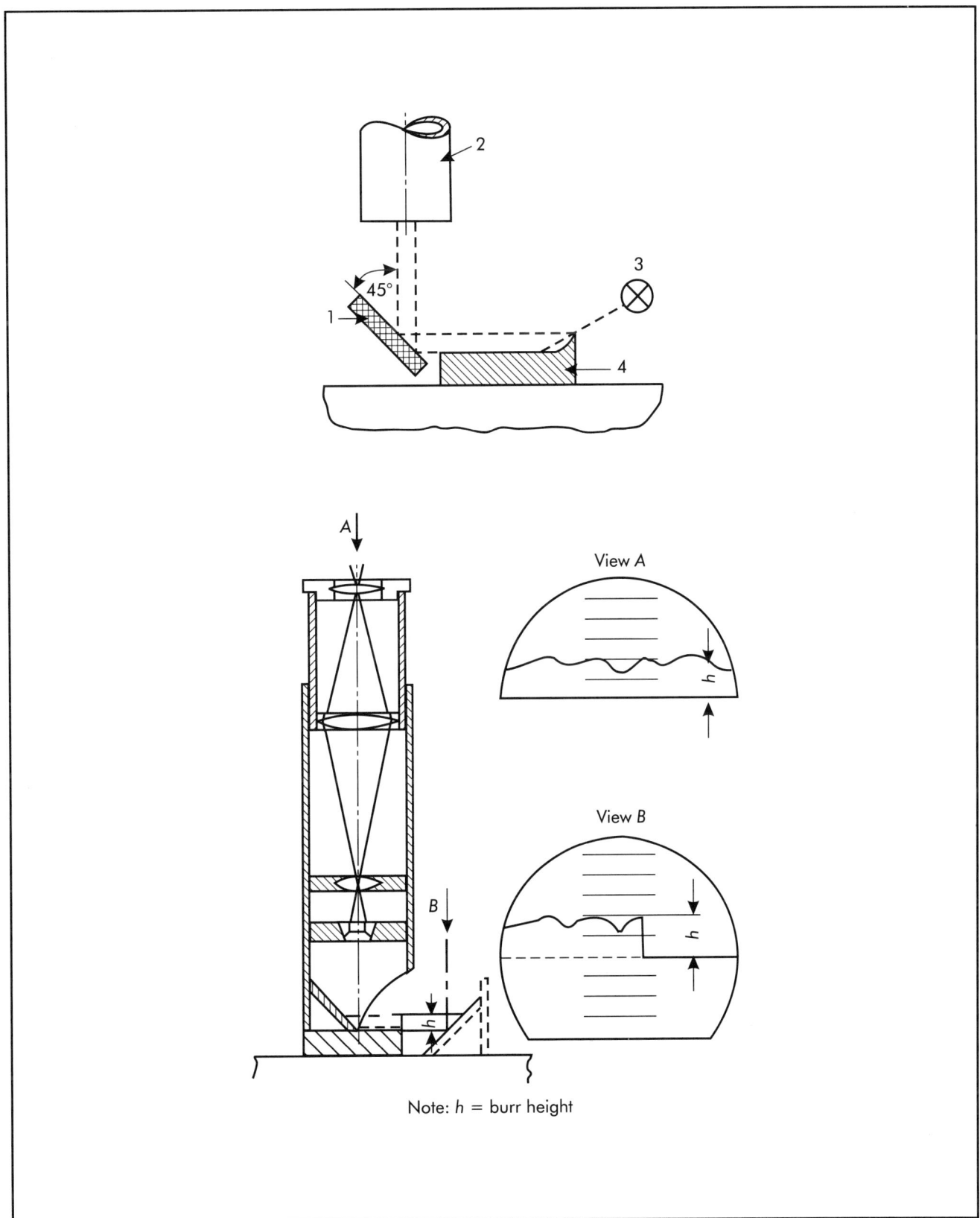

Figure 34-20. Burr measurement by handheld projection micrometer (Przyklenk and Schlatter 1987; Wentzel and Mehlhorn 1964).

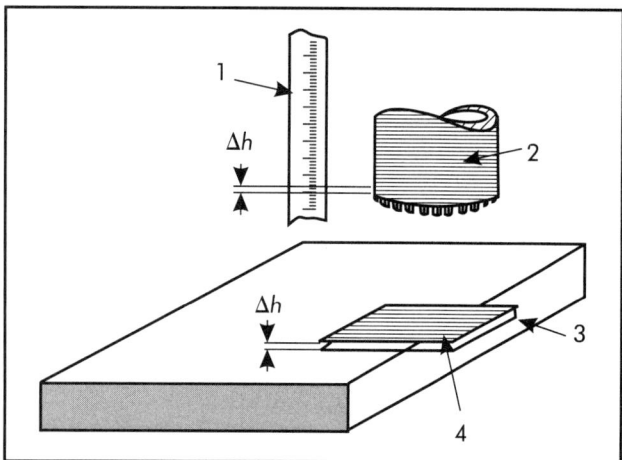

Figure 34-21. By focusing on the part, then the top of the burr, height can be measured (Przyklenk and Schlatter 1987).

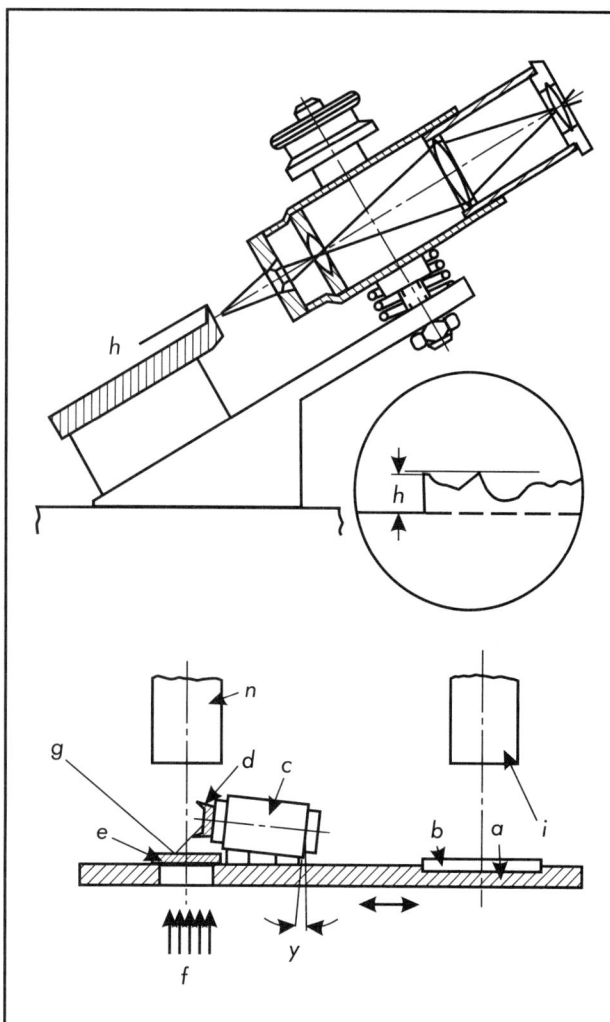

Figure 34-22. Special-design optical comparator for burrs (Wentzel and Mehlhorn 1964; Pahl 1963; Seidenberg 1967a, b).

light-section approach for measuring thin flanges in the center of a part. It works similarly on the exterior of parts.

Diode Line-scan Camera

Figure 34-26 illustrates light projection using a diode line-scan camera. In this case, the face of the burr, rather than its height, is seen. In most applications involving burrs, the burr is slightly triangular, or rectangular and vertical. Thus, the view closely represents burr height. This approach is not widely used. There is another approach, laser-diode analysis, discussed in literature (Flores 1995).

Blur Circle Technique

When using optics, an image is clear when it is well-focused. If unfocused, the image is blurred. A clear point shows as a circle with a diameter, even though the boundary is unclear. This circle is called a blur circle (Ko et al. 1988). The size of the circle is determined by the focus length and distance between the object and lens. Because the brightness at the boundaries is unclear, a modeling process is necessary for brightness to determine the size of the blur circle. The amount of illumination affects the measurement ability of this approach. The measuring range is limited to a few micrometers of height.

Electron-beam Approaches

Large companies and universities have access to sophisticated analytical instruments that can detect items and features as small as a molecule or a series of atoms. That may be too much power for shop use, but machines that use electron beams to "photograph" provide very clear evidence of edge quality when you can afford that level of analysis. Many electron-beam approaches can detect oxides and thin films of contamination as well.

Scanning Electron Microscopes

Scanning electron microscopes (SEMs) have been widely used by researchers to detect and measure burrs. These machines typically contain a built-in scale that prints on photos to show burr measurements. If parts are small enough to put in the SEM chamber, the measurement can be nondestructive. In addition, a reference part of known size can be placed in the picture area to provide additional size scale. This technique can measure to 39.37-μin. (1-μm) levels, while having a range of 0.04 in. (1.0 mm), or more. It provides height, width, and profile measurements if edges can be located appropriately (Ko et al. 1988). SEMs photograph at magnifications from 50–500,000×.

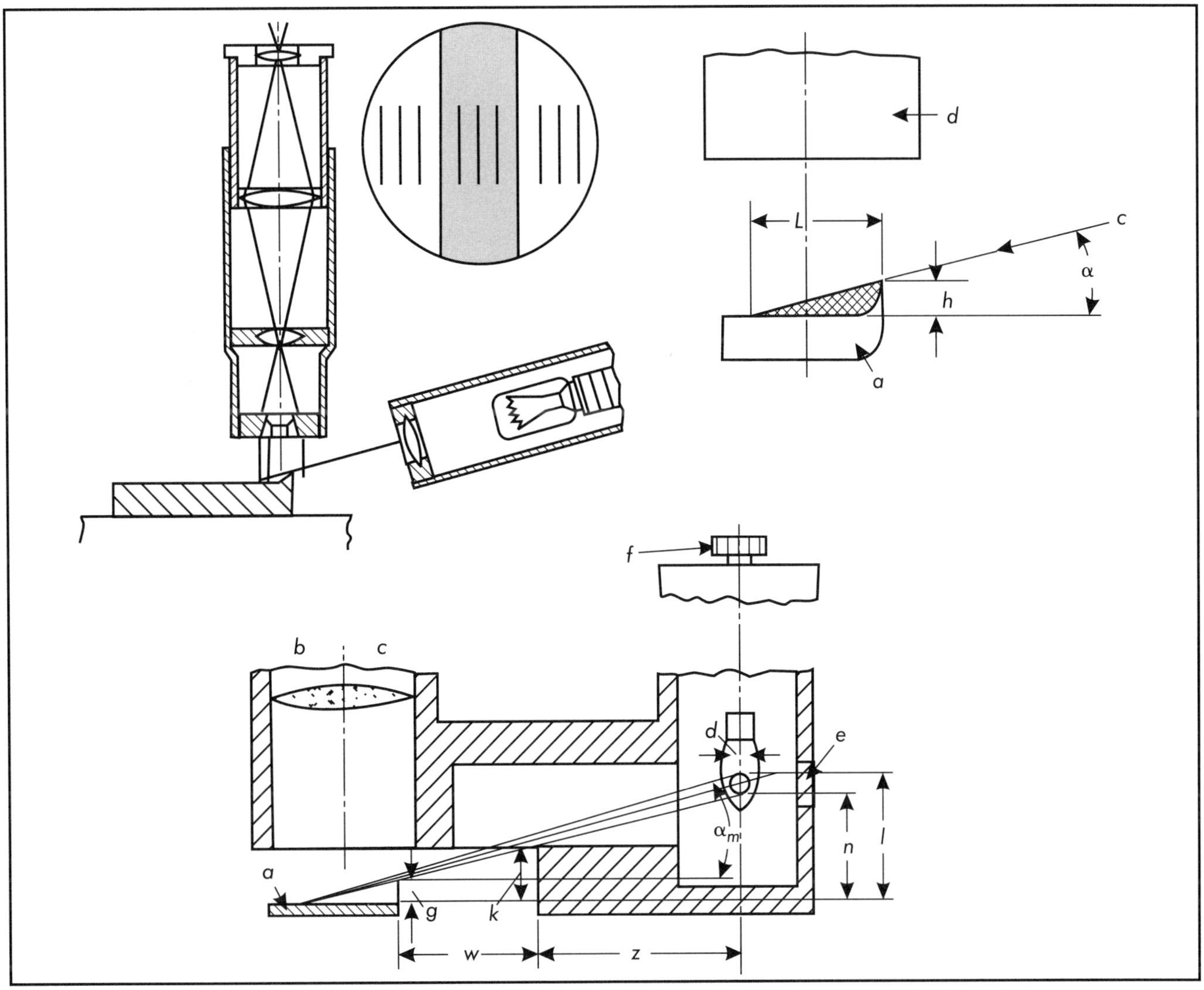

Figure 34-23. Another optical approach used for burr measurement (Wentzel and Mehlhorn 1964; Seidenberg 1967a, b).

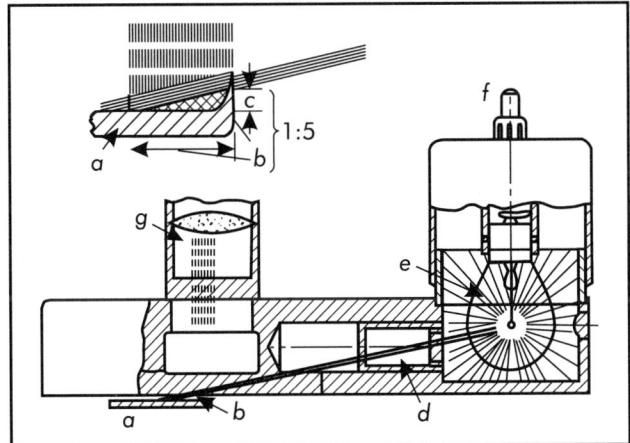

Figure 34-24. Principle of burr shadow length (Von Obering 1955; Wentzel and Mehlhorn 1964; Pahl 1963).

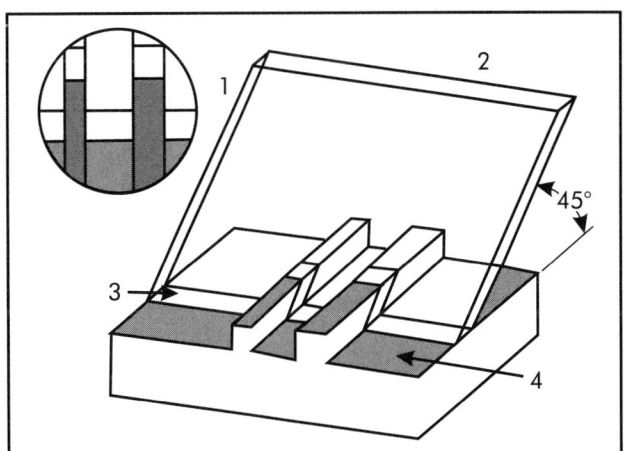

Figure 34-25. Principle of light-section method (Przyklenk and Schlatter 1987).

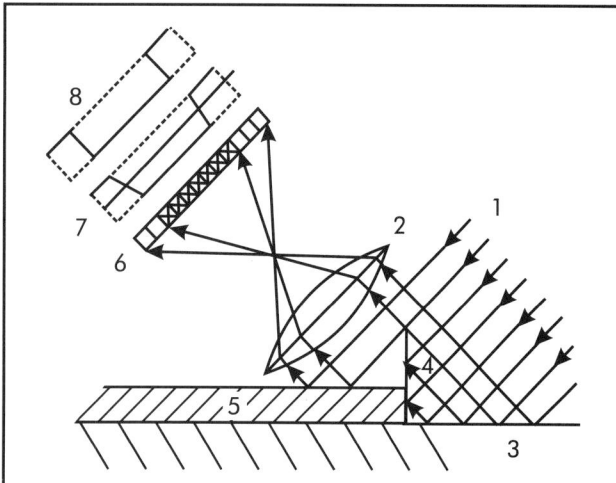

Figure 34-26. Principle of diode-line-scan camera (Przyklenk and Schlatter 1987).

A 500× magnification will provide clear edge-quality viewing. Conventional light-beam cameras do not have enough depth of field to clearly show edge quality at high magnification. The threaded photos illustrated in Chapter 32 are examples of SEM photos.

Small plants do not normally employ SEMs for production use, but the photos are so clear that when others cannot see what you see it is worth paying a university or analytical lab for a few photographs to show others. Some large companies use them to document edge quality.

Mechanically Gaged Methods

While several of the previous measuring/detecting means could be classified as mechanical, that term is reserved to define the industrial mechanical movement measuring devices. All of the following examples have widespread commercial uses.

Plug Gage for Burr Height Measurement

Figure 34-27 shows a plug gage, which is used to verify that burr height does not exceed a certain dimension. The gage is smaller than the diameter of the hole into which it is inserted. It is inserted until it stops on the gage-holder ledge (0.035 in. [0.89 mm]), which indicates the desired depth or burr height. This is fast, easy, and inexpensive; and, it can be used in many applications other than holes. By grinding a step in the end of the gage, two values of burr height can be detected with the same gage (go/no go).

Micrometers

Micrometers are used to measure burr height, but many burrs are mashed in the attempt to obtain an accurate height value. Such use also wears the micrometer anvils.

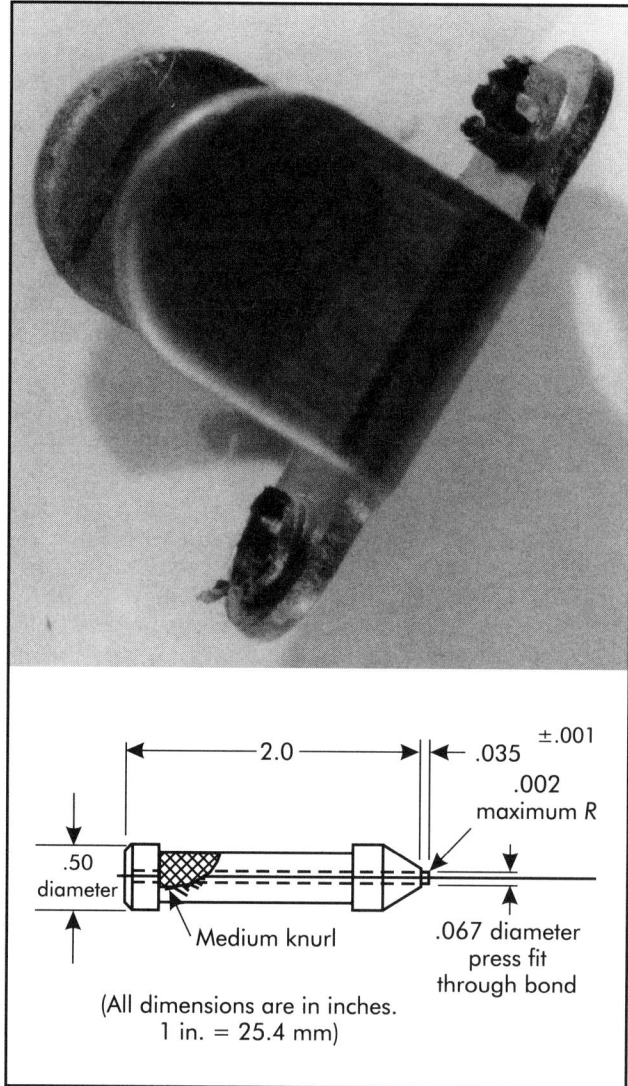

Figure 34-27. Burr height plug gage used for sheet-metal parts (Gillespie and Monteiro 1977).

Height Gage and Indicator

Standard height gages and indicators provide a reasonable estimate of the height of strong burrs (Figure 34-28). They are inadequate for precision studies since they also bend the burr during measurement.

Bench-type Dial Gage

For sheet-metal parts, one researcher used a bench-type, dial-gage micrometer with constant load on the movable anvil to provide reproducible measurements on laminated materials (Figure 34-29).

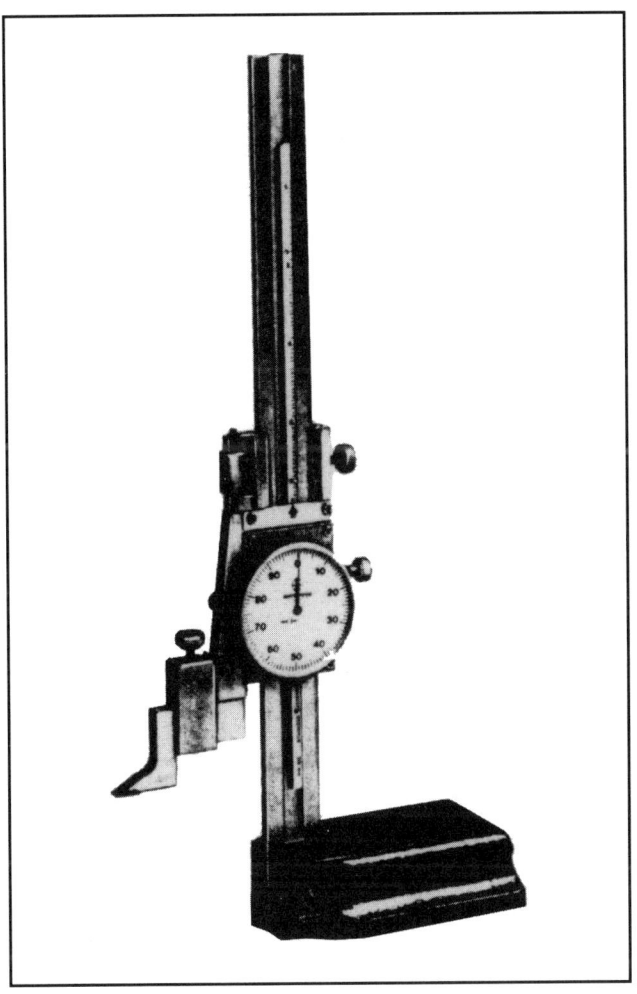

Figure 34-28. Height gage and indicator for burr-height measurements.

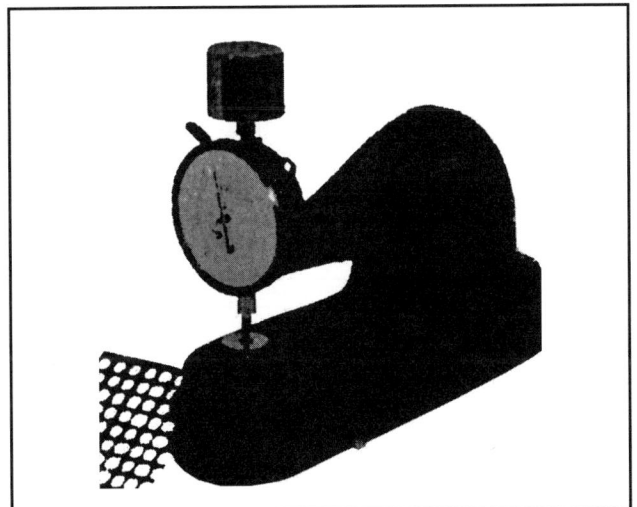

Figure 34-29. Bench-type dial gage for burr-height measurement (Hamill 1950).

Stacked Part Gage

One researcher developed a dial-indicator gage to accommodate a stack of sheet-metal parts (Figure 34-30). The overall height of five parts was the total of the part thickness plus the burr heights. This assumes, of course, that parts are flat enough to prevent them from bowing during the press operation. By measuring five parts at a time, the required time was reduced and a better average of burr height was obtained.

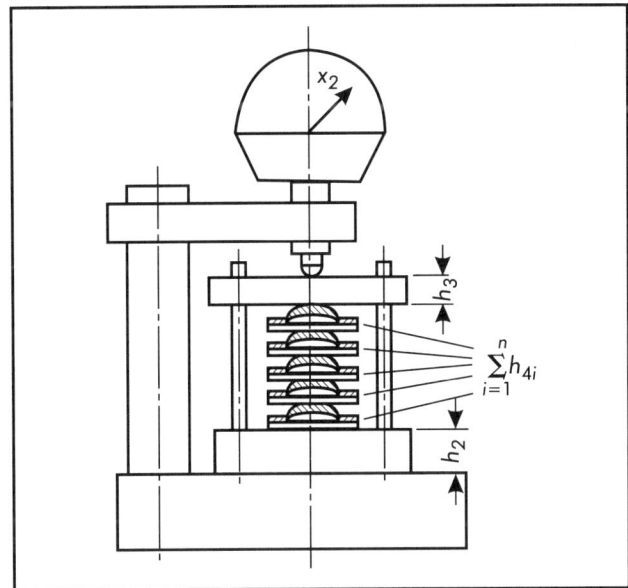

Figure 34-30. Gage for checking the height of stacks of burr-laden parts (Seidenberg 1967a, b).

Combination Approaches

Figure 34-31 shows a combination technique that provides burr height and thickness calculations. Height measurement is provided by the dial indicator, while thickness is determined by focusing a toolmaker's microscope on the flat surface and moving laterally.

Functional Testing

Functional approaches ignore common measurements and are based on how the edge or part is used. When fluid must flow in a smooth stream, watch how the flow comes off of a finished part. If cut hands are the issue, then use a device that simulates cut hands.

Fluid-flow Trajectories

When drilled holes must be free of burrs, some users force fluid through the hole, then monitor the trajectory of the fluid to determine if burrs or sharp edges

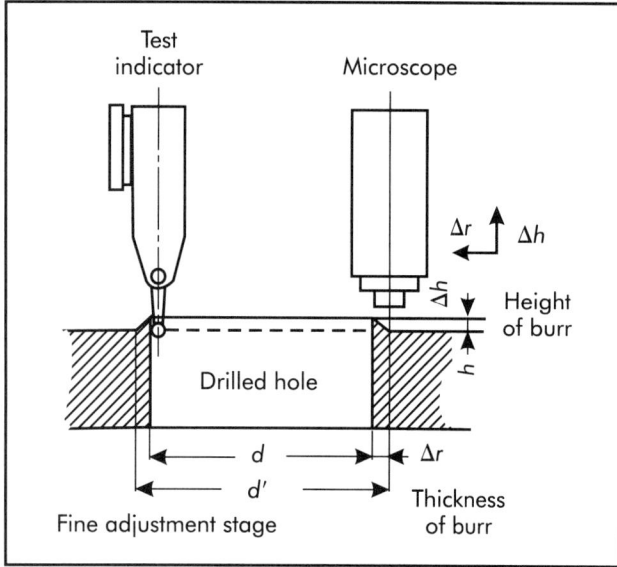

Figure 34-31. Measuring method for burr thickness and height (Zaima et al. 1975).

exist at hole exits (Figure 34-32). Any variation from a smooth radius on the hole disrupts the path of the fluid. An actual size is not provided, but the presence of burrs and the edge quality are obvious. It is quick, encompasses the entire hole contour, and can be used with multiple holes per part.

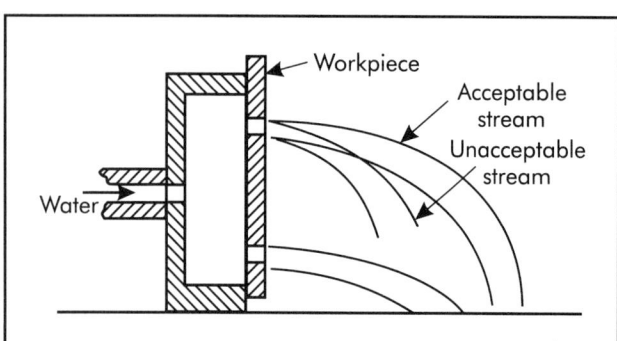

Figure 34-32. Burrs disrupt water flow from holes (Gillespie 1981).

Sharp-edge Tester

Figure 34-33 illustrates a sharp-edge tester that simulates adult finger skin (Underwriters Laboratories 1995). Using a soft-backed probe, the edge in question is run over the tool. If the edge is too sharp, the adhesive tape, which simulates finger skin, is cut. This simple-to-use tool is one of a few that mimics nature. U.S. Patent 3,931,732 describes another sharp-edge tester (Von Heitlinger 1976).

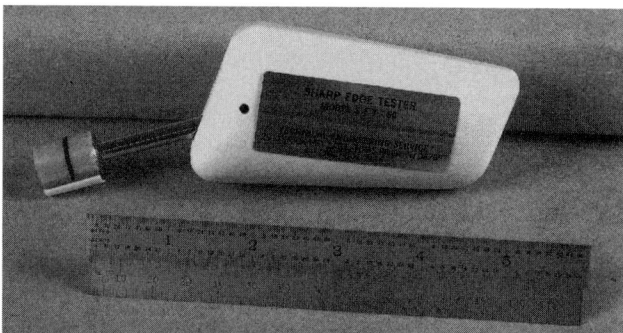

Figure 34-33. Underwriters Laboratories sharp-edge tester (Bogue 1977).

Profilometer Methods

Profilometers trace the profile of a part. A stylus is dragged over the edge of a part and a printout of the contour at the edge is produced. If a burr is present, a large jump in the profile at the edge appears.

Mechanical Stylus Profilometer

Surface-finish profilometers are used to measure small excursions from the nominal condition (millionths of an inch [μm] difference from the average). The success of this approach depends on the cross-section and height of the burr. Burr height will be reasonably accurate if the burr is not bent over and does not bend during the tracing operation. This method also provides burr thickness if the burr is not bent over. In some instances, the side of the profilometer stylus can run against, rather than over, the burr. The biggest concern with using the profilometer is to protect it from damage that could affect its calibration and accuracy. A fall from the part edge could damage the stylus.

One researcher designed a special glass stylus that provides minimum contact with the sides of the burr as it is being measured (Figure 34-34). It also allows measurements to be taken without removing cutting oil or cleaning solution from the part, which speeds testing.

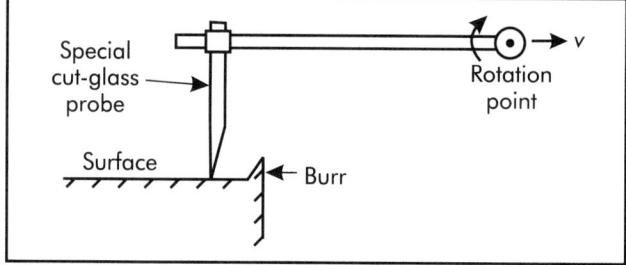

Figure 34-34. Principle of special glass probe for profilometer (Riis 1979).

Flat-end Form Coder

A flat-ended probe has been studied for use with surface-profile machines to measure burr height on sheet-metal parts (Ko et al. 1988). The results provide relative height measurement; but, as with all contact measurements, the burr will bend under pressure. In addition, the flat end can only capture maximum height—not the actual burr profile. For the study reported, the probe had a 0.008-in. (0.20-mm) diameter and was used to measure micro-burrs on the edge of 0.016-in. (0.41-mm) holes.

Coordinate Measuring Machine

Coordinating measuring machines (CMMs) have been used to measure burrs, but the results have been unsatisfactory because they tend to bend burrs. CMMs are fast and can be programmed to follow any nominal path. They can be used on unusually stiff burrs. However, the ragged edges of the burrs rapidly wear probe tips.

Three-dimensional Laser Profiling

Phase-measuring interferometry and vertical-scanning-interference microscopy provide three-dimensional images of part surfaces or two-dimensional plots with key summary data (see Figure 34-35). They provide images of the burrs at part edges. If the burrs are unbent, both height and thickness values can be obtained and documented for some length of edge, not just at one point on an edge. They are expensive for measuring burrs, but provide noncontact results and can detect (phase-measuring interferometry only) variations as fine as 0.000000004 in. (0.1 nm or 0.0001 μm). Vertical-scanning-interference microscopy can record height variations of several millimeters. These machines provide data capture, bitmapped and tagged image file (TIF) images, and a variety of plotting capabilities.

Laser Confocal Measurement

Laser measurements in the micro-inch (nanometer or micrometer) accuracy range work by at least one of two mechanisms: confocal and interferometry. In the confocal method, the light beam is focused on a surface, causing the reflected light to converge to a pinhole. The degree of convergence determines the resolution of height measurement. For example, a laser wave with a length of 27.6 μin. (670 nm or 0.7 μm) can produce a beam spot of 275.6 μin. (7 μm). This allows a height resolution of 7.9 μin. (0.2 μm) and a measuring range of ±0.039 in. (±1 mm). While this provides an accurate measurement, it is difficult to measure the actual profile of the burr because of diffused reflection at the tip of the burr (Ko et al. 1988).

Destructive Methods

Some burrs are too difficult to see because of their location on the part. Others can be seen, but are hard to explain to new workers. By cutting these parts in half near the burr, it is easier to visualize what the edge or burr actually looks like. All of these techniques destroy the part.

Metallographic Section

Accurate evaluation of burr properties requires destructive measurements (Figure 34-36). When drilling burrs, a thin saw is used to cut away a little less than one half of the hole, then the sample is potted in a plastic matrix using metallurgical mount techniques. After solidifying, the exposed surface is lapped to reveal the true cross-section surface. This is a time-consuming approach, but it enables users to see the true burr profile, measure both height and thickness, and obtain microhardness as well. Grain structures, flow lines, and other properties can be viewed with this approach. However, it destroys the part. Burr sizes are measured with toolmaker's microscopes, reflection comparators, or scanning-electron microscopes.

Metallographic Taper Section

Slicing burrs perpendicular to axis provides an accurate representation of dimensions. However, by slicing at a known angle, a skewed but taller image of the burr can be obtained. This is called taper sectioning, as shown in Figure 34-37. An apparent magnification ratio of 10:1 can be produced by mounting the specimen at an angle of 5°44'21" instead of the usual 90° angle. This technique is widely used for measuring scratch contours and raised metal at scratch edges.

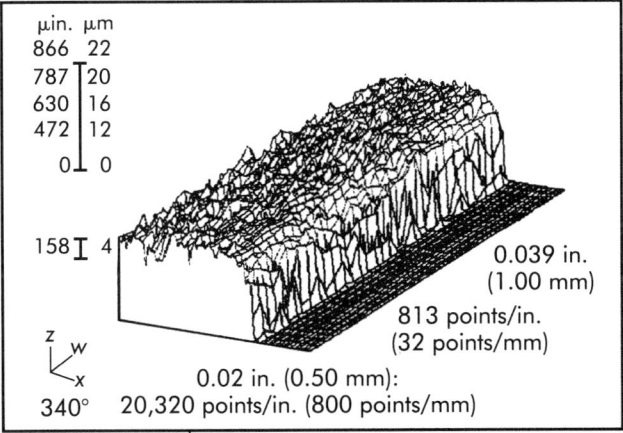

Figure 34-35. Three-dimensional perspective of a burr showing burr height and thickness (Flores 1995).

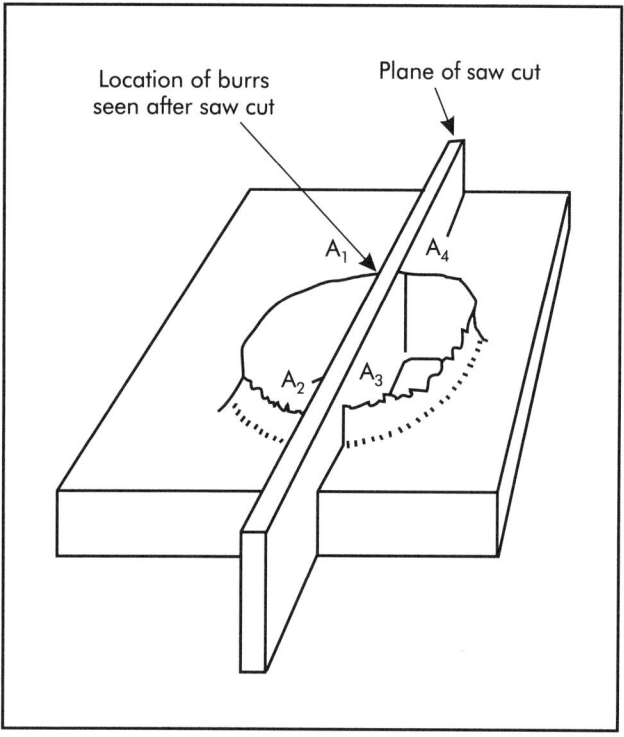

Figure 34-36. Destructive cross-sectioning to obtain burr properties (Gillespie 1974).

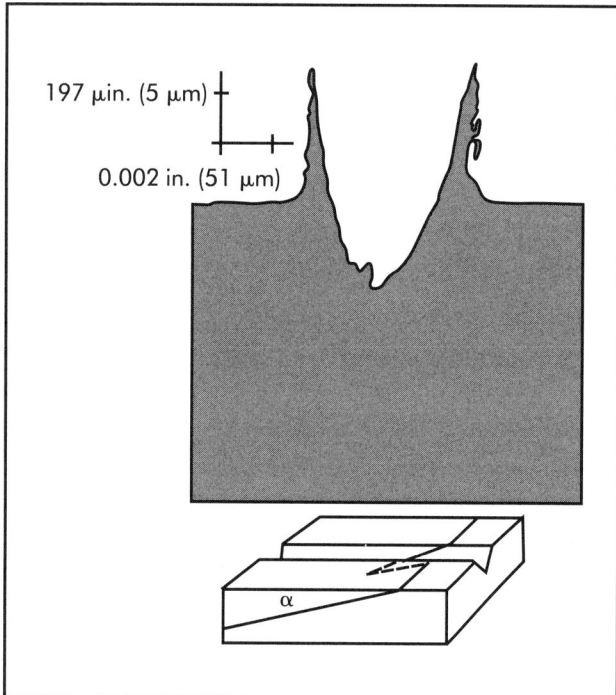

Figure 34-37. Taper sectioning concept. A relatively shallow scratch and burr-like projection are magnified in height by grinding the mount at a small angle (α) (American Society for Metals 1985).

Replica Methods

Plastic replica materials, such as room-temperature vulcanizing (RTV) elastomer, can be used to mold an image of the edge. The replica then can be sliced in half with a razor blade, X-Acto® knife, or similar sharp, thin blade. The resulting surface can be viewed with a microscope or comparator. Burr height and thickness can be measured, and the entire cross-sectional profile can be seen. This is nondestructive, but it may leave traces of the plastic material on part surfaces. It is somewhat less accurate than measuring the actual burrs. In some instances, it may be impossible to obtain a true cross-section, especially if the burr is bent over to the extent that the plastic cannot get under the bent burr.

Electrical Methods

Electrical or electronic gages are some of the newer approaches to burr measurement. These methods are not widely used but have their place in unique situations.

Electrical Contact

Figure 34-38 illustrates a micrometer head mounted to a base. When the thimble touches the burr, contact is signaled by either a lighted bulb or sound. This provides a more accurate measurement of height than the conventional mechanical micrometer or height gage since it does not bend the burr tip.

Leaf Spring

Figure 34-39 illustrates a leaf-spring device for measuring burr height. It has been used in Germany

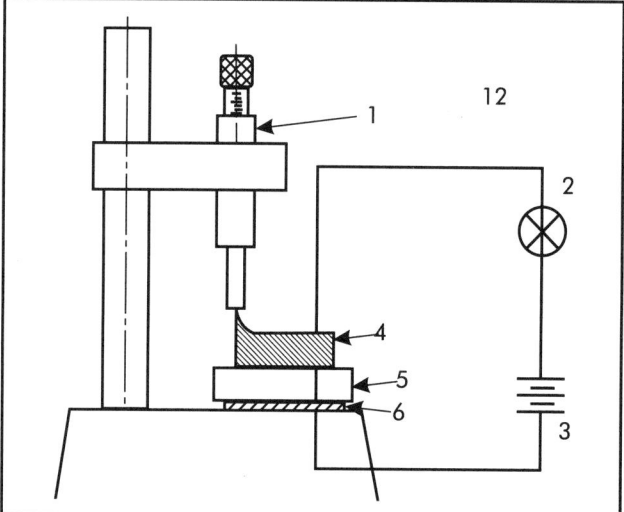

Figure 34-38. Burr height measured by micrometer with electrical contact signal (Przyklenk and Schlatter 1987).

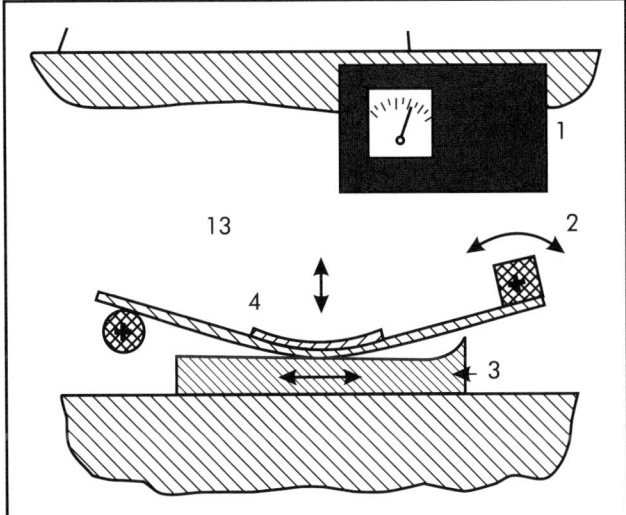

Figure 34-39. Leaf-spring and strain-gage burr height measurement (Przyklenk and Schlatter 1987).

but may not offer any advantage over other approaches.

Inductive Methods

As illustrated in Figure 34-40, it is possible to stack the parts and measure induction to obtain some measure of burr height while keeping parts spaced (Wentzel and Mehlhorn 1964). This is only practical in a few applications, and does not directly measure height. It provides a relative measurement and must be compared to a standard to obtain an estimate of burr size. One user noted that when he used this method many years ago it was not very sensitive.

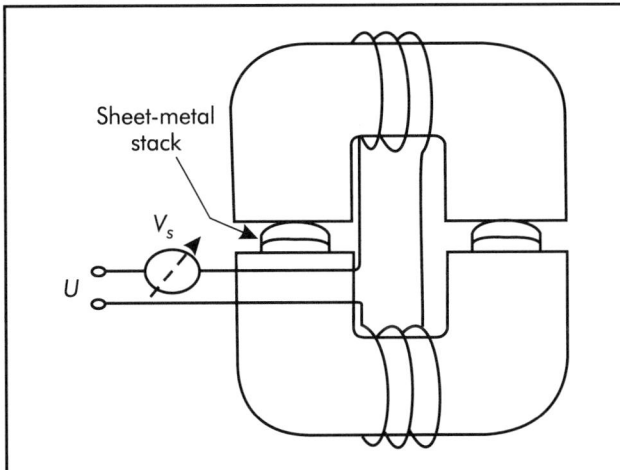

Figure 34-40. Inductive burr measurement concept (Wentzel and Mehlhorn 1964).

Capacitive Methods

One of the newest approaches for providing relative burr size involves a capacitive surface-finish analyzer (Figure 34-41). A unique signal is generated for each material, part surface finish, and configuration when the capacitive head is moved over the edge of a burr-laden part. With calibration for this specific part design, researchers and shop personnel can identify when the burr is larger or smaller than normal. This is a fast and efficient technique for high-production runs and for laboratory research. Results from different part geometries and materials cannot be compared directly.

When the burrs are shorter, a different signal is produced (Lee et al. 1996; Elbestawi et al. 1991). A rough correlation can be established between the burr size and signal. While it is not as precise as some of the other approaches, it is much faster—providing size information while scanning at several inches (millimeters) per minute. This makes it one of the fastest devices available. Users must set their own standard, and each part requires a different adjustment.

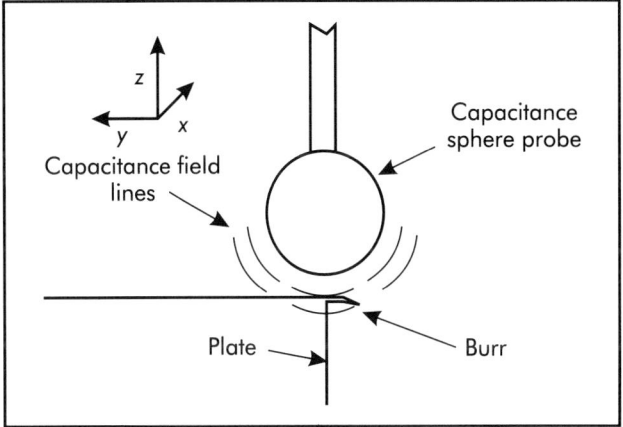

Figure 34-41. Schematic for capacitance burr measurements (Jones and Furness 1997).

Force Feedback Methods

Robotic deburring studies have used force feedback (Figure 34-42) to determine the size of the burr being removed, as well as the point at which the burr has been removed (Elbestawi et al. 1991; Stango 1997). Since deburring force involves the amount of metal removed in a given time, it is more a measure of burr volume than of height or thickness. This approach is used to provide information to the robot, which then makes decisions. Robot logic determines only that the burr is gone. Not concerned with height or thickness, it uses force to determine the amount of metal removed.

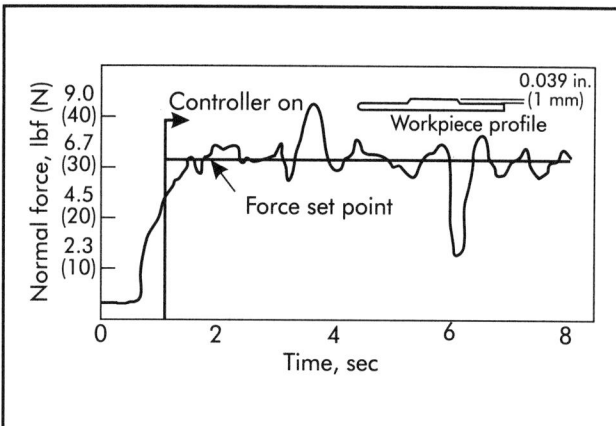

Figure 34-42. Force feedback as used in brush deburring (Elbestawi et al. 1991; Stango 1997).

In one approach, a series of master curves are generated to document force measurements for specific brushing conditions, part material, and burr size. Higher forces for a given set of brushing conditions indicate a thick or tall burr, or both (Stango 1997).

Acoustic Methods

At least one company produces an acoustic monitor for detecting wear in metalworking dies (Westinghouse Electric Corporation 1976). Noise is generated and correlated to burr size as dies wear. This is one of the few applications that measures burrs at the point of generation. Some researchers have performed additional studies of burr measurements through acoustic emission for milling and other operations (Dornfeld 1984, 1991).

Combinations of Inspection Approaches

Many users find that the best solution involves combining two or more inspection techniques. In the most common situations, inspectors visually look for burrs and, if they think a small burr might be present, use a toothpick as a probe. Others use the naked eye until a question is raised, then resort to a 4× magnifying glass or 10× microscope.

Consistency in Use

The measurement and detection of burrs can suffer from a lack of control. Specific inspection approaches may change with time and practice, which means those approaches may not be reflected accurately in documentation. "Oh, we switched to using a number 3 pencil point because that is all the office had," is an example of subtle changes that can occur unless someone monitors for consistency and compliance.

TRAINING WORKERS AND INSPECTORS

A well-trained staff is key to consistent success in burr detection and measurement. Well-trained workers greatly lighten the workload of inspectors.

The Course

Descriptions of programs that train inspectors for burr detection are not found in the literature. The material in this book provides much of the needed material. Training course outlines include the:

- objective of the course;
- definition of the skill measurement approach;
- testing of abilities or comprehension, and
- statement of company or plant requirements.

Visual Ability

Not all people can adequately see or detect burrs. Companies should periodically review staff and equipment capabilities because eyes change over time; optics develop films of dirt, oil, and dust, which affect visibility; and an individual's level of quality expectation can change over time.

Feedback

Inspectors need and generally receive immediate feedback when they fail to find burrs on edges. There are times, however, when the product requires immediate action but feedback is not provided. The problem will repeat itself unless participants understand the issue.

Parts

If parts are expensive, they are difficult to obtain for training purposes. It may be difficult, also, to replicate areas that are challenging to inspect or occasional, unusual conditions. One company trained 100 people to hand deburr and inspect, and found that several thousand parts were required to attain the desired quality level. Photographs, line-art sketches, or videos of the issue can be saved for training purposes.

The buddy system of having an experienced inspector check over the work of a new inspector is the most common approach for reviewing deburring work. Real production parts are used since no parts are risked with this approach.

Pseudo-burrs

For tests and training, it is frequently more convenient to manufacture a pseudo-burr because burr size, shape, and location can be controlled.

One of the most persistent problems with any form of testing and inspection is obtaining enough test samples to provide necessary information. Any test or inspection can be performed better by using a controlled sample. In the case of burrs, it is desirable to control the size, shape, hardness, and position of the burr during training. It is a challenge, and frequently difficult, to provide realistic burr variation and sizes. In the past, because of this problem trainers and researchers ignored burr variation in training and studies. In fact, to improve the accuracy of metal removal rate studies in deburring, many researchers required parts to be burr-free before they began their study of finishing parameters (Dornfeld 1991).

Shapes

Figure 34-43 illustrates several types of pseudo-burr shapes (Gillespie 1975). In (a), the burr is actually a flange that purposely has not been machined away. Using this approach, it is possible to use precision miniature lathes to produce flanges only 0.001-in. (25-μm) thick. The diameter of these flanges can be controlled to within 0.0002 in. (5 μm) of any size desired. This approach creates a rather conveniently produced and accurately controlled pseudo-burr.

The advantage of a cylindrical sample is that it is consistent with the basic configuration used in many parts and requires minimal inspection time for training. Stock loss, scratches, and other damage from deburring can be readily seen.

Figure 34-43b illustrates another method of fabricating a pseudo-burr. In this case, the burr is parallel to the axis of the workpiece.

It is easy to produce a triangular-shaped pseudo-burr by using a tapered-form tool. Rectangular cross-sections and long, narrow burrs can be produced. It is also possible to weld or braze a thin shim of any shape desired to almost any workpiece material (Figure 34-43c). The convenience and accessibility of shim, or feeler, stock in several different materials and tempers make this a convenient approach for use with a resistance welder. It is important to have a continuous weld on this type of pseudo-burr or the shim may tear loose or its response may not be as accurate as that of real burrs.

To provide more variation for training inspectors, put the parts through the normal deburring process, then search for the burr. Pseudo-burrs left after deburring can be very small if they began as small burrs. In some instances, the pseudo-burr would be beaten over; in others, the burr largely (but not completely) would be removed from the edge. One advantage of using parts after finishing operations is that

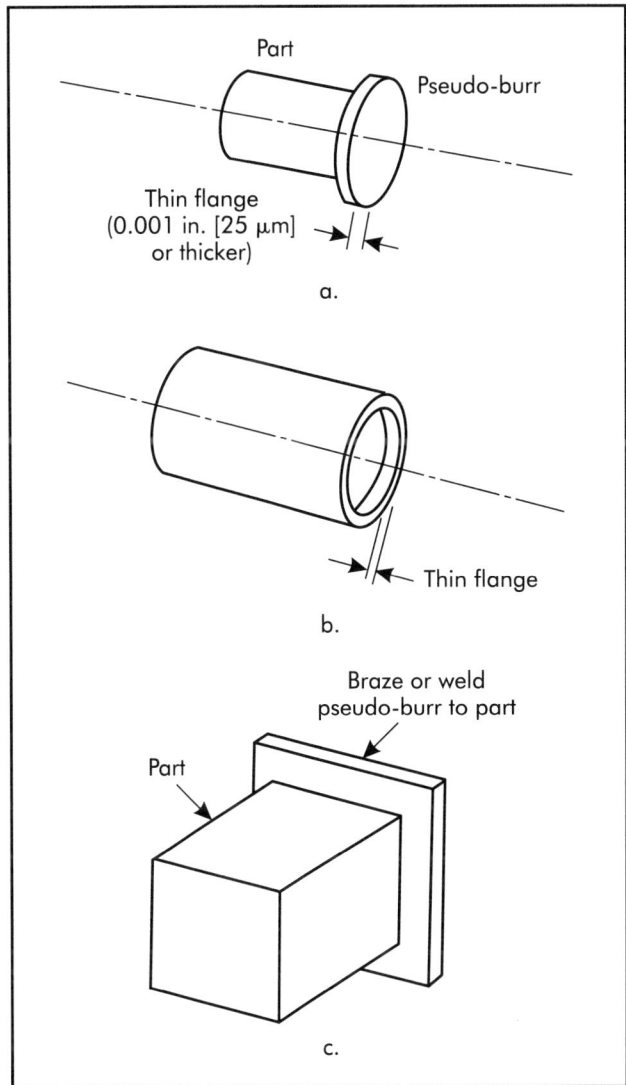

Figure 34-43. (a) Flange-style pseudo-burr; (b) trepan-style pseudo-burr; (c) shim-stock-style pseudo-burr. (Courtesy Honeywell)

they display a more typical surface. Sometimes, the surface appearance can mask small burrs.

Limitations

The problem with artificial burrs is that they do not necessarily represent the actual shape or size of the burr to be removed from production parts. For example, extensive studies have shown that a wide variety of burr shapes is possible with just one or two operations, such as drilling and contour milling (Schäfer 1975). Several of these shapes are also shown and discussed elsewhere (Gillespie 1978, 1996).

In some instances, a vibratory machine will beat over burrs that are 0.002-in. (51-μm) thick by 0.125-

in. (3.18-mm) high rather than remove them. One user found that vibratory finishing beat the material flush with the surface, and the thin burr left was only 0.0001-in. (2.5-μm) thick, but resulted in an interference fit in the next assembly. The burr could not be detected by unaided eye or fingernail, but it was evident when a microscope was used.

Pseudo-burrs can be effective and economical for initial training and helping to solve deburring problems, but the pseudo-burr must be based on realistic conditions.

CALIBRATION OF BURR INSPECTION TECHNIQUES

How do you calibrate the effectiveness of an inspector's ability to detect burrs? It is important to calibrate people and equipment, as noted in the following sections.

Pseudo-burrs

Pseudo-burrs can be used to provide a standard calibration artifact. They are simple to make, remake, and control to exact conditions.

Gage Calibration

Burr inspection devices should be calibrated if they are used to take measurements. There are no special requirements for calibration, but users may be unaware that some lens and microscope combinations produce magnification ratios different from what the reticule says because of additional optical aids within the microscope.

Calibration with Customer Expectation and Practice

Calibration usually implies a hardware calibration, but it is equally important to calibrate part quality with customer expectation to ensure consistency. Typically, the consumer needs to know how the parts are being inspected. When inspection techniques are changed, a recalibration should be provided.

As previously stated, companies should periodically review the capabilities of equipment and staff because eyes change over years, optics develop films of dirt, oil, and dust that affect visibility, and people's level of quality expectation can change over time.

WHICH EDGE CHARACTERISTICS SHOULD BE MEASURED?

Burr height is often the measurement taken in deburring. Sometimes it may be the appropriate variable to measure, but in many instances it is not.

Users must decide which edge characteristic they need to monitor. Is it an edge chamfer, an edge radius or blend, absence of sharpness, complete removal of sharp material, removal of all material at an edge, burr height, or burr thickness? In some cases, measuring burr height because it is more convenient may not yield enough information to make the best decision.

Figure 34-44 illustrates the problem of using maximum burr length for a milling burr in stainless steel. Choosing the maximum height, shown on the left side of the illustration, will not yield the same result as using a burr dimension that has been straightened back into the basic path that the cutter teeth used to generate the burr (right-hand view).

In an unusual variation, one researcher found that burr measurements were more repeatable on the slug that was thrown away in a blanking operation than the burr on the part (Wang et al. 1970). So, the study relied on the scrap piece, rather than the part, to determine die wear and burr trends. Also, it was much faster since the scrap was smaller and easier to handle.

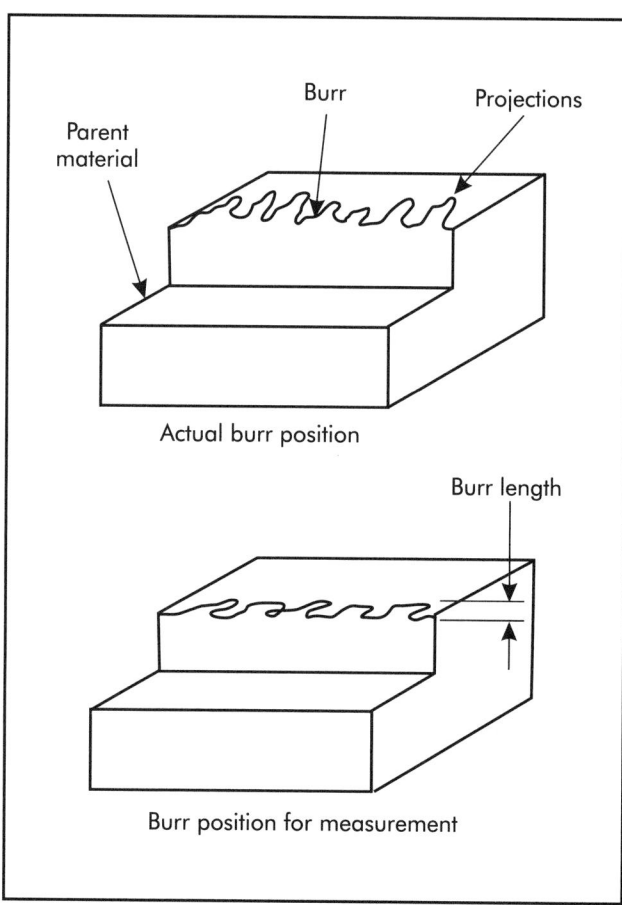

Figure 34-44. Effective burr length for end-milling burrs.

ONE MEASUREMENT IS NOT ENOUGH

The final point in burr measurement is that burrs are often so irregular that a single measurement per part results in tremendous swings in measured results (Wang et al. 1970; Kienzle and Kienzle 1958). Measuring the burr at the same location on each part provides more consistent results, but may mask large variations around the contour. For blanking burrs, an early researcher found that average burr height produces much more repeatable results (50 times more repeatable) than using the maximum burr height (Wang et al. 1970).

MEASUREMENT CAPABILITY STUDIES

Some knowledge of measuring capability is expected in today's environment. This necessitates a simple gage repeatability study. The approach is the same as for any measurement study (Gillespie 1994; Wilson 1989).

DOCUMENTATION

Where does your company keep information about burr measurements or deburring? Where are your standards for deburring? Are they written?

Results

Most plants have failed to document their deburring or burr-rejection efforts. Solutions may be frustrating because the real cause of the problem has not been defined. Many companies find themselves working on the wrong issue because the problem to be attacked was not documented by fact.

Training

Quality assurance requires documenting answers to the following questions:

- Which individuals have been trained?
- What version of the training program was used?
- What are the training and skill requirements?
- When was the training performed?

Calibration

A record of all devices in calibration should be available for all quality systems.

Inspection Performed

A note that parts were inspected for edge conditions should be included in all related inspection records.

THE BURR INSPECTOR

Burr inspection can be a challenging task. Normally, it is one of few areas of compliment and acknowledgment of excellence. The operating staff can do much to build pride in the job, acknowledge excellence, demonstrate performance, and the importance of the function.

Burr inspection can be simple and straightforward, or it can involve some of the most sophisticated equipment and training. Whatever the application, documenting expectations and approaches is critical for consistent results.

Improving manual deburring, or any other form of deburring, requires an understanding of the edge before and after deburring. As seen by previous examples in this chapter, the tools used can be very simple and low-cost. Measurement, accept/reject criteria, training, and communication will make any operation more efficient and more effective.

REFERENCES

American Machinist. 1973. "Practical Ideas: Gaging the Sharp Edge of a Hole." *American Machinist,* (August 6), p. 58.

American Society for Metals. 1985. "Taper Sectioning." *Metals Handbook,* 9th ed. Metals Park, OH: American Society for Metals (ASM), pp. 450–451.

Bell, Clarence L. and Blaine Kearsley. 1963. "The Effects of Drill-point Geometry, Feed, and Speed on Cutting Forces and Burrs." Course Paper for Tool and Manufacturing Engineering (TME). Logan, UT: Utah State University, pp. 187–188.

Bogue, R. J. 1977. "U.L. Sharp-edge Tester." Technical Paper MR77-469. Dearborn, MI: Society of Manufacturing Engineers (SME).

Dornfeld, David. 1984. "Acoustic-emission Monitoring and Analysis in Manufacturing." Proceedings of the symposium at 1984 ASME Winter Annual Meeting. ASME-PED. Vol. 14. New York: American Society of Manufacturing Engineers (ASME).

——. 1991. "Monitoring the Machining Process by Means of Acoustic-emission Sensors, Acoustic Emission: Current Practice and Future Directions." ASTM Special Technical Publication (1077). Edited by W. Sachse, J. Roget, and K. Yamaguchi. Philadelphia, PA: American Society for Testing and Materials (ASTM).

Elbestawi, M. A., K. M. Yuen, A. K. Srivastave, and H. Dai. 1991. "Adaptive Force Control for Robotic Disc Grinding." *Annals of the CIRP* 40 (1). Paris, France: International Institution for Production Engineering Research, pp. 391–394.

Flores, Gerhard. 1995. "Mechanical Deburring: Process, Tools, Machines, and Applications." Technical Paper MR95-272. Dearborn, MI: Society of Manufacturing Engineers (SME).

Gillespie, LaRoux K. 2000. *Guide to Deburring, Deflashing, and Trimming Equipment, Services and Supplies.* Kansas City, MO: Deburring Technology International, Inc.

———. 1998. "An Integrated International Standard for Burrs and Edge Conditions." WBTC-STD 14. Draft. Worldwide Burr Technology Committee. Kansas City, MO: Deburring Technology International.

———. 1994. "Characterization of Burrs and Deburring." Paper TR94-1. Kansas City, MO: Deburring Technology International.

———. 1981. "Progress in the Battle of the Burr." Technical Paper MRR81-07. Dearborn, MI: Society of Manufacturing Engineers (SME).

———. 1979. "How to Look for Burrs: A Training Manual for Precision Hand Deburring." Part 1. Report BDX-613-2245 (November). Kansas City, MO: Bendix. (Available from National Technical Information Service [NTIS].)

———. 1978. *Advances in Deburring.* Dearborn, MI: Society of Manufacturing Engineers (SME), pp. 37–56.

———. 1975. "Effects of Measurement Technique and Experimental Design in the Analysis of Burrs." Technical Paper MR75-985. Dearborn, MI: Society of Manufacturing Engineers (SME).

———. 1974. "The Measurement of Burrs." Technical Paper MR74-993. Dearborn, MI: Society of Manufacturing Engineers (SME).

Gillespie, LaRoux K., ed. 1996. "Standard Terminology for Researchers of Burrs and Edge Finishing." WBTC STD-02. Draft. Kansas City, MO: Deburring Technology International.

Gillespie, LaRoux K. and Arnold J. Monteiro. 1977. "Deburring Case Histories." Technical Paper MR77-06. Dearborn, MI: Society of Manufacturing Engineers (SME).

Hamill, A. T. 1950. "Die Life Test Cuts Cost of Laminations." *American Machinist,* November 27: pp. 100–102.

Jones, Steven D. and Richard J. Furness. 1997. "An Experimental Study of Burr Formation for Face Milling 356 Aluminum." *Transactions of the NAMRI/SME,* 25. Dearborn, MI: Society of Manufacturing Engineers (SME), pp. 183–188.

Kienzle, Otto and Werner Kienzle. 1958. "Tool Wear in the Cutting of Thin Gauge Steel Sheets." (Translation of *Stahl und Eisen,* 78 [12]. June 12: pp. 820–828.) Research Report 22 (May 1). Dearborn, MI: American Society of Tool and Manufacturing Engineers (now SME).

Ko, Sung-Lim, Jing-Koo Lee, and Gun-Bae Jun. 1988. "Measurement Technology for Micro Burrs." Proceedings of the 5th International Deburring and Surface Finishing Conference. San Francisco, CA (September 29–30). Kansas City, MO: Deburring Technology International, pp. 314–326.

Lee, S. H., D. S. Park, and D. A. Dornfeld. 1996. "Burr-size Measurements Using a Capacitance Sensor." Proceedings of the Second S. M. Wu Symposium on Manufacturing Science. Ann Arbor, MI. Dearborn, MI: Society of Manufacturing Engineers (SME), pp. 31–36.

Pahl, E. 1963. "En Messungen Zur Ermittlung des Schnittgrates." *VDI-Berichte,* 76: pp. 33–35.

Przyklenk, K. and M. Schlatter. 1987. *Entgraten von Werkstucken aus Aluminum.* Dusseldorf, Germany: Aluminum-Verlag.

Riis, Arne. 1979. "Etseafgratning: Almen Procesteknik, Afdelingen for Mekanish Tekologi (AMT) Danmarks Tekniske Hooskole." AMT Publication AP.79.04-B. Afdelingen for Mekanish Tekologi.

Seidenberg, Harald. 1967a. "Neues Verfahren zur Messung der Grathole an Schnittelen." *Werkstatt und Betrieb,* 100 (6): pp. 801–805.

———. 1967b. Stanzereitechnik: schnittgrad und messverfahren. *Werkstatt und Betrieb,* 100 (6): pp. 471–477.

Shäfer, Fredrich. 1975. *Entgraten.* Mainz, Germany: Krausskopf-Verlag.

Sorrels, John R. and Robert E. Berger. 1974. "An Inspection Procedure for Detecting Hazardous Edges." Report NBSIR 74-428 (April). Washington, DC: National Bureau of Standards.

Stango, Robert J. 1997. "Robotic Brush Deburring." Proceedings of the ASME International Mechancial Engineering Congress." Transactions of the ASME, November. New York: American Society of Mechanical Engineers (ASME).

Underwriters Laboratories. 1995. *Catalog of Standards.* Northbrook, IL: Underwriters Laboratories (UL), pp. 16–17.

———. 1998. "Test for Sharpness of Edges of Electrical Equipment." Bulletin on Subject 1439. Northbrook, IL: Underwriters Laboratories.

Von Heitlinger, Eugene. 1976. Sharp-edge Tester. U.S. Patent 3,931,732. Chicago, IL: Schwinn Bicycle Co.

Von Obering, Hans Peter. 1955. "Stanzereitechnik: Die Fertigung Formgeschliffener Schnittwerkzeuge." *Werkstatt und Betrieb,* 88 (1): pp. 27–31.

Wang, K. K., K. Taraman, and S. M. Wu. 1970. "An Analysis of Punching Variables by Two-level Fractional Factorial Design." *Journal of Engineering for Industry, Transactions of the ASME,* Series B (May). New York: American Society of Mechanical Engineers (ASME).

Wentzel, H., and H. Mehlhorn. 1964. "Erkenntnisse uber Gratbildung und Verschleill von Schnittwerkzeugen." *Fertigungstechnik und Betrieb,* 14 (April): pp. 236–240.

Westinghouse Electric Corporation. 1976. "Acoustical Die Monitor." Brochure Catalog 17-221 C WE A (June 7). Sunnyvale, CA: Westinghouse Electric Corporation.

Wilson, Mario-Parez. 1989. *Machine/Process Capability Study.* Scottsdale, AZ: Advanced Systems Consultants.

Zaima, Shigeo, Yoshio Hasegawa, and Akiyasu Yuki. 1975. "Burr Drilling Aluminum and Prevention of It." Technical Paper MR75-480. Dearborn, MI: Society of Manufacturing Engineers (SME).

BIBLIOGRAPHY

Asai, S., T. Taguchi, K. Horio, T. Kasai, and A. Kobayashi. 1990. "Measuring the Very Small Cutting-edge Radius for a Diamond Tool Using a New Kind of SEM Having Two Detectors." *CIRP Annals*. 39 (1) Paris, France: International Institution for Production Engineering Research, pp. 85–88.

Buchmann, K. 1961. "Beitrag zur Verschleissbeurteilung Beim Schneiden von Stahlfeinblechen. Disseration. Hanover, Germany: TH Hanover.

Lam, Anselm. 1993. "Burr Detection and Measurement Using a Machine Vision System." MS Report, Mechanical Engineering Department. Berkeley, CA: University of California.

Lam, Anselm and Jaramporn Hassamontr. 1993/1994. "Burr Detection and Measurement Using a Machine Vision System." LMA Research Report, Mechanical Engineering Department. Berkeley, CA: University of California, pp. 43–45.

Moore, A. J. W. 1948. "A Refined Metallography Technique for the Examination of Surface Contours and Surface Structure of Metals." *Metallurgia*, vol. 38, no. 224 (June): pp. 71–75.

Owens, Damon C. 1991. "Burr Detection and Measurement Using a Machine Vision Metrology." MS Report, Mechanical Engineering Department. Berkeley, CA: University of California.

Siemens. 1975. "Prufeinrichtung fur die Mechanische Messung Dunner und Labiler Metallteile." Brochure. Munich, Germany: Siemens AG.

Vander Voort, George F. 1989. *Applied Metallography*, 136. New York: Van Nostrand Reinhold Company, pp. 278–279.

Yeh, Po-ting 1991. "A Machine Vision System Used for Detection of Burr Heights." MS Report, Mechanical Engineering Department. Berkeley, CA: University of California.

Index

A

abrasive cork tools, 201-202 (Table 16-1)
abrasive cotton-fiber products, 205 (Figure 17-1), 207-208
    resin-bonded tools, 205 (Figure 17-1)
    sizes and shapes, 183-185 (Table 13-1), 205-207 (Table 17-1)
    usage, 208-209 (Figures 17-2 through 17-4)
abrasive rubber tools, 211-213 (Figures 18-1 through 18-3)
    characteristics, 211-213
    comparisons, 214-219 (Figures 18-4 through 18-8)
    limitations, 214
    shank configurations, 212 (Figure 18-2), 214
    sizes, 213-216 (Tables 18-1, 18-2)
    use, 216, 220 (Figure 18-9)
abrasive wood tools, 197-198 (Figures 15-1, 15-2, Table 15-1)
air motors (dental), 286-287 (Figure 23-5)
automation and mechanization, 341
    automation to replace manual deburring, 346-349 (Figures 29-7 through 29-9)
    mechanization to replace manual deburring, 341-347 (Figures 29-1 through 29-6)

B

back pain, 366-367 (Table 31-4)
ball-end probes, 326, 328 (Figure 26-34)
ballizing, 271-272 (Figures 21-1 through 21-3)
bench grinders, 301-302 (Figure 24-10)
bench motors, 299-301 (Figures 24-1 through 24-7)
bench-top chamfering machines, 305-307 (Figures 25-1 through 25-4)
bonded abrasive tools, 221
    environmental, health, and safety issues, 233, 235, 238 (Figure 19-25)
    materials, 231-234 (Tables 19-3, 19-4, Figures 19-13 through 19-15)
    shapes and sizes, 221-231 (Tables 19-1, 19-2, Figures 19-1 through 19-12)
    use, 232-233 (Figures 19-16 through 19-24)
brushes, 239
    aggressiveness, 246-248 (Figure 20-7)
    applications, 257-262 (Figures 20-12 through 20-19)
    combined process deburring, 266
    equipment, 252, 255-257 (Figures 20-9 through 20-11)
    fill material, 243-246 (Figures 20-5, 20-6)
    nontraditional brushing, 264-266 (Table 20-8, Figures 20-20 through 20-24)
    power requirements, 249-251 (Tables 20-1, 20-2, Figure 20-8)
    selection, 251-255 (Tables 20-4 through 20-6)
    side effects, 263-264
    tool wear, 248-249
    troubleshooting, 263 (Table 20-7)
    types, 239-243 (Table 20-1, Figures 20-1 through 20-4)
buffers (industrial), 302-303 (Figure 24-12)
burr
    and edge
        standards, 25, 435
        terminology, 39-40 (Figure 3-8)
    inspection, 435-456 (Table 34-1, Figures 34-1 through 34-42), 458-459 (Figure 34-44)
    location, 394-399 (Figures 32-16 through 32-19)
    size, 7, 19, 60, 391
    technology fundamentals, 2
burs (rotary), 125 (Table 10-1)
    applications, 51 (Figures 4-3 through 4-6), 144-155 (Tables 10-5 through 10-7, Figures 10-29 through 42)

materials, 142, 146-148 (Figures 10-26 through 10-28)
styles, 125-137 (Tables 10-2, 10-3, Figures 10-1 through 10-21)
tooth design, 137-146 (Table 10-4, Figures 10-22 through 10-25)

C

calibration of burr inspection techniques, 458-459 (Figure 34-44)
carpal tunnel syndrome, 359, 363, 365 (Figure 31-2)
chairs, 330
chamfer mikes, 326-327 (Figure 26-33)
chamfering machines, 290, 293 (Figures 23-15, 23-16), 305-307 (Figures 25-1 through 25-4)
chamfering tools and countersinks, 115-123 (Table 9-1, Figures 9-1 through 9-14)
chippers and scalers, 290, 293 (Figures 23-17, 23-18)
combined process deburring, 266
common practices, 43-47
 flash removal, 64, 66
 intersecting holes, 48-56 (Figures 4-1 through 4-15, Tables 4-1, 4-2)
 lathe parts, 66 (Figure 4-35)
 management, 46-47
 operator, 47-68 (Figures 4-1 through 4-15, Tables 4-1, 4-2)
 threads, 56-65 (Figures 4-16 through 4-34, Tables 4-1, 4-2)
cork tools, 201-202 (Table 16-1)
costs, 71-81 (Tables 5-1 through 5-4, 6-1, Figure 5-1), 85-89 (Tables 6-4 through 6-6, Figure 6-1), 358, 371-376 (Tables 32-1 through 32-5), 402-405 (Tables 32-12 through 32-17)
countersinks and chamfering tools, 115-123 (Table 9-1, Figures 9-1 through 9-14)
cryogenic probes, 327, 330 (Figure 26-40)
cumulative trauma disorders, 359, 363

D

data-driven solutions, 383-405 (Tables 32-6 through 32-17, Figures 32-2 through 32-22), 409-410 (Figures 32-23 through 32-30)
 deburring repeatability tests, 384
 deburring tool capabilities, 397-400 (Figures 32-20, 32-21)
 effect of burr location, 394-399 (Figures 32-16 through 32-19)
 effects of burr size and part geometry, 391
 operational cost data, 402-405 (Tables 32-12 through 32-17)
 problem areas, 405-409 (Figures)
deburring, 1-22 (Figures 1-1, 1-2)
 burr size, 7

burr technology fundamentals, 2
combined process, 266
costs, 71-81 (Tables 5-1 through 5-4, 6-1, Figure 5-1), 358, 402-405 (Tables 32-12 through 32-17)
 calculation, 85-89 (Tables 6-4 through 6-6, Figure 6-1)
 increases, 82
 reducing, 82-84 (Tables 6-2, 6-3), 371-376 (Tables 32-1 through 32-5)
 time and motion studies, 82
 tooling, 85-86
 validating, 84-85
 versus part finishing, 81-82
economics, 425
effectiveness, 4-7 (Table 1-1)
equipment allocation, 382-383
intersecting holes, 48-56 (Figures 4-1 through 4-15, Tables 4-1, 4-2)
method selection, 12-21 (Tables 2-1 through 2-4, Figure 2-1)
 burr size, 19
 examples, 21-22
 external edges, 20
 holes, 20-21
 limitations, 19
operation efficiency, 371-372, 375, 377-381 (Figure 32-1), 383-405 (Tables 32-6 through 32-17, Figures 32-2 through 32-22), 409-420 (Figures 32-30 through 31-50)
 coordinating people, 377-381
 effect of burr location, 394-399 (Figures 32-16 through 32-19)
 effects of burr size and part geometry, 391
 operator mistakes, 378-381 (Figure 32-1)
 problem identification and solution, 371-372, 375, 377, 405-409 (Figures 32-23 through 32-29)
people, 8
repeatability tests, 384
tool capabilities, 397-400 (Figures 32-20, 32-21)
versus edge finishing, 9-10
dental air motors, 286-287 (Figure 23-5)
dictionaries, 38
drills (portable), 290, 292 (Figure 23-14)

E

edge requirements and standards, 25-29 (Figure 3-1, Table 3-1), 435
 burr and edge terminology, 39-40 (Figure 3-8)
 dictionaries, 38
 existing, 28
 issues and problems, 25-26
 reasons to adopt, 26, 28
 sample, 29-38 (Figures 3-2 through 3-7, Tables 3-2 through 3-4)

edgers and chamfering machines, 290, 293 (Figures 23-15, 23-16)
electron-beam inspection, 448-450
environmental, health, and safety issues, 191 (Figure 14-6), 195, 233, 235, 238 (Figure 19-25), 266-267, 351, 355-356
   back pain, 366-367 (Table 31-4)
   carpal tunnel syndrome, 359, 363, 365 (Figure 31-2)
   cumulative trauma disorders, 359, 363
   files, 354
   injury statistics, 355 (Table 30-2)
   knives, 351, 354 (Figure 30-1)
   legal liabilities, 358
   medical and insurance costs, 358
   motorized operations, 354-355
   preventing injuries, 359-365 (Tables 31-1 through 31-3, Figure 31-1)
   training, 355-356
equipment allocation, 382-383
ergonomics, 357
   back pain, 366-367 (Table 31-4)
   carpal tunnel syndrome, 359, 363, 365 (Figure 31-2)
   cost of, 357-359
   cumulative trauma disorders, 359, 363
   data, 357
   hand-tool design, 363-366 (Figures 31-3 through 31-5)
   high stress, 366-367 (Table 31-5)
   illumination, 366
   improving productivity, 358
   legal liabilities, 358
   medical and insurance costs, 358
   noise, 367-369 (Tables 31-6, 31-7, Figure 31-6)
   preventing injuries, 359-365 (Tables 31-1 through 31-3, Figure 31-1)
   red flags, 358-359
exhaust systems, 327-328, 330 (Figure 26-41)

F

files, 161, 354
   cleaning and care, 172
   glossary of terms, 173-175
   selecting and using, 169-172 (Tables 12-3, 12-4, Figures 12-11 through 12-13)
   types, 161-169 (Tables 2-1, 2-2, Figures 12-1 through 12-10)
finishing machines, 306-307 (Figures 25-1, 25-2, 25-5, 25-6)
finishing versus deburring, 81-82
flash removal, 64, 66
flexible-positioning arm, 330, 332 (Figure 26-43)

G

gages, 450-452 (Figures 34-27 through 34-31), 458
gloves (protective), 328, 331 (Figure 26-42)

H

hand deburring, 1-22 (Figures 1-1, 1-2)
   burr technology fundamentals, 2
   effectiveness, 4-8 (Table 1-1, Figures 1-4, 1-5)
   method selection, 12-21 (Tables 2-1 through 2-4, Figure 2-1)
      burr size, 19
      examples, 21-22
      external edges, 20
      holes, 20-21
      limitations, 19
   people, 8
   reasons for, 2-3, 11-12
   when to use, 3
handheld motors, 283-290 (Table 23-1, Figures 23-1 through 23-20)
   chippers and scalers, 290, 293 (Figures 23-17, 23-18)
   dental air motors, 286-287 (Figure 23-5)
   portable drills, 290, 292 (Figure 23-14)
   portable edgers and chamfering machines, 290, 293 (Figures 23-15, 23-16)
   portable sanders, 288-291 (Figures 23-9, 23-10)
   use, 290-298 (Figures 23-19 through 23-23)
hand-operated mechanized machines, 305-311 (Figures 25-1 through 25-15, Table 25-12)
hand stones, 187
   composition, 187 (Figure 14-1)
   environmental, health, and safety issues, 191 (Figure 14-6), 195
   grit size, 189-190 (Tables 14-1, 14-2)
   nomenclature, 187-189
   shapes and sizes, 190-194 (Tables 14-3 through 14-5, Figures 14-2 through 14-4)
   stone bonds, 189
   stoning oil, 191
   use, 191 (Figure 14-5), 194
hand tools
   design, 363-366 (Figures 31-3 through 31-5)
   special, 275-281 (Figures 22-1 through 22-13)
health, environmental, and safety issues, 191 (Figure 14-6), 195, 233, 235, 238 (Figure 19-25), 266-267, 351, 355-356
heat considerations, 330
holes, 20-21, 48-56 (Figures 4-1 through 4-15, Tables 4-1, 4-2)
hot-wire tools for thermoplastic parts, 335
   hot-tool concept, 103 (Figure 7-14), 335-336 (Figures 28-1, 28-2)

suitable plastic materials, 335-339 (Figures 28-3, 28-4)

I

illumination, 366
industrial buffers, 302-303 (Figure 24-12)
injury prevention, 359-365 (Tables 31-1 through 31-3, Figure 31-1)
injury statistics, 355 (Table 30-2)
inspecting for burrs and sharp edges, 435-439 (Table 34-1, Figures 34-1 through 34-3), 456
   acoustic methods, 456
   burr and edge standards, 435
   calibration, 458-459 (Figure 34-44)
   capacitive methods, 455 (Figure 34-41)
   consistency, 456
   destructive methods, 453-454 (Figures 34-36, 34-37)
   electrical methods, 454-455 (Figures 34-38, 34-39)
   electron-beam, 448-450
   force feedback methods, 455-456 (Figure 34-42)
   functional testing, 451-452 (Figures 34-32, 34-33)
   importance of clean parts, 445-446
   inductive methods, 455 (Figure 34-40)
   measuring microscopes, 446-448 (Figures 34-19 through 34-21)
   mechanically gaged methods, 450-452 (Figures 34-27 through 34-31)
   profilometer methods, 452-453 (Figures 34-34, 34-35)
   replica methods, 454
   structured lighting methods, 446
   visual methods, 439-445 (Figures 34-4 through 34-18)
intersecting holes, 48-56 (Figures 4-1 through 4-15, Tables 4-1, 4-2)

K

knives, 91, 105-106 (Table 7-3, Figures 7-22, 7-23), 351, 354 (Figure 30-1)
   blade materials, 101
   styles, 91-101 (Tables 7-1, 7-2, Figures 7-1 through 7-17)
   using, 101-105 (Figures 7-18 through 7-21)

L

lathes, 66 (Figure 4-35), 301-302 (Figures 24-8, 24-9)
lights, 324-326 (Figures 26-24 through 26-31), 366, 446

M

machining and blanking process, 7-8 (Figures 1-4, 1-5)

magnifiers, 313-320 (Figures 26-1 through 26-16)
manual deburring costs, 81 (Table 6-1)
   calculation, 85-89 (Tables 6-4 through 6-6, Figure 6-1)
   increases, 82
   reducing, 82-84 (Tables 6-2, 6-3)
   time and motion studies, 82
   tooling, 85-86
   validating, 84-85
   versus part finishing, 81-82
measurement capability studies, 459
measuring
   devices for sharp edges, 326
   microscopes, 446-448 (Figures 34-19 through 34-21)
   surface finish, 327
mechanization and automation, 341
   automation to replace manual deburring, 346-349 (Figures 29-7 through 29-9)
   mechanization to replace manual deburring, 341-347 (Figures 29-1 through 29-6)
microburrs, 60
microscopes, 319-324 (Figures 26-17 through 26-23), 446-448 (Figures 34-19 through 34-21)
mirrors, 326 (Figure 26-32)
motion and time studies, 82
motorized operations, 354-355
motors
   chippers and scalers, 290, 293 (Figures 23-17, 23-18)
   dental air motors, 286-287 (Figure 23-5)
   handheld, 283-290 (Table 23-1, Figures 23-1 through 23-20)
   pedestal, 299, 302 (Figure 24-11)
   portable drills, 290, 292 (Figure 23-14)
   portable edgers and chamfering machines, 290, 293 (Figures 23-15, 23-16)
   portable sanders, 288-291 (Figures 23-9, 23-10)
   use, 290-298 (Figures 23-19 through 23-23)
mounted points, 177
   applications for, 181 (Table 13-1), 183-186 (Tables 13-2, 13-3)
   hardness of, 178, 180-181
   materials in, 178
   types, 177-182 (Figures 13-1 through 13-5)

N

noise, 367-369 (Tables 31-6, 31-7, Figure 31-6)

O

operation efficiency, 358, 371, 410-420 (Figures 32-31 through 32-50)
   coordinating people, 377-381

data-driven solutions, 371-372, 375, 377, 383-405 (Tables 32-6 through 32-17, Figures 32-2 through 32-22), 409-410 (Figure 32-30)
  cost, 371-376 (Tables 32-1 through 32-5), 402-405 (Tables 32-12 through 32-17)
  deburring repeatability tests, 384
  deburring tool capabilities, 397-400 (Figures 32-20, 32-21)
  effect of burr location, 394-399 (Figures 32-16 through 32-19)
  effects of burr size and part geometry, 391
  typical problem areas, 405-409 (Figures 32-23 through 32-29)
equipment allocation, 382-383
operator mistakes, 378-381 (Figure 32-1)
problem identification and solution, 371-372, 375, 377

P

part finishing versus deburring, 81-82
part geometry and burr size, 391
pedestal motors, 299, 302 (Figure 24-11)
portable edgers and chamfering machines, 290, 293 (Figures 23-15, 23-16)
portable sanders, 288-291 (Figures 23-9, 23-10)
probes (cryogenic), 327, 330 (Figure 26-40)
problem identification and solutions, 371-372, 375, 377
productivity (improving), 358, 371, 410-420 (Figures 32-31 through 32-50)
profilometer, 452-453 (Figures 34-34, 34-35)
protective gloves, 328, 331 (Figure 26-42)
pseudo-burrs, 458

Q

quality requirements, 425

R

reamers, 157-159 (Figures 11-1 through 11-3)
reducing deburring costs, 82-84 (Tables 6-2, 6-3), 371-376 (Tables 32-1 through 32-5)
resin-bonded tools, 205 (Figure 17-1)
rotary burs, 125 (Table 10-1)
  applications, 51 (Figures 4-3 through 4-6), 144-155 (Tables 10-5 through 10-7, Figures 10-29 through 10-42)
  materials, 142, 146-148 (Figures 10-26 through 10-28)
  styles, 125-137 (Tables 10-2, 10-3, Figures 10-1 through 10-21)
  tooth design, 137-146 (Table 10-4, Figures 10-22 through 10-25)

S

safety, environmental, and health issues, 191 (Figure 14-6), 195, 233, 235, 238 (Figure 19-25), 266-267, 351, 355-356
sanders, 288-291 (Figures 23-9, 23-10), 307-310 (Figures 25-7 through 25-12, Table 25-12)
scalers and chippers, 290, 293 (Figures 23-17, 23-18)
scrapers, 109-113 (Table 8-1, Figures 8-1 through 8-8)
sharp-edge measuring devices, 326
special hand tools, 275-281 (Figures 22-1 through 22-13)
speed lathes, 301-302 (Figures 24-8, 24-9)
standards (burr and edge), 25-38 (Figures 3-2 through 3-7, Tables 3-2 through 3-4), 435
support devices, 313 (Table 26-1)
surface-finish-measuring machines, 327
Swiss file terms, 173-175

T

thermoplastic parts (hot-wire tools), 103 (Figure 7-14), 335-339 (Figures 28-1 through 28-4)
threads, 56-65 (Figures 4-16 through 4-34, Tables 4-1, 4-2)
time and motion studies, 82
tool
  capabilities, 397-400 (Figures 32-20, 32-21)
  comparisons, 214-219 (Figures 18-4 through 18-8)
  costs, 85-86
  design, 363-366 (Figures 31-3 through 31-5)
  holders, 327-330 (Figures 26-35 through 26-39)
  materials, 142, 146-148 (Figures 10-26 through 10-28), 178, 231-234 (Tables 19-3, 19-4, Figures 19-13 through 19-15)
  that beat down edges, 333-334 (Figures 27-1 through 27-3)
  tube-end finishing, 121-123 (Figure 9-12)
  use, 198 (Table 15-1), 216, 220 (Figure 18-9), 232-233 (Figures 19-16 through 19-24), 290-298 (Figures 23-19 through 23-23)
training, 423
  case history, 427-433 (Tables 33-2, 33-3, Figures 33-1, 33-2)
  deburring economics, 425
  elements, 423-424 (Table 33-1)
  goals, 423-424
  hands-on, 424-425
  identifying tools, 424
  parts, 424
  people, 424
  plant expectations, 34 (Figures 3-6, 3-7), 425 (Figure 33-1)
  quality requirements, 425
  safety, 355-356

workers and inspectors, 355-356, 456-458 (Figure 34-43)
tube-end finishing tools, 121-123 (Figure 9-12)

V

visual inspection methods, 439-445 (Figures 34-4 through 34-18)

W

wood tools, 197 (Figures 15-1, 15-2, Table 15-1)
workbenches, 327
workers and inspectors, 355-356, 456-458 (Figure 34-43)